ANNALS

OF THE

ASTRONOMICAL OBSERVATORY OF HARVARD COLLEGE.

VOL. IV. PART I.

CATALOGUE OF POLAR AND CLOCK STARS.

PRINTED FROM THE STURGIS FUND.

CAMBRIDGE:
WELCH, BIGELOW, AND COMPANY,
PRINTERS TO THE UNIVERSITY.
1863.

CATALOGUE

OF

STANDARD POLAR AND CLOCK STARS,

FOR

THE REDUCTION OF OBSERVATIONS IN RIGHT ASCENSION.

BY

T. H. SAFFORD, A.A.S.,

ASSISTANT AT THE OBSERVATORY OF HARVARD COLLEGE.

FROM THE MEMOIRS OF THE AMERICAN ACADEMY, NEW SERIES, VOL. VIII.

CAMBRIDGE:
WELCH, BIGELOW, AND COMPANY,
PRINTERS TO THE UNIVERSITY.
1863.

INTRODUCTORY NOTE.

THE following Memoir contains the positions of standard stars originally prepared for the reduction of meridian observations of right ascension made at the Observatory of Harvard College, to which purpose they have been applied since May, 1862. It was communicated by the author to the American Academy, September 9th, 1862.

The expense of publication in its present form, as a part of the Transactions of the Observatory, has been supplied from the income of a permanent fund given by Hon. William Sturgis of Boston, and held in trust by J. Ingersoll Bowditch and William Sturgis Hooper.

G. P. BOND,

Director of the Observatory of Harvard College.

A CATALOGUE

OF

STANDARD POLAR AND CLOCK STARS, FOR THE REDUCTION OF OBSERVATIONS IN RIGHT ASCENSION; WITH A DISCUSSION OF THEIR POSITIONS.

By TRUMAN HENRY SAFFORD.

(*Communicated September* 9, 1862.)

The object of the present Catalogue is to furnish (additional to the very accurate positions of the Berlin Jahrbuch) the means of reducing observations of right ascension with greater convenience and accuracy, and in a somewhat more systematic manner, than by the aid of former catalogues. It is divided primarily into two catalogues, the one of "polar stars," the other of "clock-stars," so called; with a supplementary list of stars which for large instruments are not commonly used in either capacity, but merely as tests of the accuracy of observation and reduction.

Excepting 25 "polar stars," all the stars here given are found either in the Nautical Almanac Catalogue, or in that of the Berlin Jahrbuch.

This Catalogue contains, then, in the first place, the 47 stars of whose apparent places an ephemeris is given in the Jahrbuch; 2 of these being "polar stars," 36 "clock-stars," and 9 of the supplementary list.

In the second place, our Catalogue contains the 25 "polar stars" — stars within 10° of the north pole — mentioned above.

In the third place, all the stars of the Nautical Almanac Catalogue are here given, except α Columbæ and δ Sculptoris, ζ Ursæ Minoris, η Draconis, and γ Cephei, and also the stars below 45° of south declination.

The mean positions in this Catalogue of the 47 stars are derived from Wolfers's *Tabulae Reductionum*,* without further investigation. It will be often convenient to

* Tabulae Reductionum Observationum Astronomicarum ab anno 1860, usque ad 1880. Auctore J. Ph. Wolfers. Additur, Tabulae Regiomontanae ab anno 1850, usque ad 1860, ab Ill. Zech continuatae. Berolini, 1858.

have them in the form here presented, which differs from that adopted in those tables. The secular variations given for these stars are the same which Wolfers used, and which are taken from the *Tabulae Regiomontanae.*

Computation of the Mean Places of 25 *Polar Stars.*

In this discussion the following catalogues have been employed; to the name of each catalogue has been prefixed the abridgment by which it is cited in the course of the work: —

Br. The Fundamenta Astronomiae (full title given in Vol. VIII. of the Memoirs of the American Academy, Pt. I. p. 299).

Gr. Groombridge (full title, *loco modo citato*).

Str. 1815. F. G. W. Struve, Observationes Astronomicae institutae in Specula Universitatis caesareae Dorpatensis. Vol. I. Pars II. Dorpati. 1817. 4to.

Str. 1824. Stellarum Fixarum imprimis duplicium et multiplicium Positiones Mediae pro Epocha 1830.0 deductae ex observationibus meridianis annis 1822 ad 1843 in Specula Dorpatensi institutis; auctore F. G. W. Struve. Petropoli ex typographia academica, 1852. Fol. The positions which I have used are a few for 1824.0 from the Introduction, page xxxxviii. *et seqq.*

Schw. Schwerd's Beobachtungen von Circumpolarsternen in mittleren Positionen 1828.0. Von Wilhelm Oeltzen. Aus dem X. B[de] der Denkschriften der Mathem.-Naturw. Classe der K. Akademie der Wissenschaften. . . . . Wien, 1856. 4to.

Arg. DLX Stellarum Fixarum Positiones Mediae ineunte Anno 1830. . . . [auctore] Argelandro. . . . Helsingforsiae, 1835. 4to.

Pond. A Catalogue of 1112 Stars reduced from Observations made at the Royal Observatory at Greenwich from the Years 1816 to 1833. London. . . . . 1833 Fol.

Hend. Astronomical Observations made at the Royal Observatory, Edinburgh, by Thomas Henderson. . . . . . Vol. III. for the Year 1837 [to V. for 1839] Edinburgh. 1840 [– 1843]. 4to.

[The same] reduced and edited by his successor, Charles Piazzi Smyth. Vol VI. for the Year 1840 [to X. for 1844 – 7].

Rob. Places of 5345 Stars observed from 1828 to 1854 at the Armagh Observatory by Rev. T. R. Robinson, D. D. . . . . . Dublin, 1859. 8vo.

Airy. Airy's "Twelve-Year Catalogue," "Six-Year Catalogue," "Greenwich Observa tions" from 1854–1858. (For full titles see Vol. VIII. of the Memoirs of the American Academy, pp. 300, 301.)

The "Greenwich Observations" for 1859 and 1860 have come to hand since the main part of these calculations were made.

Joh. The Radcliffe Catalogue of 6317 Stars chiefly Circumpolar reduced to the Epoch 1845.0, formed from the Observations made at the Radcliffe Observatory under the Superintendence of Manuel John Johnson, M. A., late Radcliffe Observer. With Introduction by the Rev. Robert Main, M. A., Radcliffe Observer. Oxford. 1860. 8vo.

Car. A Catalogue of 3735 Circumpolar Stars observed at Redhill in the years 1854, 1855, and 1856, and reduced to Mean Positions for 1855.0, by Richard Christopher Carrington. London. 1857. Fol.

LeV. Annales de l'Observatoire Impérial de Paris publiées par U. J. Le Verrier, Directeur de l'Observatoire. Observations. Tome XII. [XIII.] Paris. 1860, [1861]. 4to.

Tomes XIV., XV. were received later.

I found, when too late to use them, that Airy's Cambridge Observations contained two or three places which should have been inserted.

No use was made of the *Histoire Céleste*, or Fedorenko's Catalogue; nor of Piazzi, nor Groombridge's right-ascensions. Any material from these sources is of uncertain accuracy; or would require insecure systematic corrections.

The method by which mean places for 1855, and proper motions referred to the ecliptic of the same year have been found, is as follows: The positions of the Fundamenta, where such existed, were reduced from 1755 to 1855 by the use of Struve and Peters's data. In some cases where there was no observed declination for 1755, a value for 1755 referred to the ecliptic of 1855 was assumed, and by a slight modification of the method given in the introduction (p. ix.) to the *Tabulae Regiomontanae* (see, also, Wolfers's *Tabulae Reductionum*, pp. lii., liii. of the Introduction) was made to serve the same purpose. For if (see the place just cited) δ' be supposed known, we shall have ([32]) $x = \frac{\delta' - (d' + d)}{\beta + \alpha}$. In this case the declination for 1755 corresponding to that assumed for the ecliptic of 1855, will be

$$\delta = d' - d + (\beta - \alpha)\, x.$$

The indirect form of Bohnenberger's method which Bessel has given, and which has been transferred by Wolfers, is greatly preferable to the more direct one. It needs a little more care, however, in the computation.

We have now positions for 1755, referred to the ecliptic of 1855, which agree with *all* the data of the Fundamenta referring to these stars. We have also some modern observations referred with more or less accuracy to the ecliptic of the same epoch, or which we can easily refer to that ecliptic approximately. These give us the means of obtaining a formula which shall be of the form

$$\alpha = \alpha^\circ + h^\circ\,(t - 1755)\,, \qquad \delta = \delta^\circ + i^\circ\,(t - 1755)\,.$$

Where α° δ° are (Bessel's) Bradley's positions of 1755, in cases where they are complete, referred to the co-ordinate-planes of 1855 ; or, in case the Fundamenta position of 1755 is incomplete, α° is the AR. for 1755 referred to the ecliptic and equator of 1855, by means of δ°, as explained above ; δ° being here arbitrary within certain small limits: where again h° i° are *assumed* proper motions referred to the ecliptic and equator of 1855, but *so* assumed as to nearly represent more modern observations. For stars not in the Fundamenta, α° δ° are both arbitrary.

In other words, if all observed positions are referred to 1855 by precession alone the formulæ

$$\alpha = \alpha^\circ + \text{etc.}, \qquad \delta = \delta^\circ + \text{etc.},$$

would represent those taken from the Fundamenta *exactly*, and all others *nearly*. But as certain positions are exceptionally incomplete, and others so as a rule, we shall not attempt so to refer them, but shall proceed in the opposite way. We shall compute theoretical places of each star for the epochs of the catalogues. This will be done most conveniently by computing places for epochs at equal intervals, which can then be tested, and afterwards interpolated ; thus saving labor, and securing accuracy.

Table I. contains the assumed values of α° δ° h° i°. For the stars contained in Bradley, the values of α° and δ° have been verified by Mr. A. Hall, now Aide at the U. S. Naval Observatory, Washington ; who has recomputed the positions referred to the ecliptic of 1755 from them.

From these values positions for 1810, 1830, 1850, 1870 were computed, and in some cases for 1820, 1840, 1860 ; using Struve and Peters's constants. The positions for 1815 were computed independently, and also interpolated as a check. The proper motions (h' i'), and secular variation (in AR.) were computed for 1830 and 1850 the former by formulæ which will afterwards be cited with some modifications from Bessel, and the latter by the formula,

$$0''.0284 + 0''.1950\,[\text{tg}\,\delta'^2 + \tfrac{1}{2}]\sin 2\,\alpha' + 0''.4481\sin(\alpha' + 91^\circ\,6')\,\text{tg}\,\delta' + 0.01944\,[h'\,\text{tg}\,\delta'\cos\alpha' + i'\sec\delta'^2\sin\alpha']$$

The secular variation in declination is

$$-\,0''.1950\;\text{tg}\,\delta'\,\sin\alpha'^2 + 0''.4481\,\cos\,(\alpha' + 91^\circ\,6') - 0''.01944\,h'\,\sin\alpha'.$$

This, however, it was not necessary to compute, as existing catalogues furnished by interpolation sufficiently accurate values for our immediate purpose, which was to test the computed right-ascensions and declinations.

Table II. contains the comparison of the theoretical and observed positions; the observed positions having been corrected so as to correspond to the same proper motions as had been previously computed.

In certain cases it was necessary to form the "observed" positions by combining the results of several years' observations. For Henderson's and Le Verrier's AR. this was done by means of the annual variations derived from our computed places: for the Greenwich observations, by the annual variations given in the separate volumes. In either case care was taken to have the "mean epoch" adopted correspond very precisely to the mean of the times of observation.

The computed right-ascension and declination of a star for any time depend strictly each upon four elements (besides precession); upon the AR. and proper motion in AR. for some fixed epoch, and also the Decl. and P. M. in Decl. for the same epoch.

Let α, δ, be the star's right-ascension and declination at the time $t + 1855$, referred to the co-ordinate planes of 1855; α' and δ' being the right-ascension and declination for the same time, referred to its own co-ordinate-planes.

Then, if we vary the assumed AR. for 1855 by η and the assumed proper motion in AR. for n years by η'; the assumed Dec. for 1855 by ι, and the assumed proper motion in Dec. for n years by ι'; we shall have

$$\varDelta\alpha = \eta + \frac{\eta' t}{n}, \qquad \varDelta\delta = \iota + \frac{\iota' t}{n}.$$

[In our calculations we shall make $n = 30$.]

Again (*Fundamenta Astronomiae*, p. 304 (15)).

$$\varDelta\alpha' = \varDelta\alpha\,(\cos\theta + \sin\theta\, tg\,\delta' \cos(\alpha' + \lambda' - z')) + \frac{\varDelta\delta}{\cos\delta}\,\frac{\sin\theta\sin(\alpha' + \lambda' - z')}{\cos\delta'},$$

$$\varDelta\delta' = -\varDelta\alpha\sin\theta\sin(\alpha' + \lambda' - z') + \frac{\varDelta\delta}{\cos\delta}\cos\delta'\,(\cos\theta + \sin\theta\, tg\,\delta'\cos(\alpha' + \lambda' - z')).$$

The notation here used is that of the Fundamenta.

Let

$$\sin\theta\sin(\alpha' + \lambda' - z') = m\sin\xi,$$

$$\cos\theta\cos\delta' + \sin\theta\sin\delta'\cos(\alpha' + \lambda' - z') = m\cos\xi,$$

taking the quadrant of ξ, so that m shall be positive.

We shall readily find (*Fundamenta*, p. 300 (9))

$$\sin\delta = \cos\theta \sin\delta' - \sin\theta \cos\delta' \cos(\alpha' + \lambda' - z'),$$

$$\sin\delta^2 + m^2 = \sin\delta^2 + m^2 \sin\xi^2 + m^2 \cos\xi^2 = 1.$$

Hence,

$$m^2 = \cos\delta^2 \,;\ m \text{ (always positive)} = \cos\delta,$$

$$\sin\xi = \frac{\sin\theta \sin(\alpha' + \lambda' - z')}{\cos\delta},$$

and

$$\varDelta\alpha' \cos\delta' = \varDelta\alpha \cos\delta \cos\xi + \varDelta\delta \sin\xi,$$

$$\varDelta\delta' = -\varDelta\alpha \cos\delta \sin\xi + \varDelta\delta \cos\xi.$$

Hence, again,

$$\varDelta\alpha' \cos\delta' = \left(\eta + \frac{\eta' t}{30}\right) \cos\delta \cos\xi + \left(\iota + \frac{\iota' t}{30}\right) \sin\xi,$$

$$\varDelta\delta' = -\left(\eta + \frac{\eta' t}{30}\right) \cos\delta \sin\xi + \left(\iota + \frac{\iota' t}{30}\right) \cos\xi.$$

If η .η', ι ι' be of the nature of corrections to the assumed place, and $\varDelta\alpha' \cos\delta'$, $\varDelta\delta'$ be of an *opposite* character, — that is, be "computed — observed," — we must necessarily change the sign of one term; or, what is the same, transfer $\varDelta\alpha' \cos\delta'$, $\varDelta\delta'$ to the right members of their proper equations.

We have now the form of the equations of condition; perfectly rigorous. It is, however, obvious that in most cases perfect rigor is not necessary. For example, for stars near the equator no one would think of considering the declinations of 1755 as so erroneous that a perfectly accurate reduction of AR. to 1855 could not be effected by its means.

The rigorous course has here been followed for the four stars for which the value of sin ξ was the largest. These stars are λ Ursæ Minoris (Bode), 51 Cephei Hevelii, 4 Ursæ Minoris (Bode) and 20 Ursæ Minoris (Bode).

For the remaining stars of our polar list, a value of ι, ι' was deduced from the observed declinations, assuming that η, η' were $= 0$. Substituting these values in the equations derived from the right-ascensions observed, the values of η, η' were now obtained; employing in both cases the method of least squares, as will be further explained in its place.

These values of η, η' were now substituted in the declination equations, and a new set of values of ι, ι', differing by very small amounts from the former, were obtained.

The effect of these corrections was in no case over 0''.01, when substituted in the right-ascension equations; so that the former values of η, η' were retained.

This process is equivalent simply to this: the computed declinations derived from our final theory are virtually employed to reduce the observed right-ascensions to 1855.0, and the resulting right-ascensions treated by the method of least squares give the final values of right-ascension and proper motion.

It remains to explain the details of the application of the method of least squares. For the four stars treated rigorously we have

$$\frac{\eta}{15} = w, \qquad \frac{\eta'}{15} = x, \qquad \iota = y, \qquad \iota' = z.$$

The assumed weights are denoted by ω for right-ascension, by ω' for declination. These were assigned in a somewhat arbitrary manner; bearing in mind the magnifying power of the various instruments employed, and in one or two instances the less than usual care bestowed on observation and reduction. It is possible that the weight which has been assigned is in some cases too large or too small; thus Bradley's observations have received a weight which, though small in itself, is yet rather larger than the probable error as given in the Fundamenta would seem to warrant. This, however, if an error, is but a venial one, as the Fundamenta are justly considered as entitled to a very large proportional weight.

In all cases where many observations have been made at the same observatory and with the same instrument, less than the proportional weight has been given. The maximum weight has been put at 3, while the minimum for a single observation has been $\frac{1}{8}$.

We shall now make in AR.

$$a = \sqrt{\omega}\; 15 \cos \delta \cos \xi,$$

$$b = a \frac{t - 1855}{30} = \frac{a\, t}{30} \text{ [if } t \text{ be counted from 1855]},$$

$$c = \sqrt{\omega} \sin \xi,$$

$$d = c \frac{t - 1855}{30} = \frac{c\, t}{30} \text{ [if } t \text{ be counted from 1855]},$$

$$n = \sqrt{\omega}\; \Delta \alpha' \cos \delta' \text{ [taking } \Delta \alpha' \text{ in the sense } c - o].$$

And in declination

$$a = - \sqrt{\omega'}\; 15 \cos \delta \sin \xi,$$

$$b = \frac{a\,t}{30},$$

$$c = \sqrt{\omega'} \cos \xi,$$

$$d = \frac{c\,t}{30},$$

$$n = \sqrt{\omega'}\, \varDelta\, \delta';\ (c - o).$$

The equations of equal precision for these four stars will be of the form

$$0 = n + a\,w + b\,x + c\,y + d\,z,$$

whose solution by least squares is a mere matter of mechanical computation, according to Encke's form in the Jahrbuch for 1836.

For the polar stars excepting these four we shall put

$$x = \eta \cos \delta, \qquad x' = \iota,$$

$$y = \eta' \cos \delta, \qquad y' = \iota',$$

and the quantities x y will be eliminated from the declination equations, and x' y' from those in right-ascension by the process mentioned above. We have, then, making

$$a = \sqrt{\omega} \cos \xi, \text{ or } \sqrt{\omega'} \cos \xi, \qquad \varDelta\, n\ (\text{AR.}) = \sqrt{\omega} \left[\iota + \frac{\iota' t}{30}\right] \sin \xi,$$

$$b = a\,\frac{t}{30}, \qquad \varDelta\, n\ (\text{Dec.}) = -\sqrt{\omega'} \left[\eta + \frac{\eta' t}{30}\right] \sin \xi \cos \delta',$$

the following equations of equal precision in right-ascension

$$0 = n + \varDelta\, n + a\,x + b\,y = n' + a\,x + b\,y,$$

and in declination

$$0 = n + \varDelta\, n + a\,x' + b\,y' = n' + a\,x' + b\,y'.$$

The process of forming these is, — 1st, assume in declination $\varDelta\, n = 0$, and form and solve the declination-equations; 2d, substitute these values of x', y' or ι, ι' in the equation for $\varDelta\, n$ in AR., and form and solve the right-ascension equations. This gives the means of correcting the declination-equations. These last corrections were shown by trial to have no perceptible effect on the equations for AR., and the process of approximation ceased here.

Tables III. and IV. contain the formation and solution of the equations. In order to add new equations in future, we must, after determining the weights, compute for the epoch in question the theoretical AR. and Dec. derived from the numbers in Table I.

Comparing these with the new observations, as in Table II., the quantities $\Delta\alpha'$, $\Delta\alpha'\cos\delta'$, $\Delta\delta'$ will be readily formed. If the star be one of the four strictly treated, the equations can be at once formed, by the formulæ for that purpose. Then the quantities $a\,n$, $b\,n$, $c\,n$, $d\,n$, etc., as also $a\,a$, $a\,b$, $a\,c$, etc., being computed for each new equation, and added in the proper place; the final equations in Table III. Part I. being

$$0 = [a\,n] + [a\,a]\,w + [a\,s]\,x + [a\,c]\,y + [a\,d]\,t,$$
$$0 = [b\,n] + [a\,b]\,w + [b\,b]\,x + [b\,c]\,y + [b\,d]\,z,$$
$$\text{etc.,} \qquad \text{etc.,}$$

will give at once new final equations which can be solved.

For the 21 other stars, the corrections in declination must first be computed, according to the values of x' and y' given in Table III. Part II.; after this is done, the quantity $\Delta\,n$ for the new equations in AR. can be formed, and their solution will present no difficulty. It will in general make no difference whether the new values of x and y, or those already given in Table III. Part II. be used, to obtain Δn for declination. In either case there will be no difficulty in computing Δn for the new observations, and thus forming the new equations according to the precepts for the old.

An extreme case may occur, where it will be found that Δn for AR. would differ by a sensible amount when computed according to the new values of x' and y'. This, however, will not happen for a good many years.

It remains merely to explain Tables V., VI., which contain the calculations of AR. for the clock-stars. (Table VI., Supplement, contains the positions of the more northern stars not often used as clock-stars.)

The method here pursued was, to form positions for the years 1838 – 40, 1850, 1854 – 1858, and in some few cases for 1859, depending on the places of the Fundamenta for 1755, and the "Greenwich Results" for 1855. The positions of the Fundamenta were assumed as correct systematically, because the mean difference between the former reduction and Le Verrier's and Peters's new reductions of the 36 stars called "fundamental" is almost precisely zero; and so far as the casual errors of Bessel's places are concerned, it seemed possible to diminish their influence in great part by using comparatively late modern observations.

One of the objects of the observations now in progress at Cambridge, for the reduction of which these tables are designed, is to deduce new places of the fundamental stars, and to afford materials for a thorough examination of the relative systematic errors, whatever they may be, of the fundamental catalogue employed, so that for our purpose it is best to have a catalogue which is symmetrical as far as may be; which would not

2

be the case if we made a doubtful application of the method of least squares. The method here employed is the best I could devise for bringing down the relative errors of near stars.

The form of reduction proposed, and to some extent actually carried out, will enable the following things to be done with great ease: firstly, to reduce (if desired) the observations of each star so as to depend directly on the *Tabulae Reductionum*, and, secondly, to re-compute from our own observations the places of all the stars of the present list, so as to depend solely on our own observations, so far as our instrumental means will permit.

The positions derived from the Greenwich results for 1855, compared with Bessel's for 1755, will, assuming the correctness of the latter, require the addition of $x\,\frac{(t-1755)}{100}$, where x is a correction applicable to the places for 1855.

The positions given in the Edinburgh observations of 1838, 1839, 1840 were compared then with our calculated places, and would give in the mean the quantity

$$0.84\,x = \left(\frac{1839-1755}{100}\right)x\,,$$

after correcting them, to reduce to the equinox of Wolfers. See *Tabulae Reductionum*, p. xliii., at bottom, where we find

$$\text{N. C.} - \text{Henderson} = +\,0^{s}.038 \text{ (for 1830)}.$$

Reducing this to 1839, the mean epoch of the observations used

$$\text{N. C.} - \text{Henderson} = 0^{s}.043 \text{ (for 1835)}.$$

The number $+\,0^{s}.044$ was used instead; the change will be insensible.

The Edinburgh observations after 1840 were not used. The same catalogue had been employed to reduce them that was also employed at Greenwich for the same year, and the resulting numbers (it was feared) were not precisely homogeneous with those of the former years. The few observations made at Edinburgh on the new list of 47 stars introduced in 1857 into the Nautical Almanac were omitted. The Greenwich catalogue for 1850, depending, as is well known, upon the two instruments employed at Greenwich, — the old Transit and the new Transit-Circle, — was also compared. The Greenwich observations from 1854 to 1858 were compared; after the work was nearly complete, the volume for 1859 was received, and used for a few stars inadequately observed in the preceding years.

The Paris observations of 1856 and 1857 were also used. To my great regret, the volumes for 1858 and 1859 did not come to hand till after the completion of the computations. The corrections for equinox here used were

Gr. C. 1850	+ 0s.017,
Gr. Obs. 1854 — 1859	+ 0s.027,
Paris	+ 0s.056.

The mean of the corrected deviations of these four catalogues from the computed places, gives the correction x° for the mean t° of the dates. This was reduced to 1855 by the formula $x = x^\circ \frac{100}{t^\circ - 1755}$, the correction x was applied to the assumed places for 1855 (the Greenwich results 1855), and the right-ascensions of Table VI. formed. The annual variation given here is that which results from the combination of the mean annual variation derived from these right-ascensions compared with Bessel's Bradley, with the secular variation of the last column; which had been computed.

For the 47 stars given by Wolfers the positions are derived from the materials of his work, and are here computed for convenience' sake.

TABLE I.

Star's Name.	α°	h°	δ°	i°
43 Cephei Hevelii	12 24 6.64	+0.95	85 28 36.24	
Groombridge 527	36 47 36.39	+0.40	80 49 47.09	—0.10
323 Cephei (Bode)	49 48 50.00	+2.00	86 10 45.30	—0.05
Groombridge 750	58 6 11.20		85 9 51.60	
64 Camelopardali (Bode) . .	78 59 48.00		85 6 24.30	
51 Cephei Hevelii	97 46 45.00		87 15 9.00	
4 Ursæ Minoris (Bode) . .	105 55 0.00		89 1 49.00	
25 Camelopardali Hevelii . .	105 4 25.50		82 40 32.10	
156 Camelopardali (Bode) .	115 24 0.00		84 27 39.00	
1 Draconis Hevelii	139 0 30.00		81 57 38.50	
30 Camelopardali Hevelii . .	153 14 51.00		83 17 32.50	
202 Camelopardali (Bode) .	170 22 50.00	—0.75	81 55 29.10	
5 Ursæ Minoris (Bode) . .	183 0 50.42	+4.70	87 14 31.09	
6 Ursæ Minoris (Bode) . .	183 38 16.24	—1.75	88 30 5.72	+0.08
214 Camelopardali (Bode) .	200 12 12.50	—1.30	85 30 45.00	
20 Ursæ Minoris (Bode) . .	211 17 0.00		86 27 5.00	
57 Ursæ Minoris (Bode) . .	231 30 0.00		87 46 50.00	
62 Ursæ Minoris (Bode) . .	239 42 0.00		83 22 42.00	
ε Ursæ Minoris	255 14 30.66	+0.15	82 16 6.56	
λ Ursæ Minoris (Bode) . . .	302 9 38.49	—1.17	88 52 31.29	+0.01
74 Draconis	309 24 14.64	+0.40	80 34 28.12	+0.26
119 Cephei (Bode)	321 14 53.97	+0.06	83 38 36.08	—0.04
34 Cephei Hevelii	341 58 45.22		82 22 58.20	+0.05
39 Cephei Hevelii	351 55 24.42	+1.15	86 30 26.79	
309 Cephei (Bode)	358 11 50.56	+0.25	85 53 56.80	

TABLE II.

4 Ursæ Minoris (Bode). = Gr. 1119.
[sec δ' for 1815 = 61.2 ; for 1855 = 59.1.]

	$\Delta\alpha'$	$\Delta\alpha' \cos\delta'$	N	ω	$\Delta\delta'$	N	ω'
Gr. 1807					—0.″22	9	½
Str. 1815	+ 9.″35	+0.″15	15	1			
Schw. 1828 . . .	—22.85	—0.37	2	¼	+0.91	2	¼
Pond, 1830 . . .	[—120.55]		4		—0.36	10	½
Hend. 1839 . . .	+ 1.25	+0.02	10	1	—0.72	10	1
Rob. 1844	+12.15	+0.20	30	2	—0.44	11	1
Joh. 1849, 1850 . .	+ 6.95	+0.12	70	3	—0.43	43	3
Airy, 1857	+43.80	+0.73	5	½	—0.50	5	½
Car. 1855	+ 6.00	+0.10	27	1½	—0.50	27	1½

λ Ursæ Minoris (Bode).
[sec δ' for 1815 = 45.8 ; for 1855 = 51.0.]

	$\Delta\alpha'$	$\Delta\alpha' \cos\delta'$	N	ω	$\Delta\delta'$	N	ω'
Br. 1755	0.″00	0.″00	3	¼	0.″00	3	¼
Gr. 1807					+0.84	19	1
Str. 1815	+19.02	+0.42	4	½			
Schw. 1828 . . .	+17.48	+0.36	22	1	+0.01	12	1
Arg. 1830	+10.80	+0.22	17	1	+0.14	17	2
Pond, 1830 . . .	+10.80	+0.22	6	¼	+0.24	10	½
Airy, 1840 . . .	—14.11	—0.28	25	2	+0.03	77	3
Rob. 1843, 1852 . .	+17.75	+0.36	36	2	—0.42	5	1
Hend. 1840 . . .	— 5.05	—0.10	26	2	—0.01	132	3
Joh. 1843, 1847 . .	+ 5.43	+0.11	111	3	+0.42	158	3
Airy, 1844	—13.50	—0.27	15	1	+0.14	56	3
Airy, 1851	— 9.97	—0.20	49	2	—0.10	106	3
Airy, 1857	— 3.79	—0.07	63	3	+0.16	97	3
LeV. 1857 . . .	+11.79	+0.23	50	3	—0.25	59	3

6 Ursæ Minoris (Bode). = Gr. 1884.
[sec δ' for 1815 = 45 ; for 1855 = 38.4.]

	$\Delta\alpha'$	$\Delta\alpha' \cos\delta'$	N	ω	$\Delta\delta'$	N	ω'
Br. 1752	0.″00	0.″00	4	¼	0.″00	4	¼
Gr. 1807					—0.47	6	½
Str. 1815	+18.55	+0.41	19	1			
Bessel, 1815 . . .					+0.02	24	1
Schw. 1828 . . .	+24.04	+0.56	11	¾	+0.26	10	¾
Arg. 1830	+24.84	+0.58	16	1	+0.44	10	1
Hend. 1839 . . .	+42.09	+1.03	10	1	—0.26	10	1
Airy, 1840	—23.98	—0.56	5	½			
Rob. 1839	+26.02	+0.64	59	3	—0.22	18	1½
Airy, 1842	+ 8.53	+0.21	4	½	+0.91	7	1
Joh. 1848	—11.54	—0.29	63	3	+0.48	67	3
Airy, 1851	— 8.28	—0.21	3	½	—0.74	3	½
Car. 1855	—10.26	—0.27	15	1	+0.52	15	1

57 φ Ursæ Minoris (Bode). = Gr. 2283.

[sec δ′ for 1815 = 27.4; for 1855 = 25.8.]

	Δα′	Δα′ cos δ′	N	ω	Δδ′	N	ω′
Gr. 1808					+1″.58	8	½
Str. 1815	+11″.17	+0″.41	13	1			
Schw. 1828 . . .	+ 8.89	+0.33	5	½	+0.85	5	½
Hend. 1839 . . .	+13.18	+0.50	10	1	+0.14	10	1
Joh. 1850	—19.13	—0.73	47	3	—0.54	11	1
Airy, 1851	+ 6.75	+0.26	5	½	+0.40	4	½
Car. 1855	+ 4.50	+0.17	12	1	—0.60	12	1

5 Ursæ Minoris (Bode). = Gr. 1871.

[sec δ′ for 1815 = 22.6; for 1855 = 20.8.]

	Δα′	Δα′ cos δ′	N	ω	Δδ′	N	ω′
Br. 1751	0″.00	0″.00	2	⅕			
Gr. 1807					—1″.49	7	½
Str. 1815	— 8.37	—0.37	17	1			
Schw. 1828* . . .	+ 4.57	+0.21	2	¼	—0.78	2	¼
Arg. 1830	+10.25	+0.47	19	1½	+0.61	19	2
Airy, 1840	+ 1.12	+0.05	6	½	—0.08	12	1
Joh. 1850	+ 1.15	+0.05	28	2	—0.43	8	1
Airy, 1851	+ 6.08	+0.29	4	½	—0.04	5	½
Car. 1855	+17.92	+0.86	5	½	—0.41	5	½

51 Cephei Hevelii.

[sec δ′ for 1815 = 21.0; for 1855 = 20.9.]

	Δα′	Δα′ cos δ′	N	ω	Δδ′	N	ω′
Gr. 1807					—1″.48	11	½
Str. 1815	— 6″.57	—0″.31	14	1			
Schw. 1828 . . .	— 7.24	—0.35	17	1	—0.63	14	1
Pond, 1830† . . .	— 2.37	—0.11	23	1	+0.70	20	1
Hend. 1842 . . .	+ 6.20	+0.30	29	2	—0.31	108	3
Airy, 1840	+ 7.67	+0.37	17	1	+0.49	78	3
Airy, 1843	+16.47	+0.78	33	2	—0.30	40	2
Joh. 1847	— 0.42	—0.02	157	3	—0.37	122	3
Airy, 1850	+ 8.08	+0.39	50	2	+0.28	66	2½
Airy, 1857	+10.67	+0.51	100	3	+0.56	185	3
LeV. 1857 . . .	—18.54	—0.88	3	¼	+1.46	1	
Car. 1855	+ 5.70	+0.27	28	2	+0.28	28	2

20 Ursæ Minoris (Bode). = Gr. 2099.

[sec δ′ for 1815 = 17.0; for 1855 = 16.1.]

	Δα′	Δα′ cos δ′	N	ω	Δδ′	N	ω′
Gr.					+0″.72	5	½
Str. 1815	—10″.54	—0″.62	12	1			
Schw. 1828* . . .	— 8.33	—0.50	3	⅓	+1.62	3	⅓
Joh. 1850, 1852 . .	— 2.43	—0.15	20	2	—0.21	7	¾
Car. 1855	+25.50	+1.59	4	½	—2.20	4	½

* See Carrington.

† Declination derived from observations of 1833 alone. Those of 1831, which differ in the mean 3″.11 from those of 1833, are excluded.

39 Cephei Hevelii.

[sec δ' for 1815 = 15.4 ; for 1855 = 16.4.]

	$\varDelta\alpha'$	$\varDelta\alpha' \cos\delta'$	N	ω	$\varDelta\delta'$	N	ω'
Br. 1755	0″.00	0″.00	4	¼	0″.00	5	¼
Gr. 1808					+1.71	6	½
Str. 1815	+21.53	+1.40	3	⅕			
Schw. 1828 . . .	+ 1.48	+0.09	7	½	+0.12	6	½
Hend. 1839 . . .	+ 4.24	+0.26	11	1	—0.19	10	1
Airy, 1840	—12.68	—0.79	5	½			
Rob. 1839, 1842 . .	+18.49	+1.16	13	1	+0.55	5	½
Airy, 1842	— 0.04	0.00	5	½	—0.55	5	½
Joh. 1850	+ 0.01	0.00	38	2½	+0.95	31	2
Airy, 1851, 1852 .	— 6.53	—0.40	5	½	+0.71	10	1
Car. 1855	— 3.08	—0.19	21	2	—0.81	21	½

323 Cephei (Bode). = Gr. 642.

[sec δ' for 1815 = 14.4 ; for 1855 = 15.0.]

	$\varDelta\alpha'$	$\varDelta\alpha' \cos\delta'$	N	ω	$\varDelta\delta'$	N	ω'
Gr. 1807					+0″.23	6	½
Str. 1815	+25″.68	+1″.78	19	1			
Schw. 1828	— 0.54	—0.04	8	½	—0.48	7	¾
Airy, 1840							
Hend. 1839 . . .	+ 2.71	+0.18	10	1	—0.65	10	1
Airy, 1845	— 6.54	—0.46	3	¼	—0.43	5	½
Joh. 1845, 1850 . .	+ 8.39	+0.57	50	3	+0.39	14	1¼
Airy, 1851	+ 4.35	+0.29	5	½	+0.34	4	½
Car. 1855	+ 1.00	+0.07	5	½	—0.50	5	½

43 Cephei Hevelii = 2 Ursæ Minoris. = Gr. 177.

[sec δ' for 1815 = 12.1 ; for 1855 = 12.7.]

	$\varDelta\alpha'$	$\varDelta\alpha' \cos\delta'$	N	ω	$\varDelta\delta'$	N	ω'
Br. 1755	0″.00	0″.00	8	⅜	0″.00	5	⅓
Gr. 1807					+0.24	7	½
Str. 1815	+ 5.28	+0.42	6	⅝			
Bess. 1815					—0.45	28	1
Schw. 1828 . . .	+ 1.28	+0.10	3	¼	—0.66	2	¼
Pond, 1830 . . .	— 3.78	—0.31	10	½	+0.52	12	⅔
Airy, 1837	— 5.98	—0.48	5	½			
Hend. 1839 . . .	— 2.83	—0.23	7	¾	+0.01	16	1½
Rob. 1837, 1852 . .	+ 2.37	+0.19	8	¾	+1.57	2	½
Airy, 1845					+0.66	7	¾
Joh. 1849, 1851 . .	— 1.60	—0.12	24	2	+0.65	12	1
Airy, 1851	—12.01	—0.95	4	½	—0.24	4	½
Car. 1855	— 1.86	—0.15	3	¼	—0.26	3	¼

309 Cephei (Bode). = Gr. 4193.

[sec δ' for 1815 = 13.3 ; for 1855 = 14.0.]

	$\Delta\alpha'$	$\Delta\alpha' \cos\delta'$	N	ω	$\Delta\delta'$	N	ω'
Br. 1755	0″.00	0″.00	2	$\frac{1}{8}$			
Gr. 1808					—0″.34	6	$\frac{1}{2}$
Str. 1815	+12.20	+0.91	3	$\frac{1}{4}$			
Bess. 1815					—1.21	27	1
Schw. 1828 . . .	+13.45	+1.00	2	$\frac{1}{4}$	+0.27	2	$\frac{1}{4}$
Rob. 1844, 1846 . .	— 0.87	—0.06	5	$\frac{1}{2}$	+1.10	5	$\frac{1}{2}$
Joh. 1849	— 8.88	—0.64	29	2	+0.44	18	$1\frac{1}{2}$
Airy, 1851	—11.62	—0.84	8	1	—0.03	10	1
Car. 1855	—12.94	—0.92	4	$\frac{1}{2}$	—1.80	4	$\frac{1}{2}$

214 Camelopardali (Bode). = Gr. 2007.

[sec δ' for 1815 = 13.4 ; for 1855 = 12.8.]

	$\Delta\alpha'$	$\Delta\alpha' \cos\delta'$	N	ω	$\Delta\delta'$	N	ω'
Gr. 1808					+0″.86	6	$\frac{1}{2}$
Str. 1815	+3″.51	+0″.26	4	$\frac{1}{2}$			
Schw. 1828 . . .	—1.17	—0.09	2	$\frac{1}{4}$	—1.32	2	$\frac{1}{4}$
Hend. 1839 . . .	+3.63	+0.28	10	1	—0.42	10	1
Airy, 1845 . . .					—0.59	5	$\frac{1}{2}$
Joh. 1849, 1850 . .	—3.16	—0.24	29	$2\frac{1}{2}$	+1.40	22	2
Airy, 1850, 1851 .	—1.51	—0.12	8	$\frac{3}{4}$	—0.62	7	$\frac{3}{4}$
Car. 1855	+1.00	+0.08	3	$\frac{1}{4}$	—1.80	3	$\frac{1}{4}$

Gr. 750.

[sec δ' for 1815 = 11.6 ; for 1855 =11.9.]

	$\Delta\alpha'$	$\Delta\alpha' \cos\delta'$	N	ω	$\Delta\delta'$	N	ω'
Gr. 1808					+1″.66	6	$\frac{1}{2}$
Str. 1815	+3″.12	+0″.28	5	$\frac{1}{2}$			
Schw. 1828 . . .	+1.75	+0.14	2	$\frac{1}{4}$	—0.62	2	$\frac{1}{4}$
Pond. 1830 . . .	—1.67	—0.14	11	$\frac{3}{4}$	+1.06	13	$\frac{3}{4}$
Hend. 1839 . . .	—2.52	—0.21	10	1	—0.47	10	1
Joh. 1848, 1852 . .	+3.00	+0.25	43	3	—0.59	11	1
Airy, 1857	—3.25	—0.27	5	$\frac{1}{2}$	+0.08	7	1
Car. 1855	—5.30	—0.45	16	$1\frac{1}{2}$	—0.90	16	1

156 Camelopardali (Bode). = Gr. 1359.

[sec δ' for 1815 = 10.5 ; for 1855 = 10.4.]

	$\Delta\alpha'$	$\Delta\alpha' \cos\delta'$	N	ω	$\Delta\delta'$	N	ω'
Gr. 1808					+0″.96	7	$\frac{1}{2}$
Str. 1815	+ 3″.57	+0″.34	3	$\frac{1}{4}$			
Schw. 1828 . . .	+ 4.05	+0.39	7	$\frac{3}{4}$	+0.44	7	$\frac{3}{4}$
Hend. 1844 . . .	+14.26	+1.36	5	$\frac{3}{4}$			
Joh. 1850, 1848 . .	+ 1.54	+0.15	38	$2\frac{1}{2}$	+0.50	23	2
Airy, 1844	+ 7.09	+0.68	4	$\frac{1}{2}$	—0.11	3	$\frac{1}{4}$
Carr. 1855	—10.50	—1.01	3	$\frac{1}{4}$	—0.40	3	$\frac{1}{4}$

64 Camelopardali (Bode). = Gr. 944.
[sec δ′ for 1815 = 11.6 ; for 1855 = 11.7.]

	Δα′	Δα′ cos δ′	N	ω	Δδ′	N	ω′
Gr. 1807					+0″.14	6	½
Str. 1815	+14″.75	+1″.27	8	¾			
Schw. 1828 . . .	+ 0.53	+0.05	5	½	+1.02	5	½
Hend. 1839 . . .	+ 3.90	+0.33	10	1	—0.58	11	1
Rob. 1850, 1851 . .	— 5.26	—0.47	2	¼	+1.66	5	½
Joh. 1848, 1846 . .	— 1.12	—0.10	33	2½	—0.09	18	1½
Airy, 1844	+ 5.18	+0.44	4	½	+0.02	5	½
Airy, 1851	+ 4.95	+0.42	3	½	—0.14	3	½
Car. 1855	0.00	0.00	14	1¼	—0.70	14	1

119 Cephei (Bode).*
[sec δ′ for 1815 = 8.8 ; for 1855 = 9.0.]

	Δα′	Δα′ cos δ′	N	ω	Δδ′	N	ω′
Br. 1755	0″.00	0″.00	2	⅛	0″.00	1	⅛
Gr. 1808					+2.59	7	½
Str. 1815	+10.42	+1.18	1	⅛			
Schw. 1828 . . .	+ 5.58	+0.63	2	¼	+0.57	2	¼
Rob. 1846, 1849 . .	— 7.22	—0.81	5	½	+0.13	5	½
Joh. 1851, 1848 . .	— 1.31	—0.15	5	½	+0.65	3	½
Car. 1855	— 6.03	—0.67	3	¼	—1.52	3	¼

2315 Groombr. = 62 Ursæ Minoris (B).
[sec δ′ for 1815 = 8.8 ; for 1855 = 8.7.]

	Δα′	Δα′ cos δ′	N	ω	Δδ′	N	ω′
Gr. 1807					+0″.74	6	½
Str. 1815	+ 1″.64	+0″.19	6	½			
Schw. 1828 . . .	+13.48	+1.55	2	¼	—0.31	2	¼
Airy, 1847					+1.43	2	½
Joh. 1851, 1850 . .	— 3.18	—0.37	4	½	+0.68	6	½
Airy, 1850	— 4.75	—0.55	2	¼			
Car. 1855	— 6.00	—0.69	3	¼	—1.10	3	¼

30 Camelopardali Hevelii. = Gr. 1633 = P. X. 22.
[sec δ′ for 1855 = 8.8 ; for 1855 = 8.6.]

	Δα′	Δα′ cos δ′	N	ω	Δδ′	N	ω′
Gr. 1807					+0″.34	6	½
Str. 1815	—13″.31	—1″.51	2	¼			
Schw. 1828 . . .	+ 8.89	+1.01	4	½	+0.83	4	½
Rob. 1844 1854 . .	+15.86	+1.83	5	½	+0.85	5	½
Joh. 1845	+ 8.45	+0.98	7	¾	+0.52	6	¾
Airy, 1847					—0.64	9	1
Airy, 1850	+13.24	+1.54	9	1	+0.17	20	2
Car. 1855	+18.00	+2.09	3	¼	—1.10	3	¼

* 13″.2, or one division, has been subtracted from the declination in the *Fundamenta Astronomiae.* The correction seems necessary.

25 Camelopardali Hevelii. = Gr. 1259.

[sec δ′ for 1815 = 7.9 ; for 1855 = 7.84.]

	$\Delta\alpha'$	$\Delta\alpha' \cos\delta'$	N	ω	$\Delta\delta'$	N	ω'
Gr. 1808					—0″.54	6	½
Str. 1815	— 0″.59	—0″.07	3	¼			
Str. 1824	+ 0.31	+0.04	14	1½	—0.21	14	1
Schw. 1828 . . .	— 0.24	—0.03	19	1¼	+0.08	17	1½
Pond, 1830 . . .	—10.65	—1.36	10	¾	+0.10	11	¾
Airy, 1840	— 0.80	—0.10	5	½	—0.18	16	1½
Rob. 1831, 1852 . .	— 7.85	—1.00	2	¼	+2.28	5	½
Airy, 1844	+ 5.19	+0.66	4	½	—0.13	3	¼
Joh. 1843 . . .	— 9.52	—1.22	5	½	+0.98	7	¾
Airy, 1849					—0.63	5	½
Car. 1855	—12.00	—1.53	3	¼	0.00	3	¼
Airy, 1858					+1.41	2	¼

34 Cephei Hevelii.

[sec δ′ for 1815 = 7.34 ; for 1855 = 7.55.]

	$\Delta\alpha'$	$\Delta\alpha' \cos\delta'$	N	ω	$\Delta\delta'$	N	ω'
Br. 1755	0″.00	0″.00	1	⅛	0″.00	3	¼
Gr. 1808					+0.53	6	½
Str. 1815	+14.68	+2.00	1	¼			
Schw. 1828 . . .	+ 5.52	+0.74	9	⅚	—0.26	8	¾
Rob. 1842, 1843 . .	— 1.51	—0.20	10	1	—0.04	10	1
Airy, 1839	— 6.01	—0.81	6	½			
Airy, 1846					—0.36	3	⅖
Joh. 1847, 1844 .	+ 1.34	+0.18	4	½	+0.97	6	½
Car. 1855	— 5.78	—0.77	3	¼	—1.50	3	¼

ε Ursæ Minoris.

[sec δ′ for 1815 = 7.49 ; for 1855 = 7.43.]

	$\Delta\alpha'$	$\Delta\alpha' \cos\delta'$	N	ω	$\Delta\delta'$	N	ω'
Br. 1755	0″.00	0″.00	5	⅓	0″.00	5	⅓
Gr. 1807					+0.36	16	1
Str. 1815	+ 2.71	+0.36	38	2			
Str. 1824	— 1.71	—0.23	15	2	—0.19	15	2
Schw. 1828 . . .	+ 2.58	+0.35	29	1	—0.32	18	1
Pond, 1830 . . .	+ 1.62	+0.22	9	½	—0.19	12	½
Hend. 1841 . . .	— 0.87	—0.12	19	1½	+0.11	99	3
Airy, 1840	— 4.84	—0.65	28	2	—0.21	102	3
Rob. 1828, 1836 . .	+18.78	+2.50	2	¼	—0.11	4	½
Peters, 1840 . . .	— 4.09	—0.55		3			
Airy, 1845	— 8.19	—1.10	22	1½	+0.09	84	3
Joh. 1842, 1843 .	+ 1.19	+0.16	19	1½	—0.11	24	3
Airy, 1851	— 3.96	—0.53	40	2½	—0.11	64	3
Car. 1855	— 0.84	—0.11	3	¼	—1.14	3	¼
Airy, 1857	+ 1.16	+0.16	45	3	+0.14	72	3
LeV. 1857 . . .	+ 5.43	+0.73	16	2	—0.28	38	2

202 Camelopardali (Bode).

[sec δ′ for 1815 = 7.3 ; for 1855 = 7.1.]

	Δα′	Δα′ cos δ′	N	ω	Δδ′	N	ω′
Gr. 1807					+1″.00	6	½
Str. 1815	+0″.41	+0″.06	3	⅓			
Schw. 1828 . . .	+5.13	+0.70	7	⅔	+0.05	5	½
Joh. 1844, 1846 .	+3.53	+0.50	5	½	+0.78	5	½
Car. 1855	—1.00	—0.14	3	¼	—1.00	3	¼

1 Draconis Hevelii. = Gr. 1537.*

[sec δ′ for 1815 = 7.3 ; for 1855 = 6.95.]

	Δα′	Δα′ cos δ′	N	ω	Δδ′	N	ω′
Gr. 1810					+0″.38	6	½
Str. 1815	— 1″.90	—0″.26	6	¾			
Schw. 1828 . . .	— 0.26	—0.04	57	3	+0.18	46	3
Pond, 1830 . . .	— 6.50	—0.91	14	¾	+0.57	20	1
Peters, 1840 . . .	— 0.53	—0.08	16	2			
Airy, 1840	— 4.88	—0.69	5	½	+0.16	20	2
Airy, 1845	— 7.13	—1.01	6	½	+0.55	7	¾
Joh. 1845, 1850 . .	— 2.33	—0.33	5	½	+0.24	17	1½
Airy, 1851	— 0.23	—0.03	7	¾	+0.09	1	¼
Car. 1855	—10.50	—1.51	3	¼	+0.30	3	¼

Gr. 527.

[sec δ′ for 1815 = 6.15 ; for 1855 = 6.26.]

	Δα′	Δα′ cos δ′	N	ω	Δδ′	N	ω′
Br. 1755	0″.00	0″.00	1	⅛			
Gr. 1807					+0″.56	6	½
Str. 1815	— 3.48	—0.57	11	1			
Schw. 1828 . . .	+ 1.38	+0.23	7	¾	+0.14	7	7/10
Rob. 1853, 1850 . .	— 0.82	—0.13	3	⅜	+0.42	5	½
Airy, 1841					—1.35	2	3/10
Airy, 1847	+12.56	+2.01	2	¼			
Joh. 1844, 1846 . .	— 0.60	—0.10	3	¼	—0.31	5	½
Airy, 1849, 1850 .	+ 4.17	+0.67	2	¼	—0.27	9	1

74 Draconis.

[sec δ′ for 1815 = 6.02 ; for 1855 = 6.11.]

	Δα′	Δα′ cos δ′	N	ω	Δδ′	N	ω′
Br. 1755	0″.00	0″.00	2	¼	0″.00	4	3/10
Gr. 1808					+0.99	6	½
Str.† 1815	[+19.59]	+0.76	1				
Bess. 1818	+ 2.41	+0.40	2	⅓			
Schw. 1828 . . .	+13.26	+2.19	3	¼	+2.12	3	3/10
Arg. 1830	+ 7.06	+1.17	9	1	+1.76	9	1
Rob. 1853, 1842 . .	+ 2.33	+0.38	1	⅛	+3.58	7	⅘
Joh. 1850, 1846 . .	+11.71	+1.91	5	½	+3.50	7	⅗

* I have not been able to verify Robinson's declination from his observations.

† Probably one second of time in error (see Argelander).

TABLE III. Part 1.*

4 Ursæ Minoris (Bode) = Gr. 1119.

$0 = +0.2970 + 0.5950\,w - 0.2076\,x - 0.0284\,y + 0.0181\,z$	$w = -0^{s}.683$
$0 = -0.0520 - 0.2076\,w + 0.1674\,x + 0.0180\,y - 0.0107\,z$	$x = -0^{s}.576$
$0 = -3.5740 - 0.0284\,w + 0.0180\,x + 8.2660\,y - 2.7550\,z$	$y = +0''.515$
$0 = +0.8430 + 0.0181\,w - 0.0107\,x - 2.7550\,y + 2.3150\,z$	$z = +0''.252$

λ Ursæ Minoris (Bode).

$0 = +0.2000 + 1.8200\,w - 0.6210\,x - 0.0770\,y + 0.0660\,z$	$w = -0^{s}.070$
$0 = -0.1070 - 0.6210\,w + 0.6027\,x + 0.0650\,y - 0.0880\,z$	$x = +0^{s}.144$
$0 = +1.9550 - 0.0770\,w + 0.0650\,x + 24.4830\,y - 8.6580\,z$	$y = +0''.017$
$0 = -2.1960 + 0.0660\,w - 0.0880\,x - 8.6580\,y + 8.5800\,z$	$z = +0''.276$

57 φ Ursæ Minoris (Bode) = Gr. 2283.

$0 = -0.5390 + 2.3600\,w - 0.9560\,x + 0.0440\,y - 0.0260\,z$	$w = +0^{s}.642$
$0 = -0.2750 - 0.9560\,w + 0.8580\,x - 0.0260\,y + 0.0270\,z$	$x = +1^{s}.007$
$0 = +0.4250 + 0.0440\,w - 0.0260\,x + 4.5040\,y - 2.0130\,z$	$y = +0''.551$
$0 = -1.6060 - 0.0260\,w + 0.0270\,x - 2.0130\,y + 1.8700\,z$	$z = +1''.446$

51 Cephei Hevelii.

$0 = +2.6900 + 9.3030\,w - 3.1160\,x + 0.0410\,y - 0.0010\,z$	$w = -0^{s}.595$
$0 = +0.2230 - 3.1160\,w + 2.2790\,x - 0.0040\,y - 0.0130\,z$	$x = -0^{s}.916$
$0 = +1.1060 + 0.0410\,w - 0.0040\,x + 21.0020\,y - 6.9660\,z$	$y = -0''.295$
$0 = +1.2060 - 0.0010\,w - 0.0130\,x - 6.9660\,y + 4.4620\,z$	$z = -0''.734$

TABLE III. Part 2.†

6 Ursæ Minoris B.

Equation	Solution	Equation	Solution
$0 = +2.99 + 12.50\,x - 7.02\,y$	$x = -0.089$	$0 = +2.34 + 11.50\,x' - 7.20\,y'$	$x' = -0.324$
$0 = -2.63 - 7.02\,x + 7.51\,y$	$y = +0.268$	$0 = -0.72 - 7.20\,x' + 8.32\,y'$	$y' = -0.194$

5 Ursæ Minoris B.

Equation	Solution	Equation	Solution
$0 = +1.09 + 6.45\,x - 4.14\,y$	$x = -0.285$	$0 = -0.45 + 5.75\,x' - 3.41\,y'$	$x' = -0.007$
$0 = -0.20 - 4.14\,x + 5.43\,y$	$y = -0.181$	$0 = +0.41 - 3.41\,x' + 3.00\,y'$	$y' = -0.144$

39 Cephei Hevelii.

Equation	Solution	Equation	Solution
$0 = +0.83 + 9.00\,x - 3.63\,y$	$x = +0.004$	$0 = +2.12 + 7.75\,x' - 3.46\,y'$	$x' = -0.176$
$0 = -1.04 - 3.63\,x + 4.48\,y$	$y = +0.234$	$0 = -1.66 - 3.46\,x' + 4.82\,y'$	$y' = +0.219$

20 Ursæ Minoris B.

Equation	Solution	Equation	Solution
$0 = -0.32 + 3.83\,x - 1.64\,y$	$x = -0.233$	$0 = -0.36 + 2.08\,x' - 1.16\,y'$	$x' = +1.05$
$0 = +0.98 - 1.64\,x + 1.83\,y$	$y = -0.74$	$0 = -1.01 - 1.16\,x' + 1.41\,y'$	$y' = +1.58$

323 Cephei B.

Equation	Solution	Equation	Solution
$0 = +3.73 + 6.75\,x - 2.97\,y$	$x = -0.174$	$0 = -0.63 + 5.00\,x' - 2.63\,y'$	$x' = +0.146$
$0 = -2.73 - 2.97\,x + 2.58\,y$	$y = +0.86$	$0 = +0.29 - 2.63\,x' + 2.34\,y'$	$y' = +0.039$

309 Cephei B.

Equation	Solution	Equation	Solution
$0 = -2.14 + 4.62\,x - 1.70\,y$	$x = +0.68$	$0 = -1.03 + 5.25\,x' - 2.92\,y'$	$x' = -0.116$
$0 = -0.12 - 1.70\,x + 2.15\,y$	$y = +0.59$	$0 = +1.53 - 2.92\,x' + 3.33\,y'$	$y' = -0.561$

* In this portion of Table III. w denotes the correction in *time* of the assumed AR., and x 30 times the correction in time of the assumed annual proper motion; y in like manner denotes the correction in space of the assumed declination, and z 30 times the correction in space of its assumed annual proper motion; — all assumed values holding good for 1855.

† In this portion of Table III. x denotes the correction of the assumed right ascension for 1855, expressed in seconds of space and multiplied by the cosine of the declination; y denotes the correction of the assumed proper motion (for 1855) in AR., multiplied by 30 and by the cosine of the declination, — likewise, in space; x' and y' denote the same quantities as y and z in the first part of this table.

214 Camelopardali B.			
$0 = -0.29 + 5.25x - 1.91y$	$x = +0.202$	$0 = +1.27 + 5.25x' - 2.24y'$	$x' = -0.238$
$0 = -0.20 - 1.91x + 1.45y$	$y = +0.404$	$0 = -0.55 - 2.24x' + 1.85y'$	$y' = +0.011$

43 Cephei Hevelii. (2 Ursæ Minoris.)			
$0 = -0.92 + 6.50x - 4.22y$	$x = +0.246$	$0 = +0.57 + 8.00x' - 6.24y'$	$x' = -0.27$
$0 = -0.01 - 4.22x + 6.53y$	$y = +0.160$	$0 = +0.53 - 6.24x' + 8.73y'$	$y' = -0.25$

Gr. 750.			
$0 = -0.20 + 7.50x - 2.02y$	$x = +0.085$	$0 = -0.41 + 5.50x' - 2.32y'$	$x' = +0.585$
$0 = -0.25 - 2.02x + 1.94y$	$y = +0.215$	$0 = -1.43 - 2.32x' + 2.29y'$	$y' = +1.203$

64 Camelopardali B.			
$0 = +1.38 + 7.25x - 2.83y$	$x = +0.139$	$0 = -0.13 + 6.00x' - 2.52y'$	$x' = +0.157$
$0 = -1.48 - 2.83x + 2.21y$	$y = +0.847$	$0 = -0.29 - 2.52x' + 2.14y'$	$y' = +0.319$

156 Camelopardali B.			
$0 = +1.83 + 5.00x - 1.83y$	$x = -0.21$	$0 = +1.65 + 3.75x' - 2.06y'$	$x' = -0.22$
$0 = -0.91 - 1.83x + 1.26y$	$y = +0.41$	$0 = -1.26 - 2.06x' + 2.01y'$	$y' = +0.40$

30 Camelopardali H.			
$0 = +3.68 + 3.25x - 1.37y$	$x = -1.92$	$0 = +0.82 + 5.50x' - 2.10y'$	$x' = -0.01$
$0 = -0.75 - 1.37x + 1.01y$	$y = -1.86$	$0 = -0.69 - 2.10x' + 1.86y'$	$y' = +0.36$

119 Cephei B.			
$0 = -0.37 + 1.75x - 1.02y$	$x = +0.38$	$0 = +1.44 + 2.12x' - 1.64y'$	$x' = -0.13$
$0 = -0.13 - 1.02x + 1.86y$	$y = +0.28$	$0 = -2.25 - 1.64x' + 2.88y'$	$y' = +0.71$

62 Ursæ Minoris B.			
$0 = -0.03 + 1.75x - 0.99y$	$x = +0.45$	$0 = +1.08 + 2.00x' - 1.23y'$	$x' = -0.43$
$0 = -0.40 - 0.99x + 1.11y$	$y = +0.76$	$0 = -0.79 - 1.23x' + 1.50y'$	$y' = +0.18$

25 Camelopardali H.			
$0 = -1.96 + 5.75x - 4.46y$	$x = +0.86$	$0 = +1.23 + 7.75x' - 5.09y'$	$x' = -0.77$
$0 = +1.14 - 4.46x + 4.02y$	$y = +0.67$	$0 = +0.45 - 5.09x' + 4.70y'$	$y' = -0.93$

34 Cephei H.			
$0 = +0.46 + 3.37x - 2.20y$	$x = +0.09$	$0 = +0.03 + 3.65x' - 3.00y'$	$x' = +0.10$
$0 = -0.73 - 2.20x + 2.67y$	$y = +0.34$	$0 = -0.33 - 3.00x' + 4.80y'$	$y' = +0.13$

ε Ursæ Minoris.			
$0 = -2.91 + 23.33x - 11.69y$	$x = +0.16$	$0 = -1.60 + 25.58x' - 11.49y'$	$x' = +0.08$
$0 = +0.91 - 11.69x + 12.80y$	$y = +0.08$	$0 = +0.47 - 11.49x' + 12.00y'$	$y' = +0.04$

202 Camelopardali B.			
$0 = +0.72 + 1.75x - 1.22y$	$x = -0.29$	$0 = +0.66 + 1.75x' - 1.39y'$	$x' = +0.14$
$0 = -0.56 - 1.22x + 1.20y$	$y = +0.17$	$0 = -0.93 - 1.39x' + 1.71y'$	$y' = +0.66$

1 Draconis H.			
$0 = -2.45 + 9.00x - 6.01y$	$x = +0.48$	$0 = -0.28 + 9.25x' - 5.88y'$	$x' = -0.30$
$0 = +1.39 - 6.01x + 5.03y$	$y = +0.29$	$0 = +0.18 - 5.88x' + 5.02y'$	$y' = 0.00$

Gr. 527.			
$0 = +0.19 + 3.00x - 2.64y$	$x = -0.45$	$0 = -0.24 + 3.50x' - 1.96y'$	$x' = +0.35$
$0 = +0.47 - 2.64x + 3.77y$	$y = -0.44$	$0 = -0.31 - 1.96x' + 1.98y'$	$y' = +0.51$

74 Draconis.			
$0 = +2.99 + 2.62x - 2.64y$	$x = -1.71$	$0 = +7.75 + 3.50x' - 3.41y'$	$x' = -3.42$
$0 = -1.86 - 2.64x + 4.72y$	$y = -0.56$	$0 = -4.71 - 3.41x' + 5.61y'$	$y' = -1.24$

TABLE IV.

	AR.	log a	log b	log c	log d	μ	ν	Dec.	μ'	Mag.
	5 Ursæ Minoris B.									
	h m s					s	s	° ′ ″	″	
1855	12 12 34.30	0.1904	0.1404n	0.1409n	8.8806n	+0.306	+0.001	87 14 31.08	—0.005	6
1865	12 52.79	0.1886	0.1317n	0.1322n	8.8824n	+0.300	+0.001	11 10.79		
	6 Ursæ Minoris B.									
1855	12 14 21.19	*	0.4061n	0.4062n	9.2035n	—0.094	—0.007	88 30 13.40	+0.074	6
1865	14 19.56		0.3903n	0.3904n	9.1869n	—0.091	—0.006	26 53.97		
	43 Cephei Hevelii.									
1855	0 49 42.98	0.8267	9.9154	9.9167	9.2599	+0.068	—0.006	85 28 35.97	—0.001	4.5
1865	50 51.40	0.8349	9.9202	9.9215	9.2749	+0.068	—0.006	31 51.70		
	Polaris † (from Wolfers).									
1860	1 8 2.65							88 33 47.12		2
1865	9 38.10							35 22.91		
1870	11 16.92							36 58.48		
	214 Camelopardali B.									
1855	13 20 40.34	0.4474n	9.9017n	9.9030n	9.4680n	—0.075	—0.005	85 30 44.76	0.000	7.8
1865	20 12.08	0.4315n	9.8969n	9.8982n	9.4605n	—0.074	—0.005	27 36.43		
	20 Ursæ Minoris B. = Gr. 2099.									
1855	14 5 7.75	0.9098n	9.9632n	9.9641n	9.7477n	—0.027	—0.013	86 27 6.05	+0.053	7.8
1865	3 47.52	0.8959n	9.9590n	9.9598n	9.7577n	—0.027	—0.012	24 14.87		
	Gr. 527.									
1855	2 27 12.90	0.9048	9.5192	9.5248	9.3989	+0.021	—0.003	80 49 37.44	—0.083	6
1865	28 33.75	0.9083	9.5194	9.5249	9.4043	+0.020	—0.003	52 16.84		
	323 Cephei B.									
1855	3 19 28.49	1.2642	9.8083	9.8093	9.8835	+0.162	—0.016	86 10 40.45	—0.049	6
1865	22 35.44	1.2717	9.8053	9.8062	9.8924	+0.164	—0.016	12 48.17		
	57 Ursæ Minoris B.									
1855	15 26 0.64	1.3789n	0.0297n	0.0300n	0.1295n	+0.034	—0.040	87 46 50.55	+0.048	7
1865	22 5.61	1.3642n	0.0319n	0.0322n	0.1165n	+0.032	—0.039	44 44.85		
	Gr. 750.									
1855	3 52 24.81	1.2171	9.6195	9.6211	9.8270	+0.006	—0.009	85 9 52.18	+0.040	7
1865	55 10.61	1.2220	9.6136	9.6152	9.8329	+0.006	—0.009	11 37.53		

* a (number) $-0^s.1314$ for 1855, $-0^s.0107$ for 1865.

† Apparent Places from the *Berliner Astronomisches Jahrbuch.*

62 Ursæ Minoris B.

	AR.	log a	log b	log c	log d	μ	ν	Dec.	μ'	Mag.
	h m s					s	s	° ′ ″	″	
1855	15 58 48.26	0.8371n	9.4620n	9.4649n	9.6982n	+0.015	—0.004	83 22 41.57	+0.006	7.8
1865	57 40.04	0.8326n	9.4638n	9.4667n	9.6951n	+0.015	—0.004	21 0.01		

ε Ursæ Minoris.

	AR.	log a	log b	log c	log d	μ	ν	Dec.	μ'	Mag.
1855	17 0 59.13	0.8097n	9.0971n	9.1011n	9.6805n	+0.011	—0.001	82 16 6.64	+0.001	4.5
1865	16 59 54.87	0.8076n	9.1039n	9.1079n	9.6792n	+0.011	—0.001	15 15.13		

64 Camelopardali B.

	AR.	log a	log b	log c	log d	μ	ν	Dec.	μ'	Mag.
1855	5 15 59.31	1.2649	*	*	9.8849	+0.022	—0.004	85 6 24.46	+0.011	6
1865	19 3.91	1.2665			9.8869	+0.022	—0.003	7 1.52		

δ Ursæ Minoris.†

	AR.	log a	log b	log c	log d	μ	ν	Dec.	μ'	Mag.
1860	18 17 30.16							86 36 6.24		4.5
1865	15 53.34							36 13.63		
1870	14 16.41							36 20.33		

51 Cephei Hevelii.

	AR.	log a	log b	log c	log d	μ	ν	Dec.	μ'	Mag.
1855	6 31 6.40	1.4868	‡	‡	0.1392	—0.031	+0.008	87 15 8.70	—0.024	5
1865	36 12.01	1.4843			0.1365	—0.031	+0.009	14 39.12		

25 Camelopardali Hevelii.

	AR.	log a	log b	log c	log d	μ	ν	Dec.	μ'	Mag.
1855	7 0 18.15	1.1178	9.1300n	9.1336n	9.7032	+0.012	+0.002	82 40 31.33	—0.031	5
1865	2 29.19	1.1162	9.1442n	9.1478n	9.7012	+0.011	+0.002	39 37.93		

4 Ursæ Minoris B.

	AR.	log a	log b	log c	log d	μ	ν	Dec.	μ'	Mag.
1855	7 3 39.32	1.8979	0.0335n	0.0335n	0.5785	—0.019	+0.120	89 1 49.52	+0.008	7
1865	16 36.81	1.8833	0.1040n	0.1040n	0.5633	—0.017	+0.140	0 49.16		
1861		1.8893	0.0781n	0.0781n	0.5696	—0.018	+0.132			
1862		1.8879	0.0848n	0.0848n	0.5681	—0.018	+0.134			
1863		1.8864	0.0913n	0.0913n	0.5665	—0.018	+0.136			
1864		1.8848	0.0977n	0.0977n	0.5649	—0.017	+0.138			

156 Camelopardali B.

	AR.	log a	log b	log c	log d	μ	ν	Dec.	μ'	Mag.
1855	7 41 35.85	1.1911	9.4696n	9.4716n	9.7951	+0.009	—0.005	84 27 38.78	+0.013	6
1865	44 10.61	1.1876	9.4778n	9.4799n	9.7909	+0.009	—0.005	26 11.81		

* b (number) +.1486 for 1855; +.1386 for 1865. Log c —Log b = +.0016.

† Apparent Places from the *Berliner Astronomisches Jahrbuch.*

‡ b (number) —.1880 for 1855; —.2179 for 1865. Log c —Log b = +.0005.

	AR.	log a	log b	log c	log d	μ	ν	Dec.	μ'	Mag.
	λ Ursæ Minoris B.									
	h m s					s	s	° ′ ″	″	
1855	20 8 30.70	1.7374n	0.2568	0.2569	0.4590n	—0.073	+0.148	88 52 32.31	+0.015	6.7
1865	19 59 8.86	1.7604n	0.2386	0.2386	0.4808n	—0.077	+0.151	54 15.64		
1861		1.7512n	0.2461	0.2462	0.4721n	—0.075	+0.150			
1862		1.7535n	0.2443	0.2444	0.4743n	—0.076	+0.150			
1863		1.7558n	0.2424	0.2425	0.4765n	—0.076	+0.150			
1864		1.7581n	0.2405	0.2405	0.4786n	—0.077	+0.151			
	74 Draconis.									
1855	20 37 38.95	0.4990n	9.4068	9.4127	9.4979n	+0.019	+0.002	80 34 50.70	+0.217	6.7
1865	37 7.40	0.5038n	9.4073	9.4131	9.5004n	+0.019	+0.002	37 0.01		
	1 Draconis Hevelii.									
1855	9 16 2.23	0.9676	9.5518n	9.5561n	9.4951	+0.005	+0.004	81 57 38.20	0.000	4.5
1865	17 34.69	0.9638	9.5520n	9.5563n	9.4895	+0.005	+0.004	55 6.36		
	30 Camelopardali Hevelii.									
1855	10 12 58.30	0.9133	9.7043n	9.7073n	9.4099	—0.036	+0.005	83 17 32.49	+0.012	5
1865	14 19.36	0.9081	9.7023n	9.7054n	9.4016	—0.036	+0.005	14 33.26		
	119 Cephei (Bode).									
1855	21 25 0.23	0.6473n	9.6690	9.6716	9.5761n	+0.010	+0.004	83 38 31.95	—0.016	7
1865	24 15.52	0.6552n	9.6708	9.6735	9.5808n	+0.010	+0.004	41 7.98		
	34 Cephei Hevelii.									
1855	22 47 55.06	*	9.6759	9.6797	9.1920n	+0.006	+0.001	82 23 3.30	+0.054	5
1865	47 54.79		9.6790	9.6828	9.1951n	+0.006	+0.001	26 14.56		
	202 Camelopardali (Bode).									
1855	11 21 26.20	0.6675	9.6658n	9.6701n	8.9003	—0.047	+0.002	81 55 29.24	+0.022	6
1865	22 12.02	0.6635	9.6631n	9.6675n	8.8887	—0.047	+0.002	52 11.68		
	39 Cephei Hevelii.									
1855	23 27 49.30	†	0.0340	0.0349	9.1851n	+0.085	+0.002	86 30 26.61	+0.007	6
1865	27 49.98		0.0410	0.0418	9.1919n	+0.086	+0.002	33 45.27		
	309 Cephei (Bode).									
1855	23 52 49.67	0.3959	9.9682	9.9693	8.4649n	+0.035	—0.001	85 53 56.68	—0.019	7
1865	53 15.04	0.4006	9.9742	9.9752	8.4444n	+0.035	—0.001	57 16.96		

* a (number) —0s 0215 for 1855; —0s.0435 for 1865.

† a (number) +0s.0061 for 1855; —0s.0418 for 1865.

TABLE V.

	Assumed AR. 1855.		Deviation.	
12 Ceti,	h m s 0 22 38.34	H.		
		Ai.	+39	+56
		LeV.		
		Ai.	+21	+48
		1853		+52
		1855		+53
β Ceti,	0 36 18.53	H.	—36	+ 8
		Ai.	+28	+45
		LeV.	+12	+68
		Ai.	— 4	+23
		1850		+36
		1855		+38
ε Piscium,	0 55 25.27	H.		
		Ai.	+16	+33
		LeV.	—80	—24
		Ai.	—13	+14
		1854		+ 8
		1855		+ 8
θ Ceti,	1 16 46.56	H.	—37	+ 7
		Ai.	+31	+48
		LeV.	+10	+66
		Ai.	— 8	+19
		1850		+35
		1855		+37
η Piscium,	1 23 43.82	H.		
		Ai.	—56	—39
		LeV.	—66	—10
		Ai.	— 5	+22
		1854		— 9
		1855		— 9
ν Piscium,	1 33 53.34	H.		
	[wt. = ½]	Ai.	—21	— 4
		LeV.	+30	+86
		Ai.	—25	+ 2
		1854		+16
		1855		+16
β Arietis,	h m s 1 46 38.21	H.		
		Ai.	+64	+81
		LeV.	+32	+88
		Ai.	+51	+78
		1854		+82
		1855		+83
67 Ceti,	2 9 45.20	H.		
		Ai.	—59	—42
		LeV.		
		Ai.	—25	+ 2
		1853		—22
		1855		—22
ξ² Ceti,	2 20 27.25	H.		
		Ai.	+ 2	+19
		LeV.	—54	+ 2
		Ai.	— 8	+19
		1854		+13
		1855		+13
γ Ceti,	2 35 47.45	H.	—44	0
		Ai.	+20	+37
		LeV.	—19	+37
		Ai.	+ 6	+33
		1850		+27
		1855		+28
δ Arietis,	3 3 20.66	H.		
		Ai.	+11	+28
		LeV.	—45	+11
		Ai.	+11	+38
		1854		+26
		1855		+26
η Tauri,	3 38 52.34	H.	—53	— 9
		Ai.	—22	— 5
		LeV.	—73	—17
		Ai.	+ 4	+31
		1850		0
		1855		0

	Assumed AR. 1855.		Deviation.			Assumed AR. 1855.		Deviation.	
	h m s					h m s			
γ Eridani,	3 51 15.94	H.	—15	+ 29	α Leporis,	5 26 20.20	H.		
		Ai.	+21	+ 38			Ai.	—12	+ 5
	[wt. = ½]	LeV.	+50	+106		[wt. = ½]	LeV.	—57	— 1
		Ai.	— 6	+ 21		[wt. = ½]	Ai.	—55	—28
		1850		+ 40			1853		— 5
		1855		+ 42			1855		— 5
ο Eridani,*	4 4 47.37	H.			ε Orionis,	5 28 51.42	H.	— 5	+39
		Ai.	+31	+48			Ai.	+25	+42
		LeV.					LeV.	—13	+43
		Ai.	—15	+12			Ai.	—22	+ 5
		1853		+30			1850		+32
		1855		+31			1855		+34
ε Tauri,	4 20 9.26	H.			ν Orionis,*	5 59 17.56	H.		
		Ai.	+ 4	+21			Ai.		
		LeV.	— 6	50			LeV.	+53	+109
		Ai.	— 2	25			Ai.	+32	+ 59
		1854		+32			1859		+ 84
		1855		+32			1855		+ 81
ι Aurigæ,	4 47 33.42	H.			μ Geminor.	6 14 11.34	H.	—31	+13
		Ai.	—13	+ 4			Ai.	—25	— 8
		LeV.	—91	—35			LeV.	—97	—41
		Ai.	— 7	+20			Ai.	—48	—21
		1854		— 4			1850		—14
		1855		— 4			1855		—15
ε Leporis,*	4 59 19.43	H.			γ Geminor.	6 29 20.15	H.		
		Ai.	+ 7	+24			Ai.	—29	—12
		LeV.					LeV.	—70	—14
		Ai.	0	+27			Ai.	—72	—45
		1853		+25			1854		—24
		1855		+26			1855		—24
δ Orionis,	5 24 36.03	H.	—14	+30	ε Canis Maj.	6 52 55.68	H.		
		Ai.	+ 7	+24			Ai.	+29	+46
		LeV.	—41	+15			LeV.	—16	+40
		Ai.	—19	+ 8			Ai.	0	+27
		1850		+19			1854		+38
		1855		+20			1855		+38

* The Greenwich observations for 1859 - 60, and those made at Paris in 1858 - 59, were included.

	Assumed AR. 1855.	Deviation.				Assumed AR. 1855.	Deviation.		
γ Canis Maj.	h m s 6 57 11.92	H.			ι Ursæ Maj.	h m s 8 49 15.67	H.	— 66	—22
		Ai.	+ 9	+26			Ai.	—100	—83
		LeV.					LeV.	— 59	— 3
		Ai.	+ 2	+29			Ai.	— 45	—18
		1853		+27			1850		—31
		1855		+28			1855		—33
δ Geminorum,	7 11 27.68	H.	— 91	—47	83 Cancri,	9 10 53.01	H.		
		Ai.	— 54	—37			Ai.	—22	— 5
		LeV.	—152	—96			LeV.		
		Ai.	— 73	—46			Ai.	—22	+ 5
		1850		—56			1853		0
		1855		—59			1855		0
χ Geminorum, [6 Cancri,]	7 54 36.38	H.			θ Ursæ Maj.	9 23 8.06	H.	—103	—59
		Ai.	+86	+103			Ai.	— 18	— 1
		LeV.	—96	— 40			LeV.		
		Ai.	+ 5	+ 32			Ai.	— 4	+23
		1853		+ 32			1848		—12
		1855		+ 33			1855		—13
ι Navis, [15 Argus,]	8 1 22.17	H.			ε Leonis,	9 37 36.82	H.	—85	—41
		Ai.	+13	+30			Ai.	— 9	+ 8
		LeV.	—20	+36			LeV.	—66	—10
		Ai.	+15	+42			Ai.	+10	+37
		1854		+36			1850		— 2
		1855		+36			1855		— 2
η Cancri,	8 24 19.06	H.			π Leonis,	9 52 32.88	H.		
		Ai.	+ 8	+25			Ai.	—29	—12
		LeV.					LeV.	— 2	+54
		Ai.	+11	+38			Ai.	+ 4	+31
		1853		+31			1854		+24
		1855		+32			1855		+24
ε Hydræ,	8 39 5.66	H.	—31	+13	γ Leonis, preceding,	10 11 58.54*	H.		
		Ai.	+14	+31			Ai.	—156	—139
		LeV.	— 3	+53			LeV.	—223	—167
		Ai.	— 7	+20			Ai.	—175	—148
		1850		+29			1854		—151
		1855		+31			1855		—152

* The value for the mean of the two stars was used as an assumed AR. The Greenwich observations for 1859 were included.

	Assumed AR. 1855.	Deviation.				Assumed AR. 1855.	Deviation.		
	h m s					h m s			
ϱ Leonis,	10 25 10.40	H.			ε Corvi,*	12 2 40.41	H.		
		Ai.	+17	+34			Ai.	—13	+ 4
		LeV.	—27	+29			LeV.		
		Ai.	— 6	+21			Ai.	+21	+48
		1854		+28			1853		+26
		1855		+28			1855		+26
l Leonis,	10 41 37.97	H.			η Virginis,	12 12 29.29	H.		
		Ai.	—25	— 8			Ai.	— 9	+ 8
		LeV.					LeV.	+11	+67
		Ai.	—20	+ 7			Ai.	+30	+57
		1853		0			1854		+44
		1855		0					
χ Leonis,	10 57 32.13	H.			β Corvi,	12 26 46.65	H.		
		Ai.	+ 9	+26			Ai.	+69	+86
		LeV.	—42	+14			LeV.	+24	+80
		Ai.	—19	+ 8			Ai.	+48	+75
		1854		+16			1854		+81
		1855		+16			1855		+82
δ Leonis,	11 6 23.49	H.	+17	+61	γ Virginis,* [mean of two stars,]	12 34 18.90	H.		
		Ai.	+25	+42			Ai.		
		LeV.	—42	+14			LeV.		
		Ai.	+ 3	+30			Ai.	—2	+25
		1850		+37			1857		+25
		1855		+39			1855		+25
δ Crateris,	11 12 5.63	H.	—13	+31	12 Canum,*	12 49 14.36	H.	—61	—17
		Ai.	+51	+68			Ai.	—41	—24
		LeV.	—24	+32			LeV.	—95	—39
		Ai.	+21	+48			Ai.	—16	+11
		1850		+45			1850		—17
		1855		+47			1855		—18
υ Leonis,	11 29 31.51	H.			θ Virginis,*	13 2 26.76	H.		
		Ai.	+ 6	+23			Ai.	—17	0
		LeV.	—28	+28			LeV.	—21	+35
		Ai.	—12	+15			Ai.	—17	+10
		1854		+22			1854		+15
		1855		+22			1855		+15

* Observations made at Greenwich in 1859 included.

	Assumed AR. 1855.		Deviation.			Assumed AR. 1855.		Deviation.	
	h m s					h m s			
ζ Virginis,	13 27 18.44	H.			β Libræ,	15 9 12.58	H.	—52	— 8
		Ai.	+18	+35			Ai.	—32	—15
		LeV.	—37	+19			LeV.	—60	— 4
		Ai.	+17	+44			Ai.	—35	— 8
		1854		+33			1850		— 9
		1855		+33			1855		— 9
η Boötis,	13 47 46.84	H.	— 9	+35	β Scorpii,	15 57 0.78	H.	—103	—59
		Ai.	+22	+39			Ai.	— 79	—62
		LeV.	—28	+28			LeV.	— 96	—40
		Ai.	+ 2	+29			Ai.	— 65	—38
		1850		+33			1850		—50
		1855		+35			1855		—52
τ Virginis,	13 54 16.20	H.			δ Ophiuchi,	16 6 45.01	H.	—59	—15
		Ai.	—44	—27			Ai.	+16	+33
		LeV.	—42	+14			LeV.	+ 3	+59
		Ai.	—19	+ 8			Ai.	+ 1	+28
		1850		— 2			1850		+26
		1855		— 2			1855		+27
ϱ Boötis,	14 25 34.87	H.			ζ Herculis,	16 35 49.28	H.		
		Ai.	—39	—22			Ai.	— 41	—24
		LeV.				[wt. = ½]	LeV.	—121	—65
		Ai.	—21	+ 6			Ai.	— 37	—10
		1853		— 8			1854		—27
		1855		— 8			1855		—27
ε Boötis, following,	14 38 39.30	H.	—48	— 4	κ Ophiuchi,	16 50 48.40	H.		
		Ai.	0	+17			Ai.	+19	+36
		LeV.	—73	—17			LeV.	—17	+39
		Ai.	+ 2	+29			Ai.	— 4	+23
		1850		+ 6			1854		+33
		1855		+ 7			1855		+33
ψ Boötis,	14 58 14.02	H.			θ Ophiuchi,	17 13 6.49	H.		
		Ai.	+12	+29			Ai.	—26	— 9
		LeV.					LeV.	—57	— 1
		Ai.	—18	+ 9			Ai.	+10	+37
		1853		+19			1854		+ 9
		1855		+19			1855		+ 9

	Assumed AR. 1855.		Deviation.	
	h m s			
β Draconis,	17 27 9.58	Ai.	— 64	— 47
	[wt. = ½]	H.	— 37	+ 7
	[wt. = ½]	LeV.	—167	—111
		Ai.	+ 4	+ 31
		1850		— 23
		1855		— 24
μ Herculis,	17 40 47.18	H.		
		Ai.	—46	—29
	[wt. = ½]	LeV.	—87	—31
		Ai.	—34	— 7
		1854		—21
		1855		—21
μ¹ Sagittarii,	18 5 5.52	H.		
		Ai.	— 6	+11
		LeV.	+21	+77
		Ai.	+ 5	+32
		1854		+40
		1855		+40
β Lyræ,	18 44 43.65	H.	— 55	—11
		Ai.	— 26	— 9
		LeV.	—106	—50
		Ai.	+ 19	+46
		1850		— 6
		1855		— 6
ζ Aquilæ,	18 58 44.77	H.	—123	—79
		Ai.	— 24	— 7
		LeV.	— 59	— 3
		Ai.	0	+27
		1850		—15
		1855		—16
ω Aquilæ,	19 11 0.63	H.		
		Ai.	—37	—20
	[wt. = ½]	LeV.	—49	+ 7
		Ai.	— 4	+23
		1854		+ 3
		1854		+ 3

	Assumed AR. 1855.		Deviation.	
	h m s			
δ Aquilæ,	19 18 11.18	H.	— 6	+38
		Ai.	+32	+49
		LeV.	+22	+78
		Ai.	+12	+39
		1850		+51
		1855		+53
h² Sagittarii,*	19 27 52.83	H.		
		Ai.		
		LeV.	—59	— 3
		Ai.	—92	—65
		1856		—34
		1855		—34
ϱ Capricorni,	20 20 35.23	H.		
	(4 obs.)	Ai.	—136	—119
	(2 obs.)	LeV.	+ 53	+ 3
	(13 obs.)	Ai.	—105	— 78
		1855		— 78
32 Vulpeculæ,	20 48 22.87	H.		
		Ai.	0	+17
		LeV.	—53	+ 3
		Ai.	+44	+71
		1854		+30
		1855		+31
61 Cygni, [mean of two stars,]	21 0 24.75	H.	—85	—41
		Ai.	+ 2	+15
		LeV.	—28	+28
		Ai.	+66	+93
		1850		+24
		1855		+25
ζ Cygni,	21 6 46.03	H.	—77	—33
		Ai.	—19	— 2
		LeV.	—55	+ 1
		Ai.	— 6	+21
		1850		— 3
		1855		— 3

* Greenwich observations of 1859 included.

	Assumed AR. 1855.		Deviation.	
	h m s			
β Aquarii,	21 23 55.35	H.	—13	+31
		Ai.	+11	+28
		LeV.	0	+56
		Ai.	— 8	+19
		1850		+34
		1855		+35
ε Pegasi,	21 37 3.85	H.	—23	+21
		Ai.	— 9	+ 8
		LeV.	—46	+10
		Ai.	+18	+45
		1850		+21
		1855		+22
16 Pegasi,	21 46 28.03	H.		
		Ai.	—41	—24
		LeV.		
		Ai.	—20	+ 7
		1853		— 9
		1855		— 9
θ Aquarii,	22 9 10.75	H.		
		Ai.	— 1	+16
		LeV.	—41	+15
		Ai.	+ 2	+29
		1854		+20
		1855		+20
η Aquarii,	22 27 54.24	H.		
		Ai.	— 2	+15
		LeV.	—19	+37
		Ai.	+ 9	+36
		1854		+29
		1855		+30
	h m s			
ζ Pegasi,	22 34 13.90	H.	—96	—52
		Ai.	—32	—15
		LeV.	—81	—25
		Ai.	—10	+17
		1850		—19
		1855		—20
γ Piscium,	23 9 38.92	H.		
		Ai.	+37	+54
		LeV.	— 9	+47
		Ai.	+15	+42
		1854		+48
		1855		+48
κ¹ Piscium,	23 19 29.94	H.		
		Ai.	+26	+43
		LeV.	+37	+93
		Ai.	— 3	+24
		1854		+53
		1855		+54
ι Piscium,	23 32 29.55	H.	+92	+136
		Ai.	+86	+103
		LeV.	+13	+ 69
		Ai.	+29	+ 56
		1850		+ 91
		1855		+ 96
ω Piscium,	23 51 52.06	H.		
		Ai.	— 3	+14
		LeV.	—81	—25
		Ai.	—41	—14
		1854		— 8
		1855		— 8

NOTE. — The columns headed "Deviation" contain the difference "Observation less Computation," expressed in thousandths of seconds; the numbers in the first of these columns being the direct result from the Catalogues, and those in the second, the same corrected by the quantities on page 11.

TABLE VI.

	AR. 1855.0	Annual Variation.	Secular Variation.		AR. 1855.0	Annual Variation.	Secular Variation.
	h m s	s	s		h m s	s	s
α Andromedæ,	0 0 54.010	+3.08397	+0.01770	ε Leonis,	9 37 36.818	+3.42179	—0.01798
γ Pegasi,	5 46.457	3.08076	+0.00970	π Leonis,	52 32.904	3.17822	—0.00819
12 Ceti,	22 38.393	3.06042	+0.00059	α Leonis,	10 0 38.763	3.20353	—0.01020
β Ceti,	36 18.568	3.01391	—0.00580	γ Leonis, prec.	11 58.388	3.31951	—0.01515
ε Piscium,	55 25.278	3.10944	+0.00852	ρ Leonis,	25 10.428	3.16739	—0.00817
θ Ceti,	1 16 46.597	+2.99669	+0.00158	l Leonis,	10 41 37.970	+3.16068	—0.00831
η Piscium,	23 43.811	3.19707	+0.01392	χ Leonis,	57 32.146	3.09982	—0.00581
ν Piscium,	33 53.356	3.11423	+0.00893	δ Leonis,	11 6 23.529	3.20499	—0.01354
β Arietis,	46 38.293	3.29597	+0.01809	δ Crateris,	12 5.677	2.99470	+0.00613
α Arietis,	59 0.500	3.36392	+0.02010	υ Leonis,	29 31.532	3.07136	+0.00008
67 Ceti,	2 9 45.178	+2.98604	+0.00479	β Leonis,	11 41 39.669	+3.06661	—0.00770
ξ² Ceti,	20 27.263	3.17892	+0.01153	β Virginis,	43 8.533	3.12472	—0.00060
γ Ceti,	35 47.478	3.10049	+0.00928	ε Corvi,	12 2 40.436	3.07399	+0.01379
α Ceti,	54 42.231	3.12678	+0.00970	η Virginis,	12 29.334	3.06671	+0.00273
δ Arietis,	3 3 20.686	3.41691	+0.01717	β Corvi,	26 46.732	3.13101	+0.01603
η Tauri,	3 38 52.340	+3.55008	+0.01789	γ Virginis, med.	12 34 18.925	+3.03755	+0.00403
γ Eridani,	51 15.982	2.79480	+0.00475	12 Canum Ven.	49 14.342	2.81928	—0.01561
o′ Eridani,	4 4 47.401	2.92074	+0.00600	θ Virginis,	13 2 26.775	3.09894	+0.00758
ε Tauri,	20 9.292	3.49347	+0.01244	α Virginis,	17 33.567	3.14988	+0.01120
α Tauri,	27 36.289	3.43414	+0.01090	ζ Virginis,	27 18.473	3.05219	+0.00611
ι Aurigæ,	4 47 33.416	+3.89371	+0.01523	η Boötis,	13 47 46.875	+2.85875	—0.00066
ε Leporis,	59 19.456	2.53580	+0.00342	τ Virginis,	54 16.198	3.04728	+0.00634
α Aurigæ,	5 5 59.061	4.42031	+0.01840	α Boötis,	14 9 2.949	2.73349	+0.00120
β Orionis,	7 34.294	2.88069	+0.00430	ρ Boötis,	25 34.862	2.58811	—0.00169
β Tauri,	17 7.747	3.78683	+0.00890	ε Boötis, seq.	38 39.307	2.62017	—0.00005
δ Orionis,	5 24 36.050	+3.06375	+0.00410	α¹ Libræ,	14 42 40.403	+3.30308	+0.01560
α Leporis,	26 20.195	2.64574	+0.00319	α² Libræ,	42 51.833	3.30414	+0.01550
ε Orionis,	28 51.454	3.04138	+0.00381	ψ Boötis,	58 14.039	2.57055	+0.00113
α Orionis,	47 19.409	3.24714	+0.00310	β Libræ,	15 9 12.571	3.21778	+0.01188
ν Orionis,	59 17.641	3.42628	+0.00227	α Coronæ,	28 33.018	2.53833	+0.00240
μ Geminorum,	6 14 11.325	+3.63290	+0.00023	α Serpentis,	15 37 7.734	+2.94935	+0.00630
γ Geminorum,	29 20.126	3.46792	—0.00094	β Scorpii,	57 0.728	3.47595	+0.01455
α Canis Maj.*	38 45.452	2.64508	+0.00040	δ Ophiuchi,	16 6 45.037	3.13560	+0.00846
ε Canis Maj.	52 55.718	2.35827	+0.00164	α Scorpii,	20 31.399	3.66500	+0.01570
γ Canis Maj.	57 11.948	2.71602	+0.00067	ζ Herculis,	35 49.253	2.26165	+0.00319
δ Geminorum,	7 11 27.621	+3.59256	—0.00676	κ Ophiuchi,	16 50 48.433	+2.83429	+0.00458
α Geminorum,†	25 20.245	3.84131	—0.01250	α Herculis,	17 8 2.283	2.73289	+0.00370
α Canis Min.	31 42.628	3.14698	—0.00440	θ Ophiuchi,	13 6.499	3.67702	+0.00871
β Geminorum,	36 26.276	3.68356	—0.01220	α Ophiuchi,	28 12.321	2.78117	+0.00350
χ Geminorum,	54 36.413	3.69912	—0.01429	μ Herculis,	40 47.159	2.34272	+0.00345
ι Navis,	8 1 22.206	+2.55499	+0.00107	μ¹ Sagittarii,	18 5 5.560	+3.58524	+0.00154
η Cancri,	24 19.092	3.48165	—0.01281	α Lyræ,	32 1.781	2.03091	+0.00160
ε Hydræ,	39 5.691	3.18498	—0.00692	β Lyræ,	44 43.644	2.21283	+0.00172
83 Cancri,	9 10 53.010	3.35922	—0.01337	ζ Aquilæ,	58 44.754	2.75318	+0.00061
α Hydræ,	20 27.725	+2.94942	—0.00150	ω Aquilæ,	19 11 0.633	+2.81459	—0.00003

* Peters's correction is yet to be added.

† Mean of the two stars.

	AR. 1855.0	Annual Variation.	Secular Variation.		AR. 1855.0	Annual Variation.	Secular Variation.
	h m s	s	s		h m s	s	s
δ Aquilæ	19 18 11.233	+3.02489	—0.00153	ε Pegasi,	21 37 3.872	+2.94836	—0.00059
h^2 Sagittarii,	27 52.796	3.65940	—0.00965	16 Pegasi,	46 28.021	2.72578	+0.00513
γ Aquilæ,	39 21.991	2.85281	—0.00080	α Aquarii,	58 20.150	3.08414	—0.00430
α Aquilæ,	43 42.531	2.92881	—0.00140	θ Aquarii,	22 9 10.770	3.17178	—0.00771
β Aquilæ,	48 11.470	2.94775	—0.00150	η Aquarii,	27 54.270	3.08381	—0.00324
$α^1$ Capricorni,	20 9 36.510	+3.33175	—0.00810	ζ Pegasi,	22 34 13.880	+2.98752	+0.00211
$α^2$ Capricorni,	10 0.414	3.33500	—0.00810	α Piscis Austr.	49 37.775	3.33287	—0.02170
ϱ Capricorni,	20 35.152	3.42960	—0.01118	α Pegasi,	57 32.469	2.98307	+0.00530
α Cygni,	36 29.380	2.04250	+0.00240	γ Piscium,	23 9 38.968	3.10795	+0.00029
32 Vulpeculæ,	48 22.901	2.55446	+0.00264	$κ^1$ Piscium,	19 29.994	3.07564	—0.00021
61 Cygni, med.	21 0 24.775	+2.67784	+0.00392	ι Piscium,	23 32 29.646	+3.08413	+0.00281
ζ Cygni,	6 46.027	2.54826	+0.00384	ω Piscium,	51 52.052	+3.07735	+0.00445
β Aquarii,	23 55.385	3.16430	—0.00711				

TABLE VI. Supplement.

	AR. 1855.0	Annual Variation.	Secular Variation.		AR. 1855.0	Annual Variation.	Secular Variation.
	h m s	s	s		h m s	s	s
α Cassiopeiæ,	0 32 18.284	+3.35312	+0.05440	η Ursæ Maj.	13 41 49.377	+2.37462	—0.01050
α Persei,	3 13 59.591	4.24123	+0.04850	β Ursæ Min.	14 51 10.547	—0.26640	+0.10660
ι Ursæ Maj.	8 49 15.637	4.14858	—0.04433	β Draconis,	17 27 9.556	+1.35102	+0.00545
θ Ursæ Maj.	9 23 8.047	4.06080	—0.05615	γ Draconis,	17 53 14.502	+1.39321	+0.00350
α Ursæ Maj.	10 54 44.415	3.77579	—0.08380	α Cephei,	21 15 6.946	+1.43846	—0.00640
γ Ursæ Maj.	11 46 11.102	+3.19678	—0.04440	β Cephei,	21 26 46.341	+0.80464	—0.03300

NOTES AND ERRATA.

In Table II., λ Ursæ Minoris B., $\Delta\delta'$ (Rob.) *for* —0″.42 *read* —0″.29.
51 Cephei H., *for* Airy, 1845, *read* Airy, 1840, 1839.
for $\Delta\alpha'$ + 7″.67 *read* — 7″.67.
for $\Delta\alpha'\cos\delta'$ + 0 37 *read* —0.37
323 Cephei B., *dele* Airy, 1840.

Table III. is not affected by these errata.

The order of some of the dates in Table II. is reversed in printing: that is, the date for Declination is before that for A.R.

It should perhaps have been stated, that the logarithms a, b, c, d in Table IV. are in the original order of Bessel; the English have arbitrarily changed a, b, c, d to c, d, a, b, and the same inconvenient alteration has been made in the American Ephemeris. Log a is the log of the annual precession in A.R.; μ, μ' are proper motion; ν is $-\frac{1}{2}\frac{d^2\alpha}{dt^2}$.

In reducing to apparent place for a fraction τ of the year 1865 + n, we interpolate log a, b, c, d for the time 1865 + n + τ, and we shall have (α° being A.R. for 1865),

$$\alpha' - \alpha^\circ = (A+n)\,a + Bb + Cc + Dd + (\tau+n)\,\mu + (\tau+n)^2\,\nu.$$

This includes all the sensible terms except the very small ones depending on $\odot + \Omega$, $\odot - \Omega$, if the tables for Log A, B, etc., include them, as do O. Struve's *Tabulæ Quantitatum Besselianarum*.

ANNALS

OF THE

ASTRONOMICAL OBSERVATORY OF HARVARD COLLEGE.

VOL. IV. PART II.

OBSERVATIONS IN RIGHT ASCENSION OF 505 STARS.

PRINTED FROM THE STURGIS FUND.

CAMBRIDGE:
PRESS OF JOHN WILSON AND SON.
1878.

RIGHT ASCENSIONS OF 505 STARS

DETERMINED WITH THE

AST TRANSIT CIRCLE AT THE OBSERVATORY OF HARVARD COLLEGE.

GEORGE PHILLIPS BOND, A.M., DIRECTOR.

1862–1865.

BY

TRUMAN HENRY SAFFORD, PH.D.,

FIELD MEMORIAL PROFESSOR OF ASTRONOMY IN WILLIAMS COLLEGE.

CAMBRIDGE:

PRESS OF JOHN WILSON AND SON.

1878.

INTRODUCTION.

Within the past few years, it has become evident that, besides the fundamental stars proper (by which term is here understood the Catalogue of 36 principal stars selected by Maskelyne), and the two principal polar stars α and δ Ursæ Minoris, positions of many others of very great precision are needed for the various uses of astronomy.

The selection and observation of these additional half-fundamental or secondary stars have been differently made by different observers. The present series of right-ascensions is intended as a modest contribution to this object. When it was begun, Professor G. P. Bond, under whose direction the larger part of the observations were made, as well as myself, who planned the series in detail, had but little expectation that the observations would prove of very high quality; as the instrument was of an antique construction, and the collimation must be determined by reversals on a distant mark. Moreover, the clock was not so regular in its rate as was desirable.

It was therefore thought best to observe the stars contained in the chief ephemerides, and other bright stars, in order to ascertain by comparison with their right-ascensions, elsewhere determined, what were the individual peculiarities of the instrument.

But as the work progressed, and various discrepancies between the observations and some of the principal ephemerides became manifest, it was almost invariably found that the observations were the more accurate of the two; and a pretty thorough discussion of results from time to time showed clearly enough that considerable improvement might be made by such observations in the right-ascensions of almost any star save the few most frequently and regularly observed elsewhere.

For example, the predicted right-ascension of Polaris itself was corrected by these observations; and substantially the same correction as they give was found to result from other modern series, was at once employed in the European determinations of longitude, and after some hesitation in those of our own Coast Survey. So was it with other stars whose right-ascensions exhibited considerable discrepancies from the results of previous observation: in such cases, our results were confirmed upon the whole, as will be manifest upon an examination of the details given at the time in Professor Bond's reports. Although the individual results do not agree quite so well as those made by the most modern and powerful instruments, they are by no means unusually discrepant. Their individual accuracy is about the same as that of Argelander's late right-ascensions, and rather greater than that of Struve's; and it is hoped that proportionate care will be seen to have been taken in eliminating constant error.

The one chief weakness of the instrument is its want of collimators; and it is so located that it would be rather difficult to put them up; and it would not be possible to set one upon the other without raising the instrument from its Y[s], as it has no cube, but is constructed in an entirely different manner.

Consequently the greatest care was taken to ascertain its remaining errors, and to trace and measure any large sudden variations in the value of $(n + c)$ as ascertained by polar star observations, of which a very great many were made. *Per contra*, the very pains so taken has resulted in a catalogue of right-ascensions of stars within 10° of the pole, of considerable accuracy (as tested in all possible ways), and of quite unusual completeness; and, as the "working list" seems to have fallen under the notice of the officers of the Coast Survey, portions of it have been incorporated with their Field Catalogue lately published. Moreover, as the clock-rate was rather variable, especial pains were taken to observe the fundamental stars proper with great regularity, and thus determine the clock-correction with entire certainty for any other observations. Of course, under these circumstances these stars are not considered as determined, and no results are set down for their right-ascensions, which are used as taken from Wolfers' "Tabulæ Reductionum," with necessary corrections for Sirius, Castor, and Procyon.

In another point of view, the resulting positions are more entirely independent. The right-ascension of Polaris is determined from about 175 consecutive transits above and below pole; those of the other principal polars, 51 Cephei H. and ϵ, δ, λ Ursæ Minoris, by comparison with Polaris, and as far as possible at double transits; no right-ascension being set down for either of these stars, save on days when Polaris was observed. The right-ascensions of the 22 auxiliary polars, whose places from preceding observations are investigated in the first part of this volume, are also determined by the present observations: the former positions were used for the determination of the instrumental errors only till September, 1864.

Again, care was taken to observe the principal circumpolars to about 35° of polar distance, both above and below pole; thus affording a tolerable check upon the differences of right-ascension of fundamental stars twelve hours apart, and on the general accuracy of the whole work. Of course, this control is not so valuable as in higher latitudes.

The working list includes:—

1. All the stars of the "Nautical Almanac" visible at Cambridge, latitude 42° 23′. I was never able to see *α* Gruis, which it would have been quite useless to observe, as its altitude above the horizon is due almost entirely to refraction; and stars of the altitude of *α* Phœnicis (between 4° and 5°) could be but ill-determined.

2. The additional 22 polars mentioned above. The basis of this list was Argelander's, mentioned in Volume I. of the Bonn Observations, and Carrington's set of standard circumpolars. One object I had in view was the detection, if it existed, of a periodic error of rather complex form in Carrington's right-ascensions, first suspected by Professor Peirce.

3. Many other bright stars, partly from the "Connaissance des Temps," or the American Ephemeris, added from time to time; and such of Argelander's "Gemeinschäftliche Sterne" as were not already in the working list, and could be conveniently observed before I left Cambridge in the spring of 1866.

4. Various stars, selected chiefly from the "Uranometria Nova," as needing observation; some Flamsteed stars, for instance, which are not in Bradley, and are apt to be overlooked on that account; some other stars of the 5th and 6th magnitudes, which are not in the British Association Catalogue at all; and a few stars of suspected or sure proper motion, which other observers had not noticed.

In all this portion, too, of the series, I paid especial regard to the region about the north pole.

This regular list, in which, excluding fundamentals, are over 500 stars, was observed with much persistency from 1862 to 1865 inclusive. The observations, however, were, as to myself, much interrupted by equatorial work, and also by a series of experiments at determining declinations with the meridian circle, at the instance of Colonel Graham, State Superintendent of the United States Lake Survey. The results obtained for him proved useful and sufficiently accurate: they are contained in my catalogue of 2018 stars now in press, to be published by the War Department. But this experiment showed that for declinations a new meridian circle was necessary. Some small stars were observed in right-ascension at odd times: it is, of course, possible that one or the other of them may prove interesting on account of proper motion, when the great zones now in progress are completed. They were originally observed in pursuance of a design of mine to render more accurate the equatorial zones already published, as well as others then contemplated and actually begun.

The observers were, besides myself: —

During a portion of 1862, Professor A. Hall, U. S. N., then assistant at the Harvard College Observatory. He became Aide at the U. S. Naval Observatory that summer.

During a part of 1864, Mr. W. A. Rogers, assistant professor at the Harvard College Observatory, then teacher of mathematics in Alfred Academy, New York. For the remainder of the series, Mr. A. McConnel was of very great help: during the whole interval, I gave a large portion of my time to the series, and especially to those parts of it upon which the fundamental determinations depend.

The instrument, the East Meridian Circle, is sufficiently described elsewhere in these Annals. (See Vol. I.) At the time of these observations, the clock (Bond 312) was provided with two mercurial pendulums, connected by a fine wire, and beating together; an experimental arrangement, which was afterwards done away with. This peculiarity of the clock probably was in part the cause of its irregularities. As it stood, not in the observing room, but in a room generally heated in winter, the rate should be more uniform during a day than that of a transit clock with a mercurial pendulum usually is: I have, however, not depended to any great extent upon the daily rate, but have divided the fundamental stars into such groups that this dependence is not necessary.

The observations were *all* chronographically registered. Much trouble was found with the close polar stars, especially Polaris; the object-glass defined well; but the position of the observing room, next the warmed room in which is the clock, is quite unfavorable, and the shutters are not so wide as I should make them from my experience. It happens very frequently at Cambridge that the east wind carries the currents of heated air from the dwelling-house across the opening of the East Transit Room.

I am inclined to think that an observatory ought to be chiefly made up of detached buildings, or at least that great care should be taken with respect to the air-currents from the apartments of the observers and computers. This was the arrangement I carried out for the meridian circle room at Chicago. The chronograph employed was Bond's Spring Governor, and satisfactory in all respects. The sheets were read off only to tenths of seconds.

The system of wires was once changed during the series; but from some unknown cause the intervals seemed to vary from time to time. The change was made about November 10, 1865, or very nearly at the end. I then took occasion to replace wires in the original scores, as graduated by Simms in 1847: these, of course, were at wide intervals, and seven in number. Of these, the middle one only had been used from 1862 to 1865, and four additional ones at equatorial intervals of about 3^s; so that most of the present series was made upon a single close group of 5, and a very small portion near the end upon 11 wires, 5 close and 6 others more distant. The former arrangement is better adapted to large masses of work, the latter is more accurate for single observations. The wire intervals, as used, are given in Table A. The wire A. is that one of the close series of 5 which a star above pole first passes when the lamp is east; M. denotes the mean of these five, to which the few observations made after November 14, 1865, are also reduced.

The steps of reduction are the following, after the chronograph sheets are read off, and the means of wires of complete observations taken: —

1. Reduction of broken transits. These were, fortunately, few in number, save for stars near the north pole of the heavens.

2. Computation of instrumental corrections. The collimation having been determined from time to time by reversal on a distant mark, the value of n is ascertained from the computed right-ascensions of circumpolars. At first the 27 of Part I. were all used, but with very different weights, and afterwards only Polaris, ϵ, δ, λ Ursæ Minoris and 51 Cephei Hevelii. The values of $n + c$ were combined into periods, sometimes of considerable length, where they seemed unchanged. After September 2, 1864, the right-ascensions given in Part I. were corrected by provisional quantities, nearly identical with the final results.

3. Computation of clock-corrections. Such stars only, aside from the 36 Maskelyne's fundamentals, are used as appeared indispensable for a sufficient clock-correction; and no observation thus used furnishes an observed right-ascension; so that the resulting right-ascensions are independently based upon Wolfers' system. Supplementary catalogue places are mostly taken from Part I. of this volume, with the help of the "Berlin Jahrbuch," and "Nautical Almanac" for the reductions to apparent right-ascension.

The two trifling incongruities thus introduced — first, that Wolfers' right-ascensions, and those resulting from my own computations on pages 24 to 30 of Part I., are not quite in accordance, save upon the average, and are systematically a little different; and, second, that the "Nautical Almanac" apparent places do not contain the term $(0^s.0078 + 0^s.0034 \sin \alpha \tan \delta) \sin (\odot - \Gamma)$, — will have the effect of slightly altering the systematic correction to our results, and especially the terms

$$m \sin \alpha + n \cos \alpha.$$

But, as these terms of the systematic correction arise in part from the combination of unnoticed causes producing very small effects, this change will be of no consequence.

In point of fact, I do not know any series of modern observations now extant which is entirely free from suspicion of error of this kind, save, perhaps, the Pulcova observations about 1845: in most cases, the correction is trifling, and the present series (it is to be hoped) will prove nearly free from such errors, to which my attention was called many years ago, and before these observations were begun. (See my paper "On the Positions of the Radcliffe Catalogue.")

In Table B., I give a comparison between the right-ascensions of time-stars in Part I. of this volume, as carried back to 1845, with the Pulcova right-ascensions for that date.

4. Computation of apparent observed right-ascensions.

The instrumental corrections, $n \tan \delta + c \sec \delta$, from $-40°$ to $+70°$ of declination, were tabulated to 3 decimal places for convenient intervals, so that they could be very easily set down for each star to hundredths of seconds. In Table C., I give a specimen of this work.

The clock-corrections for each date were divided into two groups, if the observations were sufficiently numerous, and there was any suspicion of change of rate in the clock. On a good many dates there were observations by two observers: in this case, those of the one were divided into two groups by those of the other. Here the rate used for reducing the observations of the latter was derived from the two groups of the former observer. In other cases, the clock-corrections, including Bessel's m, were obtained by interpolation between consecutive groups, allowing for personal equation only when the groups were more than twelve hours apart; otherwise, comparing the same observer's results only.

The apparent observed right-ascensions were formed by summing up

$$T + \Delta t + m + n \tan \delta + c \sec \delta,$$

where but three separate numerical terms are added, unless the star be beyond 70° of north declination; T, $\Delta t + m$, and $n \tan \delta + c \sec \delta$.

5. Computation of mean right-ascensions for 1865.0.

For stars contained in published ephemerides, the "Berliner Jahrbuch," the "Nautical Almanac," the "Connaissance des Temps," and for 1865 the "American Ephemeris," and Dr. von Asten's calculations in No. 1578 of the "Astronomische Nachrichten," the discrepancies from these ephemerides were at once set down and applied to the mean places for 1865. These positions need a trifling correction, so far as they are taken from the "Nautical Almanac" and "Connaissance des Temps," on account of the small term of nutation previously mentioned; whose amount with changed sign is set down in Table D. with the arguments, *dies reductus*, and annual precession in right-ascension.

Besides this correction, the error in the "Nautical Almanac's" annual variations was taken into account where necessary. The "Connaissance des Temps" gives its mean right-ascensions only to hundredths. I, therefore, took the means of the three values for 1863, 1864, 1865, and added a correct annual variation.

For stars not in the ephemerides the reductions were effected by the data of Wolfers' "Tabulæ Reductionum;" and for verification (in part), by those of the "American Ephemeris." During 1862, 1863, and 1864, I was obliged in using these latter to allow for the known error in fixing the beginning of the year. Those given for 1865 are correct, the previous practice having been to follow the example set by the English "Nautical Almanac" before 1857.

In employing the formula

$$\alpha' - \alpha = f + g \sin (G + \alpha) \tan \delta + h \sin (H + \alpha) \sec \delta$$

I found it well to compute f, g, G, &c., directly from the logs. A, B, C, D, given in the "Berlin Jahrbuch," and arrange them for immediate reduction to 1865.0. Ten-day ephemerides of the reductions to 1865.0 were computed by their help, for stars observed on closely following dates.

6. For Polaris, as before stated, a special discussion was made, including only the consecutive transits. The clock-correction $\Delta t + m$ having been applied to the clock time of transit, the mean of the two values of $T + \Delta t + m$ given by the two transits at opposite culminations is taken, omitting the hours and minutes, thus giving the seconds of observed right-ascensions, and compared immediately with the "Berlin Jahrbuch." The resulting mean correction was considered to hold good for 1864.0, and brought up to 1865 by the annual variation, as given in the "American Ephemeris Tables," which offer a result very closely equal to that in my paper on the right-ascensions of Polaris. I did not think it wise to apply a correction for personal equation varying with declination; but few observations of Polaris were made otherwise than by Mr. McConnel and myself, and our joint result as given is very nearly equal to the arithmetical mean between our separate determinations.

With regard to the remaining 26 polars of Part I., I intended at first to adopt a different method. I had hoped that the right-ascensions of δ Ursæ Minoris and 51 Cephei H. might be obtained by double transits, or by comparison with double transits of Polaris, or in default of these by single transits of Polaris on the same day. But these observations were not made often enough in connection, and the instrument seemed to vary during the night's work more than from night to night; so that I could not make sure that the right-ascensions thus obtained represented any thing real. I preferred then to sacrifice the independent determinations of these stars, as well as of λ Ursæ Minoris, and employ the computed right-ascensions of these to determine instrumental error. Fortunately, the change which the introduction of the best values of the right-ascensions of 51 Cephei H. δ, and λ, Ursæ Minoris into my instrumental corrections would have made, is not very considerable; and, with the exception of these three stars, the results of the observations here presented are substantially independent. The three stars' right-ascensions are in parentheses in the catalogue. The right-ascensions of all the polars, as finally given, depend upon the same values of n and c as the remaining stars.

The body of the work comprises the following divisions:—

I. Table of Individual Clock-corrections, from Computed Right-ascensions, with Means of Instrumental Corrections (pp. 1–23).

In case I had foreseen that the original instrumental corrections would not have been modified, I should have given them individually. I will here give a specimen for the longest period, viz. from December 9, 1864, to January 25, 1865, both dates inclusive. The computation was so arranged that $\Delta t + m$ and c being supposed known, n and c would be obtained from each polar star by the formula

$$n + c = [n \tan \delta + c \sec \delta + n (\sec \delta - \tan \delta)] \cos \delta,$$

where $n \tan \delta + c \sec \delta$ is at once given $= a - T - (\Delta t + m)$ and $n (\sec \delta - \tan \delta)$ never exceeds $0.0888\, n$ or about $0^s.04$ for a star in $80°$ declination.

Date.	Star.	$n+c$
		s.
1864 Dec. 9	34 Cephei H.	—0.174
	202 Camelop. B. s.p.	—0.095
	39 Cephei H.	—0.215
	5 Ursæ Minoris B. s.p.	—0.002
	Polaris	—0.070
14	Polaris	—0.071
	214 Camelop. B. s.p.	—0.061
20	5 Ursæ Minoris B. s.p.	—0.012
	43 Cephei H.	—0.005
	Polaris	—0.029
	20 Ursæ Minoris B. s.p.	—0.128
30	39 Cephei H.	—0.169
	5 Ursæ Minoris B. s.p.	+0.011
	6 Ursæ Minoris B. s.p.	—0.084
	43 Cephei H.	—0.095
	Polaris	—0.119
	20 Ursæ Minoris B. s.p.	—0.112
1865 Jan. 2	64 Camelop. B.	—0.054
	51 Cephei H.	—0.100
4	57 Ursæ Minoris B.	—0.012
	Gr. 750.	—0.064
	64 Camelop. B.	—0.080
11	Polaris	—0.140
	Gr. 750.	—0.105
	51 Cephei H.	—0.023
12	43 Cephei H.	—0.119
	Polaris	—0.108
	214 Camelop. B. s.p.	—0.067
	323 Cephei B.	—0.074
16	Polaris	—0.074
	4 Ursæ Minoris B.	—0.057
18	Polaris	—0.078
	62 Ursæ Minoris B. s.p.	—0.042
1865 Jan. 20	Polaris	—0.103
	57 Ursæ Minoris B. s.p.	—0.068
	Gr. 750.	—0.034
24	Gr. 527.	—0.006
	δ Ursæ Minoris s.p.	—0.058
25	Polaris	—0.161
	62 Ursæ Minoris B. s.p.	—0.156
	δ Ursæ Minoris	—0.161
	51 Cephei H.	—0.102

Mean for separate days: —

	s.	
1864 Dec. 9	—0.111	5
14	—0.066	2
20	—0.044	4
30	—0.095	6
1865 Jan. 2	—0.077	2
4	—0.052	3
11	—0.089	3
12	—0.092	4
16	—0.066	2
18	—0.060	2
20	—0.068	3
24	—0.032	2
25	—0.120	4

Adopted Means —0.080 ± 0.0052 (s.)
Probable error of one star ± 0.034
Probable error of one date ± 0.019 2.6 stars.

The observations of Polaris during the period give $n + c = -0^s.088 \pm 0^s.008$; probable error of one observation $\pm 0.^s0235$.

It will be seen, from these numbers, that the variations during an evening were fully comparable with those from night to night; and so it was during the whole series. The individual clock-corrections in this table were obtained by the use of the "Berlin Jahrbuch," wherever stars enough of the original Maskelyne's catalogue were observed.* The nine northern stars were used only in extreme cases.

The "Nautical Almanac" stars, employing the mean right-ascensions of Part I. of this volume, were often used; and no "Nautical Almanac" star, when thus used, was considered as having its right-ascension determined on the same evening.

* The right-ascension of Sirius was corrected by $-0^s.08$; that of Procyon, by $-0^s.01$ in 1863, by $-0^s.02$ in 1864, by $-0^s.04$ in 1865; that of Castor, by $+0^s.08$.

II. Mean Values of Clock-corrections (pp. 24–29).

These are so taken that the injurious effects of the variable clock rate may be so far as possible avoided; of course, taking care to use each observer's clock-corrections in reducing his own observations only. The stars whose right-ascensions are determined are usually between two groups of clock-stars of the same date, or else very near one such group.

III. Results. Mean Right-ascensions for 1865 (pp. 30–108).

These do not include proper motion, save for stars of which ephemerides were at hand and for η Cassiopeiæ. For stars found in the "Nautical Almanac" and other ephemerides, the deviations (Obs$^{d}_{.}$ — Comp$^{d}_{.}$) from those ephemerides are in nearly all cases given; and the resulting right-ascensions obtained by means of the mean right-ascensions for 1865 as given in the Ephemeris.

For the "Connaissance des Temps," this value is not always that given in that work for 1865, but sometimes a mean from the values for 1862, 1863, 1864, and 1865, brought up to 1865 with more correct annual variation. These inequalities are all eliminated by the column $\Delta\alpha$ of the catalogue. For the stars of my Polar Catalogue (in Part I. of this volume) the deviations are from the tables of pages 121 to 143.

IV. General Catalogue (pp. 109–120).

The stars' names are usually the same as in the previous division, and commonly agree with Argelander's "Uranometria Nova," but frequently differ from those in the British Association Catalogue. The precessions and secular variations are computed from Bessel's elements, using Argelander's precepts and tables in the seventh volume of the Bonn Observations. I have utilized, as far as they go, the values for 1875 of those quantities given in the "Vierteljahrsschrift der astronomischen Gesellschaft," Volume IV., pages 324 to 349 inclusive, save for stars within 10° of the pole, for which I have (with much trouble) computed declinations for 1865, exact at least to seconds, and when needful more closely. In cases where the third term, according to Tiele, is inserted in the "Vierteljahrsschrift," I have used six one-hundredths of it as the secular variation of the secular variation.

From this it will be seen, by referring to the seventh volume of the Bonn Observations, pages 147, 148, that these secular variations are geometrical only, and do not contain the additional terms

$$0.01944\ (h'\ \mathrm{tg}\ \delta'\ \cos\alpha' + i'\ \sec\delta'^2\ \sin\alpha')\ \text{and}\ -0.01944\ h'\ \sin\alpha',$$

depending upon proper motion and mentioned on page 4 of Part I.

The secular variations on pages 31, 32 of that Part, contain one-half of the first of these terms, and hold good for $1805 = \frac{1}{2}(1755 + 1855)$ and for Struve's elements: hence they are slightly different from the values here given.

The proper motions are derived from the following authorities, reduced in all cases to Bessel's precession: —

A. Argelander, Volume VII. of the Bonn Observations, or the Vierteljahrsschrift.

N. Newcomb, American Ephemeris for 1880, Table VIII.

M. Mädler, Dorpat Observations, Volume XIV.

Ast. Von Asten, Vierteljahrsschrift, Volume IV. as before cited.

Wg. Wagner, Observations de Poulcova, Volume III.

W. Wolfers, Tabulæ Reductionum.

R. Rogers, Volume X. of these Annals.

S. My own investigations; provisional only.

When proper motion is not given, it is usually very small: a few cases need farther study, but the epoch can always be readily obtained.

The declinations are mostly reliable to 0′.1: sometimes not, especially for southern stars, as I had no good modern catalogue at hand. For high northern stars, more accurate values were used in computing precessions, &c.

The column $\Delta\alpha$ contains the corrections applied, due to proper motion; or for stars contained in ephemerides, to differences of proper motion and to periodic terms neglected (see page ix.). Where this column is vacant, neither correction was necessary.

V. Tables for the Stars of the Polar Catalogue (pp. 121–143).

These are in the usual form as given by Bessel and others after him, save for the terms depending on 2 ☾: for these, the argument *in days* is obtained by subtracting from the *dies reductus* as corrected for the star the date just preceding it in the table. In these tables, I have omitted the terms depending upon $\odot + \Omega$: the former, which alone is sensible, varies between $+ 0^s.14$ and $- 0^s.14$ for λ Ursæ Minoris, and between $+ 0^s.02$ and $- 0^s.02$ for 51 Cephei H.; both according to the Star Tables of the "American Ephemeris."

This term is substantially eliminated in the mean of observations above and below pole, and is of very slight consequence in all cases.

TABLE A.

Wire Intervals, Illumination East.

Period.	Dates included.	A to M.	B to M.	C to M.	D to M.	E to M.
		s.	s.	s.	s.	s.
I	1862 March 11 — May 23.	+6.700	+3.524	—0.303	—3.248	—6.673
II	1862 May 27 — 1863 March 12.	6.739	3.393	0.295	3.101	6.736
III	1863 March 23 — July 11.	6.753	3.428	0.336	3.108	6.736
IV	1863 July 20 — Oct. 1.	6.659	3.317	0.255	2.925	6.794
V	1863 Oct. 22 — 1864 Feb. 19.	6.681	3.326	0.238	2.989	6.783
VI	1864 Feb. 23 — 1864 June 17.	6.651	3.356	0.241	2.935	6.836
VII	1864 June 21 — 1864 Sept. 30.	6.683	3.402	0.340	2.913	6.831
VIII	1864 Oct. 5 — 1864 Nov. 16.	6.663	3.381	0.301	2.941	6.803
IX	1864 Nov. 25 — 1865 April 26.	6.663	3.354	0.265	2.913	6.839
X	1865 May 3 — June 13.	6.667	3.384	0.313	2.899	6.839
XI	1865 June 19 — Nov. 11.	6.619	3.385	—0.316	2.869	6.819
XII	1865 Nov. 14 — 1866 Jan. 4.	+6.858	+3.360	+0.078	—3.500	—6.797

TABLE B.

Comparison of Time Stars in Part I. with the Pulcova Catalogue for 1845.

Star's Name.	Wolfers *minus* Pulcova.	Safford *minus* Pulcova.	Star's Name.	Wolfers *minus* Pulcova.	Safford *minus* Pulcova.
	s.	s.		s.	s.
α Andromedæ	+0.011		α Virginis	+0.013	
γ Pegasi	+0.055		ζ Virginis		+0.026
ε Piscium		—0.013	η Bootis		+0.034
θ Ceti		—0.018	τ Virginis		+0.027
η Piscium		—0.001	α Bootis	+0.017	
β Arietis		+0.040	ρ Bootis		+0.050
α Arietis		+0.019	ε Bootis		+0.049
ξ^2 Ceti		+0.026	α^1 Libræ	+0.003	
γ Ceti		+0.019	α^2 Libræ	+0.051	
α Ceti	+0.018		β Libræ		+0.037
η Tauri		+0.018	α Coronæ	+0.046	
γ Eridani		+0.046	α Serpentis	+0.041	
ε Tauri		+0.040	β Scorpii		+0.045
α Tauri	+0.064		δ Ophiuchi		+0.037
ι Aurigæ		+0.073	α Scorpii	+0.030	
α Aurigæ	+0.053		ζ Herculis		+0.012
β Orionis	+0.041		κ Ophiuchi		+0.081
β Tauri	+0.040		α Herculis	+0.065	
δ Orionis		+0.060	α Ophiuchi	+0.050	
ε Orionis		+0.041	μ Herculis		+0.057
α Orionis	+0.077		α Lyræ	+0.006	
μ Geminorum		+0.050	β Lyræ		+0.024
γ Geminorum	+0.016		ζ Aquilæ		+0.029
δ Geminorum		+0.034	δ Aquilæ		+0.013
α Geminorum	[—0.035]		γ Aquilæ	+0.026	
β Geminorum	+0.042		α Aquilæ	+0.029	
ι Navis		+0.056	β Aquilæ	+0.043	
ε Hydræ		+0.052	α^1 Capricorni	+0.042	
α Hydræ	+0.041		α^2 Capricorni	+0.028	
ε Leonis		+0.029	α Cygni	+0.011	
α Leonis	+0.042		ζ Cygni		+0.038
γ Leonis		+0.097	β Aquarii		—0.013
ρ Leonis		+0.059	ε Pegasi		+0.007
δ Leonis		+0.040	α Aquarii	+0.042	
δ Crateris		+0.022	η Aquarii		+0.004
β Leonis	+0.029		ζ Pegasi		+0.002
β Virginis	+0.038		α Pegasi	+0.045	
η Virginis		+0.047	γ Piscium		+0.052
γ Virginis		+0.046	ι Piscium		+0.031
12 Canum Ven.		+0.026	ω Piscium		+0.020

TABLE C.

SPECIMEN OF INSTRUMENTAL CORRECTIONS, 1864 DECEMBER 9 TO 1865 JANUARY 25.

$n = +\overset{s.}{0}.229$ δ	$c = -\overset{s.}{0}.309$ $n \tan \delta + c \sec \delta$	
—40°	$-\overset{s.}{0}.596$	
		107
—30°	—0.489	
		77
—20°	—0.412	
		58
—10°	—0.354	
		45
0°	—0.309	
		35
+10°	—0.274	
		29
+20°	—0.245	
		20
+30°	—0.225	
		13
+40°	—0.212	
		4
+45°	—0.208	
		0
+50°	—0.208	
		4
+55°	—0.212	
		9
+60°	—0.221	
		6
+62°	—0.227	
		8
+64°	—0.235	
		10
+66°	—0.245	
		13
+68°	—0.258	
		16
+70°	—0.274	

TABLE D.

Correction for Neglected Terms of Nutation.

Arguments: Annual Precession at Top, Reduced Mean Date at Side.

	1s	2s	3s	4s	5s	6s	7s	8s	9s	10s
	s.	s.	s.	s.	s.	s.	s.	s.	s.	s.
Jan. 0	.000	.000	—.001	—.001	—.001	—.001	—.001	—.002	—.002	—.002
30	—.002	—.003	—.005	—.006	—.008	—.010	—.011	—.013	—.014	—.016
Mar. 1	—.003	—.005	—.008	—.010	—.013	—.016	—.018	—.021	—.023	—.026
31	—.003	—.006	—.009	—.012	—.014	—.017	—.020	—.023	—.026	—.029
April 30	—.002	—.001	—.007	—.010	—.012	—.015	—.017	—.020	—.022	—.025
May 30	—.001	—.003	—.004	—.006	—.007	—.008	—.010	—.011	—.013	—.014
June 29	.000	.000	.000	.000	.000	.000	.000	.000	.000	.000
July 29	+.001	+.003	+.004	+.006	+.007	+.008	+.010	+.011	+.013	+.014
Aug. 28	+.002	+.005	+.007	+.010	+.012	+.015	+.017	+.020	+.022	+.025
Sept. 27	+.003	+.006	+.009	+.012	+.015	+.017	+.020	+.023	+.026	+.029
Oct. 27	+.003	+.005	+.008	+.010	+.013	+.016	+.018	+.021	+.023	+.026
Nov. 26	+.002	+.003	+.005	+.006	+.008	+.010	+.011	+.013	+.014	+.016
Dec. 26	.000	.000	.000	+.001	+.001	+.001	+.001	+.001	+.001	+.001

These corrections are to be added to mean right-ascensions obtained by the use of the "Nautical Almanac," or subtracted from the apparent right-ascensions of that work.

TABLE OF INDIVIDUAL CLOCK CORRECTIONS

FROM COMPUTED RIGHT ASCENSIONS,

WITH MEANS OF INSTRUMENTAL CORRECTIONS.

Date.	Name of Star.	Value of $\Delta t+m$	Date.	Name of Star.	Value of $\Delta t+m$	Date.	Name of Star.	Value of $\Delta t+m$
1862.		s.	1862.		s.	1862.		s.
March 11	α Canis Maj.	5.06	April 4	α Tauri	4.96	April 24	β Leonis	14.46
H.	α Geminorum	5.19	H.	α Hydræ	4.64	S.	β Virginis	14.44
	β Geminorum	4.95		α Leonis	4.53			
				χ Leonis	4.35*	April 24	12 Canum	14.39*
March 24	β Orionis	32.57		δ Leonis	4.54*	H.	α Virginis	14.32
S.	β Tauri	32.50					ζ Virginis	14.31*
	α Geminorum	32.22	April 7	η Cancri	56.17*			
	α Canis Min.	32.28	S.	ε Hydræ	56.20*	April 25	α Hydræ	12.76
	β Geminorum	32.16		α Hydræ	56.07	S.	α Leonis	12.61
				β Leonis	55.88		α Virginis	12.39
March 26	β Tauri	27.34		β Virginis	55.84			
S.	α Orionis	27.35		γ Virginis	55.89*	April 26	l Leonis	10.09*
	α Canis Maj.	27.20				H.	χ Leonis	10.08*
	β Leonis	26.70	April 12	α Leonis	43.77		β Virginis	9.97
	β Virginis	26.62	H.	ρ Leonis	43.74*		α Virginis	9.83
	α Virginis	26.55		δ Crateris	43.73*			
				β Leonis	43.58	April 29	η Virginis	2.75*
March 27	η Cancri	24.45*				S.	γ Virginis	2.76*
H.	α Hydræ	24.41	April 16	α Leonis	35.11		α Virginis	2.76
	ε Leonis	24.40*	H.	β Leonis	34.97			
	ρ Leonis	24.22*		β Virginis	34.90	April 30	l Leonis	0.78*
				α Virginis	34.80	H.	β Leonis	0.81
March 28	α Tauri	22.17					β Virginis	0.73
S.	β Orionis	22.11	April 18	α Leonis	30.60		α Virginis	0.66
	β Tauri	22.04	H.	ρ Leonis	30.61*			
	α Orionis	22.09		l Leonis	30.46*	May 7	l Leonis	44 39*
	α Canis Min.	21.95		χ Leonis	30.43*	S.	β Leonis	44.23
	β Geminorum	21.97					β Virginis	44.12
	α Hydræ	21.74	April 19	α Hydræ	28.47		τ Virginis	43.98*
			S.	ε Leonis	28.45*			
March 29	α Hydræ	19.31		l Leonis	28.23*	May 14	α Geminorum	28.91
H.	α Leonis	19.25		δ Leonis	28.26*	H.	α Canis Min.	28.84
	β Leonis	19.06	April 23				β Geminorum	28.87
	β Virginis	19.00	S.	β Virginis	17.34		α Virginis	28.18

Date	n	c	Date	n	c
	s.	s.		s.	s.
March 11	$n = +0.341$	$c = +0.050$	April 19 to 9.50	$n = +0.258$	$c = +0.029$
24–27	+0.293	+0.086	19 after 9.50	+0.258	+0.072
28	+0 278	+0.074	24	+0 008	+0.144
29–April 4	+0.344	−0.080	25	+0.216	+0.144
April 7	+0.195	+0.146	26 to 30	+0.193	−0.047
12	+0.289	−0.053	May 7	+0.234	−0.047
16–18	+0.168	−0.032	14, 15	+0.259	+0.032

Date.	Name of Star.	Value of $\Delta t + m$
1862.		s.
May 14 H.	α Bootis	28.04
	α^1 Libræ	27.97
	α^2 Libræ	27.98
	α Coronæ	27.94
	α Serpentis	27.90
May 15 S.	β Leonis	25.84
	β Virginis	25.82
May 17 S.	α Virginis	21.46
	α^1 Libræ	21.21
	α^2 Libræ	21.17
May 23 S.	α Leonis	7.27
	α Virginis	6.96
	α Bootis	7.03
	α^1 Libræ	6.79
	α^2 Libræ	6.74
	α Coronæ	6.81
	α Serpentis	6.63
	α Scorpii	6.66
May 28 H.	η Bootis	54.94*
	α Bootis	54.91
	ρ Bootis	54.87*
June 5 H.	η Bootis	34.51*
	τ Virginis	34.48*
June 11 H.	α Bootis	18.86
	α^1 Libræ	18.82
	α^2 Libræ	18.83
	α Serpentis	18.66
June 16 S.	α Virginis	5.77
	α Bootis	5.66
	α Coronæ	5.60
	α Serpentis	5.46
	α Scorpii	5.37
June 17 H.	α Virginis	3.50
	α Bootis	3.45
	α^1 Libræ	3.38
	α^2 Libræ	3.38
	α Coronæ	3.38
	α Serpentis	3.28

Date.	Name of Star.	Value of $\Delta t + m$
1862.		s.
July 3 H.	α Coronæ	21.56
	α Serpentis	21.57
	α Scorpii	21.54
July 18 S.	α Virginis	45.21
	α Coronæ	44.99
	α Serpentis	44.92
	α Scorpii	44.92
	γ Aquilæ	44.49
	α Aquilæ	44.50
	β Aquilæ	44.40
	α^2 Capricorni	44.48
July 22 S.	α Coronæ	34.57
	α Serpentis	34.44
	α Scorpii	34.45
	α Herculis	34.33
	α Ophiuchi	34.32
July 26 S.	α Ophiuchi	23.38
	γ Aquilæ	23.27
	α Aquilæ	23.20
	β Aquilæ	23.22
July 28 S.	α Bootis	18.98
Aug. 1	β Scorpii	5.86*
	ζ Herculis	5.92*
	κ Ophiuchi	5.91*
Aug. 6	α Bootis	52.12
	α Herculis	51.83
	α Ophiuchi	51 82
	ω Aquilæ	51.46*
	δ Aquilæ	51.32*
Aug. 11	α Herculis	37.38
	α Ophiuchi	37.34
	γ Aquilæ	37.12
	β Aquilæ	37.06
Aug. 16	α Ophiuchi	24.60
Aug. 20	α Bootis	15.69
	γ Aquilæ	15.13
	β Aquilæ	15.15

Date.	Name of Star.	Value of $\Delta t + m$
1862.		s.
Aug. 21	α Virginis	13.27
	α Bootis	13.23
	α Herculis	13.00
	γ Aquilæ	12.68
	β Aquilæ	12.69
	α^1 Capricorni	12.58
	α^2 Capricorni	12.61
Aug. 29	α Bootis	53.65
	β Lyræ	53.10*
Sept. 2	α Ophiuchi	44.28
	γ Aquilæ	44.03
	α^1 Capricorni	43.91
	α^2 Capricorni	43.87
Sept. 4	α Herculis	39.38
	γ Aquilæ	39.14
	β Aquilæ	39.13
	α Aquarii	38.91
	α Piscis Aus.	38.74
	α Pegasi	38.85
Sept. 8	κ Ophiuchi	30.62*
	α Herculis	30.61
	μ Sagittarii	30.46*
	ρ Capricorni	30.19*
	61 Cygni	30.09*
Sept. 9	α Herculis	27.95
	θ Ophiuchi	27.96*
Sept. 10	α Bootis	25.75
	α Herculis	25.57
	α Ophiuchi	25.50
	α^1 Capricorni	25.22
	α^2 Capricorni	25.23
	α Aquarii	24.85
	α Piscis Aus.	24.77
Sept. 11	α Herculis	23.16
	α Ophiuchi	23.09
	γ Aquilæ	22.86
	β Aquilæ	22.88
	α^1 Capricorni	22.75
	α^2 Capricorni	22.70

Date		n	c
		s.	s.
May	17, 24	$n = +0.159$	$c = -0.080$
	23–June 17	+0.186	+0.184
July	18–22	+0.278	+0.140
	26	+0.255	−0.140
Aug.	6	+0.167	−0.150
	11	+0.129	−0.150

Date		n	c
		s.	s.
Aug.	20	$n = +0.163$	$c = -0.150$
	21 to 17.28	+0.174	−0.150
	21 (17.28) –Sept. 4	+0.287	+0.108
Sept.	8–9	+0.215	−0.125
	10–27 (Polaris sp.)	+0.285	−0.125

Date.	Name of Star.	Value of $\Delta t+m$
1862.		s.
Sept. 15	α Herculis	13.91
	α Ophiuchi	13.89
Sept. 16	α Scorpii	11.74
	α Herculis	11.71
	α Ophiuchi	11.77
	γ Aquilæ	11.47
Sept. 17	α Herculis	9.68
	α Ophiuchi	9.58
	α Orionis	8.44
	α Canis Maj.	8.29
Sept. 18	α Herculis	7.46
	μ Herculis	7.46*
Sept. 21	α Canis Maj.	56.89
	α Leonis	55.94
Sept. 22	ζ Herculis	55.09*
Sept. 23	α Herculis	52.70
	α Ophiuchi	52.62
	α Aquarii	52.10
	α Piscis Aus	52.15
	α Pegasi	52.08
	α Andromedæ	52.03
	γ Pegasi	51.94
Sept. 25	α Herculis	48.32
	α Ophiuchi	48.12
	α Lyræ	48.11
	α Cygni	47.99
	32 Vulpeculæ	47.86*
Sept. 27	μ Sagittarii	43.16*
Sept. 29	α Ophiuchi	39.00
	α^1 Capricorni	38.76
	α^2 Capricorni	38.82
	α Cygni	38.81
	α Andromedæ	38.52
	γ Pegasi	38.55
Oct. 5	α Tauri	24.53
	α Aurigæ	24.59
	β Orionis	24.53

Date.	Name of Star.	Value of $\Delta t+m$
1862.		s.
Oct. 5	α Orionis	24.44
	α Canis Maj.	23.97
	α Geminorum	24.13
	α Canis Min.	24.08
	β Geminorum	24.11
	α Leonis	23.79
	γ Leonis	23.93*
	δ Leonis	23.74*
Oct. 6	θ Ophiuchi	23.24
	α Ophiuchi	23.19
Oct. 7	α Ophiuchi	21.33
	α^1 Capricorni	21.22
	α^2 Capricorni	21.13
	α Cygni	21.15
	α Aquarii	21.09
	α Canis Maj.	20.21
	α Geminorum	20.20
	α Canis Min.	20.03
	β Geminorum	20.19
Oct. 8	α Lyræ	19.32
	α^1 Capricorni	19.22
	α^2 Capricorni	19.23
	α Cygni	19.07
Oct. 9	α Lyræ	17.41
Oct. 14	β Leonis	3.20
Oct. 16	α Herculis	59.55
	α Ophiuchi	59.59
Oct. 18	α Ophiuchi	54.40
	α Lyræ	54.37
	α Aquilæ	54.13
	β Aquilæ	54.13
Oct. 30	α Piscis Aus.	22.01
	α Pegasi	22.17
Nov. 1	α Aquilæ	17.47
Nov. 3	γ Aquilæ	13.20
	α Aquilæ	13.34
	β Aquilæ	13.24

Date.	Name of Star.	Value of $\Delta t+m$
1862.		s.
Nov. 3	α Aquarii	12.92
	α Andromedæ	12.93
	γ Pegasi	12.95
Nov. 4	32 Vulpeculæ	10.62*
	ζ Cygni	10.60*
	α Aquarii	10.47
	α Piscis Aus.	10.33
	α Pegasi	10.41
Nov. 10	γ Pegasi	53.02
	12 Ceti	52.85*
	β Ceti	52.83*
	α Bootis	51.42
Nov. 13	α Aquilæ	45.85
	β Aquilæ	45.84
	α^1 Capricorni	45.76
	α^2 Capricorni	45.80
Dec. 6	ε Pegasi	38.86*
	16 Pegasi	38.85*
Dec. 8	β Ceti	32.97*
	ε Piscium	32.87*
	β Arietis	32.87
	α Arietis	32.77
Dec. 10	ε Piscium	27.35*
	θ Ceti	27.34*
	α Arietis	27.38
	α Ceti	27.22
	ε Corvi	26.06*
	α Virginis	25.98
	ε Bootis	25.87*
Dec. 11	α Aquarii	25.27
	α Piscis Aus.	25.10
	α Pegasi	25.13
	ε Tauri	24.52*
	α Tauri	24.59*
Dec. 12	α Piscis Aus.	22.97
	α Pegasi	23.04
Dec. 16	β Ceti	11.85*
	ε Piscium	11.86*

Date		n	c
Sept.	27 (18.04)–29	$n = +$ 0.259 s.	$c = +$ 0.110 s.
Oct.	5, 6	+ 0.326	+ 0.110
	7, 8 (to 20.30)	+ 0.203	+ 0.110
	8 (20.30), 9	+ 0.113	− 0.075
	18	+ 0.316	− 0.075
	30	+ 0.293	− 0.075

Date		n	c
Nov.	3	$n = +$ 0.233 s.	$c = +$ 0.045 s.
	4	+ 0.371	+ 0.045
	10	+ 0.336	+ 0.045
Dec.	8	+ 0.287	+ 0.045
	12–16	+ 0.267	+ 0.012

Date.	Name of Star.	Value of $\Delta t+m$
1862.		s.
Dec. 16	β Arietis . .	11.90*
	α Arietis . .	11.84
	ξ^2 Ceti . . .	11.80*
Dec. 20	α Pegasi . .	1.23
	α Ceti . .	0.89
	δ Arietis . .	1.14*
	α Persei . .	1.15*
Dec. 24	η Piscium . .	50.50*
Dec. 27	β Arietis . .	42.67*
	α Arietis . .	42.62
Dec. 28	α Andromedæ	39.97
	γ Pegasi . .	40.01
	β Leonis . .	38.48
	β Virginis . .	38.42
	α Virginis . .	38.34
Dec. 29	α Pegasi . .	37.69
	γ Pegasi . .	37.48
1863.		
Jan. 3	α Andromedæ	25.76
	γ Pegasi . .	25.85
Jan. 5	β Ceti . . .	20.62*
	η Tauri . .	20.36*
	α Tauri . .	20.34
Jan. 16	α Lyræ . .	52.04
Jan. 17	α Pegasi . .	51.55
	α Andromedæ	51.45
	γ Pegasi . .	51.39
Jan. 19	ε Piscium . .	46.00*
	α Tauri . .	45.70
	α Aurigæ . .	45.69
Jan. 30	α Arietis . .	20.42
Jan. 31	α Aurigæ . .	17.82
	β Orionis . .	17.58
1863.		s.
Feb. 2	α Ceti . . .	13.15
	β Orionis . .	13.02
	α Orionis . .	12.90
	α Canis Maj. .	12.77
Feb. 7	α Ceti . . .	59.71
	β Geminorum	59.59
	ε Hydræ . .	59.40*
	83 Cancri . .	59.34*
Feb. 10	α Arietis . .	54.22
Feb. 14	η Tauri . .	45.45*
	γ Eridani . .	45.52*
	o^1 Eridani . .	45.57*
Feb. 16	β Arietis . .	41.50*
	γ Ceti . . .	41.31*
	α Ceti . . .	41.31
	α Canis Min. .	41.00
Feb. 17	γ Eridani . .	39.21*
Feb. 18	α Arietis . .	37.45
Feb. 20	α Orionis . .	33.79
	ν Orionis . .	33.74*
	δ Geminorum	33.62*
	α Coronæ . .	32.88
	α Serpentis .	32.90
Feb. 21	α Geminorum	31.75
	β Geminorum	31.90
Feb. 23	β Arietis . .	27.59*
	α Ceti . . .	27.63
	α Orionis . .	27.50
	γ Canis Maj. .	27.29*
Feb. 25	α Hydræ . .	23.38
	ε Leonis . .	23.45
March 2	α Aurigæ . .	14.08
	β Orionis . .	14.06
	α Geminorum	14.10
	β Geminorum	13.97
	α Hydræ . .	13.83
1863.		s.
March 5	γ Geminorum	5.17*
	ε Canis Maj. .	5.08*
	γ Canis Maj. .	4.89*
	η Cancri . .	4.80*
March 6	ρ Leonis . .	1.46*
	l Leonis . .	1.40*
	β Leonis . .	1.41
	β Virginis . .	1.37
March 9	α Aurigæ . .	52.67
	β Orionis . .	52.68
	α Geminorum	52.34
	α Canis Min. .	52.27
	β Geminorum	52.30
	l Leonis . .	51.80*
	χ Leonis . .	51.90*
March 11	α Canis Maj. .	46.89
	α Geminorum	46.84
	α Canis Min. .	46.78
	β Geminorum	46.65
	α Hydræ . .	46.67
March 23	χ Geminorum	4.32*
March 27	α Hydræ . .	51.96
March 30	ε Hydræ . .	42.62*
	α Hydræ . .	42.58
April 1	ε Hydræ . .	36.31*
	83 Cancri . .	36.12*
	α Hydræ . .	36.19
April 6	β Geminorum	19.76
April 9	α Canis Maj. .	11.73
	α Hydræ . .	11.54
	α Leonis . .	11.44
	β Leonis . .	11.23
	β Virginis . .	11.15
	α Virginis . .	11.06
April 18	α Leonis . .	45.02
	γ Leonis . .	44.92*

Date	n	c
	s.	s.
Dec. 20	$n = +0.169$	$c = +0.012$
24	$+0.370$	$+0.012$
27	$+0.254$	$+0.012$
28–29	$+0.233$	$+0.012$
1863.		
Jan. 3–Feb. 2	$+0.259$	-0.042
Feb. 7	$+0.304$	-0.042
Feb. 10–16	$n = +0.262$	$c = -0.042$
17	$+0.295$	-0.042
18–21	$+0.201$	-0.042
23–March 6 (11.15)	$+0.261$	-0.042
March 6 (11.15)–11	$+0.292$	$+0.092$
23–April 9	$+0.192$	$+0.092$
April 18–May 15	$+0.003$	$+0.092$

Date.	Name of Star.	Value of $\Delta t+m$
1863.		s.
April 21	α Canis Maj. .	35.55
	α Hydræ . .	35.29
April 22	α Hydræ . .	32.36
	θ Virginis. .	31.87
April 23	α Hydræ . .	29.51
April 27	α Canis Maj. .	17.18
May 1	α Virginis. .	4.83
May 2	γ Leonis . .	2.79*
May 9	χ Leonis . .	41.12*
	α Virginis. .	40.65
May 15	ψ Bootis . .	21.63*
May 18	α Serpentis .	12.35
May 19	χ Leonis . .	10.12*
	δ Leonis . .	10.29*
	α Bootis . .	9.77
May 23	β Virginis. .	1.06
	α Bootis . .	0.72
May 25	α Virginis. .	54.71
	α Coronæ . .	54.60
	α Serpentis .	54.35
May 27	B A C 6651**	47.71
June 1	ε Virginis **	36.37
June 2	α Bootis . .	33.76
	α Coronæ . .	33.73
June 3	α^2 Libræ . .	30.96
June 4	β Libræ . .	27.85*
June 5	32 Vulpeculæ.	24.27*
	ζ Cygni . .	24.20*
June 11	ζ Virginis. .	10.00*

Date.	Name of Star.	Value of $\Delta t+m$
1863.		s.
June 17	α Ophiuchi .	50.31
June 23	32 Vulpeculæ	32.56*
	ε Pegasi . .	32.54*
June 27	ψ Bootis . .	22.07*
	α Coronæ . .	22.05
June 29	16 Pegasi . .	16.24*
June 30	α Pegasi . .	13.82
July 11	α Scorpii . .	44.29
	α Herculis .	44.29
	α Ophiuchi .	44.23
July 20	α Serpentis .	18.24
	δ Ophiuchi .	18.13*
July 21	α Andromedæ	13.96
	γ Pegasi . .	13.97
July 22	β Libræ . .	12.12*
	α Scorpii . .	11.97
	α Andromedæ	10.99
	γ Pegasi . .	10.92
July 23	β Lyræ. . .	8.61*
	δ Aquilæ . .	8.59*
July 24	α Lyræ . .	5.84
July 29	α Lyræ . .	53.99
	16 Pegasi . .	53.54*
	α Aquarii . .	53.49
July 31	α Ophiuchi .	48.91
	α Aquilæ . .	48.61
	β Aquilæ . .	48.70
Aug. 3	α Herculis .	41.79
	α Ophiuchi .	41.74
	α^1 Capricorni	41.59
	α^2 Capricorni	41.63
Aug. 4	α^1 Capricorni	39.20
	α^2 Capricorni	39.29

Date.	Name of Star.	Value of $\Delta t+m$
1863.		s.
Aug. 6	α Andromedæ	33.66
	γ Pegasi . .	33.85
Aug. 10	α Ophiuchi .	52.37
	γ Pegasi . .	51.58
Aug. 12	α Ophiuchi .	47.19
	β Aquilæ . .	46.97
	α^1 Capricorni	46.96
	α^2 Capricorni	46.89
	16 Pegasi . .	46.70
Aug. 14	γ Aquilæ . .	40.73
	β Aquilæ . .	40.78
	α^1 Capricorni	40.71
	α^2 Capricorni	40.71
	ζ Cygni . .	40.62*
Aug. 15	μ Herculis .	38.33*
	β Lyræ . .	38.15*
Aug. 17	γ Aquilæ . .	31.80
	β Aquilæ . .	31.81
	α^2 Capricorni	31.71
Aug. 18	β Aquilæ . .	28.83
	α^1 Capricorni	28.79
	α^2 Capricorni	28.81
	α Cygni . .	28.87
	32 Vulpeculæ	28.87*
Aug. 19	γ Aquilæ . .	26.27
	β Aquilæ . .	26.25
	α^2 Capricorni	26.21
Aug. 22	α Ophiuchi .	18.37
	ω Aquilæ . .	18.04*
Aug. 24	α Bootis . .	13.56
	α Herculis .	13.31
	α Ophiuchi .	13.22
Aug. 30	α Andromedæ	53.75
	γ Pegasi . .	53.82
Aug. 31	θ Cygni .	51.22**

Dates	n	c
	s.	s.
May 18–June 4	$n = +0.079$	$c = -0.122$
June 5	$+0.047$	$+0.167$
6–17	-0.001	$+0.399$
23–July 11	-0.026	$+0.399$
July 20	$+0.104$	$+0.399$

Dates	n	c
	s.	s.
July 21–24	$n = -0.076$	$c = +0.399$
28–Aug. 3	-0.014	$+0.399$
Aug. 4–12	$+0.031$	-0.482
14–Sept. 4	-0.018	$+0.412$

Date.	Name of Star.	Value of $\Delta t+m$
1863.		s.
Sept. 2	α Ophiuchi .	44.75
	α Cygni . .	44.32
Sept. 4	ζ Cygni . .	37.27*
	α Aquarii .	37.29
Sept. 5	β Lyræ . . .	34.47*
	γ Aquilæ . .	34.38
Sept. 7	α Cygni . .	27.90
Sept. 9	α Cassiopeiæ.	21.19*
	ε Piscium . .	20.91*
	θ Ceti . . .	20.85*
Sept. 10	ζ Aquilæ . .	17.82*
	ξ² Ceti . . .	16.71*
Sept. 14	α Cygni . .	3.61
Sept. 15	β Lyræ . .	0.85*
	α¹ Capricorni	0.64
	α² Capricorni	0.62
	32 Vulpeculæ	0.49*
Sept. 16	α Arietis . .	56.94
Sept. 17	32 Vulpeculæ	54.53*
	ζ Cygni . .	54.54*
	α Pegasi . .	54.29
Sept. 20	α Leonis . .	42.49
Sept. 21	β Lyræ . .	41.20*
	β Aquilæ . .	41.03
Sept. 22	α Aquilæ . .	37.43
	α² Capricorni.	37.30
	α Cygni . .	37.02
	α Piscis Aus..	36.73
	κ Piscium . .	36.67*
	ι Piscium . .	36.61*
Sept. 23	α Aquilæ . .	33.90
	α¹ Capricorni	33.91
	α² Capricorni	33.77
	α Cygni . .	33.61

Date.	Name of Star.	Value of $\Delta t+m$
1863.		s.
Sept. 24	α Piscis Aus..	29.36
	α Andromedæ	29.35
	γ Pegasi . .	29.28
Sept. 28	α¹ Capricorni	15.72
	α² Capricorni	15.63
	α Cygni . .	15.60
	α Piscis Aus..	15.24
	ω Piscium . .	14.96*
	ε Piscium . .	14.96*
Sept. 29	α Aquarii . .	11.74
	α Piscis Aus..	11.58
	α Pegasi . .	11.49
	γ Pegasi . .	11.47
Sept. 30	μ Herculis .	8.90*
	γ Draconis .	8.99*
	α Pegasi . .	8.05
	γ Pegasi . .	7.95
	ε Piscium .	7.76*
Oct. 1	β Lyræ . . .	5.10*
	ϱ Capricorni .	4.81*
Oct. 22	α Piscis Aus. .	48.54
	γ Piscium . .	48.38*
	κ Piscium . .	48.30*
Oct. 23	γ Aquilæ . .	44.99
Oct. 28	α¹ Capricorni	27.51
	α² Capricorni	27.45
	α Piscis Aus..	27.11
	α Pegasi . .	27.17
	θ Ceti . . .	26.60*
	ν Piscium . .	26.60*
Oct. 29	α¹ Capricorni	24.03
	α² Capricorni	24.09
	12 Ceti . . .	23.49*
Nov. 2	α Aquarii . .	10.49
	ω Piscium . .	10.11*
	12 Ceti . . .	10.16*
	β Ceti . . .	10.09*

Date.	Name of Star.	Value of $\Delta t+m$
1863.		s.
Nov. 3	α Aquarii . .	7.22
	α Andromedæ	7.01
	γ Pegasi . .	7.07
	β Arietis . .	6.80*
	α Arietis . .	6.89
Nov. 4	α¹ Capricorni	4.09
	α² Capricorni	4.05
	α Cygni . .	3.95
	α Andromedæ	3.48
	η Piscium . .	3.38*
	α Arietis . .	3.43
	ξ² Ceti . . .	3.22*
Nov. 19	32 Vulpeculæ	23.85*
	ζ Cygni . .	23.79*
	α Virginis . .	21.86
Nov. 22	α Virginis . .	12.68
Nov. 23	α Ceti . . .	11.21
	η Tauri . . .	11.11*
	γ Eridani . .	11.11*
Nov. 25	ε Pegasi . .	6.57*
	16 Pegasi . .	6.55*
	α Aquarii . .	6.56
Nov. 26	γ Piscium . .	3.40*
	κ Piscium . .	3.57*
	θ Ceti . . .	3.25*
Nov. 30	12 Ceti . .	58.43*
	α Arietis . .	58.15
	α Virginis . .	56.37
Dec. 1	α Cygni . .	55.42
	ε Pegasi . .	55.33*
	16 Pegasi . .	55.26*
Dec. 6	α Virginis . .	38.93
	α Bootis . .	38.58
Dec. 7	α Aquarii . .	37.88
	ε Piscium . .	37.70*
Dec. 9	α Pegasi . .	32.44

Date	n (s.)	c (s.)	Date	n (s.)	c (s.)
Sept. 5–10	+ 0.192	− 0.428	Nov. 3	+ 0.318	− 0.430
14–17	+ 0.109	+ 0.390	4	+ 0.244	− 0.430
21–22	+ 0.241	+ 0.390	19–26	+ 0.327	− 0.430
23–Oct. 1	+ 0.276	+ 0.390	29–Dec. 2	+ 0.386	− 0.430
Oct. 22, 23	+ 0.143	+ 0.395	Dec. 6–7	+ 0.446	− 0.430
28–Nov. 2	+ 0.396	− 0.430	9–15	+ 0.367	− 0.430

Date.	Name of Star.	Value of $\Delta t + m$
1863.		s.
Dec. 9	ν Piscium . .	32.05*
	α Arietis . .	32.09
Dec. 10	β Aquarii . .	29.45*
	α Piscis Aus. .	29.29
	ε Piscium . .	29.18*
Dec. 15	α Piscis Aus. .	15.60
	α Pegasi . .	15.64
	α Arietis . .	15.30
	α Ceti . . .	15.19
	α Tauri . .	15.03
	ε Orionis . .	14.84*
Dec. 21	ε Piscium . .	55.60*
	α Arietis . .	55.64
	α Ceti . . .	55.40
	α Tauri . .	55.12
Dec. 22	ξ² Ceti . . .	52.49*
	γ Ceti . . .	52.41*
Dec. 23	α Piscis Aus. .	49.97
	α Pegasi . .	49.89
	α Arietis . .	49.44
	ξ² Ceti . . .	49.33*
Dec. 26	α Andromedæ	39.99
	γ Pegasi . .	39.96
	η Tauri . .	39.48*
	γ Eridani . .	39.50*
	α Tauri . .	39.48
Dec. 29	θ Ceti . . .	29.82*
	η Piscium . .	29.70*
	β Arietis . .	29.72*
	α Arietis . .	29.75
	α Tauri . .	29.42
	α Aurigæ . .	29.26
	β Orionis . .	29.32
	β Tauri . .	29.33
	α Canis Maj. .	29.17
Dec. 30	η Piscium . .	27.08*
	α Arietis . .	26.94
	ξ² Ceti . . .	26.79*
	γ Eridani . .	26.58*

Date.	Name of Star.	Value of $\Delta t + m$
1864.		s.
Jan. 1	θ Ceti . . .	20.16*
	η Piscium . .	20.29*
Jan. 2	α Pegasi . .	16.80
Jan. 5	α Arietis . .	7.21
	α Ceti . . .	7.14
Jan. 6	ω Piscium . .	4.54*
	γ Pegasi . .	4.51
	θ Ceti . .	4.35*
	η Piscium . .	4.25*
Jan. 8	γ Virginis . .	55.79*
	α Virginis. .	55.64
	η Bootis . .	55.53*
Jan. 9	α Andromedæ	54.23
	γ Pegasi . .	54.41
	α Arietis . .	54.12
	α Ceti . . .	53 87
Jan. 10	ε Piscium . .	50.92*
	θ Ceti . . .	50.78*
Jan. 11	ε Piscium . .	46.99*
	θ Ceti . . .	46.84*
	α Ceti . . .	46.65
	α Tauri . .	46.60
	α Orionis . .	46.45
Jan. 12	θ Ceti . . .	44.32*
	γ Eridani . .	43 97*
	o¹ Eridani . .	43.98*
Jan. 14	α Andromedæ	38.56
	γ Pegasi . .	38.58
	α Arietis . .	38.41
	α Canis Maj. .	37.78
	α Canis Min. .	37.75
	β Geminorum	37.72
Jan. 16	β Ceti . . .	32.42*
	ε Piscium . .	32.41*
	θ Ceti . . .	32.36*
	γ Geminorum.	31.63*
	α Canis Maj. .	31.48

Date.	Name of Star.	Value of $\Delta t + m$
1864.		s.
Jan. 21	ε Piscium . .	16.14*
	θ Ceti . . .	16.04*
	η Piscium . .	16.10*
	α Arietis . .	16.07
	γ Eridani . .	15.89*
	o¹ Eridani . .	15.84*
Jan. 25	δ Arietis . .	5.34*
	ν Orionis . .	5.11*
	γ Geminorum.	5.12
	α Canis Maj. .	4.96
Jan. 27	α Andromedæ	0.57
	γ Pegasi . .	0.66
	α Canis Maj. .	0.05
	α Geminorum	59.95
	α Canis Min. .	59.94
	β Geminorum	59.86
Feb. 4	η Piscium . .	39.52*
Feb. 5	α Andromedæ	37.01
	α Arietis . .	36.73
	β Orionis . .	36.84
	α Canis Maj. .	36.70
Feb. 8	ξ² Ceti . . .	29.43*
	α Canis Maj. .	29.03
	ε Canis Maj. .	29.04*
Feb. 10	β Arietis . .	24.45*
	α Ceti . . .	24.10
	δ Arietis . .	24.19*
Feb. 12	α Arietis . .	19.29
	α Ceti . . .	19.05
	α Tauri . .	19.04
Feb. 18	α Tauri . .	3.03
	α Aurigæ . .	3.15
	β Orionis . .	3.13
	α Orionis . .	3.06
Feb. 19	γ Ceti . . .	0.75*
	β Orionis . .	0.45
	α Orionis . .	0.26

Date		n (s.)	c (s.)
Dec.	21–22	$n = +0.477$	$c = -0.430$
	23	+0.394	−0.430
	26–29	+0.277	+0.343
	30–Jan. 10, 1864	+0.230	+0.343
Jan.	11–12	+0.291	+0.343
	14	+0.239	+0.343
	16	+0.247	−0.329

Date		n (s.)	c (s.)
Jan.	21–27	$n = +0.210$	$c = -0.329$
Feb.	4–5 (1.58)	+0.181	−0.329
	5 (1.58) –12	+0.223	−0.015
	13	+0.266	−0.015
	18	+0.363	−0.015
	19	+0.506	−0.015

Date.	Name of Star.	Value of $\Delta t+m$
1864.		s.
Feb. 23	α Arietis . .	49.31
	α Ceti . . .	49.33
	α Geminorum	48.88
	α Canis Min. .	48.97
	β Geminorum	48.99
Feb. 24	α Geminorum	46.85
	α Canis Min. .	46.76
	β Geminorum	46.78
Feb. 25	α Canis Min. .	44.41
	β Geminorum	44.44
Feb. 26	α Canis Maj. .	42.07
	α Canis Min. .	41.96
Feb. 29	α Aurigæ . .	34.37
	β Orionis . .	34.42
March 3	α Arietis . .	27.64
	α Ceti . . .	27.69
	α Geminorum	27.18
	α Canis Min. .	27.17
	β Geminorum	27.19
March 7	β Geminorum	16.93
	α Hydræ . .	16.95
March 8	α Arietis . .	15.17
March 9	α Canis Maj. .	12.47
	α Geminorum	12.43
	α Canis Min. .	12.48
	β Geminorum	12.40
March 12	α Aurigæ . .	5.43
	β Orionis . .	5.46
	β Tauri . .	5.34
	83 Cancri . .	4.99*
	α Hydræ . .	5.07
March 14	α Tauri . .	0.35
	π Leonis . .	59.79*
	α Leonis . .	59.82
March 16	α Hydræ . .	55.40

Date.	Name of Star.	Value of $\Delta t+m$
1864.		s.
March 17 S.	α Leporis . .	53.49*
	ε Orionis* . .	53.28
March 17 R.	83 Cancri . .	53.06*
March 19 S.	α Orionis . .	48.60
	α Canis Maj. .	48.56
March 19 R.	α Geminorum	48.54
	β Geminorum	48.55
	α Hydræ . .	48.55
March 21 S.	α Canis Maj. .	27.26
	ε Canis Maj. .	27.17*
	ε Leonis . .	27.04*
	π Leonis . .	27.03*
March 21 R.	α Geminorum	27.28
	α Canis Min. .	27.28
	β Geminorum	27 36
	α Hydræ . .	27.07
March 24 S.	α Geminorum	21.33
	α Canis Min. .	21.35
	β Geminorum	21.32
	β Leonis . .	21.01
	β Virginis . .	20.96
March 24 R.	83 Cancri . .	21.22*
	θ Ursæ Maj. .	21.20*
April 7 R.	χ Geminorum	51.69*
	η Cancri . .	51.80*
	ε Hydræ . .	51.77*
	ε Leonis . .	51.64*
April 8 S.	η Cancri . .	49.88*
	π Leonis . .	49.67*
	α Leonis . .	49.66
April 9 R.	ε Hydræ . .	48.22*
	83 Cancri . .	48.06*
	α Hydræ . .	48.02
	l Leonis . .	47.94*
	χ Leonis . .	47.84*

Date.	Name of Star.	Value of $\Delta t+m$
		s.
April 15 S.	α Hydræ . .	34.22
	α Leonis . .	34.07
April 16 R.	α Hydræ . .	32.26
	α Leonis . .	32.10
April 21 S.	α Leonis . .	22.24
	α Virginis . .	21.96
	α¹ Libræ . .	21.76
	α² Libræ . .	21.78
April 21 R.	β Leonis . .	22.23
	β Corvi . .	22.02*
April 22 S.	α Hydræ . .	20.34
	β Leonis . .	20.12
	β Virginis . .	20.07
	α Virginis . .	20.01
April 27 S.	α Hydræ . .	11.26
	ε Leonis . .	11.25*
April 29 S.	α Virginis . .	7.35
	α Bootis . .	7.38
	α¹ Libræ . .	7.22
	α² Libræ . .	7.24
	α Coronæ . .	7.32
April 30 R.	α Leonis . .	6.08
	β Virginis . .	5.97
	α Virginis . .	5.93
May 2 S.	α Geminorum	2.31
	β Geminorum	2.29
May 3 S.	β Geminorum	0.53
	ε Hydræ . .	0.27*
	γ Leonis . .	0.28*
	α Virginis . .	0.07
	α Bootis . .	59.95
	α¹ Libræ . .	59.81
	α² Libræ . .	59.90
May 4 S.	α Leonis . .	58.46
	γ Leonis . .	58.38*
	θ Virginis . .	58.12*
	α Bootis . .	58.08

Dates	n (s.)	c (s.)
Feb. 23	$n = +0.415$	$c = -0.015$
24–March 8	$+0.314$	-0.015
March 9–19	$+0.230$	$+0.095$
21–24	$+0.173$	$+0.396$
April 6–27	$n = +0.114$	$c = +0.396$
29	$+0.057$	-0.401
30–May 4	$+0.135$	-0.401

Date.	Name of Star.	Value of $\Delta t+m$
1864.		s.
May 5 S.	α Leonis	56.53
	α Bootis	56.25
	α^1 Libræ	56.12
	α^2 Libræ	56.19
May 6 S.	α Leonis	55.09
	γ Leonis	55.05*
	β Virginis	54.95
	α Virginis	55.06
May 9 R.	ε Corvi	50.77*
	α Virginis	50.59
	η Bootis	50.57*
	τ Virginis	50.53*
May 10 S.	α^1 Libræ	49.08
	α^2 Libræ	49.14
	α Coronæ	49.07
May 18 R.	β Leonis	35.96
	ε Corvi	35.94*
	ζ Virginis	35.83*
	α Bootis	35.64
May 18 S.	α Virginis	35.71
	α^1 Libræ	35.70
	α^2 Libræ	35.64
May 19 S.	α Virginis	34.04
	α Bootis	33.90
	α^1 Libræ	33.62
	α^2 Libræ	33.82
	α Coronæ	33.83
	α Serpentis	33.71
May 19 R.	β Leonis	34.11
	β Virginis	34.07
May 22 R.	ε Corvi	29.81*
	η Virginis	29.84*
	α Bootis	29.73
	α^1 Libræ	29.45
	α^2 Libræ	29.61
	α Coronæ	29.77
	α Serpentis	29.55
	α Scorpii	29.28
May 23 S.	α Leonis	28.19

Date.	Name of Star.	Value of $\Delta t+m$
1864.		s.
May 23 S.	β Leonis	28.14
	β Virginis	28.08
May 27 S.	β Virginis	21.75
	α Virginis	21.64
	α Bootis	21.69
	α^1 Libræ	21.48
	α^2 Libræ	21.59
	α Coronæ	21.59
	α Serpentis	21.52
May 28 R.	γ Virginis	20.35*
	12 Canum	20.56*
	α Virginis	20.21
	α Bootis	20.33
	ψ Bootis	20.18*
	β Libræ	20.09*
May 29 R.	α Virginis	18.49
	α Bootis	18.53
	α^1 Libræ	18.35
	α^2 Libræ	18.38
	α Coronæ	18.28
	α Serpentis	18.17
May 31 S.	α Virginis	15.62
	α^1 Libræ	15.56
	α^2 Libræ	15.64
	α Coronæ	15.68
	α Scorpii	15.50
June 2 R.	γ Virginis	12.85*
	12 Canum	12.86*
	α Virginis	12.69
	α Bootis	12.76
June 2 S.	ρ Bootis	12.68*
	ε Bootis	12.64*
	α Coronæ	12.52
	α Serpentis	12.46
June 3 S.	α Arietis	10.62
June 4 S.	α^1 Libræ	9.97
	α^2 Libræ	10.06
	α Coronæ	10.09

Date.	Name of Star.	Value of $\Delta t+m$
1864.		s.
June 4	α Serpentis	9.98
June 6 S.	ε Bootis	7.61*
	ψ Bootis	7.63*
	α Serpentis	7.53
	δ Ophiuchi	7.40*
June 6 R.	γ Aquilæ	7.33
	α^1 Capricorni	7.39
	α Cygni	7.22
June 7 S.	β Leonis	6.38
	α Virginis	6.22
	α Coronæ	6.22
	α Serpentis	6.14
	α Scorpii	6.03
June 7 R.	τ Virginis	6.23*
	α^1 Libræ	6.29
	α^2 Libræ	6.24
	ψ Bootis	6.36*
June 8 S.	α Virginis	4.92
	α^1 Libræ	4.81
	α^2 Libræ	4.78
June 8 R.	α Coronæ	4.86
	α Serpentis	4.80
	α Scorpii	4.67
June 10 S.	α Virginis	1.57
	α Bootis	1.52
	α^1 Libræ	1.43
	α^2 Libræ	1.50
June 10 M.	α Scorpii	1.30
	α Herculis	1.41
June 13 S.	β Leonis	56.91
	α Virginis	56.77
	α^1 Libræ	56.66
	α^2 Libræ	56.74
	α Coronæ	56.77
	α Scorpii	56.63
June 13 R.	β Scorpii	56.82*
	δ Ophiuchi	56.73*

Date	n	c
May 5–19	$n = +$ 0.066 s.	$c = -$ 0.401 s.
22, 23	− 0.102	+ 0.355
27	− 0.039	+ 0.355
28	− 0.152	+ 0.355
29	− 0.084	+ 0.355
May 31–June 2	$n = -$ 0.107 s.	$c = +$ 0.355 s.
June 3–6	+ 0.046	+ 0.355
7–9	+ 0.059	− 0.394
10–13	+ 0.133	− 0.394

Date.	Name of Star.	Value of $\Delta t+m$
1864.		s.
June 14	β Leonis . .	55.90
S.	α Virginis .	55.83
	α Coronæ . .	55.84
	α Serpentis .	55.73
	α Scorpii . .	55.70
	α Herculis .	55.72
	α Ophiuchi .	55.72
June 14	ε Bootis . .	55.97*
R.	α^1 Libræ . .	55.84
	α^2 Libræ . .	55.89
	β Lyræ . .	55.77*
June 16	α Bootis . .	53.45
S.	α^1 Libræ . .	53.45
	α^2 Libræ . .	53.51
	α Coronæ . .	53 37
	α Serpentis .	53.36
	α Scorpii . .	53.11
	α Herculis . .	53.22
June 17	α Virginis . .	52.10
S.	α^2 Libræ . .	51.87
	α Coronæ . .	51.97
	α Serpentis .	51.89
	α Scorpii . .	51.67
	α Herculis .	51.77
June 18	α Virginis . .	50.84
M.	α Bootis . .	50.89
	α^1 Libræ . .	50.78
	α^2 Libræ . .	50.86
June 18	α Coronæ . .	50.80
R.	α Serpentis .	50.78
	δ Ophiuchi .	50.68*
	κ Ophiuchi .	50.72*
June 19	α Virginis . .	49.49
R.	ζ Virginis . .	49.59*
June 21	α^1 Libræ . .	46.76
S.	α^2 Libræ . .	46.76
	α Coronæ . .	46.75
	α Serpentis .	46.73
	α Scorpii . .	46.58
	α Herculis . .	46.70

Date.	Name of Star.	Value of $\Delta t+m$
1864.		s.
June 21	α Ophiuchi .	46.60
S.	α Lyræ . .	46.60
June 21	τ Virginis . .	46.84*
R.		
June 22	α^2 Libræ . .	45.43
R.	α Coronæ . .	45.52
	α Serpentis .	45.52
	α Herculis .	45.50
	α Ophiuchi .	45.35
June 23	α Virginis . .	44.38
R.	τ Virginis . .	44.31*
June 23	α Bootis . .	44.27
S.	α Lyræ . . .	44.03
June 24	α Bootis . .	43.00
S.	α^1 Libræ . .	42.87
	α^2 Libræ . .	42.89
	α Ophiuchi .	42.67
June 24	α Coronæ . .	43.13
M.	α Serpentis .	43.08
	α Scorpii . .	43.05
June 26	ζ Herculis .	41.00*
R.		
June 27	α Virginis . .	39.68
R.	α Coronæ . .	39.71
	α Serpentis .	39.37
June 27	ρ Bootis . .	39.63*
M.	ε Bootis . .	39.64*
June 28	α Virginis . .	37.46
S.	α^1 Libræ . .	37.33
	α^2 Libræ . .	37.33
	α Coronæ . .	37.37
	α Serpentis .	37.38
	α Lyræ . . .	37.08
June 28	ρ Bootis . .	37.72*
M.		

Date.	Name of Star.	Value of $\Delta t+m$
1864.		s.
June 29	α Virginis . .	35.91
S.	α^1 Libræ . .	35.69
	α^2 Libræ . .	35.79
	α Coronæ . .	35.70
	α Serpentis .	35.70
June 29	η Bootis . .	36.10*
M.	α Bootis . .	36.12
June 30	α^1 Libræ . .	34.06
S.	α^2 Libræ . .	34.00
	α Scorpii . .	33.90
June 30	α Coronæ . .	34.05
R.	α Serpentis .	34.09
July 3	α Virginis. .	28.93
R.	α Coronæ . .	28.82
	α Serpentis .	28.72
	α Scorpii . .	28.79
	α Herculis .	28.80
	α Ophiuchi .	28.65
	α Lyræ . .	28.61
July 4	α Virginis. .	27.53
R.		
July 4	α Bootis . .	27.36
S.	α^1 Libræ . .	27.33
	α^2 Libræ . .	27.31
	α Coronæ . .	27.25
	α Serpentis .	27.31
	α Scorpii . .	27.15
	α Herculis. .	27.24
	α Ophiuchi .	27.16
July 5	α^1 Libræ . .	25.84
S.	α^2 Libræ . .	25.87
	α Coronæ . .	25.86
	α Serpentis .	25.80
	α Lyræ. . .	25.58
	γ Aquilæ . .	25.56
July 6	β Leonis . .	24.50
S.		

Date	n (s.)	c (s.)
June 14	$n = +0.079$	$c = -0.394$
16–19	$+0.043$	-0.394
21	$+0.094$	-0.326
June 22–27	$n = +0.032$	$c = -0.326$
28–July 5	$+0.079$	$+0.316$
July 6	-0.037	$+0.316$

Date.	Name of Star.	Value of $\Delta t+m$	Date.	Name of Star.	Value of $\Delta t+m$	Date.	Name of Star.	Value of $\Delta t+m$
1864.		s.	1864.		s.	1864.		s.
July 6	α Coronæ	24.37	July 15	α Serpentis	11.46	July 26	β Aquilæ	54.98
R.	α Serpentis	24.38	S.	α Scorpii	11.31	S.	α^1 Capricorni	54.94
				α Herculis	11.34		α^2 Capricorni	54.89
July 8	α Virginis	21.21		α Ophiuchi	11.32			
S.	α^2 Libræ	21.13				July 27	α Scorpii	54.07
	α Coronæ	20.99	July 16	β Scorpii	9.98*	S.	κ Ophiuchi	54.14*
	α Serpentis	20.99	R.	α Herculis	10.08			
	α Scorpii	20.85		α Ophiuchi	9.98	July 28	α Ophiuchi	52.62
				δ Aquilæ	9.67*	S.	α Lyræ	52.51
July 9	α Virginis	19.40					γ Aquilæ	52.50
S.	α^2 Libræ	19.32	July 18	ζ Herculis	7.12*		α^1 Capricorni	52.51
	α Coronæ	19.30	R.	α Herculis.	7.08		α^2 Capricorni	52.54
	α Serpentis	19.28		α Ophiuchi	6.97			
	γ Aquilæ	19.04		γ Aquilæ	6.88	July 29	α Ophiuchi	51.14
						S.	γ Aquilæ	51.09
July 10	ζ Herculis	17.98*	July 19	α Bootis	5.73		α Aquilæ	51.13
R.	κ Ophiuchi	17.91*	S.	α^2 Libræ	5.76		β Aquilæ	51.02
							α^1 Capricorni	51.11
July 12	α Virginis	15.82	July 19	α Scorpii	5.64		α^2 Capricorni	50.98
S.	α^2 Libræ	15.81	R.	κ Ophiuchi	5.68*		α Cygni	50.97
	α Coronæ	15.82		μ Herculis	5.65*			
	α Serpentis	15.86		μ Sagittarii	5.60*	July 30	α Herculis	49.78
				δ Aquilæ	5.55*	S.	α Ophiuchi	49.69
July 12	α Scorpii	15.71					α Lyræ	49.61
R.	α Herculis	15.82	July 20	α Herculis	4.31			
	α Ophiuchi	15.76	S.	α Ophiuchi	4.18	Aug. 1	α Serpentis	47.32
	μ Sagittarii	15.65*		γ Aquilæ	4.12	S.	α Scorpii	47.14
	δ Aquilæ	15.59*		α Aquilæ	4.17		α Herculis	47.20
				β Aquilæ	4.12		α Lyræ	47.09
July 13	α Coronæ	14.53					α Aquilæ	47.07
S.	α Serpentis	14.53	July 22	α Coronæ	0.90		β Aquilæ	47.07
	α Scorpii	14.34	S.	α Serpentis	0.91		α^1 Capricorni	47.06
	α Herculis	14.39		α Herculis	0.86		α^2 Capricorni	47.03
	α Ophiuchi	14.28		α Ophiuchi	0.71		α Cygni	46.97
	α Lyræ	14.20						
	α Aurigæ	13.48	July 23	α Herculis	59.29	Aug. 5	α Serpentis	40.42
	β Orionis	13.28	R.	α Ophiuchi	59.27	S.	α Scorpii	40.25
	α Canis Maj.	13.34		α Lyræ	59.14			
						Aug. 8	α Scorpii	36.24
July 14	α Virginis	13.04	July 26	α Virginis	55.46	S.	α Herculis	36.25
S.	α Coronæ	12.94	S.	α Bootis	55.37		α Ophiuchi	36.21
	α Serpentis	12.97		α Coronæ	55.24		α Lyræ	36.05
	α Scorpii	12.86		α Serpentis	55.30		γ Aquilæ	36.07
	α Ophiuchi	12.78		α Scorpii	55.18		β Aquilæ	36 09
	γ Aquilæ	12.78		α Lyræ	55.04		α^1 Capricorni	35.91
July 15				γ Aquilæ	55.07		α^2 Capricorni	35.94
S.	α Coronæ	11.50		α Aquilæ	55.05			

	n	c		n	c
	s.	s.		s.	s.
July 8–10	$+0.080$	$+0.316$	July 29–30	$+0.102$	$+0.338$
12–16	$+0.071$	-0.371	Aug. 1	$+0.060$	$+0.338$
18–20	$+0.110$	-0.371	8	$+0.110$	$+0.338$
22–28	$+0.158$	$+0.338$			

Date.	Name of Star.	Value of $\Delta t+m$
1864.		s.
Sept. 2	α Lyræ	59.53
S.	γ Aquilæ	59.30
	α Aquilæ	59.27
	β Aquilæ	59.36
	α¹ Capricorni	59.28
	α² Capricorni	59.18
Sept. 6	α Scorpii	52.77
S.	α Lyræ	52.72
	γ Aquilæ	52.61
	β Aquilæ	52.65
	α¹ Capricorni	52.62
	α² Capricorni	52.61
Sept. 7	α Scorpii	51.02
S.	ε Pegasi	50.65*
	16 Pegasi	50.50*
	α Aquarii	50.51
Sept. 9	α Lyræ	47.20
S.	α Aquilæ	47.14
	β Aquilæ	47.05
	α² Capricorni	47.01
Sept. 10	α Scorpii	45.64
S.	α Herculis	45.61
	α Lyræ	45.44
	α Piscis Aus.	45.22
	α Pegasi	45.25
Sept. 15	α Coronæ	35.32
S.	β Aquarii	34.85*
Sept. 16	α Herculis	33.26
S.	α Ophiuchi	33.21
	γ Aquilæ	33.03
	β Aquilæ	33.02
	α¹ Capricorni	32.94
	α² Capricorni	32.84
	α Piscis Aus.	32.54
	α Pegasi	32.70
Sept. 17	α Ophiuchi	31.19
S.	α Lyræ	31.16
	γ Aquilæ	31.02
	β Aquilæ	31.08
	α Aquarii	30.87

Date.	Name of Star.	Value of $\Delta t+m$
1864.		s.
Sept. 17	α Piscis Aus.	30.80
S.	α Pegasi	30.86
Sept. 22	α Ophiuchi	22.93
S.	α Lyræ	22.97
Sept. 26	α Lyræ	14.65
S.	β Lyræ	14.77*
	α Aquarii	14.56
	α Piscis Aus.	14.21
	α Pegasi	14.38
	α Andromedæ	14.29
Sept. 27	α Lyræ	13.09
S.	γ Aquilæ	13.01
	α Aquilæ	13.06
	β Aquilæ	12.92
	α Piscis Aus.	12.71
	α Pegasi	12.78
Sept. 29	16 Pegasi	9.07*
S.	α Aquarii	9.23
Sept. 30	α Piscis Aus.	7.62
S.	α Pegasi	7.69
	γ Pegasi	7.40
Oct. 4	α Lyræ	59.17
S.		
Oct. 5	α Lyræ	57.68
S.	β Lyræ	57.65*
	α¹ Capricorni	57.36
	α² Capricorni	57.36
Oct. 7	α Ophiuchi	53.75
S.	α Lyræ	53.79
	γ Aquilæ	53.68
	β Aquilæ	53.62
	α Aquarii	53.53
	α Piscis Aus.	53.31
	α Pegasi	53.52
	γ Pegasi	53.46
	ε Piscium	53.39*
Oct. 8	α Ophiuchi	51.95
S.	α Lyræ	51.99

Date.	Name of Star.	Value of $\Delta t+m$
1864.		s.
Oct. 10	α Cygni	49.28
S.	32 Vulpeculæ	49.27*
	ζ Cygni	49.33*
	α Aquarii	49.15
	ν Piscium	48.94*
	β Arietis	49.06*
	α Arietis	48.93
Oct. 11	α Aquarii	47.88
S.	α Piscis Aus.	47.71
	α Pegasi	47.91
	α Andromedæ	47.84
	γ Pegasi	47.84
	ε Piscium	47.81*
Oct. 17	α Lyræ	40.03
S.	γ Aquilæ	39.97
	β Aquilæ	39.97
	α Aquarii	39.59
	α Piscis Aus.	39.54
	α Pegasi	39.82
	α Andromedæ	39.80
	γ Pegasi	39.75
	α Arietis	39.65
Oct. 18	ζ Cygni	38.49*
S.	β Aquarii	38.30*
Oct. 19	α Lyræ	37.21
S.	γ Aquilæ	37.03
	β Aquilæ	37.01
	α¹ Capricorni.	36.91
Oct. 21	γ Aquilæ	32.70
S.	β Aquilæ	32.72
	α¹ Capricorni.	32.68
	α² Capricorni.	32.66
	α Piscis Aus.	32.39
	α Pegasi	32.52
Oct. 24	α Lyræ	26.39
S.	β Lyræ	26.39*
Oct. 25	α Lyræ	24.75
S.	γ Aquilæ	24.69
	β Aquilæ	24.78
	α¹ Capricorni.	24.65

Date	n (s.)	c (s.)
Sept. 2	$n = +0.141$	$c = +0.338$
6	+0.320	−0.360
7	+0.399	−0.360
9	+0.328	−0.360
10–17 (17.51)	+0.223	+0.275
17 (17.51) –27	+0.322	−0.040

Date	n (s.)	c (s.)
Sept. 29–30	$n = +0.256$	$c = -0.040$
Oct. 5	+0.196	+0.023
7–8	+0.139	+0.023
10	+0.268	+0.023
11–26	+0.309	+0.023

Date.	Name of Star.	Value of $\Delta t+m$	Date.	Name of Star.	Value of $\Delta t+m$	Date.	Name of Star.	Value of $\Delta t+m$
1864.		s.	1864.		s.	1864.		s.
Oct. 25	α^2 Capricorni	24.57	Nov. 10	ξ^2 Ceti . . .	0.28*	Nov. 19	α Persei . .	42.32*
S.	α Cygni . .	24.66	S.	γ Ceti . . .	0.19*	S.		
	α Piscis Aus. .	24.41						
	α Pegasi . .	24.42	Nov. 10	ε Pegasi . .	0.84*	Nov. 19	α Andromedæ	42.65
	α Andromedæ	24.48	M.	α Piscis Aus. .	0.75	M.	γ Pegasi . .	42.83
	γ Pegasi . .	24.48		α Pegasi . .	0.62		α Arietis . .	42.56
				$\varkappa$ Piscium . .	0.59*		ξ^2 Ceti . . .	42.37*
Oct. 26	α Lyræ . .	23.49						
S.	γ Aquilæ . .	23.33	Nov. 11	ζ Cygni . .	58.50*	Nov. 21	α Bootis . .	37.14
	β Aquilæ . .	23.39	S.	β Aquarii . .	58.31*	S.		
	α^1 Capricorni	23.33		α Arietis . .	57.83			
	α^2 Capricorni	23.37		67 Ceti. . .	57.83*	Nov. 22	γ Aquilæ . .	36.83
	α Pegasi . .	23.31		ξ^2 Ceti . . .	58.08*	S.	α Aquilæ . .	36.87
	α Andromedæ	23.15					α Piscis Aus. .	36.54
	γ Pegasi . .	23.25	Nov. 11	ε Pegasi . .	58.56*		α Pegasi . .	36.67
	θ Ceti . . .	22.99*	M.	16 Pegasi . .	58.51*			
				α Aquarii . .	58.37	Nov. 23	γ Pegasi . .	34.97
Nov. 1	α^1 Capricorni	14.11		α Piscis Aus. .	58.36	S.	12 Ceti. . .	34.96*
S.	α^2 Capricorni	13.97		α Pegasi . .	58.38		β Ceti . .	34.88*
	α Cygni . .	14.04		ω Piscium . .	58.27*		η Tauri . .	34.80*
	θ Ceti . . .	13.66*					γ Eridani . .	34.92*
	α Arietis . .	13.94	Nov. 14	α Aquarii . .	51.69			
			S.	α Piscis Aus. .	51.69	Nov. 23	α Arietis . .	35.14
Nov. 2	α Aquilæ . .	13.12		α Pegasi . .	51.66	M.	ξ^2 Ceti . . .	34.91*
S.	β Aquilæ . .	12.92		α Andromedæ	51.57		γ Ceti . . .	34.95*
	α^1 Capricorni	12.99		γ Pegasi . .	51.62			
	α^2 Capricorni	12.95		α Arietis . .	51.47	Nov. 25	γ Aquilæ . .	31.11
	ε Piscium . .	12.89*				S.	α Aquilæ . .	31.22
	θ Ceti . . .	12.80*	Nov. 16	α Lyræ. . .	48.29		β Aquilæ . .	31.14
			S.	γ Aquilæ . .	48.12		α Aquarii . .	31.10
Nov. 3	α Aquilæ . .	12.13		α Aquarii . .	48.03		β Ceti . . .	30.90*
S.	β Aquilæ . .	12.04		θ Ceti . . .	47.51*			
	α^1 Capricorni	11.99		η Piscium . .	47.56*	Nov. 25	ν Piscium . .	31.04*
	α^2 Capricorni	12.01				M.	α Arietis . .	31.06
	α Cygni . .	12.01	Nov. 16	12 Ceti . .	47.66*		o^1 Eridani . .	30.97*
			M.	β Ceti . . .	47.60*		ε Tauri . .	30.93*
Nov. 4	α Cygni . .	10.57		ε Piscium . .	47.71*			
S.	α Piscis Aus. .	10.28		α Arietis . .	47.52	Nov. 29	α Aquarii . .	23.73
	α Pegasi . .	10.40		α Ceti . . .	47.47	S.	α Piscis Aus. .	23.56
	α Andromedæ	10.30					ι Piscium . .	23.58*
	γ Pegasi . .	10.44	Nov. 19	α Lyræ . .	42.88			
			S.	α Aquilæ . .	42.93	Nov. 29	α Andromedæ	23.60
Nov. 10	ϱ Capricorni .	0.69*		β Aquilæ . .	42.85	M.	γ Pegasi . .	23.68
S.	α Cygni . .	0.77		α Piscis Aus. .	42.63		η Piscium . .	23.49*
	α Andromedæ	0.43		α Pegasi . .	42.69			
	α Arietis . .	0.23		α Ceti . . .	42.33	Nov. 30	α Aquarii . .	22.20
	67 Ceti. . .	0.22*		δ Arietis . .	42.37*	M.	α Piscis Aus. .	22.25

Oct. 31–Nov. 4 $n = +0^s.282$ $c = +0^s.023$
Nov. 10 $+0.277$ $+0.277$
11–14 $+0.114$ $+0.277$
15–19 $+0.232$ $+0.277$

Nov. 22–23 $n = +0^s.158$ $c = +0^s.277$
25 $+0.264$ $+0.277$
29 $+0.215$ $+0.277$
30–Dec. 1 $+0.170$ $+0.277$

Date.	Name of Star.	Value of $\Delta t+m$
1864.		s.
Nov. 30	α Pegasi	22.21
M.	α Andromedæ	21.97
	γ Pegasi	22.13
	γ Ceti	21.86*
	α Ceti	21.82
	δ Arietis	21.78*
Dec. 1	α Aquarii	20.19
S.	α Piscis Aus.	20.11
	α Pegasi	20.07
	γ Pegasi	19.88
Dec. 5	α Andromedæ	12.80
M.	γ Pegasi	12.74
Dec. 8	α Aquilæ	8.00
S.	α Cygni	7.89
	α Andromedæ	7.72
	γ Pegasi	7.62
	α Arietis	7.53
Dec. 8	α Aquarii	7.95
M.	α Piscis Aus.	7.97
	α Pegasi	7.94
Dec. 9	α Aquarii	5.63
S.	α Piscis Aus.	5.54
	α Pegasi	5.58
Dec. 9	α Andromedæ	5.77
M.	γ Pegasi	5.66
	α Arietis	5.68
	ξ² Ceti	5.66*
Dec. 13	α Cygni	55.84
S.		
Dec. 14	θ Ceti	54.61*
S.	α Arietis	54.52
	α Ceti	54.51
Dec. 20	α Cygni	42.36
S.	α Andromedæ	42.11
	γ Pegasi	42.21
Dec. 20	θ Ceti	42.29*
M.	η Piscium	42.27*

Date.	Name of Star.	Value of $\Delta t+m$
1864.		s.
Dec. 20	α Arietis	42.39
M.	α Ceti	42.14
Dec. 30	α Pegasi	22.15
M.	α Andromedæ	22.27
	γ Pegasi	22.21
	α Arietis	22.21
	α Ceti	22.08
	α Persei	22.13*
1865.		
Jan. 2	α Aurigæ	37.12
M.	β Orionis	36.61
	β Tauri	36.87
	α Orionis	36.65
Jan. 3	α Andromedæ	35.63
M.	γ Pegasi	35.68
Jan. 3	β Ceti	35.36*
S.	ε Piscium	35.42*
	ν Piscium	35.57*
	α Arietis	35.62
Jan. 4	η Tauri	33.93*
M.	γ Eridani	33.85*
	α Tauri	33.94
	α Aurigæ	34.12
	β Orionis	34.07
	β Tauri	34.05
	α Orionis	34.00
Jan. 11	α Pegasi	23.56
M.	α Andromedæ	23.71
	γ Pegasi	23.77
	α Tauri	23.69
	α Aurigæ	23.88
	β Orionis	23.62
	α Orionis	23.73
	α Canis Maj.	23.54
Jan. 11	ν Piscium	23.54*
S.	β Arietis	23.51*

Date.	Name of Star.	Value of $\Delta t+m$
1865.		s.
Jan. 12	α Arietis	22.93
S.	α Ceti	22.83
	α Tauri	22.84
Jan. 12	η Tauri	22.89*
M.	γ Eridani	22.93*
	o¹ Eridani	22.90*
	ε Tauri	22.89*
Jan. 13	γ Ceti	21.97*
M.	α Ceti	22.09
	α Tauri	22.10
	α Aurigæ	22.27
	β Orionis	22.13
Jan. 16	γ Pegasi	18.14
M.	β Ceti	18.00*
	α Ceti	17.80
	α Tauri	17.89
	α Aurigæ	17.91
	β Orionis	17.78
	β Tauri	17.85
	α Orionis	17.85
	α Canis Maj.	17.82
	α Geminorum	18.01
	α Canis Min.	17.97
	β Geminorum.	17.95
Jan. 16	η Piscium	17.80*
S.	ν Piscium	17.85*
	α Arietis	17.90
Jan. 18	α Andromedæ	15.72
M.	γ Pegasi	15.74
	α Tauri	15.54
	α Aurigæ	15.67
	β Orionis	15.34
	β Tauri	15.62
	α Orionis	15.61
Jan. 18	η Piscium	15.38*
S.	ν Piscium	15.41*
	α Arietis	15.41
	ξ² Ceti	15.38*
Jan. 20	α Andromedæ	12.76
M.	α Arietis	12.77

Dec. 5–8 $n = +0.103^s$ $c = +0.277^s$

Dec. 9–Jan. 25, 1865 $n = +0.229^s$ $c = -0.309^s$

Date.	Name of Star.	Value of $\Delta t + m$
1865.		s.
Jan. 20	α Ceti . . .	12.79
M.	α Tauri . .	12.80
	α Aurigæ . .	12.94
	β Orionis . .	12.71
	β Tauri . .	12.79
	α Orionis . .	12.72
	α Canis Maj. .	12.58
Jan. 24	α Arietis . .	5.67
S.	α Ceti . . .	5.59
	α Orionis . .	5.50
	α Canis Maj. .	5.25
Jan. 24	α Tauri . .	5.78
M.	α Aurigæ . .	5.76
	β Orionis . .	5.53
Jan. 25	β Ceti . . .	4.53*
M.	β Arietis . .	4.58*
	α Arietis . .	4.75
	α Tauri . .	4.69
	α Aurigæ . .	4.92
	β Orionis . .	4.58
	β Tauri . .	4.65
	α Orionis . .	4.75
	α Canis Maj. .	4.62
Jan. 26	β Arietis . .	3.41*
S.	α Arietis . .	3.44
	ξ² Ceti . .	3.31*
	α Tauri . .	3.36
	α Aurigæ . .	3.30
	β Orionis . .	3.36
	β Tauri . .	3.37
Jan. 27	β Arietis . .	1.98*
M.	α Arietis . .	2.10
	α Ceti . . .	1.86
	α Tauri . .	1.91
	α Aurigæ . .	1.90
	β Orionis . .	1.91
	β Tauri . .	2.04
	α Orionis . .	1.89
	α Canis Maj. .	1.80
Jan. 28	α Arietis . .	0.93
S.	α Ceti . . .	0.77

Date.	Name of Star.	Value of $\Delta t + m$
1865.		s.
Jan. 28	α Tauri . .	0.79
S.	μ Geminorum	0.79*
Jan. 28	ι Aurigæ . .	0.93*
M.	α Aurigæ . .	1.09
	β Tauri . .	0.99
Jan. 30	α Arietis . .	59.00
M.	γ Ceti . . .	58.85*
	α Tauri . .	58.79
	α Aurigæ . .	58.88
	β Orionis . .	58.96
	α Orionis . .	58.89
	α Canis Maj. .	58.83
	α Geminorum	58.78
	α Canis Min. .	58.81
	β Geminorum	58.87
Feb. 1	η Tauri . .	56.47*
S.	γ Eridani . .	56.38*
	o¹ Eridani . .	56.41*
	ε Tauri . .	56.47*
	χ Geminorum	56.26*
	ι Navis . .	55.96*
Feb. 1	α Tauri . .	56.60
M.	α Aurigæ . .	56.80
	β Orionis . .	56.62
	β Tauri . .	56.51
	α Orionis . .	56.62
	α Canis Maj. .	56.45
	α Geminorum	56.44
	α Canis Min. .	56.40
	β Geminorum	56.46
Feb. 2	α Arietis . .	54.71
S.	α Tauri . .	54.53
	α Aurigæ . .	54.49
	β Orionis . .	54.46
	α Canis Maj. .	54.23
	α Geminorum	54.11
	α Canis Min. .	54.19
	β Geminorum	54.14
Feb. 2	ε Orionis . .	54.47*
M.	α Orionis . .	54.53
	ν Orionis . .	54.41*

Date.	Name of Star.	Value of $\Delta t + m$
1865.		s.
Feb. 3	α Aurigæ . .	53.01
M.	β Orionis . .	52.89
	β Tauri . .	52.95
Feb. 9	α Arietis . .	43.02
S.	α Ceti . . .	42.79
	α Aurigæ . .	42.96
	β Orionis . .	42.87
	α Orionis . .	42.96
	α Canis Maj. .	42.80
	α Geminorum	42.84
	α Canis Min. .	42.81
	β Geminorum	42.87
Feb. 9	ε Tauri . .	43.10*
M.	α Tauri . .	43.19
	ι Aurigæ . .	43.17*
Feb. 11	α Arietis . .	40.11
S.	α Ceti . . .	40.07
	α Tauri . .	40.20
	α Aurigæ . .	40.05
	β Orionis . .	40.02
	β Tauri . .	40.07
	α Orionis . .	40.19
Feb. 13	α Ceti . . .	35.70
M.	α Tauri . .	35.96
	α Aurigæ . .	36.03
	β Orionis . .	35.90
	β Tauri . .	35.87
	α Orionis . .	35.98
Feb. 14	α Arietis . .	35.17
S.	α Ceti . . .	35.01
	α Canis Min. .	35.00
	β Geminorum	34.94
Feb. 14	α Aurigæ . .	35.36
M.	β Orionis . .	35.27
	α Orionis . .	35.28
	α Canis Maj. .	35.23
Feb. 21	α Arietis . .	22.90
S.	α Ceti . . .	22.79
	α Aurigæ . .	22.69
	β Orionis . .	22.71

		n	c
		s.	s.
Jan.	26–30	$n = +0.175$	$c = +0.281$
Feb.	1	$+0.097$	$+0.281$
	2, 3	$+0.171$	$+0.281$

		n	c
		s.	s.
Feb.	9–11	$n = +0.137$	$c = +0.281$
	13	$+0.071$	$+0.281$
	14–22	$+0.184$	$+0.281$

Date.	Name of Star.	Value of $\Delta t+m$	Date.	Name of Star.	Value of $\Delta t+m$	Date.	Name of Star.	Value of $\Delta t+m$
1865.		s.	1865.		s.	1865.		s.
Feb. 21	α Orionis	22.74	March 6	β Geminorum	6.40	March 18	α Canis Maj.	55.93
S.	α Geminorum	22.60	M.	α Hydræ	6.31	S.	α Canis Min.	55.88
	α Canis Min.	22.60					β Geminorum	55.94
	β Geminorum	22.56	March 6	α Aurigæ	6.38			
			S.	β Orionis	6.36	March 20	α Tauri	54.49
Feb. 21	γ Geminorum	22.84*				M.	α Aurigæ	54.61
M.	α Canis Maj.	22.77	March 7	α Aurigæ	5.61		β Orionis	54.55
	δ Geminorum	22.73*	S.	β Orionis	5.54		β Tauri	54.48
							α Orionis	54.41
Feb. 22	α Orionis	21.88	March 8	η Cancri	4.88*		α Ophiuchi	54.21
M.			M.	ε Hydræ	4.84*		α Lyræ	54.12
Feb. 24	α Arietis	19.26	March 10	ε Hydræ	2.88*	March 21	α Aurigæ	53.88
M.	α Aurigæ	19.32	M.	83 Cancri	2.80*	S.	β Orionis	53.84
	β Orionis	19.32		α Leonis	2.86		β Tauri	53.80
	β Tauri	19.24		l Leonis	2.69*			
	α Orionis	19.28		χ Leonis	2.73*	March 23	α Hydræ	51.30
	α Canis Maj.	19.23				S.	α Leonis	51.28
	α Geminorum	19.14	March 13	α Tauri	0.25			
	α Canis Min.	19.36	M.	α Aurigæ	0.40	March 25	α Canis Maj.	49.21
	β Geminorum	19.30		β Orionis	0.23	S.	α Hydræ	49.04
				β Tauri	0.25		α Leonis	48.99
Feb. 25	α Arietis	18.14		α Canis Maj.	0.30		β Leonis	48.99
S.	α Ceti	17.99		α Geminorum	0.18		β Virginis	48.94
	α Tauri	17.92		α Canis Min.	0.32			
	α Aurigæ	18.01		β Geminorum	0.10	March 25	α Geminorum	49.20
	β Orionis	18.03				M.	ι Ursæ Maj.	49.12*
	α Canis Maj.	17.86	March 16	α Aurigæ	58.25			
	α Geminorum	17.84	M.	β Orionis	58.27	March 27	α Aurigæ	47.45
				β Tauri	58.17	M.	β Orionis	47.45
Feb. 27	α Arietis	14.59					β Tauri	47.31
M.	α Aurigæ	14.72	March 17	α Tauri	57.05		α Orionis	47.42
	β Orionis	14.67	M.	α Aurigæ	57.17		α Geminorum	47.36
	β Tauri	14.53		β Orionis	56.97		α Canis Min.	47.35
				β Tauri	57.02		β Geminorum	47.42
March 4	α Canis Maj.	10.02		α Canis Maj.	57.05		α Hydræ	47.30
S.	α Geminorum	9.95		α Canis Min.	56.97		α Leonis	47.17
	α Canis Min.	10.03		β Geminorum	57.00		χ Leonis	47.10*
	β Geminorum	9.99		α Hydræ	56.87		δ Leonis	47.28*
	π Leonis	9.91*		α Leonis	56.83			
	α Leonis	9.94		γ Leonis	56.81*	March 28	α Orionis	46.76
						M.	α Geminorum	46.77
March 6	ε Tauri	6.28*	March 18	α Orionis	56.28		α Canis Min.	46.79
M.	α Tauri	6.42	M.	ν Orionis	56.19*		β Geminorum	46.71
	α Canis Maj.	6.47		η Cancri	56.00*		α Hydræ	46.59
	α Geminorum	6.40		ε Hydræ	56.07*		α Leonis	46.66
	α Canis Min.	6.40						

	n (s.)	c (s.)		n (s.)	c (s.)
Feb. 24	$n = +0.130$	$c = +0.281$	March 17–28	$n = +0.059$	$c = +0.316$
25–March 13	$+0.175$	$+0.316$			

Date.	Name of Star.	Value of $\Delta t+m$	Date.	Name of Star.	Value of $\Delta t+m$	Date.	Name of Star.	Value of $\Delta t+m$
1865.		s.	1865.		s.	1865.		s.
April 1	α Orionis	43.48	April 11	π Leonis	36.83*	April 17	12 Canum	29.90*
S.			S.	α Leonis	36.66	S.	α Virginis	29.90
				ρ Leonis	36.74*			
April 3	α Tauri	42.09				April 18	β Geminorum	28.90
M.	α Aurigæ	41.95	April 12	γ Leonis	35.89*	S.	β Leonis	28.72
	β Orionis	42.24	M.	ε Leonis	35.83*		β Virginis	28.75
	β Tauri	42.07		δ Leonis	35.91*		α Virginis	28.73
	α Orionis	42.16		δ Crateris	35.96*			
	α Canis Maj.	42.35				April 19	α Aurigæ	28.12
	α Hydræ	42.30	April 13	α Aurigæ	34.43	M.	β Orionis	28.29
	α Leonis	42.29	M.	β Orionis	34.55		α Orionis	28.22
				β Tauri	34.47		α Canis Maj.	28.22
April 3	α Geminorum	42.14		α Orionis	34.46		α Geminorum	28.22
S.	α Canis Min.	42.13		α Canis Maj.	34.43		α Canis Min.	28.23
	β Geminorum	42.11		α Geminorum	34.40		β Geminorum	28.21
				α Canis Min.	34.38		α Hydræ	28.10
April 4	α Orionis	41.59		α Hydræ	34.18		β Leonis	28.00
S.	α Canis Maj.	41.55		α Leonis	34.15		β Virginis	28.04
	α Canis Min.	41.57		β Leonis	34.05			
	β Geminorum	41.49		β Virginis	34.08	April 19	ε Leonis	28.02*
	β Corvi	41.43*		α Virginis	34.04	S.	μ Leonis	28.05**
April 4	χ Geminorum	41.60*	April 14	α Orionis	32.99	April 24	β Orionis	22.31
M.	ι Navis	41.55*	S.	α Canis Maj.	33.03	M.	β Tauri	22.25
	η Cancri	41.51*		α Hydræ	32.86		α Orionis	22.30
				α Leonis	32.82		α Canis Maj.	22.24
April 5	α Aurigæ	41.18		β Leonis	32.83		α Geminorum	22.38
M.	β Orionis	41.09		β Virginis	32.87		α Canis Min.	22.36
	β Tauri	41.16		α Virginis	32.80		β Geminorum	22.37
	α Orionis	41.19					α Leonis	22.34
	α Hydræ	40.73	April 17	α Aurigæ	30.17		β Leonis	22.32
	α Leonis	40.76	S.				β Virginis	22.23
	β Leonis	40.68					α Virginis	22.16
	β Virginis	40.66	April 17	β Orionis	30.17			
			M.	β Tauri	30.21	April 25	α Orionis	21.51
April 5	α Canis Min.	40.98		α Orionis	30.24	M.	α Canis Maj.	21.46
S.	β Geminorum	40.80		α Canis Maj.	30.17		α Geminorum	21.42
				α Geminorum	30.21		α Canis Min.	21.40
April 8	α Aurigæ	38.92		α Canis Min.	30.23		β Geminorum	21.44
S.	β Tauri	38.83		β Geminorum	30.18		α Hydræ	21.32
	β Leonis	38.57		α Hydræ	30.10		α Leonis	21.41
	η Virginis	38.48*		α Leonis	30.16		β Leonis	21.47
				β Leonis	30.09		β Virginis	21.38
April 8	ε Leonis	38.81*		β Virginis	30.02		α Virginis	21.38
M.	α Leonis	38.78		β Lyræ	29.74*			
	δ Leonis	38.81*	April 17			April 26	α Orionis	20.94
	δ Crateris	38.69*	S.	γ Virginis	29.97*	M.	α Canis Maj.	20.87

April 1–4 $n = +\overset{s.}{0.206}$ $c = -\overset{s.}{0.351}$
5–8 $+0.004$ -0.351

April 11–26 $n = +\overset{s.}{0.067}$ $c = -\overset{s.}{0.351}$

Date.	Name of Star.	Value of $\Delta t+m$
1865.		s.
April 26 M.	α Geminorum	20.87
	α Canis Min. .	20.82
	β Geminorum	20.88
	α Hydræ . .	20.78
	α Leonis . .	20.81
May 3 M.	α Leonis . .	15.23
	β Virginis . .	15.16
	α Virginis . .	15.19
	α Bootis . .	15.15
May 10 S.	α Geminorum	9.47
	α Canis Min. .	9.51
	α Leonis . .	9.41
	β Leonis . .	9.25
	β Virginis . .	9.22
May 12 M.	α Virginis . .	7.59
May 13 M.	α Geminorum	7.13
	α Canis Min. .	7.06
May 13 S.	β Geminorum	7.07
	β Virginis . .	6.87
	α Virginis . .	6.75
May 15 M.	α Geminorum	5.61
	α Canis Min. .	5.61
	β Geminorum	5.68
	α Hydræ . .	5.63
	α Leonis . .	5.60
	β Leonis . .	5.50
	β Virginis . .	5.53
	α Virginis . .	5.47
	α Bootis . .	5.44
	α^1 Libræ . .	5.31
	α^2 Libræ . .	5.33
May 17 M.	α Geminorum	4.41
	α Canis Min. .	4.44
	β Geminorum	4.34
May 18 S.	β Leonis . .	3.25
	β Virginis . .	3.21
	α Virginis . .	3.20
	α Bootis . .	3.09

Date.	Name of Star.	Value of $\Delta t+m$
1865.		s.
May 18 S.	α^2 Libræ . .	3.00
	α Coronæ . .	3.13
	α Serpentis .	3.02
May 23 M.	β Leonis . .	59.48
	β Virginis . .	59.38
	α Bootis . .	59.36
May 23 S.	θ Virginis . .	59.28*
	α Virginis . .	59.21
May 24 S.	α Leonis . .	58.32
	β Leonis . .	58.26
	β Virginis . .	58.21
	α Virginis . .	58.12
	α Bootis . .	58.16
May 25 M.	α Leonis . .	57.10
	β Leonis . .	57.08
	β Virginis . .	57.10
	α Virginis . .	57.02
	α Bootis . .	57.01
	α^1 Libræ . .	56.93
	α^2 Libræ . .	57.01
May 26 M.	α Geminorum	56.20
	α Canis Min. .	56.24
	β Geminorum.	56.19
	β Leonis . .	56.06
	β Virginis . .	56.05
	α Bootis . .	56.23
	α^1 Libræ . .	56.14
	α^2 Libræ . .	56.09
May 26 S.	θ Virginis . .	56.08*
	α Virginis . .	56.19
May 30 S.	α Virginis . .	52.14
	α^1 Libræ . .	52.06
	α^2 Libræ . .	52.13
	α Coronæ . .	52.15
May 31 S.	α Virginis . .	51.45
	ζ Virginis . .	51.41*
June 1 S.	α Virginis . .	50.51
	α Bootis . .	50.56

Date.	Name of Star.	Value of $\Delta t+m$
1865.		s.
June 1 S.	α^1 Libræ . .	50.69
	α^2 Libræ . .	50.61
June 2 S.	η Virginis . .	49.74*
	θ Virginis . .	49.83*
	α Virginis . .	49.88
	α Bootis . .	49.77
	α^1 Libræ . .	49.71
	α^2 Libræ . .	49.83
June 3 S.	γ Leonis . .	49.00*
	α Virginis . .	48.99
	α Bootis . .	48.95
	α^1 Libræ . .	48.89
	α^2 Libræ . .	48.88
	α Coronæ . .	49.01
	α Serpentis .	48.88
June 5 S.	α Virginis . .	47.94
	α Bootis . .	47.82
	ε Bootis . .	47.92*
June 6 S.	α Leonis . .	46.93
	α Virginis . .	46.75
	α Coronæ . .	46.64
	α Serpentis .	46.61
June 7 S.	η Virginis . .	45.76*
	θ Virginis . .	45.81*
	α Virginis . .	45.80
June 8 S.	α Leonis . .	44.89
	θ Virginis . .	44.89*
	α Virginis . .	44.82
June 12 S.	α Bootis . .	40.82
	α Coronæ . .	40.85
	α Serpentis .	40.77
June 13 S.	α Leonis . .	40.14
	α Bootis . .	40.07
	α^1 Libræ . .	40.14
	α^2 Libræ . .	40.07
	α Coronæ . .	40.13
	α Serpentis .	40.12
June 19 M.	α Leonis . .	35.15

May 1–18 $n = +\overset{s.}{0.027}$ $c = +\overset{s.}{0.304}$
23–26 -0.008 -0.344

May 29–June 29 $n = +\overset{s.}{0.045}$ $c = -\overset{s.}{0.344}$

Date.	Name of Star.	Value of $\Delta t+m$	Date.	Name of Star.	Value of $\Delta t+m$	Date.	Name of Star.	Value of $\Delta t+m$
1865.		s.	1865.		s.	1865.		s.
June 19	α Virginis	35.04	June 29	α Bootis	27.18	July 13	α Serpentis	34.00
M.	α Bootis	35.04	S.	α¹ Libræ	27.16	S.	α Scorpii	33.87
	α¹ Libræ	34.94		α² Libræ	27.20			
	α² Libræ	34.97		α Coronæ	27.09	July 14	α Virginis	33.45
				α Serpentis	27.10	M.	α² Libræ	33.40
June 20	α Leonis	33.84		α Scorpii	27.09		α Coronæ	33.48
M.							α Serpentis	33.39
			July 3	α Virginis	13.14		α Scorpii	33.41
June 20	α Virginis	33.79	M.	α Bootis	13.20		α Herculis	33.34
S.	α Bootis	33.69		α¹ Libræ	13.14		α Ophiuchi	33.34
				α² Libræ	13.05			
June 21	α Bootis	33.16		α Coronæ	13.17	July 15	α Coronæ	32.97
M.	α¹ Libræ	33.14		α Serpentis	13.02	M.	α Serpentis	32.92
	α² Libræ	33.15		α Scorpii	13.18		α Scorpii	32.96
	α Serpentis	33.13		α Herculis	13.02		α Herculis	32.81
							α Ophiuchi	32.96
June 23	α Virginis	31.44	July 5	α Leonis	36.87			
M.	α Bootis	31.30	M.	α Virginis	36.80	July 17	α Coronæ	31.92
	α¹ Libræ	31.53				M.	α Serpentis	31.88
	α² Libræ	31.52	July 5	α Coronæ	36.57		α Scorpii	31.73
	α Herculis	31.37	S.	α Serpentis	36.61		α Herculis	31.81
	α Ophiuchi	31.33		α Scorpii	36.58		α Ophiuchi	31.84
				α Herculis	36.66		α Lyræ	31.77
June 24	α Virginis	30.67		α Ophiuchi	36.59			
M.	α Bootis	30.59				July 18	α Virginis	31.17
			July 8	α Virginis	35.87	S.		
June 24	α¹ Libræ	30.40	S.	α Bootis	35.84			
S.	α² Libræ	30.49				July 18	α Coronæ	31.39
	α Coronæ	30.56	July 8	α² Libræ	35.90	M.	α Serpentis	31.21
	α Serpentis	30.48	M.	α Coronæ	35.89		α Scorpii	31.16
				α Serpentis	36.07		α Herculis	31-30
June 27	α Leonis	29.46		α Scorpii	35.90		α Ophiuchi	31.25
M.	α Virginis	29.36		α Herculis	35.88			
	α Bootis	29.50				July 20	α Virginis	30.42
	α¹ Libræ	29.41	July 10	α Coronæ	35.14	M.		
	α² Libræ	29.30	M.	α Serpentis	35.23			
	α Serpentis	29.25				July 20	α Coronæ	30.42
	α Scorpii	29.31	July 11	α Coronæ	34.75	S.	α Serpentis	30.31
	α Herculis	29.26	S.	α Serpentis	34.60		α Scorpii	30.20
				α Scorpii	34.57		α Herculis	30.33
June 28	α Bootis	28.09					α Ophiuchi	30.34
S.	α¹ Libræ	27.98	July 12	α Serpentis	34.53		α Lyræ	30.44
	α² Libræ	27.99	M.					
	α Coronæ	28.04				July 21	α Virginis	30.07
	α Serpentis	27.98	July 13	α Virginis	34.06	M.	α Coronæ	30.13
June 29			S.	α² Libræ	33.97		α Serpentis	30.05
S.	α Virginis	27.16		α Coronæ	33.96			

July 8–11 $n = -\overset{s.}{0.003}$ $c = +\overset{s.}{0.286}$
13–15 $+0.061$ $+0.286$

July 17–26 $n = -\overset{s.}{0.025}$ $c = -\overset{s.}{0.286}$

Date.	Name of Star.	Value of $\Delta t + m$
1865.		s.
July 22	α Virginis	29.50
S.	α Bootis	29.57
	α Herculis	29.59
	α Ophiuchi	29.55
	α Lyræ	29.67
July 22	α² Libræ	29.67
M.	α Coronæ	29.75
	α Serpentis	29.70
	α Scorpii	29.60
July 24	α Virginis	28.95
M.	α Bootis	28.98
	α Coronæ	28.98
	α Serpentis	28.96
	α Ophiuchi	28.86
	α Lyræ	28.94
	γ Aquilæ	28.92
	α Aquilæ	28.99
	β Aquilæ	28.95
July 26	α Serpentis	28.50
M.	α Scorpii	28.47
	α Herculis	28.54
	α Ophiuchi	28.43
	α Lyræ	28.60
	γ Aquilæ	28.54
	α Aquilæ	28.62
	β Aquilæ	28.54
July 27	α Virginis	27.83
S.	α Bootis	27.80
	α Lyræ	27.75
	δ Aquilæ	27.56*
	h² Sagittarii	27.61*
July 27	α Coronæ	27.99
M.	α Serpentis	27.89
	α Scorpii	27.77
	α Herculis	27.85
	α Ophiuchi	27.78
July 28	α Serpentis	27.54
M.	α Scorpii	27.50
	α Ophiuchi	27.51
	α Lyræ	27.62
	γ Aquilæ	27.56

Date.	Name of Star.	Value of $\Delta t + m$
1865.		s.
July 28	β Aquilæ	27.50
M.		
July 29	α Scorpii	27.42
S.	α Ophiuchi	27.31
July 31	α Virginis	26.36
M.	α Bootis	26.42
	α Serpentis	26.29
	α Scorpii	26.10
	α Herculis	26.25
	α Ophiuchi	26.19
	α Lyræ	26.26
	γ Aquilæ	26.13
Aug. 1	α Bootis	25.64
S.	α Scorpii	25.59
	α Herculis	25.65
	α Ophiuchi	25.60
Aug. 2	α Scorpii	25.54
M.	α Herculis	25.48
	α Ophiuchi	25.47
	α Lyræ	25.50
	γ Aquilæ	25.53
	β Aquilæ	25.42
Aug. 4	α Herculis	24.86
	α Ophiuchi	24.87
	α Lyræ	24.96
	γ Aquilæ	24.77
	β Aquilæ	24.82
Aug. 8	α Scorpii	23.22
M.	α Herculis	23.21
	α Ophiuchi	23.28
	γ Aquilæ	23.27
	α Aquilæ	23.22
	β Aquilæ	23.26
Aug. 9	α Lyræ	22.71
M.	γ Aquilæ	22.67
	β Aquilæ	22.56
Aug. 10	α Virginis	21.97
M.	α Bootis	21.79

Date.	Name of Star.	Value of $\Delta t + m$
1865.		s.
Aug. 12	α Virginis	21.08
M.	α Herculis	21.15
	α Ophiuchi	21.09
	α Lyræ	21.25
	γ Aquilæ	20.94
	β Aquilæ	20.95
Aug. 14	α Virginis	19.76
	α Bootis	19.68
	α Herculis	19.74
	α Ophiuchi	19.62
	α Lyræ	19.75
	γ Aquilæ	19.53
	α Aquilæ	19.68
Aug. 15	α Ophiuchi	19.35
M.	α Lyræ	19.54
	γ Aquilæ	19.38
	α Aquilæ	19.37
	β Aquilæ	19.32
Aug. 16	α Herculis	18.95
M.		
Aug. 18	α Lyræ	17.60
M.	γ Aquilæ	17.32
	α Aquilæ	17.52
	β Aquilæ	17.49
Aug. 19	α Herculis	16.80
S.	α Ophiuchi	16.67
	α Lyræ	16.74
Aug. 21	α Bootis	15.96
M.	α Serpentis	16.04
	α Ophiuchi	16.00
Aug. 23	α Lyræ	14.49
M.	γ Aquilæ	14.20
	α Aquilæ	14.17
	β Aquilæ	14.21
Aug. 24	α Lyræ	13.08
M.	γ Aquilæ	12.90
	α Aquilæ	12.94
	β Aquilæ	12.85

July 27–28 $n = +0^s.023$ $c = -0^s.220$ Aug. 18–23 $n = +0^s.147$ $c = -0^s.220$

July 29–Aug. 16 $n = +0.063$ $c = -0.220$ Aug. 24–25 $n = +0.192$ $c = -0.220$

Date.	Name of Star.	Value of $\Delta t+m$
1865.		s.
Aug. 25 M.	α Virginis	12.03
	α Ophiuchi	12.03
	α Lyræ	12.06
	γ Aquilæ	11.99
	α Aquilæ	12.05
	β Aquilæ	12.05
Aug. 26 S.	α Lyræ	11.33
	γ Aquilæ	11.17
Aug. 28 M.	α Herculis	9.99
	α Ophiuchi	9.97
	α Lyræ	9.83
	γ Aquilæ	9.93
	α Aquilæ	9.88
	β Aquilæ	9.81
Aug. 29 S.	α Coronæ	9.17
	α Serpentis	9.10
	α Lyræ	9.04
Aug. 31 S.	α Lyræ	7.26
	α Cygni	7.27
Sept. 2 S.	α Lyræ	7.04
	γ Aquilæ	6.95
	β Aquilæ	6.92
	α^1 Capricorni	6.99
	α^2 Capricorni	6.96
Sept. 4 M.	α Ophiuchi	5.86
	α Lyræ	5.86
	γ Aquilæ	5.99
	β Aquilæ	5.85
Sept. 5 S.	γ Aquilæ	5.02
	β Aquilæ	5.00
	α^1 Capricorni	5.06
	α^2 Capricorni	5.11
Sept. 6 S.	α Ophiuchi	4.66
	μ Herculis	4.48*
	μ Sagittarii	4.50*
Sept. 6 M.	α Lyræ	4.62
	γ Aquilæ	4.54
	β Aquilæ	4.50

Date.	Name of Star.	Value of $\Delta t+m$
1865.		s.
Sept. 6 M.	α^1 Capricorni	4.51
	α^2 Capricorni	4.40
Sept. 7 S.	α Scorpii	3.73
	α Lyræ	3.84
	γ Aquilæ	3.77
	α Aquilæ	3.82
	α Cygni	3.58
Sept. 9 S.	γ Aquilæ	2.20
	β Aquilæ	2.10
	α^1 Capricorni	1.97
	α^2 Capricorni	1.97
Sept. 11 M.	γ Aquilæ	0.59
	α Aquilæ	0.69
	β Aquilæ	0.56
	α^1 Capricorni	0.49
	α^2 Capricorni	0.50
	α Aquarii	0.54
	α Piscis Aus.	0.48
	α Pegasi	0.56
Sept. 12 M.	α Lyræ	0.33
	γ Aquilæ	0.16
	α Aquilæ	0.25
	β Aquilæ	0.04
	α^1 Capricorni	0.06
	α^2 Capricorni	59.99
	α Cygni	0.15
	α Piscis Aus.	0.18
	α Pegasi	59.99
Sept. 13 M.	α Lyræ	59.58
	γ Aquilæ	59.41
	α Aquilæ	59.47
	β Aquilæ	59.46
	α^1 Capricorni	59.44
	α^2 Capricorni	59.46
	α Aquarii	59.35
	α Piscis Aus.	59.29
	α Pegasi	59.32
	α Andromedæ	59.24
	γ Pegasi	59.29
Sept. 14 S.	γ Aquilæ	58.51
	α^1 Capricorni	58.53

Date.	Name of Star.	Value of $\Delta t+m$
1865.		s.
Sept. 14 S.	α^2 Capricorni	58.47
Sept. 15 M.	α Lyræ	58.40
	γ Aquilæ	58.34
	α Aquilæ	58.41
	β Aquilæ	58.32
	α^1 Capricorni	58.33
	α^2 Capricorni	58.37
	α Aquarii	58.38
Sept. 16 S.	α Lyræ	57.44
	α^1 Capricorni	57.28
	α^2 Capricorni	57.32
	α Aquarii	57.24
Sept. 19 M.	α Herculis	55.16
	α Ophiuchi	55.06
	α Lyræ	55.20
	γ Aquilæ	55.10
	α Aquilæ	55.23
	β Aquilæ	54.97
	α^1 Capricorni	54.98
	α^2 Capricorni	55.08
	α Aquarii	54.85
Sept. 21 S.	α^2 Capricorni	52.99
Sept. 22 M.	α Lyræ	52.69
Sept. 23 S.	α Lyræ	51.38
	α Cygni	51.19
	α Pegasi	51.05
Sept. 25 M.	α Lyræ	49.29
	γ Aquilæ	49.34
	α Aquilæ	49.35
	β Aquilæ	49.32
	α^1 Capricorni	49.26
	α^2 Capricorni	49.19
	α Piscis Aus.	48.91
	α Pegasi	49.17
	γ Pegasi	49.23
Sept. 26 S.	α Lyræ	48.26
	α Aquarii	47.81

	n	c
	s.	s.
Aug. 26–Sept. 5	$n = +0.137$	$c = +0.210$
Sept. 6–12	$+0.060$	$+0.210$
13–15	$+0.130$	$+0.220$
Sept. 16	$n = +0.150$	$c = +0.230$
19–21	$+0.206$	$+0.237$
22–26	$+0.175$	-0.275

Date.	Name of Star.	Value of $\Delta t+m$
1865.		s.
Sept. 26	θ Aquarii . .	47.73*
S.	12 Ceti . .	47.95*
Sept. 26	γ Aquilæ . .	48.20
M.	α Aquilæ . .	48.23
	β Aquilæ . .	48.22
	α^1 Capricorni	48.18
	α^2 Capricorni	48.13
	α Cygni . .	48.27
Sept. 27	α Ophiuchi .	47.04
M.	α Lyræ . .	46.89
	γ Aquilæ . .	46.83
	α Aquilæ . .	46.85
	β Aquilæ . .	46.82
	α^1 Capricorni	46.80
	α^2 Capricorni	46.81
	α Aquarii . .	46.56
	α Piscis Aus. .	46.58
	α Pegasi . .	46.63
	α Andromedæ	46.48
	γ Pegasi . .	46.63
Sept. 29	γ Aquilæ . .	44.36
M.	α Aquilæ . .	44.42
	β Aquilæ . .	44.33
	α^1 Capricorni	44.22
	α^2 Capricorni	44.29
	α Piscis Aus. .	44.25
	α Pegasi . .	44.18
	α Andromedæ	44.17
	γ Pegasi . .	44.18
Sept. 30	α Lyræ . .	43.41
S.	α Aquarii . .	43.29
	α Pegasi . .	43.30
	α Andromedæ	43.27
	γ Pegasi . .	43.30
	θ Ceti . . .	43.13*
Oct. 3	α Herculis .	40.58
M.		
Oct. 3	α Lyræ . .	40.47
S.	α^1 Capricorni	40.36
	α^2 Capricorni	40.34
	α Piscis Aus. .	40.25
1865.		s.
Oct. 3	α Pegasi . .	40.41
S.	κ Piscium . .	40.29*
Oct. 5	ε Delphini .	38.49**
S.	λ Cygni . .	38.59**
	32 Vulpeculæ	38.57*
Oct. 6	α Lyræ . .	37.97
M.	α^2 Capricorni	37.80
	α Aquarii . .	37.95
	α Piscis Aus. .	37.74
	α Pegasi . .	37.72
	α Andromedæ	37.83
	γ Pegasi . .	37.75
Oct. 7	α Aquarii . .	37.04
S.	α Piscis Aus. .	36.97
	α Pegasi . .	36.82
	α Andromedæ	36.93
	γ Pegasi . .	37.11
Oct. 9	α Lyræ . .	35.51
M.	γ Aquilæ . .	35.36
	β Aquilæ . .	35.37
	α^1 Capricorni	35.36
	α^2 Capricorni	35.34
	α Aquarii . .	35.27
	α Piscis Aus. .	35.23
Oct. 10	α Lyræ . .	34.59
S.	ε Pegasi . .	34.47*
	α Aquarii . .	34.50
Oct. 10	γ Aquilæ . .	34.56
M.	α Aquilæ . .	34.62
	β Aquilæ . .	34.59
Oct. 11	γ Aquilæ . .	34.04
M.	α Aquilæ . .	34.22
	β Aquilæ . .	34.05
	α Aquarii . .	33.95
	α Piscis Aus. .	34.03
	α Pegasi . .	34.07
Oct. 12	α Aquarii . .	33.22
M.	α Piscis Aus. .	33.34
	α Pegasi . .	33.44
1865.		s.
Oct. 13	α Lyræ .	32.85
M.	γ Aquilæ . .	32.80
	α Aquilæ . .	32.87
	β Aquilæ . .	32.72
	α Aquarii . .	32.77
	α Piscis Aus. .	32.67
	α Pegasi . .	32.64
	α Andromedæ	32.62
	γ Pegasi .	32.55
Oct. 16	α Aquarii . .	30.97
M.	α Piscis Aus. .	30 86
	α Pegasi . .	30.96
	α Andromedæ	30.91
	γ Pegasi . .	30.94
Oct. 17	α Lyræ . .	30.55
S.	γ Aquilæ . .	30.53
	α Piscis Aus. .	30.39
	α Pegasi . .	30.49
	γ Pegasi . .	30.26
Oct. 19	γ Aquilæ . .	29.05
S.	ζ Cygni . .	28.99*
Oct. 20	α Lyræ . .	28.46
S.		
Oct. 23	α Lyræ . .	26.47
M.	α Aquilæ . .	26.45
	β Aquilæ . .	26.40
	α^1 Capricorni	26.31
	α^2 Capricorni	26.31
	α Aquarii . .	26.43
	α Piscis Aus. .	26.36
	α Pegasi . .	26.33
	α Andromedæ	26.37
	γ Pegasi . .	26.48
Oct. 24	α Piscis Aus. .	26.08
S.	α Pegasi . .	26.00
	α Andromedæ	26.11
	γ Pegasi . .	26.07
Oct. 25	α Lyræ . .	25.92
M.	γ Aquilæ . .	25.83
	α Aquilæ . .	25.91

	n	c		n	c
	s.	s.		s.	s.
Sept. 27	$n = +0.281$	$c = -0.275$	Oct. 5	$n = +0.365$	$c = -0.275$
28–30	$+0.207$	-0.275	6–20	$+0.334$	$+0.234$
Oct. 2–3	$+0.255$	-0.275	Oct. 23–Nov. 11	$+0.332$	-0.246

Date.	Name of Star.	Value of $\Delta t+m$
1865.		s.
Oct. 25 M.	β Aquilæ . .	25.87
	α^1 Capricorni	25.82
	α^2 Capricorni	25.83
	α Aquarii . .	25.83
	α Piscis Aus. .	25.74
	α Pegasi . .	25.87
	α Andromedæ	25.90
	γ Pegasi . .	25.81
Oct. 26 S.	γ Aquilæ . .	24.95
	β Aquilæ . .	25.01
Oct. 30 M.	α Lyræ . .	22.07
	γ Aquilæ . .	22.09
	β Aquilæ . .	22.13
	α^1 Capricorni	21.96
	α^2 Capricorni	22.15
	α Aquarii . .	22.02
	α Piscis Aus. .	21.93
	α Pegasi . .	21.99
	α Andromedæ	22.03
	γ Pegasi . .	22.08
Nov. 1 S.	γ Aquilæ . .	21.05
	α^2 Capricorni	21.00
	α Andromedæ	20.92
	α Arietis . .	20.91
Nov. 7 S.	γ Aquilæ . .	15.90
	β Aquilæ . .	15.86
	α^2 Capricorni	15.86
	γ Pegasi . .	15.43
Nov. 9 S.	α Piscis Aus. .	13.68
	α Pegasi . .	13.77
Nov. 11 M.	γ Aquilæ . .	12.00
	α Aquilæ . .	12.20
	β Aquilæ . .	12.12
	α Piscis Aus. .	12.17
	α Pegasi . .	12.36
	γ Pegasi . .	12.16
Nov. 14 M.	α Cygni . .	9.52
	α Piscis Aus. .	9.55
	α Pegasi . .	9.56
	α Andromedæ	9.55
1865.		s.
Nov. 14 M.	γ Pegasi . .	9.60
	θ Ceti . . .	9.46*
	η Piscium . .	9.44*
Nov. 15 S.	α Cygni . .	8.67
	32 Vulpeculæ	8.72*
	ζ Cygni . .	8.78*
	α Pegasi . .	8.80
	γ Pegasi . .	8.80
Nov. 16 M.	ζ Cygni . .	8.05*
	α Aquarii . .	8.08
	α Andromedæ	8.07
	γ Pegasi . .	8.05
Nov. 16 S.	θ Aquarii . .	7.85*
	η Aquarii . .	7.82*
	α Piscis Aus. .	7.79
Nov. 17 M.	α Piscis Aus. .	7.05
	α Pegasi . .	7.06
	α Virginis . .	6.40
Nov. 18 S.	γ Aquilæ . .	6.06
	α Aquilæ . .	6.14
Nov. 27 M.	η Piscium . .	56.17*
	α Virginis . .	55.91
	α Bootis . .	55.83
Nov. 28 M.	α Cygni . .	55.85
	α Aquarii . .	55.78
	α Arietis . .	55.82
Dec. 1 M.	α Aquarii . .	53.76
	α Piscis Aus. .	53.76
	α Pegasi . .	53.76
	α Andromedæ	53.83
	γ Pegasi . .	53.76
	α Arietis . .	53.82
	ξ^2 Ceti . . .	53.69*
Dec. 2 S.	α Piscis Aus. .	52.73
	α Pegasi . .	52.83
Dec. 5 S.	α Andromedæ	50.63
	γ Pegasi . .	50.63
1865.		s.
Dec. 6 S.	α Cygni . .	49.34
	α Andromedæ	49.24
	γ Pegasi . .	49.21
Dec. 8 S.	ε Pegasi . .	44.24*
	16 Pegasi . .	44.18*
	α Andromedæ	44.11
	γ Pegasi . .	44.13
Dec. 11 S.	α Andromedæ	49.20
	γ Pegasi . .	49.14
Dec. 21 M.	α Piscis Aus. .	36.06
	α Pegasi . .	35.99
	α Andromedæ	36.06
	γ Pegasi . .	35.98
	α Virginis . .	35.87
Dec. 22 M.	α Piscis Aus. .	35.93
	α Pegasi . .	35.95
	γ Pegasi . .	36.00
	α Arietis . .	36.02
	α Virginis . .	35.74
Dec. 23 M.	α Piscis Aus. .	35.52
	α Pegasi . .	35.56
	α Andromedæ	35.63
	γ Pegasi . .	35.63
	α Arietis . .	35.74
Dec. 25 M.	α Arietis . .	33.66
	α Ceti . . .	33.71
	α Virginis . .	33.83
	α Bootis . .	33.88
Dec. 29 M.	α Andromedæ	32.74
	γ Pegasi . .	32.85
	α Arietis .	32.88
	α Ceti . . .	32.83
1866.		
Jan. 4 S.	α Andromedæ	25.51
	ε Piscium . .	25.60*
M.	α Virginis . .	24.50

Dates	n	c
	s.	s.
Nov. 14–15	+ 0.350	+ 0.320
16	+ 0.250	+ 0.320
17–18	+ 0.200	− 0.350
Nov. 27–Dec. 6	+ 0.270	− 0.350
Dec. 21–23	+ 0.350	− 0.350
25–Jan. 4, 1866	+ 0.210	− 0.460

MEAN VALUES OF CLOCK CORRECTIONS.

Date.	Observer.	Sidereal Hour.	Mean $\Delta t+m$.	No. Stars.	Date.	Observer.	Sidereal Hour.	Mean $\Delta t+m$.	No. Stars.	Date.	Observer.	Sidereal Hour.	Mean $\Delta t+m$.
1862.		h. m.	s.		1862.		h. m.	s.		1862.		h. m.	s.
March 11	H.	7. 8	5.005	2	July 3	H.	15.49	21.557	3	Sept. 25		20.43	47.925
24	S.	5.12	32.535	2	18	S.	15.11	45.010	4	27	S.	18. 6	43.16
		7.31	32.220	3			19.51	44.468	4	29	S.	19.36	38.848
26	S.	5.54	27.297	3	22	S.	16.25	34.422	5			0. 4	38.535
		12.14	26.623	3	26	S.	19.10	23.268	4	Oct. 5	S.	4.54	24.550
27	H.	9.39	24.370	4	28	S.	14. 9	18.980	1			7. 0	24.146
28	S.	5.10	22.102	4	Aug. 1	S.	16.28	5.897	3			10.27	23.820
		8. 9	21.887	3	6	S.	16.16	51.923	3	6	S.	17.21	23.215
29	H.	9.41	19.280	2			19.15	51.390	2	7	S.	20. 5	21.184
		11.42	19.030	2	11	S.	17.18	37.360	2		S.	7.18	20.158
April 4	H.	4.28	4.960	1			19.44	37.090	2	8	S.	19.52	19.210
		10.22	4.515	4	16	S.	17.29	24.600	1	9	S.	18.32	17.410
7	S.	8.48	56.147	3	20	S.	14. 9	15.690	1	14	S.	11.42	3.200
		12. 0	55.870	3			19.44	15.140	2	16	S.	17.18	59.570
12	H.	10.48	43.705	4	21	S.	14.52	13.167	3	18	S.	18.53	54.258
16	H.	11.41	34.945	4			19.57	12.640	4	30	S.	22.54	22.090
18	H.	10.31	30.525	4	29	S.	14. 9	53.650	1	Nov. 1	S.	19.44	17.470
19	S.	10.11	28.352	4			18.45	53.100	1	3	S.	20.18	13.175
23	S.	11.43	17.340	1	Sept. 2	S.	17.29	44.280	1			0. 4	12.940
24	S.	11.42	14.450	2			20. 0	43.937	3	4	S.	21.57	10.486
	H.	13.11	14.340	3	4	S.	18.52	39.217	3	10	S.	0.22	52.900
25	S.	10.52	12.587	3			22.36	38.833	3			14. 9	51.420
26	H.	11.40	9.922	4	8	S.	17.22	30.563	3	13	S.	19.58	45.812
29	S.	12.41	2.753	3			20.41	30.140	2	Dec. 6	S.	21.42	38.855
30	H.	11.51	0.745	4	9	S.	17.11	27.955	2	8	S.	1.20	32.870
May 7	S.	12. 0	44.180	4	10	S.	16.16	25.607	3	10	S.	1.47	27.322
14	H.	7.31	28.873	3			21.17	25.018	4			13.20	25.970
		14.40	28.002	6	11	S.	17.18	23.125	2	11	S.	22.36	25.167
15	S.	11.42	25.830	2			19.57	22.798	4			4.24	24.555
17	S.	14.15	21.280	3	15	S.	17.18	13.900	2	12	S.	22.54	23.005
23	S.	11.40	7.115	2	16	S.	17.39	11.672	4	16	S.	0.46	11.855
		15.10	6.777	6	17	S.	17.18	9.630	2			2. 2	11.880
28	H.	14. 8	54.907	3			6.13	8.365	2	20	S.	22.58	1.230
June 5	H.	13.51	34.495	2	18	S.	17.25	7.460	2			3. 4	1.060
11	H.	14.48	18.792	4	21	S.	8.20	56.415	2	24	S.	1.24	50.500
16	S.	13.44	5.715	2	22	S.	16.36	55.090	1	27	S.	1.53	42.645
		15.49	5.477	3	23	S.	17.18	52.660	2	28	S.	0. 4	39.990
17	H.	13.44	3.475	2			23.11	52.060	5			12.15	38.413
		15. 8	3.355	4	25	S.	17.43	48.183	3	29	S.	23.32	37.585

Date.		Observer.	Sidereal Hour.	Mean $\Delta t+m$.	No. Stars.
1863.			h. m.	s.	
Jan.	1	S.	22.54	30.680	2
	2	S.	0. 4	27.995	2
	3	S.	0. 4	25.805	2
	5	S.	0.37	20.620	1
			4. 4	20.350	2
	16	S.	18.32	52.040	1
	17	S.	23.42	51.463	3
	19	S.	0.56	46.000	1
			4.57	45.695	2
	30	S.	1.59	20.420	1
	31	S.	5. 7	17.700	2
Feb.	2	S.	2.55	13.150	1
		S.	5.52	12.897	3
	7	S.	2.55	59.710	1
			8.29	59.443	3
	10	S.	1.59	54.220	1
	14	S.	3.52	45.513	3
	16	S.	2.26	41.373	3
			7.32	41.000	1
	17	S.	3.52	39.210	1
	18	S.	1.59	37.450	1
	20	S.	6.20	33.717	3
			15.33	32.890	2
	21	S.	7.31	31.825	2
	23	S.	2.21	27.610	2
			6.23	27.395	2
	25	S.	9.29	23.415	2
March	2	S.	5. 7	14.070	2
			8. 8	13.967	3
	5	S.	7.11	4.985	4
	6	S.	11. 8	1.410	4
	9	S.	6.34	52.452	5
			10.50	51.850	2
	11	S.	7.43	46.766	5
	23	S.	7.55	4.320	1
	27	S.	9.21	51.960	1
	30	S.	9. 0	42.600	2
April	1	S.	9. 4	36.207	3
	6	S.	7.37	19.760	1
	9	S.	8.40	11.570	3
			12.15	11.147	3
	18	S.	10. 7	44.970	2
	21	S.	8. 0	35.420	2
	22	S.	11.12	32.115	2
	23	S.	9.21	29.510	1
	27	S.	6.39	17.180	1
May	1	S.	13.18	4.830	1
	2	S.	10.12	2.790	1
	9	S.	12. 8	40.885	2
	15	S.	14.59	21.630	1
	18	S.	15.38	12.350	1
	19	S.	12. 5	10.060	3
	23	S.	12.57	0.890	2
	25	S.	14.48	54.553	3

Date.		Observer.	Sidereal Hour.	Mean $\Delta t+m$.	No. Stars.
1863.			h. m.	s.	
May	27	S.	19.19	47.71	1
June	1	S.	12.55	36.37	1
	2	S.	14.49	33.745	2
	3	S.	14.43	30.960	1
	4	S.	15.10	27.850	1
	5	S.	20.58	24.235	2
	11	S.	13.28	10.000	1
	17	S.	17.29	50.310	1
	23	S.	21.13	32.550	2
	27	S.	15.14	22.060	2
	29	S.	21.47	16.240	1
	30	S.	22.58	13.820	1
July	11	S.	16.59	44.270	3
	20	S.	15.52	18.185	2
	21	S.	0. 4	13.965	2
	22	S.	15.45	12.045	2
			0. 4	10.955	2
	23	S.	19. 2	8.600	2
	24	S.	18.32	5.840	1
	29	S.	20.46	53.673	3
	31	S.	18.59	48.740	3
Aug.	3	S.	18.44	41.688	4
	4	S.	20.10	39.245	2
	6	S.	0. 4	33.755	2
	10	S.	17.29	52.370	1
			0. 6	51.880	1
	12	S.	17.29	47.190	1
			20.29	46.880	4
	14	S.	20.11	40.710	5
	15	S.	18.13	38.240	2
	17	S.	19.53	31.773	3
	18	S.	20.19	28.834	5
	19	S.	19.53	26.243	3
	22	S.	18.20	18.205	2
	24	S.	16.15	13.363	3
	30	S.	0. 4	53.785	2
	31	S.	19.23	51.220	2
Sept.	2	S.	19. 3	44.535	2
	4	S.	21.33	37.280	2
	5	S.	19.12	34.425	2
	7	S.	20.37	27.900	1
	9	S.	0.55	20.983	3
	10	S.	18.59	17.820	1
			2.21	16.710	1
	14	S.	20.37	3.610	1
	15	S.	19.59	0.650	4
	16	S.	2. 0	56.940	1
	17	S.	21.38	54.453	3
	20	S.	10. 1	42.490	1
	21	S.	19.17	41.115	2
	22	S.	20.11	37.250	3
			23.14	36.670	3
	23	S.	20.10	33.798	4
	24	S.	23.39	29.330	3

Date.		Observer.	Sidereal Hour.	Mean $\Delta t+m$.	No. Stars.
1863.			h. m.	s.	
Sept.	28	S.	20.19	15.650	3
			23.53	15.053	3
	29	S.	22.58	11.570	4
	30	S.	17.47	8.945	2
			0. 0	7.920	3
Oct.	1	S.	19.33	4.955	2
	22	S.	23. 7	48.407	3
	23	S.	19.40	44.990	1
	28	S.	21.32	27.310	4
			1.26	26.600	2
	29	S.	21.35	23.870	3
Nov.	2	S.	23.43	10.212	4
	3	S.	0.23	6.998	5
	4	S.	21.15	3.892	4
			1.54	3.343	3
	19	S.	20.58	23.820	2
			13.18	21.860	1
	22	S.	13.18	12.680	1
	23	S.	3.29	11.143	3
	25	S.	21.48	6.560	3
	26	S.	23.56	3.407	3
	30	S.	1.11	58.290	2
			13.18	56.370	1
Dec.	1	S.	21.20	55.337	3
	6	S.	13.44	38.755	2
	7	S.	23.27	37.790	2
	9	S.	0.51	32.193	3
	10	S.	23. 3	29.307	3
	15	S.	22.54	15.620	2
			3.43	15.090	4
	21	S.	1.27	55.620	2
			3.41	55.260	2
	22	S.	2.29	52.450	2
	23	S.	22.54	49.930	2
			2.10	49.385	2
	26	S.	0. 4	39.975	2
			3.59	39.487	3
	29	S.	1.37	29.748	4
			5.20	29.300	5
	30	S.	1.42	27.010	2
			3. 6	26.685	2

Date.	Observer.	Sidereal Hour.	Mean $\Delta t+m$.	No. Stars.	Date.	Observer.	Sidereal Hour.	Mean $\Delta t+m$.	No. Stars.	Date.	Observer.	Sidereal Hour.	Mean $\Delta t+m$.	No. Stars.
1864.		h. m.	s.		1864.		h. m.	s.		1864.		h. m.	s.	
Jan. 1	S.	1.21	20.225	2	March 24	S.	7.32	21.333	3	June 10	S.	14.13	1.505	4
2	S.	22.58	16.800	1		R.	9.18	21.210	2		M.	16.45	1.355	2
		2.27	7.175	2		S.	11.43	20.985	2	13	S.	12.30	56.840	2
6	S.	0.40	4.412	4	April 7	R.	8.39	51.725	4		S.	15. 8	56.700	4
8	S.	13.14	55.653	3	8	S.	9.26	49.737	3		R.	16. 2	56.775	2
9	S.	1.16	54.158	4	9	R.	9. 4	48.100	3	14	S.	12.30	55.865	2
10	S.	1. 7	50.850	2		R.	10.50	47.890	2		R.	14.42	55.900	3
11	S.	1.43	46.827	3	15	S.	9.41	34.145	2		S.	16.25	55.742	5
		5. 8	46.525	2	16	R.	9.41	32.180	2		R.	18.45	55.770	1
12	S.	3. 5	44.090	3	21	S.	10. 1	22.240	1	16	S.	14.32	53.470	3
14	S.	0.42	38.517	3		R.	12. 5	22.125	2		S.	16. 9	53.265	4
		7.16	37.750	3		S.	14.15	21.833	3	17	S.	14.30	51.980	3
16	S.	0.57	32.397	3	22	S.	9.21	20.340	1		S.	16.22	51.777	3
		6.35	31.555	2		S.	12.15	20.067	3	18	M.	14.13	50.842	4
21	S.	1.24	16.088	4	27	S.	9.29	11.255	2		R.	16. 1	50.745	4
		3.58	15.865	2	29	S.	14.29	7.302	5	19	R.	13.23	49.540	2
25	S.	3. 4	5.340	1	30	R.	11.41	5.993	3	21	R.	13.55	46.840	1
		6.23	5.063	3	May 2	S.	7.31	2.300	2		S.	15. 8	46.750	4
27	S.	0. 4	0.615	2	3	S.	8.50	0.360	3		S.	17.23	46.620	4
		7.18	59.950	4		S.	14.13	59.932	4	22	R.	15.17	45.490	3
Feb. 4	S.	1.24	39.520	1	4	S.	10. 7	58.420	2		R.	17.19	45.425	2
5	S.	1. 0	36.870	2		S.	13.36	58.100	2	23	R.	13.18	44.380	1
		5.54	36.770	2	5	S.	10. 1	56.530	1		S.	16.21	44.150	2
8	S.	2.21	29.430	1		S.	14.32	56.187	3	24	S.	14.32	42.920	3
		6.46	29.035	2	6	S.	10. 7	55.070	2		M.	15.49	43.087	3
10	S.	2.35	24.247	3		S.	12.31	55.005	2		S.	17.29	42.670	1
12	S.	3. 8	19.127	3	9	R.	12.41	50.680	2	26	R.	16.36	41.000	1
18	S.	5. 8	3.092	4		R.	13.51	50.550	2	27	M.	14.33	39.635	2
19	S.	2.36	0.770	1	10	S.	14.58	49.097	3		R.	14.48	39.587	3
		5.28	0.355	2	18	R.	11.53	35.950	2	28	M.	14.26	37.720	1
23	S.	2.27	49.320	2		R.	13.49	35.735	2		S.	14.46	37.374	5
		7.32	48.947	3		S.	14.15	35.683	3		S.	18.32	37.080	1
24	S.	7.32	46.797	3	19	R.	11.43	34.090	2	29	M.	13.59	36.110	2
25	S.	7.35	44.425	2		S.	14.40	33.820	6		S.	14.46	35.758	5
26	S.	7. 6	42.015	2	22	R.	12. 8	29.825	2	30	S.	15.16	33.987	3
29	S.	5. 7	34.395	2		R.	15.11	29.565	6		R.	15.33	34.070	2
March 3	S.	2.27	27.665	2	23	S.	11. 9	28.137	3	July 3	R.	14.48	28.823	3
		7.32	27.180	3	27	S.	12.31	21.695	2		R.	17.23	28.712	4
7	S.	8.29	16.940	2		S.	14.57	21.574	2	4	R.	13.18	27.530	1
8	S.	2. 0	15.170	1	28	R.	12.54	20.373	3		S.	14.56	27.312	5
9	S.	7.19	12.445	4		R.	14.46	20.200	3		S.	16.59	27.183	3
12	S.	5.11	5.410	3	29	R.	14.40	18.367	6	5	S.	15. 8	25.842	4
		9.16	5.030	2	31	S.	14.55	15.600	5		S.	19. 6	25.570	2
14	S.	4.28	0.350	1	June 2	R.	13.13	12.790	4	6	S.	11.42	24.500	1
		9.57	59.805	2		S.	15. 3	12.575	4		R.	15.33	24.375	2
16	S.	9.21	55.400	1	3	S.	2. 0	10.620	1	8	S.	14. 1	21.170	2
17	S.	5.28	53.385	2	4	S.	15. 8	10.025	4		S.	15.49	20.943	3
	R.	9.11	53.060	1	6	S.	15.21	7.542	4	9	S.	14.47	19.325	4
19	S.	6.14	48.580	2		R.	20. 9	7.313	3		S.	19.40	19.040	1
	R.	7.31	48.545	2	7	S.	12.30	6.300	2	10	R.	16.44	17.945	2
	R.	9.21	48.550	1		R.	14.35	6.280	4	12	S.	14.47	15.828	4
21	S.	6.46	27.215	2		S.	15.49	6.130	3		R.	16.59	15.763	3
	R.	7.58	27.248	4	8	S.	14.15	4.837	3		R.	18.42	15.620	2
	S.	9.46	27.035	2		R.	15.49	4.779	3	13	S.	16.46	14.378	6

Date.	Observer.	Sidereal Hour.	Mean $\Delta t+m$.	No. Stars.	Date.	Observer.	Sidereal Hour.	Mean $\Delta t+m$.	No. Stars.	Date.	Observer.	Sidereal Hour.	Mean $\Delta t+m$.	No. Stars.
1864.		h. m.	s.		1864.		h. m.	s.		1864.		h. m.	s.	
July 13	S.	5.38	13.367	3	Oct. 18	S.	21.16	38.395	2	Dec. 20	S.	0. 4	42.160	3
14	S.	14.48	12.983	3	19	S.	19.33	37.040	4		M.	1.54	42.272	4
	S.	17.50	12.807	3	21	S.	19.57	32.690	4	30	M.	23.42	22.210	3
15	S.	16.25	11.386	5		S.	22.54	32.455	2		M.	2.43	22.140	3
16	R.	17.28	9.928	4	24	S.	18.39	26.390	2					
18	R.	17.43	7.012	4	25	S.	19.50	24.683	6					
19	S.	14.26	5.745	2		S.	23.29	24.448	4					
	R.	17.39	5.624	5	26	S.	19.40	23.382	5	**1865.**				
20	S.	18.46	4.180	5		S.	0. 6	23.175	4	Jan. 2	M.	5.20	36.812	4
22	S.	16.26	0.845	4	Nov. 1	S.	20.19	14.040	3	3	M.	0. 4	35.655	2
23	R.	17.43	59.233	3		S.	1.38	13.800	2		S.	1.17	35.492	4
26	S.	14.59	55.310	5	2	S.	19.58	12.995	4	4	M.	4. 0	33.907	3
	S.	19.41	54.995	6		S.	1. 7	12.845	2		M.	5.20	34.060	4
27	S.	16.36	54.105	2	3	S.	20. 6	12.036	5	11	M.	23.42	23.680	3
28	S.	19.12	52.536	5	4	S.	22. 8	10.417	3		M.	5.26	23.692	5
29	S.	19.40	51.063	7		S.	0. 4	10.370	2		S.	1.41	23.525	2
30	S.	17.43	49.693	3	10	S.	20.29	0.730	2	12	S.	3. 7	22.867	3
Aug. 1	S.	16.22	47.220	3		M.	22.41	0.700	4		M.	3.59	22.902	4
	S.	19.50	47.048	6		S.	1.49	0.270	5	13	M.	2.46	22.030	2
5	S.	15.59	40.335	2	11	S.	21.16	58.405	2		M.	4.54	22.167	3
8	S.	17.23	36.188	4		M.	22.31	58.408	6	16	M.	1.13	17.980	3
	S.	19.57	36.002	4		S.	2.10	57.913	3		S.	1.39	17.850	3
Sept. 2	S.	19.41	59.320	6	14	S.	23.39	51.617	6		M.	6. 7	17.892	9
6	S.	17.27	52.745	2	16	S.	20. 3	48.147	3	18	M.	0. 4	15.730	2
	S.	19.57	52.622	4		M.	0.39	47.657	3		S.	1.50	15.395	4
7	S.	16.21	51.020	1		S.	1.21	47.535	2		M.	5.10	15.556	5
	S.	21.48	50.553	3		M.	2.27	47.495	2	20	M.	1.38	12.773	3
9	S.	19.34	47.100	4	19	S.	20.47	42.796	5		M.	5.25	12.756	6
10	S.	17.21	45.563	3		M.	1. 7	42.602	4	24	S.	2.27	5.630	2
	S.	22.54	45.235	2		S.	3. 5	42.340	3		M.	4.54	5.690	3
15	S.	15.29	35.320	1	21	S.	14. 9	37.140	1		S.	6.14	5.375	2
	S.	21.24	34.850	1	22	S.	21.18	36.728	4	25	M.	1.28	4.620	3
16	S.	18.32	33.130	4	23	S.	0.22	34.937	3		M.	5.25	4.702	6
	S.	21.32	32.755	4		M.	2.19	35.000	3	26	S.	2. 3	3.387	3
17	S.	18.34	31.123	3		S.	3.45	34.860	4		S.	5. 0	3.347	4
	S.	22.36	30.843	3	25	S.	19.44	31.157	3	27	M.	2.14	1.980	3
22	S.	18. 1	22.950	2		S.	23.18	31.000	2		M.	5.25	1.908	6
26	S.	18.39	14.710	2		M.	1.47	31.050	2	28	S.	2.27	0.850	2
	S.	22.57	14.360	4		M.	4.13	30.950	2		S.	5. 4	1.003	3
27	S.	19.26	13.020	4	29	S.	22.47	23.623	3		S.	5.21	0.790	2
	S.	22.54	12.745	2		M.	0.30	23.590	3	30	M.	2.18	58.925	2
29	S.	21.53	9.150	2	30	M.	23.10	22.152	5		M.	6.13	58.851	8
30	S.	23.18	7.570	3		M.	2.52	21.820	3	Feb. 1	S.	3.59	56.432	4
Oct. 4	S.	18.32	59.170	1	Dec. 1	S.	22.58	20.062	4		M.	5.22	56.630	5
5	S.	19.25	57.512	4	5	M.	0. 4	12.770	2		M.	7.19	56.437	4
7	S.	18.52	53.710	4	8	S.	20.10	7.945	2		S.	7.59	56.110	2
	S.	23.22	53.442	5		M.	22.36	7.953	3	2	S.	4.11	54.547	4
8	S.	18. 1	51.970	2		S.	0.42	7.623	3		M.	5.46	54.470	3
10	S.	21. 8	49.258	4	9	S.	22.36	5.583	3		S.	7.19	54.167	4
	S.	1.47	48.977	3		M.	0. 4	5.715	2	3	M.	5.11	52.950	3
11	S.	22.36	47.833	3		M.	2.10	5.670	2	9	S.	2.27	52.905	2
	S.	0.21	47.830	3	13	S.	20.37	55.840	1		M.	4.32	43.153	3
17	S.	19.20	39.990	3	14	S.	2. 4	54.547	3		S.	6.28	42.873	7
	S.	23.39	39.692	6	20	S.	20.37	42.360	1	11	S.	2.27	40.090	2

Date.	Observer.	Sidereal Hour.	Mean $\Delta t + m$.	No. Stars.	Date.	Observer.	Sidereal Hour.	Mean $\Delta t + m$.	No. Stars.	Date.	Observer.	Sidereal Hour.	Mean $\Delta t + m$.	No. Stars.
1865.		*h. m.*	*s.*		1865.		*h. m.*	*s.*		1865.		*h. m.*	*s.*	
Feb. 11	S.	5.10	40.106	5	April 4	S.	12.27	41.430	1	May 26	M.	9.12	56.148	5
13	M.	3.42	35.830	2	5	M.	5.20	41.155	4		S.	13.11	56.135	2
	M.	5.20	35.945	4		S.	7.35	40.890	2		M.	14.32	56.153	3
14	S.	2.27	35.090	2		M.	10.42	40.708	4	30	S.	14.33	52.120	4
	M.	5.41	35.285	4	8	S.	5.12	38.875	2	31	S.	13.23	51.430	2
	S.	7.35	34.970	2		M.	10.30	38.772	4	June 1	S.	14.14	50.592	4
21	S.	2.27	22.845	2		S.	11.58	38.525	2	2	S.	12.51	49.817	3
	S.	5.21	22.713	3	11	S.	10. 7	36.743	3		S.	14.32	49.770	3
	M.	6.47	22.780	3	12	M.	10.48	35.892	4	3	S.	12.33	48.980	3
	S.	7.32	22.587	3	13	M.	5.36	34.468	5		S.	15. 8	48.915	4
22	M.	5.48	21.880	1		M.	8.35	34.278	4	5	S.	14. 2	47.893	3
24	M.	4.23	19.285	4		M.	12.15	34.057	3	6	S.	10. 1	46.930	1
	M.	7. 0	19.262	5	14	S.	6.14	33.005	2		S.	14.28	46.667	3
25	S.	3. 8	18.016	3		S.	9.41	32.840	2	7	S.	12.51	45.790	3
	S.	6. 5	17.935	4		S.	12.22	32.833	3	8	S.	12. 7	44.867	3
27	M.	1.59	14.590	1	17	S.	5. 7	30.170	1	12	S.	15. 5	40.813	3
	M.	5.11	14.640	3		M.	5.43	30.198	4	13	S.	10. 1	40.140	1
March 4	S.	7.19	9.998	4		M.	7.59	30.180	4		S.	14.57	40.106	5
	S.	9.57	9.925	2		M.	11. 9	30.090	3	19	M.	10. 1	35.150	1
6	M.	4.24	6.350	2		S.	12.54	29.923	3		M.	14.14	34.997	4
	S.	5. 7	6.370	2		M.	18.45	29.740	1	20	M.	10. 1	33.840	1
	M.	7.42	6.396	5	18	S.	7.37	28.900	1		S.	13.44	33.740	2
7	S.	5. 7	5.575	2		S.	12.15	28.733	3	21	M.	14.48	33.145	4
8	M.	8.32	4.860	2	19	M.	6.28	28.216	7	23	M.	14.14	31.447	4
10	M.	8.56	2.840	2		S.	9.42	28.035	2		M.	17.19	31.350	2
	M.	10.34	2.760	3		M.	10.56	28.047	3	24	M.	13.44	30.630	2
13	M.	5. 0	0.282	4	24	M.	5.43	22.275	4		S.	15. 8	30.482	4
	M.	7.19	0.225	4		M.	8. 9	22.363	4	27	M.	12.30	29.440	3
16	M.	5.11	58.230	3		M.	12.22	22.237	3		M.	15.43	29.306	5
17	M.	5. 0	57.052	4	25	M.	6.14	21.485	2	28	S.	14.57	28.016	5
	M.	7.16	57.007	3		M.	8.23	21.398	5	29	S.	14.14	27.175	4
	M.	9.52	56.837	3		M.	12.15	21.410	3		S.	15.49	27.093	3
18	M.	5.54	56.235	2	26	M.	6.14	20.905	2	July 3	M.	14.14	13.132	4
	S.	7.16	55.917	3		M.	7.32	20.857	3		M.	16. 9	13.097	4
	M.	8.32	56.035	2		M.	9.41	20.795	2	5	M.	11.40	36.835	2
20	M.	5.10	54.508	5	May 3	M.	10.52	15.195	2		S.	16.25	36.602	5
	M.	18. 1	54.165	2		M.	13.44	15.170	2	8	S.	13.44	35.855	2
21	S.	5.11	53.840	3	10	S.	7.29	9.490	2		M.	15.52	35.928	5
23	S.	9.41	51.290	2		S.	11. 9	9.293	3	10	M.	15.33	35.185	2
25	S.	8. 0	49.125	2	12	M.	13.18	7.590	1	11	S.	15.49	34.640	3
	M.	8. 8	49.160	2	13	M.	7.29	7.095	2	12	M.	15.38	34.530	1
	S.	11. 9	48.973	3		S.	7.37	7.070	1	13	S.	14. 1	34.015	2
27	M.	5.20	47.407	4		S.	12.31	6.810	2		S.	15.49	33.943	3
	M.	7.32	47.377	3	15	M.	8.23	5.626	5	14	M.	14.47	33.430	4
	M.	10.22	47.212	4		M.	13.23	5.430	6		M.	16.59	33.363	3
28	M.	5.48	46.760	1	17	M.	7.32	4.397	3	15	M.	16.25	32.924	5
	M.	8.23	46.704	5	18	S.	12.15	3.220	3	17	M.	17.29	31.825	6
April 1	S.	5.48	43.480	1		S.	15. 0	3.060	4	18	S.	13.18	31.170	1
3	M.	5.23	42.143	6	23	M.	12.32	59.407	3		M.	16.25	31.262	5
	S.	7.32	42.127	3		S.	13.11	59.245	2	20	M.	13.18	30.420	1
	M.	9.41	42.295	2	24	S.	11. 9	58.263	3		S.	16.46	30.340	6
4	S.	6.14	41.570	2		S.	13.44	58.140	2	21	M.	14.48	30.083	3
	S.	7.35	41.530	2	25	M.	11. 9	57.093	3	22	S.	13.44	29.535	2
	M.	8.11	41.553	3		M.	14.14	56.992	4		M.	15.33	29.680	4

Date.	Observer.	Sidereal Hour.	Mean $\Delta t+m$.	No. Stars.
		h. m.	*s.*	
1865.				
July 22	S.	17.43	29.603	3
24	M.	14.39	28.968	4
	M.	19. 3	28.932	5
26	M.	15.59	28.485	2
	M.	18.44	28.545	6
27	S.	13.44	27.815	2
	M.	16.25	27.856	5
	S.	19. 7	27.640	3
28	M.	16.29	27.517	3
	M.	19.20	27.560	3
29	S.	16.55	27.365	2
31	M.	14.52	26.292	4
	M.	18.12	26.208	4
Aug. 1	S.	16.17	25.620	4
2	M.	18.10	25.490	6
4	M.	18.32	24.856	5
8	M.	16.59	23.237	3
	M.	19.44	23.250	3
9	M.	19.20	22.647	3
10	M.	13.44	21.880	2
12	M.	13.18	21.080	1
	M.	18.32	21.076	5
14	M.	13.44	19.720	2
	M.	18.31	19.664	5
15	M.	19. 3	19.392	5
16	M.	17. 8	18.950	1
18	M.	19.26	17.482	4
19	S.	17.43	16.737	3
21	M.	15.45	16.000	3
23	M.	19.26	14.268	4
24	M.	19.26	12.942	4
25	M.	15.23	12.030	2
	M.	19.26	12.038	4
26	S.	19. 6	11.250	2
28	M.	18.44	9.902	6
29	S.	16.33	9.103	3
31	S.	19.35	7.265	2
Sept. 2	S.	19.40	6.972	5
4	M	18.52	5.890	4
5	S.	19.57	5.048	4
6	S.	17.45	4.547	3
	M	19.40	4.514	5
7	S.	16.21	3.730	1
	S.	19.40	3.752	4
9	S.	19.57	2.060	4
11	M.	19.55	0.566	5
	M.	22.36	0.527	3
12	M.	19.49	0.140	7
1865.				
Sept. 12	M.	22.54	0.085	2
13	M.	19.41	59.470	6
	M.	23.11	59.298	5
14	S.	20. 0	58.503	3
15	M.	20. 1	58.364	7
16	S.	20.13	57.320	4
19	M.	17.43	55.140	3
	M.	20.15	55.035	6
21	S.	20.11	52.990	1
22	M.	18.32	52.690	1
23	S.	20.42	51.207	3
25	M.	19.41	49.292	6
	M.	23.18	49.103	3
26	S.	18.32	48.260	1
	M.	20. 2	48.205	6
	S.	22.50	47.830	3
27	M.	19.22	46.863	7
	M	23.11	46.576	5
29	M.	19.55	44.324	5
	M.	23.29	44.195	4
30	S.	20.16	43.350	2
	S.	0. 6	43.250	4
Oct. 3	M.	17. 8	40.580	1
	S.	19.38	40.390	3
	S.	23. 3	40.317	3
5	S.	20.39	38.550	3
6	M.	20.14	37.907	3
	M.	23.29	37.760	4
7	S.	23.11	36.974	5
9	M.	19.40	35.388	5
	M.	22.25	35.250	2
	M.	19.44	34.590	3
10	S.	20.43	34.520	3
11	M.	19.44	34.103	3
	M.	22.36	34.017	3
12	M.	22.36	33.333	3
13	M.	19.26	32.810	4
	M.	23.11	32.650	5
16	M.	23.11	30.928	5
17	S.	19. 6	30.540	2
	S.	23.18	30.380	3
19	S.	20.24	29.020	2
20	S.	18.32	28.460	1
23	M.	19.41	26.388	5
	M.	23.11	26.394	5
24	S.	23.29	26.065	4
25	M.	19.41	25.863	6
	M.	23.11	25.830	5
1865.				
Oct. 26	S.	19.44	24.980	2
30	M.	19.40	22.080	5
	M.	23.11	22.010	5
Nov. 1	S.	22.28	20.970	4
7	S.	19.53	15.873	3
	S.	0. 6	15.430	1
9	S.	22.54	13.725	2
11	M.	19.44	12.107	3
	M.	23.18	12.230	3
14	M.	22. 8	9.543	3
	M.	0.42	9.512	4
15	S.	20.51	8.723	3
	S.	23.32	8.800	2
16	M.	21.33	8.065	2
	M.	0. 4	8.060	2
	S.	22.29	7.820	3
17	M.	22.54	7.055	2
	M.	13.18	6.400	1
18	S.	19.42	6.100	2
27	M.	1.24	56.170	1
	M.	13.44	55.870	2
28	M.	21.18	55.815	2
	M.	2. 0	55.820	1
Dec. 1	M.	22.36	53.760	3
	M.	1. 7	53.775	4
2	S.	22.54	52.780	2
5	S.	0. 4	50.630	2
6	S.	22.55	49.263	3
8	S.	21.42	44.210	2
	S.	0. 4	44.120	2
11	S.	0. 4	49.170	2
21	M.	23.29	36.022	4
	M.	13.18	35.870	1
22	M.	23.59	35.975	4
	M.	13.18	35.740	1
23	M.	22.54	35.540	2
	M.	0.42	35.667	3
25	M.	2.27	33.685	2
	M.	13.44	33.855	2
29	M.	1.16	32.825	4
1866.				
Jan. 4	S.	0.29	25.555	2
	M.	13.18	24.500	1

RESULTS. MEAN RIGHT ASCENSIONS FOR 1865.0.

PIAZZI. 0.19.

Date.	Observer.	A.R. 1865.0.
		h. m. s.
1865 Sept. 13	M.	0 8 10.97
29	M.	10.84
Oct. 13	M.	10.91
16	M.	10.90
17	S.	11.01
23	M.	10.81
Nov. 11	M.	10.94
14	M.	10.89
Dec. 1	M.	10.90
5	S.	10.98
1866 Jan. 4	S.	10.88

Mean A.R. 1865.0 **0ʰ 8ᵐ 10ˢ.912**

PIAZZI. 0.38.

Date.	Observer.	A.R. 1865.0.
		h. m. s.
1865 Sept. 13	M.	0 11 35.72
27	M.	35.44
Oct. 13	M.	35.55
16	M.	35.58
17	S.	35.79
23	M.	35.46
Nov. 14	M.	35.59
Dec. 1	M.	35.55
5	S.	35.62
29	M.	35.67
1866 Jan. 4	S.	35.61

Mean A.R. 1865.0 **0ʰ 11ᵐ 35ˢ.598**

ι SCULPTORIS = PI. 0.50 = B. A. C. 72.

Date.	Observer.	A.R. 1865.0.
		s.
1865 Sept. 13	M.	+ .06
27	M.	— .04
29	M.	+ .05
30	S.	+ .03
Oct. 13	M.	+ .11
16	M.	+ .15
1865 Oct. 23	M.	— .14
30	M.	+ .07
Nov. 11	M.	+ .16
14	M.	— .01
Dec. 1	M.	+ .07
Mean		+ .046
v. Asten		44.00

Mean A.R. 1865.0 **0ʰ 14ᵐ 44ˢ.046**

PIAZZI. 0.56.

Date.	Observer.	A.R. 1865.0.
		h. m. s.
1865 Sept. 27	M.	0 16 13.44
29	M.	13.38
Oct. 13	M.	13.50
16	M.	13.48
23	M.	13.25
30	M.	13.42
Nov. 11	M.	13.49
14	M.	13.43
Dec. 1	M.	13.50
5	S.	13.41
29	M.	13.44

Mean A.R. 1865.0 **0ʰ 16ᵐ 13ˢ.431**

A. Oe. 326.

Date.	Observer.	A.R. 1865.0.
		h. m. s.
1865 Sept. 27	M.	0 18 22.43
Oct. 13	M.	22.37
16	M.	22.47
23	M.	22.34
Dec. 1	M.	22.51
5	S.	22.52
29	M.	22.62

Mean A.R. 1865.0 **0ʰ 18ᵐ 22ˢ.466**

α PHŒNICIS. C. des T.

Date.	Observer.	A.R. 1865.0.
		s.
1862 Sept. 23	S.	+ .25
Nov. 10	S.	— .01
1864 Oct. 7	S.	+ .42
17	S.	+ .59
Nov. 14	S.	+ .59
19	M.	+ .25
30	M.	+ .02
Dec. 8	S.	+ .53
Mean		0.33
C. des T.		36.15

Mean A.R. 1865.0 **0ʰ 19ᵐ 36ˢ.48**

12 CETI. N.A.

Date.	Observer.	A.R. 1865.0.
		s.
1862 Nov. 3		+ .06
Dec. 28		— .02
1863 Sept. 24		— .07
29		— .01
30		.00
Nov. 4		— .08
Dec. 26		+ .06
1864 Jan. 9		+ .15
Sept. 30		— .10
Oct. 7		+ .06
17		+ .07
Nov. 4		+ .03
10		.00
14		+ .08
19		+ .04
30		+ .09
Dec. 9		— .26
30		+ .13
Mean		+0.013
N.A.		8.972

Mean A.R. 1865.0 **0ʰ 23ᵐ 8ˢ.985**

Pi. **0.91** = B. A. C. 115.

Date.	Observer.	A.R. 1865.0.
		h. m. s.
1865 Sept. 27	M.	0 23 37.61
Oct. 13	M.	37.77
16	M.	37.72
Nov. 1	S.	37.62
14	M.	37.52
Dec. 1	M.	37.64
29	M.	37.57

Mean A.R. 1865.0 **0^h 23^m $37^s.636$**

Pi. **0.100.**

Date.	Observer.	A.R. 1865.0.
		h. m. s.
1865 Sept. 27	M.	0 25 17.07
29	M.	17.16
Oct. 13	M.	17.08
16	M.	17.17
23	M.	17.35
Nov. 1	S.	17.18
Dec. 5	S.	17.19
29	M.	17.13
1866 Jan. 4	S.	17.10

Mean A.R. 1865.0 **0^h 25^m $17^s.159$**

Pi. **0.109** = B. A. C. 135.

Date.	Observer.	A.R. 1865.0.
		h. m. s.
1865 Sept. 27	M.	0 27 0.22
29	M.	0.38
Oct. 13	M.	0.24
16	M.	0.39
23	M.	0.24
Nov. 1	S.	0.26
14	M.	0.20
Dec. 1	M.	0.25
29	M.	0.21
1866 Jan. 4	S.	0.23

Mean A.R. 1865.0 **0^h 27^m $0^s.262$**

Pi. **0.122.**

Date.	Observer.	A.R. 1865.0.
		h. m. s.
1865 Sept. 27	M.	0 29 11.93
29	M.	12.00
Oct. 13	M.	11.98
16	M.	12.04
23	M.	12.02
Nov. 1	S.	12.13
14	M.	12.00
1865 Dec. 1	M.	0 29 12.10
1866 Jan. 4	S.	12.10

Mean A.R. 1865.0 **0^h 29^m $12^s.033$**

ζ Cassiopeiæ.

Date.	Observer.	A.R. 1865.0.
		h. m. s.
1862 Nov. 3	S.	0 29 27.86
10	S.	27.93
Dec. 8	S.	27.88
1863 Nov. 2	S.	27.78
1864 Jan. 14	S.	27.97
27	S.	27.91
Nov. 10	S.	27.83
16	M.	27.92
23	S.	27.81
25	S.	28.04
29	M.	27.82
30	M.	28.08
Dec. 5	M.	27.92
20	S.	27.77
1865 Jan. 3	S.	27.98
11	M.	27.82
20	M.	28.03

Mean A.R. 1865.0 **0^h 29^m $27^s.903$**

Bradley 48.

Date.	Observer.	A.R. 1865.0.
		h. m. s.
1863 Nov. 4	S.	0 29 42.83
26	S.	42.72
30	S.	43.45
Dec. 26	S.	42.96
1864 Jan. 6	S.	43.01
9	S.	42.78
Sept. 30	S.	42.87
Oct. 17	S.	42.74
Dec. 8	S.	42.96

Mean A.R. 1865.0 **0^h 29^m $42^s.924$**

Sub Polo.

Date.	Observer.	A.R. 1865.0.
		h. m. s.
1864 Jan. 8	S.	0 29 43.96
Apr. 22	S.	42.80
29	R.	43.18
30	R.	42.87
May 4	S.	42.96
6	S.	42.78
1864 May 9	R.	0 29 42.85
18	R.	42.78
22	R.	43.16
23	S.	43.40
27	S.	42.95
29	R.	42.77
June 2	R.	42.87

Mean A.R. 1865.0 **0^h 29^m $43^s.025$**

Pi. **0.131** = B. A. C. 161.

Date.	Observer.	A.R. 1865.0.
		h. m. s.
1865 Sept. 27	M.	0 30 33.58
29	M.	33.65
Oct. 13	M.	33.58
16	M.	33.51
23	M.	33.45
Nov. 1	S.	33.70
14	M.	33.57
Dec. 1	M.	33.64
29	M.	33.68
1866 Jan. 4	S.	33.49

Mean A.R. 1865.0 **0^h 30^m $33^s.585$**

α Cassiopeiæ.

Date.	Observer.	A.R. 1865.0.
		s.
1862 Nov. 3	S.	+ .16
10	S.	— .06
Dec. 29	S.	— .05
1863 Jan. 1	S.	— .08
3	S.	— .05
Nov. 2	S.	— .02
4	S.	— .17
26	S.	— .05
Dec. 26	S.	.00
1864 Jan. 9	S.	— .11
16	S.	— .15
27	S.	— .04
Feb. 5	S.	— .18
Sept. 30	S.	+ .02
Nov. 16	M.	.00
29	M.	— .09
30	M.	+ .05
Dec. 30	M.	+ .05
1865 Jan. 11	M.	— .15
20	M.	— .05
Mean		— .048
B. J.		51.842

Mean A.R. 1865.0 **0^h 32^m $51^s.794$**

Sub Polo.

Date			Observer.	A.R. 1865.0.
				s.
1862	April	7	S.	— .21
	Dec.	10	S.	— .08
1864	April	21	S.	— .07
		22	S.	+ .41
		30	R.	— .12
	May	3	S.	+ .09
		4	S.	+ .06
		6	S.	+ .21
		9	R.	+ .08
		18	R.	.00
		22	R.	— .13
		28	R.	— .18
		29	R.	— .29
	Mean			— .018
	B. J.			51.842

Mean A.R. 1865.0
0h 32m 51s.824

Pi. 0.142.

Date			Observer.	A.R. 1865.0.
				h. m. s.
1865	Sept.	27	M.	0 32 55.16
		29	M.	55.17
	Oct.	13	M.	55.15
		16	M.	55.17
		23	M.	55.25
	Nov.	1	S.	55.34
		14	M.	55.11
	Dec.	1	M.	55.23
		29	M.	55.13
1866	Jan.	4	S.	55.15

Mean A.R. 1865.0
0h 32m 55s.186

Pi. 0.152 = B. A. C. 185.

Date			Observer.	A.R. 1865.0.
				h. m. s.
1865	Oct.	23	M.	0 35 25.51
	Nov.	1	S.	25.64
	Dec.	1	M.	25.55
1866	Jan.	4	S.	25.43

Mean A.R. 1865.0
0h 35m 25s.532

β Ceti.

Date			Observer.	A.R. 1865.0.
				s.
1863	Jan.	17	S.	+ .21
	Nov.	30	S.	+ .09
	Dec.	26	S.	+ .22
1864	Jan.	6	S.	+ .07
1864	Jan.	9	S.	+ .18
		27	S.	+ .10
	Sept.	30	S.	+ .10
	Oct.	17	S.	+ .14
	Nov.	10	S.	+ .07
		14	S.	+ .28
		16	M.	+ .13
		30	M.	+ .15
	Dec.	5	M.	+ .29
		30	M.	+ .24
1865	Jan.	11	M.	+ .14
	Mean			+ .161
	N. A.			48.591

Mean A.R. 1865.0
0h 36m 48s.752

A. 0.681.

Date			Observer.	A.R. 1865.0.
				h. m. s.
1865	Sept.	27	M.	0 37 6.29
	Oct.	16	M.	6.52
		23	M.	6.50
	Dec.	1	M.	6.49
1866	Jan.	4	S.	6.38

Mean A.R. 1865.0
0h 37m 6s.436

Pi. 0.174.

Date			Observer.	A.R. 1865.0.
				h. m. s.
1865	Sept.	27	M.	0 38 56.74
	Oct.	13	M.	56.83
		16	M.	56.83
		23	M.	56.80
	Nov.	14	M.	56.72
	Dec.	1	M.	56.76
1866	Jan.	4	S.	56.71

Mean A.R. 1865.0
0h 38m 56s.770

η Cassiopeiæ.

Date			Observer.	A.R. 1865.0.
				h. m. s.
1863	Nov.	3	S.	0 40 57.06
		4	S.	56.92
1865	Sept.	29	M.	56.88
	Oct.	13	M.	56.92
		16	M.	57.02
		23	M.	56.84
	Nov.	14	M.	56.97
		15	S.	57.17
	Dec.	1	M.	57.02

Mean A.R. 1865.0
0h 40m 56s.978

Sub Polo.

Date			Observer.	A.R. 1865.0.
				h. m. s.
1865	May	24	S.	0 40 57.27
		25	M.	57.06
		26	M.	57.24
		30	S.	57.58
	June	2	S.	57.26
		6	S.	57.38

Mean A.R. 1865.0
0h 40m 57s.298

[*Note.*—The proper motion 0s.140 yearly was allowed in the reduction to 1875.0.]

δ Piscium.

Date			Observer.	A.R. 1865.0.
				h. m. s.
1862	Nov.	10	S.	0 41 40.88
	Dec.	8	S.	40.89
		10	S.	40.92
		16	S.	40.75
1863	Jan.	5	S.	40.88
1864	Jan.	16	S.	40.81
		27	S.	40.97
	Sept.	30	S.	40.86
	Oct.	11	S.	40.90
		17	S.	40.90
		26	S.	40.82
	Nov.	1	S.	40.84
		10	S.	40.78
		14	S.	40.89
		16	M.	40.89
		23	S.	40.83
		25	S.	40.91
		29	M.	40.84
		30	M.	40.88
	Dec.	5	M.	41.10
		14	S.	40.85
		30	M.	40.99
1865	Jan.	16	M.	40.74
		18	M.	40.90

Mean A.R. 1865.0
0h 41m 40s.876

Bradley. 74.

Date			Observer.	A.R. 1865.0.
				h. m. s.
1862	Dec.	8	S.	0 42 27.71
		10	S.	27.68
1863	Sept.	28	S.	27.35
		30	S.	27.97
	Oct.	29	S.	27.42
	Nov.	2	S.	27.99
		26	S.	27.55

Date.	Observer.	A.R. 1865.0.
		h. m. s.
1863 Dec. 26	S.	0 42 27.78
1864 Jan. 6	S.	28.08
9	S.	27.96
16	S.	28.31
Sept. 30	S.	28.04
Nov. 1	S.	28.24

Mean A.R. 1865.0 **$0^h\ 42^m\ 27^s.853$**

Sub Polo.

Date.	Observer.	A.R. 1865.0.
		h. m. s.
1864 Jan. 8	S.	0 42 28.41
April 21	S.	27.23
22	S.	28.28
29	S.	28.01
May 4	S.	27.96
6	S.	28.17
9	R.	27.68
18	R.	27.56
22	R.	28.11
23	S.	28.33
28	R.	27.92
29	R.	27.67
June 4	S.	28.04

Mean A.R. 1865.0 **$0^h\ 42^m\ 27^s.952$**

A. Oe. 799.

Date.	Observer.	A.R. 1865.0.
		h. m. s.
1865 Sept. 27	M.	0 43 13.61
29	M.	13.55
Oct. 13	M.	13.67
16	M.	13.55
23	M.	13.62
Nov. 1	S.	13.77
14	M.	13.56
Dec. 1	M.	13.81

Mean A.R. 1865.0 **$0^h\ 43^m\ 13^s.642$**

Pi. 0.210.

Date.	Observer.	A.R. 1865.0.
		h. m. s.
1865 Sept. 27	M.	0 44 31.71
Oct. 16	M.	31.71
23	M.	31.52
25	M.	31.83
Nov. 1	S.	31.66
14	M.	31.66
Dec. 1	M.	31.80
29	M.	31.80

Mean A.R. 1865.0 **$0^h\ 44^m\ 31^s.711$**

γ Cassiopeiæ.

Date.	Observer.	A.R. 1865.0.
		h. m. s.
1862 Dec. 8	S.	0 48 34.88
10	S.	35.00
1863 Jan. 3	S.	34.80
5	S.	34.99
Aug. 6	S.	34.62
Sept. 9	S.	34.88
1864 Jan. 14	S.	35.14
Feb. 5	S.	34.72
Dec. 30	M.	34.95

Mean A.R. 1865.0 **$0^h\ 48^m\ 34^s.887$**

Sub Polo.

Date.	Observer.	A.R. 1865.0.
		h. m. s.
1864 April 22	S.	0 48 35.01
29	S.	35.04
May 3	S.	35.19
4	S.	35.08
28	R.	34.86
1865 April 17	S.	35.07
18	S.	35.06

Mean A.R. 1865.0 **$0^h\ 48^m\ 35^s.044$**

μ Andromedæ.

Date.	Observer.	A.R. 1865.0.
		h. m. s.
1863 Nov. 3	S.	0 49 16.26
4	S.	16.00
1864 Jan. 6	S.	16.11
9	S.	16.22
11	S.	16.31
16	S.	16.15
27	S.	16.12
Sept. 30	S.	16.15
Oct. 17	S.	16.05
26	S.	16.14
Nov. 1	S.	16.20
10	S.	16.25
16	M.	16.15
23	S.	16.22
25	S.	16.18
Dec. 20	S.	16.01
1865 Jan. 16	M.	16.21
20	M.	16.25

Mean A.R. 1865.0 **$0^h\ 49^m\ 16^s.166$**

Pi. 0.242.

Date.	Observer.	A.R. 1865.0.
		h. m. s.
1865 Sept. 27	M.	0 50 50.72
29	M.	50.74
Oct. 13	M.	50.76
16	M.	50.76
23	M.	50.71
25	M.	50.68
Nov. 14	M.	50.75
Dec. 1	M.	50.83
29	M.	50.81

Mean A.R. 1865.0 $0^h\ 50^m\ 50^s.751$

43 Cephei H.

Date.	Observer.	A.R. 1865.0.
		s.
1862 Nov. 10	S.	—0.13
Dec. 8	S.	+1.61
10	S.	+0.03
16	S.	+0.73
29	S.	+0.95
1863 Jan. 3	S.	—0.23
5	S.	+0.10
Sept. 28	S.	—0.53
Oct. 29	S.	—0.53
Nov. 2	S.	+0.30
26	S.	+0.25
30	S.	+0.06
Dec. 7	S.	—0.11
10	S.	—0.59
23	S.	+0.06
26	S.	+0.04
1864 Jan. 9	S.	+0.51
11	S.	+0.67
12	S.	—0.19
14	S.	+0.21
16	S.	+0.77
21	S.	+0.44
27	S.	+0.48
1864 Oct. 17	S.	—0.24
26	S.	+0.41
Nov. 1	S.	+0.33
10	S.	+0.93
23	S.	+0.18
30	M.	—0.20
Dec. 8	S.	+0.55
20	S.	—0.76
30	M.	+0.33
1865 Jan. 12	S.	+0.63
Oct. 24	S.	—0.78
Nov. 1	S.	+0.15
15	S.	+1.79
1866 Jan. 4	S.	+0.05
Mean		+0.224
T.H.S.		51.40

Mean A.R. 1865.0 **$0^h\ 50^m\ 51^s.624$**

SUB POLO.

Date.	Observer.	A.R. 1865.0.
		s.
1862 April 7	S.	—0.09
25	S.	—0.74
29	S.	—0.15
30	H.	—0.37
May 7	S.	+1.46
17	S.	0.00
23	S.	—0.70
Dec. 10	S.	—0.15
29	S.	+0.58
1863 June 1	S.	+0.13
1864 Jan. 8	S.	+0.57
April 21	S.	+0.54
22	S.	—0.18
29	S.	—0.32
May 3	S.	—0.14
4	S.	—0.22
6	S.	—0.19
10	S.	—0.47
19	S.	+1.00
22	R.	+0.40
23	S.	+0.60
29	R.	—0.36
31	S.	+0.24
June 2	R.	+0.32
4	S.	0.00
10	S.	—0.23
14	S.	+0.49
1865 May 18	S.	—0.70
23	M.	—0.04
25	M.	—0.47
26	S.	+0.20
30	S.	+0.79
June 1	S.	+0.52
8	S.	—0.24
Mean		+0.061
T.H.S.		51.40

Mean A.R. 1865.0
0ʰ 50ᵐ 51ˢ.461

α SCULPTORIS = PI. 0.250.

Date.	Observer.	A.R. 1865.0.
		h. m. s.
1865 Sept. 29	M.	0 52 6.10
30	S.	5.98
Oct. 13	M.	6.14
16	M.	6.02
23	M.	5.72
25	M.	5.89
Dec. 1	M.	6.09
29	M.	6.01

Mean A.R. 1865.0
0ʰ 52ᵐ 5ˢ.994

PI. 0.258.

Date.	Observer.	A.R. 1865.0.
		h. m. s.
1865 Sept. 27	M.	0 54 26.46
30	S.	26.42
Oct. 13	M.	26.38
16	M.	26.37
23	M.	26.27
25	M.	26.27
Nov. 1	S.	26.32
14	M.	26.35
Dec. 29	M.	26.40

Mean A.R. 1865.0
0ʰ 54ᵐ 26ˢ.360

ε PISCIUM. N.A.

Date.	Observer.	A.R. 1865.0.
		s.
1863 Nov. 3	S.	.00
4	S.	— .04
Dec. 9	S.	— .01
15	S.	.00
23	S.	+ .02
26	S.	+ .07
1864 Jan. 6	S.	— .07
9	S.	+ .07
Oct. 17	S.	— .05
Nov. 1	S.	— .04
23	S.	— .09
29	M.	+ .02
30	M.	+ .01
Dec. 9	M.	+ .10
30	M.	+ .13
1865 Jan. 16	M.	— .05
18	M.	+ .23
Mean		+ .018
N.A.		56.358

Mean A.R. 1865.0
0ʰ 55ᵐ 56ˢ.376

PI. 0.265 = σ SCULPTORIS.

Date.	Observer.	A.R. 1865.0.
		h. m. s.
1865 Sept. 27	M.	0 55 59.46
29	M.	59.51
30	S.	59.61
Oct. 13	M.	59.46
16	M.	59.55
23	M.	59.53
25	M.	59.54
Nov. 14	M.	59.49
Dec. 1	M.	59.55

Mean A.R. 1865.0
0ʰ 55ᵐ 59ˢ.522

A. Oe. 1092 = 70* HEIS CASSIOPEIÆ.

Date.	Observer.	A.R. 1865.0.
		h. m. s.
1865 Sept. 27	M.	0 58 49.59
29	M.	49.72
30	S.	49.70
Oct. 13	M.	49.67
16	M.	49.69
23	M.	49.51
25	M.	49.54
Nov. 1	S.	49.45
14	M.	49.67

Mean A.R. 1865.0
0ʰ 58ᵐ 49ˢ.616

44 CEPHEI H.

Date.	Observer.	A.R. 1865.0.
		h. m. s.
1862 Dec. 8	S.	1 0 44.65
10	S.	43.93
1863 Jan. 19	S.	44.36
Sept. 28	S.	43.77
30	S.	43.74
Nov. 30	S.	43.62
Dec. 7	S.	44.05
9	S.	44.11
15	S.	43.86
21	S.	43.43
23	S.	43.91
26	S.	44.29
1864 Jan. 9	S.	43.78
11	S.	43.73
21	S.	44.00

Mean A.R. 1865.0
1ʰ 0ᵐ 43ˢ.949

SUB POLO.

Date.	Observer.	A.R. 1865.0.
		h. m. s.
1864 Jan. 8	S.	1 0 44.54
April 21	S.	43.91
22	S.	44.08
29	S.	43.88
May 3	S.	43.93
4	S.	44.03
9	R.	43.85
18	S.	43.98
19	S.	44.21
23	S.	44.04
28	R.	44.01
29	R.	43.28
June 2	R.	43.52

Mean A.R. 1865.0
1ʰ 0ᵐ 43ˢ.943

Date.	Observer.	A.R. 1865.0.
β Andromedæ.		C. des T.
		s.
1862 Dec. 8	S.	—0.04
10	S.	+0.14
1863 Jan. 1	S.	—0.02
2	S.	+0.13
3	S.	—0.07
19	S.	+0.18
Aug. 10	S.	—0.06
Sept. 9	S.	+0.04
Nov. 30	S.	—0.08
Dec. 23	S.	+0.04
26	S.	—0.05
1864 Jan. 6	S.	—0.04
9	S.	—0.06
10	S.	—0.01
11	S.	—0.06
14	S.	+0.09
16	S.	+0.11
21	S.	+0.13
27	S.	—0.04
Feb. 5	S.	—0.07
Nov. 30	M.	—0.08
Dec. 20	M.	—0.23
30	M.	+0.05
1865 Jan. 11	M.	—0.04
16	M.	—0.14
18	M.	—0.10
Nov. 28	M.	—0.13
Mean		—0.015
C. des T.		10.904
Mean A.R. 1865.0		**1h 2m 10s.889**
Polaris.		
		s.
1862 Mar. 26, 26	S.	—2.25
April 7, 7	S.	+0.13
25, 25	S.	+0.30
May 7, 7	S.	+1.00
13, 14	S.H.	—1.16
14, 14	H.S.	—1.32
Nov. 10, 10	S.	+0.42
Dec. 10, 10	S.	—0.34
28, 29	S.	+1.68
1863 April 9, 9	S.	+2.37
20, 21	S.	+2.30
21, 21	S.	+1.05
22, 22	S.	0.00
Sept. 9, 10	S.	+0.39
28, 28	S.	+2.91
28, 29	S.	+2.91
30, 30	S.	+2.74
Oct. 22, 22	S.	+2.68

Date.	Observer.	A.R. 1865.0.
		s.
1863 Oct. 22, 23	S.	+1.80
28, 28	S.	—1.21
28, 29	S.	+1.84
Nov. 3, 3	S.	+0.36
3, 4	S.	+1.37
22, 23	S.	—2.09
26, 26	S.	+1.14
29, 30	S.	+0.00
30, 30	S.	+1.48
Dec. 6, 7	S.	—0.26
1864 Jan. 8, 9	S.	+1.77
Apr. 21, 21	S.	+0.77
21, 22	S.	—0.28
29, 29	S.	+0.42
May 5, 5	S.	+3.65
5, 6	S.	+2.66
9, 9	R.S.	+0.62
9, 10	S.	+0.85
22, 22	R.S.	+0.46
22, 23	S.	+1.16
27, 27	S.	+0.88
27, 28	S.R.	—2.02
June 3, 4	S.	—1.53
7, 7	S.R.	+0.78
7, 8	R.S.	+1.63
10, 10	S.	—1.14
28, 28	S.R.	+0.52
28, 29	R.S.	+0.26
Oct. 7, 7	S.	—1.42
9, 10	S.	—0.42
10, 10	S.	—0.27
10, 11	S.	—1.89
16, 17	S.	—0.02
20, 21	S.	—0.42
25, 25	S.	+1.02
25, 26	S.	+1.28
Oct. 31, Nov. 1	S.	+0.48
1, 1	S.	+1.99
1, 2	S.	+0.31
11, 11	S.	+2.94
14, 14	S.	+3.32
15, 16	S.	+0.72
25, 25	S.	+1.68
1865 Apr. 13, 13	S.	+2.63
13, 14	S.	+2.17
16, 17	S.	+0.38
24, 24	S.	+1.80
24, 25	S.	+2.22
May 3, 3	S.	—0.25
23, 23	S.	+0.14
23, 24	S.	+1.50
25, 25	S.	+1.96
25, 26	S.	+1.74
29, 30	S.	+0.55
June 22, 23	S.	—0.12

Date.	Observer.	A.R. 1865.0.
		s.
1865 June 23, 24	S.	+0.52
27, 27	S.	+0.47
29, 29	S.	+1.41
July 3, 3	S.	+1.24
4, 5	S.	+1.29
9, 10	S.	+0.78
14, 14	S.	+1.10
14, 15	S.	+0.52
20, 20	S.	+0.51
20, 21	S.	+0.34
24, 24	S.	+1.55
30, 31	S.	+1.24
Aug. 1, 1	S.	+1.09
1, 2	S.	+0.75
28, 28	M.	+1.63
Sept. 13, 13	M.	—0.23
27, 27	M.	+1.26
29, 29	M.	+1.29
30, 30	S.	+1.25
Oct. 16, 16	M.	+0.53
24, 24	S.	—0.64
24, 25	S.M.	+0.79
Nov. 14, 14	M.	+0.96
16, 16	M.	+0.65
17, 17	M.	—1.04
27, 27	M.	+2.82
Dec. 21, 21	M.	+1.66
21, 22	M.	+0.04
22, 22	M.	+1.80
22, 23	M.	+1.64
Mean Deviations:		
Spring & Summer (53)		+0.748
Autumn & Winter (50)		+0.866
Final Mean		+0.805
B.J.		38.104
Mean A.R. 1865.0		**1h 9m 38s.909**
ψ Cassiopeiæ.		
		h. m. s.
1863 Oct. 29	S.	1 16 26.73
Dec. 15	S.	26.15
26	S.	26.13
1864 Jan. 6	S.	26.20
9	S.	26.15
10	S.	25.93
11	S.	26.21
12	S.	26.30
14	S.	26.32
16	S.	26.18
21	S.	26.08
27	S.	26.23
Sept. 30	S.	26.16
Oct. 10	S.	26.13

Date.	Observer.	A.R. 1865.0.
		h. m. s.
1864 Dec. 9	M.	1 16 26.09
1865 Jan. 12	S.	26.18
16	M.	25.81

Mean A.R. 1865.0 **1h 16m 26s.175**

Sub Polo.

Date.	Observer.	A.R. 1865.0.
		h. m. s.
1863 May 25	S.	1 16 26.48
1864 April 29	S.	26.36
May 3	S.	26.24
9	R.	26.30
18	S.	26.25
27	S.	26.35
28	R.	26.10
29	R.	26.17
31	S.	26.18
June 2	S.	25.99
8	S.	26.56
13	S.	26.31
14	S.	26.17

Mean A.R. 1865.0 **1h 16m 26s.266**

θ Ceti. N.A.

Date.	Observer.	A.R. 1865.0.
		s.
1862 Dec. 16	S.	+.06
1863 Jan. 3	S.	+.02
Nov. 4	S.	+.10
Dec. 23	S.	+.10
26	S.	+.25
1864 Jan. 9	S.	+.21
14	S.	+.16
16	S.	+.06
21	S.	+.14
27	S.	+.13
Nov. 23	S.	+.13
30	M.	+.19
Dec. 8	S.	+.20
1865 Jan. 18	M.	+.17
Mean		+.137
N.A.		16.485

Mean A.R. 1865.0 **1h 17m 16s.622**

η Piscium N.A.

Date.	Observer.	A.R. 1865.0.
		s.
1862 Dec. 10	S.	+.07
16	S.	+.05
1863 Dec. 23	S.	+.18
26	S.	+.12
1864 Jan. 9	S.	+.09
1864 14	S.	+.42
27	S.	.00
Nov. 11	S.	+.07
14	S.	+.23
30	M.	+.24
Dec. 8	S.	+.08
9	M.	+.10
14	S.	+.11
Mean		+.122
N.A.		15.713

Mean A.R. 1865.0 **1h 24m 15s.835**

υ Persei (51 Andromedæ).

Date.	Observer.	A.R. 1865.0.
		h. m. s.
1862 Dec. 8	S.	1 29 43.26
10	S.	43.25
16	S.	43.10
1863 Jan. 1	S.	43.27
Oct. 28	S.	43.08
Dec. 9	S.	43.13
23	S.	43.09
29	S.	43.22
1864 Jan. 6	S.	43.35
27	S.	43.25
Oct. 10	S.	43.23
Nov. 1	S.	43.18
11	S.	43.19
16	S.	43.16
23	S.	43.32
25	M.	43.50
30	M.	43.31
Dec. 8	S.	43.12
20	M.	43.23
1865 Jan. 20	M.	43.20
27	M.	43.37

Mean A.R. 1865.0 **1h 29m 43s.229**

ν Piscium N.A.

Date.	Observer.	A.R. 1865.0.
		s.
1862 Dec. 8	S.	+0.11
10	S.	0.05
27	S.	0.12
1863 Dec. 23	S.	0.10
1864 Jan. 9	S.	0.23
27	S.	0.08
Nov. 11	S.	0.09
14	S.	0.22
16	S.	0.16
30	M.	0.10
Dec. 8	S.	0.16
20	M.	0.09
1865 Jan. 24	S.	+0.07
Mean		+0.122
N.A.		24.433

Mean A.R. 1865.0 **1h 34m 24s.555**

τ Ceti.

Date.	Observer.	A.R. 1865.0.
		h. m. s.
1862 Dec. 8	S.	1 37 48.09
10	S.	48.02
16	S.	47.97
1863 Oct. 28	S.	47.81
4	S.	48.03
Dec. 23	S.	47.78
29	S.	47.96
30	S.	47.92
1864. Jan. 14	S.	48.03
27	S.	47.92
Oct. 16	S.	48.00
Nov. 11	S.	47.89
14	S.	47.79
25	S.	47.90
Dec. 8	S.	48.00
1865 Jan. 3	S.	48.00
11	S.	47.86
16	S.	47.86
18	S.	47.86

Mean A.R. 1865.0 **1h 37m 47s.93]**

χ Ceti.

Date.	Observer.	A.R. 1865.0.
		h. m. s.
1862 Dec. 8	S.	1 42 57.29
10	S.	57.35
16	S.	57.34
1863 Oct. 28	S.	57.24
Nov. 3	S.	57.39
4	S.	57.34
Dec. 21	S.	57.21
23	S.	57.29
29	S.	57.40
30	S.	57.37
1864 Jan. 21	S.	57.39
27	S.	57.28
Oct. 10	S.	57.43
Nov. 1	S.	57.41
11	S.	57.38
14	S.	57.32
25	M.	57.39
30	M.	57.38
Dec. 8	S.	57.41
14	S.	57.29
1865 Jan. 3	S.	57.26
11	S.	57.30
28	S.	57.65

Mean A.R. 1865.0 **1h 42m 57s.353**

ω Cassiopeiæ. Sub Polo only.

Date.	Observer.	A.R. 1865.0.
		h. m. s.
1865 May 25	M.	1 45 33.74
26	M.	33.85
June 2	S.	33.73
3	S.	33.81
5	S.	34.06
19	M.	33.69

Mean A.R. 1865.0 1h 45m 33s.813

β Arietis.

Date.	Observer.	A.R. 1865.0.
		s.
1864 Jan. 27	S.	+0.09
Feb. 5	S.	+0.01
March 3	S.	+0.24
Nov. 11	S.	—0.05
14	S.	+0.01
19	M.	+0.01
30	M.	+0.09
Dec. 8	S.	+0.14
20	M.	+0.02
1865 Jan. 20	M.	+0.04
24	S.	+0.05
Feb. 11	S.	+0.05
Mean		+0.058
N. A.		11.181

Mean A.R. 1865.0 1h 47m 11s.239

***A* Cassiopeiæ.** Sub Polo only.

Date.	Observer.	A.R. 1865.0.
		h. m. s.
1865 May 26	M.	1 50 55.83
June 2	S.	55.55
3	S.	55.62
5	S.	55.81
20	S.	55.39

Mean A.R. 1865.0 1h 50m 55s.640

50 Cassiopeiæ.

Date.	Observer.	A.R. 1865.0.
		h. m. s.
1863 Oct. 28	S.	1 51 58.15
Nov. 4	S.	58.21
Dec. 29	S.	57.97
30	S.	57.90
1864 Jan. 6	S.	57.97
21	S.	58.05
27	S.	58.13
Oct. 10	S.	58.11
Nov. 1	S.	58.37
19	M.	58.02
1865 Jan. 16	S.	58.42
18	S.	57.97
26	S.	58.10

Mean A.R. 1865.0 1h 51m 58s.105

50 Cassiopeiæ. Sub Polo.

Date.	Observer.	A.R. 1865.0.
		h. m. s.
1864 April 29	S.	1 51 58.41
May 3	S.	58.30
4	S.	58.32
9	R.	58.28
18	R.	58.31
19	S.	58.81
27	S.	58.26
28	R.	58.03
29	R.	57.87
June 7	R.	58.16
8	S.	58.45
16	S.	58.32

Mean A.R. 1865.0 1h 51m 58s.293

α Piscium. Med.

Date.	Observer.	A.R. 1865.0.
		h. m. s.
1863 Dec. 23	S.	1 55 3.82
29	S.	3.73
30	S.	3.81
1864 Jan. 21	S.	3.89
27	S.	3.71
Nov. 1	S.	3.80
1865 Jan. 12	S.	3.80
16	S.	3.80

Mean A.R. 1865.0 1h 55m 3s.795

α Piscium. Foll.

Date.	Observer.	A.R. 1865.0.
		h. m. s.
1862 Dec. 10	S.	1 55 3.81
1864 Oct. 10	S.	3.86
Nov. 14	S.	3.90
25	M.	3.97
30	M.	3.92
Dec. 14	S.	3.86
20	M.	3.85
1865 Jan. 3	S.	3.88
18	S.	3.86

Mean A.R. 1865.0 1h 55m 3s.879

α Piscium. Point not stated.

Date.	Observer.	A.R. 1865.0.
		h. m. s.
1863 Nov. 4	S.	1 55 3.83
1864 Nov. 16	M.	3.87
19	M.	3.86
23	M.	3.86
Dec. 8	S.	3.90

Mean A.R. 1865.0 1h 55m 3s.864

γ Andromedæ.

Date.	Observer.	A.R. 1865.0.
		h. m. s.
1862 Dec. 8	S.	1 55 37.45
16	S.	37.49
1863 Feb. 16	S.	37.45
23	S.	37.52
Dec. 21	S.	37.46
1864 Feb. 5	S.	37.39
12	S.	37.54
19	S.	37.57
Nov. 11	S.	37.43
1865 Jan. 20	M.	37.51
24	S.	37.45
25	M.	37.55
26	S.	37.70
27	M.	37.54
30	M.	37.47
Feb. 2	S.	37.55
9	S.	37.29
11	S.	37.45
14	S.	37.46
21	S.	37.37

Mean A.R. 1865.0 1h 55m 37s.482

Groombridge 454. Sub Polo only.

Date.	Observer.	A.R. 1865.0.
		h. m. s.
1865 May 26	M.	2 1 0.28
June 3	S.	0 59.80
5	S.	59.81
19	M.	59.83
20	S.	59.60

Mean A.R. 1865.0 2h 0m 59s.864

67 Ceti.

Date.	Observer.	A.R. 1865.0.
		s.
1862 Dec. 10	S.	+0.16
1863 Dec. 29	S.	+0.09
1864 Jan. 9	S.	+0.31
14	S.	+0.13

Date.	Observer.	A.R. 1865.0.
		s.
1864 Jan. 16	M.	+0.14
23	M.	+0.19
25	M.	+0.28
30	M.	+0.17
Dec. 8	S.	+0.28
9	M.	+0.07
14	S.	+0.05
20	M.	+0.07
30	M.	+0.22
1865 Jan. 11	M.	+0.09
16	M.	+0.09
18	S.	+0.05
24	S.	+0.07
25	M.	+0.18
27	M.	+0.23
Mean		+0.151
N.A.		14.992

Mean A.R. 1865.0 **2ʰ 10ᵐ 15ˢ.143**

o Ceti.

Date.	Observer.	A.R. 1865.0.
		s.
1862 Dec. 10	S.	+0.01
16	S.	—0.01
1863 Sept. 10	S.	—0.22
Nov. 4	S.	—0.05
Dec. 29	S.	—0.08
30	S.	+0.02
1864 Jan. 9	S.	+0.19
Nov. 10	S.	—0.02
11	S.	+0.06
Dec. 20	M.	—0.21
1865 Jan. 11	M.	+0.08
16	M.	—0.12
18	S.	—0.13
24	S.	—0.09
25	M.	+0.10
27	M.	+0.08
28	S.	—0.23
30	M.	—0.09
Mean		—0.039
C. des T.		31.854

Mean A.R. 1865.0 **2ʰ 12ᵐ 31ˢ.815**

ι Cassiopeiæ. C. des T.

Date.	Observer.	A.R. 1865.0.
		s.
1863 Feb. 2	S.	+0.06
16	S.	+0.34
Nov. 4	S.	—0.26
Dec. 22	S.	+0.54
23	S.	+0.18
29	S.	+0.12
30	S.	+0.14
1864 Jan. 9	S.	+0.25
Feb. 8	S.	+0.32
Nov. 11	S.	+0.08
16	M.	+0.21
19	M.	+0.13
23	M.	+0.17
25	M.	—0.24
30	M.	+0.20
Dec. 9	M.	—0.23
Mean		+0.126
C. des T.		58.876

Mean A.R. 1865.0 **2ʰ 17ᵐ 59ˢ.002**

ι Cassiopeiæ. Sub Polo.

Date.	Observer.	A.R. 1865.0.
		s.
1863 May 23	S.	+0.30
1864 April 29	S.	+0.12
May 3	S.	+0.44
19	S.	+0.74
22	R.	+0.09
27	R.	+0.28
28	R.	+0.21
29	R.	+0.36
June 7	R.	—0.05
8	S.	+0.67
18	M.	+0.34
21	S.	+0.37
Mean		+0.322
C. des T.		58.876

Mean A.R. 1865.0 **2ʰ 17ᵐ 59ˢ.198**

ξ² Ceti. N.A.

Date.	Observer.	A.R. 1865.0.
		s.
1862 Dec. 10	S.	+0.10
1863 Nov. 4	S.	+0.08
1864 Jan. 5	S.	+0.19
9	S.	+0.10
Feb. 5	S.	—0.13
Nov. 16	M.	—0.04
25	M.	+0.03
30	M.	+0.03
Dec. 9	M.	+0.03
14	S.	+0.02
20	M.	+0.10
30	M.	+0.11
1865 Jan. 24	S.	+0.04
26	S.	+0.10
Mean		+0.054
N.A.		59.026

Mean A.R. 1865.0 **2ʰ 20ᵐ 59ˢ.080**

36 H. Cassiopeiæ.

Date.	Observer.	A.R. 1865.0.
		h. m. s.
1863 Nov. 4	S.	2 25 15.86
Dec. 9	S.	16.07
22	S.	16.12
23	S.	15.92
29	S.	15.92
30	S.	16.18
1864 Jan. 9	S.	15.77
Feb. 5	S.	16.08
8	S.	15.68
Nov. 16	M.	15.93
19	M.	15.86
23	M.	15.67
30	M.	15.98

Mean A.R. 1865.0 **2ʰ 25ᵐ 15ˢ.926**

36 H. Cassiopeiæ. Sub Polo.

Date.	Observer.	A.R. 1865.0.
		h. m. s.
1863 May 23	S.	2 25 15.73
1864 April 21	S.	15.99
29	S.	16.36
May 3	S.	16.15
19	S.	16.27
22	R.	16.19
27	R.	16.04
28	R.	15.97
29	R.	16.06
31	S.	16.06
June 6	S.	16.08

Mean A.R. 1865.0 **2ʰ 25ᵐ 16ˢ.082**

Groombridge 527.

Date.	Observer.	A.R. 1865.0.
		s.
1862 Dec. 10	S.	—0.24
16	S.	—0.08
1863 Sept. 9	S.	—0.27
Nov. 4	S.	—0.35
30	S.	—0.40
Dec. 9	S.	+0.05
21	S.	—0.09
23	S.	—0.10
29	S.	—0.29
30	S.	—0.28
1864 Jan. 9	S.	—0.44
Feb. 8	S.	+0.72
Nov. 30	M.	—0.46
1865 Dec. 1	M.	—0.28
Mean		—0.179
T.H.S.		33.75

Mean A.R. 1865.0 **2ʰ 28ᵐ 33ˢ.571**

Sub Polo.

Date	Observer.	A.R. 1865.0.
		s.
1862 May 14	H.	—0.13
17	S.	+0.07
23	S.	—0.24
June 16	S.	—0.24
1864 April 21	S.	—0.41
29	S.	—0.18
May 3	S.	—0.07
5	S.	—0.10
10	S.	+0.08
19	S.	—0.22
27	S.	—0.19
28	R.	—0.04
31	S.	—0.10
June 2	S.	—0.11
6	S.	—0.50
27	M.	—0.70
Mean		—0.192
T.H.S.		33.75

Mean A.R. 1865.0
2h 28m 33s.558

ε Ceti.

Date	Observer.	A.R. 1865.0.
		h. m. s.
1862 Dec. 10	S.	2 33 2.11
1863 Feb. 2	S.	2.22
Dec. 29	S.	2.12
1864 Jan. 9	S.	2.11
11	S.	2.15
Nov. 10	S.	2.12
16	M.	2.12
19	M.	2.10
23	M.	2.18
25	M.	2.25
30	M.	2.14
Dec. 14	S.	2.11
30	M.	2.27
1865 Jan. 12	S.	2.15
13	M.	2.15
16	S.	2.00
18	M.	2.23
24	S.	2.15
26	S.	2.18
28	S.	2.14
30	M.	2.16
Feb. 2	S.	2.29

Mean A.R. 1865.0
2h 33m 2s.157

γ Ceti. N.A.

Date	Observer.	A.R. 1865.0.
		s.
1862 Dec. 10	S.	+.08
1863 Feb. 2	S.	+.12
Dec. 29	S.	+.08
1864 Jan. 9	S.	+.15
11	S.	+.02
Nov. 10	S.	+.06
19	M.	+.12
25	M.	+.06
30	M.	+.06
Dec. 14	M.	+.03
30	S.	+.14
1865 Jan. 12	M.	+.12
13	S.	+.15
16	M.	+.10
18	M.	+.07
24	M.	+.13
26	S.	+.15
28	S.	+.17
Feb. 9	S.	+.08
11	S.	+.17
Mean		+.103
N.A.		18.406

Mean A.R. 1865.0
2h 36m 18s.509

41 Arietis. C des T.

Date	Observer.	A.R. 1865.0.
		s.
1862 Dec. 10	S.	+.02
20	S.	+.01
1863 Feb. 2	S.	+.01
16	S.	+.05
Dec. 22	S.	+.04
1864 Jan. 9	S.	+.05
11	S.	—.02
Feb. 10	S.	+.01
12	S.	+.09
Nov. 19	S.	+.03
23	M.	+.10
25	M.	+.14
30	M.	+.11
Dec. 14	S.	—.02
30	M.	+.07
1865 Jan. 13	M.	+.11
16	M.	+.08
18	M.	+.27
Mean		+.064
C. des T.		2.595

Mean A.R. 1865.0
2h 42m 2s.659

η Eridani.

Date	Observer.	A.R. 1865.0.
		h. m. s.
1862 Dec. 10		2 49 50.02
16		50.05
1863 Dec. 29		50.04
1864 Jan. 9		50.05
11		50.06
25		50.15
Feb. 10		49.97
Nov. 16		50.00
19		50.00
23		50.01
25		49.95
30		50.04
Dec. 14		49.98
20		50.11
30		50.10
1865 Jan. 12		50.04
13		50.06
16		49.86
18		50.09
24		50.08

Mean A.R. 1865.0
2h 49m 50s.033

Groombridge 580. Sub Polo.

Date	Observer.	A.R. 1865.0.
		h. m. s.
1865 May 25	M.	2 51 1.10
26	M.	1.72
June 2	S.	1.45
3	S.	1.44
19	M.	1.32
21	M.	0.94
27	M.	0.87
July 3	M.	0.53

Mean A.R. 1865.0
2h 51m 1s.171.

ε Arietis.

Date	Observer.	A.R. 1865.0.
		h. m. s.
1862 Dec. 10	S.	2 51 29.79
16	S.	29.84
1864 Jan. 11	S.	29.92
Nov. 16	M.	29.78
19	S.	29.87
25	M.	29.88
30	M.	29.96
Dec. 14	S.	29.81
1865 June 12	S.	29.94
18	M.	30.00
20	M.	29.80

Date.	Observer.	A.R. 1865.0.
		h. m. s.
1865 June 24	S.	2 51 29.85
28	S.	29.80
Feb. 9	S.	29.76

Mean A.R. 1865.0
2h 51m 29s.857

β Persei. C. des T.

Date.	Observer.	A.R. 1865.0.
		s.
1863 Feb. 16	S.	+.07
Dec. 29	S.	+.11
1864 Jan. 5	S.	+.04
9	S.	+.15
11	S.	+.12
25	S.	+.15
Feb. 10	S.	+.12
12	S.	+.25
Nov. 16	M.	—.08
19	S.	+.15
23	S.	+.08
30	M.	+.17
1865 Jan. 12	S.	+.15
13	M.	—.05
16	M.	+.21
18	M.	+.16
20	M.	—.01
24	S.	+.07
27	M.	+.09
Mean		+.103
C. des T.		23.625

Mean A.R. 1865.0
2h 59m 23s.728

48 H. Cephei. Sub Polo.

Date.	Observer.	A.R. 1865.0.
		h. m. s.
1865 June 3	S.	3 3 19.27
21	M.	18.90
24	S.	19.07
27	M.	18.81
29	S.	19.17
July 3	M.	18.73
8	M.	19.31

Mean A.R. 1865.0
3h 3m 19s.037

δ Arietis. N.A.

Date.	Observer.	A.R. 1865.0.
		s.
1863 Feb. 16		.00
1864 Jan. 9		+.02
11		+.11
Feb. 12		+.01
1864 Nov. 16	M.	—.01
30	M.	+.08
Dec. 30	M.	+.06
1865 Jan. 12	S.	+.07
13	M.	+.09
16	M.	+.11
18	M.	+.27
24	S.	+.04
27	M.	+.10
28	S.	+.14
Feb. 9	S.	+.01
11	S.	+.03
14	S.	+.08
Mean		+ .071
N.A.		54.810

Mean A.R. 1865.0
3h 3m 54s.881

12 Eridani.

Date.	Observer.	A.R. 1865.0.
		h. m. s.
1864 Jan. 9	S.	3 6 20.37
11	S.	20.34
Feb. 10	S.	20.21
12	S.	20.37
Nov. 16	M.	20.32
19	S.	20.28
1865 Jan. 12	S.	20.28
16	M.	20.31
18	M.	20.39
20	M.	20.34
24	S.	20.29
27	M.	20.34
28	S.	20.35
Feb. 9	S.	20.31
13	M.	20.40
14	S.	20.38

Mean A.R. 1865.0
3h 6m 20s.330

ζ Eridani.

Date.	Observer.	A.R. 1865.0.
		h. m. s.
1862 Dec. 16	S.	3 9 16.75
1863 Nov. 23	S.	16.61
Dec. 29	S.	16.65
1864 Jan. 5	S.	16.75
9	S.	16.65
11	S.	16.63
25	S.	16.69
Feb. 10	S.	16.60
12	S.	16.68
Nov. 16	M.	16.82
19	S.	16.68
1864 Nov. 23	S.	3 9 16.76
25	M.	16.63
30	M.	16.68
1865 Jan. 12	S.	16.68
13	M.	16.78
16	M.	16.82
18	M.	16.76
20	M.	16.80
24	S.	16.68
27	M.	16.78

Mean A.R. 1865.0
3h 9m 16s.709

α Persei. B.J.

Date.	Observer.	A.R. 1865.0.
		s.
1862 Dec. 16	S.	—.07
1863 Jan. 5	S.	—.10
Nov. 23	S.	—.11
Dec. 29	S.	—.09
1864 Jan. 5	S.	—.17
11	S.	—.20
21	S.	—.37
25	S.	—.06
Feb. 10	S.	—.08
Nov. 16	M.	—.16
25	S.	—.09
Dec. 20	M.	—.21
1865 Jan. 13	M.	—.17
16	M.	—.26
18	M.	—.11
20	M.	—.28
27	M.	—.18
Feb. 13	M.	—.34
14	S.	—.01
24	M.	—.03
25	S.	—.08
27	M.	—.17
Mean		—0.152
B. J.		42.028

Mean A.R. 1865.0
3h 14m 41s.876

ο Tauri.

Date.	Observer.	A.R. 1865.0.
		h. m. s.
1862 Dec. 16	S.	3 17 33.07
1863 Jan. 5	S.	33.14
Nov. 23	S.	33.00
1864 Jan. 5	S.	33.01
9	S.	33.04
11	S.	33.15
21	S.	33.12

Date.	Observer.	A.R. 1865.0.
		h. m. s.
1864 Feb. 10	S.	3 17 33.09
12	S.	33.14
Nov. 16	M.	32.94
19	S.	33.05
23	S.	33.06
25	M.	33.03
Dec. 20	M.	33.05
30	M.	33.05
1865 Jan. 4	M.	33.01
11	M.	33.22
12	S.	33.11
16	M.	33.21
18	M.	33.19

Mean A.R. 1865.0 **$3^h\ 17^m\ 33^s.084$**

Gr. 642 = 323 Cephei B.

Date.	Observer.	A.R. 1865.0.
		s.
1863 Jan. 5	S.	+0.62
Feb. 16	S.	+0.48
Dec. 15	S.	+1.75
21	S.	−0.11
26	S.	+0.49
29	S.	+0.70
1864 Jan. 9	S.	−0.78
21	S.	−0.56
Feb. 13	S.	+0.26
1865 Jan. 12	S.	+0.06
Mean		+0.291
T.H.S.		35.44

Mean A.R. 1865.0 **$3^h\ 22^m\ 35^s.731$**

Sub Polo.

Date.	Observer.	A.R. 1865.0.
		s.
1862 May 23	S.	−0.55
1863 May 25	S.	−0.74
June 4	S.	+0.89
16	S.	−1.24
1864 April 29	S.	+0.52
May 10	S.	+0.09
19	S.	+1.18
22	R.	+1.92
27	S.	−0.14
28	R.	+0.43
31	S.	+0.29
June 2	S.	−0.45
6	S.	−0.82
8	S.	−0.06
July 9	S.	+0.11
12	S.	+0.20
14	S.	−0.27
		s.
1865 June 27	M.	+0.24
29	S.	+0.73
July 3	M.	−0.43
14	M.	+0.43
Mean		+0.111
T.H.S.		35.44

Mean A.R. 1865.0 **$3^h\ 22^m\ 35^s.551$**

ε Eridani.

Date.	Observer.	A.R. 1865.0.
		h. m. s.
1863 Dec. 26	S.	3 26 34.41
1864 Jan. 11	S.	34.41
21	S.	34.28
25	S.	34.38
27	S.	34.41
Feb. 5	S.	34.30
Nov. 16	S.	34.36
23	S.	34.45
25	S.	34.38
Dec. 30	S.	34.35
1865 Jan. 4	S.	34.32
11	S.	34.36
12	S.	34.29
13	S.	34.32
16	S.	34.48
18	S.	34.40
20	S.	34.34
25	S.	34.42
27	S.	34.42
Feb. 9	S.	34.34
13	S.	34.30
14	S.	34.32

Mean A.R. 1865.0 **$3^h\ 26^m\ 34^s.365$**

6 H. Camelopardali = Gr. 716.

Date.	Observer.	A.R. 1865.0.
		h. m. s.
1863 Feb. 10	S.	3 30 28.09
Dec. 15	S.	28.31
21	S.	28.20
26	S.	28.22
1864 Jan. 25	S.	28.18
27	S.	28.11
Feb. 5	S.	28.05
13	S.	28.19
Nov. 23	S.	28.06
1865 Feb. 14	S.	28.10

Mean A.R. 1865.0 **$3^h\ 30^m\ 28^s.151$**

Sub Polo.

Date.	Observer.	A.R. 1865.0.
		h. m. s.
1864 May 19	S.	3 30 28.44
22	R.	28.20
27	S.	28.12
31	S.	28.18
June 6	S.	28.16
7	R.	28.20
8	R.	28.15
13	S.	28.21
16	S.	28.35
17	S.	28.21
18	R.	28.07
21	S.	28.18
27	R.	28.01
28	S.	28.22
29	S.	28.34
July 3	R.	28.15
12	S.	28.29

Mean A.R. 1865.0 **$3^h\ 30^m\ 28^s.205$**

δ Persei. C. des T.

Date.	Observer.	A.R. 1865.0.
		s.
1863 Jan. 5	S.	+.13
Nov. 23	S.	+.21
Dec. 21	S.	+.04
26	S.	+.04
1864 Jan. 11	S.	+.04
21	S.	−.03
27	S.	+.03
Feb. 5	S.	+.10
Nov. 23	S.	+.33
25	M.	+.11
1865 Jan. 11	M.	+.10
13	M.	+.05
16	M.	+.02
18	M.	+.23
20	M.	+.12
27	M.	+.21
Feb. 1	S.	+.04
13	M.	−.07
14	S.	+.22
21	S.	+.01
25	S.	+.08
27	M.	+.14
Mean		+.098
C. des. T		19.395

Mean A.R. 1865.0 **$3^h\ 33^s\ 19^m.493$**

δ Eridani.

Date.	Observer.	A.R. 1865.0.
		h. m. s.
1863 Jan. 5	S.	3 36 46.96
Dec. 26	S.	46.99
1864 Jan. 11	S.	49.07
21	S.	47.00
27	S.	46.86
Feb. 5	S.	46.96
Nov. 23	S.	47.14
25	M.	47.04
1865 Jan. 4	M.	47.05
11	M.	47.10
12	M.	46.96
13	M.	46.91
16	M.	46.94
18	M.	47.02
20	M.	47.06
24	M.	47.15
25	M.	47.08
27	M.	47.11
Feb. 1	S.	46.96
13	M.	46.99
14	S.	47.05
21	S.	46.92
25	S.	46.98

Mean A.R. 1865.0 **3h 36m 47s.013**

η Tauri. N.A.

Date.	Observer.	A.R. 1865.0.
		s.
1862 Dec. 11	S.	+0.09
1863 Dec. 21	S.	—0.11
1864 Jan. 11	S.	+0.10
27	S.	+0.06
Feb. 5	S.	+0.05
Nov. 25	S.	+0.07
1865 Jan. 13	M.	+0.09
16	M.	+0.11
18	M.	+0.12
20	M.	+0.26
24	M.	+0.15
25	M.	+0.06
27	M.	+0.14
Mean		+0.092
N.A.		27.784

Mean A.R. 1865.0 **3h 39m 27s.876**

ζ Persei.

Date.	Observer.	A.R. 1865.0.
		h. m. s.
1862 Dec. 11	S.	3 45 39.12
1863 Jan. 5	S.	39.08
Feb. 14	S.	39.14
		h. m. s.
1863 Feb. 16	S.	3 45 39.16
Nov. 23	S.	39.15
Dec. 26	S.	39.17
1864 Jan. 5	S.	38.88
11	S.	39.17
27	S.	39.07
Nov. 23	S.	39.18
1865 Jan. 4	M.	39.14
11	M.	39.13
12	M.	39.02
13	M.	39.17
18	M.	39.16
20	M.	39.05
25	M.	39.05
27	M.	39.12
Feb. 1	S.	38.99
2	S.	39.26
14	S.	39.12

Mean A.R. 1865.0 **3h 45m 39s.111**

49 H. Cephei. Sub Polo only.

Date.	Observer.	A.R. 1865.0.
		h. m. s.
1865 June 27	M.	3 47 37.12
28	S.	37.44
29	S.	37.26
July 3	M.	36.51
5	S.	37.22
8	M.	36.99
11	S.	37.45
13	S.	37.05
14	M.	36.98
15	M.	36.87
17	M.	36.84
18	M.	37.02
22	M.	37.05

Mean A.R. 1865.0 **3h 47m 37s.062**

γ Eridani N.A.

Date.	Observer.	A.R. 1865.0.
		s.
1862 Dec. 11	S.	+0.04
1863 Jan. 5	S.	+0.12
19	S.	+0.15
Feb. 16	S.	+0.12
Dec. 15	S.	+0.13
21	S.	—0.09
29	S.	+0.14
1864 Jan. 11	S.	0.00
27	S.	+0.09
Feb. 18	S.	+0.16
Nov. 25	M.	+0.06
		s.
1865 Jan. 11	M.	+0.2
13	M.	+0.2
18	M.	+0.2
20	M.	+0.1
24	M.	+0.1
25	M.	+0.1
27	M.	+0.2
30	M.	+0.2
Mean		+0.13
		43.82

Mean A.R. 1865.0 **3h 51m 43s.96**

Groombridge 750.

Date.	Observer.	A.R. 1865.0.
		s.
1863 Jan. 5	S.	+0.2
19	S.	+0.6
Feb. 16	S.	—0.0
Nov. 23	S.	—0.3
Dec. 15	S.	—0.0
21	S.	—0.4
26	S.	+0.7
29	S.	—0.2
30	S.	+0.0
1864 Jan. 11	S.	+0.6
12	S.	+0.7
Nov. 25	M.	—0.4
1865 Jan. 11	M.	+0.3
20	M.	—0.4
27	M.	+1.4
Mean		+0.20
T.H.S.		10.6

Mean A.R. 1865.0 **3h 55m 10s.81**

Sub Polo.

Date.	Observer.	A.R. 1865.0.
		s.
1863 May 25	S.	+0.5
1864 April 21	S.	+0.1
May 19	S.	+0.3
22	R.	—1.4
27	S.	+0.2
31	S.	—0.0
June 2	S.	+0.3
4	S.	+0.3
7	S.	+0.7
10	S.	+0.4
13	S.	—0.3
14	S.	+0.3
16	S.	—0.5
17	S.	—0.5
21	S.	+0.2

Date.	Observer.	A.R. 1865.0.
		s.
1865 July 3	R.	—0.19
8	S.	+0.83
9	S.	+0.19
12	S.	+0.12
1865 July 13	S.	+0.22
Mean		+0.096
T.H.S.		10.61
Mean A.R. 1865.0		**3h 55m 10s.706**

GROOMBRIDGE 766. Sub Polo only.

Date.	Observer.	A.R. 1865.0.
		h. m. s.
1865 July 3	M.	3 57 14.00
5	S.	15.31
8	M.	14.61
11	S.	15.23
17	M.	14.58
15	M.	14.33
17	M.	14.35
18	M.	14.32
22	M.	14.54
29	S.	14.73
Mean A.R. 1865.0		**3h 57m 14s.600**

GROOMBRIDGE 774. Sub Polo only.

Date.	Observer.	A.R. 1865.0.
		h. m. s.
1865 July 3	M.	4 0 36.34
5	S.	37.63
8	M.	36.78
11	S.	37.57
14	M.	36.39
15	M.	36.54
17	M.	36.78
18	M.	36.69
22	M.	36.51
29	S.	36.98
Mean A.R. 1865.0		**4h 0m 36s.821**

GROOMBRIDGE 779.

Date.	Observer.	A.R. 1865.0.
		h. m. s.
1863 Jan. 19	S.	4 3 42.97
Feb. 14	S.	43.29
Dec. 15	S.	42.77
29	S.	42.78
30	S.	42.48
1864 Jan. 11	S.	42.39
12	S.	42.69
21	S.	42.63
27	S.	43.54
Mean A.R. 1865.0		**4h 3m 42s.838**

SUB POLO.

Date.	Observer.	A.R. 1865.0.
		h. m. s.
1863 Feb. 20	S.	4 3 43.65
July 20	S.	41.62
1864 April 21	S.	42.92
May 22	R.	42.10
31	S.	42.71
June 2	S.	42.74
10	S.	43.13
13	R.	42.85
14	S.	43.05
16	S.	43.29
18	R.	42.80
21	S.	42.94
24	M.	43.45
30	R.	42.59
July 16	R.	43.19
26	S.	42.82
Mean A.R. 1865.0		**4h 3m 42s.866**

o^1 ERIDANI. N.A.

Date.	Observer.	A.R. 1865.0.
		s.
1363 Jan. 5	S.	+0.02
19	S.	+0.14
Dec. 15	S.	—0.03
29	S.	+0.06
1864 Jan. 27	S.	+0.12
Feb. 18	S.	+0.16
1865 Jan. 4	S.	+0.02
13	M.	+0.20
18	M.	+0.41
24	M.	+0.15
25	M.	+0.15
26	S.	+0.26
27	M.	+0.29
30	M.	+0.24
Feb. 1	S.	+0.03
Mean		+.148
		16.594
Mean A.R. 1865.0		**4h 5m 16s.742**

γ TAURI. C. des T.

Date.	Observer.	A.R. 1865.0.
		s.
1862 Dec. 11	S.	—.02
1863 Jan. 5	S.	—.07
19	S.	+.09
Dec. 15	S.	—.04
19	S.	—.06
1864 Jan. 11	M.	—.07
27	M.	—.11
1865 Jan. 11	M.	+.03
12	M.	—.02
13	M.	.00
16	M.	—.07
20	M.	.00
25	M.	—.02
26	S.	—.01
27	M.	+.04
30	M.	—.03
Feb. 1	S.	—.05
2	S.	+.14
9	M.	+.06
14	S.	—.05
24	M.	—.07
25	S.	.00
Mean		—.015
C. des T.		6.852
Mean A.R. 1865.0		**4h 12m 6s.837**

δ TAURI.

Date.	Observer.	A.R. 1865.0.
		h. m. s.
1863 Jan. 5	S.	4 15 9.29
19	S.	9.26
Dec. 15	S.	9.14
29	S.	9.05
1864 Jan. 11	S.	9.14
Nov. 25	M.	9.08
1865 Jan. 4	M.	9.16
11	M.	9.24
12	M.	9.03
13	M.	9.18
16	M.	9.17
20	M.	9.20
24	M.	9.23
25	M.	9.17
26	S.	9.12
27	M.	9.16
30	M.	9.21
Feb. 1	S.	9.12
2	S.	9.24
9	M.	9.22
11	S.	9.15
Mean A.R. 1865.0		**4h 15m 9s.170**

64 TAURI.

Date.	Observer.	A.R. 1865.0.
		h. m. s.
1862 Dec. 11	S.	4 16 19.02
1863 Jan. 5	S.	18.99
19	S.	19.05
Dec. 15	S.	19.02
29	S.	18.97

Date.			Observer.	A.R. 1865.0.
				h. m. s.
1864	Jan.	11	S.	4 16 19.03
1865	Jan.	4	M.	19.07
		11	M.	19.04
		12	M.	18.93
		13	M.	19.14
		16	M.	19.00
		20	M.	19.05
		24	M.	18.98
		25	M.	18.99
		26	S.	19.10
		27	M.	18.96
		30	M.	19.04
	Feb.	1	S.	18.98
		2	S.	19.10
		9	M.	19.07
		11	S.	18.96

Mean A.R. 1865.0
4^h 16^m $19^s.023$

ε Tauri. N.A.

Date.			Observer.	A.R. 1865.0.
				s.
1863	Jan.	5	S.	—.05
		19	S.	+.09
	Dec.	15	S.	+.12
1864	Feb.	18	S.	+.11
1865	Jan.	4	M.	+.02
		11	M.	+.16
		13	M.	+.04
		16	M.	+.04
		20	M.	.00
		24	M.	+.05
		26	S.	+.11
		27	M.	+.07
	Feb.	1	S.	.00
		2	S.	+.19
		11	S.	+.04
		25	S.	+.06
	Mean			+.066
				44.161

Mean A.R. 1865.0
4^h 20^m $44^s.227$

53 Eridani.

Date.			Observer.	A.R. 1865.0.
				h. m. s.
1862	Oct.	5	S.	4 36 10.11
1863	Jan.	19	S.	59.98
	Dec.	29	S.	59.99
1864	Jan.	11	S.	59.93
	Feb.	12	S.	59.95
		18	S.	59.76
1865	Jan.	4	M.	59.96
		11	M.	60.12
				h. m. s.
1865	Jan.	13	M.	4 32 60.02
		16	M.	60.19
		18	M.	60.13
		20	M.	60.01
		24	M.	59.91
		26	S.	60.07
		27	M.	60.21
		30	M.	60.02
	Feb.	1	M.	59.96
		2	S.	60.00
		9	M.	59.90
		11	S.	60.18
		13	M.	59.86

Mean A.R. 1865.0
4^h 32^m $0^s.012$

50 Cephei H.

Date.			Observer.	A.R. 1865.0.
				h. m. s.
1863	Dec.	26	S.	4 35 12.92
		29	S.	13.07
1864	Jan.	11	S.	13.38
	Feb.	18	S.	12.76
1865	Jan.	26	S.	13.40
	Feb.	1	M.	12.75
		2	S.	13.30
		11	S.	13.53
		13	M.	12.41
		25	S.	13.42

Mean A.R. 1865.0
4^h 35^m $13^s.094$

50 Cephei H.

Date.			Observer.	A.R. 1865.0.
				h. m. s.
1863	July	22	S.	4 35 13.11
1864	May	31	S.	12.94
	Jan.	8	S.	13.33
		10	S.	13.64
		13	S.	13.35
		14	S.	12.98
		16	S.	13.25
		17	S.	13.42
		24	S.	13.64
	July	3	R.	12.75
		12	R.	12.81
		18	R.	13.35
		19	R.	13.28
		26	S.	13.51
	Aug.	1	S.	13.30

Mean A.R. 1865.0
4^h 35^m $13^s.244$

μ Eridani.

Date.			Observer.	A.R. 1865
				h. m.
1862	Oct.	5	S.	4 38 45
1863	Jan.	19	S.	45
	Dec.	26	S.	45
1865	Jan.	4	M.	45
		11	M.	45
		13	M.	45
		16	M.	45
		18	M.	45
		20	M.	45
		26	S.	45
		27	M.	45
		30	M.	45
	Feb.	1	M.	45
		2	S.	45
		9	M.	45
		11	S.	45
		13	M.	45
		27	M.	45

Mean A.R. 1865.0
4^h 38^m $45^s.3$

9 Camelopardali.

Date.			Observer.	A.R. 1865
				h. m.
1863	Jan.	19	S.	4 40 39
	Dec.	26	S.	39
1864	Jan.	11	S.	39
1865	Jan.	13	M.	38
		26	S.	38
	Feb.	2	S.	39
		11	S.	39
		25	S.	38

Mean A.R. 1865.0
4^h 40^m $39^s.$

Sub Polo.

Date.			Observer.	A.R. 1865
				h. m.
1864	June	8	R.	4 40 38
		10	M.	39
		14	S.	39
		16	S.	39
	July	3	R.	38
		10	R.	38
		16	R.	39
		18	R.	39
		19	R.	38

Mean A.R. 1865.0
4^h 40^m $39^s.$

ι Aurigæ. N.A.

Date.	Observer.	A.R. 1865.0.
		s.
1862 Oct. 5	S.	+0.29
1865 Jan. 4	M.	.04
13	M.	.11
16	M.	.19
18	M.	+.11
20	M.	—.02
24	M.	—.01
26	S.	+.04
27	M.	+.04
30	M.	+.06
Feb. 1	M.	+.12
2	S.	+.19
3	M.	+.09
9	M.	.00
11	S.	+.05
13	M.	—.06
14	M.	+.02
21	S.	—.02
25	S.	+.05
27	M.	.00
Mean		+.065
		12.340

Mean A.R. 1865.0 $4^h\ 48^m\ 12^s.405$

10 Camelopardali. Sub Polo. C. des T.

Date.	Observer.	A.R. 1865.0.
		s.
1865 June 23	M.	+0.19
27	M.	—0.20
July 3	M.	—0.22
5	S.	—0.16
8	M.	—0.11
11	S.	—0.01
14	M.	—0.09
17	M.	—0.11
22	M.	—0.16
26	M.	—0.23
28	M.	—0.11
31	M.	—0.03
Mean		—0.103
C. des T.		25.30

Mean A.R. 1865.0 $4^h\ 51^m\ 25^s.197$

ι Tauri.

Date.	Observer.	A.R. 1865.0.
		h. m. s.
1862 Oct. 5	S.	4 55 1.73
1863 Jan. 19	S.	1.82
1864 Feb. 19	S.	1.64
29	S.	2.00
		h. m. s.
1865 Jan. 2	M.	4 55 1.94
4	M.	1.85
11	M.	1.79
13	M.	1.78
16	M.	2.14
20	M.	1.65
24	M.	1.47
25	M.	1.69
26	S.	1.79
27	M.	1.83
28	M.	1.56
30	M.	1.78
Feb. 1	M.	1.85
2	S.	1.80
3	M.	1.82
9	S.	1.57

Mean A.R. 1865.0 $4^h\ 55^m\ 1^s.775$

Radcliffe 1377. S.P.

Date.	Observer.	A.R. 1865.0.
		h. m. s.
1865 June 23	M.	4 58 19.74
July 15	M.	19.59
17	M.	20.37
22	M.	19.11
29	S.	20.09
31	M.	19.95

Mean A.R. 1865.0 $4^h\ 58^m\ 19^s.808$

ε Leporis. N.A.

Date.	Observer.	A.R. 1865.0.
		s.
1865 Jan. 2	M.	+0.11
4	M.	0.26
11	M.	0.20
16	M.	0.41
18	M.	0.43
20	M.	0.19
24	M.	0.24
25	M.	0.24
26	S.	0.23
27	M.	0.30
28	M.	0.39
30	M.	0.29
Feb. 1	M.	0.31
2	S.	0.19
3	M.	0.32
9	S.	0.26
13	M.	0.30
14	M.	0.21
25	S.	0.33
27	M.	0.29
		s.
Mean		+0.275
N.A.		44.732

Mean A.R. 1865.0 $4^h\ 59^m\ 45^s.007$

τ Orionis.

Date.	Observer.	A.R. 1865.0.
		h. m. s.
1862 Oct. 5	S.	5 11 3.18
1863 Jan. 31	S.	3.21
Mar. 2	S.	3.16
9	S.	3.20
Dec. 29	S.	3.13
1864 Feb. 18	S.	3.04
Mar. 12	S.	3.42
1865 Jan. 2	M.	3.55
4	M.	3.28
11	M.	3.28
16	M.	3.59
18	M.	3.48
20	M.	3.26
26	S.	3.34
27	M.	3.34
30	M.	3.31
Feb. 1	M.	3.29
3	M.	3.04
9	S.	3.12
11	S.	3.33
13	M.	3.15
14	M.	3.57

Mean A.R. 1865.0 $5^h\ 11^m\ 3^s.285$

64 Camelopardali B.

Date.	Observer.	A.R. 1865.0.
		s.
1863 Jan. 19	S.	—0.29
Mar. 2	S.	+0.71
9	S.	+0.52
Dec. 15	S.	+0.71
29	S.	+0.76
1864 March 3	S.	+0.34
1865 Feb. 24	M.	—0.09
Mar. 6	M.	+0.07
Mean		+0.341
T.H.S.		3.91

Mean A.R. 1865.0 $5^h\ 19^m\ 4^s.251$

SUB POLO.

Date.	Observer.	A.R. 1865.0.
		s.
1862 Aug. 21	S.	—0.35
1863 Aug. 24	S.	—0.11
1864 June 10	S.	—0.78
14	S.	+0.25
16	S.	+0.05
21	S.	—0.43
July 3	R.	—0.98
4	S.	+0.21
12	R.	+0.05
14	S.	+0.35
19	R.	+0.14
1865 July 5	S.	—0.31
14	M.	—0.85
15	M.	+0.74
Mean		—0.144
T.H.S.		3.91

Mean A.R. 1865.0 **5h 19m 3s.766**

GR. 966 = 74 CAMELOP. BODE.

Date.	Observer.	A.R. 1865.0.
		h. m. s.
1863 Feb. 2	S.	5 21 41.79
Dec. 29	S.	41.49
1865 Mar. 6	S.	41.57
7	S.	41.85

Mean A.R. 1865.0 **5h 21m 41s.675**

SUB POLO.

Date.	Observer.	A.R. 1865.0.
		h. m. s.
1864 June 21	S.	5 21 41.87
1865 June 23	M.	41.74
July 5	S.	41.60
14	M.	41.52
15	M.	41.58
17	M.	41.54
18	M.	41.48
20	S.	41.58
31	M.	41.55

Mean A.R. 1865.0 **5h 21m 41s.607**

β LEPORIS.

Date.	Observer.	A.R. 1865.0.
		h. m. s.
1862 Oct. 5	S.	5 22 27.97
1864 Mar. 2	S.	27.78
1865 Jan. 4	S.	27.95
1865 Jan. 11	S.	5 22 27.76
16	S.	27.85
18	S.	27.94
20	M.	27.71
25	M.	27.80
27	M.	27.86
28	M.	27.80
30	M.	27.86
Feb. 1	M.	27.86
3	M.	27.86
9	S.	27.81
11	S.	27.90
14	M.	27.83
24	M.	27.87
27	M.	27.92
Mar. 13	M.	27.79

Mean A.R. 1865.0 **5h 22m 27s.848**

δ ORIONIS. N.A.

Date.	Observer.	A.R. 1865.0.
		s.
1862 Mar. 24	S.	+0.02
26	S.	+0.03
Oct. 5	S.	+0.11
1863 Feb. 2	S.	0.00
Mar. 9	S.	+0.07
Dec. 15	S.	+0.16
29	S.	+0.09
1864 Feb. 19	S.	—0.05
Mar. 3	S.	—0.02
12	S.	+0.10
1865 Jan. 2	M.	+0.31
4	M.	+0.09
11	M.	+0.28
16	M.	+0.03
18	M.	+0.19
20	M.	+0.11
25	M.	+0.14
27	M.	+0.18
28	M.	+0.13
30	M.	+0.11
Feb. 1	M.	+0.11
3	M.	+0.08
13	M.	+0.04
Mar. 6	S.	+0.05
Mean		+0.098
N.A.		6.655

Mean A.R. 1865.0 **5h 25m 6s.753**

α LEPORIS. N.A.

Date.	Observer.	A.R. 1865.0.
		s.
1862 Mar. 24	S.	+0.04
26	S.	+0.09
28	S.	+0.05
1863 Feb. 2	S.	+0.01
Mar. 9	S.	+0.06
Dec. 15	S.	—0.04
29	S.	+0.01
1864 Mar. 3	S.	—0.12
12	S.	+0.08
1865 Jan. 2	S.	+0.26
11	M.	+0.05
16	M.	+0.21
18	M.	+0.37
25	M.	+0.15
27	M.	+0.23
28	M.	+0.12
30	M.	+0.06
Feb. 1	M.	—0.02
3	M.	+0.11
13	M.	+0.08
Mar. 6	S.	+0.16
Mean		+0.093
		46.626

Mean A. R. 1865.0 **5h 26m 46s.719**

ε ORIONIS. N.A.

Date.	Observer.	A.R. 1865.0.
		s.
1862 Mar. 24	S.	+.10
26	S.	+.08
28	S.	+.13
Oct. 5	S.	+.20
1863 Feb. 2	S.	.00
23	S.	+.07
Dec. 15	S.	+.12
29	S.	—.02
1864 Feb. 5	S.	+.14
Mar. 3	S.	—.02
12	S.	+.16
1865 Jan. 2	M.	+.35
4	M.	+.30
11	M.	+.15
16	M.	+.04
20	M.	+.06
25	M.	+.20
27	M.	+.08
28	M.	+.06
30	M.	+.05
Feb. 1	M.	+.32

Date.		Observer.	A.R. 1865.0.
			s.
Feb.	3	M.	+.23
	13	M.	+.16
Mar.	6	S.	+.02
Mean			+ .124
N.A.			21.808

Mean A.R. 1865.0
5^h 29^m $21^s.932$

α Columbæ. N.A.

Date.		Observer.	A.R. 1865.0.
			s.
1862 Mar.	24	S.	+.01
Oct.	5	S.	—.28
1863 Feb.	20	S.	+.06
	23	S.	—.12
1864 Mar.	3	S.	—.19
1865 Jan.	2	M.	—.05
	4	M.	—.07
	16	M.	—.04
	18	M.	+.14
	20	M.	—.06
	25	M.	—.06
	27	M.	+.05
	30	M.	—.04
Feb.	1	M.	+.10
	2	M.	.00
	9	S.	+.08
	11	S.	+.11
	13	M.	—.12
	14	M.	—.02
	21	S.	—.10
Mean			—0.030
N.A.			45.832

Mean A.R. 1865.0
5^h 34^m $45^s.802$

λ Orionis.

Date.		Observer.	A.R. 1865.0.
			h. m. s.
1862 Oct.	5	S.	5 41 21.39
1863 Feb.	2	S.	21.31
	20	S.	21.41
	23	S.	21.33
1865 Jan.	2	M.	21.57
	4	M.	21.55
	11	M.	21.43
	16	M.	21.47
	20	M.	21.34
	25	M.	21.52
	27	M.	21.40
	30	M.	21.47
Feb.	1	M.	21.67
	2	M.	21.36
	11	S.	21.41

Date.		Observer.	A.R. 1865.0.
			h. m. s.
Feb.	13	M.	5 41 21.36
	14	M.	21.50
	21	S.	21.25
	22	M.	21.45
	24	M.	21.38
Mar.	28	M.	21.38
April	8	S.	21.39

Mean A.R. 1865.0
5^h 41^m $21^s.425$

θ Aurigæ.

Date.		Observer.	A.R. 1865.0.
			h. m. s.
1862 Mar.	26	S.	5 50 31.01
	28	S.	30.97
Sept.	17	S.	30.88
1863 Feb.	2	S.	30.99
1864 Mar.	19	S.	31.02
1865 Jan.	2	S.	31.13
	4	M.	31.10
	11	M.	30.91
	16	M.	30.98
	20	M.	30.80
	25	M.	31.01
	27	M.	30.98
	30	M.	31.03
Feb.	1	M.	31.18
	9	S.	30.98
	11	S.	30.96
	14	M.	31.12
	21	S.	31.00
	22	M.	31.16
	24	M.	31.09
Mar.	18	M.	31.08

Mean A.R. 1865.0
5^h 50^m $31^s.018$

1 Geminorum.

Date.		Observer.	A.R. 1865.0.
			h. m. s.
1862 Sept.	17	S.	5 55 54.86
Oct.	5	S.	54.80
1863 Feb.	20	S.	54.89
1864 Jan.	11	S.	54.92
	25	S.	54.87
Feb.	23	S.	54.94
1865 Jan.	2	M.	55.08
	11	M.	54.93
	16	M.	55.00
	20	M.	54.96
	25	M.	54.95
	27	M.	54.92
	30	M.	54.93

Date.		Observer.	A.R. 1865.0.
			h. m. s.
Feb.	1	M.	5 55 54.96
	2	M.	54.88
	3	M.	54.97
	9	S.	54.89
	11	S.	54.97
	14	M.	55.00
	21	S.	54.90
	22	M.	55.10
	24	M.	54.87
Mar.	6	M.	54.92
	18	M.	55.00

Mean A.R. 1865.0
5^h 55^m $54^s.938$

ν Orionis. N.A.

Date.		Observer.	A.R. 1865.0.
			s.
1862 Sept.	17	S.	—0.11
Oct.	5	S.	—0.06
Dec.	29	S.	+0.06
1864 Jan.	14	S.	+0.26
Feb.	23	S.	+0.04
1865 Jan.	2	S.	+0.25
	11	M.	+0.03
	20	M.	+0.09
	24	S.	—0.03
	25	M.	+0.11
	30	M.	+0.10
Feb.	1	M.	+0.15
	3	M.	+0.04
	9	S.	—0.03
	11	S.	+0.20
	13	M.	+0.05
	14	M.	+0.19
	24	M.	+0.10
March	6	S.	0.00
Mean			+0.076
N.A.			51.856

Mean A.R. 1865.0
5^h 59^m $51^s.932$

η Geminorum.

Date.		Observer.	A.R. 1865.0.
			h. m. s.
1862 Mar.	26	S.	6 6 43.74
	28	S.	43.71
Sept.	12	S.	43.62
Oct.	5	S.	43.74
1863 Feb.	2	S.	43.75
	20	S.	43.73
1864 Jan.	11	S.	43.81
	25	S.	43.67
1865 Jan.	2	M.	43.97
	11	M.	43.71

Date.		Observer.	A.R. 1865.0.
			h. m. s.
1865 Jan.	20	M.	6 6 43.79
	27	M.	43.69
	28	S.	43.80
	30	M.	43.74
Feb.	1	M.	43.80
	3	M.	43.79
	11	S.	43.81
	13	M.	43.80
	14	M.	43.86
Mar.	18	S.	43.74

Mean A.R. 1865.0
6h 6m 43s.764

ν Columbæ.

Date.		Observer.	A.R. 1865.0.
			h. m. s.
1862 Oct.	5	S.	6 11 45.00
1865 Jan.	11	M.	45.19
	16	M.	45.02
	20	M.	44.98
	25	M.	44.98
	27	M.	45.11
	30	M.	45.07
Feb.	1	M.	44.97
	9	S.	45.10
	13	M.	45.29
	21	S.	45.09
	24	M.	45.17
Mar.	6	M.	44.95
	13	M.	45.09
	18	S.	45.21

Mean A.R. 1865.0
6h 11m 45s.081

μ Geminorum. N.A.

Date.		Observer.	A.R. 1865.0.
			s.
1865 Jan.	2	M.	+0.25
	11	M.	+0.01
	16	M.	+0.10
	20	M.	+0.09
	24	S.	0.00
	25	M.	+0.03
	27	M.	—0.01
Feb.	1	M.	+0.14
	11	S.	+0.14
	13	M.	—0.06
	14	M.	+0.09
	21	S.	+0.01
	24	M.	+0.07
Mar.	6	M.	+0.05
	13	M.	+0.05
	28	M.	+0.06
			s.
1865 April	3	M.	+0.14
	4	S.	+0.03
	5	M.	—0.05
	14	S.	+0.02
Mean			+0.058
N.A.			47.581

Mean A.R. 1865.0
6h 14m 47s.639

β Canis Majoris. C. des T.

Date.		Observer.	A.R. 1865.0.
			s.
1865 Jan.	2	M.	+0.26
	11	M.	+0.01
	20	M.	+0.02
	28	S.	—0.06
	30	M.	—0.09
Feb.	9	S.	—0.07
	11	M.	—0.07
	13	M.	—0.01
	14	M.	—0.07
	21	M.	+0.03
	24	M.	—0.04
Mar.	6	M.	+0.16
	13	M.	—0.08
	18	S.	—0.13
	27	M.	—0.13
	28	M.	—0.10
April	3	M.	—0.28
	4	S.	—0.07
	5	M.	—0.07
	14	S.	—0.01
	17	M.	—0.04
	19	M.	—0.08
	24	M.	—0.03
	25	M.	—0.02
Mean			—0.040
C. des T.			45.44

Mean A.R. 1865.0
6h 16m 45s.400

ν Geminorum.

Date.		Observer.	A.R. 1865.0.
			h. m. s.
1862 Mar.	26	S.	6 20 56.87
Oct.	5	S.	56.72
	7	S.	56.73
1863 Feb.	2	S.	56.86
1864 Jan.	11	S.	56.88
	25	S.	56.87
	27	S.	56.81
Feb.	26	S.	56.84
Mar.	9	S.	56.94
			h. m. s.
1864 Mar.	19	S.	6 20 56.94
	24	S.	56.82
1865 Jan.	2	M.	56.93
	11	M.	56.79
	16	M.	56.93
	20	M.	56.98
	25	M.	56.81
	27	M.	56.86
	30	M.	56.87
Feb.	1	M.	56.87
	9	S.	56.86
	13	M.	56.95
	14	M.	56.97
	21	M.	56.88
Mar.	18	S.	56.93

Mean A.R. 1865.0
6h 20m 56s.871

23 Camelopardali H.

Date.		Observer.	A.R. 1865.0.
			h. m. s.
1863 Feb.	2	S.	6 23 8.24
Mar.	2	S.	8.57
Mar.	5	S.	8.54
Dec.	29	S.	8.16
1864 Jan.	11	S.	7.84
	14	S.	8.71
	16	S.	8.49
	25	S.	8.36
	27	S.	7.97
Feb.	23	S.	8.60
	26	S.	8.47
Mar.	19	S.	8.40
	21	S.	8.17
1865 Jan.	2	M.	8.50
	11	M.	7.63

Mean A.R. 1865.0
6h 23m 8s.310

Sub Polo.

Date.		Observer.	A.R. 1865.0.
			h. m. s.
1863 July	23	S.	6 23 8.30
	29	S.	8.45
Aug.	3	S.	7.98
	10	S.	8.73
	12	S.	8.76
	15	S.	8.97
	17	S.	8.35
	24	S.	8.31
Sept.	5	S.	8.31
	10	S.	8.17
	30	S.	8.47

Date.		Observer.	A.R. 1865.0.
			h. m. s.
1864 June	14	S.	6 23 8.66
	21	S.	8.61
	23	S.	8.76
	28	S.	8.56
July	3	R.	8.07
	12	R.	8.06

Mean A.R. 1865.0 **6ʰ 23ᵐ 8ˢ.442**

γ Geminorum. N.A.

Date.		Observer.	A.R. 1865.0.
			s.
1862 Sept.	17	S.	—0.07
	21	S.	—0.06
Oct.	5	S.	—0.17
	7	S.	—0.12
1863 Feb.	2	S.	—0.03
1864 Jan.	14	S.	—0.05
Feb.	26	S.	+0.10
Mar.	9	S.	+0.05
	19	S.	+0.08
	24	S.	+0.03
1865 Jan.	11	M.	+0.05
	16	M.	—0.02
	20	M.	+0.06
	25	M.	—0.09
	27	M.	+0.01
Feb.	3	M.	+0.06
	13	M.	+0.10
	14	M.	+0.17
	24	M.	+0.18
	25	S.	—0.09
Mar.	4	S.	+0.08
	6	M.	+0.06
	13	M.	—0.01
	20	S.	+0.07
Mean			+0.016
N.A.			54.770

Mean A.R. 1865.0 **6ʰ 29ᵐ 54ˢ.786**

51 H. Cephei.

Date.		Observer.	A.R. 1865.0.
			s.
1862 Mar.	26	S.	+1.66
	28	S.	+0.11
1863 Feb.	16	S.	+0.04
	23	S.	—1.42
Mar.	2	S.	+1.41
Dec.	29	S.	—1.03
1864 Jan.	11	S.	—2.03
	14	S.	—0.25
	16	S.	—0.65
	27	S.	—0.22
1864 Feb.	8	S.	—0.22
	23	S.	+1.01
1865 Jan.	11	M.	—1.18
	25	M.	+0.64
	27	M.	—1.32
	30	M.	+0.31
Feb.	14	M.	+2.94
	24	M.	+0.56
Mar.	6	M.	+0.19
Mean			+0.029
T.H.S.			12.01

Mean A.R. 1865.0 **6ʰ 36ᵐ 12ˢ.039**

Sub Polo.

Date.		Observer.	A.R. 1865.0.
			s.
1862 July	18	S.	+1.58
1863 Aug.	19	S.	+0.34
	24	S.	—0.36
Sept.	10	S.	+0.35
	21	S.	+0.68
	22	S.	+0.79
	23	S.	—0.05
	28	S.	+0.32
	30	S.	—0.05
1864 June	14	R.	—0.27
	21	S.	+0.11
	23	S.	—0.40
	28	S.	+1.34
July	3	R.	—1.62
	5	S.	—0.44
	9	S.	—0.49
	13	S.	—0.01
	14	S.	—0.94
	15	S.	—0.66
Oct.	5	S.	—0.20
	7	S.	—0.59
	8	S.	—0.20
1865 July	20	S.	—0.20
	22	S.	—0.47
	24	M.	—1.07
	27	S.	—0.56
	31	M.	—0.23
Aug.	2	M.	—0.01
	12	M.	+0.33
	14	M.	—0.15
	23	M.	—0.17
	25	M.	—0.14
Sept.	16	S.	+0.63
	27	M.	—0.09
	29	M.	—0.06
	30	S.	—0.80
1865 Oct.	3	S.	—0.31
	10	S.	—0.51
	23	M.	—1.30
	25	M.	+0.35
Mean			—0.138
T.H.S.			12.01

Mean A.R. 1865.0 **6ʰ 36ᵐ 11ˢ.872**

ε Geminorum.

Date.		Observer.	A.R. 1865.0.
			h. m. s.
1862 Sept.	21	S.	6 35 37.72
1865 Jan.	20	M.	37.56
	25	M.	37.52
	27	M.	37.44
Feb.	2	M.	37.50
	9	S.	37.42
	13	M.	37.49
	14	M.	37.66
	21	M.	37.58
	24	M.	37.74
	25	S.	37.52
Mar.	4	S.	37.52
	6	M.	37.52
	13	M.	37.77
April	4	S.	37.56
	13	M.	37.58
	14	S.	37.54
	17	M.	37.51
	19	M.	37.47

Mean A.R. 1865.0 **6ʰ 35ᵐ 37ˢ.559**

24 Camelopardali H. Sub Polo only.

Date.		Observer.	A.R. 1865.0.
			h. m. s.
1865 July	27	S.	6 40 20.12
	28	M.	19.86
Aug.	2	M.	19.80
	4	M.	19.94
	12	M.	19.63
	18	M.	19.64
	26	S.	19.86
	29	S.	20.20
	31	S.	19.92
Sept.	4	M.	20.03

Mean A.R. 1865.0 **6ʰ 40ᵐ 19ˢ.900**

κ Canis Majoris.

Date.	Observer.	A.R. 1865.0.
		h. m. s.
1862 Mar. 11	H.	6 44 48.07
Oct. 5	S.	47.91
7	S.	47.92
1863 Feb. 2	S.	47.97
21	S.	48.05
Mar. 2	S.	47.99
5	S.	47.76
11	S.	47.91
1864 Jan. 27	S.	47.97
Mar. 21	S.	47.89
24	S.	48.10
1865 Jan. 2	M.	48.25
11	M.	48.02
Feb. 1	M.	48.18
21	M.	48.04
24	M.	48.13
25	M.	48.01
Mar. 4	S.	48.11
6	M.	48.05
13	M.	47.86

Mean A.R. 1865.0
$6^h\ 44^m\ 48^s.010$

θ Canis Majoris.

Date.	Observer.	A.R. 1865.0.
		h. m. s.
1862 Oct. 5	S.	6 47 55.20
7	S.	55.13
1863 Feb. 21	S.	55.17
Mar. 2	S.	55.09
5	S.	55.05
11	S.	55.13
1864 Feb. 26	S.	55.19
Mar. 9	S.	55.12
21	S.	55.00
24	S.	54.91
1865 Jan. 11	M.	55.25
Feb. 21	M.	55.17
24	M.	55.26
25	S.	55.10
Mar, 4	S.	55.15
6	M.	55.02
13	M.	55.11
18	S.	55.07
25	S.	55.08

Mean A.R. 1865.0
$6^h\ 47^m\ 55^s.116$

ε Canis Majoris. N.A.

Date.	Observer.	A.R. 1865.0.
		s.
1862 Mar. 11	H.	+.07
24	S.	+.06
Oct. 5	S.	+.04
7	S.	—.08
1863 Mar. 2	S.	+.05
5	S.	—.06
11	S.	+.06
1864 Jan. 27	S.	+.12
Feb. 26	S.	+.05
Mar. 3	S.	—.08
9	S.	+.01
24	S.	+.04
1865 Jan. 11	M.	+.05
30	M.	+.11
Feb. 1	M.	+.20
2	S.	+.10
14	M.	+.06
21	M.	—.01
24	M.	+.10
Mar. 4	S.	+.25
6	M.	+.03
Mean		+0.056
N.A.		19.237

Mean A.R. 1865.0
$6^h\ 53^m\ 19^s.293$

ζ Geminorum.

Date.	Observer.	A.R. 1865.0.
		h. m. s.
1862 Mar. 11	H.	6 56 6.22
24	S.	6.16
Oct. 5	S.	5.96
7	S.	6.01
1863 Feb. 21	S.	6.18
Mar. 2	S.	6.04
5	S.	6.09
11	S.	6.23
1864 Jan. 14	S.	6.09
27	S.	5.99
Feb. 25	S.	6.05
26	S.	6.12
Mar. 3	S.	6.02
9	S.	6.11
24	S.	6.00
1865 Jan. 11	M.	5.98
16	M.	6.15
30	M.	6.14
Feb. 1	M.	6.17
2	S.	6.06

Mean A.R. 1865.0
$6^h\ 56^m\ 6^s.088$

γ Canis Majoris. N.A.

Date.	Observer.	A.R. 1865.0.
		s.
1862 Mar. 24	S.	—.01
Oct. 7	S.	—.04
1863 Feb. 21	S.	+.14
Mar. 2	S.	+.26
1864 Jan. 27	S.	+.07
Feb. 25	S.	—.06
26	S.	+.01
Mar. 3	S.	—.11
9	S.	—.02
24	S.	—.03
1865 Jan. 16	M.	+.22
30	M.	+.11
Feb. 1	M.	+.11
2	M.	—.04
9	S.	+.15
21	M.	+.06
24	M.	—.01
25	S.	—.10
Mar. 4	S.	+.02
6	M.	+.09
13	M.	—.10
27	M.	+.12
28	M.	+.04
Mean		+.038
N. A.		39.080

Mean A.R. 1865.0
$6^h\ 57^m\ 39^s.118$

25 Camel. H.

Date.	Observer.	A.R. 1865.0.
		s.
1862 Mar. 11	H.	—0.19
Oct. 7	S.	—0.25
1863 Feb. 20	S.	—0.53
21	S.	—0.14
23	S.	—0.51
Mar. 2	S.	+0.16
5	S.	—0.48
11	S.	+0.26
1865 Mar. 18	S.	—0.12
27	M.	—0.30
28	M.	—0.38
Mean		—0.225
T.H.S.		29.19

Mean A.R. 1865.0
$7^h\ 2^m\ 28^s.965$

Sub Polo.

Date.	Observer.	A.R. 1865.0.
		s.
1862 July 26	S.	—0.21
Aug. 6	S.	—0.17
11	S.	—0.14

Date.	Observer.	A.R. 1865.0.
		s.
1862 Aug. 20	S.	+0.35
Sept. 2	S.	+0.04
16	S.	—0.37
Oct. 8	S.	—0.92
18	S.	—0.72
1863 July 23	S.	—0.26
29	S.	+0.27
31	S.	+0.04
Aug. 17	S.	—0.58
1864 Oct. 7	S.	—0.41
17	S.	+0.06
24	S.	+0.36
Mean		—0.177
T.H.S.		29.19

Mean A. R. 1865.0 **7h 2m 29s.013**

δ Canis Majoris.

Date.	Observer.	A.R. 1865.0.
		h. m. s.
1862 Oct. 5	S.	7 2 54.19
1864 Feb. 29	S.	54.20
Mar. 9	S.	54.30
1865 Jan. 16	M.	54.29
30	M.	54.13
Feb. 2	S.	54.28
21	M.	54.19
24	M.	54.21
25	S.	54.13
Mar. 6	M.	54.34
13	M.	54.18
April 5	S.	54.12
13	M.	54.18
24	M.	54.32
25	M.	54.30

Mean A.R. 1865.0 **7h 2m 54s.224**

λ Geminorum.

Date.	Observer.	A.R. 1865.0.
		h. m. s.
1862 Mar. 24	S.	7 10 20.22
Oct. 5	S.	19.98
1864 Jan. 14	S.	20.15
27	S.	20.14
Feb. 25	S.	20.18
26	S.	20.04
Mar. 3	S.	19.94
19	R.	20.03
21	R.	20.01
1865 Jan. 16	M.	20.06
30	M.	20.07
1865 Feb. 1	M.	7 10 20.14
21	M.	19.96
24	M.	20.07
Mar. 4	S.	19.99
6	M.	19.95
13	M.	20.03
27	M.	20.08
28	M.	20.03
April 3	S.	19.99
5	S.	19.96

Mean A.R. 1865.0 **7h 10m 20s.049**

4 Ursæ Minoris B. (Decl. 89° 1′).

Date.	Observer.	A.R. 1865.0.
		s.
1863 Feb. 7	S.	+2.04
21	S.	+5.44
Mar. 2	S.	—0.40
5	S.	—0.72
9	S.	—3.60
11	S.	+0.71
1864 Mar. 9	S.	—0.95
1865 Jan. 16	M.	—2.17
Feb. 24	M.	+0.24
25	S.	—2.76
Mar. 6	M.	+2.66
27	M.	+0.48
28	M.	—1.30
Mean		—0.025
T.H.S.		36.81

Mean A.R. 1865.0 **7h 16m 36s.785**

Sub Polo.

Date.	Observer.	A.R. 1865.0.
		s.
1862 July 26	S.	—1.68
Aug. 11	S.	—0.57
20	S.	—0.55
21	S.	—3.53
Sept. 4	S.	+0.88
16	S.	—4.10
1863 May 27	S.	—3.92
June 5	S.	+0.36
July 23	S.	—1.67
28	S.	+1.99
29	S.	—1.43
Aug. 3	S.	—3.07
12	S.	—0.83
14	S.	—0.52
1863 Sept. 10	S.	—2.37
14	S.	+2.49
15	S.	—2.42
21	S.	—1.20
23	S.	+0.94
1864 Sept. 17	S.	—1.93
Mean		—1.156
T.H.S.		36.81

Mean A.R. 1865.0 **7h 16m 35s.654**

δ Geminorum N.A.

Date.	Observer.	A.R. 1865.0.
		s.
1862 Mar. 11	H.	+0.14
24	S.	+0.04
Oct. 5	S.	—0.02
7	S.	—0.05
1863 Feb. 20	S.	+0.04
1864 Jan. 14	S.	+0.14
27	S.	+0.10
Feb. 25	S.	+0.12
26	S.	+0.03
Mar. 24	S.	—0.11
1865 Jan. 30	M.	+0.02
Feb. 1	M.	+0.13
2	S.	—0.02
Mar. 4	S.	+0.08
6	M.	+0.09
13	M.	—0.06
28	M.	+0.03
April 3	S.	+0.02
5	S.	—0.11
13	M.	+0.02
Mean		+0.032
N.A.		3.519

Mean A.R. 1865.0 **7h 12m 3s.551**

Piazzi VII. 67.

Date.	Observer.	A.R. 1865.0.
		h. m. s.
1864 Jan. 14	S.	7 16 48.57
Feb. 25	S.	48.18
26	S.	48.14
Mar. 3	S.	48.24
24	S.	47.98
1865 Feb. 21	M.	47.93
April 5	S.	47.94

Mean A.R. 1865.0 **7h 16m 48s.140**

Sub Polo.

Date.	Observer.	A.R. 1865.0.
		h. m. s.
1864 July 14	S.	7 16 47.86
Sept. 6	S.	48.62
9	S.	48.28
10	S.	48.19
Oct. 7	S.	48.09
1865 July 24	M.	47.98
28	M.	48.33
Aug. 2	M.	48.26
Sept. 4	M.	48.75
5	S.	48.24
6	M.	48.09

Mean A.R. 1865.0
7ʰ 16ᵐ 48ˢ.245

η Canis Majoris.

Date.	Observer.	A.R. 1865.0.
		h. m. s.
1862 Mar. 11	H.	7 18 45.48
Oct. 7	S.	45.26
1864 Jan. 14	S.	45.49
27	S.	45.28
Feb. 26	S.	45.36
Mar. 3	S.	45.27
19	R.	45.34
21	R.	45.42
24	S.	45.27
1865 Jan. 30	M.	45.53
Feb. 1	M.	45.65
2	S.	45.41
21	M.	45.51
24	M.	45.52
Mar. 4	S.	45.46
27	M.	45.40
April 5	S.	45.36

Mean A.R. 1865.0
7ʰ 18ᵐ 45ˢ.412

28 Camelopardali H. Sub Polo only.

Date.	Observer.	A.R. 1865.0.
		h. m. s.
1865 July 28	M.	7 33 48.94
31	M.	48.73
Aug. 2	M.	48.27
4	M.	48.75
8	M.	48.75
9	M.	48.40
12	M.	49.09
26	S.	48.40
Sept. 2	S.	47.61
4	M.	48.62

Mean A.R. 1865.0
7ʰ 33ᵐ 48ˢ.556

156 Camelopardali B.

Date.	Observer.	A.R. 1865.0.
		s.
1862 Mar. 11	H.	—0.66
28	S.	+0.19
1863 Feb. 7	S.	+0.48
21	S.	—0.17
Mar. 2	S.	+0.11
1864 Mar. 3	S.	+0.15
7	S.	+0.40
1865 April 4	S.	+0.10
Mean		+0.075
T.H.S.		10.61

Mean A.R. 1865.0
7ʰ 44ᵐ 10ˢ.685

Sub Polo.

Date.	Observer.	A.R. 1865.0.
		s.
1862 Aug. 11	S.	+0.31
20	S.	—0.11
Sept. 4	S.	—0.31
11	S.	+0.02
16	S.	+0.63
1863 Aug. 12	S.	+0.55
14	S.	+0.70
17	S.	—0.99
Sept 21	S.	—0.03
Oct. 23	S.	+0.62
1864 Jan. 6	R.	—0.46
Sept. 6	S.	+0.16
16	S.	+0.37
17	S.	—0.43
Oct. 17	S.	+0.11
21	S.	+0.09
25	S.	+0.13
26	S.	—0.11
1865 Oct. 17	S.	+0.19
30	M.	+0.06
Nov. 1	S.	—0.50
Mean		+0.048
T.H.S.		10.61

Mean A.R. 1865.0
7ʰ 44ᵐ 10ˢ.658

Piazzi. VII. 187.

Date.	Observer.	A.R. 1865.0.
		h. m. s.
1865 July 28	M.	7 43 25.16
31	M.	24.77
Aug. 2	M.	24.84
4	M.	24.65
9	M.	24.35
12	M.	24.89
26	S.	24.71
1865 Sept. 2	S.	7 43 24.65
4	M.	25.14
6	M.	24.43

Mean A.R. 1865.0
7ʰ 43ᵐ 24ˢ.759

ξ Argus. C. des T.

Date.	Observer.	A.R. 1865.0.
		s.
1864 Feb. 23	S.	—0.27
Mar. 19	R.	+0.11
21	R.	+0.03
1865 Feb. 24	M.	—0.02
Mar. 4	S.	+0.06
6	M.	+0.12
17	M.	+0.15
27	M.	+0.01
28	M.	+0.08
April 3	S.	—0.22
4	S.	—0.05
5	M.	+0.23
24	M.	+0.13
Mean		+0.028
C. des T.		37.006

Mean A.R. 1865.0
7ʰ 43ᵐ 37ˢ.034

χ Geminorum N.A.

Date.	Observer.	A.R. 1865.0.
		s.
1862 Mar. 11	H.	+.08
1863 Feb. 7	S.	—.04
1864 Feb. 23	S.	—.01
25	S.	—.07
26	S.	+.04
Mar. 3	S.	—.10
7	S.	.00
19	R.	+.15
21	R.	+.06
24	S.	+.01
1865 Feb. 24	M.	+.08
Mar. 17	M.	—.05
27	M.	+.05
28	M.	—.01
April 3	M.	.00
5	M.	.00
13	M.	—.05
Mean		+0.008
N.A.		13.423

Mean A.R. 1865.0
7ʰ 55ᵐ 13ˢ.431

ζ Argus.

Date.	Observer.	A.R. 1865.0.
		h. m. s.
1864 Feb. 23	S.	7 58 50.26
26	S.	50.45
Mar. 24	S.	50.53
1865 Feb. 21	S.	50.45
April 5	M.	50.48

Mean A.R. 1865.0 **7h 58m 50s.434**

ι Navis. N.A.

Date.	Observer.	A.R. 1865.0.
		s.
1863 Feb. 7	S.	+.06
1864 Mar. 19	R.	+.17
21	R.	+.16
24	S.	+.09
1865 Feb. 21	S.	+.06
24	M.	+.03
Mar. 28	M.	+.13
April 3	M.	—.02
5	M.	+.10
13	M.	+.06
Mean		+0.084
N.A.		47.705

Mean A.R. 1865.0 **8h 1m 47s.789**

β Cancri. C. des T.

Date.	Observer.	A.R. 1865.0.
		s.
1862 Mar. 28	S.	+.07
1863 Feb. 7	S.	—.12
Mar. 5	S.	+.01
11	S.	+.10
April 1	S.	—.01
1864 Feb. 26	S.	—.06
Mar. 3	S.	—.01
7	S.	—.03
21	R.	—.01
April 7	R.	—.12
1865 Feb. 24	M.	+.06
Mar. 6	M.	+.03
17	M	+.05
27	M.	+.02
April 3	M.	—.07
4	M.	—.05
5	M.	+.01
13	M.	—.05
24	M.	+.03
25	M.	+.05
Mean		—.005
C. des T.		11.587

Mean A.R. 1865.0 **8h 9m 11s.582**

Gr. 1418.

Date.	Observer.	A.R. 1865.0.
		h. m. s.
1863 Mar. 5	S.	8 15 32.98
11	S.	34.15
April 1	S.	34.52
1864 Feb. 26	S.	34.08
Mar. 3	S.	34.44
7	S.	34.77
21	R.	35.43
April 8	S.	35.65
1865 Mar. 4	S.	33.19
17	M.	34.13
April 5	M.	33.99

Mean A.R. 1865.0 **8h 15m 34s.303**

Gr. 1418. Sub Polo.

Date.	Observer.	A.R. 1865.0.
		h. m. s.
1863 July 29	S.	8 15 34.73
Aug. 14	S.	34.60
Sept. 24	S.	34.68
28	S.	34.77
Oct. 28	S.	34.38
29	S.	33.95
1864 June 6	R.	33.86
Sept. 16	S.	32.58
Oct. 25	S.	34.47
26	S.	33.60
Nov. 2	S.	34.42
3	S.	34.42

Mean A.R. 1865.0 **8h 15m 34s.205**

o Ursæ Majoris.

Date.	Observer.	A.R. 1865.0.
		h. m. s.
1862 Mar. 11	S.	8 19 1.47
28	S.	1.22
April 4	S.	1.42
7	S.	1.37
1863 Feb. 7	S.	1.29
Mar. 11	S.	1.39
April 1	S.	1.18
1864 Feb. 26	S.	1.47
Mar. 7	S.	1.47
19	R.	1.54
21	R.	1.53
April 7	R.	1.36
8	S.	1.39

Mean A.R. 1865.0 **8h 19m 1s.393**

o Ursæ Majoris. Sub Polo.

Date.	Observer.	A.R. 1865.0.
		h. m. s.
1864 June 6	R.	8 19 1.19
Sept. 6	S.	1.57
16	S.	1.78
Oct. 5	S.	1.39
25	S.	1.36
26	S.	1.27
Nov. 1	S.	1.29
2	S.	1.44
1865 Oct. 9	M.	1.31

Mean A.R. 1865.0 **8h 19m 1s.400**

A. Ursæ Majoris.

Date.	Observer.	A.R. 1865.0.
		h. m. s.
1862 Mar. 27	S.	8 22 28.91
29	S.	28.86
April 4	S.	28.85
7	S.	28.70
1863 Feb. 7	S.	28.73
Mar. 5	S.	28.73
11	S.	28.83
30	S.	29.06
April 1	S.	28.83
1864 Feb. 26	S.	29.09
Mar. 12	S.	28.96
19	R.	28.39
21	R.	28.97
April 7	R.	28.97
8	S.	28.76

Mean A.R. 1865.0 **8h 22m 28s.843**

A. Ursae Majoris. Sub Polo.

Date.	Observer.	A.R. 1865.0.
		h. m. s.
1863 Oct. 1	S.	8 22 28.63
1864 Sept. 6	S.	29.01
16	S.	28.71
Oct. 5	S.	28.68
10	S.	28.99
25	S.	28.92
26	S.	28.81
Nov. 1	S.	28.56
2	S.	29.00
3	S.	29.04
10	S.	28.98
1865 Oct. 9	M.	28.63

Mean A.R. 1865.0 **8h 22m 28s.830**

η Cancri. N.A.

Date.	Observer.	A.R. 1865.0.
		s.
1862 Mar. 29	H.	+.22
April 4	H.	+.08
Feb. 7	S.	+.01
1863 Mar. 11	S.	+.01
1864 Feb. 26	S.	+.02
Mar. 7	S.	+.05
19	R.	.00
21	R.	+.12
1865 Feb. 24	M.	+.17
Mar. 6	M.	+.08
17	M.	+.07
27	M.	+.08
April 3	M.	+.02
5	M.	+.06
8	S.	+.06
17	M.	+.07
Mean		+0.070
N.A.		53.848

Mean A.R. 1865.0 8h 24m 53s.918

δ Cancri.

Date.	Observer.	A.R. 1865.0.
		h. m. s.
1862 Mar. 27	S.	8 37 0.59
1864 Feb. 26	S.	0.57
Mar. 19	R.	0.67
April 7	R.	0.52
9	R.	0.55
1865 Mar. 6	M.	0.65
10	M.	0.59
17	M.	0.60
25	M.	0.62
27	M.	0.64
April 4	M.	0.60
5	M.	0.75
8	S.	0.56
13	M.	0.63
17	M.	0.63

Mean A.R. 1865.0 8h 37m 0s.611

ε Hydræ. N.A.

Date.	Observer.	A.R. 1865.0.
		s.
1862 April 4	H.	+.13
1863 Mar. 2	S.	+.12
April 9	S.	.00
1864 Mar. 3	S.	+.02
1864 Mar. 12	S.	+.03
19	R.	—.06
24	S.	+.09
April 15	S.	—.08
1865 Mar. 4	S.	+.05
6	M.	.00
17	M.	+.01
27	M.	+.07
Mean		+.029
N.A.		37.513

Mean A.R. 1865.0 8h 39m 37s.542

6 Ursæ Majoris.

Date.	Observer.	A.R. 1865.0.
		h. m. s.
1863 Feb. 7	S.	8 45 0.62
Mar. 2	S.	0.80
9	S.	0.91
30	S.	0.70
April 9	S.	0.77
1864 Mar. 7	S.	0.88
12	S.	0.93
24	S.	0.82
April 7	R.	0.84
9	R.	0.79
15	S.	0.83
1865 Mar. 10	M.	0.89

Mean A.R. 1865.0 8h 45m 0s.815

6 Ursæ Majoris. Sub Polo.

Date.	Observer.	A.R. 1865.0.
		h. m. s.
1863 Aug. 14	S.	8 45 0.75
Sept. 17	S.	0.84
Nov. 19	S.	0.93
1864 June 6	R.	0.75
Sept. 6	S.	0.96
15	S.	0.87
16	S.	0.88
Oct. 10	S.	1.07
Nov. 1	S.	0.81
10	S.	1.01

Mean A.R. 1865.0 8h 45m 0s.887

R. C. 2218.

Date.	Observer.	A.R. 1865.0.
		h. m. s.
1865 Sept. 11	M.	8 46 36.43
13	M.	36.80
14	S.	37.47
1865 Sept. 15	M.	8 46 37.15
19	M.	36.90
23	S.	37.43
25	M.	36.32
26	M.	37.47
27	M.	36.86

Mean A.R. 1865.0 8h 46m 36s.981

ι Ursæ Majoris. N.A.

Date.	Observer.	A.R. 1865.0.
		s.
1862 Mar. 28	S.	+0.09
April 4	H.	+0.19
7	S.	+0.14
1863 Feb. 7	S.	+0.08
Mar. 2	S.	+0.04
30	S.	+0.08
April 1	S.	—0.07
9	S.	+0.17
21	S.	+0.02
1864 Mar. 12	S.	+0.21
19	R.	+0.01
21	R.	+0.16
April 7	R.	+0.12
9	R.	+0.11
27	S.	+0.20
May 3	S.	+0.24
1865 March 4	S.	+0.13
10	M.	+0.17
17	M.	+0.09
27	M.	+0.21
April 3	M.	+0.16
4	M.	+0.02
5	M.	+0.18
Mean		+0.120
N.A.		56.921

Mean A. R. 1865.0 8h 49m 57s.041

σ² Ursæ Majoris.

Date.	Observer.	A.R. 1865.0.
		h. m. s.
1862 Mar. 29	H.	8 58 28.23
April 4	H.	28.29
7	S.	28.01
1863 Feb. 7	S.	28.15
Mar. 2	S.	28.22
30	S.	28.43
April 1	S.	28.07
9	S.	28.44
2	S.	28.05

Date.	Observer.	A.R. 1865.0.
		h. m. s.
1864 Mar. 7	S.	8 58 28.23
12	S.	28.37
17	R.	28.26
19	R.	27.96
21	R.	28.45
24	R.	28.35
April 7	R.	28.34

Mean A.R. 1865.0
8h 58m 28s.241

σ² Ursæ Majoris. Sub Polo.

Date.	Observer.	A.R. 1865.0.
		h. m. s.
1862 Sept. 25	S.	8 58 28.37
Oct. 7	S.	28.25
1863 Aug. 14	S.	28.28
Sept. 30	S.	28.01
Nov. 19	S.	28.44
1864 Sept. 10	S.	28.24
16	S.	28.11
Oct. 7	S.	28.21
10	S.	28.68
Nov. 3	S.	28.78

Mean A.R. 1865.0
8h 58m 28s.337

κ Cancri.

Date.	Observer.	A.R. 1865.0.
		h. m. s.
1862 Mar. 28	S.	9 0 26.06
29	H.	26.09
April 4	H.	26.01
7	S.	25.95
1863 Feb. 7	S.	25.92
Mar. 2	S.	26.08
30	S.	25.97
April 1	S.	26.00
9	S.	26.01
21	S.	25.84
1864 Mar. 7	S.	25.96
12	S.	26.05
21	R.	26.16
24	R.	25.99
April 7	R.	25.91
9	R.	26.09
16	R.	25.99
1865 Mar. 6	M.	26.10
10	M.	25.95
17	M.	26.01
27	M.	26.05
28	M.	26.16
April 3	M.	26.08

Mean A.R. 1865.0
9h 0m 26s.019

λ Argus.

Date.	Observer.	A.R. 1865.0.
		h. m. s.
1862 Mar. 28	S.	9 3 2.13
29	S.	2.07
April 4	S.	1.97
1863 Feb. 7	S.	2.00
April 1	S.	1.94
9	S.	1.97
1864 Mar. 12	S.	2.05
14	S.	1.98
24	R.	1.96
April 7	R.	2.07
1865 Mar. 6	M.	2.10
10	M.	1.98
17	M.	1.84
23	M.	1.97
27	M.	1.97
28	M.	1.82
April 3	M.	1.62
5	M.	2.25
13	M.	1.94
14	S.	2.09
17	M.	2.22

Mean A.R. 1865.0
9h 3m 1s.997

θ Hydræ.

Date.	Observer.	A.R. 1865.0.
		h. m. s.
1862 Mar. 29	H.	9 7 20.45
April 4	H.	20.44
7	S.	20.30
1863 Feb. 7	S.	20.36
April 9	S.	20.26
1864 Mar. 14	S.	20.31
17	R.	20.32
19	R.	20.43
21	R.	20.38
24	R.	20.36
April 7	R.	20.30
16	R.	20.32
27	S.	20.40
1865 Mar. 6	M.	20.46
10	M.	20.33
17	M.	20.35
27	M.	20.42
28	M.	20.42
April 3	M.	20.34
5	M.	20.49
13	M.	20.40
14	M.	20.32

Mean A.R. 1865.0
9h 7m 20s.371

83 Cancri. N.A.

Date.	Observer.	A.R. 1865.0.
		s.
1862 April 4	H.	+0.13
7	S.	+0.09
1863 Mar. 9	S.	+0.18
April 9	S.	+0.02
21	S.	+0.08
22	S.	+0.16
1864 Mar. 19	R.	+0.28
21	R.	+0.24
April 16	R.	+0.20
1865 Mar. 6	M.	+0.26
17	M.	+0.19
23	S.	+0.18
25	S.	+0.20
27	M.	+0.22
April 17	M.	+0.19
19	M.	+0.08
Mean		+0.169
N.A.		26.457

Mean A.R. 1865.0
9h 11m 26s.626

1 Drac. H.

Date.	Observer.	A.R. 1865.0.
		s.
1862 April 25	S.	+0.08
1863 Mar. 2	S.	0.00
9	S.	—0.40
April 9	S.	—0.07
21	S.	—0.52
22	S.	—0.38
1864 Mar. 24	R.	—0.60
April 7	R.	—0.24
22	S.	—0.24
1865 Mar. 6	M.	—0.55
25	S.	—0.07
April 3	M.	—0.56
13	M.	—0.07
14	S.	+0.13
17	M.	—0.16
25	M.	—0.10
Mean		—0.234
T.H.S.		34.69

Mean A.R. 1865.0
9h 17m 34s.456

Sub Polo.

Date.	Observer.	A.R. 1865.0.
		s.
1862 Sept. 10	S.	—0.24
23	S.	—0.22
29	S.	—0.33
Oct. 7	S.	—0.19
8	S.	—0.53

Date.	Observer.	A.R. 1865.0.
		s.
1863 Nov. 4	S.	—0.09
19	S.	—0.38
1864 June 6	R.	—0.46
Sept. 15	S.	—0.44
16	S.	—0.25
Oct. 7	S.	—0.75
10	S.	+0.34
17	S.	—0.04
Nov. 10	S.	+0.48
11	S.	+0.44
16	S.	+0.04
25	S.	—0.06
1865 Sept. 13	M.	—0.39
27	M.	+0.27
29	M.	—0.76
Oct. 11	M.	—0.06
Nov. 28	M.	+0.03
Mean		—0.163
T.H.S.		34.69

Mean A.R. 1865.0
9h 17m 34s.527

θ Ursæ Majoris. N.A.

Date.	Observer.	A.R. 1865.0.
		s.
1862 Mar. 27	H.	+0.18
29	H.	+0.16
April 4	H.	+0.09
1864 Mar. 7	S.	+0.13
19	R.	+0.19
21	R.	+0.20
April 7	R.	+0.10
9	R.	+0.07
15	S.	+0.12
16	R.	+0.20
1865 Mar. 4	S.	+0.28
17	M.	+0.09
April 14	S.	+0.04
17	M.	+0.29
19	M.	+0.01
25	M.	+0.12
26	M.	+0.15
Mean		+0.142
N.A.		48.554

Mean A.R. 1865.0
9h 23m 48s.696

Gr. 1564.

Date.	Observer.	A.R. 1865.0.
		h. m. s.
1863 Feb. 7	S.	9 30 38.23
April 9	S.	38.26
1864 Mar. 14	S.	38.24
21	S.	38.32
24	R.	38.12
April 7	R.	38.37
9	R.	38.36
15	S.	38.39
16	R.	38.28
27	S.	38.44
1865 Mar. 10	M.	38.19
17	M.	38.87
April 3	M.	38.12
5	M.	38.00
13	M.	38.20

Mean A.R. 1865.0
9h 30m 38s.293

Gr. 1564. Sub Polo.

Date.	Observer.	A.R. 1865.0.
		h. m. s.
1863 Sept. 17	S.	9 30 38.34
Nov. 25	S.	38.33
Dec. 1	S.	38.40
1864 Sept. 16	S.	38.03
Oct. 7	S.	38.08
10	S.	38.53
Nov. 16	S.	38.27

Mean A.R. 1865.0
9h 30m 38s.283

o Leonis.

Date.	Observer.	A.R. 1865.0.
		h. m. s.
1862 Mar. 27	H.	9 33 56.68
29	H.	56.66
April 4	H.	56.67
7	S.	56.56
19	S.	56.68
25	S.	56.69
1863 Feb. 7	S.	56.58
25	S.	56.62
April 9	S.	56.71
1864 Mar. 7	S.	56.56
14	S.	56.67
21	S.	56.54
April 7	R.	56.61
9	R.	56.57
16	R.	56.60
27	S.	56.59
1865 Mar. 10	M.	56.60
17	M.	56.68
27	M.	56.68
28	M.	56.68
April 3	M.	56.64

Mean A.R. 1865.0
9h 33m 56s.632

ε Leonis. N.A.

Date.	Observer.	A.R. 1865.0.
		s.
1862 Mar. 27	H.	+.02
29	H.	+.03
April 4	H.	+.06
7	S.	—.04
19	S.	.00
25	S.	+.15
1863 April 9	S.	—.03
1864 Mar. 14	S.	+.06
19	R.	—.01
April 9	R.	—.09
15	S.	—.02
16	R.	—.06
21	S.	—.08
1865 Mar. 10	M.	.00
17	M.	+.02
27	M.	+.11
28	M.	+.12
April 3	M.	+.06
5	M.	+.11
May 24	S.	—.00
Mean		+.020
N.A.		10.984

Mean A.R. 1865.0
9h 38m 11s.004

υ Ursæ Majoris.

Date.	Observer.	A.R. 1865.0.
		h. m. s.
1862 Mar. 29	H.	9 41 21.87
April 4	H.	21.90
7	S.	21.79
1863 Mar. 9	S.	21.73
1864 Mar. 14	S.	22.01
19	R.	22.17
April 9	R.	21.93
15	S.	21.84
16	R.	22.05
27	S.	21.95
1865 Mar. 10	M.	21.94

Mean A.R. 1865.0
9h 41m 21s.925

υ Ursæ Majoris. Sub Polo.

Date.	Observer.	A.R. 1865.0.
		h. m. s.
1863 Oct. 7	S.	9 41 21.96
Nov. 25	S.	21.94
1864 Sept. 7	S.	21.84
10	S.	21.90
16	S.	21.65
17	S.	21.98

Date.	Observer.	A.R. 1865.0.
		h. m. s.
1864 Oct. 7	S.	9 41 21.70
10	S.	22.12
17	S.	21.94

Mean A.R. 1865.0
9ʰ 41ᵐ 21ˢ.892

μ Leonis. C. des T.

Date.	Observer.	A.R. 1865.0.
		s.
1862 Mar. 27	H.	+0.12
29	H.	—0.02
April 4	H.	+0.04
1864 Mar. 14	S.	—0.06
19	R.	—0.06
21	S.	—0.17
April 7	R.	—0.18
9	R.	—0.13
16	R.	—0.19
1865 Mar. 10	M.	—0.02
17	M.	—0.05
23	S.	+0.04
27	M.	+0.05
28	M.	+0.08
April 5	M.	+0.03
8	M.	—0.02
11	S.	—0.09
13	M.	+0.10
14	S.	—0.02
May 24	S.	+0.05
Mean		—0.025
C. des T.		4.83

Mean A.R. 1865.0
9ʰ 45ᵐ 4ˢ.805

π Leonis. N.A.

Date.	Observer.	A.R. 1865.0.
		s.
1862 Mar. 29	H.	+.08
April 4	H.	+.07
25	S.	+.13
1864 Mar. 19	R.	+.07
April 9	R.	—.01
16	R.	+.03
1865 Mar. 10	M.	+.16
17	M.	+.05
23	S.	+.05
25	S.	+.02
27	M.	+.08
28	M.	+.04
April 5	M.	+.01
8	M.	+.07
13	M.	+.11
14	S.	.00
Mean		.060
N.A.		4.633

Mean A.R. 1865.0
9ʰ 53ᵐ 4ˢ.693

λ Hydræ.

Date.	Observer.	A.R. 1865.0.
		h. m. s.
1862 Mar. 29	H.	10 4 0.48
April 4	H.	0.58
7	S.	0.46
12	H.	0.56
16	H.	0.51
18	H.	0.48
25	S.	0.51
1863 May 2	S.	0.37
1864 Mar. 14	S.	0.49
19	R.	0.46
April 9	R.	0.52
16	R.	0.44
21	S.	0.44
May 3	S.	0.52
4	S.	0.44
5	S.	0.43
6	S.	0.46
1865 Mar. 4	S.	0.47
10	M.	0.43
17	M.	0.45
23	S.	0.45
27	M.	0.45

Mean A.R. 1865.0
10ʰ 4ᵐ 0ˢ.473

λ Ursæ Majoris.

Date.	Observer.	A.R. 1865.0.
		h. m. s.
1862 April 4	H.	10 8 56.64
7	S.	56.66
12	H.	56.63
16	H.	56.58
18	H.	56.72
25	S.	56.79
1863 April 9	S.	56.54
18	S.	56.62
May 2	S.	56.71
1864 Mar. 19	R.	56.61
May 3	S.	56.78
4	S.	56.88
5	S.	56.69
6	S.	56.55
1865 Mar. 10	M.	56.69
17	M.	56.67
23	S.	56.66
25	S.	56.69
27	M.	56.59
28	M.	56.72
April 5	M.	56.53
8	M.	56.60

Mean A.R. 1865.0
10ʰ 8ᵐ 56ˢ.661

γ Leonis. N.A.

Date.	Observer.	A.R. 1865.0.
		s.
1862 April 7	S.	—0.03
19	S.	+0.06
Oct. 5	S.	—0.05
7	S.	0.00
1865 Mar. 10	M.	+0.03
23	S.	+0.09
25	S.	+0.02
27	M.	+0.02
April 5	M.	+0.01
8	M.	—0.07
13	M.	+0.16
17	M.	+0.05
18	S.	+0.03
19	M.	—0.08
May 10	S.	—0.02
June 6	S.	+0.05
8	S.	+0.06
13	S.	—0.03
Mean		+0.017
N.A.		31.538

Mean A.R. 1865.0
10ʰ 12ᵐ 31ˢ.555

30 Camelopardali H.

Date.	Observer.	A.R. 1865.0.
		s.
1862 April 7	S.	—0.78
16	H.	0.00
28	S.	+0.84
1863 Mar. 9	S.	—0.27
May 3	S.	—0.28
5	S.	—0.23
6	S.	+0.03
1865 Mar. 25	S.	+0.22
April 3	M.	—0.41
13	M.	+0.12
14	S.	+0.18
17	M.	—0.42
24	M.	—0.31
25	M.	—0.59
Mean		—0.136
T.H.S.		19.36

Mean A.R. 1865.0
10ʰ 14ᵐ 19ˢ.224

30 Camelop. H. Sub Polo.

Date.	Observer.	A.R. 1865.0.
		s.
1862 Sept. 10	S.	—0.66
23	S.	—0.60
Oct. 7	S.	—0.03
Nov. 3	S.	+0.48

Date.			Observer.	A.R. 1865.0.
				s.
1862	Dec.	11	S.	+0.21
1863	Sept.	29	S.	+0.32
	Nov.	3	S.	+0.62
	Dec.	7	S.	+1.38
		10	S.	+0.12
1864	Sept.	16	S.	—0.03
	Oct.	7	S.	+0.28
		11	S.	—0.08
	Mean			+0.168
	T.H.S.			19.36

Mean A.R. 1865.0
10^h 14^m 19^s.528

μ Hydræ.

Date.			Observer.	A.R. 1865.0.
				h. m. s.
1862	Mar.	27	H.	10 19 33.80
		29	H.	33.76
	April	4	H.	33.81
		12	H.	33.91
		16	H.	33.80
		18	H.	33.84
		25	S.	33.84
1863	Mar.	6	S.	33.95
	April	9	S.	33.68
1864	April	9	R.	33.83
	May	3	S.	33.77
		6	S.	33.79
1865	Mar.	10	M.	33.81
		23	S.	33.81
		27	M.	33.77
	April	3	M.	33.74
		5	M.	33.78
		8	M.	33.80
		13	M.	33.77
		17	M.	33.73
		18	S.	33.77
		19	M.	33.78

Mean A.R. 1865.0
10^h 19^m 33^s.797

ϱ Leonis N.A.

Date.			Observer.	A.R. 1865.0.
				s.
1862	Mar.	29	H.	+.04
	April	4	H.	+.06
		16	H.	+.04
		25	S.	+.12
1863	April	9	S.	—.03
1864	April	16	R.	—.10
	May	3	S.	+.06
		6	S.	+.03
1865	Mar.	10	M.	+.04
		23	S.	—.01
				s.
1865	Mar.	25	S.	+.02
		27	M.	+.07
	April	3	M.	+.09
		5	M.	—.05
		8	M.	—.05
		13	M.	+.06
	Mean			+0.024
	N.A.			42.058

Mean A.R. 1865.0
10^h 25^m 42^s.082

Bradley 1458 = 193 Camelopardali B.

Date.			Observer.	A.R. 1865.0.
				h. m. s.
1863	Mar.	6	S.	10 29 59.48
	April	9	S.	58.94
1864	April	30	R.	58.61
	May	3	S.	59.16
		6	S.	59.07
1865	Mar.	10	M.	59.46
		25	S.	59.10
		27	M.	59.24
	April	5	M.	58.47
		11	S.	59.08
		12	M.	59.92
		17	M.	58.52
		24	M.	58.88
		25	M.	59.01

Mean A.R. 1865.0
10^h 29^m 59^s.067

Bradley 1458 = 193 Camelopar. B. Sub Polo.

Date.			Observer.	A.R. 1865.0.
				h. m. s.
1863	Nov.	3	S.	10 29 59.23
1864	Sept.	10	S.	59.24
		16	S.	59.02
		17	S.	59.59
	Nov.	11	M.	59.22
		14	S.	59.90

Mean A.R. 1865.0
10^h 29^m 59^s.367

l Leonis. N.A.

Date.			Observer.	A.R. 1865.0.
				s.
1862	Mar.	29	H.	+0.04
	April	4	H.	+0.14
		7	S.	+0.05
		12	H.	+0.09
		16	H.	+0 10
1863	April	9	S.	0
1864	April	16	R.	+0
		30	R.	+0
	May	6	S.	+0
1865	Mar.	27	M.	+0
	April	3	M.	+0
		5	M.	+0
		8	M.	+0
		12	M.	+0
	Mean			+0.
	N.A.			9.

Mean A.R. 1865.0
10^h 42^m 9^s.

β Ursæ Majoris.

Date.			Observer.	A.R. 1865.0.
				h. m.
1862	Mar.	29	H.	10 53 40
	April	4	H.	40
		7	S.	40
		12	H.	40
		16	H.	40
		18	H.	40
		19	S.	40
		25	S.	40
		30	H.	40
	May	7	S.	40
	Oct.	5	S.	40
1863	May	9	S.	40
		19	S.	40
1864	April	9	R.	40
		16	R.	40
		30	R.	40

Mean A.R. 1865.0
10^h 53^m 40^s.

β Ursæ Majoris. Sub Po

Date.			Observer.	A.R. 1865.0.
				h. m.
1863	Sept.	30	S.	10 53 40
	Dec.	9	S.	40
		23	S.	40
1864	Sept.	10	S.	40
		16	S.	40
		26	S.	40
	Oct.	7	S.	40
		25	S.	40
		26	S.	40

Mean A.R. 1865.0
10^h 53^m 40^s.

α Ursæ Majoris. B.J.

Date.	Observer.	A.R. 1865.0.
		s.
1862 Mar. 29	H.	0.00
April 4	H.	+0.11
7	S.	+0.08
12	H.	+0.23
16	H.	—0.01
18	H.	+0.23
26	H.	+0.07
30	H.	+0.08
Oct. 5	S.	+0.16
1863 Mar. 6	S.	+0.28
May 9	S.	+0.19
19	S.	+0.16
1864 April 9	R.	+0.15
16	R.	—0.05
30	R.	+0.09
1865 Mar. 10	M.	+0.35
27	M.	+0.10
April 3	M.	+0.13
5	M.	0.00
8	M.	—0.03
Mean		+0.120
B.J.		22.131

Mean A.R. 1865.0 **10ʰ 55ᵐ 22ˢ.251**

α Ursæ Majoris. Sub Polo.

Date.	Observer.	A.R. 1865.0.
		s.
1863 Jan. 1	S.	+0.04
Sept. 17	S.	+0.08
30	S.	—0.03
Dec. 9	S.	+0.24
15	S.	+0.08
23	S.	+0.40
1864 Sept. 10	S.	+0.07
16	S.	+0.16
26	S.	+0.21
27	S.	+0.34
30	S.	+0.26
Oct. 7	S.	+0.24
17	S.	—0.01
25	S.	+0.12
26	S.	+0.01
Nov. 10	M.	+0.36
11	M.	—0.34
Mean		+0.131
B. J.		22.131

Mean A.R. 1865.0 **10ʰ 55ᵐ 22ˢ.262**

χ Leonis. N.A.

Date.	Observer.	A.R. 1865.0.
		s.
1862 Mar. 26	H.	+.05
29	H.	+.06
April 7	S.	+.04
12	H.	+.10
16	H.	+.04
24	H.	+.07
25	S.	+.13
May 7	S.	+.05
1864 April 16	R.	+.08
30	R.	+.02
May 6	S.	+.07
19	S.	+.14
April 3	M.	+.06
5	M.	+.12
8	M.	+.07
12	M.	+.03
13	M.	—.03
May 3	M.	—.15
10	S.	+.08
Mean		+.054
N.A.		3.107

Mean A.R. 1865.0 **10ʰ 58ᵐ 3ˢ.161**

ψ Ursæ Majoris.

Date.	Observer.	A.R. 1865.0.
		h. m. s.
1862 Mar. 26	S.	11 2 3.74
29	H.	3.68
April 4	H.	3.77
12	H.	3.82
16	H.	3.74
19	S.	3.77
26	H.	3.79
30	H.	3.72
1863 May 19	S.	3.76
1864 April 16	R.	3.68
30	R.	3.71
1865 Mar. 25	S.	3.85
27	M.	3.75
April 3	M.	3.75
5	M.	3.79
8	M.	3.78
13	M.	3.74
14	S.	3.67
17	M.	3.80

Mean A.R. 1865.0 **11ʰ 2ᵐ 3ˢ.753**

δ Leonis. N.A.

Date.	Observer.	A.R. 1865.0.
		s.
1862 Mar. 26	S.	+.05
29	H.	—.03
April 25	S.	+.09
26	H.	—.04
30	H.	—.02
1863 Mar. 6	S.	+.04
1864 May 6	S.	+.08
June 7	S.	—.05
13	S.	+.06
1865 Mar. 25	S.	+.04
April 3	M.	+.09
5	M.	—.09
13	M.	+.13
May 3	M.	—.03
Mean		+0.023
N.A.		55.499

Mean A.R. 1865.0 **11ʰ 6ᵐ 55ˢ.522**

δ Crateris. N.A.

Date.	Observer.	A.R. 1865.0.
		s.
1862 Mar. 26	S.	+.08
April 7	S.	+.04
16	H.	+.08
25	S.	+.20
26	H.	+.08
30	H.	+.06
May 7	S.	+.08
15	S.	—.04
1863 Mar. 6	S.	+.01
1864 April 16	R.	—.14
22	S.	+.08
May 6	S.	+.13
19	R.	+.08
1865 Mar. 25	S.	+.01
April 3	M.	—.01
5	M.	+.15
8	M.	—.05
13	M.	+.16
14	S.	.00
17	M.	+.20
Mean		+.060
N.A.		35.529

Mean A.R. 1865.0 **11ʰ 12ᵐ 35ˢ.598**

γ Crateris.

Date.	Observer.	A.R. 1865.0.
		h. m. s.
1862 Mar. 26	S.	11 18 8.46
29	H.	8.42

Date.	Observer.	A.R. 1865.0.
		h. m. s.
1862 April 7	S.	11 18 8.47
12	H.	8.47
16	H.	8.48
26	H.	8.46
30	H.	8.39
May 7	S.	8.48
15	S.	8.36
Dec. 28	S.	8.64
1864 April 16	R.	8.32
22	S.	8.40
May 6	S.	8.50
19	R.	8.42
1865 Mar. 25	S.	8.40
April 5	M.	8.40
8	M.	8.53
12	M.	8.51
13	M.	8.40
14	S.	8.40
17	M.	8.42
19	M.	8.40

Mean A.R. 1865.0
$11^h\ 18^m\ 8^s.442$

τ Leonis.

Date.	Observer.	A.R. 1865.0.
		h. m. s.
1862 April 7	S.	11 20 59.62
12	H.	59.75
16	H.	59.68
1864 April 16	R.	59.56
22	S.	59.69
May 19	R.	59.64
1865 Mar. 25	S.	59.67
April 3	M.	59.68
8	S.	59.71
13	M.	59.70
14	S.	59.65
17	M.	59.70
18	S.	59.74
19	M.	59.59
24	M.	59.68
May 3	M.	59.61
13	M.	59.63
15	S.	59.72
24	M.	59.68
25	M.	59.69

Mean A.R. 1865.0
$11^h\ 20^m\ 59^s.670$

202 Camelopardali. Bode.

Date.	Observer.	A.R. 1865.0.
		s.
1862 Mar. 26	S.	—0.75
April 7	S.	—0.74
30	H.	—0.46
1862 May 7	S.	—0.17
15	S.	—0.33
Dec. 28	S.	+0.23
1863 May 9	S.	—0.39
1864 May 6	S.	+0.12
19	R.	—0.46
Mean		—0.328
T.H.S.		12.02

Mean A.R. 1865.0
$11^h\ 22^m\ 11^s.692$

202 Camelop. B. Sub Polo.

Date.	Observer.	A.R. 1865.0.
		s.
1862 Sept. 23	S.	—0.63
Dec. 28	S.	—0.05
29	S.	+0.07
1863 Sept. 22	S.	+0.18
28	S.	—0.59
30	S.	—0.61
Oct. 29	S.	—0.49
Nov. 2	S.	+0.35
3	S.	—0.30
4	S.	+0.07
Dec. 10	S.	+0.15
21	S.	—0.28
23	S.	+0.28
26	S.	+0.01
1864 Sept. 16	S.	—0.53
30	S.	—0.53
Nov. 10	M.	+0.08
11	M.	—0.43
14	S.	+0.11
Dec. 9	S.	—0.28
1865 Sept. 30	S.	—0.16
Oct. 3	S.	+0.18
Mean		—0.155
T.H.S.		12.02

Mean A.R. 1865.0
$11^h\ 22^m\ 11^s.865$

λ Draconis. Sub Polo. A.E.

Date.	Observer.	A.R. 1865.0.
		s.
1865 Sept. 13	M.	+0.42
25	M.	—0.07
Oct. 6	M.	—0.06
13	M.	—0.06
16	M.	+0.18
23	M.	+0.14
25	M.	+0.10
Nov. 11	M.	+0.07
14	M.	—0.05
Dec. 22	M.	—0.02
Mean		+0.065
A.E.		21.132

Mean A.R. 1865.0
$11^h\ 23^m\ 21^s.197$

υ Leonis. N.A.

Date.	Observer.	A.R. 1865.0.
		s.
1864 April 22	S.	—.07
1865 April 3	M.	—.05
5	M.	+.12
13	M.	+.06
14	S.	+.03
17	M.	+.06
18	S.	+.01
24	M.	.00
25	M.	+.03
May 3	M.	—.09
10	S.	—.01
13	S.	—.02
15	M.	+.04
18	S.	.00
24	S.	+.02
Mean		+0.009
N.A.		2.214

Mean A.R. 1865.0
$11^h\ 30^m\ 2^s.223$

χ Ursæ Majoris.

Date.	Observer.	A.R. 1865.0.
		h. m. s.
1862 Mar. 26	S.	11 38 54.67
29	H.	54.59
April 16	H.	54.66
26	H.	54.60
30	H.	54.62
May 7	S.	54.60
15	S.	54.70
Dec. 28	S.	54.61
1863 March 6	S.	54.71
1864 April 21	R.	54.66
22	S.	54.69
May 6	S.	54.78
19	R.	54.63
23	S.	54.58
1865 Mar. 25	S.	54.55
April 5	M.	54.53
8	S.	54.48
13	M.	54.74
17	M.	54.69
18	S.	54.68
19	M.	54.54

Mean A.R. 1865.0
$11^h\ 38^m\ 54^s.634$

γ Ursæ Majoris. B.J.

Date.	Observer.	A.R. 1865.0.
		s.
1862 Mar. 26	S.	—0.13
29	H.	—0.11
April 7	S.	+0.03
12	H.	—0.06
16	H.	—0.12
26	H.	—0.17
30	H.	—0.19
May 7	S.	—0.03
15	S.	—0.08
Dec. 28	S.	0.00
1863 May 23	S.	+0.03
1864 April 21	R.	—0.07
22	S.	—0.20
30	R.	—0.08
May 6	S.	+0.09
18	R.	—0.19
27	S.	—0.20
29	R.	—0.16
1865 April 24	M.	—0.04
25	M	—0.04
May 3	M.	—0.05
15	M.	0.00
23	M.	—0.07
25	M.	—0.08
Mean		—0.080
B.J.		43.048

Mean A.R. 1865.0
$11^h\ 46^m\ 42^s.968$

Gr. 1845.

Date.	Observer.	A.R. 1865.0.
		h. m. s.
1865 Sept. 13	M.	11 53 11.42
25	M.	11.03
29	M.	10.62
Oct. 6	M.	11.14
16	M	11.26
Nov. 1	S.	11.03
11	M.	11.75
Dec. 21	M.	10.28
22	M.	10.59
23	M.	10.93

Mean A.R. 1865.0
$11^h\ 53^m\ 11^s.005$

π Virginis.

Date.	Observer.	A.R. 1865.0.
		h. m. s.
1862 April 23	S.	11 53 57.40
26	H.	57.26
May 7	S.	57.28
1864 May 6	S.	11 53 57.43
19	R.	57.23
23	S.	57.29
27	S.	57.26
29	R.	57.18
1865 April 13	M.	57.33
14	S.	57.30
17	M.	57.30
18	S.	57.34
24	M.	57.27
25	M.	57.34
May 3	M.	57.28
10	S.	57.32
13	S.	57.24
15	M.	57.30
23	M.	57.38
24	S.	57.29
25	M.	57.31
26	M.	57.36

Mean A.R. 1865.0
$11^h\ 53^m\ 57^s.304$

Groombridge 1850.

Date.	Observer.	A.R. 1865.0.
		h. m. s.
1865 Sept. 25	M.	11 57 53.37
Oct. 7	S.	53.80
24	S.	54.76
Nov. 1	S.	53.87
11	M.	53.52
Dec. 21	M.	51.73
22	M.	52.13
23	M.	53.32
1866 Jan. 4	S.	54.30

Mean A.R. 1865.0
$11^h\ 57^m\ 53^s.422$

α Corvi.

Date.	Observer.	A.R. 1865.0.
		h. m. s.
1865 May 23	M.	12 1 27.42
24	M.	27.39
25	S.	27.38
26	M.	27.42

Mean A.R. 1865.0
$12^h\ 1^m\ 27^s.402$

δ Corvi. N.A.

Date.	Observer.	A.R. 1865.0.
		s.
1862 Mar. 26	S.	+0.10
April 30	H.	+0.13
May 7	S.	+0.12
17	S.	—0.08
23	S.	+0.06
		s.
1862 Dec. 28	S.	+0.09
1864 April 22	S.	+0.02
30	R.	+0.09
May 23	S.	+0.11
29	R.	+0.13
1865 April 8	S.	+0.04
13	M.	+0.13
17	M.	+0.12
18	S.	+0.19
24	M.	+0.12
25	M.	+0.07
May 3	M.	+0.06
13	S.	+0.07
15	M.	+0.04
23	S.	+0.13
26	M.	+0.15
Mean		+0.090
N.A.		11.124

Mean A.R. 1865.0
$12^h\ 3^m\ 11^s.214$

207 Camelopardali B.

Date.	Observer.	A.R. 1865.0.
		h. m. s.
1865 May 10	S.	12 4 52.08
13	S.	52.02
15	M.	51.78
18	S.	52.17
23	M.	51.84
24	S.	51.86
25	M.	51.96

Mean A. R. 1865.0
$12^h\ 4^m\ 51^s.959$

Sub Polo.

Date.	Observer.	A.R. 1865.0.
		h. m. s.
1865 Sept. 13	M.	12 4 51.95
29	M.	51.54
Oct. 6	M.	51.10
13	M.	52.44
16	M.	52.26
23	M.	52.55
30	M.	51.78
1866 Jan. 4	S.	52.16

Mean A.R. 1865.0
$12^h\ 4^m\ 51^s.972$

γ Corvi.

Date.	Observer.	A.R. 1865.0.
		h. m. s.
1862 Mar. 29	H.	12 8 51.98
April 7	S.	52.02

Date.	Observer.	A.R. 1865.0.
		h. m. s.
1862 April 25	S.	12 8 52.12
30	H.	51.99
May 23	S.	52.09
Dec. 10	S.	52.14
1863 Mar. 6	S.	52.03
1864 Jan. 8	S.	52.05
Mar. 24	S.	52.01
April 21	R.	52.10
29	S.	52.10
30	R.	51.99
May 9	R.	52.01
18	R.	52.06
19	R.	52.04
22	R.	51.95
23	S.	52.07
29	R.	52.01
1865 April 8	R.	52.00
13	M.	52.01

Mean A.R. 1865.0
12ʰ 8ᵐ 52ˢ.039

2 H. Canum.

Date.	Observer.	A.R. 1865.0.
		h. m. s.
1865 May 23	M.	12 9 42.73
24	S.	42.62
25	M.	42.61
26	M.	42.70
June 2	S.	42.78

Mean A.R. 1865.0
12ʰ 9ᵐ 42ˢ.688

5 Ursæ Minoris Bode.

Date.	Observer.	A.R. 1865.0.
		s.
1862 Mar. 26	S.	—0.95
April 7	S.	0.00
16	H.	—0.27
25	S.	—0.83
May 17	S.	—0.93
Dec. 28	S.	—1.81
1863 April 9	S.	—0.39
1864 Mar. 24	S.	—0.24
April 21	R.	—0.22
29	S.	—0.60
30	R.	—0.06
May 9	R.	+0.63
27	S.	—0.61
Mean		—0.483
T.H.S.		52.79

Mean A.R. 1865.0
12ʰ 12ᵐ 52ˢ.307

5 Ursæ Min. B. Sub Polo.

Date.	Observer.	A.R. 1865.0.
		s.
1862 Nov. 10	S.	+1.37
Dec. 28	S.	+0.81
29	S.	+0.54
1863 Aug. 6	S.	+0.31
10	S.	+0.41
30	S.	+1.63
Sept. 24	S.	—0.76
28	S.	—0.57
29	S.	—1.47
30	S.	—0.54
Oct. 22	S.	—1.49
Nov. 2	S.	—0.52
3	S.	+0.53
4	S.	+0.57
26	S.	—2.01
30	S.	—1.64
Dec. 26	S.	+0.56
1864 Sept. 30	S.	+0.38
Oct. 17	S.	—0.41
Nov. 30	M.	—0.57
Dec. 9	M.	+1.79
20	S.	+1.23
30	M.	—0.41
1865 Dec. 6	S.	—1.09
Mean		—0.056
T.H.S.		52.79

Mean A.R. 1865.0
12ʰ 12ᵐ 52ˢ.734

η Virginis. N.A.

Date.	Observer.	A.R. 1865.0.
		s.
1862 Mar. 29	H.	+.04
April 26	H.	+.01
30	H.	+.11
1864 May 18	R.	+.09
19	R.	+.13
23	S.	+.10
29	R.	+.07
1865 April 13	M.	+.14
14	S.	+.05
17	M.	+.11
18	S.	+.12
24	M.	+.17
25	M.	+.23
May 3	M.	—.01
10	S.	+.07
13	S.	+.16
15	M.	+.18
18	S.	+.17
23	M.	+.13
1863 May 25	M.	+.13
Mean		+.110
N.A.		59.920

Mean A.R. 1865.0
12ʰ 13ᵐ 0ˢ.030

6 Ursæ Minoris Bode.

Date.	Observer.	A.R. 1865.0.
		s.
1862 Mar. 26	S.	+0.11
April 7	S.	+1.01
29	S.	—0.18
May 7	S.	+0.70
15	S.	—0.80
23	S.	+0.86
Dec. 10	S.	+0.74
28	S.	—1.77
1864 Jan. 8	S.	—1.32
April 22	S.	+1.93
May 6	S.	+1.58
18	R.	—0.61
22	R.	—2.07
29	R.	+0.42
Mean		+0.043
T.H.S.		19.56

Mean A.R. 1865.0
12ʰ 14ᵐ 19ˢ.603

6 Ursæ Min. B. Sub Polo.

Date.	Observer.	A.R. 1865.0.
		s.
1862 Sept. 23	S.	—1.03
Dec. 11	S.	+0.55
28	S.	+0.89
29	S.	+1.04
1863 Jan. 2	S.	+1.21
3	S.	+0.14
5	S.	+0.75
Oct. 22	S.	—0.74
Nov. 2	S.	+2.54
3	S.	+1.29
30	S.	—5.02
Dec. 26	S.	+0.48
1864 Jan. 9	S.	—0.05
Oct. 7	S.	+4.07
Nov. 29	M.	+0.50
Dec. 30	M.	+0.35
1865 Nov. 1	S.	—0.28
Mean		+0.395
T.H.S.		19.56

Mean A.R. 1865.0
12ʰ 14ᵐ 19ˢ.955

12 Comæ.

Date.			Observer.	A.R. 1865.0.
				h. m. s.
1865	May	23	M.	12 15 43.05
		24	S.	42.95
		25	M.	42.95
		26	M.	43.01
	June	1	S.	43.06
		2	S.	43.08
		7	S.	43.01

Mean A.R. 1865.0
12ʰ 15ᵐ 43ˢ.016

13 Comæ.

Date.			Observer.	A.R. 1865.0.
				h. m. s.
1865	May	24	S.	12 17 31.96
		25	M.	31.98
		26	M.	31.99
	June	1	S.	32.14
		2	S.	32.09
		7	S.	32.05

Mean A.R. 1865.0
12ʰ 17ᵐ 32ˢ.035

14 Comæ.

Date.			Observer.	A.R. 1865.0.
				h. m. s.
1865	May	23	M.	12 19 38.91
		24	S.	38.84
		25	M.	38.79
		26	M.	38.80
	June	1	S.	38.91
		2	S.	38.85

Mean A.R. 1865.0
12ʰ 19ᵐ 38ˢ.850

15 Comæ.

Date.			Observer.	A.R. 1865.0.
				h. m. s.
1862	April	26	H.	12 20 12.22
		29	S.	12.32
		30	H.	12.27
	May	14	H.	12.45
		15	S.	12.44
		23	S.	12.50
	Dec.	10	S.	12.36
		28	S.	12.35
1864	April	29	S.	12.35
		30	R.	12.26
	May	6	S.	12.44
		22	R.	12.16
		27	S.	12.35
		29	R.	12.39
1865	April	13	M.	12 20 12.45
		14	S.	12.37
		17	M.	12.37
		18	S.	12.41
		24	M.	12.34
	May	3	M.	12.38
		15	M.	12.45

Mean A.R. 1865.0
12ʰ 20ᵐ 12ˢ.363

16 Comæ.

Date.			Observer.	A.R. 1865.0.
				h. m. s.
1865	May	24	S.	12 20 14.07
	June	1	S.	14.18
		2	S.	14.12

Mean A.R. 1865.0
12ʰ 20ᵐ 14ˢ.123

17 Comæ.

Date.			Observer.	A.R. 1865.0.
				h. m. s.
1865	May	23	M.	12 22 10.11
		24	S.	10.02
		25	M.	10.05
		26	M.	10.06
		30	S.	10.04
	June	1	S.	10.22
		2	S.	9.96

Mean A.R. 1865.0
12ʰ 22ᵐ 10ˢ.066

δ Corvi. C. des T.

Date.			Observer.	A.R. 1865.0.
				s.
1862	Mar.	26	S.	—.08
	April	16	H.	—.11
		25	S.	—.07
		26	H.	—.15
		29	S.	—.04
		30	H.	—.15
	May	7	S.	—.12
		15	S.	—.11
		23	S.	—.10
	Dec.	10	S.	—.10
		28	S.	—.09
1864	Jan.	8	S.	—.14
	April	22	S.	—.07
		29	S.	—.04
		30	R.	—.15
	May	9	R.	—.11
		18	R.	—.04
		22	R.	—.01
1864	May	23	S.	—.15
		27	S.	—.16
		29	R.	—.07
	June	7	S.	—.12
	Mean			—.099
	C. des T.			53.10

Mean A.R. 1865.0
12ʰ 22ᵐ 53ˢ.001

4 Draconis.

Date.			Observer.	A.R. 1865.0.
				h. m. s.
1865	May	24	S.	12 24 10.39
		25	M.	10.58
		30	S.	10.74
	June	1	S.	11.06
		2	S.	10.75
		7	S.	10.90

Mean A.R. 1865.0
12ʰ 24ᵐ 10ˢ.737

η Corvi.

Date.			Observer.	A.R. 1865.0.
				h. m. s.
1862	Mar.	26	S.	12 25 7.13
	April	16	H.	7.08
		25	S.	7.15
		29	S.	7.05
		30	H.	7.10
	May	7	S.	7.05
		17	S.	6.97
		23	S.	7.11
	Dec.	10	S.	7.10
		28	S.	7.11
1864	Jan.		S.	7.18
	April	21	R.	7.07
		22	S.	7.01
		29	S.	7.03
		30	R.	6.95
	May	9	R.	7.00
		18	R.	7.03
		22	R.	7.07
		23	S.	7.19
		27	S.	7.04
		29	R.	7.02
1865	April	4	S.	7.20

Mean A.R. 1865.0
12ʰ 25ᵐ 7ˢ.075

5 H. Canum.

Date.			Observer.	A.R. 1865.0.
				h. m. s.
1865	May	23	M.	12 26 59.69
		24	S.	59.67

Date.	Observer.	A.R. 1865.0.
		h. m. s.
1865 May 25	M.	12 26 59.67
26	M.	59.69
30	S.	59.76
June 1	S.	59.83
2	S.	59.77

Mean A.R. 1865.0
12h 26m 59s.726

β Corvi. N.A.

Date.	Observer.	A.R. 1865.0.
		s.
1862 Mar. 26	S.	+.08
April 30	H.	+.20
May 7	S.	+.21
17	S.	+.20
23	S.	+.18
Dec. 10	S.	+.20
28	S.	+.25
1864 April 21	R.	+.19
22	S.	+.12
29	S.	+.17
30	R.	+.19
May 4	S.	+.22
9	R.	+.25
18	R.	+.33
22	R.	+.27
23	S.	+.26
27	S.	+.23
June 2	R.	+.34
10	S.	+.36
1865 April 4	S.	+.21
24	M.	+.21
May 3	M.	+.29
10	S.	+.24
15	M.	+.28
Mean		+0.228
N.A.		17.852

Mean A.R. 1865.0
12h 27m 18s.080

24 Comæ.

Date.	Observer.	A.R. 1865.0.
		h. m. s.
1865 May 24	S.	12 28 21.43
25	M.	21.43
26	M.	21.42
30	S.	21.52
June 1	S.	21.50
2	S.	21.35

Mean A.R. 1865.0
12h 28m 21s.442

γ Virginis. N.A.

Date.	Observer.	A.R. 1865.0.
		s.
1862 Mar. 26	S.	+.15
April 25	S.	+.21
26	H.	+.05
30	H.	+.04
May 14	H.	+.11
17	S.	—.07
Dec. 10	S.	+.06
1864 April 21	S.	—.01
22	S.	.00
29	S.	+.17
May 3	S.	+.09
4	S.	+.03
6	S.	+.01
22	R.	—.06
23	S.	+.07
29	R.	—.03
June 4	S.	—.01
1865 May 18	S.	+.10
June 1	S.	+.14
Mean		+.055
N.A.		49.228

Mean A.R. 1865.0
12h 34m 49s.283

ρ Virginis.

Date.	Observer.	A.R. 1865.0.
		h. m. s.
1865 May 23	M.	12 35 3.28
25	M.	3.11
26	M.	3.19
30	S.	3.21
June 2	S.	3.03

Mean A.R. 1865.0
12h 35m 3s.164

60 B. Canum.

Date.	Observer.	A.R. 1865.0.
		h. m. s.
1863 May 23	M.	12 38 46.88
24	S.	46.74
25	M.	46.87
26	M.	46.80
30	S.	46.87
June 2	S.	46.93

Mean A.R. 1865.0
12h 38m 46s.848

30 Comæ.

Date.	Observer.	A.R. 1865.0.
		h. m. s.
1865 May 23	M	12 42 42.69
24	S.	42.63
25	M.	42.63
30	S.	42.71
June 1	S.	42.71
2	S.	42.72
6	S.	42.68
7	S.	42.80
8	S.	42.76

Mean A.R. 1865.0
12h 42m 42s.703

31 Comæ.

Date.	Observer.	A.R. 1865.0.
		h. m. s.
1865 May 23	M.	12 45 7.24
24	S.	7.18
25	M.	7.22
30	S.	7.34
June 1	S.	7.40
2	S.	7.27
6	S.	7.23
7	S.	7.29
8	S.	7.25

Mean A.R. 1865.0
12h 45m 7s.269

ε Ursæ Majoris. C. des T.

Date.	Observer.	A.R. 1865.0.
		s.
1862 Mar. 26	S.	+0.05
April 7	S.	—0.19
16	H.	—0.13
26	H.	—0.16
May 17	S.	—0.11
Dec. 28	S.	—0.04
1864 Jan. 8	S.	—0.17
April 30	R.	—0.31
May 6	S.	—0.06
9	R.	—0.09
29	R.	—0.14
Mean		—0.123
C. des T.		4.94

Mean A.R. 1865.0
12h 48m 4s.817

32 Camelopardali H. foll. A.E.

Date.	Observer.	A.R. 1865.0.
		s.
1865 May 24	S.	—0.33
30	S.	—0.11
June 1	S.	+0.48
2	S.	+0.09
6	S.	+0.05

Date.	Observer.	A.R. 1865.0.
		s.
Mean		+0.036
A.E.		10.292

Mean A.R. 1865.0 12h 48m 10s.328

32 Camelopardali H. Sub Polo.

Date.	Observer.	A.R. 1865.0.
		s.
1865 Sept. 27	M.	0.00
29	M.	—0.20
30	S.	+0.28
Oct. 13	M.	—0.64
16	M.	—0.23
23	M.	—0.51
24	S.	—0.51
25	M.	—0.10
Nov. 1	S.	—0.35
14	M.	—0.48
Dec. 1	M.	—0.96
Mean		—0.364
A.E.		10.292

Mean A.R. 1865.0 12h 48m 9s.928

δ Virginis.

Date.	Observer.	A.R. 1865.0.
		h. m. s.
1864 April 30	R.	12 48 48.20
May 18	R.	48.39
19	R.	48.31
23	S.	48.22
29	R.	48.32
31	S.	48.25
June 4	S.	48.24
7	S.	48.25
10	S.	48.10
13	S.	48.31
14	S.	48.26
17	S.	48.43
1865 April 13	M.	48.27
14	S.	48.22
24	M.	48.25
May 3	M.	48.37
10	S.	48.22
15	M.	48.28
23	M.	48.33
25	M.	48.25
June 20	S.	48.25

Mean A.R. 1865.0 12h 48m 48s.272

12 Canum. N.A.

Date.	Observer.	A.R. 1865.0.
		s.
1862 April 16	H.	—.02
26	H.	+.03
May 14	H.	+.18
1864 April 29	S.	+.15
30	R.	+.01
May 3	S.	+.19
4	S.	+.07
6	S.	+.16
18	R.	+.21
19	R.	+.05
22	R.	+.04
23	S.	+.03
31	S.	+.22
June 4	S.	+.22
7	S.	+.18
10	S.	+.15
13	S.	+.10
14	S.	+.19
17	S.	+.22
1865 April 13	M.	+.15
24	M.	+.08
Mean		+.124
N.A.		42.405

Mean A.R. 1865.0 12h 49m 42s.529

ε Virginis. C. des T.

Date.	Observer.	A.R. 1865.0.
		s.
1862 Mar. 26	S.	—.02
April 16	H.	—.02
25	S.	+.06
26	H.	—.19
30	H.	—.05
May 14	H.	—.01
17	S.	—.05
23	S.	—.06
Dec. 10	S.	.00
28	S.	—.12
1864 Jan. 8	S.	—.15
April 21	S.	—.10
22	S.	—.06
29	S.	+.01
30	R.	—.12
May 3	S.	—.09
6	S.	—.04
9	R.	—.01
22	R.	—.06
23	S.	—.13
28	R.	.00
June 2	R.	—.05
8	S.	—.03
1864 June 13	S.	—.12
14	S.	—.11
17	S.	+.02
1865 June 1	S.	+0.09
7	S.	+0.02
Mean		—0.050
C. des T.		27.47

Mean A.R. 1865.0 12h 55m 27s.420

14 Canum.

Date.	Observer.	A.R. 1865.0.
		h. m. s.
1865 May 23	M.	12 59 25.46
30	S.	25.53
June 1	S.	25.72
3	S.	25.52
8	S.	25.59

Mean A.R. 1865.0 12h 59m 25s.564

θ Virginis. N.A.

Date.	Observer.	A.R. 1865.0.
		s.
1862 May 17	S.	—0.01
Dec. 28	S.	+0.03
1864 April 21	S.	—0.04
22	S.	+0.13
29	S.	+0.06
May 3	S.	+0.10
6	S.	+0.06
9	R.	+0.10
18	S.	+0.23
19	S.	+0.27
28	R.	+0.14
29	R.	+0.09
31	S.	+0.06
June 7	S.	+0.13
8	S.	+0.07
10	S.	—0.06
13	S.	+0.01
14	S.	+0.05
17	S.	+0.04
1865 June 1	S.	+0.27
3	S.	+0.06
Mean		+0.085
N.A.		57.693

Mean A.R. 1865.0 13h 2m 57s.778

43 Comæ.

Date.	Observer.	A.R. 1865.0.
		h. m. s.
1862 May 17	S.	13 5 34.46
1864 May 29	R.	34.26
June 8	S.	34.32
10	S.	34.15
13	S.	34.23
14	S.	34.28
17	S.	34.38
1865 April 13	S.	34.39
14	S.	34.30
17	S.	34.34
18	S.	34.27
24	M.	34.27
May 3	M.	34.26
June 1	S.	34.36

Mean A.R. 1865.0
13h 5m 34s.305

ι Centauri.

Date.	Observer.	A.R. 1865.0.
		h. m. s.
1862 May 17	S.	13 13 1.04
1864 May 29	R.	1.46
31	S.	1.19
June 10	S.	0.82
13	S.	1.15
14	S.	1.06
1865 May 3	M.	1.12
13	S.	1.12
15	M.	1.16
25	M.	1.15
30	S.	1.16

Mean A.R. 1865.0
13h 13m 1s.130

214 Camelopardali B.

Date.	Observer.	A.R. 1865.0.
		s.
1862 Mar. 26	S.	—1.16
April 16	H.	—0.74
25	S.	—0.62
29	S.	+0.20
May 14	H.	—1.07
17	S.	—0.19
23	S.	—0.31
Dec. 28	S.	—1.30
1863 April 9	S.	—0.14
May 1	S.	—0.13
9	S.	—0.05
June 4	S.	—0.34
6	S.	+0.16
1864 April 21	S.	—0.31
29	S.	—0.88
1864 May 3	S.	—0.79
4	S.	—0.60
5	S.	—0.85
18	S.	—0.53
27	S.	—0.29
29	R.	—0.60
31	S.	+0.33
June 2	R.	—0.35
8	S.	—0.58
10	S.	—0.21
13	S.	—0.76
1865 April 17	S.	—0.26
May 13	S.	—0.07
15	M.	+0.28
23	S.	—0.44
24	S.	—0.68
25	M.	—0.91
26	M.	—0.26
30	S.	—0.89
31	S.	+0.24
June 2	S.	—0.33
3	S.	—0.28
Mean		—0.425
T.H.S.		12.08

Mean A.R. 1865.0
13h 20m 11s.655

214 Camelop. B. Sub Polo.

Date.	Observer.	A.R. 1865.0.
		s.
1862 Dec. 10	S.	—0.62
24	S.	—1.27
1863 Jan. 2	S.	—0.51
3	S.	—0.63
Sept. 9	S.	—0.27
Oct. 28	S.	+0.02
29	S.	—0.33
Nov. 3	S.	—0.16
4	S.	—0.65
23	S.	+0.34
26	S.	+1.22
30	S.	+0.10
Dec. 9	S.	—0.61
15	S.	—0.50
21	S.	—0.77
23	S.	—0.44
26	S.	—0.27
29	S.	+0.13
30	S.	+0.04
1864 Jan. 1	S.	+0.99
6	S.	—1.59
9	S.	—0.92
10	S.	+0.57
11	S.	—0.22
1864 Jan. 12	S.	+0.12
14	S.	—0.29
16	S.	—0.44
21	S.	—0.42
Oct. 17	S.	—0.71
Nov. 11	S.	—0.26
14	S.	—0.60
16	S.	+0.68
25	M.	—0.15
30	M.	—0.69
Dec. 14	S.	—0.09
1865 Jan. 12	S.	—0.18
Nov. 1	S.	—0.34
Dec. 6	S.	+0.32
Mean		—0.247
T.H.S.		12.08

Mean A.R. 1865.0
13h 20m 11s.833

69 H. Ursæ.

Date.	Observer.	A.R. 1865.0.
		h. m. s.
1865 May 25	M.	13 23 29.43
26	M.	29.49
June 3	S.	29.61

Mean A.R. 1865.0
13h 23m 29s.510

h Virginis.

Date.	Observer.	A.R. 1865.0.
		h. m. s.
1865 May 23	S.	13 25 51.70
25	M.	51.64
26	S.	51.74
June 2	S.	51.67
3	S.	51.68
12	S.	51.61

Mean A.R. 1865.0
13h 25m 51s.673

78 o Virginis.

Date.	Observer.	A.R. 1865.0.
		h. m. s.
1865 May 23	S.	13 27 17.50
25	M.	17.50
26	M.	17.65
June 2	S.	17.61
3	S.	17.60
12	S.	17.51

Mean A.R. 1865.0
13h 27m 17s.562

ζ Virginis. N.A.

Date.	Observer.	A.R. 1865.0.
		s.
1862 April 16	H.	+0.06
29	S.	+0.03
May 14	H.	0.00
17	S.	—0.04
23	S.	+0.04
June 16	S.	+0.02
17	H.	+0.06
Dec. 28	S.	—0.14
1864 April 29	S.	+0.15
May 3	S.	+0.04
5	S.	+0.10
9	R.	0.00
28	R.	+0.08
31	S.	—0.03
June 10	S.	+0.13
13	S.	+0.02
16	S.	+0.08
17	S.	+0.07
23	R.	+0.03
July 15	S.	+0.13
Mean		+0.042
N.A.		48.972

Mean A.R. 1865.0
13h 27m 49s.014

17 H. Canum.

Date.	Observer.	A.R. 1865.0.
		h. m. s.
May 23	S.	13 28 46.02
25	M.	45.99
26	M.	46.05
June 2	S.	46.02
3	S.	46.05
12	S.	46.08
20	S.	46.11

Mean A.R. 1865.0
13h 28m 46s.046

25 Canum.

Date.	Observer.	A.R. 1865.0.
		h. m. s.
1865 May 23	S.	13 31 27.70
25	M.	27.66
26	M.	27.67
31	S.	27.72
June 2	S.	27.66
3	S.	27.69
5	S.	27.63
12	S.	27.63
19	M.	27.66
20	S.	27.72

Mean A.R. 1865.0
13h 31m 27s.674

1 H. Bootis.

Date.	Observer.	A.R. 1865.0.
		h. m. s.
1865 May 25	M.	13 22 55.21
26	M.	55.34
31	S.	55.33
3	S.	55.33
5	S.	55.29
12	S.	55.26
20	S.	55.28

Mean A.R. 1865.0
13h 32m 55s.291

i Centauri.

Date.	Observer.	A.R. 1865.0.
		h. m. s.
1865 May 25	M.	13 38 1.61
31	S.	1.53
June 2	S.	1.51
3	S.	1.56
5	S.	1.58
12	S.	1.48
13	S.	1.62

Mean A.R. 1865.0
13h 38m 1s.556

Gr. 2043.

Date.	Observer.	A.R. 1865.0.
		h. m. s.
1865 May 25	M.	13 40 28.13
31	S.	28.19
June 2	S.	28.18
3	S.	28.24
5	S.	28.15
12	S.	28.16
13	S.	28.17

Mean A.R. 1865.0
13h 40m 28s.174

21 H. Canum.

Date.	Observer.	A.R. 1865.0.
		h. m. s.
1865 May 23	S.	13 41 10.44
26	M.	10.40
31	S.	10.40
June 3	S.	10.35
5	S.	10.38
12	S.	10.35
19	M.	10.31

Mean A.R. 1865.0
13h 41m 10s.376

η Ursæ Majoris. B.J.

Date.	Observer.	A.R. 1865.0.
		s.
1862 May 7	S.	—0.02
14	H.	+0.13
17	S.	—0.07
23	S.	—0.28
June 16	S.	—0.09
17	H.	—0.01
1864 Jan. 8	S.	+0.06
April 29	S.	—0.01
May 3	S.	—0.10
9	R.	—0.12
19	S.	—0.15
27	S.	—0.06
28	R.	—0.18
29	R.	—0.02
June 7	R.	+0.04
8	S.	—0.04
16	S.	+0.03
23	R.	—0.10
29	M.	+0.01
1865 May 3	M.	+0.06
26	M.	+0.13
Mean		—0.038
B.J.		13.118

Mean A.R. 1865.0
13h 42m 13s.080

219 Camelopardali Bode.

Date.	Observer.	A.R. 1865.0.
		h. m s.
1863 June 2	S.	13 46 21.39
1864 Jan. 6	S.	21.16
8	S.	20.36
April 29	S.	20.57
May 3	S.	20.63
4	S.	20.94
9	R.	20.57
19	S.	20.66
27	S.	20.86
29	R.	20.48
June 7	R.	20.44
8	S.	20.97
16	S.	21.30

Mean A.R. 1865.0
13h 46m 20s.795

219 Camelop. B. Sub Polo.

Date.	Observer.	A.R. 1865.0.
		h. m. s.
1863 Oct. 28	S.	13 46 20.51
Nov. 4	S.	21.64
Dec. 21	S.	20.55
23	S.	21.41

Date.	Observer.	A.R. 1865.0.
		h. m. s.
1863 Dec. 29	S.	13 46 20.37
30	S.	21.19
1864 Jan. 21	S.	20.92
27	S.	20.79
Nov. 25	M.	20.72
1865 Jan. 12	S.	20.94

Mean A.R. 1865.0
13^h 46^m 20^s.904

i Draconis.

Date.	Observer.	A.R. 1865.0.
		h. m. s.
1865 May 25	M.	13 47 29.45
26	M.	29.09
June 2	S.	29.40
3	S.	29.36
5	S.	29.39
19	M.	29.38

Mean A.R. 1865.0
13^h 47^m 29^s.345

η Bootis. N.A.

Date.	Observer.	A.R. 1865.0.
		s.
1862 May 14	H.	+0.05
17	S.	—0.01
23	S.	0.00
28	H.	+0.09
June 16	S.	—0.02
17	H.	—0.01
1864 May 9	R.	+0.07
19	S.	+0.01
27	S.	+0.01
29	R.	—0.01
June 8	S.	+0.08
16	S.	+0.11
23	R.	—0.06
July 4	S.	+0.01
1865 May 3	M.	+0.13
15	M.	+0.09
Mean		+0.034
N.A.		15.381

Mean A.R. 1865.0
13^h 48^m 15^s.415

τ Virginis. N.A.

Date.	Observer.	A.R. 1865.0.
		s.
1862 May 7	S.	+0.08
14	H.	+0.12
17	S.	—0.01
23	S.	+0.12
28	H.	+0.14
June 16	S.	+0.06
17	H.	+0.07

Date.	Observer.	A.R. 1865.0.
		s.
1864 April 29	S.	+0.05
May 3	S.	+0.08
9	R.	+0.08
27	S.	+0.09
29	R.	0.00
June 2	R.	+0.18
7	R.	+0.15
8	S.	+0.15
18	M.	+0.16
23	S.	+0.12
1865 May 3	M.	+0.09
15	M.	+0.14
24	S.	+0.18
June 12	S.	+0.17
28	S.	—0.03
Mean		+0.100
N.A.		46.608

Mean A.R. 1865.0
13^h 54^m 46^s.708

θ Centauri. C des T.

Date.	Observer.	A.R. 1865.0.
		s.
1862 May 7	S.	+0.34
14	H.	+0.18
17	S.	+0.06
28	H.	(+0.53)
1864 April 29	S.	+0.16
May 3	S.	+0.28
9	R.	+0.16
18	R.	+0.28
19	S.	+0.23
27	S.	+0.35
1865 May 3	M.	+0.25
15	M.	+0.29
25	M.	+0.29
June 2	S.	+0.18
3	S.	+0.25
5	S.	+0.19
12	S.	+0.30
13	S.	+0.45
Mean		+0.249
C. des T.		44.69

Mean A.R. 1865.0
13^h 58^m 44^s.939

α Draconis. C. des T.

Date.	Observer.	A.R. 1865.0.
		s.
1862 May 14	H.	+0.21
17	S.	—0.04
28	H.	—0.10
June 16	S.	0.00
17	H.	+0.08

Date.	Observer.	A.R. 1865.0.
		s.
1863 May 23	S.	—0.12
June 2	S.	+0.32
1864 April 29	S.	+0.04
May 3	S.	+0.13
9	R.	+0.02
18	S.	+0.05
19	S.	—0.04
27	S.	+0.07
28	R.	—0.03
June 7	R.	—0.05
Mean		+0.036
C. des T.		44.02

Mean A.R. 1865.0
14^h 0^m 44^s.056

α Draconis. Sub Polo.

Date.	Observer.	A.R. 1865.0.
		s.
1863 Jan. 30		—0.11
Dec. 21		+0.43
22		+0.64
23		—0.18
30		+0.38
1864 Jan. 9		+0.01
21		+0.33
Nov. 1		—0.08
10		+0.08
11		—0.18
Mean		+0.132
C. des T.		44.02

Mean A.R. 1865.0
14^h 0^m 44^s.152

Gr. 2099 = 20 Ursæ Min. B.

Date.	Observer.	A.R. 1865.0.
		s.
1862 May 14	H.	—0.20
17	S.	—0.77
23	S.	—0.67
June 16	S.	—0.27
1864 May 29	R.	—0.44
31	S.	+0.43
June 7	R.	—1.23
16	S.	+0.72
21	R.	—0.50
1865 May 3	M.	+0.19
Mean		—0.274
T.H.S.		47.52

Mean A.R. 1865.0
14^h 3^m 47^s.246

Gr. 2099 = 20 Ursæ Min. B. Sub Polo.

Date.	Observer.	A.R. 1865.0.
		s.
1862 Dec. 8	S.	—0.46
16	S.	+0.24

Date.	Observer.	A.R. 1865.0.
		s.
1863 Sept. 9	S.	−1.03
Oct. 28	S.	−0.60
Nov. 3	S.	+0.52
4	S.	−1.20
30	S.	−0.37
Dec. 9	S.	−0.19
21	S.	+0.06
29	S.	+0.85
30	S.	+0.53
1864 Jan. 9	S.	+0.27
Dec. 20	M.	−1.09
30	M.	−0.72
Mean		−0.228
T.H.S.		47.52

Mean A.R. 1865.0
14ʰ 3ᵐ 47ˢ.292

42 H. Virginis.

Date.	Observer.	A.R. 1865.0.
		h. m. s.
1865 May 26	M.	14 5 25.85
June 1	S.	26.11
3	S.	26.00
5	S.	26.04
12	S.	25.97
13	S.	25.94
19	M.	25.89
23	M.	25.95
28	S.	25.77

Mean A.R. 1865.0
14ʰ 5ᵐ 25ˢ.947

λ Virginis.

Date.	Observer.	A.R. 1865.0.
		h. m. s.
1862 May 14	H.	14 11 48.57
17	S.	48.43
23	S.	48.62
28	H.	48.62
June 16	S.	48.48
1863 May 23	S.	48.49
1864 April 29	S.	48.59
May 3	S.	48.58
18	R.	48.56
19	S.	48.69
22	R.	48.66
27	S.	48.60
28	R.	48.72
29	R.	48.83
31	S.	48.54
June 2	R.	48.72
7	R.	48.55
16	S.	48.63
18	M.	48.61
21	S.	48.58
23	S.	48.60

Mean A.R. 1865.0
14ʰ 11ᵐ 48ˢ.603

A. Bootis.

Date.	Observer.	A.R. 1865.0.
		h. m. s.
1865 May 25	M.	14 12 17.35
26	M.	17.34
June 1	S.	17.42
3	S.	17.40
5	S.	17.36
12	S.	17.39
19	M.	17.37
20	S.	17.42
21	M.	17.40
23	M.	17.55
24	M.	17.34
27	M.	17.43
28	S.	17.34
July 3	M.	17.35

Mean A.R. 1865.0
14ʰ 12ᵐ 17ˢ.390

18 H. Bootis.

Date.	Observer.	A.R. 1865.0.
		h. m. s.
1865 May 15	M.	14 16 44.41
26	M.	44.40
June 1	S.	44.60
3	S.	44.57
12	S.	44.45
19	M.	44.43
20	S.	44.45
21	M.	44.57
27	M.	44.55
27	M.	44.45
28	S.	44.40
29	S.	44.55

Mean A.R. 1865.0
14ʰ 16ᵐ 44ˢ.478

f Bootis.

Date.	Observer.	A.R. 1865.0.
		h. m. s.
1865 May 25	M.	14 20 10.62
26	M.	10.63
June 1	S.	10.75
5	S.	10.74
13	S.	10.68
19	M.	10.69
20	S.	10.65
24	M.	10.53
27	M.	10.68
28	S.	10.60
29	S.	10.73

Mean A.R. 1865.0
14ʰ 20ᵐ 10ˢ.664

θ Bootis.

Date.	Observer.	A.R. 1865.0.
		h. m. s.
1862 May 17	S.	14 20 36.16
23	S.	36.11
28	H.	36.21
June 16	S.	35.96
17	H.	36.22
1863 May 23	S.	36.29
1864 April 21	S.	36.07
May 3	S.	36.10
19	S.	36.07
22	R.	35.93
28	R.	35.99
29	R.	35.98
31	S.	36.08
June 2	S.	36.04
6	S.	36.06
7	R.	36.12
8	S.	36.06
17	S.	36.15
1864 June 21	S.	36.04
22	R.	35.94
23	S.	36.08
24	S.	35.97
27	M.	36.15
28	M.	36.28
29	M.	36.20
July 4	S.	36.11
8	S.	36.21

Mean A.R. 1865.0
14ʰ 20ᵐ 36ˢ.096

ϱ Bootis. N.A.

Date.	Observer.	A.R. 1865.0.
		s.
1862 May 14	H.	+0.03
June 11	H.	+0.05
16	S.	−0.01
17	H.	−0.04
1864 April 21	S.	−0.02
May 19	S.	−0.03
28	R.	−0.10
June 8	S.	+0.01
16	S.	+0.06
17	S.	+0.09
18	M.	−0.09
21	S.	−0.09
22	R.	−0.12
23	S.	−0.04
24	S.	−0.11
29	S.	−0.04
30	S.	−0.01
July 4	S.	+0.03
5	S.	−0.03

Date.	Observer.	A.R. 1865.0.
		s.
1864 July 8	S.	+0.01
9	S.	—0.09
1865 May 3	S.	—0.10
Mean		—0.034
N.A.		0.714

Mean A.R. 1865.0
14h 39m 0s.680

γ Bootis.

Date.	Observer.	A.R. 1865.0.
		h. m. s.
1865 May 25	M.	14 26 38.47
26	M.	38.45
June 1	S.	38.59
3	S.	38.51
20	S.	38.49
21	M.	38.50
23	M.	38.58
24	M.	38.61
27	M.	38.53
28	S.	38.43
29	S.	38.60
July 3	M.	38.52

Mean A.R. 1865.0
14h 26m 38s.523

σ Bootis.

Date.	Observer.	A.R. 1865.0.
		h. m. s.
1864 June 13	S.	14 28 48.01
16	S.	48.21
17	S.	48.12
21	S.	48.20
24	S.	48.02
28	M.	48.26
29	S.	48.09
July 4	S.	48.12
5	S.	48.15
8	S.	48.23
1865 May 3	M.	48.18
15	M.	48.18
25	M.	48.07
30	S.	48.16
June 1	S.	48.17
3	S.	48.14
5	S.	48.16
13	S.	48.13
19	M.	48.12
20	S.	48.15
23	M.	48.10
27	M.	48.13
July 3	M.	48.18

Mean A.R. 1865.0
14h 28m 48s.143

π Bootis.

Date.	Observer.	A.R. 1865.0.
		h. m. s.
1865 May 25	M.	14 34 22.95
26	M.	22.99
June 1	S.	23.10
5	S.	23.15
13	S.	23.13
19	M.	22.90
20	S.	23.02
21	M.	22.92
23	M.	23.09
24	S.	23.04
27	M.	22.98
28	S.	22.97
29	S.	23.14
July 3	M.	22.94

Mean A.R. 1865.0
14h 34m 23s.023

μ Virginis.

Date.	Observer.	A.R. 1865.0.
		h. m. s.
1862 May 14	H.	14 35 56.92
17	S.	56.76
23	S.	56.97
28	H.	56.92
June 11	H.	56.84
16	S.	57.06
17	H.	56.98
1864 April 21	S.	56.94
29	S.	56.96
May 3	S.	56.88
5	S.	56.91
18	R.	56.94
19	S.	56.93
1864 May 27	S.	56.94
28	R.	56.98
29	R.	56.94
31	S.	56.99
June 2	S.	57.03
7	R.	56.90
8	S.	56.96
13	S.	56.88
14	R.	56.97
17	S.	56.97
18	M.	56.89
21	S.	57.03
22	R.	56.97
27	M.	57.02
28	M.	57.08
29	S.	56.88
30	S.	56.92
July 4	S.	56.90
5	S.	56.95
8	S.	56.92

Mean A.R. 1865.0
14h 35m 56s.943

ε Bootis. N.A.

Date.	Observer.	A.R. 1865.0.
		s.
1862 May 23	S.	+0.13
June 11	H.	+0.09
16	S.	+0.10
17	H.	+0.10
1863 June 3	S.	+0.05
1864 April 29	S.	+0.06
May 5	S.	+0.11
10	S.	—0.03
18	S.	+0.13
19	S.	0.00
27	S.	+0.11
28	R.	+0.11
29	R.	+0.03
June 4	S.	+0.12
8	S.	+0.09
13	S.	+0.05
18	M.	0.00
21	S.	+0.11
22	R.	—0.07
24	S.	—0.02
28	S.	+0.13
29	S.	+0.10
30	S.	+0.05
July 4	S.	+0.10
8	S.	+0.08
9	S.	+0.09
1865 May 30	S.	+0.04
Mean		+0.069
N.A.		5.406

Mean A.R. 1865.0
14h 39m 5s.475

109 Virginis.

Date.	Observer.	A.R. 1865.0.
		h. m. s.
1865 May 25	M.	14 39 25.59
26	M.	25.59
June 1	S.	25.58
13	S.	25.62
19	M.	25.51
21	M.	25.58
23	M.	25.55
24	S.	25.62
27	M.	25.61
28	S.	25.52
July 3	M.	25.52

Mean A.R. 1865.0
14h 39m 25s.572

ξ Bootis.

Date	Observer.	A.R. 1865.0.
		h. m. s.
1865 May 26	M.	14 45 9.81
June 2	S.	9.80
13	S.	9.91
19	M.	9.86
20	S.	10.11
21	M.	9.87
24	S.	9.79
27	M.	9.84
28	S.	9.81
29	S.	9.82
July 3	M.	9.80

Mean A.R. 1865.0 **$14^h\ 45^m\ 9^s.856$**

Piazzi XIV. 217 = Gr. 2164.

Date	Observer.	A.R. 1865.0.
		h. m. s.
1864 April 21	S.	14 48 1.07
29	S.	0.81
May 3	S.	1.04
10	S.	0.94
18	S.	0.97
27	S.	0.89
31	S.	0.95
June 2	S.	0.97
4	S.	1.02
6	S.	0.98
7	R.	0.84
13	R.	0.83
14	S.	1.01
18	M.	0.91
22	R.	0.94
27	M.	0.76
28	S.	1.17

Mean A.R. 1865.0 **$14^h\ 48^m\ 0^s.947$**

Piazzi XIV. 217 = Gr. 2164. Sub Polo.

Date	Observer.	A.R. 1865.0.
		h. m. s.
1863 Dec. 15	S.	14 48 0.86
21	S.	0.88
22	S.	0.96
29	S.	0.95
1864 Jan. 9	S.	1.03
25	S.	1.04
Nov. 19	M.	1.01
23	M.	1.10
25	M.	0.96
30	M.	0.87
Dec. 20	M.	0.84

Mean A.R. 1865.0 **$14^h\ 48^m\ 0^s.955$**

β Ursæ Minoris. B.J.

Date	Observer.	A.R. 1865.0.
		s.
1862 May 14	H.	+0.04
17	S.	—0.24
23	S.	+0.08
June 11	H.	+0.04
17	H.	+0.19
Dec. 10	S.	—0.09
1864 April 21	S.	+0.27
May 3	S.	+0.10
10	S.	+0.18
18	S.	+0.11
22	R.	(—0.75)
29	R.	—0.01
June 6	S.	+0.36
7	R.	—0.06
13	S.	+0.05
14	R.	+0.03
18	M.	—0.03
22	R.	—0.22
24	M.	—0.12
28	S.	+0.09
July 3	R.	—0.06
5	S.	—0.02
8	S.	+0.34
12	S.	+0.19
Mean		+0.053
B.J.		7.937

Mean A.R. 1865.0 **$14^h\ 51^m\ 7^s.990$**

β Ursæ Minoris. Sub Polo.

Date	Observer.	A.R. 1865.0.
		s.
1863 Feb. 2	S.	—0.14
16	S.	—0.05
23	S.	—0.22
Nov. 23	S.	+0.55
Dec. 15	S.	+0.28
21	S.	+0.16
22	S.	+0.26
29	S.	+0.04
1864 Jan. 9	S.	+0.14
25	S.	+0.05
Feb. 10	S.	+0.19
Dec. 20	M.	—0.19
1865 Feb. 21	S.	+0.24
24	M.	+0.32
27	M.	+0.15
Mean		+0.119
B.J.		7.937

Mean A.R. 1865.0 **$14^h\ 51^m\ 8^s.056$**

40 Bootis.

Date	Observer.	A.R. 1865.0.
		h. m. s.
1865 May 26	M.	14 54 26.33
June 19	M.	26.47
21	M.	26.42
24	S.	26.41
27	M.	26.36
28	S.	26.37
29	S.	26.51
July 3	M.	26.35
8	M.	26.33

Mean A.R. 1865.0 **$14^h\ 54^m\ 26^s.394$**

ω Bootis.

Date	Observer.	A.R. 1865.0.
		h. m. s.
1865 May 25	M.	14 56 (11.35)
26	M.	11.69
June 13	S.	11.77
19	S.	11.80
24	S.	11.75
27	M.	11.69
28	S.	11.74
29	S.	11.83
July 3	M.	11.82
8	M.	11.76

Mean A.R. 1865.0 **$14^h\ 56^m\ 11^s.761$**

20 Libræ = γ Scorpii.

Date	Observer.	A.R. 1865.0.
		h. m. s.
1862 May 14	H.	14 56 10.46
23	S.	10.58
June 11	H.	10.52
17	H.	10.55
1864 April 21	S.	10.54
29	S.	10.59
May 3	S.	10.51
10	S.	10.51
22	R.	10.66
27	S.	10.51
29	R.	10.64
June 7	R.	10.52
8	R.	10.52
18	M.	10.65
22	R.	10.64
24	M.	10.72
28	S.	10.41
July 3	R.	10.41
4	S.	10.57
5	S.	10.46
8	S.	10.39
15	S.	10.51

Mean A.R. 1865.0 **$14^h\ 56^m\ 10^s.538$**

ψ Bootis. N.A.

Date		Observer.	A.R. 1865.0.
1862 May	14	H.	s. —0.03
	23	S.	+0.06
June	11	H.	0.00
	17	H.	0.00
1864 April	21	S.	—0.04
	29	S.	—0.08
May	3	S.	—0.04
	10	S.	—0.02
	18	S.	—0.04
	22	R.	—0.23
	29	R.	—0.07
June	4	S.	—0.02
	8	R.	—0.12
	14	R.	—0.02
	17	S.	0.00
	18	M.	+0.03
	22	R.	—0.11
	24	M.	—0.05
	29	S.	—0.02
July	3	R.	—0.02
	4	S.	—0.05
	8	S.	—0.05
	9	S.	—0.05
	12	S.	+0.01
Mean			—0.040
N.A.			39.725

Mean A.R. 1865.0
14h 58m 39s.685

223 Camelopardali B = Gr. 2196.

Date		Observer.	A.R. 1865.0.
1864 June	28	S.	h. 14 m. 59 s. 40.11
July	9	S.	39.82
1865 May	25	M.	39.39
	26	M.	39.07
	30	S.	39.62
June	3	S.	40.23
	19	M.	39.71
	21	M.	39.45
	27	M.	39.81
July	3	M.	39.97

Mean A.R. 1865.0
14h 59m 39s.718

Sub Polo.

Date		Observer.	A.R. 1865.0.
1863 Dec.	21	S.	h. 14 m. 59 s. 40.56
	22	S.	39.43

Mean A.R. 1865.0
14h 59m 39s.995

c Bootis.

Date		Observer.	A.R. 1865.0.
1862 May	14	H.	h. 15 m. 1 s. 22.38
April	29	S.	22.36
May	3	S.	22.31
	10	S.	22.31
	18	S.	22.58
	10	S.	22.34
	28	R.	22.29
	29	R.	22.29
	31	S.	22.32
June	6	S.	22.28
	7	R.	22.38
	8	R.	22.28
	13	S.	22.24
	14	R.	22.35
	18	M.	22.32
	22	R.	22.28
	24	M.	22.43
	28	S.	22.26
	29	S.	22.38
July	3	R.	22.35
	4	S.	22.38
	5	S.	22.33
	8	S.	22.38
	12	S.	22.45
	14	S.	22.25

Mean A.R. 1865.0
15 1m 22s.331

β Libræ. N.A.

Date		Observer.	A.R. 1865.0.
1862 May	14	H.	s. 0.00
	23	S.	+0.22
June	11	H.	+0.09
	16	S.	+0.12
1864 April	21	S.	+0.14
	29	S.	+0.18
May	3	S.	+0.09
	10	S.	+0.19
	19	S.	+0.16
	22	R.	+0.15
	27	S.	+0.05
	29	R.	+0.14
	31	S.	—0.01
June	2	S.	+0.30
	4	S.	+0.17
	6	S.	+0.08
	8	R.	+0.05
	13	S.	+0.06
	14	R.	+0.15
	17	S.	+0.22
	18	M.	+0.12
	22	R.	+0.09
1864 June	24	M.	s. +0.24
	27	R.	+0.15
	28	S.	+0.03
	29	S.	+0.02
	30	R.	+0.14
July	3	R.	+0.11
	5	S.	+0.07
	8	S.	+0.12
	9	S.	+0.12
	12	S.	+0.12
	14	S.	+0.01
	15	S.	+0.12
1865 June	12	S.	+0.07
Mean			+0.117
N.A.			44.665

Mean A.R. 1865.0
15h 9m 44s.782

δ Bootis.

Date		Observer.	A.R. 1865.0.
1865 June	21	M.	h. 15 m. 10 s. 3.63
	24	S.	3.61
	27	M.	3.66
	28	S.	3.53
July	3	M.	3.64
	8	M.	3.56
	14	M.	3.63

Mean A.R. 1865.0
15h 10m 3s.609

5 Serpentis.

Date		Observer.	A.R. 1865.0.
1865 June	21	M.	h. 15 m. 12 s. 25.44
	24	S.	25.43
	27	M.	25.46
	28	S.	25.42
July	3	S.	25.40
	11	S.	25.45
	14	M.	25.49
	15	M.	25.51

Mean A.R. 1865.0
15h 12m 25s.450

η Coronæ.

Date		Observer.	A.R. 1865.0.
1865 June	24	S.	h. 15 m. 17 s. 37.75
	27	M.	37.64
	29	S.	37.82
July	3	M.	37.79
	8	M.	37.72
	11	S.	37.60

Date.	Observer.	A.R. 1865.0.
		h. m. s.
1865 July 13	S.	15 17 37.61
14	M.	37.66
15	M.	37.68
17	M.	37.65
18	M.	37.80
Mean A.R. 1865.0		**15h 17m 37s.702**

μ Bootis.

Date.	Observer.	A.R. 1865.0.
		h. m. s.
1862 May 14	H.	15 19 23.50
June 16	S.	23.65
1864 April 21	S.	23.61
29	S.	23.54
May 10	S.	23.46
19	S.	23.57
22	R.	23.37
31	S.	23.43
June 2	S.	23.40
4	S.	23.41
6	S.	23.43
17	S.	23.41
22	R.	23.34
24	M.	23.56
28	S.	23.56
30	R.	23.29
July 6	R.	23.42
12	S.	23.52
14	S.	23.49
1865 June 12	S.	23.49
24	S.	23.50
27	M.	23.49
28	S.	23.50
Mean A.R. 1865.0		**15h 19m 23s.476**

γ Ursæ Minoris. C. des T.

Date.	Observer.	A.R. 1865.0.
		s.
1864 May 22	R.	(—0.73)
28	R.	—0.33
29	R.	—0.43
June 22	R.	—0.24
24	M.	—0.33
27	R.	—0.07
July 3	R.	—0.10
4	S.	—0.03
6	R.	—0.12
20	S.	—0.04
1865 June 12	S.	—0.03
Mean		—0.172
C. des T.		58.16
Mean A.R. 1865.0		**15h 20m 57s.988**

γ Ursæ Minoris. Sub Polo.

Date.	Observer.	A.R. 1865.0.
		s.
1864 Feb. 10	S.	+0.03
12	S.	—0.01
Nov. 23	S.	+0.02
25	M.	—0.05
Dec. 20	M.	—0.21
1865 Jan. 27	M.	—0.60
Mean		—0.137
C. des T.		58.16
Mean A.R. 1865.0		**15h 20m 58s.023**

57 Ursæ Minoris B.

Date.	Observer.	A.R. 1865.0.
		s.
1862 May 23	S.	—3.13
June 16	S.	—1.50
1863 May 25	S.	—0.71
1864 April 21	S.	+0.70
June 4	S.	—1.49
7	S.	—0.54
13	S.	—0.64
14	R.	+1.05
16	S.	—0.34
17	S.	—0.84
21	S.	—1.44
27	R.	—0.04
28	S.	—0.57
29	S.	—0.87
July 3	R.	+0.75
4	S.	+0.32
5	S.	—2.29
8	S.	—1.12
15	S.	—1.07
Mean		—0.725
T.H.S.		5.61
Mean A.R. 1865.0		**15h 22m 4s.885**

57 Ursæ Min. B. Sub Polo.

Date.	Observer.	A.R. 1865.0.
		s.
1863 Nov. 23	S.	+0.82
Dec. 15	S.	—2.88
21	S.	—1.46
29	S.	—2.67
1864 Jan. 9	S.	—0.95
27	S.	+0.10
Feb. 5	S.	+0.31
Nov. 16	M.	—2.27
1865 Jan. 20	M.	—0.31
27	M.	—1.57
Feb. 14	S.	—0.49
Mean		—1.034
T.H.S.		5.61
Mean A.R. 1865.0		**15h 22m 4s.576**

37 Libræ.

Date.	Observer.	A.R. 1865.0.
		h. m s.
1862 June 17	H.	15 26 48.21
July 18	S.	48.05
1863 July 20	S.	48.12
1864 April 29	S.	48.22
May 19	S.	48.25
27	S.	48.21
31	S.	48.25
June 2	S.	48.26
6	S.	48.13
7	S.	48.28
13	S.	48.13
14	R.	48.33
16	S.	48.20
17	S.	48.17
21	S.	48.18
24	M.	48.26
27	R.	48.21
28	S.	48.12
29	S.	48.12
July 3	R.	48.25
4	S.	48.18
5	S.	48.12
6	R.	48.15
8	S.	48.14
9	S.	48.12
12	S.	48.22
13	R.	48.06
14	S.	48.20
Mean A.R. 1865.0		**15h 26m 48s.184**

θ Coronæ.

Date.	Observer.	A.R. 1865.0.
		h. m. s.
1865 June 21	M.	15 27 29.22
24	S.	29.21
29	S.	29.31
July 3	M.	29.20
8	M.	29.14
13	S.	29.13
14	M.	29.18
15	M.	29.25
17	M.	29.21
18	M.	29.24
Mean A.R. 1865.0		**15h 27m 29s.209**

μ Coronæ.

Date.	Observer.	h.	m.	s.
1865 June 24	S.	15	30	17.73
29	S.			17.78
July 11	S.			17.69
13	S.			17.65
15	M.			17.82
17	M.			17.70
18	M.			17.75
20	S.			17.68
21	M.			17.76
22	M.			17.56

Mean A.R. 1865.0 15ʰ 30ᵐ 17ˢ.712

φ Bootis.

Date.	Observer.	h.	m.	s.
1865 June 24	S.	15	32	58.83
27	M.			58.75
29	S.			58.92
July 3	M.			58.87
5	S.			58.99
10	M.			58.95
11	S.			58.78
13	S.			58.72
14	M.			58.85
15	M.			58.86
17	M.			58.62
18	M.			58.77
20	S.			58.81
22	M.			58.76

Mean A.R. 1865.0 15ʰ 32ᵐ 58ˢ.820

θ Ursæ Minoris.

Date.	Observer.	h.	m.	s.
1865 June 24	S.	15	35	29.27
28	S.			29.33
29	S.			29.34
July 3	M.			29.06
5	S.			29.17
8	M.			28.48
11	S.			29.16
13	S.			29.36
14	M.			29.76
15	M.			29.15
17	M.			28.82
20	M.			29.19

Mean A.R. 1865.0 15ʰ 35ᵐ 29ˢ.174

ι Serpentis.

Date.	Observer.	h.	m.	s.
1862 May 14	H.	15	35	32.00
June 11	H.			32.04
16	S.			31.98
17	H.			32.04
July 3	H.			32.01
1863 May 25	S.			32.10
July 20	S.			31.99
1864 May 19	S.			32.00
22	R.			31.90
27	S.			31.99
June 4	S.			32.06
7	S.			32.02
8	R.			31.99
13	R.			31.95
14	S.			32.03
17	S.			32.00
18	R.			32.03
21	S.			31.98
22	R.			32.00
24	M.			32.00
27	R.			31.96
28	S.			31.97
29	S.			31.97
30	R.			32.02
July 3	R.			32.11
6	R.			32.01
13	S.			31.94

Mean A.R. 1865.0 15ʰ 35ᵐ 32ˢ.003

β Serpentis.

Date.	Observer.	h.	m.	s.
1865 June 21	M.	15	39	57.52
27	M.			57.52
28	S.			57.57
29	S.			57.59
July 3	M.			57.55
5	S.			57.56
8	M.			57.52
10	M.			57,59
11	S.			57.59
11	S.			57.36
14	M.			57.59
15	M.			57.52
17	M.			57.51
18	M.			57.57
22	M.			57.42
26	M.			57.54
27	M.			57.48

Mean A.R. 1865.0 15ʰ 39ᵐ 57ˢ.529

λ Serpentis.

Date.	Observer.	h.	m.	s.
1862 May 14	H.	15	39	53.51
23	S.			53.73
June 11	H.			53.64
16	S.			53.61
17	H.			53.74
July 3	H.			53.73
18	S.			53.65
22	S.			53.64
1863 Feb. 20	S.			53.72
1864 May 19	S.			53.62
22	R.			53.59
27	S.			53.63
June 2	S.			53.68
4	S.			53.71
6	S.			53.56
8	R.			53.61
13	R.			53.56
14	S.			53.69
16	S.			53.61
17	S.			53.70
21	S.			53.66
22	R.			53.58
24	M.			53.67
28	S.			53.60
29	S.			53.51
July 3	R.			53.76
6	R.			53.54
20	S.			53.53
22	S.			53.58

Mean A.R. 1865.0 15ʰ 39ᵐ 53ˢ.633

λ Lupi. [χ Lupi.]

Date.	Observer.	h.	m.	s.
1865 June 27	M.	15	42	23.24
29	S.			23.34
July 3	M.			23.41
5	S.			23.47
8	M.			23.37
13	S.			23.27
14	M.			23.43
15	M.			23.32
17	M.			23.46
18	M.			23.49
22	M.			23.43

Mean A.R. 1865.0 15ʰ 42ᵐ 23ˢ.385

μ Serpentis.

Date.	Observer.	A.R. 1865.0.
		h. m. s.
1862 May 14	H.	15 42 34.58
23	S.	34.74
July 18	S.	34.67
22	S.	34.70
1863 Feb. 20	S.	34.73
July 22	S.	34.74
1864 May 19	S.	34.68
22	R.	34.67
June 4	S.	34.69
6	S.	34.64
8	R.	34.67
13	R.	34.73
16	S.	34.60
17	S.	34.65
21	S.	34.74
22	R.	34.68
24	M.	34.70
28	S.	34.64
29	S.	34.57
30	R.	34.73
July 3	R.	34.70
8	S.	34.63
9	S.	34.65
20	S.	34.67
22	S.	34.75
Aug. 1	S.	34.70
5	S.	34.50

Mean A.R. 1865.0
15h 42m 34s.672

δ Coronæ.

Date.	Observer.	A.R. 1865.0.
		h. m. s.
1865 June 27	M.	15 43 56.10
28	S.	56.01
29	S.	56.13
July 3	M.	56.16
5	S.	56.06
8	M.	56.04
11	S.	56.03
13	S.	55.99
14	M.	56.15
15	M.	56.06
17	M.	55.93
18	M.	56.05
22	M.	56.02
26	M.	56.01
27	M.	55.94

Mean A.R. 1865.0
15h 43m 56s.045

ε Serpentis. C. des T.

Date.	Observer.	A.R. 1865.0.
		s.
1862 May 14	H.	+0.01
23	S.	0.00
June 16	S.	—0.03
July 3	H.	+0.04
18	S.	—0.05
22	S.	—0.05
1863 Feb. 20	S.	+0.03
1864 May 19	S.	+0.13
22	R.	+0.08
27	S.	0.00
31	S.	—0.05
June 2	S.	—0.10
4	S.	—0.03
6	S.	—0.04
8	R.	—0.04
13	R.	—0.01
14	R.	+0.11
16	S.	+0.02
17	S.	+0.03
21	S.	—0.08
22	R.	—0.05
28	S.	—0.10
29	S.	—0.07
July 3	R.	+0.06
20	S.	—0.07
22	S.	+0.03
Mean		—0.009
C. des T.		5.38

Mean A.R. 1865.0
15h 44m 5s.371

λ Libræ.

Date.	Observer.	A.R. 1865.0.
		h. m. s.
1865 June 28	S.	15 45 30.10
29	S.	30.17
July 3	M.	30.30
5	S.	30.22
11	S.	30.30
13	S.	30.16
14	M.	30.27
15	M.	30.25
17	M.	30.31
18	M.	30.21
22	M.	30.19
26	M.	30.25

Mean A.R. 1865.0
15h 45m 30s.228

ζ Ursæ Minoris. N.A.

Date.	Observer.	A.R. 1865.0.
		s.
1863 Feb. 20	S.	—0.27
July 20	S.	+0.55
22	S.	—0.48
1864 April 21	S.	+0.04
May 19	S.	—0.29
27	S.	—0.24
31	S.	—0.13
June 2	S.	—0.02
8	R.	—0.29
10	S.	+0.26
18	R.	—0.44
22	R.	—0.26
30	R.	—0.46
July 3	R.	—0.08
4	S.	+0.15
20	S.	—0.29
22	S.	+0.07
26	S.	+0.03
Mean		—0.119
N.A.		57.045

Mean A.R. 1865.0
15h 48m 56s.926

ζ Ursæ Minoris. Sub Polo.

Date.	Observer.	A.R. 1865.0.
		s.
1863 Jan. 19	S.	+0.20
Feb. 14	S.	—0.43
16	S.	—0.22
Nov. 23	S.	+0.07
Dec. 15	S.	+0.04
21	S.	—0.23
26	S.	—0.07
29	S.	—0.20
30	S.	—0.01
1864 Jan. 11	S.	—0.32
12	S.	—0.21
21	S.	+0.16
27	S.	+0.05
Nov. 25	M.	—0.12
1865 Jan. 20	M.	—0.23
25	M.	—0.13
27	M.	—0.31
Feb. 14	S.	—0.11
Mean		—0.115
N.A.		57.045

Mean A.R. 1865.0
15h 48m 56s.930

γ Serpentis.

Date.	Observer.	A.R. 1865.0.
		h. m. s.
1862 May 23	S.	15 50 13.22
June 16	S.	13.03
1864 May 22	R.	13.28
June 4	S.	13.26
8	R.	13.14
13	R.	13.09
14	S.	13.07
16	S.	13.20
17	S.	13.14
18	R.	13.19
21	S.	13.14
22	R.	13.10
28	S.	13.11
29	S.	13.09
30	R.	13.16
July 3	R.	13.20
8	S.	13.10
9	S.	13.17
13	S.	13.21
14	S.	13.12
22	S.	13.26
26	S.	13.16
Aug. 1	S.	13.11

Mean A.R. 1865.0
$15^h\ 50\ 13^s.154$

π Scorpii.

Date.	Observer.	A.R. 1865.0.
		h. m. s.
1865 June 27	M.	15 50 41.46
28	S.	41.39
29	S.	41.48
July 3	M.	41.49
5	S.	41.41
8	M.	41.49
11	S.	41.52
13	S.	41.39
14	M.	41.49
15	M.	41.51
17	M.	41.56
18	M.	41.60
22	M.	41.46
26	M.	41.53
27	M.	41.52

Mean A.R. 1865.0
$15^h\ 50^m\ 41^s.487$

ε Coronæ. A.E.

Date.	Observer.	A.R. 1865.0.
		s.
1865 June 28	S.	—.10
29	S.	+.06
July 3	M.	+.06
1865 July 5	S.	—.09
8	M.	—.04
11	S.	—.08
13	S.	—.03
14	M.	—.01
15	M.	+.01
17	M.	—.12
18	M.	—.03
22	M.	—.05
26	M.	—.06
27	M.	—.15
31	M.	—.00
Mean		—0.042
A.E.		60.018

Mean A.R. 1865.0
$15^h\ 51^m\ 59^s.976$

δ Scorpii.

Date.	Observer.	A.R. 1865.0.
		h. m. s.
1862 June 16	S.	15 52 21.36
1864 May 19	S.	21.36
27	S.	21.37
31	S.	21.41
June 8	R.	21.38
10	S.	21.32
13	S.	21.31
14	S.	21.42
16	S.	21.47
17	S.	21.39
28	S.	21.44
July 4	S.	21.36
5	S.	21.35
8	S.	21.26
9	S.	21.32
12	S.	21.39
13	S.	21.32
14	S.	21.36
16	R.	21.39
20	S.	21.30
22	S.	21.39
26	S.	21.32

Mean A.R. 1865.0
$15^h\ 52^m\ 21^s.363$

β Scorpii. N.A.

Date.	Observer.	A.R. 1865.0.
		s.
1862 June 16	S.	+0.11
July 18	S.	+0.12
1864 May 22	R.	+(0.36)
27	S.	+0.14
June 8	R.	+0.02
14	S.	+0.06
1864 June 16	S.	+0.13
17	S.	+0.04
18	R.	+0.08
21	S.	+0.06
22	R.	+0.21
28	S.	+0.02
29	S.	+0.03
30	R.	+0.10
July 4	S.	+0.15
13	S.	+0.10
14	S.	+0.07
22	S.	+0.18
Aug. 5	S.	+0.14
Mean		+0.098
N.A.		35.417

Mean A.R. 1865.0
$15^h\ 57^m\ 35^s.515$

62 Ursæ Minoris B.

Date.	Observer.	A.R. 1865.0.
		s.
1862 May 23	S.	+0.68
1863 May 25	S.	+0.38
1864 April 21	S.	+0.52
May 19	S.	+0.41
27	S.	+1.21
31	S.	+0.85
June 2	S.	+0.50
4	S.	+0.98
7	S.	+0.49
10	S.	+0.40
July 8	S.	+0.34
12	S.	+0.52
1865 June 29	S.	+0.37
July 13	S.	+1.26
Mean		+0.636
T.H.S.		40.04

Mean A.R. 1865.0
$15^h\ 57^m\ 40^s.676$

62 Ursæ Min. B. Sub Polo.

Date.	Observer.	A.R. 1865.0.
		s.
1862 Dec. 11	S.	+0.93
1863 Jan. 5	S.	+0.11
19	S.	+0.73
Feb. 16	S.	+0.86
Nov. 23	S.	+0.42
Dec. 15	S.	+0.57
21	S.	+0.57
26	S.	+0.38
29	S.	+0.30
30	S.	+0.46

Date.	Observer.	A.R. 1865.0.
		s.
1864 Jan. 11	S.	+1.75
21	S.	+1.26
27	S.	+0.97
1865 Jan. 18	M.	+0.96
25	M.	—0.15
30	M.	+0.17
Mean		+0.643
T.H.S.		40.04

Mean A.R. 1865.0
15ʰ 57ᵐ 40ˢ.683

θ Draconis.

Date.	Observer.	A.R. 1865.0.
		h. m. s.
1862 July 18	S.	15 59 22.00
1864 June 6	S.	21.97
13	R.	21.79
14	S.	21.76
16	S.	21.96
17	S.	21.80
18	R.	21.79
28	S.	21.95
30	S.	21.89
July 4	S.	22.02
9	S.	21.85
14	S.	21.72
22	S.	22.01

Mean A.R. 1865.0
15ʰ 59ᵐ 21ˢ.885

θ Draconis. Sub Polo.

Date.	Observer.	A.R. 1865.0.
		h. m. s.
1863 Jan. 19	S.	15 59 21.77
1864 Feb. 5	S.	21.73
1865 Feb. 1	S.	21.95
14	S.	22.00

Mean A.R. 1865.0
15ʰ 59ᵐ 21ˢ.862

ν Scorpii.

Date.	Observer.	A.R. 1865.0.
		h. m. s.
1865 July 3	M.	16 4 9.29
5	S.	9.06
11	S.	9.19
13	S.	9.24
14	M.	9.25
15	M.	9.31
17	M.	9.31
18	M.	9.29
22	M.	9.43
27	M.	9.21
29	S.	9.05
Aug. 2	M.	8.99

Mean A.R. 1865.0
16ʰ 4ᵐ 9ˢ.218

87 Draconis B. = Gr. 2320.

Date.	Observer.	A.R. 1865.0.
		s.
1865 June 27	M.	+0.15
July 3	M.	+0.36
5	S.	+0.21
8	M.	+0.11
11	S.	+0.20
13	S.	+0.18
14	M.	+0.40
15	M.	+0.34
17	M.	+0.01
18	M.	+0.02
22	M.	+0.14
29	S.	+0.31
Mean		+0.202
A.E.		57.81

Mean A.R. 1865.0
16ʰ 5ᵐ 58ˢ.012

δ Ophiuchi. N.A.

Date.	Observer.	A.R. 1865.0.
		s.
1862 May 23	S.	+0.27
July 18	S.	+0.09
22	S.	+0.11
Sept. 16	S.	+0.02
1863 Feb. 20	S.	+0.19
July 11	S.	+0.07
1864 May 22	R.	+(0.32)
June 7	S.	+0.18
8	R.	+0.12
14	S.	+0.09
16	S.	—0.03
17	S.	+0.20
21	S.	+0.18
22	R.	+0.11
24	M.	+0.15
July 3	R.	+0.09
4	S.	+0.04
8	S.	+0.09
9	S.	+0.07
12	R.	+0.09
13	S.	+0.14
14	S.	+0.11
22	S.	+0.21
26	S.	+0.08
Aug. 5	S.	—0.02
Mean		+0.110
N.A.		16.306

Mean A.R. 1865.0
16ʰ 7ᵐ 16ˢ.416

σ Coronæ.

Date.	Observer.	A.R. 1865.0.
		h. m. s.
1865 June 27	M.	16 9 37.45
July 3	M.	37.44
8	M.	37.45
11	S.	37.41
13	S.	37.41
14	M.	37.50
15	M.	37.40
18	M.	37.43
22	M.	37.24
26	M.	37.38
29	S.	37.43

Mean A.R. 1865.0
16ʰ 9ᵐ 37ˢ.413

ε Ophiuchi.

Date.	Observer.	A.R. 1865.0.
		h. m. s.
1862 May 23	S.	16 11 10.96
July 18	S.	10.78
22	S.	10.84
Aug. 1	S.	10.72
Sept. 16	S.	10.70
1863 Feb. 20	S.	10.90
July 11	S.	10.81
22	S.	10.80
1864 May 22	R.	10.96
27	S.	10.86
31	S.	10.71
June 7	S.	10.81
8	R.	10.91
13	R.	10.91
14	S.	10.84
16	S.	10.89
17	S.	10.89
21	S.	10.98
22	R.	10.87
27	M.	10.78
30	R.	10.87
July 3	R.	10.89
4	S.	10.88
8	S.	10.91
9	S.	10.81
12	R.	10.82
13	S.	10.80
14	S.	10.83
16	R.	10.76
20	S.	10.82

Mean A.R. 1865.0
16ʰ 11ᵐ 10ˢ.844

σ Scorpii.

Date.	Observer.	A.R. 1865.0.
		h. m. s.
1865 June 27	M.	16 12 59.20
July 3	M.	59.41
5	S.	59.25
8	M.	59.29
11	S.	59.31
13	S.	59.24
14	M.	59.30
15	M.	59.32
17	M.	59.37
18	M.	59.37
22	M.	59.35
26	M.	59.37
29	S.	59.28
31	M.	59.29
Aug. 1	S.	59.20
2	M.	59.37

Mean A.R. 1865.0
$16^h\ 12^m\ 59^s.308$

19 Ursæ Minoris.

Date.	Observer.	A.R. 1865.0.
		s.
1865 June 27	M.	+0.24
July 3	M.	+0.37
5	S.	+0.39
8	M.	—0.02
11	S.	+0.22
13	S.	+0.43
14	M.	+0.29
15	M.	+0.12
17	M.	—0.09
18	M.	+0.16
22	M.	+0.03
Mean		+0.195
v. Asten		42.64

Mean A.R. 1865.0
$16^h\ 44^m\ 42^s.835$

γ Herculis.

Date.	Observer.	A.R. 1865.0.
		h. m. s.
1862 July 18	S.	16 15 57.97
22	S.	58.00
Aug. 1	S.	57.90
Sept. 16	S.	57.99
1863 June 27	S.	58.03
1864 May 22	R.	58.06
27	S.	57.96
31	S.	57.92
June 8	R.	57.95
13	R.	57.98
		h. m. s.
1864 June 14	S.	16 15 57.90
16	S.	58.01
17	S.	57.90
21	S.	58.03
24	M.	57.99
30	R.	57.98
July 3	R.	58.00
4	S.	57.96
8	S.	58.04
12	R.	57.97
15	S.	57.96
16	R.	58.05

Mean A.R. 1865.0
$16^h\ 15^m\ 57^s.980$

ξ Coronæ.

Date.	Observer.	A.R. 1865.0.
		h. m. s.
1865 June 27	M.	16 16 50.25
July 3	M.	50.34
5	S.	50.28
11	S.	50.30
13	S.	50.35
14	M.	50.36
15	M.	50.31
17	M.	50.29
18	M.	50.31
22	M.	50.34
26	M.	50.31
28	M.	50.36
29	S.	50.40
31	M.	50.35
Aug. 1	S.	50.20
2	M.	50.36

Mean A.R. 1865.0
$16^h\ 16^m\ 50^s.319$

η Draconis. N.A.

Date.	Observer.	A.R. 1865.0.
		s.
1864 May 31	S.	—0.60
June 7	S.	—0.62
8	R.	—0.80
10	S.	—0.92
13	S.	—0.69
14	S.	—0.58
16	S.	—0.63
17	S.	—0.58
21	S.	—0.59
24	S.	—0.87
30	S.	—0.59
July 3	R.	—0.75
8	S.	—0.58
12	R.	—0.95
		s.
1864 July 13	S.	—0.58
14	S.	—0.70
19	R.	—0.66
26	S.	—0.50
27	S.	—0.51
Aug. 5	S.	—0.72
8	S.	—0.66
Mean		—0.670
N.A.		10.733

Mean A.R. 1865.0
$16^h\ 22^m\ 10^s.063$

η Draconis. Sub Polo.

Date.	Observer.	A.R. 1865.0.
		s.
1863 Jan. 19	S.	—0.29
Dec. 15	S.	—0.49
29	S.	—0.78
1864 Jan. 11	S.	—0.87
Feb. 18	S.	—0.73
1865 Jan. 12	S.	—0.51
20	M.	—0.66
24	M.	—0.60
25	M.	—0.51
26	S.	—0.56
27	M.	—0.56
30	M.	—0.68
Feb. 1	S.	—0.24
9	M.	—0.64
11	S.	—0.70
25	S.	—0.83
Mean		—0.603
N.A.		10.733

Mean A.R. 1865.0
$16^h\ 22^m\ 10^s.130$

λ Ophiuchi.

Date.	Observer.	A.R. 1865.0.
		h. m. s.
1865 June 27	M.	16 24 6.41
July 3	M.	6.49
5	S.	6.35
8	M.	6.36
11	S.	6.41
13	S.	6.43
14	M.	6.52
15	M.	6.33
17	M.	6.44
18	M.	6.46
20	S.	6.41
22	M.	6.43
26	M.	6.48
28	M.	6.39
29	S.	6.43
31	M.	6.47

Date.	Observer.	A.R. 1865.0.
		h. m. s.
1865 Aug. 1	S.	16 24 6.39
2	M.	6.43

Mean A.R. 1865.0
16h 24m 6s.424

β Herculis.

Date.	Observer.	A.R. 1865.0.
		h. m. s.
1862 July 18	S.	16 24 25.11
Aug. 1	S.	25.01
Sept. 16	S.	25.02
1863 June 27	S.	25.26
July 11	S.	25.04
22	S.	25.10
1864 May 31	S.	25.00
June 8	R.	25.07
10	S.	24.87
13	S.	25.05
14	S.	25.06
16	S.	25.10
17	S.	25.04
21	S.	25.16
24	S.	24.87
July 3	R.	25.15
4	S.	25.10
8	S.	25.06
12	R.	25.05
14	S.	25.08
16	R.	25.10
18	R.	25.18
19	R.	25.06
27	S.	25.12
Aug. 8	S.	25.00

Mean A.R. 1865.0
16h 24m 25s.066

τ Scorpii.

Date.	Observer.	A.R. 1865.0.
		h. m. s.
1862 July 18	S.	16 27 28.98
Aug. 1	S.	28.81
1863 June 27	S.	29.18
July 22	S.	29.20
1864 May 31	S.	29.09
June 10	M.	28.97
13	S.	28.94
14	S.	29.03
16	S.	29.02
17	S.	29.08
21	S.	28.89
24	S.	29.04
July 3	R.	29.09
4	S.	28.98
1864 July 8	S.	16 27 28.95
12	R.	28.98
14	S.	29.21
18	R.	29.03
19	R.	29.06
27	S.	29.01
Aug. 1	S.	29.01

Mean A.R. 1865.0
16h 27m 29s.026

A. Draconis.

Date.	Observer.	A.R. 1865.0.
		s.
1865 July 5	S.	.00
13	S.	—.16
14	M.	—.20
15	M.	—.23
17	M.	—.42
18	M.	—.36
20	S.	—.15
22	M.	—.31
26	M.	—.37
28	M.	—.48
29	S.	+.08
Mean		—0.236
v. Asten		16.28 15.710

Mean A.R. 1865.0
16h 28m 15s.474

ζ Ophiuchi.

Date.	Observer.	A.R. 1865.0.
		h. m. s.
1862 July 18	S.	16 29 43.45
Aug. 1	S.	43.63
Sept. 16	S.	43.66
1863 June 27	S.	43.73
July 11	S.	43.68
1864 May 31	S.	43.62
June 8	R.	43.72
10	M.	43.52
13	S.	43.62
14	S.	(43.99)
16	S.	43.64
17	S.	43.67
21	S.	43.70
22	R.	43.79
24	S.	43.48
July 3	R	43.76
4	S.	43.72
8	S.	43.64
12	R.	43.65
14	S.	43.71
16	R.	43.63
18	R.	43.63
19	R.	43.73
		s.
1864 July 26	S.	16 29 43.65
27	S.	43.58
Aug. 1	S.	43.61
8	S.	43.58

Mean A.R. 1865.0
16h 29m 43s.646

17 Draconis.

Date.	Observer.	A.R. 1865.0.
		h. m. s.
1865 June 23	M.	16 33 2.72
27	M.	2.47
July 3	M.	2.73
5	S.	2.57
8	M.	2.56
11	S.	2.56
13	S.	2.66
14	M.	2.60
15	M.	2.54
17	M.	2.36
20	S.	2.48
22	M.	2.45

Mean A.R. 1865.0
16h 33m 2s.558

Piazzi XVI. 182 = Gr. 2372.

Date.	Observer.	A.R. 1865.0.
		h. m. s.
1863 June 27	S.	16 33 19.10
July 11	S.	19.18
22	S.	18.98
1864 May 31	S.	19.05
June 8	R.	18.76
10	M.	19.13
13	S.	19.19
14	S.	19.04
16	S.	19.42
21	S.	19.14
24	S.	19.05
July 3	R.	19.18
4	S.	19.35
8	S.	19.68
12	R.	19.38

Mean A.R. 1865.0
16h 33m 19s.175

Pi. XVI. 182. Sub Polo.

Date.	Observer.	A.R. 1865.0.
		h. m. s.
1863 Dec. 26	S.	16 33 18.98
29	S.	19.06
1864 Jan. 11	S.	19.26
Feb. 18	S.	18.89
1865 Jan. 26	S.	19.26

Date.	Observer.	A.R. 1865.0.
		h. m. s.
1865 Feb. 1	S.	16 33 19.18
2	S.	19.36
11	S.	19.31
25	S.	19.10

Mean A.R. 1865.0
16h 33m 19s.156

ζ Herculis. N.A.

Date.	Observer.	A.R. 1865.0.
		s.
1862 July 3	H.	+0.12
18	S.	+0.15
Sept. 16	S.	—0.01
1863 July 11	S.	—0.01
1864 May 22	R.	+0.07
June 8	R.	—0.01
10	S.	—0.14
17	S.	—0.06
21	S.	+0.08
22	R.	—0.01
July 4	S.	+0.08
8	S.	+0.05
14	S.	+0.05
20	S.	+0.04
23	R.	+0.06
26	S.	+0.05
30	S.	+0.05
Aug. 8	S.	+0.02
Sept. 6	S.	—0.01
7	S.	—0.01
Mean		+0.028
N.A.		11.849

Mean A.R. 1865.0
16h 36m 11s.877

Pi. XVI. 195 = Gr. 2373.

Date.	Observer.	A.R. 1865.0.
		h. m. s.
1865 June 23	M.	16 36 29.75
27	M.	29.76
July 3	M.	30.14
5	S.	30.01
8	S.	29.66
11	M.	29.81
13	S.	29.90
14	M.	29.82
15	M.	29.82
17	M.	29.66
20	S.	29.96
22	M.	29.56

Mean A.R. 1865.0
16h 36m 29s.820

η Herculis.

Date.	Observer.	A.R. 1865.0.
		h. m. s.
1862 July 18	S.	16 38 16.21
Aug. 1	S.	16.27
Sept. 16	S.	16.02
1863 July 11	S.	16.13
22	S.	16.23
1864 May 22	R.	(16.39)
June 8	R.	16.11
14	S.	16.09
16	S.	16.11
17	S.	16.12
21	S.	16.11
22	R.	16.09
24	S.	15.90
July 3	R.	16.13
4	R.	16.20
10	R.	16.15
12	R.	16.11
13	S.	16.16
14	S.	16.14
18	R.	16.15
19	R.	16.04
23	R.	16.21
Aug. 8	S.	(15.89)

Mean A.R. 1865.0
16h 38m 16s.128

ε Scorpii. C. des T.

Date.	Observer.	A.R. 1865.0.
		s.
1862 July 18	S.	+0.33
1864 May 22	R.	+0.48
31	S.	+0.32
June 10	M.	+0.15
14	S.	+0.33
17	S.	+0.08
21	S.	+0.17
24	S.	+0.12
July 4	S.	+0.20
12	R.	+0.12
13	S.	+0.24
14	S.	+0.09
15	S.	+0.44
27	S.	+0.38
30	S.	+0.25
Aug. 1	S.	+0.13
1865 June 23	M.	+0.27
27	M.	—0.04
July 3	M.	+0.21
5	S.	+0.15
8	M.	+0.23
14	M.	+0.27
Mean		+0.224
C. des T.		25.36

Mean A.R. 1865.0
16h 41m 25s.584

ζ^1 Scorpii.

Date.	Observer.	A.R. 1865.0.
		h. m. s.
1862 July 18	S.	16 44 28.73
Aug. 6	S.	28.78
1864 June 10	M.	28.77
14	S.	28.86
21	S.	28.97
July 3	R.	28.96
4	S.	28.58
12	R.	28.73
15	S.	28.94
16	R.	28.96
1865 June 23	M.	28.77
27	M.	28.84
July 3	M.	28.92
5	S.	28.80
11	S.	28.97
14	M.	28.92
15	M.	29.00
17	M.	28.94
22	M.	29.03
26	M.	28.94
28	M.	(29.24)

Mean A.R. 1865.0
16h 44m 28s.870

ζ^2 Scorpii.

Date.	Observer.	A.R. 1865.0.
		h. m. s.
1862 July 18	S.	16 45 5.39
Aug. 6	S.	5.43
1864 May 22	R.	5.82
June 10	M.	5.55
14	S.	5.68
17	S.	5.57
21	S.	5.73
July 3	R.	5.42
4	S.	5.48
12	R.	5.43
13	S.	5.65
15	S.	5.57
16	R.	5.56
30	S.	5.75
1865 June 23	M.	5.56
27	M.	5.49
July 3	M.	5.59
5	S.	5.52
11	S.	5.60
14	M.	5.68

Mean A.R. 1865.0
16h 45m 5s.574

53 Herculis.

Date.	Observer.	A.R. 1865.0.
		h. m. s.
1865 June 23	M.	16 47 51.06
27	M.	51.01
July 3	M.	51.16
5	S.	51.00
8	M.	50.97
11	S.	51.00
14	M.	51.06
15	M.	51.08
17	M.	50.87
22	M.	51.01
26	M.	51.02
28	M.	50.97
29	M.	51.01
31	M.	50.97
Aug. 1	S.	50.89
2	M.	50.94

Mean A.R. 1865.0
16ʰ 47ᵐ 51ˢ.001

ϰ Ophiuchi. N.A.

Date.	Observer.	A.R. 1865.0.
		s.
1862 July 18	S.	+0.01
Aug. 6	S.	—0.04
Sept. 16	S.	—0.13
18	S.	+0.04
25	S.	+0.00
1863 July 11	S.	0.00
22	S.	+0.10
Aug. 24	S.	+0.04
1864 June 14	S.	+0.06
17	S.	—0.02
21	S.	+0.07
22	R.	+0.06
24	S.	—0.09
July 3	R.	+0.08
4	S.	—0.03
12	R.	+0.04
14	S.	+0.05
16	R.	—0.06
18	R.	+0.01
22	S.	—0.01
23	R.	+0.04
Mean		+0.011
N.A.		16.769

Mean A.R. 1865.0
16ʰ 51ᵐ 16ˢ.780

30 Ophiuchi.

Date.	Observer.	A.R. 1865.0.
		h. m. s.
1865 June 27	M.	16 53 56.58
July 3	M.	56.75
8	M.	56.57
11	S.	56.63
14	M.	16 53 56.72
15	M.	56.70
17	M.	56.73
20	S.	56.67
22	M.	56.70
26	M.	56.63
29	S.	56.62
31	M.	56.57
Aug. 1	S.	56 55
2	M.	56.63
12	M.	56.55

Mean A.R. 1865.0
16ʰ 53ᵐ 56ˢ.640

ε Herculis. C. des T.

Date.	Observer.	A.R. 1865.0.
		s.
1862 July 18	S.	+0.01
Aug. 1	S.	—0.09
6	S.	—0.05
11	S.	—0.04
21	S.	—0.06
Sept. 8	S.	+0.12
9	S.	+0.07
16	S.	+0.06
18	S.	—0.02
25	S.	—0.01
1863 July 11	S.	—0.03
1864 June 14	S.	—0.06
16	S.	—0.09
17	S.	—0.13
22	R.	—0.07
24	S.	—0.21
July 10	R.	—0.04
12	R.	—0.05
13	S.	+0.04
16	R.	—0.07
18	R.	+0.03
19	R.	—0.11
22	S.	—0.03
23	R.	—0.19
Aug. 8	S.	+0.02
Mean		—0.040
C. des T.		7.55

Mean A.R. 1865.0
16ʰ 55ᵐ 7ˢ.510

h Draconis.

Date.	Observer.	A.R. 1865.0.
		h. m. s.
1865 June 23	M.	16 55 17.69
27	M.	17.39
July 8	M.	17.40
11	S.	17.50
14	M.	17.66
15	M.	17.49
17	M.	17.36
20	S.	17.35
22	M.	16 55 17.36
29	S.	17.40

Mean A.R. 1865.0
16ʰ 55ᵐ 17ˢ.460

ε Ursæ Minoris.

Date.	Observer.	A.R. 1865.0.
		s.
1862 July 18	S.	+0.20
Aug. 21	S.	—0.31
Sept. 10	S.	+0.14
Oct. 18	S.	+0.19
1863 Aug. 10	S.	—0.20
24	S.	—0.11
1864 June 10	M.	—0.30
14	S.	+0.25
16	S.	+0.17
17	S.	—0.32
July 3	R.	+0.42
4	S.	+0.47
12	R.	—0.07
14	S.	+0.14
19	R.	—0.04
1865 June 27	M.	—0.40
July 3	M.	+0.17
5	S.	+0.42
8	M.	—0.16
14	M.	+0.24
18	M.	—0.15
20	S.	+0.42
27	M.	—0.22
Aug. 1	S.	—0.20
2	M.	+0.10
8	M.	—0.19
12	M.	—0.21
14	M.	—0.03
Mean		+0.015
T.H.S.		54.87

Mean A. R. 1865.0
16ʰ 59ᵐ 54ˢ.885

Sub Polo.

Date.	Observer.	A.R. 1865.0.
		s.
1862 Oct. 8	S.	—0.41
1863 Jan. 19	S.	+0.19
31	S.	—0.18
Mar. 2	S.	—0.71
9	S.	+0.32
Dec. 29	S.	—0.12
1865 Feb. 21	S.	+0.29
Mar. 6	S.	+0.16
Mean		—0.058
T.H.S.		54.87

Mean A.R. 1865.0
16ʰ 59ᵐ 54ˢ.812

μ Draconis.

Date.	Observer.	A.R. 1865.0.
		h. m. s.
1865 July 18	M.	17 2 32.38
20	S.	32.38
22	S.	32.35
26	M.	32.29
27	M.	32.38
29	S.	32.56
31	M.	32.38
Aug. 1	S.	32.33
2	M.	32.42

Mean A.R. 1865.0
17ʰ 2ᵐ 32ˢ.386

η Ophiuchi.

Date.	Observer.	A.R. 1865.0.
		h. m. s.
1862 Aug. 21	S.	17 2 38.32
Sept. 10	S.	38.30
11	S.	38.33
16	S.	38.27
17	S.	38.33
18	S.	38.29
25	S.	38.28
1863 July 11	S.	38.32
Aug. 3	S.	38.27
10	S.	38.41
1864 June 10	S.	38.26
14	S.	38.35
16	S.	38.32
17	S.	38.28
July 3	R.	38.32
4	S.	38.13
10	R.	38.29
12	R.	38.25
13	S.	38.35
14	S.	38.38
15	S.	38.19
16	R.	38.21
18	R.	38.38
19	R.	38.20
30	S.	38.34
Sept. 16	S.	38.34

Mean A.R. 1865.0
17ʰ 2ᵐ 38ˢ.297

77 Ursæ Minoris B. = Gr. 2427.

Date.	Observer.	A.R. 1865.0.
		h. m. s.
1865 July 18	M.	17 5 56.01
20	S.	56.49
22	S.	56.39
26	M.	17 5 55.98
27	M.	55.87
31	M.	56.19

Mean A.R. 1865.0
17ʰ 5ᵐ 56ˢ.155

π Herculis.

Date.	Observer.	A.R. 1865.0.
		h. m. s.
1862 Aug. 6	S.	17 10 20.70
11	S.	20.77
21	S.	20.69
Sept. 9	S.	20.76
10	S.	20.83
11	S.	20.81
15	S.	20.72
16	S.	20.73
17	S.	20.84
18	S.	20.80
23	S.	20.78
25	S.	20.83
Oct. 6	S.	20.63
1864 June 10	S.	20.62
14	S.	20.79
16	S.	20.76
21	S.	20.70
22	R.	20.58
July 3	R.	20.83
4	S.	20.84
12	R.	20.79
16	R.	20.68
19	R.	20.70
30	S.	20.80

Mean A.R. 1865.0
17ʰ 10ᵐ 20ˢ.749

ξ Ophiuchi.

Date.	Observer.	A.R. 1865.0.
		h. m. s.
1862 July 22	S.	17 12 54.96
Aug. 6	S.	54.76
11	S.	54.85
21	S.	54.88
1863 Aug. 3	S.	54.88
1864 June 16	S.	54.94
21	S.	55.02
July 4	S.	55.00
13	S.	54.97
14	S.	54.88
15	S.	55.00
16	R.	54.88
18	R.	55.00
20	S.	54.97
28	S.	55.07
1864 Aug. 1	S.	17 12 54.94
1865 June 23	M.	54.90
July 5	S.	54.95
14	M.	55.04
15	M.	55.03
17	M.	55.02
18	M.	55.03
21	M.	55.05

Mean A.R. 1865.0
17ʰ 12ᵐ 54ˢ.957

θ Ophiuchi. N.A.

Date.	Observer.	A.R. 1865.0.
		s.
1862 Aug. 6	S.	−0.09
Sept. 11	S.	+0.14
15	S.	+0.14
16	S.	−0.04
17	S.	+0.20
18	S.	+0.15
23	S.	+0.11
1863 Aug. 10	S.	+0.21
1864 June 10	S.	+0.15
16	S.	+0.10
21	S.	+0.25
July 3	R.	+0.22
4	S.	+0.13
12	R.	+0.11
13	S.	+0.08
14	S.	+0.08
15	S.	+0.15
16	R.	+0.16
18	R.	+0.19
20	S.	+0.14
22	S.	+0.12
28	S.	+0.17
Mean		+'.130
N.A.		43.176

Mean A.R. 1865.0
17ʰ 13ᵐ 43ˢ.306

271 Herculis B. = Gr. 2435.
B.A.C. 5874.

Date.	Observer.	A.R. 1865.0.
		h. m. s.
1865 July 17	M.	17 17 17.89
18	M.	17.92
20	S.	17.94
24	M.	18.06
26	M.	17.87
27	M.	17.85
29	S.	17.88

Date.	Observer.	A.R. 1865.0.
		h. m. s.
1865 July 31	M.	17 17 17.70
Aug. 2	M.	17.96
4	M.	17.89

Mean A.R. 1865.0
17h 17m 17s.896

ϱ Herculis.

Date.	Observer.	A.R. 1865.0.
		h. m. s.
1865 July 17	M.	17 19 1.47
18	M.	1.50
20	S.	1.54
22	S.	1.48
24	M.	1.51
26	M.	1.48
27	M.	1.49
29	S.	1.57
31	M.	1.56
Aug. 1	S.	1.54
2	M.	1.62
28	M.	1.60

Mean A.R. 1865.0
17h 19m 1s.530

λ Herculis.

Date.	Observer.	A.R. 1865.0.
		h. m. s.
1865 July 17	M.	17 25 17.01
20	S.	17.03
22	S.	17.06
24	M.	16.96
26	M.	17.12
27	M.	16.91
28	M.	17.05
29	S.	17.00
31	M.	17.08
Aug. 1	S.	17.00
2	M.	17.10
28	M.	17.06

Mean A.R. 1865.0
17h 25m 17s.032

β Draconis. N.A.

Date.	Observer.	A.R. 1865.0.
		s.
1862 Aug. 6	S.	—0.14
11	S.	+0.12
21	S.	—0.05
Sept. 2	S.	+0.25
11	S.	+0.10
15	S.	+0.10
16	S.	—0.08
18	S.	+0.12
23	S.	+0.14
1862 Sept. 25	S.	+0.02
29	S.	+0.08
Oct. 6	S.	+0.02
7	S.	+0.08
16	S.	0.00
18	S.	+0.06
1863 June 17	S.	0.00
July 31	S.	+0.06
Aug. 22	S.	+0.11
24	S.	—0.04
1864 June 14	S.	+0.09
21	S.	+0.01
22	R.	—0.01
July 3	R.	+0.15
12	R.	—0.02
18	R.	+0.18
23	R.	+0.20
Sept. 16	S.	+0.11
17	S.	+0.07
Oct. 7	S.	+0.21
1865 June 23	M.	+0.12
Mean		+0.069
N.A.		22.945

Mean A.R. 1865.0
17h 27m 23s.014

Gr. 2456 = 50 Heis Ursæ Minoris.

Date.	Observer.	A.R. 1865.0.
		h. m. s.
1865 July 17	M.	17 29 52.81
18	M.	52.53
20	S.	52.91
22	S.	53.07
24	M.	53.05
26	M.	52.84
27	M.	52.82
28	M.	52.58
29	S.	52.84
31	M.	52.73

Mean A.R. 1865.0
17h 29m 52s.818

ι Herculis.

Date.	Observer.	A.R. 1865.0.
		h. m. s.
1865 July 17	M.	17 35 39.08
18	M.	39.22
22	S.	39.30
26	M.	39.28
27	M.	39.23
28	M.	39.35
29	S.	39.23
31	M.	39.32
1865 Aug. 4	M.	17 35 39.29
12	M.	39.13
28	M.	39.39

Mean A.R. 1865.0
17h 35m 39s.256

β Ophiuchi. C. des T.

Date.	Observer.	A.R. 1865.0.
		s.
1862 July 26	S.	+0.10
Sept. 10	S.	—0.10
18	S.	—0.02
23	S.	—0.03
25	S.	+0.03
29	S.	+0.01
Oct. 18	S.	—0.07
1863 Aug. 10	S.	—0.16
15	S.	+0.05
22	S.	—0.10
24	S.	—0.10
Sept. 2	S.	—0.02
1864 June 14	S.	0.00
16	S.	+0.04
21	S.	+0.10
22	R.	+0.01
July 3	R.	—0.02
12	R.	—0.12
13	S.	—0.04
16	R.	—0.16
18	R.	+0.03
19	R.	—0.01
22	S.	+0.07
28	S.	—0.04
Sept. 16	S.	—0.05
Mean		—0.024
C. des T.		48.30

Mean A.R. 1865.0
17h 36m 48s.276

ω Draconis. A.E.

Date.	Observer.	A.R. 1865.0.
		s.
1865 July 17	M.	—0.32
18	M.	—0.32
22	S.	—0.10
24	M.	—0.17
26	M.	—0.14
28	M.	—0.15
29	S.	—0.25
31	M.	—0.23
Aug. 28	M.	—0.02
Sept. 6	S.	0.00
Mean		—0.170
A.E.		44.68

Mean A.R. 1865.0
17h 37m 44s.510

γ Ophiuchi.

Date.	Observer.	A.R. 1865.0.
		h. m. s.
1865 July 17	M.	17 41 7.58
22	S.	7.60
26	M.	7.58
28	M.	7.58
29	S.	7.51
31	M.	7.52
Aug. 4	M.	7.39
12	M.	7.38
14	M.	7.40
15	M.	7.49
19	S.	7.46
25	M.	7.51

Mean A.R. 1865.0
17h 41m 7s.500

μ Herculis. N.A.

Date.	Observer.	A.R. 1865.0.
		s.
1862 July 26	S.	+0.14
Sept. 25	S.	+0.03
29	S.	+0.08
Oct. 18	S.	+0.03
1863 Aug. 3	S.	+0.10
10	S.	+0.12
12	S.	+0.06
24	S.	—0.02
1864 June 14	S.	+0.13
21	S.	+0.06
22	R.	+0.11
July 13	S.	+0.01
16	R.	+0.06
18	R.	+0.12
23	R.	+0.10
28	S.	+0.12
Aug. 8	S.	+0.03
Sept. 16	S.	+0.06
17	S.	0.00
22	S.	+0.09
Oct. 7	S.	+0.05
Mean		+0.070
N.A.		10.507

Mean A.R. 1865.0
17h 41m 10s.577

ψ Draconis.

Date.	Observer.	A.R. 1865.0.
		h. m. s.
1863 Aug. 3	S.	17 44 20.91
10	S.	20.62
12	S.	20.64
18	S.	20.54
22	S.	20.66
24	S.	20.59
1863 Sept. 2	S.	17 44 20.42
1864 June 14	S.	20.82
21	S.	20.95
July 3	R.	20.91
12	R.	20.50
13	S.	20.68
16	R.	20.40
18	R.	20.78
19	R.	20.59
23	R.	21.12
28	S.	20.94
Aug. 8	S.	20.75
Sept. 16	S.	20.64
17	S.	20.57

Mean A.R. 1865.0
17h 44m 20s.702

Sub Polo.

Date.	Observer.	A.R. 1865.0.
		h m. s.
1863 Feb. 2		17 44 20.75
20		21.19
23		20.76
1864 Jan. 11		20.56
1865 Feb. 2		20.62
9		20.71
11		21.08
13		21.33
22		20.97
24		20.83

Mean A.R. 1865.0
17h 44m 20s.880

f Herculis.

Date.	Observer.	A.R. 1865.0.
		h. m. s.
1865 July 22	S.	17 48 54.45
Aug. 4	M.	54.50
12	M.	54.55
14	M.	54.44
15	M.	54.33
19	S.	54.38

Mean A.R. 1865.0
17h 48m 54s.442

ξ Draconis.

Date.	Observer.	A.R. 1865.0.
		h. m. s.
1862 July 26	S.	17 51 11.63
Sept. 23	S.	11.69
29	S.	11.47
1863 Aug. 3	S.	11.75
10	S.	11.69
12	S.	11.54
1863 Aug. 14	S.	17 51 12.02
15	S.	11.73
18	S.	11.57
22	S.	11.66
1864 June 14	S.	11.72
21	S.	11.58
22	R.	11.50
July 3	R.	11.87
16	R.	11.62
19	R.	11.58
Sept. 17	S.	11.66

Mean A.R. 1865.0
17h 51m 11s.664

Sub Polo.

Date.	Observer.	A.R. 1865.0.
		h. m. s.
1862 Oct. 5	S.	17 51 11.79
1863 Feb. 20	S.	11.83
23	S.	11.76
1864 Jan. 11	S.	11.71
25	S.	11.88
Feb. 18	S.	11.67

Mean A.R. 1865.0
17h 51m 11s.773

θ Herculis.

Date.	Observer.	A.R. 1865.0.
		h. m. s.
1865 July 17	M.	17 51 37.27
22	S.	37.47
24	M.	37.47
28	M.	37.47
29	S.	37.51
31	M.	37.46
Aug. 2	M.	37.48
4	M.	37.45
8	M.	37.48
12	M.	37.45
14	M.	37.45
15	M.	37.41

Mean A.R. 1865.0
17h 51m 37s.447

γ Draconis. B.J.

Date.	Observer.	A.R. 1865.0.
		s.
1862 July 26	S.	—0.03
Sept. 11	S.	+0.01
23	S.	+0.03
29	S.	—0.09
Oct. 7	S.	—0.04
16	S.	—0.26
18	S.	—0.10

Date.	Observer.	A.R. 1865.0.
		s.
1863 Aug. 3	S.	0.00
12	S.	+0.01
15	S.	0.00
18	S.	—0.21
Sept. 4	S.	(—0.61)
28	S.	—0.09
1864 June 14	S.	—0.10
22	R.	—0.10
July 3	R.	+0.08
12	R.	—0.24
13	S.	—0.06
16	R.	—0.12
18	R.	—0.13
19	R.	—0.17
23	R.	—0.03
30	S.	—0.12
Aug. 8	S.	—0.02
Sept. 16	S.	—0.05
17	S.	—0.21
22	S.	—0.06
Oct. 7	S.	—0.11
Mean		—0.082
B.J.		28.434

Mean A.R. 1865.0
17h 53m 28s.352

35 Draconis.

Date.	Observer.	A.R. 1865.0.
		h. m. s.
1865 July 22	S.	17 55 29.67
24	M.	29.58
28	M.	29.61
31	M.	29.60
Aug. 2	M.	29.71
29	S.	29.85
31	S.	29.95
Sept. 5	S.	30.02
6	S.	29.73

Mean A.R. 1865.0
17h 55m 29s.747

γ Sagittarii. A.E.

Date.	Observer.	A.R. 1865.0.
		s.
1865 July 22	S.	+0.29
24	M.	+0.24
28	M.	+0.10
29	S.	+0.01
31	M.	+0.11
Aug. 2	M.	+0.09
4	M.	—0.03
8	M.	+0.05
12	M.	—0.11
14	M.	+0.05
Sept. 5	S.	+0.06
6	S.	+0.10

Date.	Observer.	A.R. 1865.0.
		s.
Mean		+0.080
A.E.		8.20

Mean A.R. 1865.0
17h 57m 8s.280

70 Ophiuchi. Preceding.

Date.	Observer.	A.R. 1865.0.
		h. m. s.
1862 July 26	S.	17 58 37.87
Aug. 21	S.	37.78
Sept. 8	S.	37.85
11	S.	37.91
23	S.	37.81
1863 July 24	S.	37.80
Aug. 3	S.	37.90
12	S.	37.91
15	S.	38.02
18	S.	37.88
22	S.	37.85
24	S.	37.84
1864 June 14	S.	37.96
21	S.	38.09
July 3	R.	38.05
12	R.	37.88
13	S.	37.91
14	S.	37.86
15	S.	37.92
16	R.	37.90
18	R.	37.86
19	R.	37.82
23	R.	37.76
Aug. 8	S.	37.95
Sept. 10	S.	37.96
22	S.	37.94

Mean A.R. 1865.0
17h 58m 37s.895

b Herculis.

Date.	Observer.	A.R. 1865.0.
		h. m. s.
1865 July 17	M.	18 1 54.04
22	S.	54.03
24	M.	54.11
28	M.	54.17
29	S.	54.08
31	M.	54.11
Aug. 2	M.	54.13
4	M.	54.09
8	M.	54.08
12	M.	54.09

Mean A.R. 1865.0
18h 1m 54s.093

Gr. 2517 = B.A.C. 6162.

Date.	Observer.	A.R. 1865.0.
		h. m. s.
1865 July 17	M.	18 3 24.75
22	S.	24.84
24	M.	24.89
28	M.	24.77
31	M.	24.93
Aug. 2	M.	24.87
4	M.	24.82
8	M.	24.81
12	M.	24.70
14	M.	24.77

Mean A.R. 1865.0
18h 3m 24s.815

μ Sagittarii. N.A.

Date.	Observer.	A.R. 1865.0.
		s.
1862 July 26	S.	+0.15
Aug. 6	S.	+0.11
21	S.	+0.10
Sept. 23	S.	+0.08
25	S.	+0.23
1863 July 24	S.	+0.09
Aug. 3	S.	+0.10
12	S.	+0.12
14	S.	+0.25
18	S.	+0.19
1864 June 14	S.	+0.28
21	S.	+0.25
23	S.	+0.23
July 3	R.	+0.21
12	R.	+0.12
13	S.	+0.18
14	S.	+0.15
15	S.	+0.11
16	R.	+0.23
18	R.	+0.24
20	S.	+0.17
28	S.	+0.22
30	S.	+0.14
Sept. 10	S.	+0.18
Mean		+0.172
N.A.		41.304

Mean A.R. 1865.0
18h 5m 41s.476

A Herculis.

Date.	Observer.	A.R. 1865.0.
		h. m. s.
1865 July 17	M.	18 6 49.41
22	S.	49.48
24	M.	49.49
28	M.	49.39
31	M.	49.45

Date.	Observer.	A.R. 1865.0.
		h. m. s.
1865 Aug. 2	M.	18 6 49.45
4	M.	49.43
8	M.	49.42
12	M.	49.43
14	M.	49.47

Mean A.R. 1865.0
18h 6m 49s.442

η Sagittarii.

Date.	Observer.	A.R. 1865.0.
		h. m. s.
1865 July 17	M.	18 8 29.66
28	M.	29.62
31	M.	29.58
Aug. 2	M.	29.67
4	M.	29.58
8	M.	29.66
24	M.	29.82
25	M.	29.50
29	S.	29.59
31	S.	29.58

Mean A.R. 1865.0
18h 8m 29s.626

40 Draconis.

Date.	Observer.	A.R. 1865.0.
		h. m. s.
1864 July 3	R.	18 10 8.56
12	R.	8.16

Mean A.R. 1865.0
18h 10m 8s.360

Sub Polo.

Date.	Observer.	A.R. 1865.0.
		h. m. s.
1863 Feb. 23	S.	18 10 8.05
Dec. 29	S.	8.48

Mean A.R. 1865.0
18h 10m 8s.265

41 Draconis.

Date.	Observer.	A.R. 1865.0.
		h. m. s.
1863 July 24	S.	18 10 14.17
1864 June 14	S.	14.77
21	S.	14.05
23	S.	14.51
July 5	S.	14.37
13	S.	14.47
14	S.	14.64
18	R.	14.30
19	R.	14.29
1864 July 28	S.	18 10 14.76
30	S.	14.43
Aug. 8	S.	14.18

Mean A.R. 1865.0
18h 10m 14s.412

Sub Polo.

Date.	Observer.	A.R. 1865.0.
		h. m. s.
1864 Jan. 11	S.	18 10 14.48
16	S.	14.37
25	S.	14.40
1865 Feb. 11	S.	14.40
21	S.	14.56
24	M.	14.51
Mar. 6	M.	14.87
18	S.	14.65

Mean A.R. 1865.0
18m 10h 14s.530

δ Sagittarii.

Date.	Observer.	A.R. 1865.0.
		h. m. s.
1862 July 26	S.	18 12 21.15
Aug. 6	S.	21.12
Sept. 4	S.	21.12
8	S.	21.09
23	S.	20.98
1863 Aug. 3	S.	21.04
1864 June 23	S.	21.33
July 5	S.	21.14
12	R.	21.36
13	S.	21.15
18	R.	21.30
19	R.	21.14
26	S.	21.14
28	S.	21.29
30	S.	21.22
Sept. 9	S.	21.16
10	S.	21.19
16	S.	21.30
27	S.	21.16
Oct. 5	S.	21.22

Mean A.R. 1865.0
18h 12m 21s.180

36 Draconis.

Date.	Observer.	A.R. 1865.0.
		h. m. s.
1865 July 17	M.	18 13 6.92
22	S.	7.13
24	M.	7.17
28	M.	7.09
31	M.	7.31
1865 Aug. 26	S.	18 13 7.21
31	S.	7.17
Sept. 2	S.	7.24
4	M.	7.25
5	S.	7.23
6	M.	7.02

Mean A.R. 1865.0
18h 13m 7s.158

η Serpentis.

Date.	Observer.	A.R. 1865.0.
		h. m. s.
1862 Oct. 6	S.	18 14 19.50
Sept. 4	S.	19.66
29	S.	19.71
1864 July 3	R.	19.64
12	R.	19.46
18	R.	19.66
19	R.	19.55
20	S.	19.55
23	R.	19.51
26	S.	19.56
28	S.	19.58
30	S.	19.56
Aug. 1	S.	19.62
8	S.	19.57
Sept. 9	S.	19.68
10	S.	19.61
16	S.	19.63
22	S.	19.64
27	S.	19.56
Oct. 5	S.	19.62

Mean A.R. 1865.0
18h 14m 19s.594

δ Ursæ Minoris. B.J.

Date.	Observer.	A.R. 1865.0.
		s.
1862 Sept. 23	S.	—0.08
29	S.	+0.08
Oct. 18	S.	+0.46
1863 Aug. 19	S.	+0.03
24	S.	—0.27
Sept. 5	S.	+1.08
10	S.	—0.12
28	S.	—0.81
30	S.	—0.27
1864 June 14	S.	0.00
21	S.	—0.27
20	R.	+0.19
23	S.	—0.03
28	S.	+0.34

Date.	Observer.	A.R. 1865.0.
		s.
1864 July 3	R.	+1.77
5	S.	+0.66
12	R.	—0.20
13	S.	+0.30
14	S.	+0.27
15	S.	—0.02
16	R.	+0.73
18	R.	+0.02
19	R.	+0.30
20	S.	—0.22
23	R.	+0.47
26	S.	+0.20
1865 July 18	M.	—1.00
20	S.	—0.43
22	S.	—0.27
24	M.	—1.01
27	S.	+0.09
29	S.	—0.71
31	M.	—0.22
Aug. 2	M.	+0.55
12	M.	—0.20
14	M.	+0.84
23	M.	+0.27
24	M.	—0.04
25	M.	+0.25
Sept. 13	M.	+0.77
16	S.	—0.07
27	M.	—0.06
29	M.	—0.05
30	S.	+1.15
Mean		+0.102
B.J.		53.34

Mean A.R. 1865.0
18ʰ 15ᵐ 53ˢ.442

SUB POLO.

Date.	Observer.	A.R. 1865.0.
		s.
1862 Mar. 26	S.	—0.30
1863 Feb. 23	S.	+1.31
Mar. 2	S.	—0.15
Dec. 29	S.	—0.10
1864 Jan. 11	S.	+0.69
14	S.	+0.94
16	S.	+0.46
Feb. 23	S.	+0.79
Mar. 27	S.	+0.57
1865 Jan. 25	M.	+0.31
27	M.	+1.04
30	M.	+0.05
Feb. 2	S.	+0.42
24	M.	+0.05
Mar. 6	M.	—0.76
Mean		+0.355
B.J.		53.34

Mean A.R. 1865.0
18ʰ 15ᵐ 53ˢ.695

24 URSÆ MINORIS.

Date.	Observer.	A.R. 1865.0.
		h. m. s.
1865 July 17	M.	18 20 44.59
22	S.	45.12
24	M.	45.42
26	M.	44.36
27	M.	45.14
28	M.	44.78
Aug. 4	M.	44.75
21	S.	45.85
29	S.	44.93
31	S.	45.65

Mean A.R. 1865.0
18ʰ 20ᵐ 45ˢ.059

χ DRACONIS. A.N. 1578.

Date.	Observer.	A.R. 1865.0.
		s.
1865 Sept. 2	S.	+0.25
4	M.	—0.05
5	S.	+0.07
6	M.	—0.26
16	S.	—0.02
23	S.	+0.04
25	M.	+0.14
26	S.	+0.05
27	M.	+0.05
29	M.	+0.11
Mean		+0.038
A.N.		29.19

Mean A.R. 1865.0
18ʰ 23ᵐ 29ˢ.228

42 DRACONIS.

Date.	Observer.	A.R. 1865.0.
		h. m. s.
1865 July 22	S.	18 25 35.64
24	M.	35.60
26	M.	35.59
27	M.	35.71
28	M.	35.85
31	M.	35.72
Aug. 2	M.	35.69
26	S.	35.69
29	S.	35.80

Mean A.R. 1865.0
18ʰ 25ᵐ 35ˢ.699

3 H. SCUTI. A.E.

Date.	Observer.	A.R. 1865.0.
		s.
1865 July 17	M.	+0.20
22	S.	+0.23
26	M.	+0.20
27	S.	+0.13
28	M.	+0.22
31	M.	+0.10
Aug. 2	M.	+0.10
4	M.	+0.14
9	M.	+0.12
12	M.	+0.11
26	S.	+0.00
29	S.	+0.05
Sept. 2	S.	+0.11
5	S.	+0.11
6	M.	—0.01
16	S.	+0.05
21	S.	+0.21
Mean		+0.122
A.E.		51.583

Mean A.R. 1865.0
18ʰ 27ᵐ 51ˢ.705

45 DRACONIS.

Date.	Observer.	A.R. 1865.0.
		h. m. s.
1865 July 27	S.	18 30 14.65
31	M.	14.69
Aug. 2	M.	14.70
4	M.	14.78
9	M.	14.86
12	M.	14.56
26	S.	14.83
29	S.	14.79
31	S.	14.81
Sept. 2	S.	14.87

Mean A.R. 1865.0
18ʰ 30ᵐ 14ˢ.754

110 HERCULIS.

Date.	Observer.	A.R. 1865.0.
		h. m. s.
1862 Aug. 20	S.	18 39 51.16
29	S.	51.18
Sept. 25	S.	51.14
Oct. 8	S.	51.15
18	S.	51.25
1863 July 29	S.	51.15
31	S.	51.14
Aug. 17	S.	51.28
19	S.	51.21
Sept. 10	S.	51.14

Date.	Observer.	A.R. 1865.0.
		h. m. s.
1864 June 21	S.	18 39 51.30
July 3	R.	51.30
5	S.	51.32
9	S.	51.12
13	S.	51.20
14	S.	51.25
28	S.	51.24
Aug. 1	S.	51.20
8	S.	51.14
Sept. 2	S.	51.09
6	S.	51.22
9	S.	51.24
10	S.	51.27
26	S.	51.15

Mean A.R. 1865.0
18h 39m 51s.202

β Lyræ. N.A.

Date.	Observer.	A.R. 1865.0.
		s.
1862 July 26	S.	+0.11
Aug. 11	S.	+0.05
20	S.	+0.13
Sept. 4	S.	+0.08
Oct. 8	S.	+0.19
18	S.	+0.06
1863 July 31	S.	+0.11
Aug. 12	S.	—0.03
17	S.	+0.09
Sept. 2	S.	+0.12
23	S.	+0.07
1864 June 21	S.	+0.04
July 3	R.	+0.32
5	S.	+0.12
9	S.	+0.13
13	S.	—0.03
26	S.	+0.23
Sept. 6	S.	+0.22
Oct. 17	S.	+0.14
19	S.	+0.04
25	S.	+0.14
Mean		+0.111
N.A.		5.653

Mean A.R. 1865.0
18h 45m 5s.764

ν¹ Sagittarii.

Date.	Observer.	A.R. 1865.0.
		h. m. s.
1865 July 27	S.	18 46 1.26
28	M.	1.24
31	M.	1.09
1865 Aug. 2	M.	18 46 1.16
4	M.	1.24
9	M.	1.21
12	M.	1.12
14	M.	1.11
15	M.	1.22
19	S.	1.08
23	M.	1.20
24	M.	1.26
25	M.	1.16
28	M.	1.06

Mean A.R. 1865.0
18h 46m 1s.172

σ Sagittarii.

Date.	Observer.	A.R. 1865.0.
		h. m. s.
1862 July 26	S.	18 46 53.60
Aug. 11	S.	53.59
20	S.	53.67
Sept. 4	S.	53.67
8	S.	53.51
Oct. 8	S.	53.56
18	S.	53.62
1863 July 23	S.	53.66
31	S.	53.69
Aug. 15	S.	53.82
17	S.	53.77
Sept. 5	S.	53.66
10	S.	53.57
1864 June 14	R.	53.66
July 3	R.	53.70
9	S.	53.54
13	S.	53.64
14	S.	53.59
26	S.	53.65
28	S.	53.64
Aug. 1	S.	53.62
Sept 6	S.	53.72
26	S.	53.41
Oct. 5	S.	53.63
17	S.	53.69

Mean A.R. 1865.0
18h 46m 53s.635

ν² Sagittarii.

Date.	Observer.	A.R. 1865.0.
		h. m. s.
1865 July 17	M.	18 46 57.50
27	S.	57.52
28	M.	57.48
31	M.	57.45
Aug. 2	M.	57.48
4	M.	57.43
1864 Aug. 9	M.	18 46 57.45
12	M.	57.62
14	M.	57.53
15	M.	57.57
19	S.	57.46
23	M.	57.52
24	M.	57.58
25	M.	57.45
28	M.	57.58

Mean A.R. 1865.0
18h 46m 57s.508

ο Draconis.

Date.	Observer.	A.R. 1865.0.
		h. m. s.
1865 July 17	M.	18 49 12.37
24	M.	12.42
27	S.	12.55
28	M.	12.39
Aug. 2	M.	12.41
4	M.	12.49
9	M.	12.34
12	M.	12.50
26	S.	12.48
29	S.	12.60

Mean A.R. 1865.0
18h 49m 12s.455

θ Serpentis. Pr.

Date.	Observer.	A.R. 1865.0.
		h. m. s.
1862 July 26	S.	18 49 30.55
Aug. 6	S.	30.44
11	S.	30.49
20	S.	30.56
Sept. 4	S.	30.50
8	S.	30.56
16	S.	30.47
Oct. 8	S.	30.46
18	S.	30.47
1863 July 31	S.	30.60
Aug. 15	S.	30.65
19	S.	30.50
Sept. 5	S.	30.58
10	S.	30.47
15	S.	30.54
1864 July 3	R.	30.64
5	S.	30.50
9	S.	30.51
13	S.	30.49
14	S.	30.49
26	S.	30.55
Sept. 6	S.	30.65

Mean A.R. 1865.0
18h 49m 30s.530

Date.	Observer.	A.R. 1865.0.
50 Draconis.		
		h. m. s.
1863 July 23	S.	18 50 42.58
31	S.	42.60
Aug. 15	S.	42.88
19	S.	42.69
Sept. 5	S.	42.59
10	S.	42.64
1864 July 3	R.	42.94
5	S.	42.73
26	S.	42.99
28	S.	43.22
Sept. 6	S.	42.94
Oct. 5	S.	42.73
7	S.	42.94
Mean A.R. 1865.0		**18h 50m 42s.805**
Sub Polo.		
		h. m. s.
1863 Feb. 21	S.	18 50 42.69
Mar. 2	S.	42.56
5	S.	42.96
1864 Jan. 27	S.	42.81
Feb. 26	S.	43.32
Mar. 9	S.	42.75
21	S.	42.03
24	S.	42.72
1865 Feb. 25	S.	43.09
Mar. 4	S.	43.16
Mean A.R. 1865.0		**18h 50m 42s.809**
18 B. Aquilæ = 5 Heis Aquilæ.		
		h. m. s.
1865 July 24	M.	18 52 14.43
27	S.	14.44
28	M.	14.40
Aug. 2	M	14.28
9	M.	14.41
12	M.	14.34
14	M.	14.35
15	M.	14.37
23	M.	14.36
25	M.	14.44
26	S.	14.36
29	S.	14.42
Mean A.R. 1865.0		**18h 52m 14s.383**

Date.	Observer.	A.R. 1865.0.
ε Aquilæ.		
		h. m. s.
1862 July 26	S.	18 53 29.77
Aug. 6	S.	29.61
11	S.	29.80
Sept. 4	S.	29.70
8	S.	29.80
16	S.	29.68
Oct. 8	S.	29.69
18	S.	29.82
1863 July 23	S.	29.77
31	S.	29.84
Aug. 15	S.	29.90
19	S.	29.78
Sept. 5	S.	29.84
10	S.	29.90
1864 July 5	S.	29.72
9	S.	29.74
12	R.	29.74
13	S.	29.71
14	S.	29.78
26	S.	29.80
28	S.	29.79
Aug. 8	S.	29.69
Sept. 6	S.	29.78
Oct. 5	S.	29.72
Mean A.R. 1865.0		**18h 53m 29s.765**
γ Lyræ.		
		h. m. s.
1865 July 24	M.	18 53 53.67
28	M.	53.61
31	M.	53.64
Aug. 2	M.	53.66
4	M.	53.59
9	M.	53.70
12	M.	53.71
14	M.	53.64
15	M.	53.64
23	M.	53.58
24	M.	53.74
25	M.	53.64
Mean A.R. 1865.0		**18h 53m 53s.652**
ο Sagittarii.		
		h. m. s.
1862 July 26	S.	18 56 35.63
Aug. 6	S.	35.52
11	S.	35.58
20	S.	35.60

Date.	Observer.	A.R. 1865.0.
		h. m. s.
1862 Sept. 4	S.	18 56 35.55
16	S.	35.32
Oct. 8	S.	35.45
1864 July 13	S.	35.54
14	S.	35.53
20	S.	35.61
26	S.	35.62
28	S.	35.54
Sept. 2	S.	35.54
6	S.	35.41
10	S.	35.52
17	S.	35.50
26	S.	35.45
Oct. 5	S.	35.63
7	S.	35.64
Mean A.R. 1865.0		**18h 56m 35s.536**
16 Lyræ.		
		h. m. s.
1865 July 17	M.	18 57 37.15
24	M.	37.24
28	M.	37.15
31	M.	37.08
Aug. 2	M.	37.14
4	M.	37.17
9	M.	37.13
12	M.	37.26
14	M.	37.18
15	M.	37.28
23	M.	37.18
24	M.	37.25
25	M.	37.32
Mean A.R. 1865.0		**18h 57m 37s.195**
ζ Aquilæ. N.A.		
		s.
1862 July 26	S.	+0.25
Aug. 6	S.	+0.13
11	S.	+0.23
20	S.	+0.20
Sept. 4	S.	+0.21
16	S.	+0.20
Oct. 8	S.	+0.22
1863 July 23	S.	+0.32
31	S.	+0.28
Aug. 12	S.	+0.11
14	S.	+0.18
15	S.	+0.31
17	S.	+0.07
22	S.	+0.15
24	S.	+0.14

Date.	Observer.	A.R. 1865.0.
		s.
1863 Sept. 5	S.	+0.19
15	S.	+0.17
23	S.	+0.19
1864 July 3	R.	+0.31
5	S.	+0.15
9	S.	+0.26
13	S.	+0.16
14	S.	+0.23
15	S.	+0.31
26	S.	+0.21
Aug. 8	S.	+0.07
Sept. 2	S.	+0.20
6	S.	+0.17
9	S.	+0.25
17	S.	+0.17
26	S.	+0.13
27	S.	+0.22
Oct. 5	S.	+0.15
7	S.	+0.18
17	S.	+0.26
19	S.	+0.17
25	S.	+0.20
26	S.	+0.20
Nov. 3	S.	+0.28
16	S.	+0.39
19	S.	+0.17
Mean		+0.205
N.A.		12.150

Mean A.R. 1865.0 $18^h\ 59^m\ 12^s.355$

18 Aquilæ.

Date.	Observer.	A.R. 1865.0.
		h. m. s.
1864 July 17	M.	19 0 37.53
24	M.	37.34
28	M.	37.43
31	M.	37.36
Aug. 2	M.	37.37
9	M.	37.42
12	M.	37.44
14	M.	37.28
15	M.	37.30
23	M.	37.40
24	M.	37.43
25	M.	37.43

Mean A.R. 1865.0 $19^h\ 0^m\ 37^s.394$

π Sagittarii.

Date.	Observer.	A.R. 1865.0.
		h. m. s.
1863 Aug. 12	S.	19 1 44.04
14	S.	44.13
15	S.	44.32
1863 Sept. 23	S.	19 1 44.10
1864 July 3	R.	44.13
5	S.	44.09
9	S.	44.04
13	S.	44.13
14	S.	43.95
20	S.	44.17
26	S.	44.09
28	S.	44.06
Sept. 2	S.	44.11
6	S.	44.10
9	S.	44.12
10	S.	44.06
17	S.	44.23
Oct. 5	S.	44.14
7	S.	44.16
25	S.	44.16

Mean A.R. 1865.0 $19^h\ 1^m\ 44^s.116$

ι Lyræ

Date.	Observer.	A.R. 1865.0.
		h. m. s.
1865 July 17	M.	19 2 29.08
24	M.	29.17
28	M.	29.08
31	M.	29.04
Aug. 2	M.	29.08
4	M.	28.99
9	M.	29.09
12	M.	29.17
14	M.	29.04
15	M.	29.06
18	M.	29.16

Mean A.R. 1865.0 $19^h\ 2^m\ 29^s.087$

η Lyræ.

Date.	Observer.	A.R. 1865.0.
		h. m. s.
1865 July 17	M.	19 9 9.85
24	M.	9.88
28	M.	9.80
31	M.	9.89
Aug. 2	M.	9.79
4	M.	9.72
9	M.	9.82
12	M.	9.82
14	M.	9.85
15	M.	9.89
18	M.	10.03

Mean A.R. 1865.0 $19^h\ 9^m\ 9^s.849$

ω Aquilæ. N.A.

Date.	Observer.	A.R. 1865.0.
		s.
1862 July 26	S.	+0.07
Sept. 2	S.	+0.03
1864 July 5	S.	+0.05
9	S.	+0.10
14	S.	+0.19
20	S.	+0.25
26	S.	+0.12
Sept. 2	S.	+0.19
6	S.	+0.14
9	S.	+0.13
10	S.	+0.08
17	S.	+0.08
27	S.	+0.08
Oct. 5	S.	+0.06
7	S.	+0.14
17	S.	+0.12
19	S.	+0.06
21	S.	+0.10
Mean		+0.111
N.A.		28.723

Mean A.R. 1865.0 $19^h\ 11^m\ 28^s.834$

δ Draconis. A.E.

Date.	Observer.	A.R. 1865.0.
		s.
1865 July 17	M.	—0.12
Aug. 4	M.	—0.23
14	M.	—0.03
18	M.	—0.02
28	M.	+0.05
Sept. 4	M.	—0.07
5	S.	+0.14
6	M.	—0.17
Oct. 11	M.	+0.04
30	M.	+0.04
Mean		—0.037
A.E.		30.933

Mean A.R. 1865.0 $19^h\ 12^m\ 30^s.896$

β¹ Sagittarii. (Decl. —44°.7).

Date.	Observer.	A.R. 1865.0.
		h. m. s.
1862 Aug. 6	S.	19 12 55.58
1864 July 14	S.	56.02
Sept. 2	S.	55.43
10	S.	55.69
17	S.	55.93
Oct. 7	S.	56.00

Date.	Observer.	A.R. 1865.0.
		h. m. s.
1865 Aug. 2	M.	19 12 55.97
12	M.	55.42
15	M.	55.87
23	M.	55.59
24	M.	55.86
25	M.	55.87
26	S.	55.72

Mean A.R. 1865.0
19h 12m 55s.765

59 DRACONIS.

Date.	Observer.	A.R. 1865.0.
		h. m. s.
1865 July 17		19 14 5.00
24		5.47
Aug. 4		5.23
9		5.45
14		5.53
Sept. 4		5.46
5		5.66
6		5.22
7		5.47
11		5.42
12		5.33

Mean A.R. 1865.0
19h 14m 5s.385

τ DRACONIS. A.E.

Date.	Observer.	A.R. 1865.0.
		s.
1865 July 24		—0.10
28		—0.27
31		—0.15
Aug. 4		—0.19
26		+0.19
28		+0.20
Sept. 4		—0.22
6		—0.27
7		—0.10
11		—0.23
Mean		—0.114
A.E.		7.772

Mean A.R. 1865.0
19h 18m 7s.658

δ AQUILÆ. N.A.

Date.	Observer.	A.R. 1865.0.
		s.
1862 July 26	S.	+0.05
Sept. 2	S.	0.00
Oct. 18	S.	+0.02
1863 Aug. 17	S.	(+0.30)
1864 July 5	S.	+0.03
9	S.	+0.07
1864 July 13	S.	+0.08
14	S.	+0.21
18	R.	+0.11
26	S.	+0.10
Sept. 2	S.	+0.10
6	S.	+0.15
9	S.	+0.24
10	S.	+0.15
Oct. 7	S.	+0.14
19	S.	+0.08
21	S.	+0.12
26	S.	+0.18
Nov. 3	S.	+0.19
16	S.	+0.25
19	S.	+0.08
Mean		+0.118
N.A.		41.398

Mean A.R. 1865.0
19h 18m 41s.516

π DRACONIS.

Date.	Observer.	A.R. 1865.0.
		h. m. s.
1865 Aug. 4	M.	19 19 58.02
9	M.	58.15
12	M.	58.15
14	M.	58.08
15	M.	58.00
18	M.	57.97
26	S.	58.30
28	M.	58.20
Sept. 4	M.	58.33
5	S.	58.19
6	M.	57.94

Mean A.R. 1865.0
19h 19m 58s.121

6 VULPECULÆ.

Date.	Observer.	A.R. 1865.0.
		h. m. s.
1862 July 26	S.	19 23 5.22
Aug. 21	S.	5.29
Sept. 2	S.	5.27
4	S.	5.32
16	S.	5.36
Oct. 18	S.	5.23
Nov. 4	S.	5.39
1863 July 31	S.	5.28
Aug. 14	S.	5.31
19	S.	5.38
1864 June 6	R.	5.22
July 12	R.	5.24
16	R.	5.29
18	R.	5.39
19	R.	5.26
1864 July 26	S.	19 23 5.37
Sept. 2	S.	5.44
6	S.	5.35
9	S.	5.34
Oct. 7	S.	5.32
19	S.	5.31

Mean A.R. 1865.0
19h 23m 5s.313

β CYGNI.

Date.	Observer.	A.R. 1865.0.
		h. m. s.
1862 July 26	S.	19 25 16.67
Aug. 11	S.	16.58
21	S.	16.61
Sept. 4	S.	16.63
16	S.	16.66
Oct. 18	S.	16.67
Nov. 4	S.	16.66
1863 July 31	S.	16.81
Aug. 14	S.	16.68
19	S.	16.66
Sept. 21	S.	16.69
23	S.	16.71
1864 July 12	R.	16.59
16	R.	16.56
19	R.	16.57
26	S.	16.67
Aug. 8	S.	16.64
Sept. 2	S.	16.72
6	S.	16.69
9	S.	16.72
Oct. 25	S.	16.56
Nov. 16	S.	16.78
1865 July 26	M.	16.62

Mean A.R. 1865.0
19h 25m 16s.659

ι CYGNI.

Date.	Observer.	A.R. 1865.0.
		h. m. s.
1865 July 24	M.	19 26 18.04
28	M.	18.11
31	M.	18.08
Aug. 2	M.	18.09
4	M.	18.04
8	M.	18.08
9	M.	18.12
14	M.	18.21
15	M.	17.99
23	M.	18.04
26	S.	18.15
28	M.	18.26
Sept. 2	S.	18.19

Mean A.R. 1865.0
19h 26m 18s.108

h Sagittarii. N.A.

Date.		Observer.	A.R. 1865.0.
			s.
1862 July	18	S.	+0.24
	26	S.	+0.19
Aug.	11	S.	+0.17
Sept.	4	S.	+0.22
	11	S.	+0.16
	16	S.	+0.41
Oct.	18	S.	+0.18
1863 Aug.	14	S.	+0.29
	19	S.	+0.31
1864 June	6	R.	+0.21
July	5	S.	+0.22
	9	S.	+0.23
	18	R.	+0.40
July	19	R.	+0.19
	20	S.	+0.39
	26	S.	+0.36
Sept.	2	S.	+0.36
	6	S.	+0.25
	17	S.	+0.17
	27	S.	+0.23
Oct.	19	S.	+0.23
	21	S.	+0.30
	25	S.	+0 30
	23	S.	+0.22
Mean			+0.260
N.A.			29.186

Mean A.R. 1865.0
19ʰ 28ᵐ 29ˢ.446

Gr. 2900.

Date.		Observer.	A.R. 1865.0.
			h. m. s.
1865 July	24	M.	19 29 47.84
	28	M.	47.71
	31	M.	47.64
Aug.	2	M.	47.62
	4	M.	47.68
	8	M.	48.11
	9	M.	47.90
	12	M.	47.84
	26	S.	48.50
Sept.	4	M.	48.19

Mean A.R. 1865.0
19ʰ 29ᵐ 47ˢ.903

θ Cygni.

Date.		Observer.	A.R. 1865.0.
			h. m. s.
1862 July	26	S.	19 32 48.98
Aug.	11	S.	49 22
Sept.	2	S.	49.24
	11	S.	49.31
	16	S.	49.27
1863 July	29	S.	19 32 49.15
Aug.	4	S.	49.20
	14	S.	49.06
	19	S.	49.31
Sept.	2	S.	49.01
	5	S.	49.26
	7	S.	49.25
	22	S.	49.13
1864 July	18	R.	49.33
	26	S.	49.33
Sept.	6	S.	49.39
Oct.	7	S.	49.17

Mean A.R. 1865.0
19ʰ 32ᵐ 49ˢ.212

γ Sagittæ.

Date.		Observer.	A.R. 1865.0.
			h. m. s.
1862 Aug.	21	S.	19 52 45.27
Sept.	11	S.	45.24
	16	S.	45.33
1863 Aug.	17	S.	45.19
	18	S.	45.19
1864 July	18	R.	45.32
	26	S.	45.30
Aug.	8	S.	45.15
Sept.	2	S.	45.24
	6	S.	45.29
	9	S.	45.22
	16	S.	45.11
Oct.	21	S.	45.28
Nov.	19	S.	45.16

Mean A.R. 1865.0
19ʰ 52ᵐ 45ˢ.235

λ Ursæ Minoris. T.H.S.

Date.		Observer.	A.R. 1865.0.
			s.
1862 Sept.	10	S.	—1.17
	11	S.	+3.80
	29	S.	—2.61
1863 Aug.	19	S.	—1.04
Sept.	22	S.	+1.67
	23	S.	—1.22
	24	S.	+1.76
	28	S.	—4.54
Oct.	23	S.	—2.22
	28	S.	+2.31
	29	S.	—4.80
1864 June	6	R.	—1.86
Sept.	15	S.	+1.37
	16	S.	—0.09
Oct.	21	S.	—1.56
	25	S.	—2.17
	26	S.	—1.08
1864 Nov.	2	S.	+1.29
1865 July	24	M.	+3.63
	31	M.	—2.79
Aug.	8	M.	—0.27
	12	M.	—3.19
	23	M.	+0.17
	25	M.	—1.98
Sept.	13	M.	+0.22
	27	M.	+3.66
	29	M.	—2.12
Oct.	3	S.	+1.58
	23	M.	+1.94
	25	M.	—1.00
	30	M.	+3.38
Nov.	1	S.	+1.16
Mean			—0.243
T.H.S.			8.86

Mean A.R. 1865.0
19ʰ 59ᵐ 8ˢ.617

λ Ursæ Minoris. Sub Polo.

Date.		Observer.	A.R. 1865.0.
			s.
1863 Mar.	2	S.	+2.28
	9	S.	+1.68
1864 Mar.	3	S.	+1.60
1865 Feb.	24	M.	+1.74
April	3	S.	+0.50
Mean			+1.560
T.H.S.			8.86

Mean A.R. 1865.0
19ʰ 59ᵐ 10ˢ.420

Piazzi XIX. 414.

Date.		Observer.	A.R. 1865.0.
			h. m. s.
1865 Sept.	5	S.	20 1 22.17
	6	M.	22.09
	9	S.	22.06
	16	S.	22.09
	19	M.	22.22
	26	M.	22.36
Oct.	3	S.	22.18
	6	M.	22.27
	9	M.	22.18
	30	M.	22.02

Mean A.R. 1865.0
20ʰ 1ᵐ 22ˢ.164

Piazzi XIX. 420.

Date.		Observer.	A.R. 1865.0.
			h. m. s.
1865 Sept.	14	S.	20 1 57.35
	16	S.	57.19
	19	M.	57.18

Date.	Observer.	A.R. 1865.0.
		s.
1865 Sept. 26	M.	57.36
Oct. 6	M.	57.28
9	M.	57.22
30	M.	57.14

Mean A.R. 1865.0
20h 1m 57s.246

PIAZZI XX. 6.

Date.	Observer.	A.R. 1865.0.
		h. m. s.
1865 Sept. 6	M.	20 3 54.27
9	S.	54.19
12	M.	54.24
15	M.	54.31
16	S.	54.22
19	M.	54.07
26	M.	54.42
Oct. 3	S.	54.25

Mean A.R. 1865.0
20h 3m 54s.246

PIAZZI XX. 29.

Date.	Observer.	A.R. 1865.0.
		h. m. s.
1865 Sept. 5	S.	20 6 52.02
6	M.	52.03
12	M.	52.02
15	M.	52.04
16	S.	51.97
19	M.	52.03
25	M.	51.95
26	M.	52.12
27	M.	51.88
29	M.	52.01

Mean A.R. 1865.0
20h 6m 52s.007

ϱ AQUILÆ.

Date.	Observer.	A.R. 1865.0.
		h. m. s.
1862 Oct. 7	S.	20 8 1.88
8	S.	1.92
1863 Aug. 14	S.	1.86
17	S.	1.80
Sept. 28	S.	2.05
Oct. 29	S.	1.80
1864 June 6	R.	1.83
July 29	S.	1.83
Aug. 8	S.	1.87
Sept. 2	S.	1.84
6	S.	1.94
9	S.	1.85
16	S.	1.80
		h. m. s.
1864 Oct. 19	S.	20 8 1.97
21	S.	1.78
25	S.	1.86
26	S.	1.90
Nov. 1	S.	1.94
2	S.	1.88
3	S.	1.91

Mean A.R. 1865.0
20h 8m 1s.876

κ CEPHEI. A.E.

Date.	Observer.	A.R. 1865.0.
		s.
1865 Sept. 2	S.	+0.57
5	S.	+0.29
6	M.	—0.01
7	S.	+0.33
11	M.	+0.11
12	M.	+0.14
13	M.	+0.01
15	M.	—0.05
19	M.	+0.17
25	M.	+0.14
27	M.	—0.09
29	M.	+0.34
Oct. 3	S.	+0.07
6	M.	+0.07
Mean		+0.149
A.E.		22.54

Mean A.R. 1865.0
20h 13m 22s.689

23 H. AQUILÆ.

Date.	Observer.	A.R. 1865.0.
		h. m. s.
1865 Sept. 6	M.	20 16 29.39
7	S.	29.48
11	M.	29.36
12	M.	29.42
14	S.	29.40
15	M.	29.31
19	M.	29.31
25	M.	29.29
27	M.	29.34
29	M.	29.33
Oct. 6	M.	29.42

Mean A.R. 1865.0
20h 16m 29s.368

γ CYGNI.

Date.	Observer.	A.R. 1865.0.
		h. m. s.
1862 Sept. 29	S.	20 17 23.01
Oct. 8	S.	23.11
1863 Sept. 7	S.	23.00
23	S.	22.96
1864 Sept. 6	S.	23.03
9	S.	23.04
Oct. 5	S.	22.98
Nov. 1	S.	22.99
3	S.	23.02
Dec. 8	S.	23.01
1865 Oct. 9	M.	23.09
30	M.	23.02
Nov. 1	S.	22.93

Mean A.R. 1865.0
20h 17m 23s.015

π CAPRICORNI. A.E.

Date.	Observer.	A.R. 1865.0.
		s.
1865 Sept. 5	S.	+0.06
6	M.	+0.14
7	S.	+0.15
11	M.	+0.10
12	M.	+0.21
14	S.	+0.13
15	M.	+0.03
16	S.	+0.08
19	M.	—0.02
23	S.	+0.07
25	M.	+0.11
26	M.	+0.21
27	M.	+0.06
29	M.	+0.03
Oct. 3	S.	+0.11
6	M.	+0.15
30	M.	—0.02
Nov. 1	S.	—0.02
Mean		+0.088
A.E.		35.448

Mean A.R. 1865.0
20h 19m 35s.536

ϱ CAPRICORNI. N.A.

Date.	Observer.	A.R. 1865.0.
		s.
1862 July 18	S.	+0.28
Aug. 11	S.	+0.27
Sept. 25	S.	+0.23
Oct. 7	S.	+0.29
8	S.	+0.15
1863 Aug. 14	S.	+0.23
18	S.	+0.24

Date.	Observer.	A.R. 1865.0.
		s.
1863 Sept. 24	S.	+0.14
28	S.	+0.16
1864 Aug. 8	S.	+0.28
Sept. 6	S.	+0.37
16	S.	+0.25
Oct. 5	S.	+0.34
10	S.	+0.40
25	S.	+0.34
26	S.	+0.30
Nov. 1	S.	+0.31
2	S.	+0.25
3	S.	+0.32
Mean		+0.271
N.A.		9.254

Mean A.R. 1865.0
20ʰ 21ᵐ 9ˢ.525

Piazzi XX. 174

Date.	Observer.	A.R. 1865.0.
		h. m. s.
1865 Sept. 5	S.	20 25 0.48
6	M.	0.53
7	S.	0.32
11	M.	0.49
12	M.	0.53
15	M.	0.49
23	S.	0.49
25	M.	0.53
26	M.	0.44

Mean A.R. 1865.0
20ʰ 25ᵐ 0ˢ.478

ε Delphini. A.E.

Date.	Observer.	A.R. 1865.0.
		s.
1865 Sept. 6	M.	+0.05
7	S.	+0.10
11	M.	+0.09
12	M.	+0.09
13	M.	+0.02
14	S.	+0.16
15	M.	+0.20
19	M.	—0.01
23	S.	+0.07
25	M.	+0.09
26	M.	+0.10
27	M.	+0.13
30	S.	—0.04
Oct. 3	S.	+0.13
6	M.	+0.12
Mean		+0.087
A.E.		45.761

Mean A.R. 1865.0
20ʰ 26ᵐ 45ˢ.848

θ Cephei.

Date.	Observer.	A.R. 1865.0.
		h. m. s.
1862 Sept. 8	S.	20 27 18.61
Oct. 7	S.	18.78
1863 Aug. 14	S.	18.54
18	S.	18.74
Sept. 28	S.	18.56
Nov. 4	S.	18.53
1864 June 6	R.	18.82
Sept. 6	S.	18.72
Oct. 5	S.	18.75
10	S.	18.76

Mean A.R. 1865.0
20ʰ 27ᵐ 18ˢ.681

θ Cephei. Sub Polo.

Date.	Observer.	A.R. 1865.0.
		h. m. s.
1863 Mar. 11	S.	20 27 18.76
30	S.	18.60
April 1	S.	18.64
1864 Feb. 26	S.	18.91
Mar. 7	S.	18.69
April 7	R.	18.30
1865 April 3	M.	18.86
4	M.	19.19
5	M.	18.89

Mean A.R. 1865.0
20ʰ 27ᵐ 18ˢ.760

R.C. 4894.

Date.	Observer.	A.R. 1865.0.
		h. m. s.
1865 Sept. 7	S.	20 29 33.33
11	M.	32.80
12	M.	33.07
13	M.	33.75
14	S.	33.31
15	M.	33.55
19	M.	33.03
23	S.	32.73
25	M.	33.41

Mean A.R. 1865.0
20ʰ 29ᵐ 33ˢ.220

Pi. XX. 257 = Gr. 3241.

Date.	Observer.	A.R. 1865.0.
		h. m. s.
1863 Aug. 18	S.	20 30 34.00
1864 June 6	R.	34.27
Sept. 6	S.	34.24
15	S.	34.19
16	S.	34.08
Oct. 5	S.	34.30
		h. m. s.
1864 Oct. 10	S.	20 30 34.08
25	S.	33.94
Nov. 1	S.	34.30
3	S.	33.70
4	S.	34.21
10	S.	33.84

Mean A.R. 1865.0
20ʰ 30ᵐ 34ˢ.096

Pi. XX. 257 = Gr. 3241. Sub Polo.

Date.	Observer.	A.R. 1865.0.
		h. m. s.
1863 Mar. 11	S.	20 30 33.91
1864 Feb. 26	S.	34.53
Mar. 7	S.	34.41
1865 Mar. 17	M.	34.45
27	M.	34.14
April 3	M.	34.27
4	M.	34.44
5	M.	34.28

Mean A.R. 1865.0
20ʰ 30ᵐ 34ˢ.304

1 Aquarii.

Date.	Observer.	A.R. 1865.0.
		h. m. s.
1865 Sept. 7	S.	20 32 29.88
11	M.	29.82
12	M.	29.84
13	M.	29.78
14	S.	29.74
15	M.	29.79
19	M.	29.68
23	S.	29.86
25	M.	29.77

Mean A.R. 1865.0
20ʰ 32ᵐ 29ˢ.796

α Delphini. C. des T.

Date.	Observer.	A.R. 1865.0.
		s.
1862 Sept. 8	S.	—0.11
25	S.	0.00
29	S.	+0.03
Oct. 7	S.	—0.02
8	S.	+0.02
1863 Aug. 18	S.	—0.06
1864 Aug. 8	S.	—0.12
Sept. 6	S.	+0.10
15	S.	+0.02
16	S.	—0.09
Oct. 5	S.	—0.09
25	S.	+0.02

Date.	Observer.	A.R. 1865.0.
		s.
1864 Nov. 1	S.	—0.01
2	S.	—0.06
3	S.	—0.04
10	S.	—0.16
1865 Oct. 24	S.	—0.01
Mean		—0.034
C. des T.		22.10

Mean A.R. 1865.0
20ʰ 33ᵐ 22ˢ.066

226 B. Cygni.

Date.	Observer.	A.R. 1865.0.
		h. m. s.
1865 Sept. 7	S.	20 34 37.01
11	M.	37.04
12	M.	37.02
13	M.	37.02
14	S.	37.05
15	M.	37.05
19	M.	36.84
23	S.	36.92
26	M.	36.98
27	M.	36.95

Mean A.R. 1865.0
20ʰ 34ᵐ 36ˢ.988

74 Draconis. T.H.S.

Date.	Observer.	A.R. 1865.0.
		s.
1862 Aug. 11	S.	+0.54
Sept. 8	S.	+0.35
25	S.	+0.60
Oct. 7	S.	+0.65
1863 June 5	S.	+0.83
July 29	S	+0.41
Aug. 14	S.	+0.46
Sept. 5	S.	+0.35
17	S.	+0.68
Oct. 1	S.	+0.65
Nov. 19	S.	+0.74
1864 Sept. 6	S.	+0.74
9	S.	+0.47
15	S.	+0.35
16	S.	+0.27
Oct. 5	S.	+0.32
7	S.	+0.81
1865 Sept. 11	M.	+0.50
13	M.	+0.75
25	M.	+0.71
27	M.	+0.42
29	M.	+0.68
Oct. 6	M.	+0.78
24	S.	+0.71
Mean		+0.574
T.H.S		7.40

Mean A.R. 1865.0
20ʰ 37ᵐ 7ˢ.974

74 Draconis. Sub Polo.

Date.	Observer.	A.R. 1865.0.
		s.
1862 Mar. 29	H.	+0.28
April 4	H.	—0.06
7	S.	+0.07
1863 Feb. 7	S.	+0.07
Mar. 2	S.	+0.13
11	S.	+0.07
30	S.	+0.43
April 1	S.	+0.70
9	S.	+0.42
1864 Mar. 3	S.	+0.74
7	S.	+0.52
12	S.	+0.06
24	S.	+0.55
April 15	S.	+0.15
1865 Mar. 23	S.	+0.57
Mean		+0.313
T.H.S.		7.40

Mean A.R. 1865.0
20ʰ 37ᵐ 7ˢ.713

236 B. Cygni.

Date.	Observer.	A.R. 1865.0.
		h. m. s.
1865 Sept. 13		20 40 8.79
14		8.81
23		8.70
25		8.73
26		8.75
Oct. 5		8.86
9		8.87

Mean A.R. 1865.0
20ʰ 40ᵐ 8ˢ.787

ε Cygni.

Date.	Observer.	A.R. 1865.0.
		h. m. s.
1862 Aug. 11	S.	20 40 44.84
Sept. 8	S.	44.80
25	S.	44.87
29	S.	44.84
Oct. 7	S.	44.94
8	S.	44.93
1863 Aug. 14	S.	44.92
Sept. 14	S.	44.86
17	S.	44.96
28	S.	44.89
Nov. 4	S.	44.81
19	S.	44.91
Dec. 1	S.	45.01
1864 June 6	R.	44.97
Sept. 6	S.	44.89
15	S.	45.02
16	S.	44.92
		h. m. s.
1864 Oct. 7	S.	20 40 44.97
10	S.	44.97

Mean A.R. 1865.0
20ʰ 40ᵐ 44ˢ.912

λ Cygni. A.N. 1578.

Date.	Observer.	A.R. 1865.0.
		s.
1865 Sept. 14	S.	+0.11
15	M.	+0.08
19	M.	+0.23
23	S.	+0.09
25	M.	+0.12
26	M.	+0.11
27	M.	+0.10
29	M.	+0.11
Oct. 6	M.	+0.16
9	M.	+0.14
Mean		+0.125
v. Asten		8.98

Mean A.R. 1865.0
20ʰ 42ᵐ 9ˢ.105

η Cephei.

Date.	Observer.	A.R. 1865.0.
		h. m. s.
1862 Aug. 11	S.	20 42 32.32
Sept. 8	S.	32.48
25	S.	32.20
29	S.	32.32
Oct. 7	S.	32.37
8	S.	32.34
1863 Aug. 14	S.	32.22
18	S.	32.27
Sept. 17	S.	32.42
23	S.	32.43
Nov. 19	S.	32.34
Dec. 1	S.	32.44
1864 June 6	R.	32.33
Sept. 6	S.	32.32
15	S.	32.43
16	S.	32.28

Mean A.R. 1865.0
20ʰ 42ᵐ 32ˢ.344

η Cephei. Sub Polo.

Date.	Observer.	A.R. 1865.0.
		h. m. s.
1863 Feb. 7	S.	20 42 32.26
Mar. 9	S.	32.49
11	S.	32.15
April 1	S.	32.34
1864 Mar. 7	S.	32.54
24	S.	32.08

Date.	Observer.	A.R. 1865.0.
		h. m. s.
1864 April 7	R.	20 42 32.05
9	R.	32.20
1865 Mar. 10	M.	32.23

Mean A.R. 1865.0
20h 42m 32s.260

32 Vulpeculæ. N.A.

Date.	Observer.	s.
1862 Aug. 11	S.	+0.05
Sept. 10	S.	+0.08
29	S.	+0.02
Oct. 7	S.	+0.15
8	S.	+0.11
1863 Aug. 12	S.	+0.08
14	S.	+0.08
Sept. 23	S.	+0.22
28	S.	+0.03
Nov. 4	S.	—0.09
1864 Aug. 1	S.	+0.06
8	S.	+0.01
Sept. 6	S.	+0.08
16	S.	—0.03
30	S.	+0.04
Oct. 7	S.	—0.02
Nov. 3	S.	+0.07
Mean		+0.055
N.A.		48.384

Mean A.R. 1865.0
20h 48m 48s.439

Piazzi XX. 376.

Date.	Observer.	A.R. 1865.0.
		h. m. s.
1865 Sept. 11	M.	20 48 55.07
13	M.	55.18
15	M.	55.16
25	M.	55.13
26	M.	55.16
29	M.	55.11
30	M.	54.99
Oct. 3	S.	55.05
6	M.	55.18

Mean A.R. 1865.0
20h 48m 55s.114

76 Draconis.

Date.	Observer.	A.R. 1865.0.
		h. m. s.
1863 June 5	S.	20 52 9.60
Aug. 14	S.	10.06
18	S.	10.25
Sept. 5	S.	9.68
Nov. 19	S.	10.16
1864 June 6	R.	20 52 9.74
Sept. 6	S.	10.04
16	S.	10.06
Oct. 7	S.	10.09
Nov. 3	S.	9.30

Mean A.R. 1865.0
20h 52m 9s.898

76 Draconis. Sub Polo.

Date.	Observer.	A.R. 1865.0.
		h. m. s.
1863 Feb. 7	S.	20 52 9.96
Mar. 2	S.	9.90
9	S.	10.12
11	S.	10.01
30	S.	10.04
April 1	S.	9.73
9	S.	9.80
21	S.	10.76
1864 Mar. 7	S.	10.18
12	S.	10.48
April 9	R.	9.49
15	S.	10.41

Mean A.R. 1865.0
20h 52m 10s.073

Arg. 480 = Br. 2749. A.E.

Date.	Observer.	s.
1862 Oct. 8	S.	+0.52
1865 Sept. 11	M.	0.00
13	M.	0.00
14	S.	+0.29
15	M.	+0.14
19	M.	+0.12
23	S.	+0.30
25	M.	+0.10
26	M.	—0.27
27	M.	+0.29
Mean		+0.149
A.E.		36.449

Mean A.R. 1865.0
20h 53m 36s.598

6 Equulei B.

Date.	Observer.	A.R. 1865.0.
		h. m. s.
1865 Sept. 11	M.	20 57 54.26
12	M.	54.17
13	M.	54.07
14	S.	54.14
15	M.	54.06
1864 Sept. 19	M.	20 57 54.10
23	S.	54.15
25	M.	54.16
26	M.	54.12

Mean A.R. 1865.0
20h 57m 54s.137

61 Cygni. N.A.

[*Note.*—The middle point between the components is observed.]

Date.	Observer.	s.
1862 Aug. 11	S.	+1.16
Sept. 4	S.	+1.01
10	S.	+1.03
25	S.	+1.15
29	S.	+0.99
Oct. 7	S.	+1.06
8	S.	+1.08
Nov. 4	S.	+1.01
1863 June 5	S.	+1.00
23	S.	+1.08
July 29	S.	+1.09
Aug. 3	S.	+1.22
12	S.	+1.11
14	S.	+1.08
18	S.	+1.03
Sept. 2	S.	+1.09
23	S.	+1.20
28	S.	+0.96
30	S.	+1.09
Nov. 4	S.	+0.90
19	S.	+1.06
1864 July 29	S.	+1.04
30	S.	+1.01
Sept. 6	S.	+1.05
10	S.	+1.13
16	S.	+1.01
30	S.	+1.13
Oct. 7	S.	+1.03
10	S.	+1.05
17	S.	+1.06
Nov. 1	S.	+1.03
3	S.	+0.94
10	S.	+0.98
11	S.	+1.01
19	S.	+1.02
Mean		+1.054
N.A.		50.581

Mean A.R. 1865.0
21h 0m 51s.635

PIAZZI XXI. 1.

Date.	Observer.	A.R. 1865.0.
		h. m. s.
1865 Sept. 11	M.	21 2 55.43
12	M.	55.35
13	M.	55.47
14	S.	55.39
15	M.	55.28
19	M.	55.53
23	S.	55.36
26	M.	55.39
27	M.	55.31

Mean A.R. 1865.0 **21ʰ 2ᵐ 55ˢ.390**

ζ CYGNI. N.A.

Date.	Observer.	A.R. 1865.0.
		s.
1862 Sept. 4	S.	+0.05
10	S.	—0.06
29	S.	—0.04
Oct. 7	S.	+0.02
8	S.	+0.03
1863 Sept. 23	S.	+0.20
28	S.	+0.09
Nov. 4	S.	+0.07
1864 July 29	S.	+0.19
30	S.	+0.20
Aug. 1	S.	+0.09
Sept. 6	S.	+0.09
10	S.	+0.13
16	S.	0.00
17	S.	+0.09
Oct. 7	S.	+0.21
17	S.	+0.03
Nov. 11	S.	+0.02
Dec. 8	S.	—0.01
Mean		+0.074
N.A.		11.414

Mean A.R. 1865.0 **21ʰ 7ᵐ 11ˢ.488**

114 B. CAPRICORNI.

Date.	Observer.	A.R. 1865.0.
		h. m. s.
1865 Sept. 11	M.	21 7 33.41
19	M.	33.34
26	M.	33.39
27	M.	(33.58)
29	M.	33.16
30	S.	33.32
Oct. 6	M.	33.30
9	M.	33.27
10	S.	33.25
11	M.	(33.06)

Mean A.R. 1865.0 **21ʰ 7ᵐ 33ˢ.305**

4 PISCIS AUSTRINI.

Date.	Observer.	A.R. 1865.0.
		h. m. s.
1865 Sept. 11	M.	21 9 44.85
12	M.	44.85
15	M.	44.75
19	M.	44.94
26	M.	44.91
27	M.	44.76
29	M.	44.69
30	S.	44.79
Oct. 6	M.	44.88

Mean A.R. 1865.0 **21ʰ 9ᵐ 44ˢ.824**

11 B. PEGASI.

Date.	Observer.	A.R. 1865.0.
		h. m. s.
1865 Sept. 14	S.	21 14 6.26
15	M.	6.16
27	M.	6.09
29	M.	6.07
30	S.	6.06
Oct. 3	S.	6.07
6	M.	6.12
9	M.	6.10
13	M.	6.08
30	M.	6.09

Mean A.R. 1865.0 **21ʰ 14ᵐ 6ˢ.110**

α CEPHEI. B.J.

Date.	Observer.	A.R. 1865.0.
		s.
1862 Sept. 25	S.	+0.10
29	S.	(—0.48)
1863 June 29	S.	—0.09
Aug. 12	S.	—0.08
18	S.	—0.03
Sept. 17	S.	+0.06
23	S.	+0.13
28	S.	—0.01
Nov. 4	S.	—0.28
19	S.	+0.14
Dec. 1	S.	+0.11
1867 June 6	R.	—0.05
Aug. 1	S.	+0.01
Sept. 10	S.	+0.11
15	S.	+0.18
16	S.	—0.12
17	S.	+0.03
Oct. 10	S.	—0.13
Mean		+0.005
B.J.		21.327

Mean A.R. 1865.0 **21ʰ 15ᵐ 1ˢ.332**

α CEPHEI. Sub Polo.

Date.	Observer.	A.R. 1865.0.
		h. m. s.
1862 April 7	S.	—0.32
1863 Feb. 7	S.	—0.12
Mar. 2	S.	+0.14
April 21	S.	+0.10
22	S.	+0.19
1864 Mar. 12	S.	+0.12
April 15	S.	—0.32
22	S.	+0.12
27	S.	+0.13
1865 Mar. 17	M.	—0.09
April 5	M.	—0.01
Mean		—0.005
B.J.		21.327

Mean A.R. 1865.0 **21ʰ 15ᵐ 21ˢ.322**

326 B. CYGNI.

Date.	Observer.	A.R. 1865.0.
		h. m. s.
1865 Sept. 11	M.	21 15 40.86
12	M.	40.95
14	S.	40.99
15	M	40.99
19	M.	40.98
26	M.	41.01
27	M.	40.82
29	M.	40.93
30	S.	40.89
Oct. 3	S.	40.89
10	S.	40.89

Mean A.R. 1865.0 **21ʰ 15ᵐ 40ˢ.927**

PIAZZI XXI. 120.

Date.	Observer.	A.R. 1865.0.
		h. m. s.
1865 Sept. 30	S.	21 18 34.41
Oct. 3	S.	34.57
10	S.	34.52
11	M.	34.37
13	M.	34.42
Nov. 11	M.	34.33
14	M.	34.52
16	M.	34.59
17	M.	34.58

Mean A.R. 1865.0 **21ʰ 18ᵐ 34ˢ.479**

70 Cygni.

Date.	Observer.	A.R. 1865.0.
		h. m. s.
1865 Sept. 11	M.	21 21 51.25
14	S.	51.41
15	M.	51.23
27	M.	51.30
29	M.	51.17
30	S.	51.22
Oct. 3	S.	51.30
6	M.	51.34
9	M.	51.19
12	M.	51.14

Mean A.R. 1865.0
21^h 21^m 51^s.255

119 Cephei B. T.H.S.

Date.	Observer.	A.R. 1865.0.
		s.
1862 Aug. 20	S.	+0.56
Sept. 4	S.	+0.30
10	S.	+0.08
23	S.	+0.73
29	S.	+0.77
1863 June 23	S.	+0.30
July 29	S.	+0.77
Aug. 14	S.	+0.59
18	S.	+0.80
Sept. 4	S.	+0.15
5	S.	+0.23
7	S.	+0.70
Nov. 19	S.	+0.75
25	S.	+0.67
1864 Aug. 1	S.	+0.69
Nov. 16	S.	+1.25
1865 Sept. 30	S.	+0.65
Mean		+0.588
T.H.S.		15.52

Mean A.R. 1865.0
21^h 24^m 16^s.108

119 Cephei B. Sub Polo.

Date.	Observer.	A.R. 1865.0.
		s.
1862 Mar. 28	S.	+0.73
Apr. 7	S.	+0.65
19	S.	—0.16
25	S.	+0.18
1863 Feb. 7	S.	+0.71
25	S.	+0.29
Mar. 2	S.	+0.45
9	S.	+0.79
11	S.	—0.22
27	S.	+0.95
30	S.	+0.09
		s.
1863 Apr. 9	S.	+0.63
22	S.	+0.71
23	S.	+0.13
1864 Mar. 12	S.	+0.19
Apr. 15	S.	—0.23
22	S.	+0.75
1865 Mar. 25	S.	+0.47
27	M.	+0.87
28	M.	+0.23
Apr. 13	M.	—0.02
18	S.	+0.79
Mean		+0.409
T.H.S.		15.52

Mean A.R. 1865.0
21^h 24^m 15^s.929

β Aquarii. N.A.

Date.	Observer.	A.R. 1865.0.
		s.
1862 Nov. 4	S.	—0.09
1863 Sept. 17	S.	+0.14
28	S.	—0.03
1864 Sept. 6	S.	+0.14
7	S.	+0.02
16	S.	+0.11
17	S.	+0.18
Oct. 7	S.	+0.23
10	S.	+0.21
17	S.	+0.28
Nov. 3	R.	+0.09
10	S.	+0.08
25	S.	+0.04
29	S.	+0.12
Dec. 8	M.	+0.24
20	S.	+0.25
1865 Nov. 14	M.	+0.14
Mean		+0.126
N.A.		26.930

Mean A.R. 1865.0
21^h 24^m 27^s.056

26 B Pegasi = 11 Heis Pegasi.

Date.	Observer.	A.R. 1865.0.
		h. m. s.
1865 Sept. 11	M.	21 24 37.97
13	M.	37.92
15	M.	37.87
27	M.	37.82
29	M.	37.91
Oct. 6	M.	37.91
11	M.	37.91
13	M.	37.97
23	M.	37.86
25	M.	37.94

Mean A.R. 1865.0
21^h 24^m 37^s.908

β Cephei. B.J.

Date.	Observer.	A.R. 1865.0.
		s.
1862 Sept. 14	S.	—0.22
10	S.	+0.02
23	S.	—0.03
Nov. 4	S.	—0.31
1863 Aug. 18	S.	—0.06
Sept. 17	S.	+0.04
Nov. 19	S.	—0.01
25	S.	+0.18
Dec. 1	S.	+0.25
1864 Sept. 6	S.	+0.10
7	S.	+0.13
10	S.	+0.09
16	S.	—0.07
17	S.	—0.10
Oct. 7	S.	+0.04
10	S.	+0.09
Dec. 8	M.	—0.11
Mean		+0.002
B.J.		54.371

Mean A.R. 1865.0
21^h 26^m 54^s.373

β Cephei. Sub Polo.

Date.	Observer.	A.R. 1865.0.
		s.
1863 Feb. 7	S.	+0.10
Mar. 9	S.	+0.26
1864 Mar. 12	S.	+0.18
14	S.	+0.26
24	R.	+0.17
Apr. 7	R.	—0.46
15	S.	—0.25
16	R.	—0.35
27	S.	—0.01
1865 Mar. 17	M.	+0.25
Apr. 13	M.	+0.21
17	M.	+0.11
25	M.	+0.12
Mean		+0.042
B.J.		54.371

Mean A.R. 1865.0
21^h 26^m 54^s.413

122 B Cephei = Gr. 3511.

Date.	Observer.	A.R. 1865.0.
		h. m. s.
1865 Sept. 11	M.	21 28 41.92
13	M.	41.93
15	M.	42.54
27	M.	41.55

Date.	Observer.	A.R. 1865.0.
		h. m. s.
1865 Sept. 29	M.	21 28 41.95
30	S.	42.18
Oct. 3	S.	42.50
6	M.	42.51
11	M.	42.18
12	M.	42.37

Mean A.R. 1865.0
21ʰ 28ᵐ 42ˢ.163

F 376 Aquarii = B.F. 2941.

Date.	Observer.	A.R. 1865.0.
		h. m. s.
1865 Sept. 11	M.	21 30 37.96
13	M.	37.92
15	M.	37.81
27	M.	37.86
29	M.	37.84
30	M.	37.86
Oct. 3	S.	37.76
6	M.	37.83
9	M.	37.77
11	M.	37.69
12	M.	37.78

Mean A.R. 1865.0
21ʰ 30ᵐ 37ˢ.825

γ Capricorni.

Date.	Observer.	A.R. 1865.0.
		h. m. s.
1862 Sept. 4	S.	21 32 36.45
10	S.	36.56
23	S.	36.45
Nov. 4	S.	36.70
1863 Sept. 17	S.	36.57
Nov. 25	S.	36.51
Dec. 1	S.	36.47
1864 June 6	R.	36.46
Sept. 7	S.	36.29
10	S.	36.48
16	S.	36.55
17	S.	36.52
Oct. 7	S.	36.57
10	S.	36.61
17	S.	36.67
Nov. 10	M.	36.48
16	S.	36.56
25	S.	36.50
1865 Sept. 30	S.	36.73
Oct. 3	S.	36.56
6	M.	36.52

Mean A.R. 1865.0
21ʰ 32ᵐ 36ˢ.534

ι Piscis austrini.

Date.	Observer.	A.R. 1865.0.
		h. m. s.
1865 Sept. 11	M.	21 36 54.07
13	M.	53.96
15	M.	54.05
19	M.	54.11
26	M.	54.12
27	M.	53.94
29	M.	53.99
30	S.	53.99
Oct. 3	S.	53.97
6	M.	53.91

Mean A.R. 1865.0
21ʰ 36ᵐ 54ˢ.011

ε Pegasi. N.A.

Date.	Observer.	A.R. 1865.0.
		s.
1862 Sept. 4	S.	+0.01
10	S.	+0.08
23	S.	+0.11
Oct. 7	S.	+0.04
1863 Sept. 17	S.	+0.10
1864 Sept. 6	S.	(+0.28)
10	S.	+0.01
16	S.	−0.06
17	S.	+0.07
Oct. 7	S.	+0.01
10	S.	+0.06
17	S.	+0.07
1865 Oct. 11	M.	+0.02
12	M.	−0.01
Mean		+0.039
N.A.		33.307

Mean A.R. 1865.0
21ʰ 37ᵐ 33ˢ.346

δ Capricorni. C. des T.

Date.	Observer.	A.R. 1865.0.
		s.
1862 Sept. 4	S.	+0.06
10	S.	+0.08
23	S.	+0.02
1863 Nov. 25	S.	−0.04
Dec. 1	S.	−0.08
1864 Sept. 7	S.	−0.02
10	S.	−0.02
16	S.	+0.06
17	S.	−0.01
Oct. 7	S.	+0.07
10	S.	+0.12
17	S.	−0.04
Nov. 25	S.	−0.03
Dec. 1	S.	−0.10
8	S.	−0.02
1865 Oct. 11	M.	−0.21
Mean		−0.010
C. des T.		35.200

Mean A.R. 1865.0
21ʰ 39ᵐ 35ˢ.190

θ Piscis austrini.

Date.	Observer.	A.R. 1865.0.
		h. m. s.
1865 Sept. 11	M.	21 39 48.48
13	M.	48.45
15	M.	48.46
19	M.	48.49
26	M.	48.44
27	M.	48.44
29	M.	48.51
30	S.	48.43
Oct. 3	S.	48.55
6	M.	48.51

Mean A.R. 1865.0
21ʰ 39ᵐ 48ˢ.476

388 B Cygni = Gr. 3571.

Date.	Observer.	A.R. 1865.0.
		h. m. s.
1865 Sept. 11	M.	21 42 43.46
15	M.	43.41
19	M.	43.45
26	M.	43.47
27	M.	43.36
29	M.	43.50
30	S.	43.53
Oct. 3	S.	43.54
6	M.	43.49
9	M.	43.46

Mean A.R. 1865.0
21ʰ 42ᵐ 43ˢ.467

195 Heis Cygni = R.C. 5408.

Date.	Observer.	A.R. 1865.0.
		h. m. s.
1865 Sept. 11	M.	21 44 10.94
13	M.	10.84
15	M.	10.95
19	M.	10.78
26	M.	10.75
27	M.	10.78
29	M.	10.77
30	S.	10.86
Oct. 3	S.	10.92
6	M.	10.87
9	M.	10.82

Mean A.R. 1865.0
21ʰ 44ᵐ 10ˢ.844

16 Pegasi. N.A.

Date	Observer.	A.R. 1865.0.
		s.
1862 Sept. 4	S.	+0.00
10	S.	+0.08
23	S.	+0.17
Oct. 7	S.	+0.06
Nov. 4	S.	+0.17
1863 Aug. 29	S.	—0.06
Nov. 3	S.	+0.15
1864 Sept. 10	S.	+0.04
16	S.	+0.01
Oct. 7	S.	0.00
10	S.	+0.08
11	S.	—0.06
17	S.	+0.15
Nov. 10	M.	+0.07
11	M.	+0.02
30	M.	+0.14
Dec. 8	M.	—0.04
9	S.	—0.02
1865 Nov. 9	S.	+0.03
Mean		+0.052
N.A.		55.240

Mean A.R. 1865.0
$21^h\ 46^m\ 55^s.292$

Piazzi XXI. 320.

Date	Observer.	A.R. 1865.0.
		h. m. s.
1865 Sept. 11	M.	21 47 7.35
13	M.	7.39
15	M.	7.31
19	M.	7.38
26	S.	7.57
27	M.	7.23
30	S.	7.34
Oct. 3	S.	7.53
6	M.	7.31
9	M.	7.29

Mean A.R. 1865.0
$21^h\ 47^m\ 7^s.370$

17 Pegasi.

Date	Observer.	A.R. 1865.0.
		h. m. s.
1865 Sept. 13	M.	21 50 21.40
15	M.	21.47
19	M.	21.37
26	S.	21.55
27	M.	21.33
30	S.	21.35
Oct. 3	S.	21.46
7	S.	21.35
9	M.	21.44
1865 Oct. 10	S.	21 50 21.51
11	M.	21.52

Mean A.R. 1865.0
$21^h\ 50^m\ 21^s.432$

μ Piscis austrini.

Date	Observer.	A.R. 1865.0.
		h. m. s.
1865 Sept. 11	M.	21 53 4.67
13	M.	4.63
15	M.	4.53
19	M.	4.67
26	S.	4.62
27	M.	4.56
Oct. 3	S.	4.67
6	M.	4.50
7	S.	4.58
9	M.	4.59

Mean A.R. 1865.0
$21^h\ 53^m\ 4^s.602$

Mayer 911 = B.A.C. 7665.

Date	Observer.	A.R. 1865.0.
		h. m. s.
1865 Sept. 11	M.	21 54 45.80
13	M.	45.96
15	M.	45.73
19	M.	45.95
26	S.	45.84
27	M.	45.83
30	S.	45.98
Oct. 3	S.	45.90
6	M.	45.88
7	S.	45.81

Mean A.R. 1865.0
$21^h\ 54^m\ 45^s.868$

32. Aquarii.

Date	Observer.	A.R. 1865.0.
		h. m. s.
1865 Sept. 11	M.	21 57 50.86
13	M.	50.88
15	M.	50.72
19	M.	51.01
25	M.	50.90
26	S.	50.86
27	M.	50.97
30	S.	50.85
Oct. 3	S.	50.97
6	M.	50.88
7	S.	50.79

Mean A.R. 1865.0
$21^h\ 57^m\ 50^s.881$

μ Piscis austrini.

Date	Observer.	A.R. 1865.0.
		h. m. s.
1865 Sept. 11	M.	22 0 30.16
13	M.	30.00
15	M.	30.06
19	M.	30.20
26	S.	30.07
27	M.	30.02
30	S.	30.08
Oct. 3	S.	30.15
6	M.	30.09
7	S.	30.02

Mean A.R. 1865.0
$22^h\ 0^m\ 30^s.085$

ι Pegasi.

Date	Observer.	A.R. 1865.0.
		h. m. s.
1862 Sept. 10	S.	22 0 43.68
23	S.	43.74
Oct. 7	S.	43.66
Nov. 4	S.	43.60
Dec. 11	S.	43.69
1863 Sept. 4	S.	43.52
Nov. 25	S.	43.55
Dec. 7	S.	43.83
1864 Sept. 10	S.	43.79
16	S.	43.61
17	S.	43.75
26	S.	43.48
29	S.	43.60
30	S.	43.79
Oct. 7	S.	43.71
11	S.	43.63
17	S.	43.69
Nov. 11	M.	43.71
16	S.	43.81
25	S.	43.51
30	M.	43.78
Dec. 1	S.	43.67
9	S.	43.74

Mean A.R. 1865.0
$22^h\ 0^m\ 43^s.676$

Pi. XXI. 421 = Br. 2912.

Date	Observer.	A.R. 1865.0.
		h. m. s.
1865 Sept. 11	M.	22 3 19.85
13	M.	19.83
15	M.	19.86
19	M.	19.89
26	S.	19.92
27	M.	19.87

Date.	Observer.	A.R. 1865.0.
		h. m. s.
1865 Oct. 6	M.	22 3 20.02
7	S.	19.80
9	M.	19.83
Mean A.R. 1865.0		**22h 3m 19s.874**

Pi. XXII. 19.

Date.	Observer.	A.R. 1865.0.
		h. m. s.
1865 Sept. 11	M.	22 6 9.15
13	M.	9.16
15	M.	9.18
26	S.	9.06
27	M.	9.20
Oct. 6	M.	9.09
9	M.	9.03
23	M.	9.08
25	M.	9.21
30	M.	8.93
Mean A.R. 1865.0		**22h 6m 9s.109**

ζ Cephei. C. des T.

Date.	Observer.	A.R. 1865.0.
		s.
1862 Sept. 10	S.	—0.01
23	S.	—0.14
Dec. 11	S.	—0.18
1863 June 29	S.	—0.13
30	S.	—0.02
July 22	S.	+0.06
Sept. 4	S.	—0.23
Nov. 2	S.	+0.12
3	S.	+0.09
25	S.	—0.06
Dec. 7	S.	+0.31
1864 Sept. 10	S.	+0.16
Nov. 10	M.	—0.05
11	M.	—0.12
30	M.	+0.06
Dec. 8	M.	0.00
Mean		—0.009
C. des T.		10.36
Mean A.R. 1865.0		**22h 6m 10s.351**

1 H. Lacertæ.

Date.	Observer.	A.R. 1865.0.
		h. m. s.
1865 Sept. 11	M.	22 8 5.25
13	M.	5.27
15	M.	5.23
19	M.	5.16
26	S.	5.25
1865 Oct. 6	M.	22 8 5.20
7	S.	5.23
23	M.	5.22
25	M.	5.19
Mean A.R. 1865.0		**22h 8m 5s.222**

θ Aquarii. N.A.

Date.	Observer.	A.R. 1865.0.
		s.
1862 Sept. 4	S.	+0.04
10	S.	+0.19
23	S.	+0.14
Oct. 7	S.	+0.08
Nov. 3	S.	+0.10
Dec. 11	S.	—0.03
1863 Nov. 3	S.	+0.06
Dec. 7	S.	+0.10
1864 Sept. 10	S.	+0.09
16	S.	+0.14
17	S.	+0.08
30	S.	+0.09
Oct. 11	S.	+0.12
Nov. 10	M.	—0.01
11	M.	+0.18
Nov. 25	S.	+0.09
29	S.	+0.01
30	M.	+0.13
Dec. 1	S.	—0.03
8	M.	+0.07
1865 Oct. 23	M.	+0.17
Nov. 9	S.	+0.06
Mean		+0.085
N.A.		42.445
Mean A.R. 1865.0		**22h 9m 42s.530**

Bradley 2942.

Date.	Observer.	A.R. 1865.0.
		h. m. s.
1865 Sept. 11	M.	22 10 25.36
13	M.	25.32
19	M.	25.29
27	M.	25.30
Oct. 6	M.	25.60
9	M.	25.37
23	M.	25.10
25	M.	25.54
30	M.	25.27
Nov. 1	S.	25.47
Mean A.R. 1865.0		**22h 10m 25s.362**

γ Aquarii. C. des T.

Date.	Observer.	A.R. 1865.0.
		s.
1863 Oct. 22	S.	+0.01
1864 Sept. 10	S.	+0.11
26	S.	+0.04
Oct. 17	S.	+0.23
Nov. 10	M.	+0.07
11	M.	+0.01
14	S.	—0.04
30	M.	—0.02
Dec. 1	S.	—0.05
8	M.	+0.01
9	S.	+0.08
1865 Sept. 11	M.	+0.04
Oct. 6	M.	+0.03
23	M.	+0.10
25	M.	—0.02
30	M.	—0.13
Nov. 1	S.	—0.03
9	S.	+0.01
Mean		+0.025
C. des T.		41.00
Mean A.R. 1865.0		**22h 14m 41s.025**

Piazzi XXII. 81.

Date.	Observer.	A.R. 1865.0.
		h. m. s.
1865 Sept. 11	M.	22 16 27.23
12	M.	27.38
19	M.	27.26
Oct. 6	M.	27.39
7	S.	27.38
23	M.	27.22
24	S.	27.41
25	M.	27.24
30	M.	27.07
Mean A.R. 1865.0		**22h 16m 27s.287**

3 Lacertæ.

Date.	Observer.	A.R. 1865.0.
		h. m. s.
1865 Sept. 11	M.	22 18 15.43
19	M.	15.32
29	M.	15.42
Oct. 7	S.	15.38
9	M.	15.40
11	M.	15.46
12	M.	15.38
23	M.	15.25
24	S.	15.30
25	M.	15.33
30	M.	15.24
Mean A.R. 1865.0		**22h 18m 15s.355**

LL 43886 = 18 Heis Lacertæ.

Date.	Observer.	A.R. 1865.0.
		h. m. s.
1865 Sept. 11	M.	22 21 31.80
19	M.	31.71
29	M.	31.82
Oct. 9	M.	31.85
11	M.	31.69
13	M.	31.71
23	M.	31.61
25	M.	31.71
30	M.	31.70
Nov. 9	S.	31.74

Mean A.R. 1865.0
22h 21m 31s.734

σ Aquarii.

Date.	Observer.	A.R. 1865.0.
		h. m. s.
1862 Sept. 4	S.	22 23 30.17
10	S.	30.00
23	S.	30.11
Oct. 30	S.	30.24
Nov. 4	S.	30.20
Dec. 11	S.	30.04
1863 Oct. 22	S.	30.15
1864 Sept. 10	S.	30.18
16	S.	30.03
26	S.	30.12
27	S.	30.08
30	S.	30.15
Oct. 7	S.	30.14
11	S.	30.00
17	S.	30.14
Nov. 10	M.	30.11
11	M.	30.14
14	S.	30.09
30	M.	30.12
Dec. 9	S.	30.10

Mean A.R. 1865.0
22h 23m 30s.116

32 H. Cephei.

Date.	Observer.	A.R. 1865.0.
		h. m. s.
1865 Sept. 11	M.	22 23 35.19
19	M.	35.20
29	M.	34.65
Oct. 9	M.	35.44
11	M.	35.40
12	M.	35.75
23	M.	35.03
24	S.	35.27
30	M.	34.17

Mean A.R. 1865.0
22h 23m 35s.122

Groombridge 3804.

Date.	Observer.	A.R. 1865.0.
		h. m. s.
1865 Sept. 19	M.	22 26 28.56
27	M.	28.38
29	M.	28.48
Oct. 11	M.	28.65
12	M.	28.55
13	M.	28.52
23	M.	28.58
24	S.	28.55
Nov. 1	S.	28.58
Dec. 1	M.	28.59

Mean A.R. 1865.0
22h 26m 28s.544

η Aquarii. N.A.

Date.	Observer.	A.R. 1865.0.
		s.
1862 Sept. 10	S.	+0.03
Oct. 30	S.	+0.19
Nov. 3	S.	+0.27
Dec. 11	S.	+0.08
1863 Sept. 29	S.	+0.06
Oct. 22	S.	—0.10
1864 Sept. 10	S.	+0.15
16	S.	+0.08
17	S.	—0.03
26	S.	+0.04
27	S.	+0.09
30	S.	+0.09
Oct. 11	S.	+0.09
Nov. 10	M.	+0.13
14	S.	+0.15
30	M.	+0.10
Dec. 8	M.	+0.10
9	S.	+0.14
1865 Oct. 12	M.	+0.21
13	M.	+0.08
23	M.	0.00
24	S.	+0.12
Nov. 9	S.	+0.13
Mean		+0.096
N.A.		25.053

Mean A.R. 1865.0
22h 28m 25s.149

226 B Cephei = 91 Heis Cephei. A.E.

Date.	Observer.	A.R. 1865.0.
		s.
1865 Oct. 12	M.	+0.01
13	M.	—0.47
23	M.	—0.31
24	S.	+0.11
Nov. 1	S.	+0.19
9	S.	—0.11
Mean		—0.0[illegible]
A.E.		53.5[illegible]

Mean A.R. 1865.0
22h 29m 53s.4[illegible]

ε Piscis austrini.

Date.	Observer.	A.R. 1865.0.
		h. m. s.
1865 Sept. 12	M.	22 33 11.[illegible]
27	M.	11.[illegible]
29	M.	11.[illegible]
Oct. 9	M.	11.[illegible]
11	M.	11.[illegible]
12	M.	10.[illegible]
13	M.	11.[illegible]
23	M.	11.[illegible]
24	S.	11.[illegible]
Nov. 1	S.	11.[illegible]

Mean A.R. 1865.0
22h 33m 11s.05[illegible]

Piazzi XXII. 186.

Date.	Observer.	A.R. 1865.0.
		h. m. s.
1865 Sept. 27	M	22 34 11.[illegible]
29	M.	11.[illegible]
Oct. 9	M.	11.[illegible]
11	M.	11.[illegible]
12	M.	11.[illegible]
13	M.	11.[illegible]
24	S.	11.[illegible]
Nov. 14	M.	11.[illegible]
Dec. 1	M.	11.[illegible]

Mean A.R. 1865.0
22h 34m 11s.4[illegible]

ζ Pegasi. N.A.

Date.	Observer.	A.R. 1865.0.
		s.
1862 Sept. 4	S.	+0.[illegible]
23	S.	+0.[illegible]
Nov. 3	S.	+0.[illegible]
4	S.	0.[illegible]
Dec. 12	S.	+0.[illegible]
1863 Jan. 1	S.	+0.[illegible]
Dec. 15	S.	+0.[illegible]
23	S.	+0.[illegible]
1864 Sept. 10	S.	+0.[illegible]
16	S.	0.[illegible]
17	S.	+0.[illegible]
26	S.	+0.[illegible]
27	S.	+0.[illegible]
30	S.	+0.[illegible]

Date.	Observer.	A.R. 1865.0.
		s.
1864 Oct. 7	S.	+0.15
11	S.	+0.20
26	S.	+0.19
Nov. 10	M.	+0.06
11	M.	+0.21
14	S.	+0.21
30	M.	+0.14
Dec. 1	S.	+0.17
9	S.	+0.18
Mean		+0.151
N.A.		43.661

Mean A.R. 1865.0 **22ʰ 34ᵐ 43ˢ.812**

Piazzi XXII. 195.

Date.	Observer.	A.R. 1865.0.
		h. m. s.
1865 Oct. 23	M.	22 35 17.46
Nov. 14	M.	17.56

Mean A.R. 1865.0 **22ʰ 35ᵐ 17ˢ.510**

η Pegasi.

Date.	Observer.	A.R. 1865.0.
		h. m. s.
1862 Sept. 4	S.	22 36 40.64
10	S.	40.62
23	S.	40.67
1863 Jan. 1	S.	40.60
Nov. 3	S.	40.54
Dec. 15	S.	40.58
13	S.	40.65
1864 Sept. 10	S.	40.64
16	S.	40.50
17	S.	40.43
26	S.	40.61
27	S.	40.58
Oct. 7	S.	40.70
11	S.	40.56
17	S.	40.59
21	S.	40.49
26	S.	40.57
Nov. 10	M.	40.61
11	M.	40.50
30	M.	40.71
Dec. 8	M.	40.65

Mean A.R. 1865.0 **22ʰ 36ᵐ 40ˢ.592**

13 Lacertæ.

Date.	Observer.	A.R. 1865.0.
		h. m. s.
1865 Sept. 27	M.	22 38 4.41
29	M.	4.55
Oct. 7	S.	4.58
Oct. 9	M.	22 38 4.71
11	M.	4.68
12	M.	4.55
13	M.	4.47
23	M.	4.46
24	S.	4.58
30	M.	4.40

Mean A.R. 1865.0 **22ʰ 38ᵐ 4ˢ.539**

ξ Pegasi.

Date.	Observer.	A.R. 1865.0.
		h. m. s.
1862 Sept. 4	S.	22 39 56.86
10	S.	56.96
23	S.	56.99
Nov. 3	S.	57.07
Dec. 11	S.	57.06
1863 Dec. 15	S.	56.89
1864 Sept. 10	S.	57.05
16	S.	56.95
17	S.	57.00
26	S.	56.87
27	S.	56.94
30	S.	56.97
Oct. 7	S.	56.99
11	S.	57.12
17	S.	56.97
21	S.	56.96
26	S.	56.97
Nov. 4	S.	56.99
14	S.	57.00
Dec. 8	M.	57.07
9	S.	57.07

Mean A.R. 1865.0 **22ʰ 39ᵐ 56ˢ.988**

69 Aquarii.

Date.	Observer.	A.R. 1865.0.
		h. m. s.
1865 Sept. 27	M.	22 40 32.52
29	M.	32.60
Oct. 7	S.	32.61
9	M.	32.62
11	M.	32.62
12	M.	32.50
13	M.	32.39
23	M.	32.68
24	S.	32.69
25	M.	32.59

Mean A.R. 1865.0 **22ʰ 40ᵐ 23ˢ.582**

ι Cephei. A.E.

Date.	Observer.	A.R. 1865.0.
		s.
1865 Oct. 3	S.	+0.10
7	S.	+0.25
9	M.	—0.01
12	M.	—0.01
13	M.	—0.21
17	S.	—0.04
23	M.	—0.16
25	M.	—0.04
30	M.	—0.03
Mean		—0.017
A.E. tables.		52.823

Mean A.R. 1865.0 **22ʰ 44ᵐ 52ˢ.806**

δ Aquarii.

Date.	Observer.	A.R. 1865.0.
		h. m. s.
1862 Nov. 3	S.	22 47 (29.28)
1863 Sept. 17	S.	29.00
24	S.	28.88
28	S.	28.98
1864 Sept. 26	S.	29.05
30	S.	29.00
Oct. 11	S.	28.95
17	S.	29.05
21	S.	28.96
25	S.	29.08
Nov. 4	S.	29.00
10	M.	28.97
14	S.	29.07
29	S.	29.00
30	M.	28.90
Dec. 1	S.	29.01
8	M.	29.00
1865 Sept. 12	M.	29.10
13	M.	29.10
29	M.	29.06
Oct. 3	S.	28.99
6	M.	29.04
12	M.	29.06

Mean A.R. 1865.0 **22ʰ 47ᵐ 29ˢ.011**

34 Cephei H. T.H.S.

Date.	Observer.	A.R. 1865.0.
		s.
1860 Sept. 10	S.	+0.07
23	S.	+0.58
Dec. 11	S.	+0.25
1863 Sept. 22	S.	+1.51
23	S.	+0.81
28	S.	+1.60
29	S.	+0.32

Date.	Observer.	A.R. 1865.0.
		s.
1863 Oct. 28	S.	+1.13
Dec. 9	S.	—0.07
10	S.	+0.26
15	S.	+0.39
21	S.	+1.10
23	S.	—0.02
1864 Sept. 16	S.	+0.46
30	S.	+0.99
Oct. 17	S.	+0.45
Nov. 11	M.	—0.05
1865 Oct. 7	S.	+0.96
Mean		+0.597
T.H.S.		54.79

Mean A.R. 1865.0
22ʰ 47ᵐ 55ˢ.387

34 Cephei H. Sub Polo.

Date.	Observer.	A.R. 1865.0.
		s.
1862 April 7	S.	—0.06
16	H.	+0.16
26	H.	+0.35
May 7	S.	+0.41
1863 Mar. 9	S.	+0.55
May 9	S.	+0.36
1864 April 30	R.	+0.02
May 6	S.	+0.58
10	S.	+0.83
1865 Mar. 25	S.	+0.14
April 3	M.	+0.78
13	M.	+0.27
17	M.	+0.40
21	M.	+0.46
25	M.	+0.68
May 3	M.	+0.70
Mean		+0.414
T.H.S.		54.79

Mean A.R. 1865.0
22ʰ 47ᵐ 55ˢ.204

36 Cephei H.

Date.	Observer.	A.R. 1865.0.
		h. m. s.
1865 Sept. 29	M.	22 55 21.79
30	S.	21.86
Oct. 3	S.	21.76
11	M.	21.85
13	M.	21.31
16	M.	21.06
17	S.	21.96
23	M.	21.07
24	S.	21.54
25	M.	21.46

Mean A.R. 1865.0
22ʰ 55ᵐ 21ˢ.566

β Piscium.

Date.	Observer.	A.R. 1865.0.
		h. m. s.
1862 Sept. 4	S.	22 57 0.38
23	S.	0.46
Oct. 30	S.	0.56
Dec. 20	S.	(0.19)
1863 Jan. 1	S.	0.46
Sept. 30	S.	0.43
Oct. 28	S.	0.44
Dec. 15	S.	0.39
23	S.	0.55
1864 Sept. 10	S.	0.54
26	S.	0.46
27	S.	0.38
30	S.	0.43
Oct. 7	S.	0.37
11	S.	0.39
17	S.	0.44
21	S.	0.47
25	S.	0.58
26	S.	0.42
Nov. 30	M.	0.45
Dec. 8	M.	0.41

Mean A.R. 1865.0
22ʰ 57ᵐ 0ˢ.450

118 Heis Pegasi.

Date.	Observer.	A.R. 1865.0.
		h. m. s.
1865 Sept. 11	M.	23 0 49.37
13	M.	49.39
23	S.	49.44
27	M.	49.31
Oct. 6	M.	49.40
12	M.	49.29
13	M.	49.36
23	M.	49.32
24	S.	49.39
25	M.	49.37
Nov. 1	S.	49.35

Mean A.R. 1865.0
23ʰ 0ᵐ 49ˢ.363

Piazzi XXIII. 4.

Date.	Observer.	A.R. 1865.0.
		h. m. s.
1865 Sept. 11	M.	23 4 0.39
13	M.	0.39
23	S.	0.44
25	M.	0.35
27	M.	0.41
29	M.	0.39
Oct. 6	M.	0.44
11	M.	0.33
12	M.	0.49
24	S.	0.39
25	M.	0.30

Mean A.R. 1865.0
23ʰ 4ᵐ 0ˢ.393

128 Heis Pegasi.

Date.	Observer.	A.R. 1865.0.
		h. m. s.
1865 Sept. 11	M.	23 7 57.29
13	M.	57.28
23	S.	57.33
25	M.	57.26
27	M.	57.24
29	M.	57.23
Oct. 6	M.	57.28
12	M.	57.16
13	M.	57.32
23	M.	57.16
24	S.	57.24

Mean A.R. 1865.0
23ʰ 7ᵐ 57ˢ.254

130 Heis Pegasi.

Date.	Observer.	A.R. 1865.0.
		h. m. s.
1865 Sept. 11	M.	13 9 18.95
13	M.	18.96
29	M.	18.85
Oct. 6	M.	18.94
12	M.	18.78
13	M.	18.84
23	M.	18.88
24	S.	18.90
25	M.	18.95
Nov. 1	S.	18.89
9	S.	18.94

Mean A.R. 1865.0
23ʰ 9ᵐ 18ˢ.898

γ Piscium N.A.

Date.	Observer.	A.R. 1865.0.
1862 Sept. 23	S.	+0.09
Oct. 30	S.	+0.09
Dec. 29	S.	+0.08
1863 Sept. 22	S.	—0.01
24	S.	+0.06
28	S.	+0.10
29	S.	+0.09
30	S.	+0.09
Oct. 29	S.	—0.15
Dec. 13	S.	+0.09
1864 Jan. 9	S.	+0.06
Sept. 16	S.	+0.04
17	S.	+0.12
26	S.	+0.11
Oct. 21	S.	+0.07
25	S.	+0.15
26	S.	+0.10
Nov. 4	S.	+0.11
10	M.	+0.16
11	M.	+0.26

Date.	Observer.	A.R. 1865.0.
Mean		s. +0.086
N.A.		9.993
Mean A.R. 1865.0		**23ʰ 10ᵐ 10ˢ.079**

Pi. XXIII. 36 = γ Sculptoris.

Date.	Observer.	A.R. 1865.0.
1865 Sept. 13	M.	h. 23 m. 11 s. 31.75
25	M.	31.51
29	M.	31.77
Oct. 6	M.	31.68
11	M.	31.67
12	M.	31.72
23	M.	31.72
24	S.	31.83
25	M.	31.76
Nov. 1	S.	31.67
Mean A.R. 1865.0		**23ʰ 11ᵐ 31ˢ.708**

o Cephei. A.E.

Date.	Observer.	A.R. 1865.0.
1865 Sept. 13	M.	s. —0.12
25	M.	—0.08
29	M.	—0.32
Oct. 6	M.	+0.25
12	M.	—0.25
13	M.	—0.29
16	M.	—0.10
23	M.	—0.21
24	S.	—0.19
25	M.	—0.22
Nov. 1	S.	—0.14
Mean		—0.152
A.E. tables		5.689
Mean A.R. 1865.0		**23ʰ 13ᵐ 5ˢ.537**

DM + 35°5012.

Date.	Observer.	A.R. 1865.0.
1865 Sept. 13	M.	h. 23 m. 14 s. 51.66
29	M.	51.53
Oct. 6	M.	51.63
11	M.	51.74
16	M.	51.61
23	M.	51.60
24	S.	51.57
25	M.	51.48
Nov. 1	S.	51.54
11	M.	51.48
14	M.	51.46
Mean A.R. 1865.0		**23ʰ 14ᵐ 51ˢ.573**

b Aquarii.

Date.	Observer.	A.R. 1865.0.
1862 Sept. 23	S.	h. 23 m. 15 s. 52.62
Oct. 30	S.	52.72
1863 Jan. 1	S.	52.70
Dec. 23	S.	52.68
26	S.	52.75
1864 Sept. 16	S.	52.77
17	S.	52.65
30	S.	52.73
Oct. 10	S.	52.73
21	S.	52.72
26	S.	52.70
Nov. 4	S.	52.77
10	M.	52.66
11	M.	52.77
14	S.	52.67
29	S.	52.66
Dec. 9	S.	52.65
Mean A.R. 1865.0		**23ʰ 15ᵐ 52ˢ.703**

Br. 3109 = Pi. XXIII. 71.

Date.	Observer.	A.R. 1865.0.
1865 Sept. 13	M.	h. 23 m. 17 s. 9.80
25	M.	9.58
29	M.	9.68
Oct. 12	M	9.72
13	M.	9.65
16	M.	9.77
24	S.	9.69
25	M.	9.62
Nov. 11	M.	9.60
Mean A.R. 1865.0		**23ʰ 17ᵐ 9ˢ.679**

Piazzi XXIII. 82.

Date.	Observer.	A.R. 1865.0.
1865 Sept. 13	M.	h. 23 m. 19 s. 28.47
25	M.	28.39
29	M.	28.22
Oct. 13	M.	28.40
16	M.	28.37
23	M.	28.27
24	S.	28.34
25	M.	28.35
Nov. 11	M.	28.41
14	M.	28.29
Mean A.R. 1865.0		**23ʰ 19ᵐ 28ˢ.351**

κ Piscium. N.A.

Date.	Observer.	A.R. 1865.0.
1862 Sept. 33	S.	s. +0.10
Dec. 28	S.	+0.07
1863 Sept. 29	S.	—0.04
30	S.	—0.02
Dec. 23	S.	+0.09
26	S.	+0.06
1864 Sept. 16	S.	+0.11
17	S.	+0.10
26	S.	+0.02
Oct. 10	S.	+0.10
21	S.	+0.06
25	S.	+0.10
26	S.	+0.15
Nov. 4	S.	+0.04
19	M.	+0.12
29	S.	+0.09
Dec. 1	S.	+0.06
Mean		+0.071
N.A.		0.712
Mean A.R. 1865.0		**23ʰ 20ᵐ 0ˢ.783**

128 Heis Aquarii.

Date.	Observer.	A.R. 1865.0.
1865 Sept. 13	M.	h. 23 m. 25 s. 26.16
29	M.	26.17
Oct. 6	M.	26.11
16	M.	26.16
17	S.	26.15
23	M.	26.30
24	S.	26.08
25	M.	26.26
Nov. 11	M.	26.15
14	M.	26.15
Mean A.R. 1865.0		**23ʰ 25ᵐ 26ˢ.169**

39 Cephei H. T.H.S.

Date.	Observer.	A.R. 1865.0.
1862 Sept. 23	S.	s. +1.16
29	S.	+0.29
Dec. 28	S.	+0.06
29	S.	+0.23
1863 Sept. 22	S.	+0.90
24	S.	+0.53
28	S.	+0.34
30	S.	+1.07
Oct. 22	S.	+0.56

Date.	Observer.	A.R. 1865.0.
		s.
1863 Nov. 2	S.	—0.94
3	S.	—0.06
4	S.	—0.08
23	S.	+0.25
Dec. 21	S.	+0.83
23	S.	+0.32
26	S.	—0.17
1864 Sept. 16	S.	+0.02
30	S.	—0.32
Oct. 10	S.	—0.19
11	S.	—0.80
Nov. 10	S.	+0.13
1	M.	+0.29
14	S.	+0.61
Dec. 9	M.	+2.70
30	M.	+1.42
1865 Sept. 30	S.	+0.18
Oct. 3	S.	—0.86
6	M.	+1.81
12	M.	+1.51
13	M.	+0.78
16	M.	—0.30
17	S.	+0.69
23	M.	—0.41
25	M.	—0.22
Nov. 14	M.	+0.58
16	M.	+0.40
17	M.	+0.78
Dec. 21	M.	+2.49
23	M.	+2.27
Mean		+0.483
T.H.S.		49.98

Mean A.R. 1865.0
23h 27m 50s.463

39 Cephei H. Sub Polo.

Date.	Observer.	A.R. 1865.0.
		s.
1862 Mar. 26	S.	+0.78
April 7	S.	—0.87
16	H.	+1.40
26	H.	+0.47
30	H.	—0.44
May. 7	S.	—0.38
15	S.	—0.28
Dec. 28	S.	+0.54
1863 Mar. 9	S.	+0.88
April 9	S.	+0.11
1864 April 21	R.	+0.17
22	S.	+0.64
May. 6	S.	+0.95
18	R.	—0.77
19	R.	+0.56
1865 Mar. 25	S.	—0.39
		s.
1865 April 3	M.	+1.37
13	M.	—0.24
17	M.	+0.64
24	M.	+0.62
25	M.	+0.23
May 3	M.	+0.54
15	M.	—0.32
24	S.	+0.42
Mean		+0.276
T.H.S.		49.98

Mean A.R. 1865.0
23h 27m 50s.256

Piazzi XXIII. 133.

Date.	Observer.	A.R. 1865.0.
		h. m. s.
1865 Sept. 29	M.	23 30 39.63
Oct. 24	S.	39.49
Nov. 11	M.	39.60
14	M.	39.52
16	M.	39.53
17	M.	39.49
Dec. 21	M.	39.71
23	M.	39.67

Mean A.R. 1865.0
23h 30m 39s.580

ι Piscium. N.A.

Date.	Observer.	A.R. 1865.0.
		s.
1862 Sept. 29	S.	+0.15
Nov. 3	S.	+0.13
1863 Sept. 28	S.	+0.01
1864 Jan. 9	S.	+0.04
Sept. 16	S.	—0.02
17	S.	+0.17
26	S.	+0.12
30	S.	—0.02
Oct. 10	S.	+0.11
11	S.	—0.09
Nov. 4	S.	0.00
10	S.	+0.02
11	M.	+0.12
14	S.	—0.02
19	M.	—0.03
Dec. 1	S.	—0.02
20	S.	+0.04
30	M.	+0.04
Mean		+0.042
N.A.		0.442

Mean A.R. 1865.0
23h 33m 0s.484

γ Cephei. N.A.

Date.	Observer.	A.R. 1865.0.
		s.
1862 Sept. 23	S.	+0.24
29	S.	+0.11
Dec. 28	S.	+0.09
29	S.	+0.07
1863 Jan. 2	S.	—0.03
5	S.	+0.28
17	S.	+0.66
July 22	S.	+0.34
Sept. 30	S.	+0.22
Nov. 3	S.	—0.16
4	S.	+0.08
Dec. 21	S.	+0.08
26	S.	—0.10
30	S.	—0.08
1864 Jan. 2	S.	+0.63
Mean		+0.162
N.A.		49.862

Mean A.R. 1865.0
23h 33m 50s.024

γ Cephei. Sub Polo.

Date.	Observer.	A.R. 1865.0.
		s.
1863 Mar. 6	S.	+0.14
1864 Apr. 21	R.	+0.16
22	S.	+0.13
1865 Apr. 13	M.	—0.05
14	S.	+0.31
17	M.	+0.08
24	M.	+0.14
25	M.	—0.04
May 10	S.	+0.10
13	S.	+0.51
Mean		+0.148
N.A.		49.862

Mean A.R. 1865.0
23h 33m 50s.010

Piazzi XXIII. 153.

Date.	Observer.	A.R. 1865.0.
		h. m. s.
1865 Sept. 13	M.	23 34 9.55
25	M.	9.74
29	M.	9.60
Oct. 13	M.	9.67
16	M.	9.67
24	S.	9.60
Nov. 11	M.	9.90
14	M.	9.72
Dec. 21	M.	9.73
23	M.	9.79
29	M.	9.77

Mean A.R. 1865.0
23h 34m 9s.704

368 Aquarii B.

Date.	Observer.	A.R. 1865.0.
		h. m. s.
1865 Sept. 13	M.	23 35 28.05
25	M.	28.16
29	M.	28.12
Oct. 6	M.	28.19
13	M.	28.12
16	M.	28.09
24	S.	28.13
Nov. 11	M.	28.25
14	M.	28.09
Dec. 21	M.	28.20

Mean A.R. 1865.0 $23^h\ 35^m\ 28^s.140$

979 Mayer.

Date.	Observer.	A.R. 1865.0.
		h. m. s.
1865 Sept. 13	M.	23 37 55.63
25	M.	55.57
29	M.	55.69
Oct. 6	M.	55.65
13	M.	55.66
16	M.	55.64
24	S.	55.78
1865 Nov. 11	M.	55.70
Dec. 21	M.	55.78
23	M.	55.57
29	M.	55.79

Mean A.R. 1865.0 $23^h\ 37^m\ 55^s.678$

981 Mayer.

Date.	Observer.	A.R. 1865.0.
		h. m. s.
1865 Sept. 8	M.	23 40 18.61
29	M.	18.80
Oct. 6	M.	18.81
13	M.	18.72
16	M.	18.78
24	S.	18.75
Nov. 11	M.	18.82
Dec. 21	M.	18.91
22	M.	18.83
23	M.	18.76

Mean A.R. 1865.0 $23^h\ 40^m\ 18^s.779$

δ Sculptoris = Pi. XXIII. 192 N.A.

Date.	Observer.	A.R. 1865.0.
		s.
1863 Nov. 23	S.	+0.12
Dec. 26	S.	+0.11
1864 Sept. 26	S.	+0.14
1864 Sept. 30	S.	+0.23
Oct. 11	S.	+0.23
17	S.	+0.08
Nov. 4	S.	+0.12
10	S.	+0.04
11	M.	+0.12
14	S.	+0.06
19	M.	−0.15
29	S.	−0.03
Dec. 1	S.	−0.02
20	S.	+0.07
30	M.	+0.09
Mean		+0.081
N.A.		53.345

Mean A.R. 1865.0 $23^h\ 41^m\ 53^s.426$

Pi. XXIII. 203.

Date.	Observer.	A.R. 1865.0.
		h. m. s.
1865 Sept. 13	M.	23 43 35.36
25	M.	35.49
29	M.	35.40
Oct. 6	M.	35.40
13	M.	35.43
16	M.	35.42
24	S.	35.48
Nov. 11	M.	35.52

Mean A.R. 1865.0 $23^h\ 43^m\ 35^s.438$

Bradley 3181.

Date.	Observer.	A.R. 1865.0.
		h. m. s.
1864 Sept. 30	S.	23 45 31.12
Oct 17	S.	31.16
Dec. 20	S.	30.92

Mean A.R. 1865.0 $23^h\ 45^m\ 31^s.067$

Br. 3181. Sub Polo.

Date.	Observer.	A.R. 1865.0.
		h. m. s.
1863 Mar. 6	S.	23 45 31.22
1864 April 21	R.	31.11
22	S.	31.26
30	R.	30.99
May 18	R.	31.35
19	R.	31.26
1865 April 8	S.	31.27
17	M.	31.21
18	S.	31.31
24	M.	30.79
25	M.	31.21
May 10	S.	31.65
13	S.	31.46

Mean A.R. 1865.0 $23^h\ 45^m\ 31^s.238$

Groombridge 4163. A.E.

Date.	Observer.	A.R. 1865.0.
		s.
1865 Sept. 13	M.	−0.05
25	M.	−0.24
29	M.	−0.25
Oct. 6	M.	+0.04
13	M.	+0.01
16	M.	−0.22
Nov. 1	S.	0.00
11	M.	−0.60
Dec. 22	M.	+0.19
Mean		−0.124
A.E. tables		17.849

Mean A.R. 1865.0 $23^h\ 48^m\ 17^s.725$

Bradley 3187.

Date.	Observer.	A.R. 1865.0.
		h. m. s.
1865 Sept. 13	M.	23 50 10.94
25	M.	11.44
29	M.	10.81
Oct. 6	M.	11.80
7	S.	11.41
13	M.	11.55
16	M.	10.95
24	S.	11.43
Nov. 1	S.	10.98
11	M.	10.91

Mean A.R. 1865.0 $23^h\ 50^m\ 11^s.222$

ω Piscium. N.A.

Date.	Observer.	A.R. 1865.0.
		s.
1862 Oct. 18	S.	−0.16
1863 Sept. 29	S.	+0.01
1864 Jan. 9	S.	+0.01
Oct. 11	S.	+0.03
17	S.	−0.07
Nov. 4	S.	+0.05
10	S.	−0.04
14	S.	−0.13
19	M.	+0.01
30	M.	+0.10
Dec. 5	M.	+0.01
8	M.	+0.03
9	M.	−0.08
20	S.	−0.16
30	M	+0.06
Mean		−0.022
N.A.		22.810

Mean A.R. 1865.0 $23^h\ 52^m\ 22^s.788$

309 Cephei. B. T.H.S.

Date.	Observer.	A.R. 1865.0.
		s.
1862 Sept. 23	S.	+0.81
29	S.	+0.43
Oct. 18	S.	+0.09
Nov. 3	S.	+0.27
Dec. 28	S.	+1.15
1863 Sept. 22	S.	—0.69
Nov. 4	S.	+0.71
23	S.	0.00
Dec. 26	S.	—0.09
30	S.	+0.13
1864 Oct. 17	S.	+0.76
1865 Oct. 7	S.	+1.15
24	S.	—0.75
Mean		+0.305
T.H.S.		15.04

Mean A.R. 1865.0
23ʰ 53ᵐ 15ˢ.345

309 Cephei B. Sub Polo.

Date.	Observer.	A.R. 1865.0.
		s.
1862 Mar. 26	S.	+0.91
April 7	S.	—0.08
30	H.	+0.31
May 15	S.	+0.47
Dec. 28	S.	+0.66
1863 April 9	S.	—0.07
1864 Mar. 24	S.	—0.46
April 21	R.	+0.18
30	R.	—0.24
May 9	R.	—0.03
18	R.	+0.65
22	R.	—0.25
1865 April 13	M.	—0.45
Mean		+0.123
T.H.S.		15.04

Mean A.R. 1865.0
23ʰ 53ᵐ 15ˢ.163

2 Ceti. C. des T.

Date.	Observer.	A.R. 1865.0.
		s.
1862 Sept. 23	S.	0.00
Dec. 28	S.	+0.05
1863 Nov. 2	S.	+0.03
Dec. 26	S.	—0.02
1864 Jan. 6	S.	+0.04
9	S.	+0.06
Sept. 30	S.	0.00
Oct. 11	S.	+0.10
17	S.	—0.01
25	S.	+0.15
Nov. 4	S.	+0.09
10	S.	+0.03
14	S.	+0.11
19	M.	—0.05
Dec. 5	M.	+0.03
8	M.	+0.04
Mean		+0.041
C. des T.		49.30

Mean A.R. 1865.0
23ʰ 56ᵐ 49ˢ.341

GENERAL CATALOGUE.

Number.	Star's Name.	A.R. 1865.0.	No. Obs.	Precession.	Secular Variation.	Proper Motion.	Declination.	Δα.
	h.	h. m. s.		s.	s.	s.	° ′	s.
1	Pi. 0.19	0 8 10.912	11	+ 3.0951	+0.0176		+26 32.1	
2	Pi. 0.38	11 35.598	11	3.1112	+0.0209		+30 46.1	
3	ι Sculptoris	14 44.040	11	3.0219	—0.0136	+0.0030 S.	—29 43.7	
4	Pi. 0.56	16 13.431	11	3.0426	—0.0058		—16 41.6	
5	A. Oe. 326	18 22.466	7	3.2196	+0.0484		+54 14.8	
6	α Phœnicis	19 36.48	8	2.9642	—0.0226	+0.022 St.	—43 2.4	
7	12 Ceti	23 8.991	18	3.0598	+0.0010	+0.0011 A.	— 4 42.2	+0.006
8	Pi. 0.91	23 37.636	7	3.0781	—0.0096		—24 32.1	
9	Pi. 0.100	25 17.155	9	3.0203	—0.0064	+0.0045 S.	—18 58.0	—0.004
10	Pi. 0.109	27 0.262	10	2.9789	—0.0127		—30 18.2	
11	Pi. 0.122	29 12.033	9	3.1556	+0.0192		+26 30.6	
12	ζ Cassiopeiæ	29 27.907	17	3.2915	+0.0474	+0.007 S.	+53 9.2	+0.004
13	Bradley 48	29 42.95	22	4.2630	+0.3657	—0.050 S.	+81 44.9	—0.035
14	Pi. 0.131	30 33.578	10	3.0783	+0.0050	+0.008 S.	+ 2 23.6	—0.007
15	α Cassiopeiæ	32 51.800	33	3.3520	+0.0554	+0.0082 N.	+55 47.8	+0.001
16	Pi. 0.142	32 55.186	10	3.0306	—0.0021		—11 53.1	
17	Pi. 0.152	35 25.532	4	3.0251	—0.0023	+0.004 S.	—12 32.6	
18	β Ceti	36 48.756	15	2.9984	—0.0054	+0.0172 N.	—18 43.7	+0.004
19	A. Oe. 681	37 6.436	5	3.4726	+0.0732		+61 47.3	
20	Pi. 0.174	38 56.770	7	3.0010	—0.0044		—17 9.7	
21	η Cassiopeiæ	40 57.045	15	3.4382	+0.0604	+0.140 S.	+57 5.9	
22	δ Piscium	41 40.879	24	3.1000	+0.0080	+0.0054 A.	+ 6 51.0	+0.003
23	Bradley 74	42 27.937	26	5.0692	+0.5514	+0.040 S.	+82 58.4	+0.035
24	A. Oe. 799	43 13.642	8	3.5244	+0.0730		+61 4.2	
25	Pi. 0.210	44 31.711	8	3.0248	—0.0003		—10 8.4	
26	γ Cassiopeiæ	48 34.924	16	3.5578	+0.0714	+0.0057 M.	+59 59.1	
27	μ Andromedæ	49 16.171	18	3.2918	+0.0306	+0.012 S.	+37 46.0	+0.005
28	Pi. 0.242	50 50.751	9	3.2636	+0.0262		+33 13.5	
29	43 Cephei H.	50 51.549	71	6.8357	+1.2893	+0.070 S.	+85 31.9	
30	α Sculptoris	52 5.994	8	2.8963	—0.0099		—30 5.3	
31	Pi. 0.258	54 26.354	9	3.2147	+0.0195	+0.007 S.	+24 33.9	—0.006
32	ε Piscium	55 56.379	18	3.1115	+0.0089	—0.0046 N.	+ 7 9.8	+0.001
33	σ Sculptoris	55 59.522	9	2.8666	—0.0107		—32 16.8	
34	A. Oe. 1092	58 49.616	9	+ 3.5782	+0.0624		+56 13.0	

Number.	Star's Name.	A.R. 1865.0.	No. Obs.	Precession.	Secular Variation.	Proper Motion.	Declination.	Δα.
		h. m. s.		s.	s.	s.	° ′	s.
35	44 Cephei H.	1 0 43.984	28	+ 4.8645	+ 0.3245	+0.0301 A.	+78 57.2	+0.038
36	β Andromedæ	2 10.893	27	3.3209	+ 0.0287	+0.0182 M.	+34 54.2	+0.004
37	Polaris	9 38.917	206	19.3224	+13.40	+0.119	+88 35.4	+0.008
38	ψ Cassiopeiæ	16 26.215	30	4.1235	+ 0.1199	+0.0111 A.	+67 25.4	+0.008
39	θ Ceti	17 16.623	14	3.0018	+ 0.0020	—0.0052 N.	— 8 52.8	+0.001
40	η Piscium	24 15.838	13	3.1965	+ 0.0143	+0.0029 N.	+14 38.9	+0.003
41	υ Persei	29 43.233	21	3.6363	+ 0.0484	+0.0059 M.	+47 56.6	+0.004
42	ν Piscium	34 24.559	13	3.1159	+ 0.0092	+0.000 A.	+ 4 48.2	+0.004
43	τ Ceti	37 47.846	19	2.9054	— 0.0002	—0.1176 M.	—16 39.0	—0.085
44	χ Ceti	42 57.347	23	2.9543	+ 0.0022	+0.0083 M.	—11 21.3	—0.006
45	ω Cassiopeiæ	45 33.813	6	4.5432	+ 0.1363		+68 1.2	
46	β Arietis	47 11.236	12	3.2920	+ 0.0184	+0.0071 N.	+20 8.8	—0.003
47	*A* Cassiopeiæ	50 55.640	5	4.8035	+ 0.1632		+70 15.0	
48	50 Cassiopeiæ	51 58.178	25	4.9755	+ 0.1858	—0.0122 Ast.	+71 45.9	—0.007
49	α Piscium	55 3.82	13	3.0946	+ 0.0085	+0.0053 M.	+ 2 6.7	+0.002
50	γ Andromedæ	55 37.483	20	3.6463	+ 0.0394	+0.0012 M.	+41 40.8	+0.001
51	Gr. 454	2 0 59.864	5	5.3286	+ 0.2202		+73 23.4	
52	67 Ceti	10 15.143	19	2.9820	+ 0.0051	+0.0055 A.	— 7 2.7	0.000
53	ο Ceti	12 31.807	18	3.0250	+ 0.0065	+0.0005 A.	— 3 35.5	—0.008
54	ι Cassiopeiæ	17 59.060	28	4.8368	+ 0.1307	—0.0096 M.	+66 47.6	—0.009
55	ξ² Ceti	20 59.081	14	3.1773	+ 0.0118	+0.0025 M.	+ 7 51.2	+0.001
56	36. Cassiop. H.	25 15.989	24	5.5406	+ 0.2028	—0.0125 M.	+72 13.5	
57	Gr. 527	28 33.564	30	8.0941	+ 0.6386	+0.022 S.	+80 52.3	
58	ε Ceti	33 2.160	22	2.8882	+ 0.0039	+0.0101 M.	—12 26.8	+0.003
59	γ Ceti	36 18.509	20	3.1101	+ 0.0095	—0.0091 N.	+ 2 39.9	0.000
60	41 Arietis	42 2.658	18	3.5077	+ 0.0230	+0.0037 M.	+26 42.1	—0.001
61	η Eridani	49 50.035	20	2.9209	+ 0.0053	+0.0054 M.	— 9 26.2	+0.002
62	Gr. 580	51 1.171	8	8.7639	+ 0.6533		+80 56.5	
63	ε Arietis	51 29.857	14	3.4164	+ 0.0186		+20 47.9	
64	β Persei	59 23.728	19	3.8743	+ 0.0356		+40 26.0	0.000
65	48 Cephei H.	3 3 19.027	7	7.3028	+ 0.3514	+0.0201 A.	+77 14.0	—0.010
66	δ Arietis	3 54.878	17	3.4060	+ 0.0172	+0.0123 A.	+19 12.8	—0.003
67	12 Eridani	6 20.335	16	2.5210	+ 0.0015	+0.0248 A.	—29 31.2	+0.005
68	ζ Eridani	9 16.709	21	2.9096	+ 0.0056		— 9 19.4	
69	α Persei	14 41.876	22	4.2413	+ 0.0484	+0.0043 N.	+49 22.7	0.000
70	ο Tauri	17 33.082	20	3.2235	+ 0.0116	—0.0037 M.	+ 8 33.1	—0.002
71	323 Cephei B.	22 35.609	31	18.6920	+ 3.1962	+0.167 S.	+86 12.8	
72	ε Eridani	26 34.349	22	2.8877	+ 0.0056	—0.0655 M.	— 9 55.0	—0.016
73	6 Camelop. H.	30 28.174	27	5.1357	+ 0.0899	—0.0025 A.	+62 46.3	—0.002
74	δ Persei	33 19.491	22	4.2351	+ 0.0417	+0.0047 N.	+47 21.2	—0.002
75	δ Eridani	36 47.011	23	2.8754	+ 0.0056	—0.0066 M.	—10 13.4	—0.002
76	η Tauri	39 27.875	13	3.5505	+ 0.0178	+0.0021 N.	+23 41.1	—0.001
77	ζ Persei	45 39.112	21	3.7528	+ 0.0223	+0.0020 N.	+31 28.8	+0.001
78	49 Cephei H.	47 37.062	13	9.6364	+ 0.5090	0.000 S.	+80 19.2	
79	γ Eridani	51 43.962	19	+ 2.7906	+ 0.0048	+0.0067 N.	—13 53.7	+0.001

Number.	Star's Name.	A.R. 1865.0.	No. Obs.	Precession.	Secular Variation.	Proper Motion.	Declination.	Δα.
		h. m. s.		s.	s.	s.	° ′	
80	Gr. 750	3 55 10.751	35	+16.6698	+ 1.8145	+0.0084	+85 11.6	
81	Gr. 766	57 14.600	10	13.1109	+ 1.0062		+83 28.1	s.
82	Gr. 774	4 0 36.826	10	12.5213	+ 0.8696	—0.010 S.	+83 0.3	+0.005
83	Gr. 779	3 42.856	25	10.0486	+ 0.4845		+80 29.6	
84	o^1 Eridani	5 16.741	15	2.9229	+ 0.0060	+0.0020 A.	— 7 11.5	—0.001
85	γ Tauri	12 6.836	22	3.3968	+ 0.0116	+0.0094 N.	+15 17.9	—0.004
86	δ Tauri	15 9.172	21	3.4427	+ 0.0122	+0.0067 M.	+17 13.4	+0.002
87	64 Tauri	16 19.026	21	3.4415	+ 0.0120	+0.0081 M.	+17 7.7	+0.003
88	ε Tauri	20 44.224	16	3.4858	+ 0.0122	+0.0087 N.	+18 52.7	—0.003
89	53 Eridan	32 0.011	21	2.7487	+ 0.0044	—0.0038 M.	—14 34.2	—0.001
90	50 Cephei H.	35 13.184	25	10.9052	+ 0.4167		+80 57.6	
91	μ Eridani	38 45.339	23	2.9941	+ 0.0057	+0.0056 M.	— 3 30.3	+0.001
92	9 Camelop.	40 39.047	17	5.9098	+ 0.0700		+66 6.5	
93	ι Aurigæ	48 12.405	20	3.8952	+ 0.0145	+0.0019 N.	+32 56.9	—0.004
94	10 Camelop.	51 25.200	12	5.3052	+ 0.0420		+60 14.4	+0.003
95	ι Tauri	55 1.775	20	3.5738	+ 0.0096	+0.0057 A.	+21 23.6	+0.000
96	RC. 1377.	58 19.808	6	19.5956	+ 1.1945		+85 32.4	
97	ε Leporis	59 45.003	20	2.5347	+ 0.0035	+0.0041 M.	—22 33.3	—0.004
98	τ Orionis	5 11 3.287	22	2.9110	+ 0.0041	+0.0030 M.	— 6 58.6	+0.002
99	64 Camelop. B	19 3.942	22	18.4708	+ 0.6817	+0.0246 S.	+85 7.0	
100	74 Camelop. B.	21 41.629	13	7.9726	+ 0.0792		+74 56.8	
101	β Leporis	22 27.848	19	2.5680	+ 0.0032	+0.0025 M.	—20 52.2	0.000
102	δ Orionis	25 6.753	24	3.0616	+ 0.0040	+0.0008 N.	— 0 24.1	0.000
103	α Leporis	26 46.719	21	2.6431	+ 0.0031	+0.0010 N.	—17 55.3	0.000
104	ε Orionis	29 21.932	24	3.0411	+ 0.0037	+0.0005 N.	— 1 17.5	0.000
105	α Columbæ	34 45.796	20	2.1696	+ 0.0030	+0.0024 N.	—34 8.9	—0.006
106	κ Orionis	41 21.425	22	2.8427	+ 0.0028	+0.0016 M.	— 9 43.2	0.000
107	θ Aurigæ	50 31.020	21	4.0848	+ 0.0035	+0.0059 M.	+37 12.0	+0.002
108	1 Geminorum	55 54.938	24	3.6457	+ 0.0023	—0.0009 M.	+23 16.0	0.000
109	ν Orionis	59 51.929	19	3.4237	+ 0.0018	+0.0033 A.	+14 46.9	—0.003
110	η Geminorum	6 6 43.762	20	3.6256	+ 0.0008	—0.0033 M.	+22 32.6	—0.002
111	κ Columbæ	11 45.081	15	2.1326	+ 0.0023		—35 5.9	
112	μ Geminorum	14 47.632	20	3.6257	— 0.0002	+0.0059 N.	+22 34.8	—0.007
113	β Canis Maj.	16 45.406	24	2.6405	+ 0.0018	+0.0031 M.	—17 53.5	+0.006
114	ν Geminorum	20 56.871	24	3.5632	— 0.0007		+20 17.6	
115	23 Camelop. H.	23 8.359	32	10.3905	— 0.0983	—0.0210 A.	+79 42.1	—0.021
116	γ Geminorum	29 54.784	24	3.4638	— 0.0013	+0.0039 N.	+16 30.7	—0.003
117	ε Geminorum	35 37.559	19	3.6941	— 0.0034	+0.0016 M.	+25 15.7	0.000
118	(51 Cephei H.	36 11.92[illegible])	59	30.4985	— 1.8535	—0.0274 S.	+87 14.6	
119	24 Camelop. H.	40 19.885	10	8.8369	— 0.1119	+0.0247 A.	+77 8.5	—0.015
120	κ Canis Maj.	44 48.010	20	2.2401	+ 0.0002		—32 21.3	
121	θ Canis Maj.	47 55.111	19	2.7960	+ 0.0006	—0.0067 M.	—11 52.3	—0.005
122	ε Canis Maj.	53 19.288	21	2.3560	+ 0.0015	+0 0014 N.	—28 47.4	—0.005
123	ζ Geminorum	56 6.089	20	3.5627	— 0.0049	+0.0010 M.	+20 45.9	+0.001
124	γ Canis Maj.	57 39.113	23	+ 2.7134	+ 0.0007	+0.0039 M.	—15 26.2	—0.005

Number.	Star's Name.	A.R. 1865.0.	No. Obs.	Precession.	Secular Variation.	Proper Motion.	Declination.	Δα.
		h m s.		s.	s.	s.	° ′	s.
125	25 Camelop. H.	7 2 28.993	26	+13.0670	— 0.4750	+0.013 S.	+82 39.6	
126	δ Canis Maj.	2 54.224	15	2.4326	+ 0.0013	0.0000 N.	—26 10.9	0.000
127	λ Geminorum	10 20.048	21	3.4552	— 0.0053	—0.0019 M.	+16 46.9	—0.001
128	δ Geminorum	12 3.543	20	3.5905	— 0.0071	—0.0008 N.	+22 13.7	—0.008
129	4 Ursæ Min. B.	16 36.100	33	76.4246	—27.7835	—0.009 S.	+89 0.8	
130	Gr. 1308	16 48.204	18	6.3151	— 0.0821	0.0000 Ast.	+68 44.2	
131	η Canis Maj.	18 45.411	17	2.3720	+ 0.0013	—0.0013 M.	—29 2.5	—0.001
132	28 Camelop. H.	33 48.681	10	10.4758	— 0.4240	—0.2027 A.	+80 35.7	+0.125
133	Pi. VII. 187.	43 24.759	10	9.7848	— 0.3972		+79 50.4	
134	ξ Argus	43 37.024	13	2.5223	+ 0.0010	+0.0011 M.	—24 31.4	—0.010
135	156 Camelop. B.	44 10.665	29	15.4016	— 1.2243	+0.0113 S.	+84 26.2	
136	χ Geminorum	55 13.423	17	3.6983	— 0.0146	0.000 A.	+28 10.2	—0.008
137	ζ Argus	58 50.434	5	2.1095	+ 0.0015		—39 37.5	
138	ι Navis	8 1 47.789	10	2.5598	+ 0.0014	—0.0055 N.	—23 55.0	0.000
139	β Cancri	9 11.569	20	3.2620	— 0.0069	—0.0023 M.	+ 9 36.0	—0.013
140	Gr. 1418	15 34.156	23	17.2376	— 2.1843	—0.1200 S.	+85 31.3	—0.096
141	o Ursæ Maj.	19 1.375	22	5.0660	— 0.0762	—0.0155 M.	+61 9.9	—0.020
142	A Ursæ Maj.	22 28.834	27	5.4668	— 0.1034	—0.0042 M.	+65 36.1	—0.005
143	η Cancri	24 53.911	18	3.4827	— 0.0129	—0.0022 A.	+20 53.8	—0.007
144	δ Cancri	37 0.611	15	3.4203	— 0.0123	+0.0005 M.	+18 38.9	0.000
145	ε Hydræ	39 37.533	12	3.1952	— 0.0069	—0.0116 N.	+ 6 54.7	—0.009
146	6 Ursæ Maj.	45 0.840	21	5.2378	— 0.1078		+65 7.0	
147	R. C. 2218	46 36.981	10	13.8637	— 1.7325		+84 42.9	
148	ι Ursæ Maj.	49 57.031	23	4.1879	— 0.0444	—0.0437 N.	+48 34.1	—0.010
149	σ² Ursæ Maj.	58 28.265	26	5.3888	— 0.1338	—0.0043 A.	+67 40.7	—0.005
150	κ Cancri	9 0 26.020	23	3.2579	— 0.0094	+0.0003 N.	+11 12.6	+0.001
151	λ Argus	3 1.992	21	2.2045	+ 0.0046	—0.005 St.	—42 53.3	—0.005
152	θ Hydræ	7 20.378	22	3.1169	— 0.0056	+0.0100 M.	+ 2 52.9	+0.007
153	83 Cancri	11 26.619	16	3.3671	— 0.0133	—0.0080 A.	+18 16.5	—0.007
154	1 Draconis H.	17 34.496	38	9.1988	— 0.8042	+0.0067 S.	+81 55.1	
155	θ Ursæ Maj.	23 48.684	17	4.1602	— 0.0560	—0.1057 N.	+52 17.4	—0.012
156	Gr. 1564	30 38.281	22	5.2807	— 0.1631	—0.0164 A.	+69 50.9	—0.009
157	o Leonis	33 56.620	21	3.2184	— 0.0091	—0.0098 M.	+10 30.3	—0.012
158	ε Leonis	38 10.994	20	3.4226	— 0.0178	—0.0031 N.	+24 23.7	—0.010
159	υ Ursæ Maj.	41 21.877	20	4.3705	— 0.0819	—0.0359 M.	+59 40.3	—0.040
160	μ Leonis	45 4.800	20	3.4433	— 0.0196	—0.0164 N.	+26 38.5	—0.005
161	π Leonis	53 4.684	16	3.1784	— 0 0079	—0.0024 A.	+ 8 41.4	—0.009
162	λ Hydræ	10 4 0.460	22	2.9368	+ 0.0016	—0.0118 M.	—11 41.3	—0.013
163	λ Ursæ Maj.	8 56.644	22	3.6637	— 0.0383	—0.0166 M.	+43 35.2	—0.017
164	γ Leonis pr.	12 31.549	18	3.2971	— 0.0150	+0.0214 N.	+20 31.4	—0.006
165	30 Camelop. H.	14 19.364	26	8.0904	— 0.9792	—0.0344 S.	+83 14.6	
166	μ Hydræ	19 33.789	22	2.9067	+ 0.0041	—0.0077 M.	—16 8.9	—0.008
167	ρ Leonis	25 42.073	16	3.1652	— 0.0078	+0.0010 N.	+10 0.0	—0.009
168	193 Camelop. B.	29 59.162	20	6.3494	— 0.5584	+0.020 S.	+81 7.8	+0.005
169	l Leonis	42 9.577	14	+ 3.1596	— 0.0079	+0.0010 N.	+11 15.5	—0.006

Number.	Star's Name.	A.R. 1865.0.	No. Obs.	Precession.	Secular Variation.	Proper Motion.	Declination.	Δα.
		h. m. s.		s.	s.	s.	° ′	s.
170	β Ursæ Maj.	10 53 40.445	25	+ 3.6607	— 0.0627	+0.0103 M.	+57 6.3	+0.020
171	α Ursæ Maj.	55 22.251	37	3.7849	— 0.0819	—0.0175 N.	+62 28.7	0.000
172	χ Leonis	58 3.152	19	3.1215	— 0.0055	—0.0240 A.	+ 8 3.9	—0.009
173	ψ Ursæ Maj.	11 2 3.744	19	3.4082	— 0.0366	—0.0080 M.	+45 13.8	—0.009
174	δ Leonis	6 55.513	14	3.1903	— 0.0180	+0.0113 N.	+21 15.8	—0.009
175	δ Crateris	12 35.590	20	3.0022	+ 0.0065	—0.0075 N.	—14 2.9	—0.008
176	γ Crateris	18 8.454	22	2.9969	+ 0.0083	—0.0070 R.	—16 56.6	—0.008
177	τ Leonis	20 59.670	20	3.0852	— 0.0022	+0.0015 N.	+ 3 36.0	0.000
178	202 Camelo. B.	22 11.794	31	4.6076	— 0.4133	—0.0457 S.	+81 52.2	
179	λ Draconis	23 21.197	10	3.6582	— 0.1127	—0.0098 M.	+70 4.5	0.000
180	υ Leonis	30 2.214	15	3.0707	+ 0.0005	+0.0006 N.	— 0 4.7	—0.009
181	χ Ursæ Maj.	38 54.623	21	3.2099	— 0.0357	—0.0101 M.	+48 31.7	—0.011
182	γ Ursæ Maj.	46 42.968	24	3.1792	— 0.0432	+0.0132 N.	+54 26.7	0.000
183	Groom. 1845	53 11.005	10	3.3404	— 0.2362		+81 36.4	
184	π Virginis	53 57.304	22	3.0755	— 0.0022	—0.0009 M.	+ 7 22.0	0.000
185	Groom. 1850	57 53.475	9	3.2629	— 0.5226	—0.060 S.	+80 20.1	+0.053
186	α Corvi	12 1 27.400	4	3.0747	+ 0.0154	+0.0056 R.	—23 58.5	—0.002
187	ε Corvi	3 11.208	21	3.0784	+ 0.0143	—0.0016 M.	—21 52.1	—0.006
188	207 Camelo. B.	4 51.966	15	2.8564	— 0.1898		+82 27.7	
189	γ Corvi	8 52.028	20	3.0865	+ 0.0116	—0.0093 M.	—16 47.5	—0.011
190	2 H. Canum	9 42.688	5	3.0330	— 0.0169		+33 49.0	
191	5 Ursæ Min. B.	12 52.584	37	1.5434	+ 0.0005	+0.301 S.	+87 11.2	
192	η Virginis	13 0.024	20	+ 3.0708	+ 0.0028	—0.0023 N.	+ 0 5.0	—0.006
193	6 Ursæ Min. B.	14 19.796	31	— 0.0126	+ 1.1134	—0.090 S.	+88 26.9	
194	12 Comæ Ber.	15 43.016	7	+ 3.0250	— 0.0115	—0.0009 M.	+26 35.7	0.000
195	13 Comæ Ber.	17 32.035	6	3.0192	— 0.0115	—0.0013 M.	+26 50.8	0.000
196	14 Comæ Ber.	19 38.850	6	3.0100	— 0.0120	+0.0007 M.	+28 1.0	0.000
197	15 Comæ Ber.	20 12.357	21	3.0056	— 0.0126	—0.0062 R.	+29 1.1	—0.006
198	16 Comæ Ber.	20 14.121	3	3.0093	— 0.0117	+0.0053 M.	+27 34.4	—0.002
199	17 Comæ Ber.	22 10.066	7	3.0061	— 0.0109	+0.0007 M.	+26 39.6	0.000
200	δ Corvi	22 52.988	22	3.1085	+ 0.0119	—0.0091 M.	—15 45.8	—0.013
201	4 Draconis	24 10.737	6	2.6852	— 0.0572		+69 57.0	
202	η Corvi	25 7.032	22	3.1113	+ 0.0119	—0.029 M.	—15 26.9	—0.043
203	5 H. Canum	26 59.726	7	2.9649	— 0.0150		+33 59.7	
204	β Corvi	27 18.078	24	3.1372	+ 0.0163	+0.0005 N.	—22 39.0	—0.002
205	24 Comæ Ber.	28 21.441	6	3.0137	— 0.0062	+0.0027 A.	+19 7.2	—0.001
206	γ Virginis med.	34 49.278	19	3.0734	+ 0.0044	—0.0357 M.	— 0 42.5	—0.005
207	ρ Virginis	35 3.161	5	3.0314	— 0.0016	+0.008 S.	+10 58.8	—0.003
208	60 Canum B.	38 46.848	6	2.8363	— 0.0217		+46 10.7	
209	30 Comæ Ber.	42 42.706	9	2.9376	— 0.0099	—0.0069 M.	+28 17.3	+0.003
210	31 Comæ Ber.	45 7.269	9	2.9302	— 0.0097		+28 16.6	
211	ε Ursæ Maj.	48 4.824	11	2.6472	— 0.0271	+0.0163 M.	+56 41.6	+0.007
212	32 Camelop. H.	48 10.064	16	0.3503	+ 0.2259		+84 8.8	—0.008
213	δ Virginis	48 48.267	21	3.0505	+ 0.0027	—0.0307 M.	+ 4 7.9	—0.005
214	12 Canum Ven.	49 42.529	21	+ 2.8376	— 0.0151	—0.0185 N.	+39 2.9	

Number.	Star's Name.	A.R. 1865.0.	No. Obs.	Precession.	Secular Variation.	Proper Motion.	Declination.	Δα.
		h. m. s.		s.	s.	s.	° ′	s.
215	ε Virginis	12 55 27.415	28	+ 3.0047	— 0.0006	—0.0173 M.	+11 41.1	—0.005
216	14 Canum Ven.	59 25.564	5	2.8171	— 0.0124		+36 31.3	
217	θ Virginis	13 2 57.774	21	3.1015	+ 0.0080	—0.0023 N.	— 4 49.0	—0.004
218	43 Comæ Ber.	5 34.291	14	2.8655	— 0.0078	—0.0576 M.	+28 33.8	—0.014
219	ι Centauri	13 1.130	11	+ 3.3752	+ 0.0305		—36 0.0	
220	214 Camelo. B.	20 11.745	75	— 2.7017	+ 0.9840	—0.0734 S.	+85 27.6	
221	69 Ursæ Maj.H.	23 29.513	3	+ 2.2241	— 0.0154	—0.0083 A.	+60 38.6	+0.003
222	h Virginis	25 51.674	6	3.1525	+ 0.0113		— 9 28.1	
223	78 o Virginis	27 17.562	6	3.0331	+ 0.0045		+ 4 21.3	
224	ζ Virginis	27 49.012	20	3.0701	+ 0.0066	—0.0185 N.	+ 0 5.7	
225	17 H. Canum	28 46.044	7	2.6782	— 0.0091	+0.0054 A.	+37 52.5	—0.002
226	25 Canum	31 27.674	10	2.6797	— 0.0085		+36 58.9	
227	1 H. Bootis	32 55.291	7	2.9643	+ 0.0017		+11 26.0	
228	i Centauri	38 1.571	7	3.4223	+ 0.0279	—0.0348 M.	—32 21.6	+0.015
229	Groom. 2043	40 28.174	7	2.6084	— 0.0081		+39 10.8	
230	21 H. Can.Ven.	41 10.376	7	2.6047	— 0.0080		+39 13.2	
231	η Ursæ Majoris	42 13.080	21	+ 2.3839	— 0.0102	—0.0102 N.	+49 59.3	0.000
232	219 Camelo. B.	46 20.874	23	— 2.1236	+ 0.5622	+0.040 S.	+83 25.8	+0.032
233	i Draconis	47 29.342	6	+ 1.7516	— 0.0002	+0.0060 A.	+65 23.5	—0.003
234	η Bootis	48 15.409	16	2.8606	— 0.0005	—0.0037 N.	+19 4.5	—0.006
235	τ Virginis	54 46.704	22	3.0463	+ 0.0066	+0.0019 M.	+ 2 11.9	—0.004
236	θ Centauri	58 44.929	17	3.5468	+ 0.0318	—0.0428 R.	—35 42.3	—0.010
237	α Draconis	14 0 44.100	25	+ 1.6280	+ 0.0051	—0.0054 M.	+65 1.3	+0.017
238	Groom. 2099	3 47.273	24	— 7.8685	+ 2.5087	—0.027 S.	+86 24.2	
239	42 H. Virginis	5 25.947	9	+ 3.0339	+ 0.0065		+ 3 2.8	
240	λ Virginis	11 48.603	21	3.2354	+ 0.0142		—12 44.9	
241	A Bootis	12 17.390	14	2.5383	— 0.0037		+36 8.0	
242	18 H. Bootis	16 44.478	12	2.9511	+ 0.0045		+ 9 3.8	
243	f Bootis	20 10.665	11	2.7940	+ 0.0011	—0.0022 M.	+19 50.1	+0.001
244	θ Bootis	20 36.070	27	2.0688	— 0.0023	—0.0242 N.	+52 28.6	
245	ρ Bootis	26 0.679	22	2.5937	— 0.0014	—0.0055 M.	+30 57.9	—0.001
246	γ Bootis	26 38.527	12	2.4268	— 0.0026	—0.008 S.	+38 54.0	+0.004
247	σ Bootis	28 48.143	23	2.5979	— 0.0011	+0.0175 M.	+30 20.0	0.000
248	π Bootis pr.	34 23.022	14	2.8160	+ 0.0027	+0.0020 M.	+16 59.9	—0.001
249	μ Virginis	35 56.952	33	3.1455	+ 0.0106	+0.0093 M.	— 5 4.2	+0.009
250	ε Bootis	39 5.474	27	2.6229	+ 0.0002	—0.0016 N.	+27 38.7	—0.001
251	109 Virginis	39 25.575	11	3.0340	+ 0.0075	—0.0055 M.	+ 2 27.8	+0.003
252	ξ Bootis	45 9.851	11	2.7557	+ 0.0023	+0.0114 M.	+19 39.8	—0.005
253	Gr. 2164	48 0.940	28	+ 1.5311	+ 0.0094	—0.0157 A.	+59 50.6	—0.009
254	β Ursæ Minor.	51 8.014	38	— 0.2499	+ 0.1044	—0.0054 W.	+74 42.5	
255	40 Bootis	54 26.396	9	+ 2.3026	— 0.0001	—0.0039 M.	+39 48.1	+0.002
256	γ Scorpii	56 10.535	22	3.4993	+ 0.0210	—0.0035 M.	—24 44.9	—0.003
257	ω Bootis	56 11.760	9	2.6267	+ 0.0015	+0.0015 M.	+25 32.6	—0.001
258	ψ Bootis	58 39.682	24	+ 2.5822	+ 0.0013	—0.0120 A.	+27 28.5	—0.003
259	223 Camelo. B.	59 39.744	12	— 4.6881	+ 0.7168	+0.100 S.	+83 3.8	—0.020

Number.	Star's Name.	A.R. 1865.0.	No. Obs.	Precession.	Secular Variation.	Proper Motion.	Declination.	Δα.
		h. m. s.		s.	s.	s.	° ′	s.
260	c Bootis	15 1 22.340	25	+ 2.6194	+ 0.0017	+0.0137 R.	+25 23.8	+0.009
261	β Libræ	15 9 44.780	35	3.2248	+ 0.0119	—0.0058 N.	— 8 52.9	—0.002
262	δ Bootis	15 10 3.604	7	2.4104	+ 0.0012	+0.0105 M.	+33 49.2	—0.005
263	5 Serpentis	15 12 25.437	8	3.0313	+ 0.0078	+0.025 S.	+ 2 16.8	—0.013
264	η Coronæ	15 17 37.696	11	2.4663	+ 0.0018	+0.0114 M.	+30 46.6	—0.006
265	μ Bootis	15 19 23.470	23	+ 2.2768	+ 0.0016	—0.0109 N.	+37 51.1	—0.006
266	γ Ursæ Minoris	15 20 58.002	17	— 0.1522	+ 0.0758		+72 18.9	
267	57 Urs. Min. B.	15 22 4.772	30	—23.1403	+ 7.7691	+0.031 S.	+87 44.7	
268	37 Libræ	15 26 48.199	28	+ 3.2483	+ 0.0118	+0.0211 M.	— 9 35.9	+0.017
269	θ Coronæ	15 27 29.209	10	2.4184	+ 0.0021		+31 49.0	
270	μ Coronæ	15 30 17.711	10	2.1970	+ 0.0023	+0.0018 M.	+39 27.6	—0.001
271	φ Bootis	15 32 58.816	14	+ 2.1465	+ 0.0026	+0.0067 A.	+40 47.7	—0.004
272	θ Ursæ Minoris	15 35 29.180	12	— 1.9224	+ 0.1940	—0.012 S.	+77 47.9	+0.006
273	ι Serpentis	15 35 31.998	27	+ 2.6755	+ 0.0037	—0.0049 M.	+20 6.4	—0.005
274	λ Serpentis	15 39 53.620	29	2.9214	+ 0.0061	—0.0118 M.	+ 7 46.7	—0.013
275	β Serpentis	15 39 57.525	17	2.7601	+ 0.0044	+0.0068 M.	+15 50.8	—0.004
276	λ [χ] Lupi	15 42 23.385	11	3.7931	+ 0.0240		—33 12.8	
277	μ Serpentis	15 42 34.668	27	3.1290	+ 0.0090	—0.0040 M.	— 3 0.9	—0.004
278	δ Coronæ	15 43 56.048	15	2.5188	+ 0.0030	—0.0064 M.	+26 29.0	+0.003
279	ε Serpentis	15 44 5.359	26	2.9761	+ 0.0068	+0.0094 N.	+ 4 53.2	—0.012
280	λ Libræ	15 45 30.227	12	+ 3.4708	+ 0.0153	+0.0010 M.	—19 45.7	—0.001
281	ζ Ursæ Minoris	15 48 56.927	36	— 2.3141	+ 0.2046	+0.012 Wg.	+78 12.5	—0.001
282	γ Serpentis	15 50 13.170	23	+ 2.7452	+ 0.0045	+0.0229 M.	+16 6.3	+0.016
283	π Scorpii	15 50 41.487	15	3.6151	+ 0.0181		—25 43.4	
284	ε Coronæ	15 51 59.977	15	2.4865	+ 0.0031	—0.0040 N.	+27 16.2	+0.001
285	δ Scorpii	15 52 21.363	22	3.5348	+ 0.0161	0.000 N.	—22 04.1	0.000
286	β Scorpii	15 57 35.515	18	+ 3.4769	+ 0.0143	+0.0005 N.	—19 26.0	0.000
287	62 Ursæ Min. B.	15 57 40.680	30	— 6.8010	+ 0.7200	+0.015 S.	+83 21.0	
288	θ Draconis	15 59 21.855	17	+ 1.1527	+ 0.0148	—0.038 Wg.	+ 8 55.6	—0.025
289	ν Scorpii	16 4 9.217	12	3.4761	+ 0.0137	+0.0015 M.	—19 6.4	—0.001
290	87 Draconis B.	16 5 58.013	12	0.1388	+ 0.0413	—0.010 S.	+68 10.0	+0.001
291	δ Ophiuchi	16 7 16.419	24	3.1398	+ 0.0083	—0.0021 N.	— 3 20.6	+0.003
292	σ Coronæ	16 9 37.425	11	2.2657	+ 0.0033	—0.0227 M.	+34 12.1	+0.012
293	ε Ophiuchi	16 11 10.850	30	3.1616	+ 0.0084	+0.0061 M.	— 4 21.6	+0.006
294	σ Scorpii	16 12 59.307	16	+ 3.6343	+ 0.0157		—25 15.9	
295	19 Ursæ Min.	16 14 42.834	11	— 1.8135	+ 0.1273	+0.0023 A.	+76 13.0	—0.001
296	γ Herculis	16 15 57.978	22	+ 2.6462	+ 0.0040	—0.0028 M.	+19 28.2	—0.002
297	ξ Coronæ	16 16 50.322	16	2.3432	+ 0.0034	—0.0056 M.	+31 12.4	+0.003
298	η Draconis	16 22 10.072	37	+ 0.7994	+ 0.0190	+0.0024 N.	+61 49.2	—0.009
299	λ Ophiuchi	16 24 6.424	18	3.0222	+ 0.0064		+ 2 16.9	
300	β Herculis	16 24 25.060	25	2.5823	+ 0.0038	—0.0059 M.	+21 47.1	—0.006
301	τ Scorpii	16 27 29.027	21	+ 3.7228	+ 0.0153	+0.0015 M.	—27 56.0	+0.001
302	A Draconis	16 28 15.478	11	— 0.1465	+ 0.0416	—0.0051 M.	+69 3.6	+0.004
303	ζ Ophiuchi	16 29 43.652	27	+ 3.2950	+ 0.0089	+0.0019 N.	—10 17.4	+0.001
304	17 Draconis	16 33 2.558	12	+ 1.4112	+ 0.0085		+53 11.8	

Number.	Star's Name.	A.R. 1865.0.	No. Obs.	Precession.	Secular Variation.	Proper Motion.	Declination.	Δα.
		h. m. s.		s.	s.	s.	° ′	s.
305	Pi. XVI. 182	16 33 19.168	24	— 3.4725	+ 0.1990		+78 14.9	
306	ζ Herculis	36 11.880	20	+ 2.2953	+ 0.0035	—0.0315 M.	+31 50.9	+0.003
307	Pi. XVI. 195	36 29.830	12	— 2.6647	+ 0.1428	—0.0200 A.	+77 42.8	+0.010
308	η Herculis	38 16.131	21	+ 2.0499	+ 0.0040	+0.0030 N.	+39 10.9	+0.003
309	ε Scorpii	41 25.578	22	3.9216	+ 0.0163	—0.0481 M.	—34 2.7	—0.006
310	ζ^1 Scorpii	44 28.870	20	4.2153	+ 0.0207		—42 8.1	
311	ζ^2 Scorpii	45 5.566	20	4.2160	+ 0.0206	—0.021 St.	—42 7.6	—0.008
312	53 Herculis	47 51.005	16	2.2787	+ 0.0035	—0.0081 M.	+31 55.6	+0.004
313	κ Ophiuchi	51 16.785	21	2.8552	+ 0.0046	—0.0188 N.	+ 9 35.2	+0.005
314	30 Ophiuchi	53 56.643	15	3.1609	+ 0.0062	—0.0056 M.	— 4 1.0	+0.003
315	ε Herculis	55 7.510	25	2.2957	+ 0.0034	—0.0017 M.	+31 7.6	
316	h Draconis	55 17.439	10	+ 0.2741	+ 0.0218	+0.040 S.	+65 20.5	—0.021
317	ε Ursæ Minoris	59 54.869	36	— 6.4218	+ 0.3047	+0.012 S.	+82 15.2	
318	μ Draconis	17 2 32.386	9	+ 1.2452	+ 0.0080	—0.008 S.	+54 38.9	+0.004
319	η Ophiuchi	2 38.300	26	+ 3.4314	+ 0.0075	+0.0031 M.	—15 33.3	+0.003
320	77 Ur. Min. B.	5 56.155	6	— 1.9499	+ 0.0686		+75 29.0	
321	π Herculis	10 20.749	24	+ 2.0884	+ 0.0035		+36 57.8	
322	ξ Ophiuchi	12 54.969	23	3.5724	+ 0.0075	+0.0207 M.	—20 57.9	+0.012
323	θ Ophiuchi	13 43.313	22	3.6778	+ 0.0082		—24 51.7	+0.007
324	271 Herculis B.	17 17.896	10	1.9643	+ 0.0036		+40 6.5	
325	ρ Herculis	19 1.530	12	2.0697	+ 0.0035		+37 16.3	
326	λ Herculis	25 17.032	12	2.4202	+ 0.0030		+26 12.9	
327	β Draconis	27 23.019	30	+ 1.3523	+ 0.0054	—0.0002 N.	+52 24.1	+0.005
328	Gr. 2456	29 52.818	10	— 4.6435	+ 0.0991		+80 15.1	
329	ι Herculis	35 39.256	11	+ 1.6905	+ 0.0037		+46 4.8	
330	β Ophiuchi	36 48.282	25	+ 2.9633	+ 0.0032	—0.0022 M.	+ 4 37.6	+0.006
331	ω Draconis	37 44.511	10	— 0.3632	+ 0.0141	+0.0042 A.	+68 49.2	+0.001
332	γ Ophiuchi	41 7.502	12	+ 3.0067	+ 0.0030	—0.0030 Ast.	+ 2 45.6	+0.002
333	μ Herculis	41 10.584	21	2.3684	+ 0.0028	—0.0230 N.	+27 48.1	+0.007
334	ψ Draconis	44 20.752	30	— 1.0866	+ 0.0158	—0.0018 A.	+72 12.8	—0.001
335	f Herculis	48 54.442	6	+ 1.9490	+ 0.0029		+40 2.1	
336	ξ Draconis	51 11.695	23	1.0220	+ 0.0041	+0.0139 Ast.	+56 53.7	+0.017
337	θ Herculis	51 37.447	12	2.0543	+ 0.0023		+37 16.2	
338	γ Draconis	53 28.349	27	+ 1.3906	+ 0.0033	+0.0002 N.	+51 30.4	—0.003
339	35 Draconis	55 29.740	9	— 2.7091	+ 0.0121	+0.0113 A.	+76 58.7	—0.007
340	γ Sagittarii	57 8.280	12	+ 3.8559	+ 0.0022	—0.0050 N.	—30 25.3	0.000
341	70 Ophiuchi pr.	58 37.909	26	3.0118	+ 0.0022	+0.0131 M.	+ 2 32.0	+0.014
342	b Herculis	18 1 54.097	10	2.2820	+ 0.0024	—0.0069 M.	+30 32.7	+0.004
343	Gr. 2517	3 24.815	10	1.8047	+ 0.0026		+43 26.8	
344	μ Sagittarii	5 41.482	24	3.5864	+ 0.0011	—0.0001 N.	—21 5.4	+0.006
345	A Herculis	6 49.441	10	2.2561	+ 0.0023	+0.002 S.	+31 22.4	—0.001
346	η Sagittarii	8 29.626	10	+ 4.0703	— 0.0003		—36 47.9	
347	40 Draconis	10 8.177	23	— 4.4872	— 0.0241	+0.0185 M.	+79 58.7	
348	41 Draconis	10 14.467	20	— 4.4895	— 0.0245	+0.0187 M.	+79 58.9	+0.008
349	δ Sagittarii	12 21.183	20	+ 3.8379	— 0.0004	+0.0032 M.	—29 52.9	

Number.	Star's Name.	A.R. 1865.0.	No. Obs.	Precession.	Secular Variation.	Proper Motion.	Declination.	Δα.
		h. m. s.		s.	s.	s.	° ′	s.
350	36 Draconis	18 13 7.124	11	+ 0.2912	— 0.0002	+0.0541 A.	+64 21.1	—0.034
351	η Serpentis	14 19.569	20	+ 3.1393	+ 0.0011	—0.0373 N.	— 2 55.9	—0.025
352	(δ Ursæ minoris	15 53.506)	59	—19.4039	— 0.4634	+0.029 W.	+86 36.2	
353	24 Ursæ min.	20 45.011	10	—22.1822	— 0.7778	+0.080 S.	+86 58.9	—0.048
354	χ Draconis	23 29.228	10	— 1.1915	— 0.0146	+0.1148 M.	+72 40.4	0.000
355	42 Draconis	25 35.691	9	+ 0.1584	— 0.0045	+0.0136 M.	+65 28.8	—0.008
356	3 H. Scuti	27 51.705	11	3.2653	— 0.0002	—0.0008 N.	— 8 20.1	0.000
357	45 Draconis	30 14.756	10	1.0346	— 0.0006	—0.003 M.	+56 56.6	+0.002
358	110 Herculis	39 51.205	24	2.5808	+ 0.0014	+0.0025 M.	+20 25.2	+0.003
359	β Lyræ	45 5.770	21	2.2127	+ 0.0017	+0.0012 N.	+33 12.5	+0.006
360	ν¹ Sagittarii	46 1.172	14	3.6245	— 0.0042		—22 54.5	
361	σ Sagittarii	46 53.635	25	3.7225	— 0.0052	+0.0005 N.	—26 27.7	0.000
362	ν² Sagittarii	46 57.503	15	3.6221	— 0.0043	+0.0087 M.	—22 50.2	—0.005
363	o Draconis	49 12.450	10	0.8777	— 0.0042	+0.0081 M.	+59 13.4	—0.005
364	θ Serpentis pr.	49 30.530	22	+ 2.9789	— 0.0002		+ 4 1.8	
365	50 Draconis	50 42.807	23	— 1.8913	— 0.0540		+75 16.4	
366	5 Heis Aquilæ	52 14.383	12	+ 2.6682	+ 0.0009		+17 10.9	
367	ε Aquilæ	53 29.760	24	2.7251	+ 0.0007	—0.0038 M.	+14 53.2	—0.005
368	γ Lyræ	53 53.652	12	2.2425	+ 0.0016		+32 30.4	
369	o Sagittarii	56 35.543	19	3.5930	— 0.0052	+0.0061 M.	—21 56.2	+0.007
370	16 Lyræ	57 37.195	13	1.6947	+ 0.0004		+46 44.7	
371	ζ Aquilæ	59 12.365	41	2.7567	+ 0.0005	+0.0001 N.	+13 39.9	+0.010
372	18 Aquilæ	19 0 37.394	12	2.8232	+ 0.0002		+10 51.9	
373	π Sagittarii	1 44.116	20	3.5716	— 0.0056		—21 14.1	
374	ι Lyræ	2 29.087	11	2.1393	+ 0.0014	+0.0012 A.	+35 53.4	
375	η Lyræ	9 9.849	11	2.0404	+ 0.0012		+38 54.9	
376	ω Aquilæ	11 28.841	18	2.8154	— 0.0001	0.000 A.	+11 21.3	+0.007
377	δ Draconis	12 30.896	10	0.0151	— 0.0224	+0.0203 M.	+67 25.4	0.000
378	β· Sagittarii	12 55.765	13	+ 4.3301	— 0.0193		—44 45.7	
379	59 Draconis	14 5.380	11	— 2.1420	— 0.0927	+0.008 S.	+76 20.0	—0.005
380	τ Draconis	18 7.662	10	— 1.0772	— 0.0560	—0.0323 M.	+73 6.2	+0.004
381	δ Aquilæ	18 41.531	21	+ 3.0083	— 0.0016	+0.0175 N.	+ 2 50.9	+0.007
382	π Draconis	19 58.121	11	0.3199	— 0.0185		+65 27.3	0.000
383	6 Vulpeculæ	23 5.301	21	2.5041	+ 0.0011	—0.0097 M.	+24 23.6	—0.012
384	β Cygni	25 16.659	23	2.4177	+ 0.0012		+27 40.7	0.000
385	ι Cygni	26 18.107	13	1.5111	— 0.0019	+0.0010 M.	+51 26.6	—0.001
386	h Sagittarii	28 29.454	24	+ 3.6531	— 0.0100	+0.0059 M.	—25 10.7	+0.008
387	Groomb. 2900.	29 47.902	10	— 3.4865	— 0.1973	+0.0024 A.	+79 19.8	—0.001
388	θ Cygni	32 49.209	17	+ 1.6112	— 0.0013	—0.0018 A.	+49 54.6	—0.003
389	γ Sagittæ	52 45.241	14	+ 2.6622	+ 0.0004	+0.0053 M.	+19 7.7	+0.006
390	(λ Ursæ Minoris	59 8.861)	37	—57.5734	— 29.840	—0.081 S.	+88 54.3	
391	Pi. XIX. 414	20 1 22.164	10	+ 2.7344	0.0000		+16 15.5	
392	Pi. XIX. 420	1 57.246	7	2.7346	0.0000		+16 16.4	
393	Pi. XX. 6	3 54.246	8	3.2013	— 0.0056		— 6 29.2	
394	Pi. XX. 29	6 51.937	10	+ 3.6612	— 0.0153	+0.0982 A.	—27 26.0	—0.070

Number.	Star's Name.	A.R. 1865.0.	No. Obs.	Precession.	Secular Variation.	Proper Motion.	Declination.	Δα.
		h. m. s.		*s.*	*s.*	*s.*	° ′	*s.*
395	ϱ Aquilæ	20 8 1.879	20	+ 2.7717	— 0.0003	+0.0048 M.	+14 47.3	+0.003
396	κ Cephei	13 22.693	14	— 1.8861	— 0.1639	+0.0010 A.	+77 18.2	+0.004
397	23 Aquilæ H.	16 29.368	11	+ 2.9758	— 0.0026		+ 4 54.8	
398	γ Cygni	17 23.015	13	2.1504	+ 0.0021		+39 49.6	000
399	π Capricorni	19 35.536	18	3.4410	— 0.0114	+0.0014 N.	—18 39.1	000
400	ϱ Capricorni	21 9.537	19	3.4309	— 0.0113	—0.0011 M.	—18 15.4	+ .012
401	Pi. XX. 174	25 0.464	9	3.2671	— 0.0079	+0.0194 A.	—10 18.7	— .014
402	ε Delphini	26 45.847	15	2.8655	— 0.0011	+0.0020 N.	+10 50.8	— .001
403	θ Cephei	27 18.714	19	+ 1.0117	— 0.0150	+0.0083 M.	+62 32.5	+ .008
404	R. C. 4894	29 33.262	9	— 8.3691	— 1.2582	—0.06 S.	+84 41.7	+ .042
405	Bradley 2673	30 34.168	20	— 0.2015	— 0.0664		+72 4 4	.000
406	1 Aquarii	32 29.789	9	+ 3.0707	— 0.0042	+0.0097 M.	+ 0 0.8	— .007
407	α Delphini	33 22.079	17	2.7814	0.0000	+0.0067 M.	+15 26.3	+0.013
408	226 Cygni B.	34 36.988	10	+ 2.1917	+ 0.0028		+40 6.2	
409	74 Draconis	37 7.874	39	— 3.1911	— 0.3534	+0.020 S.	+80 37.0	
410	236 Cygni B.	40 8.787	7	+ 1.9806	+ 0.0021		+46 48.5	
411	ε Cygni	40 44.951	19	2.3958	+ 0.0031	+0.0308 M.	+33 28.0	+0.039
412	λ Cygni	42 9.105	10	2.3328	+ 0.0033	+0.0003 A.	+35 59.8	
413	η Cephei	42 32.349	25	1.2165	— 0.0109	+0.0171 M.	+61 18.9	+0.023
414	32 Vulpeculæ	48 48.447	17	2.5544	+ 0.0028	+0.0007 A.	+27 32.7	+0.008
415	Pi. XX. 376	48 55.108	9	+ 3.0014	— 0.0030	+0.008 S.	+ 4 1.1	—0.006
416	76 Draconis	52 9.998	22	— 3.9065	— 0.5170	+0.0046 A.	+82 1.7	+0.005
417	Br. 2749.	53 36.601	10	— 2.4625	— 0.3062	—0.0082 A.	+80 2.6	+0.003
418	6 Equulei B.	57 54.137	9	+ 3.0419	— 0.0038		+ 1 41.3	
419	61 Cygni med.	21 0 51.647	35	2.3328	+ 0.0085	+0.3475 M.	+38 5.2	+0.012
420	Pi. XXI. 1.	2 55.386	9	2.5395	+ 0.0038	+0.006 S.	+29 39.7	—0.004
421	ζ Cygni	7 11.495	19	2.5494	+ 0.0040	—0.0007 N.	+29 40.5	+0.007
422	114 Capri. B.	7 33.305	8	3.3660	— 0.0130		—17 54.0	
423	4 Piscis Aust.	9 44.820	9	3.6522	— 0.0242	+0.0061 M.	—32 44.1	—0.004
424	11 Pegasi B.	14 6.110	16	2.7229	+ 0.0026		+21 27.4	
425	α Cephei	15 21.330	28	1.4151	— 0.0069	+0.0231 N.	+62 0.9	0.000
426	326 Cygni B.	15 40.927	11	2.5211	+ 0.0049		+32 2.4	
427	Pi. XXI. 120	18 34.479	9	2.6562	+ 0.0038		+25 35.7	
428	70 Cygni	21 51.255	10	+ 2.4405	+ 0.0060		+36 31.9	
429	119 Cephei B.	24 16.007	39	— 4.5214	— 0.8286	+0.010 S.	+83 41.1	
430	β Aquarii	24 27.063	17	+ 3.1615	— 0.0070	+0.0020 N.	— 6 9.8	+0.007
431	26 Pegasi B.	24 37.908	10	2.8996	0.0000		+11 32.8	
432	β Cephei	26 54.388	30	+ 0.7997	— 0.0340	+0.0017 W.	+69 58.1	
433	122 Cephei B.	28 42.151	10	— 1.5484	— 0.2688	+0.016 S.	+79 56.1	—0.012
434	F. 376 Aquarii	30 37.825	16	+ 3.0850	— 0.0046		— 0 59.8	
435	γ Capricorni	32 36.544	21	3.3202	— 0.0129	+0.0154 M.	—17 16.2	+0.010
436	ι Piscis Aust.	36 54.008	10	3.5909	— 0.0259	+0.004 S.	—33 38.4	—0.003
437	ε Pegasi	37 33.353	13	2.9440	— 0.0004	+0.0028 N.	+ 9 15.5	+0.007
438	δ Capricorni	39 35.200	16	3.3021	— 0.0126	+0.0173 M.	—16 44.3	+0.010
439	θ Piscis Aust.	39 48.477	10	+ 3.5418	— 0.0239	—0.0009 M.	—31 31.3	+ .001

Number.	Star's Name.	A.R. 1865.0.	No. Obs.	Precession.	Secular Variation.	Proper Motion.	Declination.	Δα.
		h. m. s.		s.	s.	s.	° ′	s.
440	388 Cygni B.	21 42 43.467	10	+ 2.4750	+ 0.0082		+38 19.8	
441	195 Heis Cygni	44 10.844	11	2.4327	+ 0.0087		+40 31.2	
442	16 Pegasi	46 55.298	19	2.7244	+ 0.0054	+0.0015 A.	+25 17.5	+0.006
443	Pi. XXI. 320	47 7.370	8	3.1338	— 0.0062		— 4 54.5	
444	17 Pegasi	50 21.432	11	2.9259	+ 0.0009		+11 26.2	.000
445	η Piscis Aust.	53 4.602	10	3.4622	— 0.0216		—29 6.0	.000
446	911 Mayeri	54 45.868	10	3.3040	— 0.0137		—18 32.9	
447	32 Aquarii	57 50.881	11	3.0894	— 0.0043		— 1 33.5	.000
448	μ Piscis Aust.	22 0 30.082	10	3.5141	— 0.0260	+0.006 S.	—33 38.7	— .004
449	ι Pegasi	0 43.691	23	2.7653	+ 0.0062	+0.020 S.	+24 41.2	+ .015
450	Br. 2912	3 19.871	9	3.1228	— 0.0056	+0.0034 M.	— 4 33.3	— .003
451	Pi. XXII. 19	6 9.101	10	3.3796	— 0.0188	+0.010 S.	—25 50.9	—0.008
452	ζ Cephei	6 10.472	27	2.0696	+ 0.0115		+57 32.2	+0.116
453	1 Lacertæ H.	8 5.219	9	2.5622	+ 0.0112	+0.004 S.	+39 2.8	—0.003
454	θ Aquarii	9 42.538	22	3.1629	— 0.0075	+0.0082 N.	— 8 27.3	+0.008
455	Br. 2942	10 25.362	10	1.1037	— 0.0262		+72 38.2	
456	γ Aquarii	14 41.023	18	3.0923	— 0.0040	+0.0089 M.	— 2 4.0	— .002
457	Pi. XXII. 81	16 27.287	9	3.1516	— 0.0070		— 7 52.5	
458	3 Lacertæ	18 15.355	11	2.3477	+ 0.0153	—0.0005 A.	+51 33.2	.000
459	18 Heis Lacer.	21 31.734	10	2.6180	+ 0.0129		+39 7.4	
460	σ Aquarii	23 30.116	20	+ 3.1808	— 0.0087		—11 22.1	.000
461	32 Cephei H.	23 35.084	9	— 3.7550	— 1.1736	+0.050 S.	+85 25.6	—0.038
462	Groomb. 3804	26 28.544	10	+ 2.6400	+ 0.0135		+39 5.2	
463	η Aquarii	28 25.157	23	3.0783	— 0.0030	+0.0062 N.	— 0 48.7	+0.008
464	226 Cephei B.	29 53.405	6	1.0860	— 0.0329		+75 31.8	.000
465	ε Piscis Aust.	33 11.053	10	3.3310	— 0.0196		—27 44.8	.000
466	ζ Pegasi	34 43.821	23	2.9841	+ 0.0024	+0.0063 N.	+10 7.7	+0.009
467	Pi. XXII. 186	34 11.401	9	2.9504	+ 0.0040	+0.020 S.	+13 50.4	—0.016
468	Pi. XXII. 195	35 17.510	2	2.9522	+ 0.0040		+13 48.7	
469	η Pegasi	36 40.592	21	2.8018	+ 0.0110		+29 31.0	.000
470	13 Lacertæ	38 4.539	10	+ 2.6627	+ 0.0160	—0.0002 A.	+41 6.7	.000
471	ξ Pegasi	39 56.999	21	2.9780	+ 0.0033	+0.015 S.	+11 28.9	+ .011
472	69 Aquarii	40 32.582	10	3.1906	— 0.0101		—14 46.0	.000
473	ι Cephei	44 52.807	9	2.1269	+ 0.0224	—0.0122 M.	+65 29.45	+ .001
474	δ Aquarii	47 29.011	22	+ 3.1944	— 0.0110		—16 32.3	
475	34 Cephei H.	47 55.301	34	— 0.0440	— 0.2202	+0.007 S.	+82 26.2	
476	36 Cephei H.	55 21.527	10	— 0.2587	— 0.2997	+0.050 S.	+83 37.4	—0.039
477	β Piscium	57 0.452	20	+ 3.0513	+ 0.0002	+0.0023 M.	+ 3 5.6	+ .002
478	118 Heis Peg.	23 0 49.363	11	2.9439	+ 0.0088		+20 24.4	
479	Pi. XXIII. 4	4 0.393	11	2.9729	+ 0.0073		+16 51.8	
480	128 Heis Peg.	7 57.254	11	2.9409	+ 0.0107		+23 22.2	
481	130 Heis Peg.	9 18.898	11	2.9401	+ 0.0112		+24 2.1	
482	γ Piscium	10 10.087	20	3.0581	+ 0.0007	+0.0493 M.	+ 2 32.7	+0.008
483	Pi. XXIII. 36	11 31.708	10	3.2546	— 0.0221		—33 12.8	
484	o Cephei	13 5.540	11	+ 2.4190	+ 0.0401	+0.011 S.	+67 22.4	+ .003

Number.	Star's Name.	A.R. 1865.0.	No. Obs.	Precession.	Secular Variation.	Proper Motion.	Declination.	Δα.
		h. m. s.		s.	s.	s.	° ′	s.
485	D.M.+35°5012	23 14 51.573	11	+ 2.8825	+ 0.0182		+35 45.7	
486	*b* Aquarii	15 52.697	17	3.1683	— 0.0123	—0.0086 M.	—20 50.2	— .006
487	Bradley 3109	17 9.665	9	2.9169	+ 0.0161	+0.018 S.	+31 47.4	— .014
488	Pi. XXIII. 82	19 28.351	10	3.1682	— 0.0132		—22 28.9	
489	κ Piscium	20 0.791	17	3.0688	+ 0.0001	+0.0070 A.	+ 0 31.0	+0.008
490	347 Aquarii B.	25 26.169	10	+ 3.1127	— 0.0062		—11 44.6	
491	39 Cephei H.	27 50.384	63	— 0.0421	— 0.5039	+0.087 S.	+86 33.8	
492	Pi. XXIII. 133	30 39.580	8	+ 3.1129	— 0.0071		—13 48.5	
493	ι Piscium	33 0.490	18	3.0575	+ 0.0031	+0.0260 N.	+ 4 53.7	+0.006
494	γ Cephei	33 50.022	25	2.4173	+ 0.0740	—0.015 S.	+76 52.7	+0.003
495	Pi. XXIII. 153	34 9.704	11	3.1040	— 0.0061		—12 25.7	
496	368 Aquarii B.	35 28.140	10	3.1124	— 0.0082		—16 11.6	
497	979 Mayeri	37 55.678	11	3.0564	+ 0.0042		+ 6 26.6	
498	981 Mayeri	40 18.779	10	3.0967	— 0.0058		—12 39.4	
499	δ Sculptoris	41 53.432	15	3.1291	— 0.0160	+0.005 S.	—28 52.6	+0.006
500	Pi. XXIII. 203	43 35.432	8	3.0968	— 0.0071	+0.008 S.	—15 9.2	—0.006
501	Bradley 3181	45 31.145	16	2.7095	+ 0.0989	+0.0787 A.	+76 51.1	—0.061
502	Groomb. 4163	48 17.725	9	2.8382	+ 0.0878		+73 39.5	
503	Bradley 3187	50 11.199	10	2.6394	+ 0.1633	+0.030 S.	+82 26.4	—0.023
504	ω Piscium	52 22.793	15	3.0662	+ 0.0049	+0.0111 N.	+ 6 7.0	+0.005
505	309 Cephei B.	53 15.254	26	2.5146	+ 0.2711	+0.036 S.	+85 57.3	
506	2 Ceti	56 49.352	16	+ 3.0774	— 0.0079	+0.0014 M.	—18 5.2	+ .011

5 Ursæ Minoris, Bode.

	Corr. of A.R., 1870.		Annual Variation.	Corr. of Dec., 1870.		Annual Variation.
	s.	s.	s.	″	″	″
Jan. 0	+ 2.03	+4.57	—0.005	—20.47	—0.04	0.000
10	+ 6.60	+4.42	—0.014	—20.51	+0.60	—0.001
20	+11.02	+4.09	—0.024	—19.91	+1.23	—0.001
30	+15.11	+3.62	—0.032	—18.68	+1.81	—0.002
Feb. 9	+18.73	+3.03	—0.039	—16.87	+2.29	—0.002
19	+21.76	+2.32	—0.046	—14.58	+2.68	—0.003
March 1	+24.08	+1.52	—0.050	—11.90	+2.97	—0.003
11	+25.60	+0.71	—0.053	— 8.93	+3.11	—0.003
21	+26.31	—0.13	—0.054	— 5.82	+3.14	—0.003
31	+26.18	—0.94	—0.052	— 2.68	+3.04	—0.003
April 10	+25.24	—1.70	—0.050	+ 0.36	+2.82	—0.003
20	+23.54	—2.38	—0.045	+ 3.18	+2.51	—0.003
30	+21.16	—2.96	—0.039	+ 5.69	+2.11	—0.003
May 10	+18.20	—3.42	—0.031	+ 7.80	+1.63	—0.002
20	+14.78	—3.77	—0.024	+ 9.43	+1.11	—0.002
30	+11.01	—4.01	—0.014	+10.54	+0.56	—0.001
June 9	+ 7.00	—4.10	—0.006	+11.10	—0.02	—0.001
19	+ 2.90	—4.10	+0.002	+11.08	—0.58	0.000
29	— 1.20	—3.99	+0.010	+10.50	—1.13	0.000
July 9	— 5.19	—3.77	+0.018	+ 9.37	—1.66	+0.001
19	— 8.96	—3.46	+0.024	+ 7.71	—2.14	+0.001
29	—12.42	—3.08	+0.029	+ 5.57	—2.57	+0.002
Aug. 8	—15.50	—2.64	+0.034	+ 3.00	—2.96	+0.002
18	—18.14	—2.12	+0.038	+ 0.04	—3.29	+0.002
28	—20.26	—1.57	+0.040	— 3.25	—3.53	+0.003
Sept. 7	—21.83	—0.97	+0.042	— 6.78	—3.73	+0.003
17	—22.80	—0.34	+0.044	—10.51	—3.83	+0.003
27	—23.14	+0.31	+0.044	—14.34	—3.87	+0.003
Oct. 7	—22.83	+0.98	+0.042	—18.21	—3.82	+0.004
17	—21.85	+1.63	+0.041	—22.03	—3.68	+0.003
27	—20.22	+2.28	+0.038	—25.71	—3.46	+0.003
Nov. 6	—17.94	+2.88	+0.036	—29.17	—3.15	+0.003
16	—15.06	+3.42	+0.030	—32.32	—2.74	+0.002
26	—11.64	+3.90	+0.024	—35.06	—2.26	+0.002
Dec. 6	— 7.74	+4.24	+0.017	—37.32	—1.71	+0.001
16	— 3.50	+4.50	+0.009	—39.03	—1.10	+0.001
26	+ 1.00	+4.58	0.000	—40.13	—0.46	0.000
36	+ 5.58		—0.009	—40.59		0.000

Add 1 day to the tabular date in Jan. and Feb. of leap years.
Add to the Reduced Date
For Upper Culmination before } Sept. 24 { + 1
after } { + 2
For Lower Culmination before } Mar. 25 { + 0.5
after } { + 1.5

6 Ursæ Minoris, Bode.

	Corr. of A.R., 1870.		Annual Variation.	Corr. of Dec., 1870.		Annual Variation.
	s.	s.	s.	″	″	″
Jan. 0	+ 3.32	+8.13	—0.011	—20.46	—0.03	0.000
10	11.45	7.85	—0.039	—20.49	+0.62	0.000
20	19.30	7.31	—0.065	—19.87	+1.24	0.000
30	26.61	6.47	—0.092	—18.63	+1.82	0.000
Feb. 9	33.08	5.40	—0.114	—16.81	+2.30	—0.001
19	38.48	4.13	—0.133	—14.51	+2.69	—0.001
March 1	42.61	2.74	—0.147	—11.82	+2.97	—0.001
11	45.35	+1.24	—0.157	— 8.85	+3.12	—0.001
21	46.59	—0.26	—0.161	— 5.73	+3.14	—0.001
31	46.33	1.72	—0.160	— 2.59	+3.04	—0.001
April 10	44.61	3.10	—0.153	+ 0.45	+2.83	—0.001
20	41.51	4.33	—0.142	+ 3.28	+2.50	—0.001
30	37.18	5.38	—0.127	+ 5.78	+2.10	—0.001
May 10	31.80	6.24	—0.108	+ 7.88	+1.63	0.000
20	25.56	6.87	—0.086	+ 9.51	+1.10	0.000
30	18.69	7.30	—0.061	+10.61	+0.55	0.000
June 9	11.39	7.49	—0.035	+11.16	—0.02	0.000
19	+ 3.90	7.48	—0.009	+11.14	—0.59	0.000
29	— 3.58	7.27	+0.018	+10.55	—1.15	0.000
July 9	10.85	6.89	+0.043	+ 9.40	—1.66	0.000
19	17.74	6.34	+0.068	+ 7.74	—2.15	0.000
29	24.08	5.67	+0.090	+ 5.59	—2.59	0.000
Aug. 8	29.75	4.82	+0.110	+ 3.00	—2.97	+0.001
18	34.57	3.93	+0.128	+ 0.03	—3.29	+0.001
28	38.50	2.92	+0.142	— 3.26	—3.54	+0.001
Sept. 7	41.42	1.85	+0.152	— 6.80	—3.72	+0.001
17	43.27	—0.73	+0.159	—10.52	—3.84	+0.001
27	44.00	+0.45	+0.161	—14.36	—3.87	+0.001
Oct. 7	43.55	1.63	+0.160	—18.23	—3.81	+0.001
17	41.92	2.80	+0.155	—22.04	—3.67	+0.001
27	39.12	3.96	+0.145	—25.71	—3.45	+0.001
Nov. 6	35.16	5.05	+0.132	—29.16	—3.12	+0.001
16	30.11	6.00	+0.114	—32.28	—2.74	+0.001
26	24.11	6.85	+0.093	—35.02	—2.25	0.000
Dec. 6	17.26	7.51	+0.070	—37.27	—1.69	0.000
16	9.75	7.95	+0.044	—38.96	—1.09	0.000
26	— 1.80	+8.12	+0.017	—40.05	—0.44	0.000
36	+ 6.32		—0.011	—40.49		0.000

Add 1 day to the tabular date in Jan. and Feb. of leap years.
Add to the Reduced Date
For Upper Culmination before } Sept. 24 { + 1
after } { + 2
For Lower Culmination before } Mar. 26 { + 0.5
after } { + 1.5

43 Cephei, Hevelii.

	Corr. of A.R., 1870.		Annual Variation.	Corr. of Dec., 1870.		Annual Variation.
	s.	s.	s.	″	″	″
Jan. 0	+ 1.83	—2.71	+0.011	+20.18	+0.46	—0.001
10	— 0.88	—2.71	+0.008	+20.64	—0.20	+0.001
20	— 3.59	—2.62	+0.004	+20.44	—0.84	+0.003
30	— 6.21	—2.42	+0.001	+19.60	—1.43	+0.004
Feb. 9	— 8.63	—2.14	—0.003	+18.17	—1.98	+0.006
19	—10.77	—1.76	—0.007	+16.19	—2.42	+0.007
March 1	—12.53	—1.32	—0.010	+13.77	—2.76	+0.009
11	—13.85	—0.81	—0.013	+11.01	—2.99	+0.009
21	—14.66	—0.30	—0.015	+ 8.02	—3.10	+0.010
31	—14.96	+0.24	—0.017	+ 4.92	—3.07	+0.010
April 10	—14.72	+0.77	—0.018	+ 1.85	—2.94	+0.010
20	—13.95	+1.25	—0.018	— 1.09	—2.69	+0.009
30	—12.70	+1.70	—0.018	— 3.78	—2.34	+0.009
May 10	—11.00	+2.07	—0.017	— 6.12	—1.93	+0.007
20	— 8.93	+2.39	—0.016	— 8.05	—1.44	+0.006
30	— 6.54	+2.63	—0.014	— 9.49	—0.92	+0.005
June 9	— 3.91	+2.78	—0.011	—10.41	—0.37	+0.003
19	— 1.13	+2.87	—0.008	—10.78	+0.20	+0.002
29	+ 1.74	+2.87	—0.004	—10.58	+0.75	0.000
July 9	+ 4.61	+2.80	—0.001	— 9.83	+1.28	—0.002
19	+ 7.41	+2.67	+0.003	— 8.55	+1.79	—0.004
29	+10.08	+2.49	+0.007	— 6.76	+2.26	—0.006
Aug. 8	+12.57	+2.24	+0.011	— 4.50	+2.66	—0.007
18	+14.81	+1.96	+0.015	— 1.84	+3.04	—0.008
28	+16.77	+1.63	+0.018	+ 1.20	+3.32	—0.009
Sept. 7	+18.40	+1.27	+0.022	+ 4.52	+3.57	—0.011
17	+19.67	+0.89	+0.025	+ 8.09	+3.72	—0.011
27	+20.56	+0.48	+0.027	+11.81	+3.81	—0.012
Oct. 7	+21.04	+0.06	+0.029	+15.62	+3.82	—0.012
17	+21.10	—0.36	+0.031	+19.44	+3.74	—0.012
27	+20.74	—0.79	+0.032	+23.18	+3.57	—0.012
Nov. 6	+19.95	—1.20	+0.033	+26.75	+3.33	—0.011
16	+18.75	—1.59	+0.033	+30.08	+2.97	—0.010
26	+17.16	—1.95	+0.032	+33.05	+2.55	—0.009
Dec. 6	+15.21	—2.26	+0.031	+35.60	+2.05	—0.007
16	+12.95	—2.49	+0.029	+37.65	+1.47	—0.006
26	+10.46	—2.65	+0.026	+39.12	+0.86	—0.004
36	+ 7.81		+0.023	+39.98		—0.003

Add 1 day to the tabular date in Jan. and Feb. of leap years.
Add to the Reduced Date
For Upper Culmination before / after } April 20 { + 1 / + 2
For Lower Culmination before / after } Oct. 4 { + 1.5 / + 2.5

214 Camelopardali, Bode.

	Corr. of A.R., 1870.		Annual Variation.	Corr. of Dec., 1870.		Annual Variation.
	s.	s.	s.	″	″	″
Jan. 0	— 3.71	+2.78	+0.007	—20.04	—1.04	—0.001
10	— 0.93	+2.83	+0.004	—21.08	—0.38	0.000
20	+ 1.90	+2.79	+0.001	—21.46	+0.29	0.000
30	+ 4.69	+2.64	—0.002	—21.17	+0.94	+0.001
Feb. 9	+ 7.33	+2.39	—0.005	—20.23	+1.54	+0.001
19	+ 9.72	+2.06	—0.007	—18.69	+2.07	+0.002
March 1	+11.78	+1.65	—0.010	—16.62	+2.50	+0.002
11	+13.43	+1.19	—0.012	—14.12	+2.84	+0.002
21	+14.62	+0.69	—0.014	—11.28	+3.04	+0.003
31	+15.31	+0.19	—0.015	— 8.24	+3.14	+0.003
April 10	+15.50	—0.33	—0.016	— 5.10	+3.10	+0.003
20	+15.17	—0.81	—0.016	— 2.00	+2.94	+0.003
30	+14.36	—1.24	—0.016	+ 0.94	+2.70	+0.003
May 10	+13.12	—1.63	—0.015	+ 3.64	+2.34	+0.002
20	+11.49	—1.97	—0.013	+ 5.98	+1.93	+0.002
30	+ 9.52	—2.24	—0.011	+ 7.91	+1.45	+0.002
June 9	+ 7.28	—2.43	—0.009	+ 9.36	+0.92	+0.001
19	+ 4.85	—2.57	—0.006	+10.28	+0.39	+0.001
29	+ 2.28	—2.62	—0.003	+10.67	—0.17	0.000
July 9	— 0.34	—2.63	+0.001	+10.50	—0.72	0.000
19	— 2.97	—2.56	+0.004	+ 9.78	—1.25	—0.001
29	— 5.53	—2.44	+0.007	+ 8.53	—1.75	—0.001
Aug. 8	— 7.97	—2.27	+0.010	+ 6.78	—2.23	—0.002
18	—10.24	—2.04	+0.013	+ 4.55	—2.65	—0.003
28	—12.28	—1.77	+0.016	+ 1.90	—3.01	—0.003
Sept. 7	—14.05	—1.46	+0.019	— 1.11	—3.34	—0.003
17	—15.51	—1.11	+0.021	— 4.45	—3.58	—0.004
27	—16.62	—0.73	+0.023	— 8.03	—3.75	—0.004
Oct. 7	—17.35	—0.31	+0.025	—11.78	—3.86	—0.004
17	—17.66	+0.11	+0.026	—15.64	—3.87	—0.004
27	—17.55	+0.56	+0.026	—19.51	—3.80	—0.004
Nov. 6	—16.99	+1.01	+0.026	—23.31	—3.64	—0.004
16	—15.98	+1.42	+0.026	—26.95	—3.36	—0.004
26	—14.56	+1.83	+0.025	—30.31	—3.01	—0.003
Dec. 6	—12.73	+2.18	+0.023	—33.32	—2.56	—0.003
16	—10.55	+2.48	+0.021	—35.88	—2.04	—0.003
26	— 8.07	+2.68	+0.019	—37.92	—1.43	—0.003
36	— 5.39		+0.016	—39.35		—0.002

Add 1 day to the tabular date in Jan. and Feb. of leap years.
Add to the Reduced Date
For Upper Culmination before / after } Oct. 11 { + 1 / + 2
For Lower Culmination before / after } April 11 { + 0.5 / + 1.5

20 Ursæ Minoris, Bode.

	Corr. of A.R., 1870.		Annual Variation.	Corr. Dec., 1870.		Annual Variation.
	s.	s.	s.	″	″	″
Jan. 0	− 8.54	+3.23	+0.022	−18.86	−1.59	−0.005
10	− 5.31	+3.44	+0.019	−20.45	−0.96	−0.003
20	− 1.87	+3.50	+0.016	−21.41	−0.29	−0.002
30	+ 1.63	+3.46	+0.011	−21.70	+0.36	0.000
Feb. 9	+ 5.09	+3.26	+0.007	−21.34	+1.02	+0.002
19	+ 8.35	+2.95	+0.002	−21.32	+1.60	+0.004
March 1	+11.30	+2.54	−0.003	−18.72	+2.12	+0.005
11	+13.84	+2.02	−0.008	−16.60	+2.54	+0.006
21	+15.86	+1.45	−0.012	−14.06	+2.85	+0.008
31	+17.31	+0.82	−0.015	−11.21	+3.05	+0.008
April 10	+18.13	+0.19	−0.018	− 8.16	+3.13	+0.009
20	+18.32	−0.44	−0.021	− 5.03	+3.09	+0.009
30	+17.88	−1.04	−0.021	− 1.94	+2.93	+0.009
May 10	+16.84	−1.60	−0.022	+ 0.99	+2.67	+0.008
20	+15.24	−2.10	−0.021	+ 3.66	+2.34	+0.007
30	+13.14	−2.53	−0.019	+ 6.00	+1.93	+0.006
June 9	+10.61	−2.87	−0.017	+ 7.93	+1.46	+0.005
19	+ 7.74	−3.15	−0.014	+ 9.39	+0.95	+0.004
29	+ 4.59	−3.33	−0.010	+10.34	+0.43	+0.002
July 9	+ 1.26	−3.44	−0.006	+10.77	−0.12	0.000
19	− 2.18	−3.46	−0.001	+10.65	−0.65	−0.002
29	− 5.64	−3.41	+0.004	+10.00	−1.18	−0.004
Aug. 8	− 9.05	−3.27	+0.009	+ 8.82	−1.68	−0.005
18	−12.32	−3.08	+0.015	+ 7.14	−2.14	−0.007
28	−15.40	−2.82	+0.020	+ 5.00	−2.57	−0.009
Sept. 7	−18.22	−2.48	+0.026	+ 2.43	−2.94	−0.010
17	−20.70	−2.09	+0.031	− 0.51	−3.27	−0.011
27	−22.79	−1.65	+0.035	− 3.78	−3.51	−0.012
Oct. 7	−24.44	−1.16	+0.040	− 7.29	−3.70	−0.014
17	−25.60	−0.62	+0.044	−10.99	−3.81	−0.014
27	−26.22	−0.06	+0.047	−14.80	−3.83	−0.015
Nov. 6	−26.28	+0.52	+0.049	−18.63	−3.76	−0.015
16	−25.76	+1.10	+0.051	−22.39	−3.59	−0.015
26	−24.66	+1.67	+0.051	−25.98	−3.32	−0.014
Dec. 6	−22.99	+2.20	+0.051	−29.30	−2.96	−0.014
16	−20.79	+2.65	+0.051	−32.26	−2.50	−0.012
26	−18.14	+3.04	+0.048	−34.76	−1.96	−0.011
36	−15.10		+0.046	−36.72		−0.010

Add 1 day to the tabular dates in Jan. and Feb. of leap years.
Add to the Reduced Date
For Upper Culmination before / after } Oct. 22 { + 1 / + 2
For Lower Culmination before / after } April 22 { + 0.5 / + 1.5

Groombridge 527.

	Corr. of A.R., 1870.		Annual Variation.	Corr. Dec., 1870.		Annual Variation.
	s.	s.	s.	″	″	″
Jan. 0	+ 4.36	−1.05	+0.006	+17.23	+1.63	−0.006
10	+ 3.31	−1.17	+0.006	+18.86	+1.04	−0.004
20	+ 2.14	−1.24	+0.006	+19.90	+0.43	−0.002
30	+ 0.90	−1.28	+0.006	+20.33	−0.20	0.000
Feb. 9	− 0.38	−1.24	+0.005	+20.13	−0.80	+0.002
19	− 1.62	−1.16	+0.004	+19.33	−1.37	+0.004
March 1	− 2.78	−1.03	+0.003	+17.96	−1.87	+0.006
11	− 3.81	−0.85	+0.002	+16.09	−2.29	+0.007
21	− 4.66	−0.65	+0.001	+13.80	−2.62	+0.008
31	− 5.31	−0.40	0.000	+11.18	−2.82	+0.009
April 10	− 5.71	−0.14	−0.001	+ 8.36	−2.93	+0.009
20	− 5.85	+0.12	−0.001	+ 5.43	−2.92	+0.010
30	− 5.73	+0.37	−0.002	+ 2.51	−2.79	+0.010
May 10	− 5.36	+0.62	−0.003	− 0.28	−2.58	+0.009
20	− 4.74	+0.85	−0.003	− 2.86	−2.28	+0.008
30	− 3.89	+1.05	−0.003	− 5.14	−1.90	+0.007
June 9	− 2.84	+1.22	−0.003	− 7.04	−1.48	+0.005
19	− 1.62	+1.34	−0.003	− 8.52	−1.01	+0.004
29	− 0.28	+1.45	−0.003	− 9.53	−0.51	+0.002
July 9	+ 1.17	+1.51	−0.003	−10.04	0.00	0.000
19	+ 2.68	+1.54	−0.002	−10.04	+0.50	−0.002
29	+ 4.22	+1.53	−0.001	− 9.54	+0.99	−0.004
Aug. 8	+ 5.75	+1.50	0.000	− 8.55	+1.47	−0.006
18	+ 7.25	+1.43	+0.001	− 7.08	+1.90	−0.008
28	+ 8.68	+1.33	+0.002	− 5.18	+2.31	−0.010
Sept. 7	+10.01	+1.22	+0.003	− 2.87	+2.67	−0.012
17	+11.23	+1.07	+0.004	− 0.20	+2.98	−0.013
27	+12 30	+0.92	+0.005	+ 2.78	+3.21	−0.015
Oct. 7	+13.22	+0.73	+0.007	+ 5.99	+3.40	−0.016
17	+13.95	+0.54	+0.008	+ 9.39	+3.52	−0.017
27	+14.49	+0.32	+0.009	+12.91	+3.56	−0.017
Nov. 6	+14.81	+0.10	+0.010	+16.47	+3.51	−0.018
16	+14.91	−0.12	+0.011	+19.98	+3.38	−0.018
26	+14.79	−0.35	+0.012	+23.36	+3.16	−0.017
Dec. 6	+14.44	−0.57	+0.012	+26.52	+2.84	−0.016
16	+13.87	−0.77	+0.013	+29.36	+2.44	−0.015
26	+13.10	−0.96	+0.014	+31.80	+1.96	−0.014
36	+12.14		+0.013	+33.76		−0.012

Add 1 day to the tabular dates in Jan. and Feb. of leap years.
Add to the Reduced Date
For Upper Culmination before / after } April 29 { + 1 / + 2
For Lower Culmination before / after } Oct. 29 { + 1.5 / + 2.5

323 Cephei, Bode.

	Corr. of A.R., 1870.		Annual Variation.	Corr. of Dec., 1870.		Annual Variation.
	s.	s.	s.	"	"	"
Jan. 0	+14.43	—2.15	+0.034	+14.60	+2.36	—0.020
10	12.28	2.56	+0.035	+16.96	+1.85	—0.016
20	9.72	2.89	+0.035	+18.81	+1.28	—0.012
30	6.83	3.10	+0.036	+20.09	+0.65	—0.008
Feb. 9	3.73	3.17	+0.033	+20.74	+0.02	—0.004
19	+ 0.56	3.11	+0.030	+20.76	—0.60	0.000
March 1	— 2.55	2.93	+0.026	+20.16	—1.19	+0.005
11	5.48	2.62	+0.022	+18.97	—1.74	+0.009
21	8.10	2.21	+0.017	+17.23	—2.19	+0.012
31	10.31	1.72	+0.012	+15.04	—2.55	+0.016
April 10	12.03	1.16	+0.006	+12.49	—2.81	+0.018
20	13.19	—0.55	+0.001	+ 9.68	—2.95	+0.019
30	13.74	+0.06	—0.004	+ 6.73	—3.00	+0.020
May 10	13.68	0.67	—0.008	+ 3.73	—2.93	+0.020
20	13.01	1.26	—0.012	+ 0.80	—2.78	+0.019
30	11.75	1.80	—0.014	— 1.98	—2.52	+0.018
June 9	9.95	2.29	—0.017	— 4.50	—2.19	+0.015
19	7.66	2.73	—0.018	— 6.69	—1.82	+0.012
29	4.93	3.07	—0.018	— 8.51	—1.38	+0.009
July 9	— 1.86	3.35	—0.017	— 9.89	—0.91	+0.004
19	+ 1.49	3.55	—0.016	—10.80	—0.43	0.000
29	5.04	3.68	—0.014	—11.23	+0.06	—0.005
Aug. 8	8.72	3.72	—0.010	—11.17	+0.56	—0.010
18	12.44	3.69	—0.006	—10.61	+1.04	—0.015
28	16.13	3.58	—0.001	— 9.57	+1.50	—0.020
Sept. 7	19.71	3.41	+0.004	— 8.07	+1.93	—0.025
17	23.12	3.17	+0.010	— 6.14	+2.34	—0.029
27	26.29	2.86	+0.016	— 3.80	+2.69	—0.033
Oct. 7	29.15	2.49	+0.022	— 1.11	+3.01	—0.037
17	31.64	2.05	+0.029	+ 1.90	+3.24	—0.041
27	33.69	1.56	+0.035	+ 5.14	+3.43	—0.043
Nov. 6	35.25	1.03	+0.042	+ 8.57	+3.52	—0.045
16	36.28	+0.47	+0.048	+12.09	+3.55	—0.046
26	36.75	—0.13	+0.053	+15.64	+3.46	—0.047
Dec. 6	36.62	0.73	+0.058	+19.10	+3.29	—0.046
16	35.89	1.30	+0.062	+22.39	+3.02	—0.045
26	34.59	—1.84	+0.065	+25.41	+2.64	—0.043
36	+32.75		+0.067	+28.05		—0.040

Add 1 day to the tabular date in Jan. and Feb. of leap years.
Add to the Reduced Date
For Upper Culmination before / after — May 13 — + 1 / + 2
For Lower Culmination before / after — Nov. 12 — + 1.5 / + 2.5

57 Ursæ Minoris, Bode.

	Corr. of A.R., 1870.		Annual Variation.	Corr. of Dec., 1870.		Annual Variation.
	s.	s.	s.	"	"	"
Jan. 0	—23.32	+3.96	+0.079	—15.14	—2.44	—0.023
10	—19.36	+4.62	+0.079	—17.58	—1.92	—0.019
20	—14.74	+5.11	+0.077	—19.50	—1.32	—0.015
30	— 9.63	+5.39	+0.073	—20.82	—0.68	—0.010
Feb. 9	— 4.24	+5.47	+0.066	—21.50	—0.02	—0.005
19	+ 1.23	+5.32	+0.058	—21.52	+0.64	0.000
March 1	+ 6.55	+4.97	+0.048	—20.88	+1.23	+0.005
11	+11.52	+4.44	+0.037	—19.65	+1.80	+0.010
21	+15.96	+3.73	+0.025	—17.85	+2.26	+0.014
31	+19.69	+2.91	+0.013	—15.59	+2.64	+0.018
April 10	+22.60	+1.98	+0.001	—12.95	+2.91	+0.021
20	+24.58	+0.99	—0.010	—10.04	+3.06	+0.023
30	+25.57	0.00	—0.020	— 6.98	+3.10	+0.024
May 10	+25.57	—1.00	—0.028	— 3.88	+3.04	+0.024
20	+24.57	—1.94	—0.035	— 0.84	+2.87	+0.023
30	+22.63	—2.81	—0.040	+ 2.03	+2.61	+0.021
June 9	+19.82	—3.61	—0.042	+ 4.64	+2.29	+0.018
19	+16.21	—4.30	—0.042	+ 6.93	+1.89	+0.014
29	+11.91	—4.87	—0.041	+ 8.82	+1.44	+0.010
July 9	+ 7.04	—5.32	—0.037	+10.26	+0.97	+0.005
19	+ 1.72	—5.65	—0.030	+11.23	+0.46	0.000
29	— 3.93	—5.85	—0.023	+11.69	—0.05	—0.005
Aug. 8	— 9.78	—5.92	—0.013	+11.64	—0.56	—0.011
18	—15.70	—5.87	—0.002	+11.08	—1.06	—0.017
28	—21.57	—5.70	+0.010	+10.02	—1.54	—0.023
Sept. 7	—27.27	—5.41	+0.023	+ 8.48	—2.00	—0.029
17	—32.68	—4.98	+0.038	+ 6.48	—2.42	—0.034
27	—37.66	—4.46	+0.053	+ 4.06	—2.79	—0.039
Oct. 7	—42.12	—3.81	+0.067	+ 1.27	—3.11	—0.043
17	—45.93	—3.06	+0.082	— 1.84	—3.37	—0.047
27	—48.99	—2.22	+0.096	— 5.21	—3.56	—0.050
Nov. 6	—51.21	—1.31	+0.110	— 8.77	—3.67	—0.052
16	—52.52	—0.34	+0.122	—12.44	—3.68	—0.054
26	—52.86	+0.67	+0.133	—16.12	—3.60	—0.054
Dec. 6	—52.19	+1.65	+0.142	—19.72	—3.42	—0.054
16	—50.54	+2.61	+0.149	—23.14	—3.13	—0.053
26	—47.93	+3.48	+0.153	—26.27	—2.73	—0.050
36	—44.45		+0.155	—29.00		—0.047

Add 1 day to the tabular date in Jan. and Feb. of leap years.
Add to the Reduced Date
For Upper Culmination before / after — Nov. 11 — + 1 / + 2
For Lower Culmination before / after — May 11 — + 0.5 / + 1.5

Groombridge 750.

	Corr. of A.R., 1870.		Annual Variation.	Corr. Dec., 1870.		Annual Variation.
	s.	s.	s.	"	"	"
Jan. 0	+12.90	−1.38	+0.020	+12.41	+2.63	−0.019
10	11.52	−1.76	+0.022	+15.04	+2.18	−0.017
20	9.76	−2.09	+0.023	+17.22	+1.65	−0.014
30	7.67	−2.31	+0.023	+18.87	+1.08	−0.011
Feb. 9	5.36	−2.45	+0.023	+19.95	+0.45	−0.007
19	2.91	−2.48	+0.022	+20.40	−0.16	−0.003
March 1	+ 0.43	−2.40	+0.021	+20.24	−0.77	+0.001
11	− 1.97	−2.23	+0.018	+19.47	−1.35	+0.004
21	4.20	−1.96	+0.016	+18.12	−1.84	+0.008
31	6.16	−1.61	+0.012	+16.28	−2.27	+0.011
April 10	7.77	−1.21	+0.009	+14.01	−2.59	+0.014
20	8.98	−0.76	+0.006	+11.42	−2.83	+0.015
30	9.74	−0.28	+0.003	+ 8.59	−2.93	+0.017
May 10	10.02	+0.20	−0.001	+ 5.66	−2.97	+0.017
20	9.82	+0.68	−0.004	+ 2.69	−2.87	+0.017
30	9.14	+1.13	−0.006	− 0.18	−2.70	+0.016
June 9	8.01	+1.55	−0.008	− 2.88	−2.44	+0.014
19	6.46	+1.93	−0.010	− 5.32	−2.13	+0.012
29	4.53	+2.24	−0.011	− 7.45	−1.75	+0.009
July 9	− 2.29	+2.52	−0.011	− 9.20	−1.33	+0.006
19	+ 0.23	+2.73	−0.011	−10.53	−0.87	+0.002
29	2.96	+2.88	−0.011	−11.40	−0.41	−0.002
Aug. 8	5.84	+2.97	−0.010	−11.81	+0.07	−0.007
18	8.81	+2.99	−0.008	−11.74	+0.54	−0.011
28	11.80	+2.97	−0.006	−11.20	+1.02	−0.016
Sept. 7	14.77	+2.88	−0.003	−10.18	+1.46	−0.020
17	17.65	+2.74	0.000	− 8.72	+1.88	−0.024
27	20.39	+2.54	+0.003	− 6.84	+2.28	−0.028
Oct. 7	22.93	+2.29	+0.007	− 4.56	+2.63	−0.032
17	25.22	+1.97	+0.012	− 1.93	+2.92	−0.035
27	27.19	+1.63	+0.016	+ 0.99	+3.17	−0.038
Nov. 6	28.82	+1.23	+0.019	+ 4.16	+3.35	−0.041
16	30.05	+0.78	+0.023	+ 7.51	+3.43	−0.042
26	30.83	+0.31	+0.027	+10.94	+3.44	−0.043
Dec. 6	31.14	−0.16	+0.031	+14.38	+3.34	−0.043
16	30.98	−0.65	+0.034	+17.72	+3.15	−0.043
26	30.33	−1.10	+0.037	+20.87	+2.86	−0.042
36	29.23		+0.039	+23.73		−0.040

Add 1 day to the tabular dates in Jan and Feb. of leap years.
Add to the Reduced Date
For Upper Culmination before / after } May 21 { + 1 / + 2
For Lower Culmination before / after } Nov. 22 { + 1.5 / + 2.5

62 Ursæ Minoris, Bode.

	Corr. of A.R., 1870.		Annual Variation.	Corr. Dec., 1870.		Annual Variation.
	s.	s.	s.	"	"	"
Jan. 0	− 9.23	+1.19	+0.008	−12.76	−2.88	−0.008
10	− 8.04	+1.46	+0.008	−15.64	−2.41	−0.007
20	− 6.58	+1.67	+0.008	−18.05	−1.86	−0.006
30	− 4.91	+1.81	+0.008	−19.91	−1.25	−0.005
Feb. 9	− 3.10	+1.90	+0.007	−21.16	−0.59	−0.003
19	− 1.20	+1.90	+0.007	−21.75	+0.07	−0.001
March 1	+ 0.70	+1.83	+0.006	−21.68	+0.72	0.000
11	+ 2.53	+1.70	+0.005	−20.96	+1.34	+0.002
21	+ 4.23	+1.51	+0.004	−19.62	+1.88	+0.003
31	+ 5.74	+1.26	+0.003	−17.74	+2.35	+0.004
April 10	+ 7.00	+0.98	+0.001	−15.39	+2.72	+0.005
20	+ 7.98	+0.66	0.000	−12.67	+2.98	+0.006
30	+ 8.64	+0.34	0.000	− 9.69	+3.12	+0.006
May 10	+ 8.98	0.00	−0.002	− 6.57	+3.17	+0.007
20	+ 8.98	−0.33	−0.003	− 3.40	+3.10	+0.007
30	+ 8.65	−0.65	−0.003	− 0.30	+2.94	+0.007
June 9	+ 8.00	−0.93	−0.003	+ 2.64	+2.70	+0.006
19	+ 7.07	−1.21	−0.004	+ 5.34	+2.37	+0.005
29	+ 5.86	−1.44	−0.004	+ 7.71	+1.99	+0.004
July 9	+ 4.42	−1.64	−0.004	+ 9.70	+1.55	+0.003
19	+ 2.78	−1.81	−0.004	+11.25	+1.09	+0.002
29	+ 0.97	−1.92	−0.003	+12.34	+0.59	0.000
Aug. 8	− 0.95	−2.00	−0.002	+12.93	+0.09	−0.002
18	− 2.95	−2.03	−0.001	+13.02	−0.43	−0.003
28	− 4.98	−2.03	0.000	+12.59	−0.93	−0.005
Sept. 7	− 7.01	−1.96	+0.001	+11.66	−1.42	−0.007
17	− 8.97	−1.87	+0.002	+10.24	−1.89	−0.009
27	−10.84	−1.72	+0.003	+ 8.35	−2.33	−0.010
Oct. 7	−12.56	−1.53	+0.005	+ 6.02	−2.72	−0.011
17	−14.09	−1.29	+0.006	+ 3.30	−3.06	−0.013
27	−15.38	−1.03	+0.008	+ 0.24	−3.34	−0.014
Nov. 6	−16.41	−0.71	+0.009	− 3.10	−3.55	−0.015
16	−17.12	−0.39	+0.011	− 6.65	−3.67	−0.015
26	−17.51	−0.03	+0.011	−10.32	−3.69	−0.006
Dec. 6	−17.54	+0.32	+0.013	−14.01	−3.61	−0.016
16	−17.22	+0.68	+0.014	−17.62	−3.42	−0.016
26	−16.54	+1.00	+0.014	−21.04	−3.11	−0.015
36	−15.54		+0.015	−24.15		−0.015

Add 1 day to the tabular dates in Jan. and Feb. of leap years.
Add to the Reduced Date
For Upper Culmination before / after } Nov. 20 { + 1 / + 2
For Lower Culmination before / after } May 21 { + 0.5 / + 1.5

ε Ursæ Minoris.

	Corr. of A.R., 1870.		Annual Variation.	Corr. of Dec., 1870.		Annual Variation.
	s.	s.	s.	″	″	″
Jan. 0	− 9.31	+0.65	+0.004	− 8.03	−3.29	−0.009
10	− 8.66	+0.92	+0.004	−11.32	−2.98	−0.008
20	− 7.74	+1.18	+0.005	−14.30	−2.54	−0.007
30	− 6.56	+1.37	+0.005	−16.84	−2.02	−0.007
Feb. 9	− 5.19	+1.51	+0.005	−18.86	−1.43	−0.005
19	− 3.68	+1.61	+0.005	−20.29	−0.79	−0.004
March 1	− 2.07	+1.63	+0.005	−21.08	−0.13	−0.003
11	− 0.44	+1.60	+0.005	−21.21	+0.53	−0.001
21	+ 1.16	+1.51	+0.004	−20.68	+1.14	0.000
31	+ 2.67	+1.36	+0.004	−19.54	+1.72	+0.001
April 10	+ 4.03	+1.17	+0.003	−17.82	+2.21	+0.003
20	+ 5.20	+0.94	+0.003	−15.61	+2.61	+0.004
30	+ 6.14	+0.69	+0.002	−13.00	+2.91	+0.005
May 10	+ 6.83	+0.42	+0.001	−10.09	+3.12	+0.005
20	+ 7.25	+0.14	0.000	− 6.97	+3.21	+0.005
30	+ 7.39	−0.14	0.000	− 3.76	+3.20	+0.006
June 9	+ 7.25	−0.43	−0.001	− 0.56	+3.11	+0.005
19	+ 6.82	−0.68	−0.001	+ 2.55	+2.92	+0.005
29	+ 6.14	−0.94	−0.002	+ 5.47	+2.66	+0.004
July 9	+ 5.20	−1.15	−0.002	+ 8.13	+2.33	+0.003
19	+ 4.05	−1.36	−0.003	+10.46	+1.96	+0.002
29	+ 2.69	−1.51	−0.003	+12.42	+1.52	+0.001
Aug. 8	+ 1.18	−1.65	−0.003	+13.94	+1.06	−0.001
18	− 0.47	−1.74	−0.003	+15.00	+0.59	−0.002
28	− 2.21	−1.80	−0.003	+15.59	+0.08	−0.004
Sept. 7	− 4.01	−1.81	−0.002	+15.67	−0.43	−0.006
17	− 5.82	−1.79	−0.002	+15.24	−0.92	−0.008
27	− 7.61	−1.72	−0.002	+14.32	−1.42	−0.009
Oct. 7	− 9.33	−1.61	−0.001	+12.90	−1.89	−0.011
17	−10.94	−1.45	0.000	+11.01	−2.32	−0.012
27	−12.39	−1.27	+0.001	+ 8.69	−2.73	−0.014
Nov. 6	−13.66	−1.03	+0.002	+ 5.96	−3.06	−0.015
16	−14.69	−0.77	+0.003	+ 2.90	−3.33	−0.016
26	−15.46	−0.47	+0.004	− 0.43	−3.51	−0.017
Dec. 6	−15.93	−0.17	+0.004	− 3.94	−3.60	−0.018
16	−16.10	+0.15	+0.005	− 7.54	−3.58	−0.018
26	−15.95	+0.46	+0.006	−11.12	−3.44	−0.018
36	−15.49		+0.007	−14.56		−0.017

Add 1 day to the tabular date in Jan. and Feb. of leap years.
Add to the Reduced Date
For Upper Culmination before } Dec. 6 { +1
after } { +2
For Lower Culmination before } June 6 { +0.5
after } { +1.5

64 Camelopardali, Bode.

	Corr. of A.R., 1870.		Annual Variation.	Corr. of Dec., 1870.		Annual Variation.
	s.	s.	s.	″	″	″
Jan. 0	+15.36	−0.48	+0.009	+ 5.84	+3.10	−0.026
10	+14.88	−0.97	+0.013	+ 8.94	+2.85	−0.025
20	+13.91	−1.41	+0.016	+11.79	+2.51	−0.023
30	+12.50	−1.79	+0.018	+14.30	+2.06	−0.021
Feb. 9	+10.71	−2.09	+0.020	+16.36	+1.54	−0.017
19	+ 8.62	−2.30	+0.021	+17.90	+0.98	−0.014
March 1	+ 6.32	−2.41	+0.022	+18.88	+0.38	−0.010
11	+ 3.91	−2.42	+0.022	+19.26	−0.24	−0.006
21	+ 1.49	−2.31	+0.021	+19.02	−0.82	−0.001
31	− 0.82	−2.12	+0.020	+18.20	−1.37	+0.003
April 10	− 2.94	−1.85	+0.018	+16.83	−1.85	+0.006
20	− 4.79	−1.50	+0.015	+14.98	−2.26	+0.010
30	− 6.29	−1.10	+0.013	+12.72	−2.58	+0.012
May 10	− 7.39	−0.67	+0.010	+10.14	−2.81	+0.014
20	− 8.06	−0.21	+0.006	+ 7.33	−2.93	+0.016
30	− 8.27	+0.26	+0.003	+ 4.40	−2.98	+0.016
June 9	− 8.01	+0.71	0.000	+ 1.42	−2.92	+0.016
19	− 7.30	+1.15	−0.003	− 1.50	−2.79	+0.015
29	− 6.15	+1.55	−0.006	− 4.29	−2.58	+0.013
July 9	− 4.60	+1.91	−0.009	− 6.87	−2.31	+0.010
19	− 2.69	+2.24	−0.011	− 9.18	−1.99	+0.007
29	− 0.45	+2.50	−0.013	−11.17	−1.62	+0.004
Aug. 8	+ 2.05	+2.73	−0.014	−12.79	−1.22	−0.001
18	+ 4.78	+2.89	−0.015	−14.01	−0.80	−0.005
28	+ 7.67	+3.00	−0.015	−14.81	−0.35	−0.010
Sept. 7	+10.67	+3.06	−0.015	−15.16	+0.10	−0.015
17	+13.73	+3.04	−0.014	−15.06	+0.57	−0.020
27	+16.77	+2.98	−0.013	−14.49	+1.01	−0.025
Oct. 7	+19.75	+2.85	−0.012	−13.48	+1.46	−0.030
17	+22.60	+2.66	−0.009	−12.02	+1.87	−0.035
27	+25.26	+2.41	−0.007	−10.15	+2.26	−0.039
Nov. 6	+27.67	+2.09	−0.004	− 7.89	+2.61	−0.043
16	+29.76	+1.73	0.000	− 5.28	+2.88	−0.048
26	+31.49	+1.29	+0.003	− 2.40	+3.10	−0.050
Dec. 6	+32.78	+0.84	+0.007	+ 0.70	+3.23	−0.052
16	+33.62	+0.33	+0.010	+ 3.93	+3.27	−0.053
26	+33.95	−0.16	+0.014	+ 7.20	+3.20	−0.053
36	+33.79		+0.017	+10.40		−0.052

Add 1 day to the tabular date in Jan. and Feb. of leap years.
Add to the Reduced Date
For Upper Culmination before } June 11 { +1
after } { +2
For Lower Culmination before } Dec. 11 { +1.5
after } { +2.5

51 Cephei Hevelii.

	Corr. of A.R., 1870.		Annual Variation.	Corr. Dec., 1870.		Annual Variation.
	s.	s.	s.	″	″	″
Jan. 0	+28.32	+0.61	-0.012	- 0.96	+3.23	-0.045
10	+28.93	-0.31	-0.002	+ 2.27	+3.19	-0.046
20	+28.62	-1.20	+0.008	+ 5.46	+3.04	-0.045
30	+27.42	-2.04	+0.018	+ 8.50	+2.78	-0.043
Feb. 9	+25.38	-2.78	+0.027	+11.28	+2.40	-0.040
19	+22.60	-3.40	+0.036	+13.68	+1.95	-0.035
March 1	+19.20	-3.87	+0.043	+15.63	+1.41	-0.030
11	+15.33	-4.16	+0.049	+17.04	+0.83	-0.024
21	+11.17	-4.28	+0.053	+17.87	+0.23	-0.017
31	+ 6.89	-4.22	+0.055	+18.10	-0.38	-0.010
April 10	+ 2.67	-4.00	+0.055	+17.72	-0.96	-0.003
20	- 1.33	-3.62	+0.053	+16.76	-1.49	+0.003
30	- 4.95	-3.12	+0.049	+15.27	-1.97	+0.009
May 10	- 8.07	-2.51	+0.045	+13.30	-2.37	+0.014
20	-10.58	-1.82	+0.038	+10.93	-2.68	+0.018
30	-12.40	-1.07	+0.030	+ 8.25	-2.92	+0.022
June 9	-13.47	-0.29	+0.022	+ 5.33	-3.06	+0.023
19	-13.76	+0.49	+0.012	+ 2.27	-3.12	+0.024
29	-13.27	+1.26	+0.002	- 0.85	-3.10	+0.024
July 9	-12.01	+1.99	-0.008	- 3.95	-3.00	+0.022
19	-10.02	+2.69	-0.018	- 6.95	-2.83	+0.018
29	- 7.33	+3.32	-0.027	- 9.78	-2.60	+0.015
Aug. 8	- 4.01	+3.88	-0.036	-12.38	-2.32	+0.009
18	- 0.13	+4.36	-0.045	-14.70	-1.97	+0.003
28	+ 4.23	+4.77	-0.053	-16.67	-1.60	-0.003
Sept. 7	+ 9.00	+5.07	-0.059	-18.27	-1.19	-0.011
17	+14.07	+5.29	-0.064	-19.46	-0.74	-0.019
27	+19.36	+5.39	-0.068	-20.20	-0.28	-0.027
Oct. 7	+24.75	+5.40	-0.071	-20.48	+0.19	-0.036
17	+30.15	+5.27	-0.072	-20.29	+0.69	-0.044
27	+35.42	+5.04	-0.072	-19.60	+1.16	-0.053
Nov. 6	+40.46	+4.67	-0.070	-18.44	+1.64	-0.060
16	+45.13	+4.19	-0.066	-16.80	+2.07	-0.068
26	+49.32	+3.58	-0.061	-14.73	+2.46	-0.074
Dec. 6	+52.90	+2.86	-0.054	-12.27	+2.79	-0.080
16	+55.76	+2.07	-0.046	- 9.48	+3.04	-0.084
26	+57.83	+1.18	-0.038	- 6.44	+3.18	-0.088
36	+59.01		-0.028	- 3.26		-0.089

Add 1 day to the tabular dates in Jan. and Feb. of leap years.
Add to the Reduced Date
For Upper Culmination before } July 1 { + 1
after } { + 2
For Lower Culmination after Jan. 0 + 1.5

25 Camelopardali Hevelii.

	Corr. of A.R., 1870.		Annual Variation.	Corr. Dec., 1870.		Annual Variation.
	s.	s.	s.	″	″	″
Jan. 0	+10.57	+0.48	-0.004	- 3.25	+3.04	-0.019
10	+11.05	+0.13	-0.002	- 0.21	+3.07	-0.020
20	+11.18	-0.22	-0.001	+ 2.86	+3.01	-0.020
30	+10.96	-0.56	+0.001	+ 5.87	+2.83	-0.020
Feb. 9	+10.40	-0.85	+0.003	+ 8.70	+2.58	-0.018
19	+ 9.55	-1.13	+0.005	+11.28	+2.09	-0.016
March 1	+ 8.42	-1.32	+0.006	+13.37	+1.66	-0.014
11	+ 7.10	-1.48	+0.007	+15.03	+1.14	-0.012
21	+ 5.62	-1.55	+0.008	+16.17	+0.56	-0.009
31	+ 4.07	-1.57	+0.008	+16.73	-0.01	-0.006
April 10	+ 2.50	-1.51	+0.009	+16.72	-0.59	-0.003
20	+ 0.99	-1.40	+0.009	+16.13	-1.13	-0.001
30	- 0.41	-1.23	+0.008	+15.00	-1.62	+0.002
May 10	- 1.64	-1.02	+0.008	+13.38	-2.05	+0.004
20	- 2.66	-0.78	+0.007	+11.33	-2.41	+0.007
30	- 3.44	-0.50	+0.006	+ 8.92	-2.69	+0.008
June 9	- 3.94	-0.22	+0.004	+ 6.23	-2.89	+0.009
19	- 4.16	+0.07	+0.003	+ 3.34	-3.00	+0.010
29	- 4.09	+0.36	+0.001	+ 0.34	-3.04	+0.010
July 9	- 3.73	+0.64	0.000	- 2.70	-3.01	+0.010
19	- 3.09	+0.90	-0.002	- 5.71	-2.90	+0.009
29	- 2.19	+1.15	-0.004	- 8.61	-2.73	+0.007
Aug. 8	- 1.04	+1.37	-0.005	-11.34	-2.50	+0.005
18	+ 0.33	+1.56	-0.007	-13.84	-2.23	+0.003
28	+ 1.89	+1.73	-0.009	-16.07	-1.90	0.000
Sept. 7	+ 3.62	+1.87	-0.010	-17.97	-1.54	-0.003
17	+ 5.49	+1.98	-0.011	-19.51	-1.13	-0.006
27	+ 7.47	+2.04	-0.012	-20.64	-0.70	-0.009
Oct. 7	+ 9.51	+2.07	-0.013	-21.34	-0.25	-0.013
17	+11.58	+2.06	-0.013	-21.59	+0.22	-0.017
27	+13.64	+2.00	-0.013	-21.37	+0.70	-0.020
Nov. 6	+15.64	+1.90	-0.013	-20.67	+1.18	-0.024
16	+17.54	+1.74	-0.013	-19.49	+1.64	-0.027
26	+19.28	+1.55	-0.012	-17.85	+2.05	-0.030
Dec. 6	+20.83	+1.30	-0.012	-15.80	+2.43	-0.033
16	+22.13	+1.01	-0.011	-13.37	+2.73	-0.035
26	+23.14	+0.70	-0.010	-10.64	+2.95	-0.036
36	+23.84		-0.008	- 7.69		-0.038

Add 1 day to the tabular dates in Jan. and Feb. of leap years.
Add to the Reduced Date
For Upper Culmination before } July 8 { + 1
after } { + 2
For Lower Culmination before } Jan. 6 { + 0.5
after } { + 1.5

4 Ursæ Minoris, Bode.

	Corr. of A.R., 1870.		Annual Variation.	Corr. of Dec., 1870.		Annual Variation.
	s.	s.	s.	″	″	″
Jan. 0	+ 76.31	+ 3.89	—0.251	— 4.80	+3.14	—0.109
10	+ 80.20	+ 1.37	—0.192	— 1.66	+3.23	—0.114
20	+ 81.57	— 1.17	—0.127	+ 1.57	+3.20	—0.116
30	+ 80.40	— 3.62	—0.057	+ 4.77	+3.04	—0.114
Feb. 9	+ 76.78	— 5.90	+0.015	+ 7.81	+2.79	—0.109
19	+ 70.88	— 7.91	+0.086	+10.60	+2.40	—0.100
March 1	+ 62.97	— 9.55	+0.152	+13.00	+1.94	—0.089
11	+ 53.42	—10.78	+0.211	+14.94	+1.40	—0.075
21	+ 42.64	—11.55	+0.262	+16.34	+0.82	—0.060
31	+ 31.09	—11.86	+0.302	+17.16	+0.21	—0.044
April 10	+ 19.23	—11.69	+0.330	+17.37	—0.39	—0.027
20	+ 7.54	—11.08	+0.344	+16.98	—0.98	—0.010
30	— 3.54	—10.09	+0.347	+16.00	—1.52	+0.006
May 10	— 13.63	— 8.74	+0.336	+14.48	—1.99	+0.021
20	— 22.37	— 7.11	+0.310	+12.49	—2.41	+0.033
30	— 29.48	— 5.26	+0.274	+10.08	—2.74	+0.044
June 9	— 34.74	— 3.28	+0.227	+ 7.34	—2.99	+0.051
19	— 38.02	— 1.19	+0.171	+ 4.35	—3.16	+0.056
29	— 39.21	+ 0.91	+0.108	+ 1.19	—3.24	+0.058
July 9	— 38.30	+ 3.00	+0.039	— 2.05	—3.25	+0.057
19	— 35.30	+ 5.02	—0.035	— 5.30	—3.18	+0.053
29	— 30.28	+ 6.91	—0.111	— 8.48	—3.03	+0.046
Aug. 8	— 23.37	+ 8.67	—0.188	—11.51	—2.82	+0.036
18	— 14.70	+10.25	—0.264	—14.33	—2.57	+0.024
28	— 4.45	+11.63	—0.337	—16.90	—2.24	+0.009
Sept. 7	+ 7.18	+12.79	—0.407	—19.14	—1.87	—0.007
17	+ 19.97	+13.69	—0.472	—21.01	—1.47	—0.025
27	+ 33.66	+14.34	—0.530	—22.48	—1.02	—0.045
Oct. 7	+ 48.00	+14.69	—0.579	—23.50	—0.54	—0.065
17	+ 62.69	+14.72	—0.619	—24.04	—0.03	—0.086
27	+ 77.41	+14.43	—0.650	—24.07	+0.47	—0.107
Nov. 6	+ 91.84	+13.78	—0.668	—23.60	+1.00	—0.127
16	+105.62	+12.77	—0.674	—22.60	+1.49	—0.147
26	+118.39	+11.41	—0.668	—21.11	+1.97	—0.165
Dec. 6	+129.80	+ 9.72	—0.649	—19.14	+2.39	—0.181
16	+139.52	+ 7.69	—0.617	—16.75	+2.74	—0.195
26	+147.21	+ 5.45	—0.575	—14.01	+3.01	—0.206
36	+152.66		—0.524	—11.00		—0.213

Add 1 day to the tabular date in Jan. and Feb. of leap years.
Add to the Reduced Date
For Upper Culmination before } July 13 { + 1
after } { + 2
For Lower Culmination before } Jan. 11 { + 0.5
after } { + 1.5

156 Camelopardali, Bode.

	Corr. of A.R., 1870.		Annual Variation.	Corr. of Dec., 1870.		Annual Variation.
	s.	s.	s.	″	″	″
Jan. 0	+13.29	+1.00	—0.011	— 6.81	+2.90	—0.021
10	+14.29	+0.54	—0.009	— 3.91	+3.07	—0.023
20	+14.83	+0.09	—0.007	— 0.84	+3.11	—0.024
30	+14.92	—0.39	—0.004	+ 2.27	+3.04	—0.024
Feb. 9	+14.53	—0.79	—0.002	+ 5.31	+2.84	—0.023
19	+13.74	—1.19	+0.001	+ 8.15	+2.54	—0.022
March 1	+12.55	—1.52	+0.003	+10.69	+2.14	—0.020
11	+11.03	—1.79	+0.006	+12.83	+1.65	—0.017
21	+ 9.24	—1.96	+0.008	+14.48	+1.12	—0.014
31	+ 7.28	—2.05	+0.010	+15.60	+0.53	—0.011
April 10	+ 5.23	—2.05	+0.011	+16.13	—0.04	—0.007
20	+ 3.18	—1.99	+0.012	+16.09	—0.63	—0.004
30	+ 1.19	—1.83	+0.012	+15.46	—1.17	—0.001
May 10	— 0.64	—1.62	+0.012	+14.29	—1.67	+0.002
20	— 2.26	—1.35	+0.011	+12.62	—2.12	+0.005
30	— 3.61	—1.04	+0.011	+10.50	—2.48	+0.008
June 9	— 4.65	—0.70	+0.009	+ 8.02	—2.77	+0.009
19	— 5.35	—0.34	+0.007	+ 5.25	—2.99	+0.011
29	— 5.69	+0.03	+0.005	+ 2.26	—3.13	+0.012
July 9	— 5.66	+0.41	+0.002	— 0.87	—3.19	+0.012
19	— 5.25	+0.75	0.000	— 4.06	—3.18	+0.011
29	— 4.50	+1.10	—0.003	— 7.24	—3.09	+0.010
Aug. 8	— 3.40	+1.43	—0.006	—10.33	—2.94	+0.009
18	— 1.97	+1.71	—0.009	—13.27	—2.73	+0.007
28	— 0.26	+1.98	—0.012	—16.00	—2.48	+0.004
Sept. 7	+ 1.72	+2.21	—0.015	—18.48	—2.14	+0.001
17	+ 3.93	+2.40	—0.017	—20.62	—1.79	—0.003
27	+ 6.33	+2.54	—0.019	—22.41	—1.37	—0.006
Oct. 7	+ 8.87	+2.63	—0.022	—23.78	—0.93	—0.010
17	+11.50	+2.69	—0.024	—24.71	—0.45	—0.015
27	+14.19	+2.66	—0.025	—25.16	+0.05	—0.019
Nov. 6	+16.85	+2.60	—0.026	—25.11	+0.57	—0.023
16	+19.45	+2.47	—0.027	—24.54	+1.08	—0.027
26	+21.92	+2.24	—0.027	—23.46	+1.57	—0.031
Dec. 6	+24.16	+1.98	—0.026	—21.89	+2.02	—0.034
16	+26.14	+1.65	—0.026	—19.87	+2.43	—0.038
26	+27.79	+1.25	—0.024	—17.44	+2.74	—0.040
36	+29.04		—0.023	—14.70		—0.042

Add 1 day to the tabular date in Jan. and Feb. of leap years.
Add to the Reduced Date
For Upper Culmination before } July 18 { + 1
after } { + 2
For Lower Culmination before } Jan. 17 { + 0.5
after } { + 1.5

	λ Ursæ Minoris, Bode.						74 Draconis.					
	Corr. of A.R., 1870.		Annual Variation.	Corr. of Dec., 1870.		Annual Variation.	Corr. of A.R., 1870.		Annual Variation.	Corr. of Dec., 1870.		Annual Variation.
	s.	s.	s.	″	″	″	s.	s.	s.	″	″	″
Jan. 0	− 67.61	− 4.71	−0.269	+ 7.27	−3.04	−0.082	− 7.13	−0.67	−0.004	+10.34	−2.89	−0.004
10	− 72.32	− 2.51	−0.243	+ 4.23	−3.22	−0.088	− 7.80	−0.43	−0.004	+ 7.45	−3.16	−0.004
20	− 74.83	− 0.19	−0.201	+ 1.01	−3.26	−0.091	− 8.23	−0.18	−0.004	+ 4.29	−3.31	−0.005
30	− 75.02	+ 2.11	−0.153	− 2.25	−3.19	−0.092	− 8.41	+0.08	−0.004	+ 0.98	−3.33	−0.005
Feb. 9	− 72.91	+ 4.31	−0.100	− 5.44	−2.98	−0.089	− 8.33	+0.33	−0.003	− 2.35	−3.22	−0.005
19	− 68.60	+ 6.29	−0.044	− 8.42	−2.67	−0.084	− 8.00	+0.58	−0.003	− 5.57	−2.99	−0.004
March 1	− 62.31	+ 7.99	+0.012	−11.09	−2.25	−0.076	− 7.42	+0.80	−0.003	− 8.56	−2.63	−0.004
11	− 54.32	+ 9.36	+0.065	−13.34	−1.76	−0.067	− 6.62	+0.99	−0.002	−11.19	−2.19	−0.004
21	− 44.96	+10.32	+0.115	−15.10	−1.18	−0.055	− 5.63	+1.13	−0.002	−13.38	−1.65	−0.003
31	− 34.64	+10.89	+0.158	−16.28	−0.58	−0.043	− 4.50	+1.26	−0.001	−15.03	−1.06	−0.003
April 10	− 23.75	+11.00	+0.194	−16.86	+0.03	−0.030	− 3.24	+1.30	0.000	−16.09	−0.44	−0.002
20	− 12.75	+10.73	+0.221	−16.83	+0.63	−0.016	− 1.94	+1.35	0.000	−16.53	+0.20	−0.001
30	− 2.02	+10.02	+0.238	−16.20	+1.22	−0.003	− 0.59	+1.31	+0.001	−16.33	+0.80	−0.001
May 10	+ 8.00	+ 9.11	+0.244	−14.98	+1.74	+0.009	+ 0.72	+1.25	+0.001	−15.53	+1.39	0.000
20	+ 17.11	+ 7.77	+0.240	−13.24	+2.20	+0.021	+ 1.97	+1.14	+0.002	−14.14	+1.93	0.000
30	+ 24.88	+ 6.23	+0.225	−11.04	+2.60	+0.029	+ 3.11	+1.00	+0.002	−12.21	+2.40	+0.001
June 9	+ 31.11	+ 4.53	+0.201	− 8.44	+2.91	+0.036	+ 4.11	+0.83	+0.002	− 9.81	+2.81	+0.002
19	+ 35.64	+ 2.70	+0.168	− 5.53	+3.14	+0.042	+ 4.94	+0.65	+0.002	− 7.00	+3.12	+0.002
29	+ 38.34	+ 0.79	+0.126	− 2.39	+3.30	+0.045	+ 5.59	+0.43	+0.002	− 3.88	+3.36	+0.002
July 9	+ 39.13	− 1.12	+0.078	+ 0.91	+3.37	+0.046	+ 6.02	+0.23	+0.002	− 0.52	+3.52	+0.002
19	+ 38.01	− 3.02	+0.025	+ 4.28	+3.35	+0.044	+ 6.25	−0.01	+0.002	+ 3.00	+3.59	+0.002
29	+ 34.99	− 4.84	−0.034	+ 7.63	+3.27	+0.040	+ 6.24	−0.22	+0.001	+ 6.59	+3.59	+0.002
Aug. 8	+ 30.15	− 6.55	−0.096	+10.90	+3.12	+0.034	+ 6.02	−0.43	+0.001	+10.18	+3.50	+0.002
18	+ 23.60	− 8.15	−0.160	+14.02	+2.90	+0.026	+ 5.59	−0.65	0.000	+13.68	+3.34	+0.002
28	+ 15.45	− 9.57	−0.224	+16.92	+2.61	+0.016	+ 4.94	−0.82	0.000	+17.02	+3.12	+0.001
Sept. 7	+ 5.88	−10.81	−0.287	+19.53	+2.28	+0.005	+ 4.12	−1.00	−0.001	+20.14	+2.83	0.000
17	− 4.93	−11.82	−0.349	+21.81	+1.89	−0.009	+ 3.12	−1.15	−0.001	+22.97	+2.47	0.000
27	− 16.75	−12.62	−0.407	+23.70	+1.46	−0.023	+ 1.97	−1.27	−0.002	+25.44	+2.05	−0.001
Oct. 7	− 29.37	−13.13	−0.460	+25.16	+0.98	−0.039	+ 0.70	−1.36	−0.003	+27.49	+1.59	−0.002
17	− 42.50	−13.38	−0.508	+26.14	+0.47	−0.055	− 0.66	−1.42	−0.004	+29.08	+1.08	−0.002
27	− 55.88	−13.31	−0.548	+26.61	−0.07	−0.071	− 2.08	−1.44	−0.004	+30.16	+0.53	−0.003
Nov. 6	− 69.19	−12.91	−0.581	+26.54	−0.59	−0.087	− 3.52	−1.44	−0.005	+30.69	−0.03	−0.004
16	− 82.10	−12.19	−0.604	+25.95	−1.15	−0.103	− 4.96	−1.38	−0.005	+30.66	−0.63	−0.005
26	− 94.29	−11.12	−0.616	+24.80	−1.66	−0.118	− 6.34	−1.30	−0.006	+30.03	−1.19	−0.005
Dec. 6	−105.41	− 9.72	−0.619	+23.14	−2.13	−0.131	− 7.64	−1.16	−0.007	+28.84	−1.74	−0.006
16	−115.13	− 8.01	−0.611	+21.01	−2.56	−0.143	− 8.80	−1.00	−0.007	+27.10	−2.24	−0.007
26	−123.14	− 6.07	−0.592	+18.45	−2.89	−0.153	− 9.80	−0.81	−0.007	+24.86	−2.68	−0.007
36	−129.21		−0.566	+15.56		−0.160	−10.61		−0.007	+22.18		−0.008

λ Ursæ Minoris, Bode.

Add 1 day to the tabular date in Jan. and Feb. of leap years.
Add to the Reduced Date
For Upper Culmination before / after } Jan. 19 { 0 / + 1
For Lower Culmination before / after } July 20 { + 0.5 / + 1.5

74 Draconis.

Add 1 day to the tabular date in Jan. and Feb. of leap years.
Add to the Reduced Date
For Upper Culmination before / after } Jan. 29 { 0 / + 1
For Lower Culmination before / after } July 31 { + 0.5 / + 1.5

1 Draconis Hevelii.

	Corr. of A.R., 1870.		Annual Variation.	Corr. Dec., 1870.		Annual Variation.
	s.	s.	s.	″	″	″
Jan. 0	+ 7.18	+1.24	−0.008	−13.84	+2.08	−0.010
10	+ 8.42	+0.98	−0.008	−11.76	+2.48	−0.011
20	+ 9.40	+0.69	−0.008	− 9.28	+2.78	−0.013
30	+10.09	+0.38	−0.007	− 6.50	+2.98	−0.014
Feb. 9	+10.47	+0.07	−0.006	− 3.52	+3.06	−0.015
19	+10.54	−0.25	−0.005	− 0.46	+3.00	−0.015
March 1	+10.29	−0.54	−0.004	+ 2.54	+2.83	−0.014
11	+ 9.75	−0.80	−0.003	+ 5.37	+2.55	−0.013
21	+ 8.95	−1.01	−0.001	+ 7.92	+2.17	−0.012
31	+ 7.94	−1.19	0.000	+10.09	+1.70	−0.011
April 10	+ 6.75	−1.29	+0.001	+11.79	+1.19	−0.009
20	+ 5.46	−1.36	+0.002	+12.98	+0.63	−0.007
30	+ 4.10	−1.38	+0.003	+13.61	+0.06	−0.005
May 10	+ 2.72	−1.32	+0.004	+13.67	−0.50	−0.003
20	+ 1.40	−1.24	+0.004	+13.17	−1.05	−0.001
30	+ 0.16	−1.11	+0.004	+12.12	−1.56	+0.001
June 9	− 0.95	−0.94	+0.005	+10.56	−2.02	+0.003
19	− 1.89	−0.76	+0.004	+ 8.54	−2.42	+0.004
29	− 2.65	−0.55	+0.004	+ 6.12	−2.75	+0.006
July 9	− 3.20	−0.33	+0.003	+ 3.37	−3.03	+0.007
19	− 3.53	−0.10	+0.003	+ 0.34	−3.22	+0.008
29	− 3.63	+0.13	+0.002	− 2.88	−3.36	+0.008
Aug. 8	− 3.50	+0.36	+0.001	− 6.24	−3.41	+0.008
18	− 3.14	+0.59	−0.001	− 9.65	−3.41	+0.008
28	− 2.55	+0.81	−0.002	−13.06	−3.32	+0.007
Sept. 7	− 1.74	+1.01	−0.003	−16.38	−3.17	+0.006
17	− 0.73	+1.21	−0.005	−19.55	−2.96	+0.005
27	+ 0.48	+1.38	−0.007	−22.51	−2.68	+0.003
Oct. 7	+ 1.86	+1.53	−0.008	−25.19	−2.34	+0.001
17	+ 3.39	+1.65	−0.009	−27.53	−1.93	−0.001
27	+ 5.04	+1.74	−0.011	−29.46	−1.47	−0.003
Nov. 6	+ 6.78	+1.81	−0.012	−30.93	−0.96	−0.005
16	+ 8.59	+1.82	−0.013	−31.89	−0.42	−0.008
26	+10.41	+1.78	−0.015	−32.31	+0.14	−0.010
Dec. 6	+12.19	+1.70	−0.015	−32.17	+0.72	−0.013
16	+13.89	+1.56	−0.016	−31.45	+1.26	−0.015
26	+15.45	+1.37	−0.016	−30.19	+1.79	−0.017
36	+16.82		−0.016	−28.40		−0.019

Add 1 day to the tabular dates in Jan. and Feb. of leap years.

Add to the Reduced Date

For Upper Culmination before / after } Aug. 11 { + 1 / + 2

For Lower Culmination before / after } Feb. 9 { + 0.5 / + 1.5

119 Cephei, Bode.

	Corr. of A.R., 1870.		Annual Variation.	Corr. Dec., 1870.		Annual Variation.
	s.	s.	s.	″	″	″
Jan. 0	− 9.03	−1.37	−0.014	+13.80	−2.40	−0.005
10	−10.40	−1.05	−0.014	+11.40	−2.79	−0.006
20	−11.45	−0.70	−0.014	+ 8.61	−3.08	−0.006
30	−12.15	−0.32	−0.013	+ 5.53	−3.24	−0.006
Feb. 9	−12.47	+0.06	−0.012	+ 2.29	−3.27	−0.006
19	−12.41	+0.44	−0.011	− 0.98	−3.17	−0.007
March 1	−11.97	−0.80	−0.010	− 4.15	−2.95	−0.006
11	−11.17	+1.13	−0.008	− 7.10	−2.61	−0.006
21	−10.04	+1.41	−0.006	− 9.71	−2.18	−0.005
31	− 8.63	+1.65	−0.003	−11.89	−1.66	−0.005
April 10	− 6.98	+1.81	−0.001	−13.55	−1.09	−0.004
20	− 5.17	+1.91	+0.001	−14.64	−0.48	−0.003
30	− 3.26	+1.94	+0.003	−15.12	+0.14	−0.002
May 10	− 1.32	+1.92	+0.005	−14.98	+0.74	−0.001
20	+ 0.60	+1.82	+0.007	−14.24	+1.31	0.000
30	+ 2.42	+1.68	+0.008	−12.93	+1.86	+0.001
June 9	+ 4.10	+1.51	+0.009	−11.07	+2.32	+0.002
19	+ 5.59	+1.24	+0.009	− 8.75	+2.74	+0.002
29	+ 6.83	+0.98	+0.009	− 6.01	+3.08	+0.003
July 9	+ 7.81	+0.69	+0.009	− 2.93	+3.34	+0.003
19	+ 8.50	+0.38	+0.009	+ 0.41	+3.53	+0.003
29	+ 8.88	+0.07	+0.008	+ 3.94	+3.63	+0.004
Aug. 8	+ 8.95	−0.26	+0.007	+ 7.57	+3.66	+0.003
18	+ 8.69	−0.55	+0.005	+11.23	+3.61	+0.003
28	+ 8.14	−0.86	+0.003	+14.84	+3.49	+0.003
Sept. 7	+ 7.28	−1.13	+0.001	+18.33	+3.29	+0.002
17	+ 6.15	−1.39	−0.001	+21.62	+3.02	+0.001
27	+ 4.76	−1.60	−0.003	+24.64	+2.68	+0.001
Oct. 7	+ 3.16	−1.80	−0.006	+27.32	+2.29	0.000
17	+ 1.36	−1.95	−0.008	+29.61	+1.82	−0.001
27	− 0.59	−2.05	−0.011	+31.43	+1.31	−0.002
Nov. 6	− 2.64	−2.11	−0.013	+32.74	+0.76	−0.003
16	− 4.75	−2.11	−0.015	+33.50	+0.17	−0.004
26	− 6.86	−2.06	−0.018	+33.67	−0.43	−0.005
Dec. 6	− 8.92	−1.94	−0.019	+33.24	−1.03	−0.006
16	−10.86	−1.77	−0.021	+32.21	−1.59	−0.008
26	−12.63	−1.53	−0.022	+30.62	−2.12	−0.008
36	−14.16		−0.022	+28.50		−0.009

Add 1 day to the tabular dates in Jan. and Feb. of leap years.

Add to the Reduced Date

For Upper Culmination before / after } Feb. 11 { + 0 / + 1

For Lower Culmination before / after } Aug. 12 { + 0.5 / + 1.5

30 Camelopardali Hevelii.

	Corr. of A.R., 1870. s.	s.	Annual Variation. s.	Corr. Dec., 1870. ″	″	Annual Variation. ″
Jan. 0	+ 6.40	+1.73	-0.007	-17.10	+1.47	-0.007
10	+ 8.13	+1.49	-0.008	-15.63	+1.98	-0.008
20	+ 9.62	+1.20	-0.008	-13.65	+2.43	-0.010
30	+10.82	+0.86	-0.008	-11.22	+2.77	-0.011
Feb. 9	+11.68	+0.51	-0.008	- 8.45	+2.99	-0.012
19	+12.19	+0.13	-0.008	- 5.46	+3.09	-0.013
March 1	+12.32	-0.23	-0.008	- 2.37	+3.07	-0.013
11	+12.09	-0.57	-0.007	+ 0.70	+2.92	-0.013
21	+11.52	-0.88	-0.006	+ 3.62	+2.65	-0.012
31	+10.64	-1.15	-0.005	+ 6.27	+2.29	-0.011
April 10	+ 9.49	-1.36	-0.004	+ 8.56	+1.85	-0.010
20	+ 8.13	-1.51	-0.003	+10.41	+1.33	-0.008
30	+ 6.62	-1.61	-0.002	+11.74	+0.78	-0.007
May 10	+ 5.01	-1.63	-0.001	+12.52	+0.22	-0.005
20	+ 3.38	-1.62	0.000	+12.74	-0.36	-0.003
30	+ 1.76	-1.54	0.000	+12.38	-0.92	-0.001
June 9	+ 0.22	-1.42	+0.001	+11.46	-1.45	-0.001
19	- 1.20	-1.25	+0.002	+10.01	-1.93	+0.002
29	- 2.45	-1.07	+0.002	+ 8.08	-2.37	+0.004
July 9	- 3.52	-0.84	+0.002	+ 5.71	-2.75	+0.005
19	- 4.36	-0.60	+0.002	+ 2.96	-3.06	+0.006
29	- 4.96	-0.35	+0.002	- 0.10	-3.31	+0.007
Aug. 8	- 5.31	-0.09	+0.001	- 3.41	-3.48	+0.008
18	- 5.40	+0.19	+0.001	- 6.89	-3.59	+0.008
28	- 5.21	+0.44	0.000	-10.48	-3.62	+0.008
Sept. 7	- 4.77	+0.72	-0.001	-14.10	-3.58	+0.008
17	- 4.05	+0.97	-0.002	-17.68	-3.47	+0.007
27	- 3.08	+1.21	-0.003	-21.15	-3.28	+0.006
Oct. 7	- 1.87	+1.44	-0.005	-24.43	-3.02	+0.005
17	- 0.43	+1.65	-0.006	-27.45	-2.67	+0.004
27	+ 1.22	+1.82	-0.008	-30.12	-2.27	+0.002
Nov. 6	+ 3.04	+1.96	-0.009	-32.39	-1.80	0.000
16	+ 5.00	+2.05	-0.011	-34.19	-1.27	-0.002
26	+ 7.05	+2.09	-0.012	-35.46	-0.69	-0.006
Dec. 6	+ 9.14	+2.07	-0.014	-36.15	-0.09	-0.006
16	+11.21	+2.00	-0.015	-36.24	+0.52	-0.008
26	+13.21	+1.85	-0.016	-35.72	+1.11	-0.010
36	+15.06		-0.017	-34.61		-0.012

Add 1 day to the tabular date in Jan. and Feb. of leap years.
Add to the Reduced Date
For Upper Culmination before / after } Aug. 25 { +1 / +2
For Lower Culmination before / after } Feb. 24 { +0.5 / +1.5

34 Cephei Hevelii.

	Corr. of A.R., 1870. s.	s.	Annual Variation. s.	Corr. Dec., 1870. ″	″	Annual Variation. ″
Jan. 0	- 4.50	-1.45	-0.006	+18.11	-1.42	0.000
10	- 5.95	-1.28	-0.007	+16.69	-1.96	0.000
20	- 7.23	-1.07	-0.008	+14.73	-2.43	+0.001
30	- 8.30	-0.80	-0.008	+12.30	-2.80	+0.001
Feb. 9	- 9.10	-0.52	-0.009	+ 9.50	-3.06	+0.001
19	- 9.62	-0.21	-0.009	+ 6.44	-3.18	+0.001
March 1	- 9.83	+0.11	-0.008	+ 3.26	-3.19	+0.001
11	- 9.72	+0.42	-0.008	+ 0.07	-3.06	+0.001
21	- 9.30	+0.71	-0.007	- 2.99	-2.81	+0.001
31	- 8.59	+0.99	-0.006	- 5.80	-2.46	0.000
April 10	- 7.60	+1.22	-0.005	- 8.26	-2.01	0.000
20	- 6.38	+1.40	-0.004	-10.27	-1.49	0.000
30	- 4.98	+1.54	-0.002	-11.76	-0.93	0.000
May 10	- 3.44	+1.63	-0.001	-12.69	-0.34	0.000
20	- 1.81	+1.66	0.000	-13.03	+0.26	0.000
30	- 0.15	+1.65	+0.002	-12.77	+0.85	0.000
June 9	+ 1.50	+1.57	+0.003	-11.92	+1.41	0.000
19	+ 3.07	+1.47	+0.004	-10.51	+1.93	0.000
29	+ 4.54	+1.33	+0.005	- 8.58	+2.41	0.000
July 9	+ 5.87	+1.14	+0.006	- 6.17	+2.81	0.000
19	+ 7.01	+0.94	+0.006	- 3.36	+3.16	-0.001
29	+ 7.95	+0.71	+0.007	- 0.20	+3.43	-0.001
Aug. 8	+ 8.66	+0.48	+0.007	+ 3.23	+3.64	-0.001
18	+ 9.14	+0.22	+0.007	+ 6.87	+3.75	-0.001
28	+ 9.36	-0.02	+0.006	+10.62	+3.80	-0.001
Sept. 7	+ 9.34	-0.27	+0.006	+14.42	+3.78	-0.001
17	+ 9.07	-0.52	+0.005	+18.20	+3.66	-0.001
27	+ 8.55	-0.75	+0.004	+21.86	+3.47	-0.001
Oct. 7	+ 7.80	-0.96	+0.003	+25.33	+3.21	-0.001
17	+ 6.84	-1.16	+0.002	+28.54	+2.86	-0.001
27	+ 5.68	-1.33	+0.001	+31.40	+2.44	0.000
Nov. 6	+ 4.35	-1.47	0.000	+33.84	+1.96	0.000
16	+ 2.88	-1.57	-0.002	+35.80	+1.42	0.000
26	+ 1.31	-1.63	-0.004	+37.22	+0.82	0.000
Dec. 6	- 0.32	-1.65	-0.005	+38.04	+0.21	0.000
16	- 1.97	-1.62	-0.006	+38.25	-0.43	0.000
26	- 3.59	-1.53	-0.007	+37.82	-1.05	0.000
36	- 5.12		-0.008	+36.77		0.000

Add 1 day to the tabular date in Jan. and Feb. of leap years.
Add to the Reduced Date
For Upper Culmination before / after } March 4 { 0 / +1
For Lower Culmination before / after } Sept. 3 { +0.5 / +1.5

202 Camelopardali Bode.

Date	Corr. of A.R., 1870. s.	s.	Annual Variation. s.	Corr. Dec., 1870. "	"	Annual Variation. "
Jan. 0	+ 2.78	+1.63	−0.003	−19.61	+0.53	−0.002
10	+ 4.41	+1.50	−0.004	−19.08	+1.15	−0.003
20	+ 5.91	+1.33	−0.005	−17.93	+1.72	−0.004
30	+ 7.24	+1.09	−0.006	−16.21	+2.21	−0.005
Feb. 9	+ 8.33	+0.85	−0.006	−14.00	+2.61	−0.006
19	+ 9.18	+0.55	−0.007	−11.39	+2.90	−0.007
March 1	+ 9.73	+0.25	−0.007	− 8.49	+3.06	−0.007
11	+ 9.98	−0.04	−0.007	− 5.43	+3.11	−0.007
21	+ 9.94	−0.33	−0.006	− 2.32	+3.01	−0.007
31	+ 9.61	−0.59	−0.006	+ 0.69	+2.81	−0.007
April 10	+ 9.02	−0.83	−0.005	+ 3.50	+2.50	−0.007
20	+ 8.19	−1.02	−0.005	+ 6.00	+2.10	−0.006
30	+ 7.17	−1.17	−0.004	+ 8.10	+1.63	−0.005
May 10	+ 6.00	−1.27	−0.003	+ 9.73	+1.11	−0.004
20	+ 4.73	−1.34	−0.002	+10.84	+0.56	−0.003
30	+ 3.39	−1.35	−0.001	+11.40	0.00	−0.002
June 9	+ 2.04	−1.33	0.000	+11.40	−0.57	−0.001
19	+ 0.71	−1.26	+0.001	+10.83	−1.12	0.000
29	− 0.55	−1.17	+0.001	+ 9.71	−1.63	+0.001
July 9	− 1.72	−1.05	+0.002	+ 8.08	−2.11	+0.002
19	− 2.77	−0.90	+0.002	+ 5.97	−2.54	+0.003
29	− 3.67	−0.73	+0.003	+ 3.43	−2.91	+0.004
Aug. 8	− 4.40	−0.54	+0.003	+ 0.52	−3.22	+0.005
18	− 4.94	−0.35	+0.003	− 2.70	−3.46	+0.005
28	− 5.29	−0.13	+0.003	− 6.16	−3.64	+0.006
Sept. 7	− 5.42	+0.08	+0.003	− 9.80	−3.75	+0.006
17	− 5.34	+0.31	+0.003	−13.55	−3.77	+0.006
27	− 5.03	+0.54	+0.002	−17.32	−3.73	+0.006
Oct. 6	− 4.49	+0.75	+0.001	−21.05	−3.59	+0.006
16	− 3.74	+0.97	+0.001	−24.64	−3.38	+0.005
26	− 2.77	+1.18	0.000	−28.02	−3.08	+0.004
Nov. 6	− 1.59	+1.34	−0.001	−31.10	−2.70	+0.004
16	− 0.25	+1.50	−0.002	−33.80	−2.24	+0.003
26	+ 1.25	+1.62	−0.003	−36.04	−1.71	+0.002
Dec. 6	+ 2.87	+1.69	−0.005	−37.75	−1.13	0.000
16	+ 4.56	+1.71	−0.006	−38.88	−0.51	−0.001
26	+ 6.27	+1.67	−0.007	−39.39	+0.13	−0.002
36	+ 7.94		−0.008	−39.26		−0.003

Add 1 day to the tabular date in Jan. and Feb. of leap years.
Add to the Reduced Date
For Upper Culmination before / after } Sept. 11 { + 1 / + 2
For Lower Culmination before / after } Mar. 13 { + 0.5 / + 1.5

39 Cephei Hevelii.

Date	Corr. of A.R., 1870. s.	s.	Annual Variation. s.	Corr. Dec., 1870. "	"	Annual Variation. "
Jan. 0	− 6.25	−3.52	−0.010	+19.60	−0.78	0.000
10	− 9.77	−3.25	−0.016	+18.82	−1.39	0.000
20	−13.02	−2.87	−0.022	+17.43	−1.94	0.000
30	−15.89	−2.37	−0.027	+15.49	−2.41	+0.001
Feb. 9	−18.26	−1.80	−0.031	+13.08	−2.78	+0.001
19	−20.06	−1.14	−0.033	+10.30	−3.03	+0.001
March 1	−21.20	−0.44	−0.035	+ 7.27	−3.15	+0.001
11	−21.64	+0.25	−0.036	+ 4.12	−3.16	+0.001
21	−21.39	+0.94	−0.036	+ 0.96	−3.02	+0.001
31	−20.45	+1.58	−0.034	− 2.06	−2.78	+0.001
April 10	−18.87	+2.16	−0.032	− 4.84	−2.43	+0.001
20	−16.71	+2.66	−0.028	− 7.27	−2.00	+0.001
30	−14.05	+3.05	−0.024	− 9.27	−1.49	+0.001
May 10	−11.00	+3.34	−0.019	−10.76	−0.94	0.000
20	− 7.66	+3.51	−0.014	−11.70	−0.36	0.000
30	− 4.15	+3.59	−0.008	−12.06	+0.22	0.000
June 9	− 0.56	+3.54	−0.003	−11.84	+0.80	0.000
19	+ 2.98	+3.42	+0.003	−11.04	+1.36	0.000
29	+ 6.40	+3.18	+0.008	− 9.68	+1.87	0.000
July 9	+ 9.58	+2.88	+0.014	− 7.81	+2.34	0.000
19	+12.46	+2.52	+0.018	− 5.47	+2.76	0.000
29	+14.98	+2.09	+0.022	− 2.71	+3.12	−0.001
Aug. 8	+17.07	+1.62	+0.025	+ 0.41	+3.40	−0.001
18	+18.69	+1.13	+0.027	+ 3.81	+3.62	−0.001
28	+19.82	+0.59	+0.029	+ 7.43	+3.77	−0.001
Sept. 7	+20.41	+0.07	+0.030	+11.20	+3.83	−0.001
17	+20.48	−0.48	+0.029	+15.03	+3.82	−0.001
27	+20.00	−1.01	+0.029	+18.85	+3.73	−0.001
Oct. 6	+18.99	−1.53	+0.027	+22.58	+3.56	−0.001
16	+17.46	−2.02	+0.024	+26.14	+3.29	−0.001
26	+15.44	−2.49	+0.021	+29.43	+2.96	−0.001
Nov. 6	+12.95	−2.88	+0.017	+32.39	+2.55	−0.001
16	+10.07	−3.22	+0.012	+34.94	+2.04	−0.001
26	+ 6.85	−3.48	+0.006	+36.98	+1.50	0.000
Dec. 6	+ 3.37	−3.64	+0.001	+38.48	+0.90	0.000
16	− 0.27	−3.70	−0.005	+39.38	+0.25	0.000
26	− 3.97	−3.63	−0.012	+39.63	−0.38	0.000
36	− 7.60		−0.018	+39.25		0.000

Add 1 day to the tabular date in Jan. and Feb. of leap years.
Add to the Reduced Date
For Upper Culmination before / after } Mar. 14 { 0 / + 1
For Lower Culmination before / after } Sept. 13 { + 0.5 / + 1.5

309 Cephei Bode.

	Corr. of A.R., 1870.		Annual Variation.	Corr. Dec., 1870.		Annual Variation.
	s.	s.	s.	″	″	″
Jan. 0	− 3.18	−3.03	−0.001	+20.07	−0.43	+0.001
10	− 6.21	−2.87	−0.005	+19.64	−1.06	+0.002
20	− 9.08	−2.60	−0.009	+18.58	−1.65	+0.002
30	−11.68	−2.24	−0.013	+16.93	−2.16	+0.003
Feb. 9	−13.92	−1.79	−0.017	+14.77	−2.58	+0.004
19	−15.71	−1.27	−0.020	+12.19	−2.90	+0.004
March 1	−16.98	−0.70	−0.023	+ 9.29	−3.09	+0.004
11	−17.68	−0.11	−0.024	+ 6.20	−3.15	+0.004
21	−17.79	+0.47	−0.025	+ 3.05	−3.09	+0.004
31	−17.32	+1.05	−0.025	− 0.04	−2.91	+0.004
April 10	−16.27	+1.57	−0.024	− 2.95	−2.61	+0.004
20	−14.70	+2.04	−0.022	− 5.56	−2.23	+0.003
30	−12.66	+2.42	−0.020	− 7.79	−1.76	+0.003
May 10	−10.24	+2.72	−0.017	− 9.55	−1.24	+0.002
20	− 7.52	+2.93	−0.014	−10.79	−0.69	+0.002
30	− 4.59	+3.06	−0.010	−11.48	−0.11	+0.001
June 9	− 1.53	+3.08	−0.006	−11.59	+0.47	+0.001
19	+ 1.55	+3.03	−0.002	−11.12	+1.03	0.000
29	+ 4.58	+2.89	+0.003	−10.09	+1.57	−0.001
July 9	+ 7.47	+2.69	+0.007	− 8.52	+2.07	−0.002
19	+10.16	+2.42	+0.011	− 6.45	+2.51	−0.002
29	+12.58	+2.10	+0.014	− 3.94	+2.91	−0.003
Aug. 8	+14.68	+1.74	+0.017	− 1.03	+3.23	−0.003
18	+16.42	+1.34	+0.020	+ 2.20	+3.50	−0.004
28	+17.76	+0.91	+0.023	+ 5.70	+3.69	−0.004
Sept. 7	+18.67	+0.46	+0.025	+ 9.39	+3.81	−0.004
17	+19.13	+0.01	+0.026	+13.20	+3.85	−0.004
27	+19.14	−0.44	+0.026	+17.05	+3.81	−0.004
Oct. 7	+18.70	−0.90	+0.026	+20.86	+3.68	−0.004
17	+17.80	−1.35	+0.026	+24.54	+3.48	−0.004
27	+16.45	−1.75	+0.024	+28.02	+3.18	−0.003
Nov. 6	+14.70	−2.14	+0.022	+31.20	+2.81	−0.003
16	+12.56	−2.48	+0.020	+34.01	+2.35	−0.002
26	+10.08	−2.76	+0.016	+36.36	+1.82	−0.002
Dec. 6	+ 7.32	−2.95	+0.013	+38.18	+1.25	−0.001
16	+ 4.37	−3.06	+0.009	+39.43	+0.61	0.000
26	+ 1.31	−3.08	+0.005	+40.04	−0.02	+0.001
36	− 1.77		0.000	+40.02		+0.001

Add 1 day to the tabular date in Jan. and Feb. of leap years.

Add to the Reduced Date

For Upper Culmination before	Mar. 21	0
after		+ 1
For Lower Culmination before	Sept. 19	+ 0.5
after		+ 1.5

TABLE FOR THE ARGUMENT OF THE TERMS DEPENDI[NG] ON THE MOON'S LONGITUDE.

1862.		1863.		1864.		1865.	
Jan.	— 7.36	Jan.	— 3.75	Jan.	— 0.14	Jan.	—10.23
	6.34		9.95		13.56		3.47
	20.04		23.65		27.26		17.17
Feb.	2.74	Feb.	6.34	Feb.	9.95		30.86
	16.43		20.04		23.65	Feb.	13.56
March	2.13	March	5.74	March	9.35		27.26
	15.83		19.44		23.05	March	12.96
	29.53	April	2.14	April	5.75		26.66
April	12.23		15.84		19.45	April	9.36
	25.92		29.53	May	3.14		23.05
May	9.62	May	13.23		16.84	May	6.75
	23.32		26.93		30.54		20.45
June	6.02	June	9.63	June	13.24	June	3.15
	19.72		23.33		26.94		16.85
July	3.42	July	7.02	July	10.63		30.54
	17.11		20.72		24.33	July	14.24
	30.81	Aug.	3.42	Aug.	7.03		27.94
Aug.	13.51		17.12		20.73	Aug.	10.64
	27.21		30.82	Sept.	3.43		24.34
Sept.	9.91	Sept.	13.52		17.12	Sept.	7.04
	23.60		27.21		30.83		20.73
Oct.	7.30	Oct.	10.91	Oct.	14.92	Oct.	4.43
	21.00		24.61		28.22		18.13
Nov.	3.70	Nov.	7.31	Nov.	10.92		31.83
	17.40		21.01		24.62	Nov.	14.53
Dec.	1.10	Dec.	4.70	Dec.	8.31		28.22
	14.79		18.40		22.01	Dec.	11.92
	28.49						25.62

Note.—**Add 1 day to the tabular date in Jan. and Feb., 1864.**

Argument in Days.	5 Ursæ Minoris B.	6 Ursæ Minoris B.	43 Cephei H.	Polaris.	214 Camelopardali B.	20 Ursæ Minoris B.	Groombridge 527.	323 Cephei B.	57 Ursæ Minoris B.	Groombridge 750.	62 Ursæ B.	ε Ursæ Minoris.	64 Camelopardali B.	δ Ursæ Minoris.	Argument in Days.
	s.	s.	s.	s.	s.	s.	s.	s.	s.	s.	s.	s.	s.	s.	
0.0	+0.12	+0.21	−0.07	−0.22	+0.07	+0.08	−0.03	−0.05	+0.09	−0.03	+0.03	+0.01	0.00	−0.01	0.0
0.5	+0.12	+0.21	−0.08	−0.24	+0.07	+0.08	−0.03	−0.06	+0.11	−0.04	+0.03	+0.02	−0.02	+0.01	0.5
1.0	+0.11	+0.20	−0.08	−0.25	+0.07	+0.09	−0.04	−0.08	+0.13	−0.06	+0.03	+0.02	−0.04	+0.03	1.0
1.5	+0.09	+0.17	−0.08	−0.24	+0.06	+0.08	−0.04	−0.09	+0.13	−0.07	+0.04	+0.03	−0.05	+0.04	1.5
2.0	+0.07	+0.14	−0.07	−0.22	+0.05	+0.08	−0.04	−0.09	+0.13	−0.07	+0.04	+0.03	−0.06	+0.06	2.0
2.5	+0.05	+0.10	−0.06	−0.19	+0.04	+0.06	−0.04	−0.09	+0.12	−0.08	+0.04	+0.03	−0.07	+0.07	2.5
3.0	+0.02	+0.05	−0.05	−0.15	+0.03	+0.05	−0.04	−0.09	+0.11	−0.08	+0.03	+0.03	−0.07	+0.08	3.0
3.5	0.00	0.00	−0.04	−0.10	+0.01	+0.03	−0.04	−0.08	+0.09	−0.07	+0.03	+0.03	−0.08	+0.08	3.5
4.0	−0.03	−0.05	−0.02	−0.05	0.00	+0.01	−0.03	−0.07	+0.07	−0.06	+0.02	+0.02	−0.07	+0.08	4.0
4.5	−0.06	−0.09	0.00	+0.06	−0.02	−0.01	−0.02	−0.05	+0.04	−0.05	+0.01	+0.02	−0.06	+0.07	4.5
5.0	−0.08	−0.13	+0.02	+0.06	−0.04	−0.03	−0.01	−0.03	+0.01	−0.03	0.00	+0.01	−0.05	+0.06	5.0
5.5	−0.10	−0.17	+0.03	+0.11	−0.05	−0.04	0.00	−0.01	−0.02	−0.02	0.00	+0.01	−0.04	+0.05	5.5
6.0	−0.11	−0.19	+0.05	+0.16	−0.06	−0.06	+0.01	+0.01	−0.05	0.00	−0.01	0.00	−0.03	+0.04	6.0
6.5	−0.12	−0.21	+0.06	+0.20	−0.07	−0.07	+0.02	+0.03	−0.08	+0.02	−0.02	−0.01	−0.01	+0.02	6.5
7.0	−0.12	−0.21	+0.07	+0.23	−0.07	−0.08	+0.03	+0.05	−0.10	+0.03	−0.03	−0.01	+0.01	0.00	7.0
7.5	−0.11	−0.21	+0.08	+0.24	−0.07	−0.09	+0.03	+0.07	−0.12	+0.05	−0.03	−0.02	+0.03	−0.02	7.5
8.0	−0.10	−0.19	+0.08	+0.25	−0.07	−0.09	+0.04	+0.08	−0.13	+0.06	−0.04	−0.02	+0.04	−0.03	8.0
8.5	−0.09	−0.16	+0.08	+0.24	−0.06	−0.08	+0.04	+0.09	−0.13	+0.07	−0.04	−0.03	+0.06	−0.05	8.5
9.0	−0.07	−0.12	+0.07	+0.21	−0.05	−0.07	+0.04	+0.09	−0.13	+0.08	−0.04	−0.03	+0.07	−0.06	9.0
9.5	−0.04	−0.08	+0.06	+0.18	−0.04	−0.06	+0.04	+0.09	−0.12	+0.08	−0.03	−0.03	+0.07	−0.07	9.5
10.0	−0.01	−0.04	+0.05	+0.14	−0.02	−0.04	+0.04	+0.09	−0.11	+0.07	−0.03	−0.03	+0.08	−0.08	10.0
10.5	+0.01	+0.01	+0.03	+0.09	−0.01	−0.03	+0.03	+0.08	−0.08	+0.07	−0.03	−0.02	+0.07	−0.08	10.5
11.0	+0.04	+0.06	+0.01	+0.03	+0.01	−0.01	+0.03	+0.06	−0.06	+0.06	−0.02	−0.02	+0.07	−0.08	11.0
11.5	+0.06	+0.11	−0.01	−0.02	+0.03	+0.01	+0.02	+0.04	−0.03	+0.05	−0.01	−0.02	+0.06	−0.07	11.5
12.0	+0.08	+0.14	−0.02	−0.08	+0.04	+0.03	+0.01	+0.02	0.00	+0.03	0.00	−0.01	+0.05	−0.06	12.0
12.5	+0.10	+0.18	−0.04	−0.13	+0.05	+0.05	0.00	0.00	+0.03	+0.01	+0.01	0.00	+0.04	−0.05	12.5
13.0	+0.11	+0.20	−0.05	−0.17	+0.06	+0.06	−0.01	−0.02	+0.06	0.00	+0.02	0.00	+0.02	−0.03	13.0
13.5	+0.12	+0.21	−0.07	−0.21	+0.07	+0.08	−0.02	−0.04	+0.08	−0.02	+0.02	+0.01	0.00	−0.01	13.5
14.0	+0.12	+0.21	−0.07	−0.23	+0.07	+0.08	−0.03	−0.06	+0.11	−0.04	+0.03	+0.01	−0.01	0.00	14.0
14.5	+0.11	+0.20	−0.08	−0.25	+0.07	+0.09	−0.04	−0.07	+0.12	−0.05	+0.03	+0.02	−0.03	+0.02	14.5
15.0	+0.10	+0.18	−0.08	−0.24	+0.06	+0.08	−0.04	−0.09	+0.13	−0.06	+0.04	+0.02	−0.05	+0.04	15.0

Argument in Days.	51 Cephei H.	25 Camelopardali H.	4 Ursæ Minoris B.	156 Camelopardali B.	λ Ursæ Minoris B.	74 Draconis.	1 Draconis H.	119 Cephei B.	30 Camelopardali H.	34 Cephei H.	202 Camelopardali B.	39 Cephei H.	309 Cephei B.	Argument in Days.
	s.	s.	s	s.	s.	s.	s.	s.	s.	s.	s.	s.	s.	
0.0	+0.03	+0.02	+0.15	+0.03	−0.15	−0.02	+0.03	−0.04	+0.05	−0.04	+0.04	−0.10	−0.08	0.0
0.5	0.00	+0.01	+0.08	+0.02	−0.09	−0.02	+0.03	−0.04	+0.04	−0.04	+0.04	−0.09	−0.08	0.5
1.0	−0.02	−0.01	0.00	.00	−0.03	−0.01	+0.02	−0.03	+0.03	−0.04	+0.03	−0.09	−0.08	1.0
1.5	−0.05	−0.02	−0.07	−0.01	+0.04	−0.01	0.00	−0.02	+0.02	−0.03	+0.02	−0.07	−0.07	1.5
2.0	−0.08	−0.03	−0.14	−0.03	+0.10	0.00	−0.01	−0.01	0.00	−0.02	+0.01	−0.06	−0.06	2.0
2.5	−0.10	−0.04	−0.21	−0.04	+0.15	0.00	−0.02	0.00	−0.01	−0.02	0.00	−0.04	−0.04	2.5
3.0	−0.11	−0.05	−0.26	−0.05	+0.21	+0.01	−0.03	+0.01	−0.02	−0.01	−0.01	−0.02	−0.02	3.0
3.5	−0.12	−0.05	−0.28	−0.06	+0.25	+0.01	−0.04	+0.02	−0.03	0.00	−0.02	+0.01	0.00	3.5
4.0	−0.12	−0.05	−0.32	−0.07	+0.27	+0.02	−0.04	+0.03	−0.04	+0.01	−0.03	+0.03	+0.02	4.0
4.5	−0.12	−0.05	−0.33	−0.07	+0.28	+0.02	−0.05	+0.04	−0.05	+0.02	−0.03	+0.05	+0.03	4.5
5.0	−0.11	−0.05	−0.32	−0.07	+0.28	+0.02	−0.05	+0.04	−0.05	+0.03	−0.04	+0.07	+0.05	5.0
5.5	−0.10	−0.04	−0.29	−0.06	+0.26	+0.03	−0.05	+0.04	−0.06	+0.04	−0.04	+0.08	+0.06	5.5
6.0	−0.08	−0.04	−0.25	−0.05	+0.23	+0.03	−0.04	+0.05	−0.05	+0.04	−0.04	+0.09	+0.08	6.0
6.5	−0.05	−0.03	−0.19	−0.04	+0.18	+0.02	−0.04	+0.04	−0.05	+0.04	−0.04	+0.10	+0.08	6.5
7.0	−0.02	−0.01	−0.13	−0.03	+0.13	+0.02	−0.03	+0.04	−0.04	+0.04	−0.04	+0.10	+0.08	7.0
7.5	−0.01	−0.00	−0.05	−0.01	+0.07	+0.02	−0.02	+0.03	−0.04	+0.04	−0.03	+0.09	+0.08	7.5
8.0	+0.03	+0.01	+0.02	0.00	+0.01	+0.01	−0.01	+0.03	−0.03	+0.04	−0.03	+0.08	+0.08	8.0
8.5	+0.06	+0.02	+0.09	+0.02	−0.06	+0.01	0.00	+0.02	−0.01	+0.03	−0.02	+0.07	+0.07	8.5
9.0	+0.08	+0.03	+0.16	+0.03	−0.12	0.00	+0.01	+0.01	−0.00	+0.02	−0.01	+0.05	+0.05	9.0
9.5	+0.10	+0.04	+0.22	+0.04	−0.17	0.00	+0.02	0.00	+0.01	+0.01	0.00	+0.03	+0.04	9.5
10.0	+0.12	+0.05	+0.27	+0.06	−0.22	−0.01	+0.03	−0.01	+0.02	0.00	+0.01	+0.01	+0.02	10.0
10.5	+0.12	+0.05	+0.31	+0.06	−0.25	−0.02	+0.04	−0.02	+0.03	−0.01	+0.02	−0.01	0.00	10.5
11.0	+0.12	+0.05	+0.32	+0.07	−0.28	−0.02	+0.04	−0.03	+0.04	−0.02	+0.03	−0.04	−0.02	11.0
11.5	+0.12	+0.05	+0.32	+0.07	−0.28	−0.02	+0.05	−0.04	+0.05	−0.02	+0.04	−0.06	−0.04	11.5
12.0	+0.11	+0.05	+0.31	+0.06	−0.27	−0.03	+0.05	−0.04	+0.05	−0.03	+0.04	−0.07	−0.06	12.0
12.5	+0.09	+0.04	+0.28	+0.06	−0.25	−0.03	+0.05	−0.05	+0.06	−0.04	+0.04	−0.09	−0.07	12.5
13.0	+0.07	+0.03	+0.23	+0.05	−0.22	−0.03	+0.04	−0.05	+0.05	−0.04	+0.04	−0.09	−0.08	13.0
13.5	+0.04	+0.02	+0.17	+0.04	−0.17	−0.02	+0.04	−0.04	+0.05	−0.04	+0.04	−0.10	−0.08	13.5
14.0	+0.02	+0.01	+0.10	+0.02	−0.11	−0.02	+0.03	−0.04	+0.04	−0.04	+0.04	−0.10	−0.08	14.0
14.5	−0.01	0.00	+0.03	+0.01	−0.05	−0.02	+0.02	−0.03	+0.03	−0.04	+0.03	−0.09	−0.08	14.5
15.0	−0.04	−0.01	−0.04	−0.01	+0.01	−0.01	+0.01	−0.02	+0.02	−0.03	+0.03	−0.08	−0.07	15.0

	5 Ursæ Min. B.		6 Ursæ Min. B.		43 Cephei H.		214 Camelop. B.	
1862.	h. m. s.	s.	h. m. s.	s.	h. m. s.	s.	h. m. s.	s.
Jan. 0	12 12 48.80	—1.16	12 14 21.76	—2.11	0 50 32.44	+0.71	13 20 20.12	—0.68
April 10	47.64	—1.15	19.65	—2.10	33.15	+0.70	19.44	—0.66
July 19	46.49	—1.16	17.55	—2.08	33.85	+0.68	18.78	—0.66
Oct. 27	45.33	—1.14	15.47	—2.05	34.53	+0.64	18.12	—0.63
Dec. 66	44.19		13.42		35.17		17.49	
1863.								
Jan. 0	12 12 46.42	—1.12	12 14 13.97	—2.01	0 50 41.83	+0.62	13 20 14.90	—0.62
April 10	45.30	—1.09	11.96	—1.95	42.45	+0.58	14.28	—0.60
July 19	44.21	—1.06	10.01	—1.86	43.03	+0.54	13.68	—0.56
Oct. 27	43.15	—1.00	8.15	—1.76	43.57	+0.48	13.12	—0.52
Dec. 66	42.15		6.39		44.05		12.60	
1864.								
Jan. 0	12 12 44.33	—0.97	12 14 6.84	—1.69	0 50 50.78	+0.45	13 20 9.98	—0.49
April 10	43.36	—0.90	5.15	—1.56	51.23	+0.39	9.49	—0.45
July 19	42.46	—0.82	3.59	—1.43	51.62	+0.33	9.04	—0.40
Oct. 27	41.64	—0.75	2.16	—1.28	51.95	+0.27	8.64	—0.34
Dec. 66	40.89		0.88		52.22		8.30	
1865.								
Jan. 0	12 12 42.98	—0.69	12 14 1.19	—1.17	0 50 59.03	+0.23	13 20 5.63	—0.31
April 10	42.29	—0.60	0.02	—1.00	59.26	+0.15	5.32	—0.25
July 19	41.69	—0.51	13 59.02	—0.81	59.41	+0.08	5.07	—0.19
Oct. 27	41.18	—0.40	58.21	—0.64	59.49	+0.02	4.88	—0.12
Dec. 66	40 78		57.57		59.51		4.76	
1862.	° ′ ″	″	° ′ ″	″	° ′ ″	″	° ′ ″	″
Jan. 0	87 12 4.00	+0.03	88 27 47.37	+0.03	85 30 59.85	—0.16	85 28 26.27	+0.26
April 10	4.03	+0.07	47.40	+0.08	59.69	—0.20	26.53	+0.31
July 19	4.10	+0.13	47.48	+0.13	59.49	—0.27	26.84	+0.37
Oct. 27	4.23	+0.19	47.61	+0.20	59.22	—0.33	27.21	+0.42
Dec. 66	4.42		47.81		58.89		27.63	
1863.								
Jan. 0	87 11 44.32	+0.23	88 27 27.65	+0.23	85 31 18.57	—0.35	85 28 8.65	+0.45
April 10	44.55	+0.28	27.88	+0.29	18.22	—0.41	9.10	+0.50
July 19	44.83	+0.34	28.17	+0.33	17.81	—0.45	9.60	+0.53
Oct. 27	45.17	+0.38	28.50	+0.39	17.36	—0.50	10.13	+0.57
Dec. 66	45.55		28.89		16.86		10.70	
1864.								
Jan. 0	87 11 25.38	+0.42	88 27 8.67	+0.42	85 31 36.59	—0.52	85 27 51.67	+0.59
April 10	25.80	+0.47	9.09	+0.46	36.07	—0.56	52.26	+0.62
July 19	26.27	+0.50	9.55	+0.51	35.51	—0.59	52.88	+0.64
Oct. 27	26.77	+0.54	10.06	+0.54	34.92	—0.61	53.52	+0.66
Dec. 66	27.31		10.60		34.31		54.18	
1865.								
Jan. 0	87 11 7.09	+0.56	88 26 50.32	+0.56	85 31 54.08	—0.63	85 27 35.12	+0.67
April 10	7.65	+0.59	50.88	+0.59	53.45	—0.64	35.79	+0.67
July 19	8.24	+0.61	51.47	+0.62	52.81	—0.66	36.46	+0.67
Oct. 27	8.85	+0.63	52.09	+0.63	52.15	—0.66	37.13	+0.67
Dec. 66	9.48		52.72		51.49		37.80	

	20 Ursæ Min. B.		Groombridge 527.		323 Cephei B.		57 Ursæ Min. B.	
1862.	h. m. s.	s.	h. m. s.	s.	h. m. s.	s.	h. m. s.	s.
Jan. 0	14 4 9.30	—0.78	2 28 11.94	+0.29	3 21 44.88	+0.57	15 23 8.15	—0.94
April 10	8.52	—0.76	12.23	+0.27	45.45	+0.52	7.21	—0.88
July 19	7.76	—0.73	12.50	+0.25	45.97	+0.47	6.33	—0.80
Oct. 27	7.03	—0.70	12.75	+0.22	46.44	+0.40	5.53	—0.72
Dec. 66	6.33		12.97		46.84		4.81	
1863.								
Jan. 0	14 3 58.60	—0.67	2 28 21.00	+0.19	3 22 5.49	+0.35	15 22 41.73	—0.66
April 10	57.93	—0.63	21.19	+0.17	5.84	+0.29	41.07	—0.57
July 19	57.30	—0.57	21.36	+0.13	6.13	+0.22	40.50	—0.47
Oct. 27	56.73	—0.52	21.49	+0.11	6.35	+0.14	40.03	—0.38
Dec. 66	56.21		21.60		6.49		39.65	
1864.								
Jan. 0	14 3 48.43	—0.47	2 28 29.67	+0.08	3 22 25.26	+0.09	15 22 16.53	—0.31
April 10	47.96	—0 41	29.75	+0.04	25.35	+0.02	16.22	—0.19
July 19	47.55	—0.35	29.79	0.00	25.37	—0.07	16.03	—0.09
Oct. 27	47.20	—0.28	29.79	—0.03	25.30	—0.13	15.94	+0.03
Dec. 66	46.92		29.76		25.17		15.97	
1865.								
Jan. 0	14 3 39.10	—0.23	2 28 37.89	—0.05	3 22 44.06	—0.19	15 21 52.79	+0.10
April 10	38.87	—0.15	37.84	—0.10	43.87	—0.26	52.89	+0.21
July 19	38.72	—0.08	37.74	—0.12	43.61	—0.33	53.10	+0.31
Oct. 27	38.64	0.00	37.62	—0.16	43.28	—0.41	53.41	+0.43
Dec. 66	38.64		37.46		42.87		53.84	
1862.	° ′ ″	″	° ′ ″	″	° ′ ″	″	° ′ ″	″
Jan. 0	86 25 0.04	+0.41	80 51 35.01	—0.49	86 12 16.02	—0.64	87 45 17.93	+0.64
April 10	0.45	+0.46	34.52	—0.53	15.38	—0.67	18.57	+0.67
July 19	0.91	+0.51	33.99	—0.58	14.71	—0.71	19.24	+0.71
Oct. 27	1.42	+0.55	33.41	—0.62	14.00	—0.74	19.95	+0.73
Dec. 66	1.97		32.79		13.26		20.68	
1863.								
Jan. 0	86 24 44.65	+0.57	80 51 48.93	—0.64	86 12 25.23	—0.75	87 45 7.79	+0.75
Apr. 10	45.22	+0.62	48.29	—0.67	24.48	—0.76	8.54	+0.76
July 19	45.84	+0.64	47.62	—0.70	23.72	—0.78	9.30	+0.77
Oct. 27	46.48	+0.66	46.92	—0.71	22.94	—0.78	10.07	+0.77
Dec. 66	47.14		46.21		22.16		10.84	
1864.								
Jan. 0	86 24 29.78	+0.68	80 52 2.37	—0.72	86 12 35.12	—0.78	87 44 57.91	+0.78
April 10	30.46	+0.69	1.65	—0.73	34.34	—0.77	58.69	+0.77
July 19	31.15	+0.71	0.92	—0.73	33.57	—0.75	59.46	+0.75
Oct. 27	31.86	+0.70	0.19	—0.73	32.82	—0.74	45 0.21	+0.72
Dec. 66	32.56		51 59.46		32.08		0.93	
1865.								
Jan. 0	86 24 15.18	+0.70	80 52 15.61	—0.71	86 12 45.00	—0.71	87 44 48.00	+0.71
April 10	15.88	+0.70	14.90	—0.71	44.29	—0.68	48.71	+0.68
July 19	16.58	+0.69	14.19	—0.68	43.61	—0.65	49.39	+0.64
Oct. 27	17.27	+0.66	13.51	—0.65	42.96	—0.59	50.03	+0.59
Dec. 66	17.93		12.86		42.37		50.62	

	Gr. 750.		62 Ursæ Min. B.		ε Ursæ Minoris.		64 Camelop. B.	
1862.	h. m. s.	s.	h. m. s.	s.	h. m. s.	s.	h. m. s.	s.
Jan. 0	3 54 26.02	+0.38	15 57 58.33	—0.26	17 0 12.02	—0.12	5 18 14.65	+0.14
April 10	26.40	+0.33	58.07	—0.24	11.90	—0.09	14.79	+0.10
July 19	26.73	+0.27	57.83	—0.22	11.81	—0.08	14.89	+0.04
Oct. 27	27.00	+0.23	57.61	—0.19	11.73	—0.06	14.93	—0.01
Dec. 66	27.23		57.42		11.67		14.92	
1863.								
Jan. 0	3 54 43.79	+0.19	15 57 50.68	—0.18	17 0 5.27	—0.05	5 18 33.40	—0.05
April 10	43.98	+0.13	50.50	—0.15	5.22	—0.02	33.35	—0.11
July 19	44.11	+0.07	50.35	—0.12	5.20	—0.01	33.24	—0.16
Oct. 27	44.18	+0.01	50.23	—0.08	5.19	+0.02	33.08	—0.22
Dec. 66	44.19		50.15		5.21		32.86	
1864.								
Jan. 0	3 55 0.84	—0.03	15 57 43.36	—0.07	16 59 58.79	+0.03	5 18 51.43	—0.26
April 10	0.81	—0.09	43.29	—0.04	58.82	+0.06	51.17	—0.30
July 19	0.72	—0.14	43.25	—0.02	58.88	+0.07	50.87	—0.35
Oct. 27	0.58	—0.21	43.23	+0.02	58.95	+0.10	50.52	—0.40
Dec. 66	0.37		43.25		59.05		50.12	
1865.								
Jan. 0	3 55 17.11	—0.24	15 57 36.45	+0.04	16 59 52.60	+0.11	5 19 8.75	—0.43
April 10	16.87	—0.30	36.49	+0.07	52.71	+0.13	8.32	—0.46
July 19	16.57	—0.36	36.56	+0.10	52.84	+0.14	7.86	—0.51
Oct. 27	16.21	—0.40	36.66	+0.13	52.98	+0.17	7.35	—0.53
Dec. 66	15.81		36.79		53.15		6.82	
1862	° ′ ″	″	° ′ ″	″	° ′ ″	″	° ′ ″	″
Jan. 0	85 11 10.35	—0.71	83 21 26.45	+0.71	82 15 28.19	+0.81	85 6 52.66	—0.83
April 10	9.64	—0.74	27.16	+0.75	29.00	+0.83	51.83	—0.84
July 19	8.90	—0.77	27.91	+0.77	29.83	+0.84	50.99	—0.85
Oct. 27	8.13	—0.78	28.68	+0.79	30.67	+0.84	50.14	—0.85
Dec. 66.	7.35		29.47		31.51		49.29	
1863.								
Jan. 0	85 11 18.10	—0.80	83 21 19.02	+0.80	82 15 26.05	+0.84	85 6 53.22	—0.84
April 10	17.30	—0.80	19.82	+0.80	26.89	+0.82	52.38	—0.83
July 19	16.50	—0.81	20.62	+0.81	27.71	+0.81	51.55	—0.80
Oct. 27	15.69	—0.80	21.43	+0.79	28.52	+0.79	50.75	—0.78
Dec. 66	14.89		22.22		29.31		49.97	
1864.								
Jan. 0	85 11 25.62	—0.79	83 21 11.77	+0.79	82 15 23.86	+0.77	85 6 53.84	—0.75
April 10	24.83	—0.77	12.56	+0.77	24.63	+0.73	53.09	—0.71
July 19	24.06	—0.75	13.33	+0.74	25.36	+0.69	52.38	—0.66
Oct. 27	23.31	—0.72	14.07	+0.71	26.05	+0.64	51.72	—0.61
Dec. 66	22.59		14.78		26.69		51.11	
1865.								
Jan. 0	85 11 33.27	—0.69	83 21 4.35	+0.69	82 15 21.29	+0.60	85 6 54.89	—0.57
April 10	32.58	—0.65	5.04	+0.65	21.89	+0.55	54.32	—0.51
July 19	31.93	—0.60	5.69	+0.59	22.44	+0.48	53.81	—0.44
Oct. 27	31.33	—0.55	6.28	+0.54	22.92	+0.42	53.37	—0.38
Dec. 66	30.78		6.82		23.34		52.99	

	51 Cephei H.		25 Camelop. H.		4 Ursæ Min. B.		156 Camelop. B.	
1862.	h. m. s.	s.	h. m. s.	s.	h. m. s.	s.	h. m. s.	s.
Jan. 0	6 34 51.09	—0.13	7 1 54.49	—0.10	7 13 13.56	—0.94	7 43 29.81	—0.23
April 10	50.96	—0.22	54.39	—0.13	12.62	—1.17	29.58	—0.28
July 19	50.74	—0.31	54.26	—0.17	11.45	—1.41	29.30	—0.32
Oct. 27	50.43	—0.40	54.09	—0.21	10.04	—1.61	28.98	—0.36
Dec. 66	50.03		53.88		8.43		28.62	
1863.								
Jan. 0	6 35 20.69	—0.45	7 2 7.05	—0.23	7 14 26.10	—1.75	7 43 44.19	—0.39
April 10	20.24	—0.54	6.82	—0.27	24.35	—1.94	43.80	—0.42
July 19	19.70	—0.62	6.55	—0.30	22.41	—2.11	43.38	—0.46
Oct. 27	19.08	—0.68	6.25	—0.33	20.30	—2.27	42.92	—0.48
Dec. 66	18.40		5.92		18.03		42.44	
1864.								
Jan. 0	6 35 49.14	—0.74	7 2 19.13	—0.35	7 15 35.64	—2.36	7 43 58.04	—0.50
April 10	48.40	—0.79	18.78	—0.37	33.28	—2.49	57.54	—0.52
July 19	47.61	—0.85	18.41	—0.39	30.79	—2.58	57.02	—0.54
Oct. 27	46.76	—0.89	18.02	—0.40	28.21	—2.66	56.48	—0.55
Dec. 66	45.87		17.62		25.55		55.93	
1865.								
Jan. 0	6 36 16.65	—0.92	7 2 30.84	—0.42	7 16 43.01	—2.70	7 44 11.53	—0.55
April 10	15.73	—0.96	30.42	—0.43	40.31	—2.73	10.98	—0.57
July 19	14.77	—0.99	29.99	—0.43	37.58	—2.74	10.41	—0.56
Oct. 27	13.78	—1.01	29.56	—0.44	34.84	—2.71	9.85	—0.55
Dec. 66	12.77		29.12		32.13		9.30	
1862.	° ′ ″	″	° ′ ″	″	° ′ ″	″	° ′ ″	″
Jan. 0	87 14 48.17	—0.85	82 39 53.04	—0.83	89 1 6.96	—0.82	84 26 35.85	—0.78
April 10	47.32	—0.84	52.21	—0.81	6.14	—0.81	35.07	—0.75
July 19	46.48	—0.82	51.40	—0.79	5.33	—0.79	34.32	—0.73
Oct. 27	45.66	—0.81	50.61	—0.77	4.54	—0.75	33.59	—0.68
Dec. 66	44.85		49.84		3.79		32.91	
1863.								
Jan. 0	87 14 42.05	—0.78	82 39 44.71	—0.74	89 0 57.74	—0.72	84 26 24.39	—0.66
Apr. 10	41.27	—0.75	43.97	—0.70	57.02	—0.68	23.73	—0.61
July 19	40.52	—0.71	43.27	—0.66	56.34	—0.64	23.12	—0.56
Oct. 27	39.81	—0.66	42.61	—0.60	55.70	—0.58	22.56	—0.50
Dec. 66	39.15		42.01		55.12		22.06	
1864.								
Jan. 0	87 14 36.25	—0.63	82 39 36.79	—0.55	89 0 48.91	—0.54	84 26 13.47	—0.46
April 10	35.62	—0.57	36.24	—0.51	48.37	—0.48	13.01	—0.39
July 19	35.05	—0.51	35.73	—0.44	47.89	—0.40	12.62	—0.33
Oct. 27	34.54	—0.44	35.29	—0.37	47.49	—0.34	12.29	—0.25
Dec. 66	34.10		34.92		47.15		12.04	
1865.								
Jan. 0	87 14 31.09	—0.40	82 39 29.62	—0.32	89 0 40.75	—0.29	84 26 3.34	—0.20
April 10	30.69	—0.32	29.30	—0.25	40.46	—0.20	3.14	—0.14
July 19	30.37	—0.25	29.05	—0.17	40.26	—0.13	3.00	—0.03
Oct. 27	30.12	—0.16	28.88	—0.09	40.13	—0.05	2.97	+0.03
Dec. 66	29.96		28.79		40.08		3.00	

	λ Ursæ Minoris.		74 Draconis.		1 Draconis, H.		119 Cephei.	
1862.	h. m. s.	s.	h. m. s.	s.	h. m. s.	s.	h. m. s.	s.
Jan. 0	20 1 39.86	+1.38	20 37 15.66	+0.22	9 17 10.46	—0.29	21 24 27.13	+0.39
April 10	41.24	+1.54	15.88	+0.22	10.17	—0.32	27.52	+0.41
July 19	42.78	+1.68	16.10	+0.22	9.85	—0.34	27.93	+0.41
Oct. 27	44.46	+1.83	16.32	+0.24	9.51	—0.36	28.34	+0.42
Dec. 66	46.29		16.56		9.15		28.76	
1863.								
Jan. 0	20 0 48.73	+1.92	20 37 13.31	+0.24	9 17 18.50	—0.38	21 24 24.13	+0.42
April 10	50.65	+2.03	13.55	+0.23	18.12	—0.39	24.55	+0.43
July 19	52.68	+2.13	13.78	+0.24	17.73	—0.41	24.98	+0.42
Oct. 27	54.81	+2.21	14.02	+0.23	17.32	—0.42	25.40	+0.42
Dec. 66	57.02		14.25		16.90		25.82	
1864.								
Jan. 0	19 59 59.05	+2.25	20 37 11.01	+0.23	9 17 26.26	—0.42	21 24 21.18	+0.41
April 10	20 0 1.30	+2.31	11.24	+0.22	25.84	—0.42	21.59	+0.39
July 19	3.61	+2.34	11.46	+0.22	25.42	—0.44	21.98	+0.38
Oct. 27	5.95	+2.34	11.68	+0.21	24.98	—0.41	22.36	+0.36
Dec. 66	8.29		11.89		24.57		22.72	
1865.								
Jan. 0	19 59 9.98	+2.34	20 37 8.65	+0.20	9 17 33.92	—0.42	21 24 18.10	+0.35
April 10	12.32	+2.32	8.85	+0.19	33.50	—0.41	18.45	+0.33
July 19	14.64	+2.28	9.04	+0.18	33.09	—0.39	18.78	+0.29
Oct. 27	16.92	+2.21	9.22	+0.16	32.70	—0.38	19.07	+0.26
Dec. 66	19.13		9.38		32.32		19.33	
1862.	° ′ ″	″	° ′ ″	″	° ′ ″	″	° ′ ″	″
Jan. 0	88 53 48.22	+0.74	80 36 25.02	+0.68	81 55 47.31	—0.58	83 40 26.07	+0.55
April 10	48.96	+0.71	25.70	+0.64	46.73	—0.54	26.62	+0.51
July 19	49.67	+0.68	26.34	+0.60	46.19	—0.48	27.13	+0.47
Oct. 27	50.35	+0.64	26.94	+0.55	45.71	—0.44.	27.60	+0.41
Dec. 66.	50.99		27.49		45.27		28.01	
1863.								
Jan. 0	88 54 0.93	+0.61	80 36 40.23	+0.51	81 55 30.21	—0.40	83 40 43.47	+0.37
April 10	1.54	+0.57	40.74	+0·47	29.81	—0.33	43.84	+0.31
July 19	2.11	+0.51	41.21	+0.40	29.48	—0.28	44.15	+0.25
Oct. 27	2.62	+0.45	41.61	+0.34	29.20	—0.21	44.40	+0.19
Dec. 66	3.07		41.95		28.99		44.59	
1864.								
Jan. 0	88 54 13.00	+0.41	80 36 54.76	+0.29	81 55 13.85	—0.17	83 41 0.12	+0.14
April 10	13.41	+0.34	55.05	+0.23	13.68	—0.09	0.26	+0.07
July 19	13.75	+0.28	55.28	+0.16	13.59	—0.03	0.33	+0.01
Oct. 27	14.03	+0.20	55.44	+0.08	13.56	+0.04	0.34	—0.07
Dec. 66	14.23		55.52		13.60		0.27	
1865.								
Jan. 0	88 54 24.19	+0.15	80 37 8.41	+0.04	81 54 58.36	+0.09	83 41 15.89	—0.11
April 10	24.34	+0.08	8.45	—0.04	58.45	+0.16	15.78	—0.18
July 19	24.42	+0.01	8.41	—0.11	58.61	+0.23	15.60	—0.25
Oct. 27	24.43	—0.07	8.30	—0.18	58.84	+0.30	15.35	—0.31
Dec. 66	24.36		8.12		59.14		15.04	

	30 Camelop. H.		34 Cephei H.		202 Camelop. B.		39 Cephei H.	
1862.	h. m. s.	s.	h. m. s.	s.	h. m. s.	s.	h. m. s.	s.
Jan. 0	10 13 58.28	—0.40	22 47 54.53	+0.40	11 22 0.24	—0.39	23 27 49.00	+0.92
April 10	57.88	—0.44	54.93	+0.41	21 59.85	—0.39	49.92	+0.94
July 19	57.44	—0.46	55.34	+0.40	59.46	—0.41	50.86	+0.93
Oct. 27	56.98	—0.48	55.74	+0.39	59.05	—0.42	51.79	+0.91
Dec. 66	56.50		56.13		58.63		52.70	
1863.								
Jan. 0	10 14 4.74	—0.48	22 47 55.97	+0.39	11 22 3.35	—0.42	23 27 52.45	+0.90
April 10	4.26	—0.50	56.36	+0.38	2.93	—0.42	53.35	+0.87
July 19	3.76	—0.50	56.74	+0.36	2.51	—0.42	54.22	+0.83
Oct. 27	3.26	—0.51	57.10	+0.35	2.09	—0.41	55.05	+0.80
Dec. 66	2.75		57.45		1.68		55.85	
1864.								
Jan. 0	10 14 11.00	—0.51	22 47 57.30	+0.33	11 22 6.38	—0.40	23 27 55.64	+0.76
April 10	10.49	—0.49	57.63	+0.31	5.98	—0.40	56.40	+0.71
July 19	10.00	—0.49	57.94	+0.28	5.58	—0.37	57.11	+0.65
Oct. 27	9.51	—0.47	58.22	+0.25	5.21	—0.37	57.76	+0.58
Dec. 66	9.04		58.47		4.84		58.34	
1865.								
Jan. 0	10 14 17.26	—0.46	22 47 58.36	+0.23	11 22 9.53	—0.35	23 27 58.21	+0.53
April 10	16.80	—0.44	58.59	+0.20	9.18	—0.32	58.74	+0.46
July 19	16.36	—0.41	58.79	+0.16	8.86	—0.29	59.20	+0.38
Oct. 27	15.95	—0.38	58.95	+0.13	8.57	—0.27	59.58	+0.29
Dec. 66	15.57		59.08		8.30		59.87	
1862.	° ′ ″	″	° ′ ″	″	° ′ ″	″	° ′ ″	″
Jan. 0	83 15 21.30	—0.41	82 25 23.46	+0.29	81 53 4.34	—0.17	86 32 52.34	+0.15
April 10	20.89	—0.35	23.75	+0.24	4.17	—0.12	52.49	+0.10
July 19	20.54	—0.31	23.99	+0.18	4.05	—0.05	52.59	+0.03
Oct. 27	20.23	—0.24	24.17	+0.12	4.00	0.00	52.62	—0.03
Dec. 66	19.99		24.29		4.00		52.59	
1863.								
Jan. 0	83 15 2.13	—0.20	82 25 43.38	+0.08	81 52 44.24	+0.04	86 33 12.47	—0.06
April 10	1.93	—0.14	43.46	+0.02	44.28	+0.11	12.41	—0.13
July 19	1.79	—0.08	43.48	—0.05	44.39	+0.16	12.28	—0.18
Oct. 27	1.71	—0.01	43.43	—0.10	44.55	+0.22	12.10	—0.25
Dec. 66	1.70		43.33		44.77		11.85	
1864.								
Jan. 0	83 14 43.76	+0.03	82 26 2.50	—0.15	81 52 24.93	+0.26	86 33 31.81	—0.28
April 10	43.79	+0.09	2.35	—0.21	25.19	+0.32	31.53	—0.33
July 19	43.88	+0.16	2.14	—0.26	25.51	+0.36	31.20	—0.39
Oct. 27	44.04	+0.23	1.88	—0.32	25.87	+0.42	30.81	—0.43
Dec. 66	44.27		1.56		26.29		30.38	
1865.								
Jan. 0	83 14 26.24	+0.26	82 26 20.79	—0.35	81 52 6.39	+0.45	86 33 50.40	—0.46
April 10	26.50	+0.32	20.44	—0.40	6.84	+0.48	49.94	—0.51
July 19	26.82	+0.39	20.04	—0.47	7.32	+0.53	49.43	—0.54
Oct. 27	27.21	+0.43	19.57	—0.51	7.85	+0.57	48.89	—0.57
Dec. 66	27.64		19.06		8.42		48.32	

	309 Cephei B.			309 Cephei B.	
1862.	h. m. s.	s.	1862.	° ′ ″	″
Jan. 0	23 53 7.56	+0.80	Jan. 0	85 56 23.69	+0.06
April 10	8.36	+0.79	Apr. 10	23.75	—0.01
July 19	9.15	+0.79	July 19	23.74	—0.05
Oct. 27	9.94	+0.76	Oct. 27	23.69	—0.12
Dec. 66	10.70		Dec. 66	23.57	
1863.			1863.		
Jan. 0	23 53 12.99	+0.75	Jan. 0	85 56 43.64	—0.16
April 10	13.74	+0.72	Apr. 10	43.48	—0.21
July 19	14.46	+0.68	July 19	43.27	—0.27
Oct. 27	15.14	+0.64	Oct. 27	43.00	—0.33
Dec. 66	15.78		Dec. 66	42.67	
1864.			1864.		
Jan. 0	23 53 18.11	+0.61	Jan. 0	85 57 2.81	—0.35
Apr. 10	18.72	+0.55	Apr. 10	2.46	—0.41
July 19	19.27	+0.50	July 19	2.05	—0.45
Oct. 27	19.77	+0.43	Oct. 27	1.60	—0.50
Dec. 66	20.20		Dec. 66	1.10	
1865.			1865.		
Jan. 0	23 53 22.61	+0.39	Jan. 0	85 57 21.30	—0.52
Apr. 10	23.00	+0.33	Apr. 10	20.78	—0.56
July 19	23.33	+0.25	July 19	20.22	—0.58
Oct. 27	23.58	+0.17	Oct. 27	19.64	—0.61
Dec. 66	23.75		Dec. 66	19.03	

ANNALS

OF THE

ASTRONOMICAL OBSERVATORY OF HARVARD COLLEGE.

VOL. V.

OBSERVATIONS ON THE GREAT NEBULA OF ORION.

PRINTED FROM THE STURGIS FUND.

CAMBRIDGE:
PRINTED AT THE RIVERSIDE PRESS.
1867.

OBSERVATIONS

UPON THE

GREAT NEBULA OF ORION.

BY THE LATE

GEORGE PHILLIPS BOND, A. M.

DIRECTOR OF THE OBSERVATORY OF HARVARD COLLEGE.

EDITED BY

TRUMAN HENRY SAFFORD, A. B.

DIRECTOR OF THE DEARBORN OBSERVATORY OF CHICAGO.

CAMBRIDGE:
PRINTED AT THE RIVERSIDE PRESS.
1867.

PREFACE.

The following work was left unfinished at the death of its author; and its completion, as far as possible, and publication, were entrusted to myself. It was necessary for me, therefore, to complete those portions which were so far advanced that the form which Professor Bond desired to give them could be distinctly recognized; and for the remainder, to give, as far as practicable, the original observations, as they stand in the records of the Observatory of Harvard College, with the amount of discussion necessary to enable astronomers to derive as much advantage as possible from them.

The portion containing the original observations of stars and the General Catalogue was in a very forward state of preparation at Professor Bond's decease; and the work of the editor upon that has been mostly one of revision. The remainder of the book has been extracted from the manuscripts, under my direction, or has been written by myself.

I have appended the charts and engraving, which were executed by Mr. Watts under the author's personal supervision; the engraving of the nebula was completed in 1864, and the edition of it printed. The drawing by Professor W. C. Bond, from which an engraving is given in Vol. III. of the New Series of the Memoirs of the American Academy, was also given to Mr. Watts to reëngrave, as the original plate did not appear to be sufficiently accurate. The differences between the two are also of some importance in respect to controverted points. It has, however, been found an extremely difficult task to represent exactly the original drawing.

I would take this opportunity to express my earnest thanks to Mrs. R. F. Bond, and to Miss Bond, for the great interest they have taken in the present work, and for their assistance in completing it, without which my task would have been much more difficult, if not impossible. My warmest thanks are also due to J. Ingersoll Bowditch, Esq., Trustee of the Sturgis Fund, for his constant interest in the book and his earnest desire that it should be creditable to its author's memory, as well as for his wise counsel and kind assistance to me in my endeavor to make it such.

TRUMAN HENRY SAFFORD.

TABLE OF CONTENTS.

SECTION V.

SECTION VI.

APPENDIX I.

APPENDIX II.

ILLUSTRATIONS.

INTRODUCTION.

In the year 1857, Prof. G. P. Bond began the observations whose results are given in these pages. During that winter and the next spring they were continued so far, that the most of the stars contained between the limits of Right Ascension of the Catalogue contained in this work and within 20′ of declination of the Trapezium were already observed. Owing, however, to the apparition of the Great Comet, (1858, V.) (and also partly from the decease of Prof. W. C. Bond in 1859,) the work of reduction was discontinued for a time, and suffered an interruption of some years. But as soon as Vol. III. of these Annals, containing the "Account of the Comet of Donati," was published, Prof. G. P. Bond returned to the Nebula of Orion, and from that time until his own death in 1865, devoted much of his time and of his gradually failing strength to this object.

The inducements to this study of the Nebula were twofold. In the first place, it had always been a favorite object of contemplation with the Great Equatorial. It will be seen, by several facts stated in the following pages, that this instrument has surpassed those previously employed upon the Nebula in some respects; especially in showing the great loop "corona Herschelii" as a continuous ring. Moreover, the recorded limits of the Nebula have been largely extended; nebulous light was traced in 1864 over an area of 2.3 square degrees; while with the smaller and less favorably situated refractor at Kazan, Liapunoff was able to trace nebulosity only in an area of 0.12 square degrees.

The early Cambridge observations upon the Nebula had been partially embodied in a memoir by Prof. W. C. Bond, contained in the Memoirs of the American Academy of Arts and Sciences, New Series, Vol. III. p. 87; entitled "Description of the Nebula about the Star θ Orionis." A somewhat strict criticism upon this by Otto Struve, then vice-director of the observatory at Pulcova, had appeared in the Bulletin of the Physico-Mathematical Class of the Imperial Academy of Sciences at St. Petersburg, also in the Monthly Notices of the Royal Astronomical Society; and this circumstance formed another inducement to investigate the positions of the stars apparently or really connected with the Nebula.

In the year 1864, Prof. Bond decided to extend his survey of the Nebula to include ι and c Orionis, and a portion of the region beyond. The Catalogue of Section III. contains, then, the positions of as many stars as could be observed with the Cambridge Refractor within the limits — $2^m\ 15^s$ to + $2^m\ 15^s$ of AR., and — 1° 30′ to + 1° 30′ of Dec. of θ Orionis. The number of square degrees contained in this area being 3.36 and the number of stars 1000,* we shall have 298 to a square degree.

The observations were made mostly by differences of AR. and Dec. For the observations of the brighter stars, the instrument was clamped; the mica scale, B, employed in the Harvard Zones † was adjusted in the focus of the instrument, so that its AR. lines were perpendicular to the apparent direction of diurnal motion, and the transits of stars across these lines were chronographically recorded; their differences of declination from the zero-point of the scale were obtained by estimation among the lines parallel to the equator.

As the width of the scale in the direction of declination is only sine 11′ × 252 French inches, the field, with the Cambridge refractor, power 141, is but 11′; which is not a sufficient extent for the zone investigated in 1857–58, 40′ in breadth. It was therefore necessary to divide this zone into four, overlapping one another in some degree. Each of these zones of 10′, also, was many times repeated, and it became desirable to equalize them by the method of least squares. Here the unknown quantities were the coördinates of each of the brighter stars, and the zero-points for the zones.

But this process requires, for the accuracy of the resulting right ascensions, that the instrument should have no constant tendency to a progressive motion in AR. Upon this point, the Harvard Zones, printed in Vol. I. Part II. of these Annals, give abundant evidence that no sensible correction is required; this evidence is given in Appendix II. 2. Appendix II. 1, contains an investigation of the relation of the instrument to the pole.

The accuracy of the declinations depends upon the correctness with which the adopted formula for the value of 1′ of scale B,

$$1'\ 0.''00 + 0.''0062\ (\theta - 67°),$$

where θ is the reading of the external thermometer, Fahrenheit, does actually represent its value for the temperatures at which the observations were made. Experiments of various kinds ‡ to deduce a correction to this formula led only to the conclusion that it was sensibly correct.

It would have been hazardous, however, to depend on this value for the much larger

* I have not included the 101 stars outside these limits.

† A description of this apparatus, and of the method of observing zones with it, is to be found in the Introduction to Vol. I. Part II. of the Annals.

‡ See Appendix II. 3.

differences of declination between θ Orionis and the stars beyond ι and c Orionis. Here the absolute positions of various stars from the catalogues were made available.

Many of the stars in and near the Nebula were too faint to observe in this manner, as in order to see the divisions of the scale with distinctness, much light was admitted into the field. For the fainter stars, another plan was adopted.

The telescope was not clamped, but left free to move in any direction.

Under these circumstances, it does not move of itself; and hence observations of differences of AR. between the fainter stars and the bright ones near them can be readily made by the help of a chronometer. After a difference of AR. has been thus observed, with a faint illumination, the declinations can be observed, also with faint illumination, but at leisure, moving the instrument occasionally towards the west. For convenience sake, the telescope was so pointed that the reading of the scale was, for the bright stars, very nearly equal to the minutes and seconds (omitting multiples of 10′) of declination relative to θ; and thus the declination of fainter stars read from the scale require very little correction besides an even 10′, to refer them at once to θ.

The arrangement of the sections is as follows: —

Section I. Part 1, contains the original observations of brighter stars within 20′ in declination of θ Orionis. They are arranged in Zones, dated and numbered, with the hour-angle τ, and temperature, θ, of the external air by Fahrenheit's thermometer. The first column contains the letter * by which the star is designated; the second, the mag-

* The notation for the stars which Prof. Bond has used, though, as he thought, rather inelegant, intending, in fact, to change it on the completion of the work, proved, however, so convenient that he determined to retain it. The following is his account of the notation for the four classes of stars, namely, the larger and smaller stars within 20′ of θ Orionis in declination, and the larger and smaller stars beyond that limit.

1st and 2d Classes. Stars about θ Orionis.

The stars in the most northern parallel were lettered in the order of their right ascensions, A, B, C, etc., through the alphabet, and when that was exhausted, the remainder were designated by AA, BB, etc. Any supplementary large stars afterwards inserted, had the same letter as the catalogue star next preceding, with a numeral subscribed, indicating its relative position in right ascension; if lettered from the star following in right ascension, the subscribed numeral had the negative sign prefixed. If the position of the supplementary star was only approximately determined, as was the case with the greater part of them, on account of their being too faint for observation in the zones, they were indicated in a similar way by the small letters of the alphabet. The stars in the three remaining principal parallels were designated in a similar way, the letters being accented once, twice, or three times respectively. The following is an example of the notation: —

A'_{-1}	$a-a^\circ = -1937.''0$	$\delta-\delta^\circ = +150.''9$
A'	-1679.7	$+106.6$
a'_1	-1375.2	$+123.5$
a'_2	-1375.2	$+539.5$
a'_3	-1253.0	$+200.5$
B'	-1245.5	$+391.8$
b'_1	-1037.0	$+569.2$

nitude, as estimated at the time, on Prof. G. P. Bond's scale, employed in the Harvard Zones; the third, the scale reading; the fourth, an approximate value, d, of a quantity which, added to the scale reading, refers the zone to a point 20′ south of θ Orionis; the fifth, the variable part of the corrections for scale-value at temperature θ, for aberration, nutation, and precession to 1857.0; the sixth, the sum of the three preceding columns. In this sixth column, then, is contained the difference of declination between the star in question and a point 20′ south of θ, involving, however, a correction to the assumed zero-point of the zone, yet to be determined, which will be constant throughout the zone. The seventh column contains the transit across the two wires, reduced to the first; the eighth, an approximate value, a, of the constant part of the reduction to θ; the ninth, the variable part of this reduction; and the tenth column (the sum of columns seven, eight, and nine) contains the difference of AR. between the star and θ, yet, however, needing a correction for constant error in the assumed value of the reduction.

Columns six and ten, therefore, will serve to form the equations of condition, in which the positions of each star, and the zero-points of the zones, (these last, quantities of no further use,) will be the unknown quantities.

Section I. Part 2, contains the assumed positions of the stars of Part 1; these were employed in forming the final equations. Part 3 of the same section contains the equations of condition, to be combined and solved by the method of least squares.

These are arranged by zones. The name of the star is used for the correction of its assumed Right Ascension; the number of the zone for the correction of Right Ascension common to all the stars of that zone. The known term of the equation is the approximate Right Ascension of each star in the zone under consideration, (extracted from Part 1 of this section,) less the assumed AR. of the star (from Part 2).

a'_1 a'_2 etc., are small stars whose approximate positions were ascertained after the completion of the zones. The supplementary stars were usually determined from neighboring zone stars with reduced illumination, by passages in right ascension over the wires of the micrometer faintly illuminated; the differences of declination being estimated by bringing the star near the edge of the scale when not bright enough to be seen between the divisions. The positions of a few of the most difficult could be derived only from mere estimates by the eye; but stars of this class, excepting in the immediate neighborhood of the trapezium, have generally been omitted altogether. Many of those retained are denoted by Greek letters, as ϕ, ψ.

3d Class. Large stars about c and ι Orionis.

The zones in these regions are designated by I–IX (around c Orionis) and X–XVIII (around ι Orionis), I being the southernmost of those around c, and X the northernmost of those around ι. These zones, however, in a few cases, overlap each other or are repetitions. The stars are denoted by A, B, C, etc., with the Roman numeral of the zone prefixed, — in other things like the zones around θ Orionis.

4th Class. The small stars in these regions are designated by small letters and a numeral subscribed, relative to the larger star next preceding or following, in the same manner as those around θ Orionis.

The known terms which have a * attached, receive half weight, as the observations were made upon but one wire for AR.

The equations in Declination are given more briefly. Here only the known term is given, as the remainder of the equation is derived immediately from the equations for AR. just preceding. All observations of Declination have equal weight.

Section I. Part 4, contains the final equations obtained by the method of least squares; there is, of course, one equation to each unknown quantity, 197 for Right Ascension, and 204 for Declination. The difference between these numbers, as well as that between corresponding equations for Right Ascension and Declination, arises from accidental omissions of one AR. wire, or of the AR. or Declination of a star in any one zone, or from omissions of Right Ascension or Declination in some entire zones.

Section I. Part 5, contains the results derived from the solution of these equations; in other words, the corrections to the approximate reduction of the separate zones, and the corrections to the approximate coördinates of the stars given in Part 2.

The solution was made in duplicate by the indirect method of successive approximations. This is the method alluded to as Gauss's, by the distinguished mathematician Claussen, in the "Briefwechsel zwischen C. F. Gauss and H. C. Schumacher," Vol. III. p. 71, and it is also employed in the Greenwich observations of each year, to determine the personal equations of the assistants.

The approximation was carried so far, that the hundredths of seconds of time in Right Ascension, and tenths of seconds of space in Declination, are perfectly secure, and the solution was made in duplicate by Prof. Bond, and by other computers.

Besides the duplicate elimination by successive approximations, the 93 unknown quantities for Right Ascension, and 99 for Declination, depending on the position of the telescope in each zone, were also directly eliminated, and the resulting 104 and 105 equations solved by inserting in each the last value, from the approximate solution, of all unknown quantities in them but the principal one of each equation. This gave new values of the corrections to the assumed coördinates of the stars, which agreed in every case within $0^{s}.01$ and $0''.1$ with the results of the approximate elimination.

Section I. Part 6, contains the observations of small stars near θ Orionis.

The names of the small stars are in the first column; the magnitudes (as before, not on the finally adopted scale) in the second; the scale readings, plus the multiple of 10′ necessary to refer the zone to a point 20′ s. of θ, in the third; in the fourth, a quantity termed Reduction, which is in fact, the constant error of pointing the telescope in the revision zones. That this Reduction is not, in the mean, equal or nearly equal to zero, arises from the fact that, in setting the telescope for these revisions, a provisional catalogue was employed in which the temperature correction of the scale reading had been omitted. The fifth column contains the variable portion of the corrections for aberration, nutation, precession, refraction, and the effect of temperature on the scale; the sixth, the declinations of the stars relatively to a point 20′ s. of θ, being, of course, the sum of the third, fourth, and fifth columns. In the seventh column, which begins the portion of these tables referring to Right Ascension, are contained the names and right ascensions of the principal stars to which the small stars are referred; the right ascensions which would be the same for successive lines are not repeated. The eighth column gives the difference of transit between the small star and the corresponding larger ones; the ninth, the correction for aberration, &c.; and finally, the small star's right ascension for 1857.0 relative to θ Orionis is given in the tenth column, whose numbers are the sums of those in the seventh, eighth, and ninth.

Section II. is devoted to the stars near c and ι Orionis. Its first Part contains observations of the larger stars in these regions. The general arrangement is similar to that of the corresponding part of Section I. The exceptions are the following: —

In Section II. Part 1, the fourth column contains the final value of the constant correction of scale reading to declination relative to θ, and the eighth, the constant term which reduces the transits across the lines of the scale expressed in clock time, to right ascensions relative to θ.

The declinations in this section are referred, not to a point 20′ s. of θ, but to θ itself, and they hold good for the equinox and equator of 1864.0.

Section II. Part 2, contains the catalogue of stars by which the observations of the preceding Part are reduced to θ. It was compiled from various sources, including the previous section and other star catalogues; some details, extracted from Prof. Bond's papers are given.

Section II. Part 3, contains the observations of small stars in the same regions, reduced to 1864.0. It is, in other respects, like Section I. Part 6, excepting that the scale reading (third column) is given without the addition of the large number of minutes necessary to reduce it to θ and that this number is incorporated in the fourth column.

The telescope was, for these zones, so set as to bring the scale readings for the principal stars closely in accordance with those in the principal zones; and Prof. Bond has therefore assumed that the constant part of this reduction was identical for the zones of Part 3 and those of Part 1. This will be seen on comparing the zones.

The results of this Section, in order to appear in the General Catalogue, are to be corrected for precession and the adopted proper motion to 1857.0.

Section III. contains the General Catalogue referred to θ Orionis. In this are incorporated the final results of Section I. Part 5, for the brighter stars near θ Orionis; of Section I. Part 7, for the small stars in the same region; and of Section II. (corrected for precession to 1857.0) for those near c and ι Orionis. To these are added the results of various direct micrometer measures and estimations from diagrams; all which are explained in the notes, as well as the materials left by Prof. Bond admit. In instances of very faint stars, Prof. Bond has given results from diagrams, and assigned to them some small weight as compared with direct observations, especially those given in Section I. Part 6, and Section II. Part 3; in all these cases the results, as finally adopted by Prof. Bond, stand in the General Catalogue, and quotations from his manuscripts are given in the notes.

The reduction of the direct micrometer measures was made, assuming one revolution of the screw $= 9.''800 + 0.''00026\ (\theta - 50^\circ)$ (θ denoting the temperature by Fahrenheit's thermometer.) This formula is given, with the evidence on which it depends, in the first volume of these Annals, Part I. p. 14. I have reduced to 0° Fahrenheit, the observations there given by the temperature coefficient ($-0.''00022$ for 1° Réaumur with a revolution $= 9.''73$) — $0.''00010$ for 1° Fahrenheit corresponding to the Pulcova experiments. This has the opposite sign to the coefficient used at Cambridge.

The results are

Temperature Fahrenheit			Value of r red. to 0°		
Temperature Fahrenheit		0°	Value of r red. to 0°		9.''771
"	"	16	"	"	9.799
"	"	61	"	"	9.807
"	"	82	"	"	9.821
			Mean		9.800

so that, on this hypothesis of the effects of temperature we shall have

$$r = 9.''800 - 0.''00010\ \theta \quad \text{instead of}$$
$$r = 9.787 + 0.00026\ \theta \quad \text{or}$$
$$\Delta r = 0.013 - 0.00036\ \theta$$

It may be safely assumed that the constant term $0.''013$ of this formula will be the maximum value of Δr for these winter observations, corresponding to a temperature 0° Fahr. $= -14^\circ.2$ Réaum. The largest distance measured in this way for the Orion observations being 15 revolutions, and this only an isolated case, we shall have $0.''195$ for the possible error arising from this source; to which is to be added fifteen times the possible error of 1 revolution which can hardly be more than $0.''02 \times 15 = 0.''30$. Hence the micrometer observations cannot be more than $0.''5$ in error from an imperfect determination of 1 revolution of the screw.

The measures upon the trapezium were reduced by myself during Prof. Bond's lifetime. It will be seen that they agree more nearly with the elder Struve's observations (Mensurae Micrometricae, p. 242, seqq.) than with Liapunoff's, (p. 44 of MM. Liapunoff and O. Struve's Memoir,) and hence negative the opinion of the latter author, that the stars b. c. d. are sensibly in motion relative to θ.

The graduation of the mica scale used in these observations was effected by this same micrometer, using this formula; and it has been found sensibly accurate for a temperature of 67° Fahr., which is about that at which the work was done.

The precessions of this part of Section III. include the proper motion of θ according to Mädler, with changed sign. This, for 100 years with unchanged sign, is $+ 2.''9$ in α and $+ 3.''2$ in δ, and the precession coefficients for AR. in the Catalogue were computed by the formula

$$-0.''029 + n \left[\sin \alpha' \left(\tan \delta' - \tan \delta\right) + 2 \sin \tfrac{1}{2} \left(\alpha' - \alpha\right) \cos \tfrac{1}{2} \left(\alpha' + \alpha\right) \tan \delta\right]$$

where $\alpha = 82° \; 3' \; 47''$, $\delta = - \; 5° \; 29' \; 14''$, the position of θ Orionis from Mädler (for 1857.0); while $\alpha' - \alpha$ and $\delta' - \delta$ are the relative right ascension and declination of any other star, in other words the quantities given in the Catalogue.

Section III. Part 2, contains the comparisons with the other large catalogues of stars in the Nebula. They have been brought together into one list, as a more ready method of exhibition. It only needs to be stated that the precessions of Part 1 have in all cases been taken into account in making the comparisons; hence arise the trifling discrepancies which are to be found between Liapunoff's and O. Struve's comparisons of the former's catalogue with the other authorities, and the same numbers as they might be derived from the present table. For the star 303 = Herschel 5, for example, Liapunoff gives L—H $= + 3.''2 \quad - 1.''5$; whereas Sect. III. Part 2 gives L—G. P. B. $= - 0.''3 \quad - 1.''8$, H—G. P. B. $= - 3.''9 \quad + 0.''5$, hence L—H $= + 3.''6 \; - \; 2.''3$. But if from Liapunoff's value we subtract 13 times the precession coefficient of Part 1, we shall find L—H $= + 3.''59 \; - \; 2.''28$, sensibly identical with the numbers just found.

The notes to both these catalogues form Part 3 of this Section. They are referred to by the asterisks of the preceding Parts, and are devoted to the following subjects: —

1. Notes to original observations.
2. Miscellaneous observations on which the positions of stars depend.
3. Difficult and doubtful identifications of stars.
4. Larger discrepancies than usual from the results of others.
5. Discrepancies in magnitude which do not probably arise from variability.

6. Prof. Bond's marginal notes to a copy of Struve's Memoir already cited.

The remainder of the work is mostly text, and therefore explains itself.

Section IV. is devoted to the discussion of magnitudes, — Part 1 being general matter, Part 2 relating to some stars probably variable.

Section V. contains (with some slight discussion by the editor) the original physical observations of Prof. Bond.

Section VI. is composed of a reprint of Prof. Bond's paper on the Spirality of the Nebula, with a few references by the editor to the original observations in the preceding Section. Prof. Bond's views on this subject were thoroughly matured, and it was therefore thought best to reprint this paper as it stands in the Proceedings of the American Academy. The present work would otherwise have been incomplete.

The Appendices are devoted, the first to Prof. W. C. Bond's observations, the second to the errors of the Equatorial, so far as they influence the micrometric observations.

The following paper on the reduction of the scale observations is extracted from Prof. Bond's manuscripts, with such notes by the editor as are necessary to extend its use to the zones observed in 1864.

REDUCTION OF THE ZONES. (1857.)

The reduction of the zones by the application of constant corrections, the same for all stars in the same zone, has referred them approximately to one zero point, viz: a point 20′ south of θ and in same AR. with it.

To the results thus obtained we must apply small corrections for differences of refraction, precession, nutation, and aberration, and for the effect of temperature on the scale. Having applied these, (the places all being referred to a common epoch,) there will remain a zero correction to be applied to all the stars of each zone, so as to bring all to the nearest mutual agreement.

The short zones of 11′ in width in Declination, and of 4^m or 5^m only in AR., by means of which we have compared the stars of the Nebula, require, for their exact reduction, the same formulae as those employed in our large zones, adapted to the declination of θ Orionis.

Let $u =$ the time of passage of a star read off from the chronograph sheet. There were commonly two passages observed, one over each right ascension wire; we will suppose the mean of the two referred to the first wire.

$\alpha =$ Right Ascension of the star referred to the mean equinox for the beginning of the year, 1857.

s = The reading of the scale (in minutes of arc).

t = The interval, expressed in minutes of time as units, from a zero point for the zone (the time of passage of θ is taken for the zero of the zone) in the passage of the particular star to be reduced.

x = The correction to be applied to the time of transit, u, of a star in the same AR. as θ (for $t=0$) and at the zero of the scale ($s=0$) to reduce u to α, so that for such a star

$$\alpha = u + x.$$

For any other star

(1)
$$\alpha = u + x + \frac{dx}{dt}t + \frac{dx}{ds}s + p's.$$

$p's$ is a small correction for error of perpendicularity of wires, which, if constant in all the zones, would show itself in difference of AR. of stars in extreme northern and southern zones.

$\frac{dx}{dt}$ includes

1st. A term $= -\frac{r}{60}$, r being the hourly rate of gain of the clock; but $\frac{r}{60}$ was less than $\frac{0^s.48}{24.60}$ or $< 0^s.00033$, and $t < 3$; hence the term of $\frac{dx}{dt}t$, depending on the clock rates, is $< 0^s.001$, and may be neglected.

2d. A term = Differential Coefficient relatively to t (or what is the same, to α) of the correction reducing the apparent AR. of a star at the commencement of the zone and zero of scale to its mean AR. for Equinox at the beginning of the year. The correction itself is as in the Nautical Almanac and other ephemerides, changing only the signs to make it the correction from apparent to mean Equinox.

(2)
$$-f - g \sin(G+\alpha) \tan\delta - h \sin(H+\alpha) \sec\delta.$$

The Differential Coefficient of this relatively to α, expressed by the proper unit, is

(3)
$$-[g \cos(G+\alpha) \tan\delta + h \cos(H+\alpha) \sec\delta] \times (\text{arc } 15' = 1^m).$$

The unit of g and h being in the Nautical Almanac seconds of *arc*, we must divide (3) by 15 to reduce to seconds of *time;* hence, since arc $15' = 15 \sin 1'$ nearly, and $\frac{15 \sin 1'}{15} = \sin 1'$, (3) becomes

(4)
$$-[g \cos(G+\alpha) \tan\delta + h \cos(H+\alpha) \sec\delta] \sin 1',$$

when the unit of t is $1^m = 15'$ and $\frac{dx}{dt}t$ is to be found in seconds of time.

3d. A term $= +p$, depending on a uniform change in the pointing of the telescope, which, however, is not supposed to exist to an amount sufficient to make $-pt$ sensible when t is $< 3^m$.

The refraction is constant, except from the minute change in the barometer and thermometer in a time less than 4 minutes, (and is therefore included in x) for all stars in the same declination, so that we have

$$(5) \qquad \frac{dx}{dt} = -\frac{r}{60} - [g \cos (G+a) \tan \delta + h \cos (H+a) \sec \delta] \sin 1' + p;$$

but p and $\frac{r}{60}$ may be assumed insensible, and therefore

$$(6) \qquad \frac{dx}{dt} = -[g \cos (G+a) \tan \delta + h \cos (H+a) \sec \delta] \sin 1',$$

g and h being expressed in seconds of arc, and $\frac{dx}{dt}$ in seconds of time.

The zones were observed between Nov. 11th and Dec. 12th, 1857.

Taking $\alpha = 5^h\ 28^m = 82°\ 00'$, for Nov. 12th, we have

$$\frac{dx}{dt} = +0^s.003. \quad \text{Maximum value of } \frac{dx}{dt}t < +0^s.009. \quad \text{Nov. 12th.}$$

For the dates between Nov. 12th and Dec. 12th, $\frac{dx}{dt}$ has a less absolute value. If we make

$$\frac{dx}{dt} = 0,$$

the maximum error will be less than $0^s.01$.

The next coefficient to be considered is $\frac{dx}{ds}$, which includes

1st. A term for the variation of $-[f + g \sin (G + \alpha) \tan \delta + h \sin (H + \alpha) \sec \delta]$ for a change in $\delta = 1'$

$$(7) \qquad -\frac{\sin 1'}{15} [g \sin (G+a) + h \sin (H+a) \sin \delta] \sec^2 \delta.$$

2d and 3d. Terms to include the correction due to the influence of refraction, which are two.

1. If the AR. wires are adjusted so as to be at right angles with the apparent diurnal motion of a star at the point of observation, then, owing to the influence of refraction, the wires will be inclined to the hour circle by a small angle,

$$\delta P = k \left(\frac{\cos \psi}{\sin \psi + \delta}\right)^2 \frac{\sin \tau}{\tan \phi} \frac{1}{\cos \delta} = \left(\frac{dq}{dt \cos \delta}\right) \text{ of Bessel, A. N. No. 69.}$$

where

$$\tan \psi = \frac{\cos \tau}{\tan \phi}. \quad \phi = \text{latitude.} \quad \tau = \text{hour angle, positive when west.}$$

Under these circumstances the AR. wires are directed to a point P in the meridian, between the pole and the zenith. AP makes the angle δP with the hour circle, and stars east of meridian north of A cross the wires too late by the interval

$$\frac{1}{15} \frac{s\,\delta P}{\cos \delta} \sin 1';$$

hence for the term of $\frac{dx}{ds}$ in question, we have

$$(8)\qquad +\frac{\sin 1'}{15}\frac{\delta P}{\cos\delta}=+k\left(\frac{\cos\psi}{\sin(\psi+\delta)}\right)^2\frac{\sin\tau}{\tan\phi\cos^2\delta}\frac{\sin 1'}{15}\text{ in seconds of time,}$$

τ and δP being positive when the star is west, and negative when east of meridian, but δP is constant for the night, so far as the influence of refraction is concerned, or till next determination of P.

2. Again, when the AR. wires have been adjusted perpendicular to the true diurnal motion, the northernmost stars transit east of meridian too late and west of meridian too early, relatively to those south of them, because the latter, being nearer the horizon, are more refracted, and therefore brought nearer to the meridian by the interval (according to Bessel's formula)

$$(9)\qquad \frac{k}{15}s\sin 1'\frac{\cos(\psi+\delta+\delta')}{\sin(\psi+\delta)\sin(\psi+\delta')}\frac{\tan\tau\sin\psi}{\cos\delta\cos\delta'},$$

from which we derive the term of $\frac{dx}{ds}$

$$(10)\qquad +k\frac{\cos(\psi+\delta+\delta')}{\sin(\psi+\delta)\sin(\psi+\delta')}\frac{\tan\tau\sin\psi}{\cos\delta\cos\delta'}\frac{\sin 1'}{15},$$

or when $\delta=\delta'$ nearly

$$k\frac{\cos(\psi+2\delta)}{\sin^2(\psi+\delta)}\frac{\tan\tau\sin\psi}{\cos^2\delta}\frac{\sin 1'}{15},$$

τ being negative when east of meridian.

From the above we have

$$\begin{aligned}
&k'_1 && \frac{dx}{ds}=-\frac{\sin 1'}{15}[g\sin(G+a)+h\sin(H+a)\sin\delta]\sec^2\delta.\\
(11)\quad &k'_3 && \qquad+\frac{\sin 1'}{15}k\left(\frac{\cos\psi}{\sin(\psi+\delta)}\right)^2\frac{\sin\tau}{\tan\phi}\sec^2\delta.\\
&k'_2 && \qquad+\frac{\sin 1'}{15}k\frac{\cos(\psi+2\delta)}{\sin^2(\psi+\delta)}\tan\tau\sin\psi\sec^2\delta.
\end{aligned}$$

For the interval covered by the zones, the first term of $\frac{dx}{ds}$ is largest on Dec. 12th $G+a=58°\ 35'$ $H+a=90°\ 4'$

	$\log g =$ 1.331		$\log h$	1.310
	$\sin G+a=$ 9.931		$\sin H+a$	0.000
+ 18″.3	1.262		$\sin\delta$	9.000
2. 0	0.310		+ 2″.0 =	0.310
20. 3	1.309			
$\sec^2\delta$	= 0.004			
$\sin 1'$	6.464			
	7.777			
15	1.176			
0ˢ.00040	6.601,			

including breadth of all four zones. $-10 \times 4 \times 0.00040 = -0^s.016$, which is the greatest error to be feared from neglect of correction from apparent to mean place.

The reduction of the differences of declination may be effected by a similar process.

Let s = The reading of the scale for a given star, which is so graduated and adjusted that s indicates the number of minutes of declination by which the star is *north* of the zero of the scale.

δ = Declination of the star referred to the mean equinox at the beginning of the year, 1857.0.

y = The correction to be applied to s, for a star at the origin of a zone * ($t = 0$) and transiting at the zero of the scale.

Then

(12) $$\delta = s + y + \frac{dy}{dt}\,t + \frac{dy}{ds}\,s + q'\,s.$$

$q's$ is a small correction for error in the assumed value of the divisions of the scale, to be ascertained from comparison of difference of declination of north and south stars.

$\frac{dy}{dt}$ includes

1st. A term for the change of the correction to be applied to the apparent declination of a star, to reduce it to the mean equinox, at the beginning of the year. This term of $\frac{dy}{dt}$ is

(13) $$[g \sin (G + a) + h \sin (H + a) \sin \delta] \times \text{arc } 15'.$$
$$\text{arc } 15' = 15 \times \sin 1' \text{ nearly.}$$

2d. A term $= + q$, depending on a uniform rate of change of refraction from change of temperature or barometer, or in the pointing of the telescope in declination, which may be neglected here.

(14) $$\frac{dy}{dt} = [g \sin (G + a) + h \sin (H + a) \sin \delta]\; 15 \sin 1'.$$

For the interval between Nov. 11th and Dec. 12th, the maximum value of the term in the brackets is 20''.3 (see p. xx), and if we neglect q, we find

$$\frac{dy}{dt} = 0''.088.$$

The greatest value of t is less than 3^m, hence the greatest value of $\frac{dy}{dt}\,t$ is $< 0''.26$.

$\frac{dy}{ds}$ includes

* See page xviii, where finally transit of θ was taken for zero of t.

1st. The variation of correction from apparent to mean equinox, depending on the change of δ, which gives the term

(15) $$+[i \sin \delta - h \cos (H+\alpha) \cos \delta] \sin 1'.$$

2d. For the effect of refraction, giving the term

(16) $$+\frac{k \sin 1'}{\sin (\psi+\delta) \sin (\psi+\delta')} = +\frac{k \sin 1'}{\sin^2 (\psi+\delta)}.$$

3d. For the influence of temperature on the scale, giving the term

(17) $$+0''.0062\ (\theta^\circ - 67^\circ).$$

Hence the complete value of $\frac{dy}{ds}$ is

(18) $$\frac{dy}{ds} = [i \sin \delta - h \cos (H+\alpha) \cos \delta] \sin 1' + \frac{k \sin 1'}{\sin^2 (\psi+\delta)} + 0''.0062\ (\theta^\circ - 67^\circ),$$

θ being the temperature Fahrenheit.

But we have for limits of zones

	$i \sin \delta =$	-0.5	
	$-h \cos (H+\alpha) \cos \delta =$	$+9.6$	
		9.1	0.959
Breadth of 4 zones $= 40'$	$40 \sin 1'$		8.066
		$0''.106$	$= 9.025$

The maximum value of $[i \sin \delta - h \cos (H+\alpha) \cos \delta] \sin 1' \times 40 = +0''.106$, so that (18) becomes

$$\frac{dy}{ds} = \frac{k \sin 1'}{\sin^2 (\psi+\delta)} + 0''.0062\ (\theta^\circ - 67^\circ).$$

But it will be still better to include 1st term as a constant, both in the correction of Right Ascension and Declination, since if we compute it for the middle of the interval between Nov. 11th and Dec. 12th, the values will be nearly correct for all the intermediate dates.

The formulae adopted will be (1) (5) (11).

In AR.,

$$\alpha = u + x + \frac{dx}{dt} t + \frac{dx}{ds} s + ps,$$

$$\frac{dx}{dt} = -[g \cos (G+\alpha) \tan \delta + h \cos (H+\alpha) \sec \delta] \sin 1',$$

for the present zones, so that the 2d term only is retained, for which we adopt the value given on p. xviii, (4).

k'_1 $$\frac{dx}{ds} = -\frac{\sin 1'}{15}\,[g \sin(G+\alpha) + h \sin(H+\alpha)\sin\delta]\sec^2\delta,$$

k'_3 $$+\frac{\sin 1'}{15}\,k\left(\frac{\cos\psi}{\sin(\psi+\delta)}\right)^2\frac{\sin\tau}{\tan\phi}\sec^2\delta,$$

k'_2 $$+\frac{\sin 1'}{15}\,k\,\frac{\cos(\psi+2\delta)}{\sin^2(\psi+\delta)}\tan\tau\sin\psi\sec^2\delta.$$

k'_3 is the correction for zero of position through refraction.

For the first term, we adopt the values as on p. xx, and for the sums of the second and third terms we have, from the same page,

		$k'_2 + k'_3$
$h = 0^h$	sum	$= 0^s.00000$
" $= 1$	"	$= 0.00075$
" $= 2$	"	$= 0.00184$
" $= 3$	"	$= 0.00401$

The sign of $k'_2 + k'_3$ is the same as that of τ, which is negative when the star is east of the meridian.

Note. k'_3 being constant for the night, should not be interpolated for each value of τ on the same night, but its value for $\tau^\circ = h$ at determination, should be used, so that a minute correction will be needed for the later observations of a night.

It will be sufficiently accurate to give $\frac{dx}{ds}$ to thousandths of a second, and we shall then obtain the following values:—

1857.		Hour angle East.		Hour angle West.	
Nov. 12	$k'_1 = -0^s.00040$	$\tau = -0^h00^m$	$k'_2 + k'_3 = -0^s.000$	$\tau = +0^h00^m$	$k'_2 + k'_3 = +0^s.000$
22	< "	0 20	.000	0 20	.000
Dec. 2	< "	0 40	.000	0 40	.000
12	< "	1 00	.001	1 00	.001
		1 20	.001	1 20	.001
		1 40	.001	1 40	.001
		2 00	.002	2 00	.002
		2 20	.002	2 20	.002
		2 40	.003	2 40	.003
		$\tau = -3\ 00$	0.004	$\tau = +3\ 00$	$+0.004$

In declination

$\delta =$ Mean Declination 1857.0,

$y =$ Constant correction,

$t =$ Interval from zero point of Zone,

$s =$ Reading of scale north of zero,

$$\delta = s + y + \frac{dy}{dt}t + \frac{dy}{ds}s + qs.$$

$$\frac{dy}{dt} = q + [g \sin(G+a) + h \sin(H+a) \sin\delta]\ 15 \sin 1'\,;\ q = 0,$$

for which we may use the constant value for the present zones, taken from p. xxi,

$$\frac{dy}{dt} = 0''.088$$

From p. xxii we have

$$d'_3 \qquad \frac{dy}{ds} = [i \sin\delta - h \cos(H+a) \cos\delta] \sin 1',$$

$$d'_2 \qquad + \frac{k \sin 1'}{\sin^2(\psi+\delta)},$$

$$d'_1 \qquad + 0''.0062\ (6^\circ - 67^\circ).$$

For the present zones we may use a constant $= 0''.002$ for the first term (p. xxii), and uniting this constant with the second term as on this page we have for the sum of the 1st and 2d terms

Hour angle East or West.		Hour angle East or West.	
$\tau = 0^h00^m$	$d'_2 = +0''.037$	$\tau = 2^h00^m$	$d'_2 = +0''.044$
0 30	.038	2 30	.049
1 00	.039	3 00	+0 .058
1 30	+0 .041		

A table for the last term, d'_1, is given in Vol. I. of these Annals, p. xi, of Part II.

On the next pages is given a summary of these results, together with those necessary for the more distant zones.

SUMMARY OF RESULTS.

$$d'_1 = +0''.0062\ (\theta^\circ - 67^\circ).$$

$$d'_2 = \frac{k \sin 1'}{\sin^2(\psi + \delta)}.$$

$$d'_3 = [i \sin \delta - h \cos (H + \alpha) \cos \delta] \sin 1'.$$

$$k'_1 = -\frac{\sin 1'}{15} [g \sin (G + \alpha) + h \sin (H + \alpha) \sin \delta] \sec^2 \delta.$$

$$k'_3 = \sin 1' \left[k \left(\frac{\cos \psi}{\sin (\psi + \delta)}\right)^2 \frac{\sin \tau}{\tan \phi} \sec^2 \delta\right].$$

$$k'_2 = \frac{\sin 1'}{15}\, k\, \frac{\cos (\psi + 2\,\delta)}{\sin^2(\psi + \delta)} \tan \tau \sin \psi \sec^2 \delta.$$

$$\frac{dx}{dt} = -[g \cos (G + \alpha) \tan \delta + h \cos (H + \alpha) \sec \delta] \sin 1'.$$

$$\frac{dy}{dt} = +[g \sin (G + \alpha) + h \sin (H + \alpha) \sin \delta]\, 15 \sin 1'.$$

$$\frac{dx}{ds} = k'_1 + k'_2 + k'_3.$$

$$\frac{dy}{ds} = d'_1 + d'_2 + d'_3.$$

TABLES OF REDUCTION.

Principal Zones.

Date.	Temp.	d'_1	$d'_1+d'_3$	k'_1	$\frac{dx}{dt}$	$\frac{dy}{dt}$
1857, Oct. 11	48°	−0″.118	−0″.113	−0ˢ.0003	+0ˢ.0051	+0″.059
Nov. 11	34	.205	.202	.0003	.0032	.063
27	33	.211	.210	.0003	.0018	.067
Dec. 1	38	.180	.179	.0003	.0014	.069
4	23	.273	.272	.0003	.0011	.070
7	40	.167	.166	.0003	.0008	.070
10	34	.205	.205	.0003	.0006	.071
12	24	.267	.267	.0003	+0.0004	.072
1864, Jan. 21	22	.279	.282	.0001	−0.0036	.028
25	35	−0.198	−0.202	−0.0001	−0.0039	+0.030

Supplementary Zones.

Date.	Temp.	d'_1	$d'_1+d'_3$	k'_1	$\frac{dx}{dt}$	$\frac{dy}{dt}$
1858, Jan. 4	45	−0.136	−0.138	−0.0004	−0.0019	+0.081
7	24	.267	.270	.0004	.0022	.082
12	32	.217	.220	.0004	.0026	.084
14	31	.223	.226	.0004	.0028	.085
20	26	.254	.257	.0004	.0032	.088
Feb. 12	16	.316	.321	.0004	.0046	.096

Date.	Temp.	d'_1	$d'_1+d'_3$	k'_1	$\frac{dx}{dt}$	$\frac{dy}{dt}$
	°	"	"	s	s	"
1858, Mar. 10	26	.254	.259	.0005	.0052	.106
13	25	.260	.265	.0005	.0052	.107
18	44	.143	.148	.0005	.0051	.108
19	32	.217	.222	.0005	.0051	.109
1864, Jan. 27	29	.236	.240	.0001	.0041	.030
28	37	.186	.190	.0001	.0042	.031
Feb. 12	36	.192	.196	.0002	.0049	.036
27	25	.260	.265	.0002	.0055	.041
29	28	.242	.247	.0002	.0055	.042
Mar. 2	23	.273	.278	.0002	.0055	.042
3	29	.236	.241	.0002	.0055	.043
12	33	.211	.216	.0002	.0056	.045
14	26	.254	.259	.0002	.0056	.045
28	32	−0.217	−0.222	−0.0002	−0.0053	+0.050

THE GREAT NEBULA OF ORION.

Letter.	Mag.	Reading of Scale.	d	D	Transit red. to 1st wire.	a	A
			Zone I. Oct. 11, 1857. $\tau = +0^h\ 32^m$ $\theta = 48°$				
		r					
D	11.12	68.75	+28′ 57.1″ +0.4″	+40′ 11.3″			
L	10.11	64.64	+0.4	39 31.0			
W	10.11	67.47	+0.5	39 58.8			
			Zone II. Oct. 11, 1857. $\tau = +0^h\ 33^m$ $\theta = 48°$				
		r					
D		67.84	+29′ 5.6″ +0.4″	+40′ 10.8″			
L		63.86	+0.4	39 31.8			
W		66.67	+0.5	39 59.4			
			Zone III. Oct. 11, 1857. $\tau = +0^h\ 51^m$ $\theta = 48°$				
		r					
A_{-1}	12	61.28	+28′ 2.6″ +0.3″	+38′ 3.4″			
G	11	62.45	+0.4	38 15.0			
H	12	59.19	+0.4	37 43.0			
Q	10	55.66	+0.4	37 8.4			
			Zone IV. Oct. 11, 1857. $\tau = +0^h\ 54^m$ $\theta = 48°$				
		r					
A_{-1}	12	61.34	+28′ 1.3″ +0.3″	+38′ 2.8″			
G	11	62.51	+0.4	38 14.3			
H	12	59.28	+0.4	37 42.7			
Q	10	55.81	+0.4	37 8.7			
V	12	63.89	+0.5	38 27.9			
			Zone V. Oct. 11, 1857. $\tau = +1^h\ 6^m$ $\theta = 48°$				
					h m s	h m s s	m s
A_{-1}					6 35 0.95	—6 36 44.38—0.00	—1 43.43
D					36 1.20	.00	—0 43.18
G					36 27.50	.00	—0 16.88
H					36 30.30	+ .01	—0 14.07
L					36 45.00	+ .01	+0 0.63
Q					37 1.30	+ .01	+0 16.93
V					37 19.65	+ .01	+0 35.28
W					37 37.80	+0.01	+0 53.43

Zone VI. Oct. 11, 1857. $\tau = 1^h\ 11^m$ $\theta = 48°$

Letter.	Mag.	Reading of Scale.	d	D	Transit red. to 1st wire.	a		A
					h m s	h m s	s	m s
A_{-1}					6 39 51.05	—6 41 34.47	—0.00	—1 43.42
D					40 51.40		+ .00	—0 43.07
G					41 17.70		.00	—0 16.77
H					41 20.40		+ .01	—0 14.06
L					41 35.05		+ .01	+0 0.59
Q					41 51.25		+ .01	+0 16.79
W					42 27.90		+0.01	+0 53.44

Zone VII. Oct. 11, 1857. $\tau = +1^h\ 15^m$ $\theta = 48°$

Letter.	Mag.	Reading of Scale.	d	D	Transit red. to 1st wire.	a		A
					h m s	h m s	s	m s
A_{-1}					6 43 18.15	—6 45 1.42	—0.00	—1 43.27
D					44 18.25		.00	—0 43.17
G					44 44.70		.00	—0 16.72
H					44 47.40		+ .01	—0 14.01
L					45 2.20		+ .01	+0 0.79
Q					45 18.30		+ .01	+0 16.89
V					45 36.75		+ .01	+0 35.34
W					45 54.85		+0.01	+0 53.44

Zone 4, Nov. 11, 1857. $\tau = +1^h\ 36^m$ $\theta = 34°$

Letter.	Mag.	Reading of Scale.	d	D	Transit red. to 1st wire.	a		A
		′ ″	′ ″ ″	′ ″	h m s	h m s	s	m s
A_{-2}	11	3 40	+29 52.9—0.7	33 32.2	7 5 5.30	—7 7 29.46	.00	—2 24.16
A_{-1}	12	8 12	—1.4	38 3.5	5 46.00		.00	—1 43.46
A	11	3 34	—0.7	33 26.2	5 58.60		.00	—1 30.86
C	12	1 30	—0.3	31 22.6	7 6 27.60		.00	—1 1.86

Zone 5, Nov. 11, 1857. $\tau = +1^h\ 48^m$ $\theta = 34°$

Letter.	Mag.	Reading of Scale.	d	D	Transit red. to 1st wire.	a		A
		′ ″	′ ″ ″	′ ″	h m s	h m s	s	m s
A_{-2}	11.12	3 39	+29 54.4—0.7	+33 32.7	7 16 44.20	—7 19 8.48	.00	—2 24.28
A_{-1}	12	8 11	—1.4	38 4.0	17 25.10		.00	—1 43.38
A	11	3 33	—0.7	33 26.7	17 37.60		.00	—1 30.88
C	12	1 29	—0.3	31 23.1	18 6.55		.00	—1 1.93
D	10.11	10 19	—1.7	40 11.7	18 25.30		+.01	—0 43.17
F	10.11	6 11	—1.0	36 4.4	18 33.35		.00	—0 35.13
G	11.12	9 50 *	—1.5	39 42.9	18 51.75		+.01	—0 16.72
K	10	1 11	—0.2	31 5.2	18 57.65		.00	—0 10.83
Q	10.11	7 15	—1.1	37 8.3	7 19 25.30		+.01	+0 16.83

Zone 6, Nov. 11, 1857. $\tau = +1^h\ 54^m$ $\theta = 34°$

Letter.	Mag.	Reading of Scale.	d	D	Transit red. to 1st wire.	a		A
		′ ″	′ ″ ″	′ ″	h m s	h m s	s	m s
A_{-2}	11.12	3 40	+29 52.4—0.7	+33 31.7	7 22 25.90	—7 24 50.13	.00	—2 24.23
A_{-1}	12.13	8 12	—1.4	38 3.0	23 6.85		.00	—1 43.28
A	11	3 36	—0.6	33 27.8	23 19.20		.00	—1 30.93
C	12	1 30	—0.3	31 22.1	23 48.30		.00	—1 1.83
D	11	10 20	—1.7	40 10.7	24 7.05		+.01	—0 43.07

* Probably 1′ 30″ too large. May have been G with error of 1′ 30″. Rejected. G. P. B.

Zone 6, Nov. 11, 1857. $\tau = +1^h\ 54^m$ $\theta = 34°$

Letter.	Mag.	Reading of Scale.	d	D	Transit red. to 1st wire.	a	A
		′ ″	′ ″ ″	′ ″	h m s	h m s s	m s
F	10.11	6 13	+29 52.4—1.0	+36 4.4	7 24 15.15	—7 24 50.13 +.01	—0 34.97
K	10	1 13	—0.2	31 5.2	24 39.25	.00	—0 10.88
Q	10.11	7 17	—1.1	37 8.3	25 7.00	+.01	+0 16.88
W	11	10 9	—1.5	39 59.9	25 43.55	+.02	+0 53.44
Y	10	6 0	—0.8	35 51.6	26 11.65	+.01	+1 21.53
Z	12.13	7 6	—1.0	36 57.4	7 26 16.40	+.01	+1 26.28

Zone 7, Nov. 11, 1857. $\tau = +2^h\ 1^m$ $\theta = 34°$

Letter.	Mag.	Reading of Scale.	d	D	Transit red. to 1st wire.	a	A
		′ ″	′ ″ ″	′ ″	h m s	h m s s	m s
A_{-2}	11.12	3 44	+29 50.6—0.7	+33 33.9	7 29 6.55	—7 31 30.84 .00	—2 24.29
A_{-1}	12.13	8 15	—1.4	38 4.2	29 47.35	.00	—1 43.49
A	11.12	3 37	—0.6	33 27.0	29 59.90	.00	—1 30.94
C	12	1 31	—0.3	31 21.3	30 28.95	.00	—1 1.89
D	11	10 20	—1.6	40 9.0	30 47.75	+.01	—0 43.08
F	10.11	6 14	—1.0	36 3.6	30 55.75	+.01	—0 35.08
K	9.10	1 16	—0.2	31 6.4	31 20.00	.00	—0 10.84
Q	10.11	7 20	—1.1	37 9.5	31 47.75	+.01	+0 16.92
W	11	10 11	—1.5	40 0.1	32 24.20	+.02	+0 53.38
Y	11	6 2	—0.8	35 51.8	32 52.25	+.01	+1 21.42
Z	12.13	7 9	—1.0	36 58.6	32 56.90	+.01	+1 26.07
BB	11.12	8 47	—1.2	38 36.4	7 33 53.10	+.02	+2 22.28

Zone 8, Nov. 11, 1857. $\tau = +2^h\ 16^m$ $\theta = 34°$

Letter.	Mag.	Reading of Scale.	d	D	Transit red. to 1st wire.	a	A
		′ ″	′ ″ ″	′ ″	h m s	h m s s	m s
C	12	1 31	+29 50.1—0.3	+31 20.8	7 43 58.80	—7 45 0.69 .00	—1 1.89
D	11	10 22	—1.6	40 10.5	44 17.45	+.01	—0 43.23
E	12	3 22	—0.5	33 11.6	44 23.40	.00	—0 37.29
F	11	6 13	—1.0	36 2.1	44 25.55	+.01	—0 35.13
K	10	1 17	—0.2	31 6.9	7 44 49.80	.00	—0 10.89

Zone 9, Nov. 11, 1857. $\tau = +2^h\ 22^m$ $\theta = 34°$

Letter.	Mag.	Reading of Scale.	d	D	Transit red. to 1st wire.	a	A
		′ ″	′ ″ ″	′ ″	h m s	h m s s	m s
C	12.13	1 28	+29 54.2—0.3	+31 21.9	7 50 42.95	—7 51 44.84 .00	—1 1.89
D	11	10 20	—1.6	40 12.6	51 1.75	+.01	—0 43.08
E	11	3 18	—0.5	33 11.7	51 7.60	.00	—0 37.24
F	10.11	6 10	—1.0	36 3.2	51 9.70	+.01	—0 35.13
K	10	1 12	—0.2	31 6.0	7 51 33.95	.00	—0 10.89

Zone 10, Nov. 11, 1857. $\tau = +2^h\ 44^m$ $\theta = 34°$

Letter.	Mag.	Reading of Scale.	d	D	Transit red. to 1st wire.	a	A
		′ ″	′ ″ ″	′ ″	h m s	h m s s	m s
C	12	1 32	+29 49.7—0.3	+31 21.4	8 12 11.45	—8 13 13.43 .00	—1 1.98
D	11	10 21	—1.6	40 9.1	12 30.35	+.02	—0 43.06
G	12	8 25	—1.3	38 13.4	12 56.60	+.02	—0 16.81
H	12	7 53	—1.2	37 41.5	12 59.30	+.02	—0 14.11
I	12	5 22	—0.8	35 10.9	13 0.40	+.01	—0 13.02
K	10	1 17	—0.2	31 6.5	8 13 2.60	.00	—0 10.83

Zone 11, Nov. 11, 1857. $\tau = +2^h\ 47^m$ $\theta = 34°$

Letter.	Mag.	Reading of Scale.	d	D	Transit red. to 1st wire.	a		A
					h m s	h m s	s	m s
C	12				8 15 44.05	—8 16 46.03	.00	—1 1.98
D	11				16 2.80		+.02	—0 43.21
G	12				16 29.00		+.02	—0 17.01
H	12				16 31.90		+.02	—0 14.11
I	12				16 32.80		+.01	—0 13.22
K	10				8 16 35.20		.00	—0 10.83

Zone 12, Nov. 11, 1857. $\tau = +2^h\ 50^m$ $\theta = 34°$

Letter.	Mag.	Reading of Scale.	d	D	Transit red. to 1st wire.	a		A
		′ ″	′ ″ ″	′ ″	h m s	h m s	s	m s
C	12	1 33	+29 49.8—0.3	+31 22.5	8 18 17.10	—8 19 18.96	.00	—1 1.86
D	11	10 23	—1.6	40 11.2	18 35.95		+.02	—0 42.99
G	12	8 27	—1.2	38 15.6	19 2.20		+.02	—0 16.74
H	12	7 55	—1.2	37 43.6	19 5.20		+.02	—0 13.74
I	12	5 20	—0.8	35 9.0	19 5.90		+.01	—0 13.05
K	10	1 16	—0.2	31 5.6	8 19 8.05		.00	—0 10.91

Zone 13, Nov. 27, 1857. $\tau = -1^h\ 9^m$ $\theta = 33°$

Letter.	Mag.	Reading of Scale.	d	D	Transit red. to 1st wire.	a		A
		′ ″	′ ″ ″	′ ″	h m s	h m s	s	m s
A_{-2}	11	3 37	+29 55.6—0.8	+33 31.8	4 17 25.80	—4 19 50.01	—.01	—2 24.22
A_{-1}	12	8 8	—1.5	38 2.1	18 6.85		—.01	—1 43.17
A	10.11	3 32	—0.7	33 26.9	18 19.10		—.01	—1 30.92
C	11.12	1 26	—0.3	31 21.3	18 48.15		.00	—1 1.86
D	12	10 17	—1.8	40 10.8	19 6.85		—.01	—0 43.17
F	10	6 10	—1.0	36 4.6	19 14.90		—.01	—0 35.12
K	9	1 11	—0.2	31 6.4	19 39.15		.00	—0 10.86
L	11	9 39	—1.6	39 33.0	19 50.70		—.01	+0 0.68
Q	11	7 13	—1.2	37 7.4	20 6.95		—.01	+0 16.93
W	10.11	10 5	—1.6	39 59.0	20 43.35		—.01	+0 53.33
Y	10	5 56	—0.9	35 50.7	4 21 11.50		—.01	+1 21.48

Zone 14, Nov. 27, 1857. $\tau = -1^h\ 3^m$ $\theta = 33°$

Letter.	Mag.	Reading of Scale.	d	D	Transit red. to 1st wire.	a		A
		′ ″	′ ″ ″	′ ″	h m s	h m s	s	m s
A_{-2}	11	3 39	+29 53.4—0.8	+33 31.6	4 23 33.05	—4 25 57.43	—.01	—2 24.39
A_{-1}	12	8 11	—1.5	38 2.9	24 14.05		—.01	—1 43.39
A	10.11	3 34	—0.7	33 26.7	24 26.60		—.01	—1 30.84
C	11.12	1 29	—0.3	31 22.1	24 55.55		.00	—1 1.88
E	12	3 18	—0.6	33 10.8	25 20.20		.00	—0 37.23
F	11	6 11	—1.1	36 3.3	25 22.35		—.01	—0 35.09
H	12	7 50	—1.3	37 42.1	25 43.45		—.01	—0 13.99
L	10	9 40	—1.6	39 31.8	25 58.15		—.01	+0 0.71
Q	10.11	7 16	—1.2	37 8.2	26 14.40		—.01	+0 16.96
W	10	10 7	—1.6	39 58.8	26 50.85		—.01	+0 53.41
Y	10	5 57	—0.9	35 49.5	4 27 18.90		—.01	+1 21.46

Zone 15, Nov. 27, 1857. $\tau = -0^h\ 49^m$ $\theta = 33°$

Letter.	Mag.	Reading of Scale.	d	D	Transit red. to 1st wire.	a		A
		′ ″	′ ″ ″	′ ″	h m s	h m s	s	m s
F	11	6 11	+29 53.7—1.0	+36 3.7	4 38 21.75	—4 38 56.71	—.01	—0 34.97
G	11.12	8 21	—1.4	38 13.3	38 39.90		—.01	—0 16.82

Zone 15, Nov. 27, 1857. $\tau = -0^h\ 49^m$ $\theta = 33°$

Letter.	Mag.	Reading of Scale.	d	D	Transit red. to 1st wire.	a		A
		′ ″	′ ″ ″	′ ″	h m s	h m s	s	m s
I	12	5 16	+29 53.7—0.9	+35 8.8	4 38 43.60	—4 38 56.71	—.01	—0 13.12
K	9	1 13	—0.2	31 6.5	38 45.80		.00	—0 10.91
N	11	1 19	—0.2	31 12.5	39 0.95		.00	+0 4.24
P	12.13	0 18	—0.1	30 11.6	39 6.45		.00	+0 9.74
R		1 13	—0.2	31 6.5	4 39 15.40		.00	+0 18.69

Zone 16, Nov. 27, 1857. $\tau = -0^h\ 41^m$ $\theta = 33°$

Letter.	Mag.	Reading of Scale.	d	D	Transit red. to 1st wire.	a		A
		′ ″	′ ″ ″	′ ″	h m s	h m s	s	m s
F	11	6 13	+29 51.2—1.1	+36 3.1	4 46 54.25	—4 47 29.32	—.01	—0 35.08
G	12	8 24	—1.4	38 13.8	47 12.70		—.01	—0 16.63
I	12	5 18	—0.9	35 8.3	47 16.25		—.01	—0 13.08
K	9	1 15	—0.2	31 6.0	47 18.40		.00	—0 10.92
N	12	1 22	—0.2	31 13.0	47 33.60		.00	+0 4.28
P	12.13	0 18	—0.1	30 9.1	47 39.10		.00	+0 9.78
R	11.12	1 14	—0.2	31 5.0	4 47 48.00		.00	+0 18.68

Zone 17, Nov. 27, 1857. $\tau = -0^h\ 38^m$ $\theta = 33°$

Letter.	Mag.	Reading of Scale.	d	D	Transit red. to 1st wire.	a		A
		′ ″	′ ″ ″	′ ″	h m s	h m s	s	m s
F	11	6 11	+29 51.2—1.0	+36 1.2	4 49 40.40	—4 50 15.47	—.01	—0 35.08
G		8 24	—1.4	38 13.8	49 58.70		—.01	—0 16.78
I		5 18	—0.9	35 8.3	50 2.40		.00	—0 13.07
K		1 15	—0.2	31 6.0	50 4.60		.00	—0 10.87
N		1 22	—0.2	31 13.0	50 19.65		.00	+0 4.18
P		0 18	—0.0	30 9.2	50 25.25		.00	+0 9.78
R		1 14	—0.2	31 5.0	4 50 34.15		.00	+0 18.68

Zone 18, Nov. 27, 1857. $\tau = -0^h\ 25^m$ $\theta = 33°$

Letter.	Mag.	Reading of Scale.	d	D	Transit red. to 1st wire.	a		A
		′ ″	′ ″ ″	′ ″	h m s	h m s	s	m s
K	9	1 13	+29 52.9—0.2	+31 5.7	5 2 37.80	—5 2 48.69	.00	—0 10.89
M	11.12	4 17	—0.7	34 9.2	2 52.70		.00	+0 4.01
Q	10.11	7 18	—1.2	37 9.7	3 5.55		—.01	+0 16.85
T	13.14	3 8	—0.5	33 0.4	3 19.20		.00	+0 30.51
V	13	8 36	—1.4	38 27.5	5 3 23.90		—.01	+0 35.20

Zone 19, Nov. 27, 1857. $\tau = -0^h\ 24^m$ $\theta = 33°$

Letter.	Mag.	Reading of Scale.	d	D	Transit red. to 1st wire.	a		A
		′ ″	′ ″ ″	′ ″	h m s	h m s	s	m s
K	9	1 11	+29 54.9—0.2	+31 5.7	5 4 23.15	—5 4 34.04	.00	—0 10.89
M	11.12	4 15	—0.7	34 9.2	4 38.05		.00	+0 4.01
Q	10.11	7 15	—1.2	37 8.7	4 50.95		—.01	+0 16.90
T	13.14	3 6	—0.5	33 0.4	5 4.80		.00	+0 30.76
V	13	8 34	—1.4	38 27.5	5 5 9.40		—.01	+0 35.35

Zone 20, Nov. 27, 1857. $\tau = -0^h\ 16^m$ $\theta = 33°$

Letter.	Mag.	Reading of Scale.	d		D	Transit red. to 1st wire.	a		A
		′ ″	′ ″	″	′ ″	h m s	h m s	s	m s
K	9	1 13	+29 53.3	—0.2	+31 6.1	5 11 50.10	—5 12 1.02	.00	—0 10.92
M	11.12	4 16		—0.7	34 8.6	12 5.10		.00	+0 4.08
O	13.14	4 54		—0.8	34 46.5	12 10.15		.00	+0 9.13
U	14	5 30		—0.9	35 22.4	12 35.50		.00	+0 34.48
V_1	14	6 30		—1.0	36 22.3	5 12 48.20		.00	+0 47.18

Zone 21, Nov. 27, 1857. $\tau = -0^h\ 14^m$ $\theta = 33°$

Letter.	Mag.	Reading of Scale.	d		D	Transit red. to 1st wire.	a		A
		′ ″	′ ″	″	′ ″	h m s	h m s	s	m s
K	9	1 9	+29 56.9	—0.2	+31 5.7	5 13 41.10	—5 13 51.96	.00	—0 10.86
M	11.12	4 13		—0.7	34 9.2	13 55.90		.00	+0 3.94
O	13.14	4 50		—0.8	34 46.1	14 0.95		.00	+0 8.99
U	14	5 26		—0.9	35 22.0	14 26.50		.00	+0 34.54
V_1	14	6 26		—1.0	36 21.9	5 14 39.10		.00	+0 47.14

Zone 22, Nov. 27, 1857. $\tau = -0^h\ 2^m$ $\theta = 33°$

Letter.	Mag.	Reading of Scale.	d		D	Transit red. to 1st wire.	a		A
		′ ″	′ ″	″	′ ″	h m s	h m s	s	m s
Q	11	7 13	+29 56.0	—1.2	+37 7.8	5 25 44.65	—5 25 27.77	—.01	+0 16.87
V	13	8 32		—1.4	38 26.6	26 2.95		—.01	+0 35.17
W	10.11	10 6		—1.6	40 0.4	26 21.20		—.01	+0 53.42
Y	10	5 56		—0.9	35 51.1	26 49.30		.00	+1 21.53
Z	12	7 0		—1.1	36 54.9	26 54.00		.00	+1 26.23
AA	12	3 13		—0.4	33 8.6	5 27 19.85		.00	+1 52.08

Zone 23, Nov. 27, 1857. $\tau = 0^h\ 0^m$ $\theta = 33°$

Letter.	Mag.	Reading of Scale.	d		D	Transit red. to 1st wire.	a		A
		′ ″	′ ″	″	′ ″	h m s	h m s	s	m s
Q	11	7 8	+30 1.9	—1.2	+37 8.7	5 28 24.00	—5 28 7.12	—.01	+0 16.87
V	13	8 26		—1.4	38 26.5	28 42.45		—.01	+0 35.32
W	10.11	9 59		—1.6	39 59.3	29 0.55		—.01	+0 53.42
Y	10	5 50		—0.9	35 51.0	29 28.60		.00	+1 21.48
Z	12	6 56		—1.1	36 56.8	29 33.20		.00	+1 26.08
AA	12	3 7		—0.4	33 8.5	5 29 59.10		.00	+1 51.98

Zone 24, Nov. 27, 1857. $\tau = +0^h\ 10^m$ $\theta = 33°$

Letter.	Mag.	Reading of Scale.	d		D	Transit red. to 1st wire.	a		A
		′ ″	′ ″	″	′ ″	h m s	h m s	s	m s
A	11	8 31	+24 56.3	—1.5	+33 25.8	5 37 2.60	—5 38 33.40	—.01	—1 30.81
B	12.13	7 56		—1.4	32 50.9	37 24.95		—.01	—1 8.46
C	12	6 26		—1.1	31 21.2	37 31.50		.00	—1 1.90
E′	12.13	—0 7		0.0	24 49.3	38 0.25		.00	—0 33.15
F′	12	—0 23		0.0	24 33.3	38 6.70		.00	—0 26.70
K	9	6 10		—1.0	31 5.3	38 22.50		.00	—0 10.90
T′	8	2 27		—0.4	27 22.9	38 47.90		.00	+0 14.50
X	13	7 0		—1.1	31 55.2	5 39 33.30		.00	+0 59.90

Zone 25, Nov. 27, 1857. $\tau=+0^h\ 13^m$ $\theta=33°$

Letter.	Mag.	Reading of Scale.	d		D	Transit red. to 1st wire.	a		A
		′ ″	′ ″	″	′ ″	h m s	h m s	s	m s
A	11	8 36	+24 52.3	—1.5	+33 26.8	5 40 53.10	—5 42 23.94	—.01	—1 30.85
B	12.13	8 1		—1.4	32 51.9	41 15.60		—.01	—1 8.35
C	12	6 31		—1.2	31 22.1	41 22.10		.00	—1 1.84
E′	12	—0 2		0.0	24 50.3	41 50.90		.00	—0 33.04
F′	11.12	—0 20		+0.1	24 32.4	41 57.20		.00	—0 26.74
K	9	6 15		—1.1	31 6.2	42 13.00		.00	—0 10.94
M	11.12	9 18		—1.6	34 8.7	42 28.00		—.01	+0 4.05
T′	8	2 32		—0.4	27 23.9	42 38.50		.00	+0 14.56
X	13.14	7 3		—1.1	31 54.2	5 43 24.15		.00	+1 0.21

Zone 26, Nov. 27, 1857. $\tau=+0^h\ 18^m$ $\theta=33°$

Letter.	Mag.	Reading of Scale.	d		D	Transit red. to 1st wire.	a		A
		′ ″	′ ″	″	′ ″	h m s	h m s	s	m s
A	11	8 33	+24 56.4	—1.5	+33 27.9	5 44 59.10	—5 46 29.97	—.01	—1 30.88
B′		1 36		—0.3	26 32.1	45 7.00		.00	—1 22.97
C	12	6 28		—1.2	31 23.2	45 28.10		.00	—1 1.87
E′		—0 4		0.0	24 52.4	45 57.00		.00	—0 32.97
F′						46 3.25		.00	—0 26.72
K	9	6 12		—1.1	31 7.3	46 19.10		.00	—0 10.87
T′	8	2 28		—0.4	27 24.0	46 44.50		.00	+0 14.53
AA	12	8 15		—1.3	33 10.1	5 48 21.95		.00	+1 51.98

Zone 27, Nov. 27, 1857. $\tau=+0^h\ 26^m$ $\theta=33°$

Letter.	Mag.	Reading of Scale.	d		D	Transit red. to 1st wire.	a		A
		′ ″	′ ″	″	′ ″	h m s	h m s	s	m s
A	11	8 33	+24 54.4	—1.5	+33 25.9	5 54 1.55	—5 55 32.39	—.01	—1 30.85
B′	12.13	1 38		—0.3	26 32.1	54 9.30		.00	—1 23.09
B	12	7 59		—1.4	32 52.0	54 24.05		—.01	—1 8.35
C	12	6 29		—1.2	31 22.2	54 30.50.		.00	—1 1.89
E	12	8 16		—1.4	33 9.0	54 55.20		—.01	—0 37.20
N	12	6 20		—1.1	31 13.3	55 36.75		.00	+0 4.36
P	13	5 18		—0.9	30 11.5	5 55 42.20		.00	+0 9.81

Zone 27 b, Nov. 27, 1857. $\tau=+0^h\ 33^m$ $\theta=33°$

Letter.	Mag.	Reading of Scale.	d		D
		′ ″	′ ″	″	′ ″
H′	12.13	3 36	+24 55.4	—0.6	+28 30.8
K	9	6 11		—1.0	31 5.4
R′	13	3 18		—0.6	28 12.8
T′	8	2 29		—0.4	27 24.0
V′	12.13	4 52		—0.8	29 46.6

Zone 27 c, Nov. 27, 1857. $\tau=+0^h\ 39^m$ $\theta=33°$

Letter.	Mag.	Reading of Scale.	d		D
		′ ″	′ ″	″	′ ″
H′		3 40	+24 52.4	—0.6	+28 31.8
K		6 15		—1.0	31 6.4
R′		3 21		—0.6	28 12.8
T′		2 32		—0.4	27 24.0
V′		4 56		—0.8	29 47.6

Zone 27 d, Nov. 27, 1857. $\tau = +0^h\ 46^m$ $\theta = 33°$

Letter.	Mag.	Reading of Scale.	d		D	Transit red. to 1st wire.	a		A
K	9	6′ 10″	+24′ 57″.4	—1″.0	+31′ 6″.4				
N	12	6 16		—1.0	31 12.4				
T′	8	2 27		—0.4	27 24.0				
v'_3	14	1 35		—0.3	26 32.1				
X	13	7 0		—1.1	31 56.3				

Zone 27 e, Nov. 27, 1857. $\tau = +0^h\ 52^m$ $\theta = 33°$

Letter.	Mag.	Reading of Scale.	d		D	Transit red. to 1st wire.	a		A
K	9	6′ 12″	+24′ 55″.4	—1″.0	+31′ 6″.4				
N	12.13	6 19		—1.1	31 13.3				
T′	8	2 29		—0.4	27 24.0				
X	14	7 0		—1.1	31 54.3				

Zone 28 a, Dec. 1, 1857. $\tau = -1^h\ 41^m$ $\theta = 38°$

Letter.	Mag.	Reading of Scale.	d		D	Transit red. to 1st wire.	a		A
						h m s	h m s	s	m s
H′	12.13	3′ 34″	+24′ 56″.3	—0″.5	+28′ 29″.8	3 47 19.00	—3 47 30.82	—.01	—0 11.83
K	9	6 9		—0.8	31 4.5	47 19.85		—.01	—0 10.98
R′	13	3 16		—0.4	28 11.9	47 40.40		—.01	+0 9.57
T′	8	2 28		—0.3	27 24.0	47 45.30		.00	+0 14.48
V′	12.13	4 50		—0.6	29 45.7	3 47 56.80		—.01	+0 25.97

Zone 28 b, Dec. 1, 1857. $\tau = -1^h\ 40^m$ $\theta = 38°$

Letter.	Mag.	Transit red. to 1st wire.	a		A
		h m s	h m s	s	m s
H′	12.13	3 48 34.00	—3 48 45.77	—.01	—0 11.78
K	9	48 34.90		—.01	—0 10.88
R′	13	48 55.20		—.01	+0 9.42
T′	8	49 0.25		.00	+0 14.48
V′	12.13	3 49 11.75		—.01	+0 25.97

Zone 29, Dec. 1, 1857. $\tau = -1^h\ 36^m$ $\theta = 38°$

Letter.	Mag.	Reading of Scale.	d		D	Transit red. to 1st wire.	a		A
						h m s	h m s	s	m s
K	9	6′ 10″	+24′ 56″.3	—0″.8	+31′ 5″.5	3 51 36.20	—3 51 47.12	—.01	—0 10.93
N	12	6 18		—0.9	31 13.4	51 51.40		—.01	+0 4.27
T′	8	2 28		—0.3	27 24.0	52 1.60		.00	+0 14.48
X	13	7 0		—0.9	31 55.4	3 52 47.05		—.01	+0 59.92

Zone 30, Dec. 1, 1857. $\tau = -1^h\ 34^m$ $\theta = 38°$

Letter.	Mag.	Reading of Scale.	d		D	Transit red. to 1st wire.	a		A
						h m s	h m s	s	m s
K	9	6′ 4″	+25′ 4″.3	—0″.8	+31′ 7″.5	3 54 19.90	—3 54 30.77	—.01	—0 10.88
N	12	6 10		—0.8	31 13.5	54 35.00		—.01	+0 4.22
T′	8	2 20		—0.3	27 24.0	54 45.25		.00	+0 14.48
X	13	6 51		—0.9	31 54.4	3 55 30.90		—.01	+1 0.12

Zone 31, Dec. 1, 1857. $\tau = -1^h\ 22^m$ $\theta = 38°$

Letter.	Mag.	Reading of Scale.	d		D	Transit red. to 1st wire.	a		A
		′ ″	′ ″	″	′ ″	h m s	h m s	s	m s
A'_{-3}	11.12	7 35	+20 0.6	—1.2	+27 34.4	4 4 18.85	—4 6 45.97	—.01	—2 27.13
A'_{-1}	10.11	2 32		—0.5	22 32.1	4 36.90		—.01	—2 9.08
A′	12	1 47		—0.4	21 47.2	4 53.95		.00	—1 52.02
B′	12	6 33		—1.0	26 32.6	5 23.00		—.01	—1 22.98
C′=C″	11					5 40.65		.00	—1 5.32
D′	12	1 6		—0.2	21 6.4	5 54.45		.00	—0 51.52
E′	11.12					6 12.90		—.01	—0 33.08
F′	11.12	4 32		—0.6	24 32.0	6 19.25		—.01	—0 26.73
H′						6 34.40		—.01	—0 11.58
M′		0 0		0.0	20 0.6	6 46.00		.00	+0 0.03
T′	8	7 23		—1.0	27 22.6	7 0.50		—.01	+0 14.52
W′	11	1 1		—0.1	21 1.5	7 28.05		.00	+0 42.08
Y′	12	3 48		—0.4	23 48.2	4 8 8.85		.00	+1 22.88

Zone 32, Dec. 1, 1857. $\tau = -1^h\ 20^m$ $\theta = 38°$

Letter.	Mag.	Reading of Scale.	d		D
		′ ″	′ ″	″	′ ″
A'_{-3}	11.12	7 37	+19 58.5	—1.2	+27 34.3
A'_{-1}	12.13	2 32		—0.5	22 30.0
A′	12.13	1 47		—0.4	21 45.1
E′		4 52		—0.7	24 49.8
F′		4 34		—0.6	24 31.9
M′		0 2		0.0	20 0.5
T′	8	7 27		—1.1	27 24.4
W′	11	1 2		—0.1	21 0.4
Y′	12	3 49		—0.4	23 47.1

Zone 33, Dec. 1, 1857. $\tau = -1^h\ 12^m$ $\theta = 38°$

Letter.	Mag.	Reading of Scale.	d		D	Transit red. to 1st wire.	a		A
		′ ″	′ ″	″	′ ″	h m s	h m s	s	m
A'_{-3}	11.12	7 38	+19 59.1	—1.2	+27 35.9	4 14 12.25	—4 16 39.47	—.01	—2 27.23
A'_{-1}	9.10	2 33		—0.5	22 31.6	14 30.25		—.01	—2 9.23
C′	11.12	0 16*		—0.1	20 15.0	15 34.15		.00	—1 5.32
D′	12	1 7		—0.2	21 5.9	15 48.00		.00	—0 51.47
E′	12	4 52		—0.7	24 50.4	16 6.40		—.01	—0 33.08
F′	11.12	4 34		—0.6	24 32.5	16 12.80		—.01	—0 26.68
M′	6	0 2		0.0	20 1.1	16 39.45		.00	—0 0.02
T′	8	7 26		—1.1	27 24.0	4 16 53.95		—.01	+0 14.47

Zone 34, Dec. 1, 1857. $\tau = -1^h\ 6^m$ $\theta = 38°$

Letter.	Mag.	Reading of Scale.	d		D	Transit red. to 1st wire.	a		A
		′ ″	′ ″	″	′ ″	h m s	h m s	s	m s
A'_{-3}	11	7 43	+19 52.9	—1.2	+27 34.7	4 20 11.05	—4 22 38.27	—.01	—2 27.23
A'_{-1}	9.10	2 38		—0.5	22 30.4	20 29.10		.00	—2 9.17
C′	11	0 21		—0.1	20 13.8	21 33.00		.00	—1 5.27
D′	11.12	1 14		—0.2	21 6.7	21 46.80		.00	—0 51.47
E′	11.12	4 58		—0.7	24 50.2	22 5.35		—.01	—0 32.93
F′	11	4 40		—0.6	24 32.3	22 11.55		—.01	—0 26.73
M′	6	0 7		0.0	19 59.9	22 38.30		.00	+0 0.03
T′	8	7 32		—1.0	27 23.9	22 52.85		—.01	+0 14.57
W′	10.11	1 7		—0.1	20 59.8	23 20.30		.00	+0 42.03
Y′	12	3 56		—0.4	23 48.5	4 24 1.05		.00	+1 22.78

* The original entry for the scale reading of this star was 0′ 36″.

Zone 35, Dec. 1, 1857. $\tau = -0^h\ 51^m$ $\theta = 38°$

Letter.	Mag.	Reading of Scale.	d		D	Transit red. to 1st wire.	a		A
						h m s	h m s	s	m s
A'_{-3}	11.12	7′ 45″	+19′ 51″.4	—1″.2	+27′ 35″.2	4 37 12.95	—4 39 40.24	—.01	—2 27.30
A'_{-2}	12	8 41		—1.4	28 31.0	37 20.20		—.01	—2 20.05
A'_{-1}	9	2 40		—0.5	22 30.9	37 31.20		.00	—2 9.04
A′	12	1 55		—0.4	21 46.0	37 48.30		.00	—1 51.94
B′	12.13	6 42		—1.0	26 32.4	4 38 17.20		—.01	—1 23.05

Zone 36, Dec. 1, 1857. $\tau = -0^h\ 49^m$ $\theta = 38°$

Letter.	Mag.	Reading of Scale.	d		D	Transit red. to 1st wire.	a		A
						h m s	h m s	s	m s
A'_{-3}	11.12	7′ 44″	+19′ 51″.1	—1″.2	+27′ 33″.9	4 38 59.90	—4 41 27.10	—.01	—2 27.21
A'_{-2}	12	8 42		—1.4	28 31.7	39 7.10		—.01	—2 20.01
A'_{-1}	9	2 41		—0.5	22 31.6	39 17.95		.00	—2 9.15
A′	12	1 57		—0.4	21 47.7	39 35.15		.00	—1 51.95
B′	12.13	6 41		—1.0	26 31.1	4 40 4.10		—.01	—1 23.01

Zone 37, Dec. 1, 1857. $\tau = -0^h\ 39^m$ $\theta = 38°$

Letter.	Mag.	Reading of Scale.	d		D	Transit red. to 1st wire.	a		A
						h m s	h m s	s	m s
C′	11	0′ 24″	+19′ 50″.1	—0″.1	+20′ 14″.0	4 48 15.45	—4 49 20.72	.00	—1 5.27
D′	12	1 17		—0.2	21 6.9	48 29.25		.00	—0 51.47
E′	11.12	5 1		—0.7	24 50.4	48 47.75		—.01	—0 32.98
F′	11	4 42		—0.6	24 31.5	48 54.05		—.01	—0 26.68
G′	12	0 17		0.0	20 7.1	49 0.25		.00	—0 20.47
H′	12	8 42		—1.2	28 30.9	49 9.10		—.01	—0 11.63
T′	8	7 35		—1.1	27 24.0	4 49 35.20		—.01	+0 14.47

Zone 38, Dec. 1, 1857. $\tau = -0^h\ 39^m$ $\theta = 38°$

Letter.	Mag.	Reading of Scale.	d		D	Transit red. to 1st wire.	a		A
						h m s	h m s	s	m s
C′	11	0′ 24″	+19′ 49″.6	—0″.1	+20′ 13″.5	4 50 10.00	—4 51 15.36	—.00	—1 5.36
D′	12	1 16		—0.2	21 5.4	50 23.90		.00	—0 51.46
E′	11.12	5 2		—0.7	24 50.9	50 42.30		—.01	—0 33.07
F′	11	4 44		—0.7	24 32.9	50 48.70		—.01	—0 26.67
G′	12	0 17		0.0	20 6.6	50 54.95		.00	—0 20.41
H′	12	8 42		—1.2	28 30.4	51 3.75		—.01	—0 11.62
T′	8	7 36		—1.1	27 24.5	4 51 29.85		—.01	+0 14.48

Zone 39, Dec. 1, 1857. $\tau = -0^h\ 23^m$ $\theta = 38°$

Letter.	Mag.	Reading of Scale.	d		D	Transit red. to 1st wire.	a		A
						h m s	h m s	s	m s
H′	12	8′ 40″	+19′ 53″.1	—1″.2	+28′ 31″.9	5 5 27.50	—5 5 39.11	—.01	—0 11.62
S′	12	10 21		—1.4	30 12.7	5 49.00		—.01	+0 9.88
T′	8	7 32		—1.1	27 24.0	5 53.60		—.01	+0 14.48
V′	12	9 57		—1.4	29 48.7	5 6 5.00		—.01	+0 25.88

Zone 40, Dec. 1, 1857. $\tau = -0^h\ 21^m$ $\theta = 38°$

Letter.	Mag.	Reading of Scale.	d		D	Transit red. to 1st wire.	a		A
						h m s	h m s	s	m s
H′	12	8′ 37″	+19′ 55″.1	—1″.2	+28′ 30″.9	5 6 51.20	—5 7 2.91	—.01	—0 11.72
S′	12	10 17		—1.4	30 10.7	7 12.80		—.01	+0 9.88
T′	8	7 30		—1.1	27 24.0	7 17.40		—.01	+0 14.48
V′	12	9 53		—1.4	29 46.7	5 7 28.90		—.01	+0 25.98

Zone 41, Dec. 1, 1857. $\tau = -0^h\ 20^m$ $\theta = 38°$

Letter.	Mag.	Reading of Scale.	d	D	Transit red. to 1st wire.	a	A
		′ ″	′ ″ ″	′ ″	h m s	h m s s	m s
H′	12	8 40	+19 53.1—1.2	28 31.9	5 8 11.50	—5 8 23.26—.01	—0 11.77
S′	12	10 22	—1.4	30 13.7	8 33.00	—.01	+0 9.73
T′	8	7 32	—1.1	27 24.0	8 37.75	—.01	+0 14.48
V′	12	9 58	—1.4	29 49.7	5 8 49.15	—.01	+0 25.88

Zone 42, Dec. 1, 1857. $\tau = +0^h\ 11^m$ $\theta = 38°$

Letter.	Mag.	Reading of Scale.	d	D	Transit red. to 1st wire.	a	A
		′ ″	′ ″ ″	′ ″	h m s	h m s s	m s
F′					5 38 36.85	—5 39 3.53+.00	—0 26.68
G′	12	0 11	+19 52.8—0.0	+20 3.8	38 43.15	.00	—0 20.38
O′	12	1 47	—0.2	21 39.6	39 4.00	.00	+0 0.47
Q′	11	1 48	—0.2	21 40.6	39 7.75	.00	+0 4.22
U′	12	3 23	—0.4	23 15.4	39 28.45	.00	+0 24.92
W′	10	1 7	—0.1	20 59.7	39 45.65	.00	+0 42.12
X′	12	0 0	+0.1	19 52.9	40 19.35	.00	+1 15.82
Y′	11	3 57	—0.5	23 49.3	5 40 26.30	.00	+1 22.77

Zone 43, Dec. 1, 1857. $\tau = +0^h\ 14^m$ $\theta = 38°$

Letter.	Mag.	Reading of Scale.	d	D	Transit red. to 1st wire.	a	A
		′ ″	′ ″ ″	′ ″	h m s	h m s s	m s
F′	11	4 42	+19 50.9—0.7	24 32.2	5 40 59.95	—5 41 26.59 .00	—0 26.64
G′	12	0 14	0.0	20 4.9	41 6.15	.00	—0 20.44
O′	12	1 46	—0.2	21 36.7	41 27.10	.00	+0 0.51
Q′	11	1 49	—0.2	21 39.7	41 30.90	.00	+0 4.31
U′	12	3 25	—0.5	23 15.4	41 51.50	.00	+0 24.91
W′	9	1 9	—0.1	20 59.8	42 8.70	.00	+0 42.11
X′	12	0 2	+0.1	19 53.0	42 42.30	.00	+1 15.71
Y′	11	3 58	—0.5	23 48.4	5 42 49.40	.00	+1 22.81

Zone 44, Dec. 1, 1857. $\tau = +0^h\ 20^m$ $\theta = 38°$

Letter.	Mag.	Reading of Scale.	d	D	Transit red. to 1st wire.	a	A
		′ ″	′ ″ ″	′ ″			
A_{-3}	11.12	7 33	+20 2.7—1.2	27 34.5			
A_{-1}	10	2 28	—0.5	22 30.2			
A′	12	1 45	—0.4	21 47.3			
B′	12	6 30	—1.0	26 31.7			
C′	11	0 12	—0.1	20 14.6			
D′	12	1 4	—0.2	21 6.5			

Zone 45, Dec. 1, 1857. $\tau = +0^h\ 25^m$ $\theta = 38°$

Letter.	Mag.	Reading of Scale.	d	D	Transit red. to 1st wire.	a	A
		′ ″	′ ″ ″	′ ″	h m s	h m s s	m s
F′	11.12	5 34	+18 58.1—0.8	24 31.3	5 52 45.95	—5 53 12.58 .00	—0 26.63
L′=U″	10	1 18	—0.1	20 16.0	53 12.25	.00	—0 0.33
T′	8	8 27	—1.1	27 24.0	53 27.00	—.01	+0 14.41
W′	9.10	2 3	—0.2	21 0.9	5 53 54.70	.00	+0 42.12

Zone 46, Dec. 1, 1857. $\tau = +0^h\ 25^m\ \theta = 38°$

Letter.	Mag.	Reading of Scale.	d		D	Transit red. to 1st wire.	a		A
		′ ″	′ ″	″	′ ″				
F′	11.12	5 48	+18 45.1	—0.8	24 32.3				
L′=U″	10	1 31		—0.1	20 16.0				
T′	8	8 39		—1.1	27 23.0				
W′	9.10	2 14		—0.2	20 58.9				

Zone 47, Dec. 1, 1857. $\tau = +0^h\ 29^m\ \theta = 38°$

Letter.	Mag.	Reading of Scale.	d		D	Transit red. to 1st wire.	a		A
		′ ″	′ ″	″	′ ″	h m s	h m s	s	m s
F′	11.12	5 27	+19 6.2	—0.8	24 32.4	5 57 25.00	—5 57 51.58	.00	—0 26.58
L′=U″	10	1 10		—0.2	20 16.0	57 51.25		.00	—0 0.33
T′	8	8 21		—1.1	27 26.1	58 6.20		—.01	+0 14.61
W′	9.10	1 56		—0.2	21 2.0	5 58 33.90		.00	+0 42.32

Zone 48, Dec. 1, 1857. $\tau = +0^h\ 32^m\ \theta = 38°$

Letter.	Mag.	Reading of Scale.	d		D	Transit red. to 1st wire.	a		A
		′ ″	′ ″	″	′ ″	h m s	h m s	s	m s
F′	11.12	5 59	+18 34.2	—0.9	24 32.3	5 59 35.35	—6 0 2.18	—.01	—0 26.84
L′=U″	10	1 42		—0.2	20 16.0	6 0 1.85		.00	—0 0.33
T′	8	8 50		—1.2	27 23.0	0 16.60		—.01	+0 14.41
W′	9.10	2 24		—0.3	20 57.9	6 0 44.15		.00	+0 41.97

Zone 49, Dec. 4, 1857. $\tau = -2^h\ 17^m\ \theta = 23°$

Letter.	Mag.	Reading of Scale.	d		D	Transit red. to 1st wire.	a		A
		′ ″	′ ″	″	′ ″	h m s	h m s	s	m s
C′	11.13*	5 13	+15 4.1	—1.2	20 15.9	3 9 50.35	—3 10 55.62	—.01	—1 5.28
D′	12	6 3		—1.4	21 5.7	10 4.15		—.02	—0 51.49
I″	12	0 45		—0.3	15 48.8	10 9.90		.00	—0 45.72
F′	11	9 32		—2.1	24 34.0	10 29.00		—.02	—0 26.64
R″	9	0 23		—0.1	15 27.0	10 49.30		.00	—0 6.32
L′=U″		5 13		—1.1	20 16.0	10 55.30		—.01	—0 0.33
I″I″	12	—0 9		0.0	14 55.1	11 30.00		.00	+0 34.38
W′	11	5 59		—1.3	21 1.8	11 37.80		—.01	+0 42.17
L″L″	9.10	4 6		—0.8	19 9.3	12 6.55		—.01	+1 10.92
Y′	11.12	8 46		—1.9	23 48.2	3 12 18.50		—.02	+1 22.86

Zone 50, Dec. 4, 1857. $\tau = -2^h\ 13^m\ \theta = 23°$

Letter.	Mag.	Reading of Scale.	d		D	Transit red. to 1st wire.	a		A
		′ ″	′ ″	″	′ ″	h m s	h m s	s	m s
C′	11.13	5 14	+15 1.1	—1.2	20 13.9	3 14 5.55	—3 15 10.72	—.01	—1 5.18
D′	12	6 7		—1.4	21 6.7	14 19.35		—.02	—0 51.39
F′	11	9 32		—2.1	24 31.0	14 44.05		—.02	—0 26.69
R″	9	0 26		—0.1	15 27.0	15 4.35		.00	—0 6.37
L′=U″		5 16		—1.1	20 16.0	15 10.40		—.01	—0 0.33
I″I″	12	—0 7		0.0	14 54.1	15 45.00		.00	+0 34.28
W′	10	6 0		—1.3	20 59.8	15 52.80		—.01	+0 42.07
L″L″	10	4 5		—0.8	19 5.3	16 21.75		—.01	+1 11.02
Y′		8 46		—1.9	23 45.2	3 16 33.60		—.02	+1 22.86

* So in the original record.

Zone 51, Dec. 4, 1857. $\tau = -1^h\ 16^m$ $\theta = 23°$

Letter.	Mag.	Reading of Scale.	d	D	Transit red. to 1st wire.	a	A
		′ ″	′ ″ ″	′ ″	h m s	h m s s	m s
C′	11	5 15	+15 0.2—1.2	20 14.0	4 11 28.75	—4 12 34.02—.01	—1 5.28
D′	11.12	6 8	—1.4	21 6.8	11 42.35	—.01	—0 51.68
I″	11.12	0 48	—0.3	15 47.9	11 48.10	.00	—0 45.92
F′	11	9 33	—2.2	24 31.0	12 7.25	—.02	—0 26.79
R″	9	0 27	—0.1	15 27.1	12 27.80	.00	—0 6.22
L′=U″	9.10	5 17	—1.2	20 16.0	12 33.70	—.01	—0 0.33
E″E″	10	3 11	—0.7	18 10.5	12 49.00	—.01	+0 14.97
I″I″	12	—0 5	0.0	14 55.2	13 8.20	.00	+0 34.18
W′	10	6 1	—1.3	20 59.9	13 15.95	—.01	+0 41.92
L″L″	9.10	4 7	—0.9	19 6.3	13 44.80	—.01	+1 10.77
Y′	11.12	8 50	—1.9	23 48.3	4 13 56.90	—.01	+1 22.87

Zone 52, Dec. 4, 1857. $\tau = -1^h\ 12^m$ $\theta = 23°$

Letter.	Mag.	Reading of Scale.	d	D	Transit red. to 1st wire.	a	A
		′ ″	′ ″ ″	′ ″	h m s	h m s s	m s
C′	11	5 9	+15 7.2—1.2	20 15.0	4 15 6.90	—4 16 12.17—.01	—1 5.28
D′	11.12	6 3	—1.4	21 8.8	15 20.70	—.01	—0 51.48
I″	11.12	0 41	—0.3	15 47.9	15 26.30	.00	—0 45.87
F′	11	9 29	—2.2	24 34.0	15 45.60	—.02	—0 26.59
R″	9	0 21	—0.1	15 28.1	16 5.75	.00	—0 6.42
L′=U″	9.10	5 10	—1.2	20 16.0	16 11.85	—.01	—0 0.33
E″E″	10	3 4	—0.7	18 10.5	16 27.20	.00	+0 15.03
I″I″	12	—0 12	0.0	14 55.2	16 46.55	.00	+0 34.38
W′	10	5 55	—1.3	21 0.9	16 54.15	—.01	+0 41.97
L″L″	10	4 1	—0.9	19 7.3	17 23.00	—.01	+1 10.82
Y′	12	8 42	—1.9	23 47.3	4 17 35.00	—.01	+1 22.82

Zone 53, Dec. 4, 1857. $\tau = -1^h\ 3^m$ $\theta = 23°$

Letter.	Mag.	Reading of Scale.	d	D	Transit red. to 1st wire.	a	A
		′ ″	′ ″ ″	′ ″	h m s	h m s s	m s
C′	11	5 12	+15 3.7—1.2	20 14.5	4 25 7.70	—4 26 12.87—.01	—1 5.18
E″	12	0 58	—0.3	16 1.4	25 16.10	.00	—0 56.77
D′	11.12	6 4	—1.4	21 6.3	25 21.35	—.01	—0 51.53
I″	11.12	0 44	—0.2	15 47.5	25 27.00	.00	—0 45.87
K″	12	3 47	—0.9	18 49.8	25 36.40	—.01	—0 36.48
P″	10.11	3 1	—0.7	18 4.0	25 56.80	—.01	—0 16.08
Q″	12	2 58	—0.7	18 1.0	4 26 2.20	—.01	—0 10.68

Zone 54, Dec. 4, 1857. $\tau = -1^h\ 1^m$ $\theta = 23°$

Letter.	Mag.	Reading of Scale.	d	D	Transit red. to 1st wire.	a	A
		′ ″	′ ″ ″	′ ″	h m s	h m s s	m s
C′	11	5 12	+15 2.9—1.2	20 13.7	4 27 18.65	—4 28 23.86—.01	—1 5.22
E″	12	0 59	—0.3	16 1.6	27 27.15	.00	—0 56.71
D′	11.12	6 5	—1.4	21 6.5	27 32.35	—.01	—0 51.52
I″	11.12	0 46	—0.2	15 48.7	27 38.00	.00	—0 45.86
K″	12	3 47	—0.9	18 49.0	27 47.35	—.01	—0 36.52
P″	10.11	3 2	—0.7	18 4.2	28 7.80	—.01	—0 16.07
Q″	12	2 59	—0.7	18 1.2	4 28 13.20	—.01	—0 10.67

Zone 55, Dec. 4, 1857. $\tau = -0^h\ 54^m$ $\theta = 23°$

Letter.	Mag.	Reading of Scale.	d	D	Transit red. to 1st wire.	a	A
		′ ″	′ ″ ″	′ ″	h m s	h m s s	m s
C′	11	5 13	+15 2.2—1.2	20 14.0	4 34 24.40	—4 35 29.70—.01	—1 5.31
H″	13	4 14	—1.0	19 15.2	34 42.85	—.01	—0 46.86
L″	13.14	3 11	—0.7	18 12.5	34 58.75	—.01	—0 30.96
P″	10.11	3 3	—0.7	18 4.5	4 35 13.60	—.01	—0 16.11

Zone 56, Dec. 4, 1857. $\tau = -0^h\ 52^m$ $\theta = 23°$

Letter.	Mag.	Reading of Scale.	d	D	Transit red. to 1st wire.	a	A
		′ ″	′ ″ ″	′ ″	h m s	h m s s	m s
C′	11	5 16	+14 59.2—1.2	20 14.0	4 35 51.30	—4 36 56.60—.01	—1 5.31
H″	13	4 17	—1.0	19 15.2	36 9.75	—.01	—0 46.86
L″	13.14	3 14	—0.7	18 12.5	36 25.85	—.01	—0 30.76
P″	10.11	3 6	—0.7	18 4.5	4 36 40.50	—.01	—0 16.11

Zone 57, Dec. 4, 1857. $\tau = -0^h\ 46^m$ $\theta = 23°$

Letter.	Mag.	Reading of Scale.	d	D	Transit red. to 1st wire.	a	A
		′ ″	′ ″ ″	′ ″	h m s	h m s s	m s
C′	11	5 15	+15 0.2—1.3	20 13.9	4 42 17.30	—4 43 22.77—.01	—1 5.48
F″	14	3 50	—0.9	18 49.3	42 33.90	—.01	—0 48.88
K″	12	3 49	—0.9	18 48.3	42 46.20	—.01	—0 36.58
G′=O″	12	5 7	—1.2	20 6.0	43 2.30	—.01	—0 20.48
L′=U″	10.11	5 17	—1.2	20 16.0	4 43 22.45	—.01	—0 0.33

Zone 58, Dec. 4, 1857. $\tau = -0^h\ 43^m$ $\theta = 23°$

Letter.	Mag.	Reading of Scale.	d	D	Transit red. to 1st wire.	a	A
		′ ″	′ ″ ″	′ ″	h m s	h m s s	m s
C′	11	5 12	+15 3.2—1.3	20 13.9	4 44 33.00	—4 45 38.32—.01	—1 5.33
F″	14	3 47	—0.9	18 49.3	44 49.40	—.01	—0 48.93
K″	12	3 46	—0.9	18 48.3	45 1.95	—.01	—0 36.38
G′	12	5 3	—1.2	20 5.0	45 18.00	—.01	—0 20.33
L′=U″	10.11	5 14	—1.2	20 16.0	4 45 38.00	—.01	—0 0.33

Zone 59, Dec. 4, 1857. $\tau = -0^h\ 28^m$ $\theta = 23°$

Letter.	Mag.	Reading of Scale.	d	D	Transit red. to 1st wire.	a	A
		′ ″	′ ″ ″	′ ″	h m s	h m s s	m s
C′	11	5 10	+15 5.1—1.2	20 13.9	4 59 1.35	—5 0 6.70—.01	—1 5.36
D″	12.13	—0 12	—0.1	14 53.0	59 5.10	.00	—1 1.60
I″	12	0 43	—0.2	15 47.9	4 59 20.95	.00	—0 45.75
R″	9	0 22	—0.1	15 27.0	5 0 0.40	.00	—0 6.30
A″A″	12	0 41	—0.2	15 45.9	0 16.85	.00	+0 10.15
C″C″	11	2 0	—0.5	17 4.6	0 18.95	.00	+0 12.25
I″I″	11	—0 10	0.0	14 55.1	5 0 41.00	.00	+0 34.30

Zone 60, Dec. 4, 1857. $\tau = -0^h\ 26^m$ $\theta = 23°$

Letter.	Mag.	Reading of Scale.	d	D	Transit red. to 1st wire.	a	A
		′ ″	′ ″ ″	′ ″	h m s	h m s s	m s
C′=C″	11	5 12	+15 2.7—1.2	20 13.5	5 1 46.80	—5 2 52.11—.01	—1 5.32
D″	12.13	—0 10	—0.1	14 52.6	1 50.60	.00	—1 1.51
I″	12	0 46	—0.2	15 48.5	2 6.15	.00	—0 45.96
K_1″	13.14	1 30	—0.4	16 32.3	2 16.15	.00	—0 35.96

Zone 60, Dec. 4, 1857. $\tau = -0^h\ 26^m$ $\theta = 23°$

Letter.	Mag.	Reading of Scale.	d		D	Transit red. to 1st wire.	a		A
		′ ″	′ ″	″	′ ″	h m s	h m s	s	m s
R″	9	0 25	+15 2.7	—0.1	15 27.6	5 2 45.90	—5 2 52.11	—.00	—0 6.21
A″A″	12	0 44		—0.2	15 46.5	3 2.20		.00	+0 10.09
C″C″	11	2 2		—0.5	17 4.2	3 4.25		.00	+0 12.14
I″I″	11	—0 8		0.0	14 54.7	5 3 26.55		.00	+0 34.44

Zone 61, Dec. 4, 1857. $\tau = -0^h\ 19^m$ $\theta = 23°$

Letter	Mag.	′ ″	′ ″	″	′ ″	h m s	h m s	s	m s
C′=C″	11	5 13	+15 1.2	—1.3	20 12.9	5 8 53.85	—5 9 59.17	—.01	—1 5.33
E″	12	1 0		—0.3	16 0.9	9 2.25		.00	—0 56.92
R″	9	0 25		—0.1	15 26.1	9 52.80		.00	—0 6.37
U″=L′	9.10	5 16		—1.2	20 16.0	9 58.85		—.01	—0 0.33
E″E″	10	3 7		—0.7	18 7.5	10 14.00		.00	+0 14.83
N″N″	11.12	0 3		+0.1	15 4.3	11 21.15		.00	+1 21.98
O″O″	11.12	—0 19		+0.1	14 42.3	5 11 31.20		.00	+1 32.03

Zone 62, Dec. 4, 1857. $\tau = -0^h\ 15^m$ $\theta = 23°$

Letter	Mag.	′ ″	′ ″	″	′ ″	h m s	h m s	s	m s
C′	11	5 14	+15 0.2	—1.3	20 12.9	5 12 14.15	—5 13 19.37	—.01	—1 5.23
E″	12	1 0		—0.3	15 59.9				
R″	9	0 26		—0.1	15 26.1	13 13.00		.00	—0 6.37
L′=U″	9.10	5 17		—1.2	20 16.0	13 19.05		—.01	—0 0.33
E″E″	10	3 9		—0.7	18 8.5	13 34.40		.00	+0 15.03
N″N″	11.12	0 3		+0.1	15 3.3	14 41.30		.00	+1 21.93
O″O″	11.12	—0 17		+0.1	14 43.3	5 14 51.60		.00	+1 32.23

Zone 63, Dec. 4, 1857. $\tau = -0^h\ 6^m$ $\theta = 23°$

Letter	Mag.	′ ″	′ ″	″	′ ″	h m s	h m s	s	m s
C′	11	5 11	+15 3.7	—1.2	20 13.5	5 21 8.90	—5 22 14.12	—.01	—1 5.23
K″	12	3 46		—0.9	18 48.8	21 37.75		—.01	—0 36.38
L″	12.13	3 9		—0.7	18 12.0	21 42.95		.00	—0 31.17
O′	12	6 36		—1.5	21 38.2	22 14.70		—.01	+0 0.57
Q′	11	6 37		—1.5	21 39.2	22 18.20		—.01	+0 4.07
L″L″	10	4 3		—0.9	19 5.8	23 24.90		.00	+1 10.78
Y′	12	8 47		—1.9	23 48.8	5 23 36.90		—.01	+1 22.77

Zone 64, Dec. 7, 1857. $\tau = -1^h\ 51^m$ $\theta = 40°$

Letter	Mag.	′ ″	′ ″	″	′ ″	h m s	h m s	s	m s
P″	11	3 3	+15 1.7	—0.4	18 4.3	3 36 45.20	—3 37 1.32	—.01	—0 16.13
Q″	12	3 1		—0.4	18 2.3	36 50.95		—.01	—0 10.38
L′=U″		5 15		—0.7	20 16.0	37 1.00		—.01	—0 0.33
C″C″	12	2 3		—0.3	17 4.4	37 13.75		.00	+0 12.43
E″E″	11	3 10		—0.4	18 11.3	3 37 16.60		—.01	+0 15.27

Zone 65, Dec. 7, 1857. $\tau = -1^h\ 50^m$ $\theta = 40°$

Letter.	Mag.	Reading of Scale.	d	D	Transit red. to 1st wire.	a	A
		′ ″	′ ″ ″	′ ″	h m s	h m s s	m s
P″	11	3 4	+15 0.7—0.4	18 4.3	3 37 55.35	—3 38 11.47—.01	—0 16.13
Q″	12	3 2	—0.4	18 2.3	38 0.95	—.01	—0 10.53
L′=U″		5 16	—0.7	20 16.0	38 11.15	—.01	—0 0.33
C″C″	11	2 4	—0.3	17 4.4	38 23.70	.00	+0 12.23
E″E″	10.11	3 9	—0.4	18 9.3	3 38 26.70	—.01	+0 15.22

Zone 66, Dec. 7, 1857. $\tau = -1^h\ 49^m$ $\theta = 40°$

Letter.	Mag.	Reading of Scale.	d	D	Transit red. to 1st wire.	a	A
					h m s	h m s s	m s
P″	11				3 38 51.25	—3 39 7.37—.01	—0 16.13
Q″	12				38 56.90	—.01	—0 10.48
L′=U″					39 7.05	—.01	—0 0.33
C″C″	11				39 19.70	.00	+0 12.33
E″E″	10.11				3 39 22.50	—.01	+0 15.12

Zone 67, Dec. 7, 1857. $\tau = -1^h\ 44^m$ $\theta = 40°$

Letter.	Mag.	Reading of Scale.	d	D	Transit red. to 1st wire.	a	A
		′ ″	′ ″ ″	′ ″	h m s	h m s s	m s
R″	10.11	0 29	+14 58.5—0.1	15 27.4	3 44 4.00	—3 44 10.30 .00	—0 6.30
Y″	6	3 27	—0.5	18 25.0	44 16.95	—.01	+0 6.64
B″B″	8	3 23	—0.4	18 21.1	44 20.35	—.01	+0 10.04
W′	10	6 2	—0.7	20 59.8	3 44 52.35	—.01	+0 42.04

Zone 68, Dec. 7, 1857. $\tau = -1^h\ 43^m$ $\theta = 40°$

Letter.	Mag.	Reading of Scale.	d	D	Transit red. to 1st wire.	a	A
		′ ″	′ ″ ″	′ ″	h m s	h m s s	m s
R″	10.11	0 22	+15 4.4—0.0	15 26.4	3 45 14.25	—3 45 20.63 .00	—0 6.38
Y″	6	3 21	—0.4	18 25.0	45 27.00	—.01	+0 6.36
B″B″	8	3 18	—0.4	18 22.0	45 30.75	—.01	+0 10.11
W′	10	5 57	—0.7	21 0.7	3 46 2.75	—.01	+0 42.11

Zone 69, Dec. 7, 1857. $\tau = -1^h\ 40^m$ $\theta = 40°$

Letter.	Mag.	Reading of Scale.	d	D
N″	10	2′ 49″	+10′ 7″.2—0″.4	12′ 55″.8
R″	10.11	5 20	—0.6	15 26.6
A″A″	13	5 40	—0.6	15 46.6
H″H″	12.13	5 7	—0.6	15 13.6
I″I″	11.12	4 46	—0.5	14 52.7
K″K″	13	5 36	—0.6	15 42.6

Zone 70, Dec. 7, 1857. $\tau = -1^h\ 38^m$ $\theta = 40°$

Letter.	Mag.	Reading of Scale.	d	D
N″	10.11	2′ 48″	+10′ 7″.4—0″.4	12′ 55″.0
R″	10.11	5 20	—0.6	15 26.8
A″A″	13	5 41	—0.6	15 47.8
H″H″	12.13	5 7	—0.6	15 13.8
I″I″	12	4 47	—0.5	14 53.9
K″K″	13	5 35	—0.6	15 41.8

Zone 71 a, Dec. 10, 1857. $\tau = -0^h\ 43^m$ $\theta = 34°$

Letter.	Mag.	Reading of Scale.	d	D	Transit red. to 1st wire.	a		A
					h m s	h m s	s	m s
R″					4 44 21.50	—4 44 27.80	.00	—0 6.30
A″A″					44 37.90		.00	+0 10.10
H″H″					44 53.70		.00	+0 25.90
I″I″					45 2.10		.00	+0 34.30
K″K″					45 21.20		.00	+0 53.40
N″N″					45 49.80		.00	+1 22.00
O″O″					4 45 59.90		.00	+1 32.10

Zone 71 b, Dec. 10, 1857. $\tau = -0^h\ 41^m$ $\theta = 34°$

Letter.	Mag.	Reading of Scale.	d	D	Transit red. to 1st wire.	a		A
		′ ″		′ ″	h m s	h m s	s	m s
R″	10	5 19	+10′ 8.4″ —0.9″	15 26.5	4 46 46.75	—4 46 53.10	.00	—0 6.35
A″A″	13	5 39	—1.0	15 46.4	47 3.00		.00	+0 9.90
H″H″	13.12	5 7	—0.8	15 14.6	47 18.90		.00	+0 25.80
I″I″	12	4 46	—0.7	14 53.7	47 27.35		.00	+0 34.25
K″K″	13	5 34	—0.9	15 41.5	47 46.50		.00	+0 53.40
N″N″	11.12	4 58	—0.7	15 5.7	48 15.15		.00	+1 22.05
O″O″	11.12	4 34	—0.7	14 41.7	4 48 25.25		.00	+1 32.15

Zone 72, Dec. 10, 1857. $\tau = -0^h\ 34^m$ $\theta = 34°$

Letter.	Mag.	Reading of Scale.	d	D	Transit red. to 1st wire.	a		A
		′ ″		′ ″	h m s	h m s	s	m s
A''_{-1}	11	4 40	+10′ 8.3″ —0.9″	14 47.4	4 53 19.40	—4 55 13.77	.00	—1 54.37
A″	12	6 3	—1.1	16 10.2	53 34.90		.00	—1 38.87
B″=B‴	12	0 10	—0.1	10 18.2	53 47.25		.00	—1 26.52
E″	12	5 54	—1.1	16 1.2	54 17.00		.00	—0 56.77
I″	12	5 40	—1.0	15 47.3	54 27.90		.00	—0 45.87
M″=M‴	11	0 0	0.0	10 8.3	54 50.00		.00	—0 23.77
N″	10.11				54 53.20		.00	—0 20.57
R″	10	5 20	—0.9	15 27.4	55 7.30		.00	—0 6.47
F″F″=L‴	12	0 9	0.0	10 17.3	55 29.25		.00	+0 15.48
I″I″	12	4 46	—0.8	14 53.5	55 48.05		.00	+0 34.28
L″L″	10	9 0	—1.4	19 6.9	56 24.70		—.01	+1 10.92
O″O″	12	4 34	—0.6	14 41.7	56 46.00		.00	+1 32.23
P″P″	12.13	1 19	—0.1	11 27.2	4 57 17.90		.00	+2 4.13

Zone 73, Dec. 10, 1857. $\tau = -0^h\ 28^m$ $\theta = 34°$

Letter.	Mag.	Reading of Scale.	d	D	Transit red. to 1st wire.	a		A
		′ ″		′ ″	h m s	h m s	s	m s
A''_{-1}	11	4 39	+10′ 9.8″ —0.9″	14 47.9	4 58 58.20	—5 0 52.63	.00	—1 54.43
A″	12	6 1	—1.1	16 9.7	59 13.90		.00	—1 38.73
B″=B‴	12	0 10	—0.1	10 19.7	59 26.00		.00	—1 26.63
E″	12	5 52	—1.1	16 0.7	4 59 55.85		.00	—0 56.78
I″	12	5 40	—1.0	15 48.8	5 0 6.80		.00	—0 45.83
M″=M‴	10.11				0 28.85		.00	—0 23.78
N″	10				0 32.00		.00	—0 20.63
R″	12	5 18	—0.9	15 26.9	0 46.20		.00	—0 6.43
F″F″	12	0 7	0.0	10 16.8	1 8.15		.00	+0 15.52
I″I″	12	4 45	—0.8	14 54.0	1 26.90		.00	+0 34.27
L″L″	10	8 58	—1.4	19 6.4	2 3.45		—.01	+1 10.81
O″O″	12	4 32	—0.6	14 41.2	2 24.90		.00	+1 32.27
P″P″	12.13	1 18	—0.1	11 27.7	5 2 56.85		.00	+2 4.22

Zone 74, Dec. 10, 1857. $\tau = -0^h\ 22^m$ $\theta = 34°$

Letter.	Mag.	Reading of Scale.	d		D	Transit red. to 1st wire.	a		A
		′ ″	′ ″	″	′ ″	h m s	h m s	s	m s
A''_{-1}	11	4 34	+10 14.1	—0.9	+14 47.2	5 4 31.40	—5 6 25.87	.00	—1 54.47
A″	12	5 57		—1.1	16 10.0	4 46.90		.00	—1 38.97
B″=B‴	11	0 7		—0.1	10 21.0	4 59.25		.00	—1 26.62
E″	12	5 48		—1.0	16 1.1	5 29.00		.00	—0 56.87
I″	12	5 35		—1.0	15 48.1	5 40.00		.00	—0 45.87
M″	10.11	—0 6		0.0	10 8.1	6 2.05		.00	—0 23.82
N″	10	2 42		—0.4	12 55.7	6 5.20		.00	—0 20.67
R″	12	5 14		—0.9	15 27.2	6 19.50		.00	—0 6.37
F″F″	12	0 3		0.0	10 17.1	6 41.35		.00	+0 15.48
I″I″	12	4 40		—0.8	14 53.3	7 0.15		.00	+0 34.28
L″L″	10	8 54		—1.4	19 6.7	7 36.75		.00	+1 10.88
O″O″	12	4 28		—0.6	14 41.5	7 58.20		.00	+1 32.33
P″P″	12.13	1 13		—0.1	11 27.0	5 8 30.10		.00	+2 4.23

Zone 75, Dec. 10, 1857. $\tau = -0^h\ 6^m$ $\theta = 34°$

Letter.	Mag.	Reading of Scale.	d		D	Transit red. to 1st wire.	a		A
		′ ″	′ ″	″	′ ″	h m s	h m s	s	m s
B″=B‴	11	0 11	+10 6.5	—0.1	+10 17.4	5 20 39.95	—5 22 6.41	.00	—1 26.46
D″	13	4 46		—0.9	14 51.6	21 4.85		.00	—1 1.56
G″	13	1 10		—0.2	11 16.3	21 18.10		.00	—0 48.31
Z″	13	2 30		—0.4	12 36.1	22 14.35		.00	+0 7.94
G″G″	13	2 7		—0.3	12 13.2	22 23.00		.00	+0 16.59
L″L″	10	9 2		—1.4	19 7.1	23 17.30		—.01	+1 10.88
M″M″=X′	12	9 47		—1.6	19 51.9	5 23 22.15		—.01	+1 15.73

Zone 76, Dec. 10, 1857. $\tau = -0^h\ 2^m$ $\theta = 34°$

Letter.	Mag.	Reading of Scale.	d		D	Transit red. to 1st wire.	a		A
		′ ″	′ ″	″	′ ″	h m s	h m s	s	m s
B″=B‴	11	0 12	+10 6.5	—0.1	+10 18.4	5 24 54.45	—5 26 20.91	.00	—1 26.46
D″	13	4 46		—0.9	14 51.6	25 19.50		.00	—1 1.41
G″	13	1 7		—0.2	11 13.3	25 32.75		.00	—0 48.16
Z″	13	2 30		—0.4	12 36.1	26 28.95		.00	+0 8.04
G″G″	13	2 6		—0.3	12 12.2	26 37.50		.00	+0 16.59
L″L″	10	9 2		—1.4	19 7.1	27 31.80		—.01	+1 10.88
M″M″=X′	12	9 48		—1.6	19 52.9	27 36.80		—.01	+1 15.88

Zone 77, Dec. 10, 1857. $\tau = +0^h\ 4^m$ $\theta = 34°$

Letter.	Mag.	Reading of Scale.	d		D	Transit red. to 1st wire.	a		A
		′ ″	′ ″	″	′ ″	h m s	h m s	s	m s
M″	11	—0 9	+10 15.7	0.0	+10 6.7	5 32 16.00	—5 32 39.77	.00	—0 23.77
N″	10	2 38		—0.5	12 53.2	32 19.00		.00	—0 20.77
U″=L′		10 2		—1.7	20 16.0	32 39.45		—.01	—0 0.33
P‴=D″D″	13	0 11		0.0	10 26.7	32 53.80		.00	+0 14.03
Q‴=F″F″	12	—0 2		0.0	10 13.7	32 55.15		.00	+0 15.38
K″K″	13	5 25		—0.8	15 39.9	5 33 33.15		.00	+0 53.38

Zone 78, Dec. 10, 1857. $\tau = +0^h\ 7^m$ $\theta = 34°$

Letter.	Mag.	Reading of Scale.	d		D	Transit red. to 1st wire.	a		A
		′ ″	′ ″	″	′ ″	h m s	h m s	s	m s
M″=M‴	11	—0 10	+10 17.7	0.0	+10 7.7	5 34 23.75	—5 34 47.38	.00	—0 23.63
N″	10	2 37		—0.5	12 54.2	34 26.80		.00	—0 20.58

Zone 78, Dec. 10, 1857. $\tau=+0^h\ 7^m$ $\theta=34°$

Letter.	Mag.	Reading of Scale.	d		D	Transit red. to 1st wire.	a		A
		′ ″	′ ″	″	′ ″	h m s	h m s	s	m s
U″=L′		10 0	+10 17.7	—1.7	+20 16.0	5 34 47.05	—5 34 47.38	.00	—0 0.33
P‴	13	0 11		0.0	10 28.7	35 1.50		.00	+0 14.12
Q‴	12	—0 2		0.0	10 15.7	35 3.00		.00	+0 15.62
K″K″	13	5 24		—0.8	15 40.9	5 35 40.90		.00	+0 53.52

Zone 79, Dec. 10, 1857. $\tau=+0^h\ 23^m$ $\theta=34°$

Letter.	Mag.	Reading of Scale.	d		D	Transit red. to 1st wire.	a		A
		′ ″	′ ″	″	′ ″	h m s	h m s	s	m s
A‴	11	3 7	+ 5 9.4	—0.6	+ 8 15.8	5 49 20.40	—5 51 31.11	.00	—2 10.71
A″₋₁	11	9 40		—1.8	14 47.6	49 36.95		—.01	—1 54.17
B″=B‴	12.11	5 9		—0.9	10 17.5	50 4.75		.00	—1 26.36
C‴	9	1 36		—0.3	6 45.1	50 10.75		.00	—1 20.36
E‴	11.10	1 4		—0.2	6 13.2	50 29.80		.00	—1 1.31
F‴	12	1 13		—0.2	6 22.2	50 32.00		.00	—0 59.11
L‴	8	3 52		—0.7	9 0.7	51 2.40		.00	—0 28.71
M‴	11	4 59		—0.9	10 7.5	51 7.35		.00	—0 23.76
P‴	13	5 21		—0.9	10 29.5	51 45.05		.00	+0 13.94
Q‴	12	5 7		—0.8	10 15.6	51 46.80		.00	+0 15.69
S‴	7	0 37		—0.1	5 46.3	52 9.70		.00	+0 38.59
V‴	6.7	—0 27		+0.1	4 42.5	52 30.75		.00	+0 59.64
N″N″	12	9 56		—1.6	15 3.8	52 53.10		—.01	+1 21.98
O″O″	12	9 33		—1.5	14 40.9	53 3.25		—.01	+1 32.13
P″P″	12.13	6 19		—0.9	11 27.5	5 53 35.15		.00	+2 4.04

Zone 80, Dec. 10, 1857. $\tau=+0^h\ 29^m$ $\theta=34°$

Letter.	Mag.	Reading of Scale.	d		D	Transit red. to 1st wire.	a		A
		′ ″	′ ″	″	′ ″	h m s	h m s	s	m s
A‴	11	3 8	+ 5 8.9	—0.6	+ 8 16.3	5 54 51.80	—5 57 2.48	.00	—2 10.68
A″₋₁	11	9 40		—1.8	14 47.1	55 8.10		.00	—1 54.38
B″=B‴	11.12	5 9		—0.9	10 17.0	55 35.95		.00	—1 26.53
C‴	9	1 36		—0.3	6 44.6	55 41.90		.00	—1 20.58
E‴	10.11	1 6		—0.2	6 14.7	56 1.00		.00	—1 1.48
F‴	12	1 14		—0.3	6 22.6	56 3.45		.00	—0 59.03
L‴	8	3 53		—0.7	9 1.2	56 33.75		.00	—0 28.73
M‴	11	4 59		—0.9	10 7.0	56 38.70		.00	—0 23.78
P‴	13	5 22		—0.9	10 30.0	57 16.35		.00	+0 13.87
Q‴	12	5 7		—0.8	10 15.1	57 18.00		.00	+0 15.52
S‴	7.8	0 36		—0.1	5 44.8	57 41.00		.00	+0 38.52
V‴	7	—0 27		+0.1	4 42.0	58 2.00		.00	+0 59.52
N″N″	12	9 57		—1.6	15 4.3	58 24.55		.00	+1 22.07
O″O″	12	9 34		—1.5	14 41.4	58 34.75		.00	+1 32.27
P″P″	12.13	6 20		—0.9	11 28.0	5 59 6.60		.00	+2 4.12

Zone 81, Dec. 10, 1857. $\tau=+0^h\ 42^m$ $\theta=34°$

Letter.	Mag.	Reading of Scale.	d		D	Transit red. to 1st wire.	a		A
		′ ″	′ ″	″	′ ″	h m s	h m s	s	m s
A‴	11	8 7	+ 0 11.5	—1.5	+ 8 17.0	6 8 17.75	—6 10 28.51	.00	—2 10.76
C‴ 1 wire	8	6 36		—1.2	6 46.3	9 7.90		.00	—1 20.61
D‴ 1 "	10.11	4 6		—0.8	4 16.7	9 8.10		.00	—1 20.41
G‴	12	4 1		—0.7	4 11.8	9 32.85		.00	—0 55.66
H‴	10	0 7		—0.1	0 18.4	9 46.60		.00	—0 41.91

Zone 81, Dec. 10, 1857. $\tau=+0^h\ 32^m$ $\theta=34°$

Letter.	Mag.	Reading of Scale.	d	D	Transit red. to 1st wire.	a		A
		′ ″	′ ″ ″	′ ″	h m s	h m s	s	m s
I‴	10	1 24	+ 0 11.5 —0.3	+ 1 35.2	6 9 47.85	—6 10 28.51	.00	—0 40.66
O‴	12	3 57	—0.7	4 7.8	10 29.15		.00	+0 0.64
R‴	12.11	2 49	—0.4	3 0.1	11 2.90		.00	+0 34.39
T‴	11	3 19	—0.4	3 30.1	11 12.00		.00	+0 43.49
U‴	11	1 20	—0.1	1 31.4	11 19.50		.00	+0 50.99
W‴	12	0 21	0.0	0 32.5	11 43.95		.00	+1 15.44
Y‴	12	3 6	—0.4	3 17.1	6 12 26.75		.00	+1 58.24

Zone 82, Dec. 10, 1857. $\tau=+0^h\ 48^m$ $\theta=34°$

Letter.	Mag.	Reading of Scale.	d	D	Transit red. to 1st wire.	a		A
		′ ″	′ ″ ″	′ ″	h m s	h m s	s	m s
A‴	11	8 9	+ 0 10.0 —1.5	+ 8 17.5	6 14 22.35	—6 16 33.10	.00	—2 10.75
C‴ 1 wire	8	6 37	—1.2	6 45.8	15 12.60		.00	—1 20.50
D‴ 1 "	10	4 5	—0.8	4 14.2	15 12.90		.00	—1 20.20
G‴ 1 "	12	4 3	—0.7	4 12.3	15 37.60		.00	—0 55.50
H‴	10	0 7	0.0	0 17.0	15 51.30		.00	—0 41.80
I‴	10	1 26	—0.3	1 35.7	15 52.65		.00	—0 40.45
O‴	12	3 58	—0.7	4 7.3	16 33.90		.00	+0 0.80
R‴	11.12	2 49	—0.4	2 58.6	17 7.60		.00	+0 34.50
T‴	11	3 20	—0.5	3 29.5	6 17 16.75		.00	+0 43.65

Zone 83, Dec. 10, 1857. $\tau=+0^h\ 55^m$ $\theta=34°$

Letter.	Mag.	Reading of Scale.	d	D	Transit red. to 1st wire.	a		A
		′ ″	′ ″ ″	′ ″	h m s	h m s	s	m s
A‴	11	8 11	+ 0 8.2 —1.5	+ 8 17.7	6 21 16.00	—6 23 26.79	.00	—2 10.79
B″=B‴	11	10 11	—1.8	10 17.4	22 0.15		.00	—1 26.64
C‴	8	6 38	—1.2	6 45.0	22 6.25		.00	—1 20.54
E‴	12	6 7	—1.1	6 14.1	22 25.35		.00	—1 1.44
L‴	8	8 55	—1.5	9 1.7	22 58.00		.00	—0 28.79
M‴	10.11	10 1	—1.7	10 7.5	23 3.00		.00	—0 23.79
Q‴	12	10 9	—1.6	10 15.6	23 42.40		.00	+0 15.61
S‴	7	5 39	—0.9	5 46.3	24 5.25		.00	+0 38.46
V‴	7	4 33	—0.7	4 40.5	24 26.20		.00	+0 59.41
Y‴	11.12	3 9	—0.4	3 16.8	6 25 24.85		.00	+1 58.06

Zone 84, Dec. 10, 1857. $\tau=+1^h\ 0^m$ $\theta=34°$

Letter.	Mag.	Reading of Scale.	d	D	Transit red. to 1st wire.	a		A
		′ ″	′ ″ ″	′ ″	h m s	h m s	s	m s
A‴	11	8 12	+ 0 6.8 —1.5	+ 8 17.3	6 26 24.05	—6 28 34.76	.00	—2 10.71
B″=B‴	11	10 12	—1.8	10 17.0	27 8.15		.00	—1 26.61
C‴	8	6 40	—1.2	6 45.6	27 14.20		.00	—1 20.56
E‴	11.12	6 8	—1.1	6 13.7	27 33.30		.00	—1 1.46
L‴	8	8 56	—1.5	9 1.3	28 6.00		.00	—0 28.76
M‴	10.11	10 2	—1.7	10 7.1	28 11.05		.00	—0 23.71
Q‴	12	10 11	—1.6	10 16.2	28 50.25		.00	+0 15.49
S‴	7	5 40	—0.9	5 45.9	29 13.30		.00	+0 38.54
V‴	6.7	4 35	—0.7	4 41.1	29 34.30		.00	+0 59.54
Y‴	11.12	3 11	—0.4	3 17.4	6 30 33.05		.00	+1 58.29

Zone 85, Dec. 12, 1857. $\tau = -0^h\ 58^m$ $\theta = 24°$

Letter.	Mag.	Reading of Scale.	d		D	Transit red. to 1st wire.	a		A
		′ ″	′ ″	″	′ ″	h m s	h m s	s	m s
M‴=M″	12	—0 8	+10 16.3	—0.0	+10 8.3	4 30 35.00	—4 30 58.72	.00	—0 23.72
N″	10.11	2 39		—0.7	12 54.6	30 38.05		.00	—0 20.67
U″=L″		10 2		—2.3	20 16.0	4 30 58.40		—.01	—0 0.33

Zone 86, Dec. 12, 1857. $\tau = -0^h\ 57^m$ $\theta = 24°$

Letter.	Mag.	Reading of Scale.	d		D	Transit red. to 1st wire.	a		A
		′ ″	′ ″	″	′ ″	h m s	h m s	s	m s
M‴=M″	12	—0 15	+10 24.3	0.0	+10 9.3	4 31 22.50	—4 31 46.22	.00	—0 23.72
N″	10.11	2 33		—0.6	12 56.7	31 25.65		.00	—0 20.57
U″=L′		9 54		—2.3	20 16.0	4 31 45.90		—.01	—0 0.33

Zone 87, Dec. 12, 1857. $\tau = -0^h\ 56^m$ $\theta = 24°$

Letter.	Mag.	Reading of Scale.	d		D	Transit red. to 1st wire.	a		A
		′ ″	′ ″	″	′ ″	h m s	h m s	s	m s
M‴=M″	12	—0 2	+10 10.3	0.0	+10 8.3	4 32 22.10	—4 32 45.92	.00	—0 23.82
N″	10.11	2 47		—0.7	12 56.6	32 25.15		.00	—0 20.77
U″=L′		10 8		—2.3	20 16.0	4 32 45.60		—.01	—0 0.33

Zone 88, Dec. 12, 1857. $\tau = -0^h\ 53^m$ $\theta = 24°$

Letter.	Mag.	Reading of Scale.	d		D	Transit red. to 1st wire.	a		A
		′ ″	′ ″	″	′ ″	h m s	h m s	s	m s
M‴=M″	12	—0 7	+10 15.3	0.0	+10 8.3	4 33 26.00	—4 33 49.67	.00	—0 23.67
N″	10.11	2 41		—0.7	12 55.6	33 29.05		.00	—0 20.62
U″=L′		10 3		—2.3	20 16.0	4 33 49.35		—.01	—0 0.33

Zone 89, Dec. 12, 1857. $\tau = -0^h\ 43^m$ $\theta = 24°$

Letter.	Mag.	Reading of Scale.	d		D	Transit red. to 1st wire.	a		A
		′ ″	′ ″	″	′ ″	h m s	h m s	s	m s
H‴	10.11	0 17	+ 0 1.4	—0.1	+ 0 18.3	4 45 45.20	—4 46 27.06	.00	—0 41.86
I‴	10.11	1 36		—0.4	1 37.0	45 46.65		.00	—0 40.41
M‴=M″	12	10 9		—2.3	10 8.1	4 46 3.35		—.01	—0 23.72

Zone 90, Dec. 12, 1857. $\tau = -0^h\ 42^m$ $\theta = 24°$

Letter.	Mag.	Reading of Scale.	d		D	Transit red. to 1st wire.	a		A
		′ ″	′ ″	″	′ ″	h m s	h m s	s	m s
H‴	10.11	0 18	+ 0 2.4	—0.1	+ 0 20.3	4 46 27.00	—4 47 8.71	.00	—0 41.71
I‴	10.11	1 36		—0.4	1 38.0	46 28.30		.00	—0 40.41
M‴=M″	12	10 8		—2.3	10 8.1	4 46 45.00		—.01	—0 23.72

Zone 91, Dec. 12, 1857. $\tau = -0^h\ 41^m$ $\theta = 24°$

Letter.	Mag.	Reading of Scale.	d		D	Transit red. to 1st wire.	a		A
		′ ″	′ ″	″	′ ″	h m s	h m s	s	m s
H‴	10.11	0 23	— 0 2.6	—0.1	+ 0 20.3	4 47 23.00	—4 48 4.61	.00	—0 41.61
I‴	10.11	1 42		—0.5	1 38.9	47 24.10		.00	—0 40.51
M‴=M″	12	10 13		—2.3	10 8.1	4 47 40.90		—.01	—0 23.72

Zone 92, Dec. 12, 1857. $\tau = -0^h\ 40^m$ $\theta = 24°$

Letter.	Mag.	Reading of Scale.	d	D	Transit red. to 1st wire.	a		A
					h m s	h m s	s	m s
H‴	10.11	0′ 19″	+ 0′ 1″.4—0″.1	+ 0′ 20″.3	4 48 26.70	—4 49 8.46	.00	—0 41.76
I‴	10.11	1 38	—0.5	1 38.9	48 27.95		.00	—0 40.51
M‴ = M″	12	10 9	—2.3	10 8.1	4 48 44.75		—.01	—0 23.72

Zone 93, Dec. 12, 1857. $\tau = -0^h\ 24^m$ $\theta = 24°$

Letter.	Mag.	Reading of Scale.	d	D	Transit red. to 1st wire.	a		A
					h m s	h m s	s	m s
I‴	10.11	1′ 33″	+ 0′ 4″.8—0″.4	+ 1′ 37″.4	5 3 54.75	—5 4 35.20	.00	—0 40.45
K‴	13	3 2	—0.7	3 6.1	3 59.95		.00	—0 35.25
N‴	13	3 43	—0.8	3 47.0	4 32.85		.00	—0 2.35
O‴	12	4 6	—0.9	4 9.9	5 4 35.90		.00	+0 0.70

Zone 94, Dec. 12, 1857. $\tau = -0^h\ 23^m$ $\theta = 24°$

Letter.	Mag.	Reading of Scale.	d	D	Transit red. to 1st wire.	a		A
					h m s	h m s	s	m s
I‴	10.11	1′ 34″	+ 0′ 4″.1—0″.4	+ 1′ 37″.7	5 5 32.40	—5 6 12.86	.00	—0 40.46
K‴	13	3 2	—0.7	3 5.4	5 37.80		.00	—0 35.06
N‴ 1 wire	13	3 43	—0.8	3 46.3	6 10.30		.00	—0 2.56
O‴	12	4 6	—0.9	4 9.2	5 6 13.60		.00	+0 0.74

Zone 95, Dec. 12, 1857. $\tau = -0^h\ 11^m$ $\theta = 24°$

Letter.	Mag.	Reading of Scale.	d	D	Transit red. to 1st wire.	a		A
					h m s	h m s	s	m s
A‴	11	8′ 15″	+ 0′ 6″.2—2″.0	+ 8′ 19″.2	5 15 23.90	—5 17 34.52	—.01	—2 10.63
D‴	11	4 10	—1.0	4 15.2	16 14.30		.00	—1 20.22
E‴	11	6 10	—1.5	6 14.7	16 33.15		.00	—1 1.37
F‴	12	6 19	—1.5	6 23.7	16 35.45		.00	—0 59.07
L‴	9	8 57	—2.1	9 1.1	17 5.75		—.01	—0 28.78
M‴ = M″	11.12	10 4	—2.3	10 7.9	17 10.80		—.01	—0 23.73
O‴	12	4 3	—0.9	4 8.3	17 35.25		.00	+0 0.73
Q‴	12	10 11	—2.3	10 14.9	17 50.05		—.01	+0 15.52
R‴	12	2 56	—0.6	3 1.6	18 8.95		.00	+0 34.43
V‴	6.7	4 36	—1.0	4 41.2	18 34.00		.00	+0 59.48
X‴	12.13	7 11	—1.5	7 15.7	19 21.20		.00	+1 46.68
Y‴	11	3 14	—0.6	3 19.6	5 19 32.65		.00	+1 58.13

Zone 96, Dec. 12, 1857. $\tau = -0^h\ 5^m$ $\theta = 24°$

Letter.	Mag.	Reading of Scale.	d	D	Transit red. to 1st wire.	a		A
					h m s	h m s	s	m s
A‴	11	8′ 15″	+ 0′ 5″.4—2″.0	+ 8′ 18″.4	5 22 21.75	—5 24 32.61	—.01	—2 10.87
D‴	11	4 12	—1.0	4 16.4	23 12.30		.00	—1 20.31
E‴	11	6 12	—1.5	6 15.9	23 31.00		.00	—1 1.61
F‴	12	6 21	—1.5	6 24.9	23 33.40		.00	—0 59.21
L‴	9	8 58	—2.1	9 1.3	24 3.85		—.01	—0 28.77
M‴ = M″	12	10 5	—2.3	10 8.1	24 8.90		—.01	—0 23.72
R‴	12	2 58	—0.6	3 2.8	25 7.00		.00	+0 34.39
V‴	6.7	4 37	—1.0	4 41.4	25 32.00		.00	+0 59.39
Y‴	11.12	3 15	—0.6	3 19.8	5 26 30.65		.00	+1 58.04

SECTION I. PART II.

APPROXIMATE POSITIONS OF BRIGHTER STARS WITHIN 20′ DEC. OF θ ORIONIS.

Star's Name.	ξ_a	η_a
	m s	′ ″
A_{-2}	—2 24.26	+33 32.3
A_{-1}	—1 43.36	+38 3.2
A	—1 30.88	+33 26.8
B	—1 8.38	+32 51.6
C	—1 1.89	+31 21.9
D	—0 43.13	+40 10.8
E	—0 37.24	+33 10.7
F	—0 35.08	+36 3.4
G	—0 16.79	+38 14.2
H	—0 14.02	+37 42.6
I	—0 13.09	+35 9.1
K	—0 10.88	+31 6.0
L	+0 0.68	+39 32.0
M	+0 4.02	+34 9.0
N	+0 4.26	+31 13.0
O	+0 9.06	+34 46.3
P=S′	+0 9.79	+30 11.2
Q	+0 16.88	+37 8.5
R	+0 18.68	+31 5.5
T	+0 30.63	+33 0.4
U	+0 34.51	+35 22.2
V	+0 35.28	+38 27.2
V_1=v_4	+0 47.16	+36 22.1
W	+0 53.41	+39 59.5
X	+1 0.06	+31 54.9
Y	+1 21.48	+35 51.0
Z	+1 26.16	+36 57.0
AA	+1 52.01	+33 9.0
BB	+2 22.28	+38 36.4
A'_{-3}	—2 27.22	+27 34.7
A'_{-2}	—2 20.03	+28 31.4
A'_{-1}	—2 9.14	+22 31.0
A′	—1 51.97	+21 46.6
B′	—1 23.02	+26 32.0
C′=C″	—1 5.29	+20 14.0
D′	—0 51.50	+21 6.4
E′	—0 33.04	+24 50.4
F′	—0 26.69	+24 32.3
G′=O″	—0 20.42	+20 5.6
H′	—0 11.67	+28 31.0
M′	+0 0.03	+20 0.5
O′	+0 0.52	+21 38.2
Q′	+0 4.20	+21 39.8
R′	+0 9.49	+28 12.5
T′	+0 14.50	+27 23.9
U′	+0 24.92	+23 15.4
V′=S	+0 25.94	+29 47.5
v'_3		+26 32.1
W′	+0 42.08	+21 0.3
X′=M″M″	+1 15.78	+19 52.8
Y′	+1 22.82	+23 48.0
A''_{-1}	—1 54.37	+14 47.4
A″	—1 38.86	+16 10.0
B″=B‴	—1 26.54	+10 18.4
D″	—1 1.52	+14 52.4
E″	—0 56.80	+16 1.0
F″	—0 48.90	+18 49.3
G″	—0 48.23	+11 14.8
H″	—0 46.86	+19 15.2
I″	—0 45.85	+15 48.1
K″	—0 36.47	+18 48.9
K''_1	—0 35.96	+16 32.3
L″	—0 30.99	+18 12.3
M″=M‴	—0 23.76	+10 7.9
N″	—0 20.64	+12 55.3
P″	—0 16.10	+18 4.3
Q″	—0 10.57	+18 1.6
R″	—0 6.35	+15 27.0
U″=L′	—0 0.33	+20 16.0
Y″	+0 6.50	+18 25.0
Z″	+0 7.99	+12 36.1
A″A″	+0 10.06	+15 46.6
B″B″	+0 10.08	+18 21.5
C″C″	+0 12.26	+17 4.4
E″E″	+0 15.07	+18 9.6
G″G″	+0 16.59	+12 12.7

Star's Name.	ξ_a	η_a
	m s	′ ″
H″H″	+0 25.85	+15 14.0
I″I″	+0 34.30	+14 54.1
K″K″	+0 53.42	+15 41.5
L″L″	+1 10.86	+19 6.7
N″N″	+1 22.01	+15 4.3
O″O″	+1 32.20	+14 41.7
P″P″	+2 4.15	+11 27.5
A‴	—2 10.73	+ 8 17.2
C‴	—1 20.52	+ 6 45.3
D‴	—1 20.28	+ 4 15.6
E‴	—1 1.43	+ 6 14.3
F‴	—0 59.09	+ 6 23.1
G‴	—0 55.60	+ 4 12.0
H‴	—0 41.79	+ 0 18.8
I‴	—0 40.50	+ 1 37.0
K‴	—0 35.15	+ 3 5.8
L‴	—0 28.75	+ 9 1.2
N‴	—0 2.45	+ 3 46.6
O‴	+0 0.72	+ 4 8.3
P‴=D″D″	+0 13.98	+10 28.9
Q‴=F″F″	+0 15.53	+10 16.0
R‴	+0 34.43	+ 3 0.4
S‴	+0 38.53	+ 5 45.8
T‴	+0 43.57	+ 3 29.8
U‴	+0 50.99	+ 1 31.4
V‴	+0 59.51	+ 4 41.5
W‴	+1 15.44	+ 0 32.5
X‴	+1 46.68	+ 7 15.7
Y‴	+1 58.16	+ 3 18.0

SECTION I. PART III.

EQUATIONS OF CONDITION.

I, Dec. only.

$+0''.5 = D - (I)$
$-1.0 = L - (I)$
$-0.7 = W - (I)$

II, Dec. only.

$0''.0 = D - (II)$
$-0.2 = L - (II)$
$-0.1 = W - (II)$

III, Dec. only.

$+0''.2 = A_{-1} - (III)$
$+0.8 = G - (III)$
$+0.4 = H - (III)$
$-0.1 = Q - (III)$

IV, Dec. only.

$-0''.4 = A_{-1} - (IV)$
$+0.1 = G - (IV)$
$+0.1 = H - (IV)$
$+0.2 = Q - (IV)$
$+0.7 = V - (IV)$

V, AR. only.

$-0^{s}.07 = A_{-1} - (V)$
$-0.05 = D - (V)$
$-0.09 = G - (V)$
$-0.05 = H - (V)$
$-0.05 = L - (V)$
$+0.05 = Q - (V)$
$0.00 = V - (V)$
$+0.02 = W - (V)$

VI, AR. only.

$-0^{s}.06 = A_{-1} - (VI)$
$+0.06 = D - (VI)$
$+0.02 = G - (VI)$
$-0.04 = H - (VI)$
$-0.09 = L - (VI)$
$-0.09 = Q - (VI)$
$+0.03 = W - (VI)$

VII, AR. only.

$+0^{s}.09 = A_{-1} - (VII)$
$-0.04 = D - (VII)$
$+0.07 = G - (VII)$
$+0^{s}.01 = H - (VII)$
$+0.11 = L - (VII)$
$+0.01 = Q - (VII)$
$+0.06 = V - (VII)$
$+0.03 = W - (VII)$

4.

$+0^{s}.10 = A_{-2} - (4) \quad -0''.1$
$-0.10 = A_{-1} - (4) \quad +0.3$
$+0.02 = A - (4) \quad -0.6$
$+0.03 = C - (4) \quad +0.7$

5.

$-0^{s}.02 = A_{-2} - (5) \quad +0''.4$
$-0.02 = A_{-1} - (5) \quad +0.8$
$0.00 = A - (5) \quad -0.1$
$-0.04 = C - (5) \quad +1.2$
$-0.04 = D - (5) \quad +0.9$
$-0.05 = F - (5) \quad +1.0$
$+0.07 = G - (5)$
$+0.05 = K - (5) \quad -0.8$
$-0.05 = Q - (5) \quad -0.2$

6.

$+0^{s}.03 = A_{-2} - (6) \quad -0''.6$
$+0.08 = A_{-1} - (6) \quad -0.2$
$-0.05 = A - (6) \quad +1.0$
$+0.06 = C - (6) \quad +0.2$
$+0.06 = D - (6) \quad -0.1$
$+0.11 = F - (6) \quad +1.0$
$0.00 = K - (6) \quad -0.8$
$0.00 = Q - (6) \quad -0.2$
$+0.03 = W - (6) \quad +0.4$
$+0.05 = Y - (6) \quad +0.6$
$+0.12 = Z - (6) \quad +0.4$

7.

$-0^{s}.03 = A_{-2} - (7) \quad +1''.6$
$-0.13 = A_{-1} - (7) \quad +1.0$
$-0.06 = A - (7) \quad +0.2$
$0.00 = C - (7) \quad -0.6$
$+0.05 = D - (7) \quad -1.8$
$0.00 = F - (7) \quad +0.2$
$+0.04 = K - (7) \quad +0.4$
$+0.04 = Q - (7) \quad +1.0$
$-0.03 = W - (7) \quad +0.6$
$-0^{s}.06 = Y - (7) \quad +0''.8$
$-0.09 = Z - (7) \quad +1.6$
$0.00 = BB - (7) \quad 0.0$

8.

$0^{s}.00 = C - (8) \quad -1''.1$
$-0.10 = D - (8) \quad -0.3$
$-0.05 = E - (8) \quad +0.9$
$-0.05 = F - (8) \quad -1.3$
$-0.01 = K - (8) \quad +0.9$

9.

$0^{s}.00 = C - (9) \quad 0''.0$
$+0.05 = D - (9) \quad +1.8$
$0.00 = E - (9) \quad +1.0$
$-0.05 = F - (9) \quad -0.2$
$-0.01 = K - (9) \quad 0.0$

10.

$-0^{s}.09 = C - (10) \quad -0''.5$
$+0.07 = D - (10) \quad -1.7$
$-0.02 = G - (10)* \quad -0.8$
$-0.09 = H - (10)* \quad -1.1$
$+0.07 = I - (10)* \quad +1.8$
$+0.05 = K - (10) \quad +0.5$

11, AR. only.

$-0^{s}.09 = C - (11)$
$-0.08 = D - (11)$
$-0.22 = G - (11)*$
$-0.09 = H - (11)*$
$-0.13 = I - (11)*$
$+0.05 = K - (11)$

12.

$+0^{s}.03 = C - (12) \quad +0''.6$
$+0.14 = D - (12) \quad +0.4$
$+0.05 = G - (12)* \quad +1.4$
$+0.28 = H - (12)* \quad +1.0$
$+0.04 = I - (12)* \quad -0.1$
$-0.03 = K - (12) \quad -0.4$

13.

$+0^{s}.04 = A_{-2} - (13) \quad -0''.5$
$+0.19 = A_{-1} - (13) \quad -1.1$

* In these cases the equation for AR. has half weight.

—0^{s}.04=A	—(13)	+0$''$.1
+0.03=C	—(13)	—0.6
—0.04=D	—(13)	0.0
—0.04=F	—(13)	+1.2
+0.02=K	—(13)	+0.4
0.00=L	—(13)	+1.0
+0.05=Q	—(13)	—1.1
—0.08=W	—(13)	—0.5
0.00=Y	—(13)	—0.3

14.

—0^{s}.13=A_{-2}	—(14)	—0$''$.7
—0.03=A_{-1}	—(14)	—0.3
+0.04=A	—(14)	—0.1
+0.01=C	—(14)	+0.2
+0.01=E	—(14)	+0.1
—0.01=F	—(14)	—0.1
+0.03=H	—(14)	—0.5
+0.03=L	—(14)	—0.2
+0.08=Q	—(14)	—0.3
0.00=W	—(14)	—0.7
—0.02=Y	—(14)	—1.5

15.

+0^{s}.11=F	—(15)	+0$''$.3
—0.03=G	—(15)	—0.9
—0.03=I	—(15)*	—0.3
—0.03=K	—(15)	+0.5
—0.02=N	—(15)	—0.5
—0.05=P	—(15)	+0.4
+0.01=R	—(15)	+1.0

16.

0^{s}.00=F	—(16)	—0$''$.3
+0.16=G	—(16)*	—0.4
+0.01=I	—(16)	—0.8
—0 04=K	—(16)	0.0
+0.02=N	—(16)	0.0
—0.01=P	—(16)	—2.1
0.00=R	—(16)	—0.5

17.

0^{s}.00=F	—(17)	—2$''$.2
+0.01=G	—(17)*	—0.4
+0.02=I	—(17)	—0.8
+0.01=K	—(17)	0.0
—0.08=N	—(17)	0.0
—0.01=P	—(17)	—2.0
0.00=R	—(17)	—0.5

18.

—0^{s}.01=K	—(18)	—0$''$.3
—0.01=M	—(18)	+0.2
—0.03=Q	—(18)	+1.2
—0.12=T	—(18)	0.0
—0.08=V	—(18)*	+0.3

19.

—0^{s}.01=K	—(19)	—0$''$.3
—0.01=M	—(19)	+0.2
+0.02=Q	—(19)	+0.2
+0.13=T	—(19)	0.0
+0.07=V	—(19)	+0.3

20.

—0^{s}.04=K	—(20)	+0$''$.1
+0.06=M	—(20)	—0.4
+0.07=O	—(20)	+0.2
—0.03=U	—(20)	+0.2
+0.02=V_1	—(20)	+0.2

21.

+0^{s}.02=K	—(21)	—0$''$.3
—0.08=M	—(21)	+0.2
—0.07=O	—(21)	—0.2
+0.03=U	—(21)	—0.2
—0.02=V_1	—(21)*	—0.2

22.

—0^{s}.01=Q	—(22)	—0$''$.7
—0.11=V	—(22)	—0.6
+0.01=W	—(22)	+0.9
+0.05=Y	—(22)	+0.1
+0.07=Z	—(22)	—2.1
+0-07=AA	—(22)	—0.4

23.

—0^{s}.01=Q	—(23)	+0$''$.2
+0.04=V	—(23)	—0.7
+0.01=W	—(23)	—0.2
0.00=Y	—(23)	0.0
—0.08=Z	—(23)	—0.2
—0.03=AA	—(23)	—0.5

24.

+0^{s}.07=A	—(24)	—1$''$.0
—0.08=B	—(24)	—0.7
—0.01=C	—(24)	—0.7
—0.11=E′	—(24)	—1.1
—0.01=F′	—(24)	+1.0
—0.02=K	—(24)	—0.7
0.00=T′	—(24)	—1.0
—0.16=X	—(24)	+0.3

25.

+0^{s}.03=A	—(25)	0$''$.0
+0.03=B	—(25)	+0.3
+0.05=C	—(25)	+0.2
0.00=E′	—(25)	—0.1
—0.05=F′	—(25)	+0.1
—0.06=K	—(25)	+0.2
+0.03=M	—(25)	—0.3
+0^{s}.06=T′	—(25)	0$''$.0
+0.15=X	—(25)	—0.7

26.

0^{s}.00=A	—(26)	+1$''$.1
+0.05=B′	—(26)	+0.1
+0.02=C	—(26)	+1.3
+0.07=E′	—(26)	+2.0
—0.03=F′	—(26)	
+0.01=K	—(26)	+1.3
+0.03=T′	—(26)	+0.1
—0.03=AA	—(26)	+1.1

27.

+0^{s}.03=A	—(27)	—0$''$.9
—0.07=B′	—(27)	+0.1
+0.03=B	—(27)	+0.4
0.00=C	—(27)	+0.3
+0.04=E	—(27)	—1.7
+0.10=N	—(27)	+0.3
+0.02=P	—(27)	+0.3

27 b, Dec. only.

—0$''$.2=H′	—(27 b)
—0·6=K	—(27 b)
+0.3=R′	—(27 b)
+0.1=T′	—(27 b)
—0.9=V′	—(27 b)

27 c, Dec. only.

+0$''$.8=H′	—(27 c)
+0.4=K	—(27 c)
+0.3=R′	—(27 c)
+0·1=T′	—(27 c)
+0.1=V′	—(27 c)

27 d, Dec. only.

+0$''$.4=K	—(27 d)
—0.6=N	—(27 d)
+0.1=T′	—(27 d)
0.0=v'_3	—(27 d)
+1.4=X	—(27 d)

27 e, Dec. only.

+0$''$.4=K	—(27 e)
+0.3=N	—(27 e)
+0.1=T′	—(27 e)
—0.6=X	—(27 e)

28 a.

—0^{s}.16=H′	—(28 a)	—1$''$.2
—0.10=K	—(28 a)	—1.5
+0.08=R′	—(28 a)	—0.6
—0.02=T′	—(28 a)	+0.1
+0.03=V′	—(28 a)	—1.8

28 b, AR. only.

s		
—0.11=H′	—(28 b)	
0.00=K	—(28 b)	
—0.07=R′	—(28 b)	
—0.02=T′	—(28 b)	
+0.03=V′	—(28 b)	

29.

s		″
—0.05=K	—(29)	—0.5
+0.01=N	—(29)	+0.4
—0.02=T′	—(29)	+0.1
—0.14=X	—(29)	+0.5

30.

s		″
0.00=K	—(30)	+1.5
—0.04=N	—(30)	+0.5
—0.02=T′	—(30)	+0.1
+0.06=X	—(30)	—0.5

31.

s		″
+0.09=A'_{-3}	—(31)	—0.3
+0.06=A'_{-1}	—(31)	+1.1
—0.05=A′	—(31)	+0.6
+0.04=B′	—(31)	+0.6
—0.03=C′	—(31)	
—0.02=D′	—(31)	0.0
—0.04=E′	—(31)	
—0.04=F′	—(31)	—0.3
+0.09=H′	—(31)	
0.00=M′	—(31)	+0.1
+0.02=T′	—(31)	—1.3
0.00=W′	—(31)	+1.2
+0.06=Y′	—(31)	+0.2

32, Dec. only.

″	
—0.4=A'_{-3}	—(32)
—1.0=A'_{-1}	—(32)
—1.5=A′	—(32)
—0.6=E′	—(32)
—0.4=F′	—(32)
0.0=M′	—(32)
+0.5=T′	—(32)
+0.1=W′	—(32)
—0.9=Y′	—(32)

33.

s		″
—0.01=A'_{-3}	—(33)	+1.2
—0.09=A'_{-1}	—(33)	+0.6
—0.03=C′	—(33)	+1.0
+0.03=D′	—(33)	—0.5
—0.04=E′	—(33)	0.0
+0.01=F′	—(33)	+0.2
—0.01=M′	—(33)	+0.6
—0.03=T′	—(33)	+0.1

34.

s		″
—0.01=A'_{-3}	—(34)	0.0
—0.03=A'_{-1}	—(34)	—0.6
+0.02=C′	—(34)	—0.2
+0.03=D′	—(34)	+0.3
+0.11=E′	—(34)	—0.2
—0.04=F′	—(34)	0.0
0.00=M′	—(34)	—0.6
+0.07=T′	—(34)	0.0
—0.05=W′	—(34)	—0.5
—0.04=Y′	—(34)	+0.5

35.

s		″
—0.08=A'_{-3}	—(35)	+0.5
—0.02=A'_{-2}	—(35)	—0.4
+0.10=A'_{-1}	—(35)	—0.1
+0.03=A′	—(35)	—0.6
—0.03=B′	—(35)	+0.4

36.

s		″
+0.01=A'_{-3}	—(36)	—0.8
+0.02=A'_{-2}	—(36)	+0.3
—0.01=A'_{-1}	—(36)	+0.6
+0.02=A′	—(36)	+1.1
+0.01=B′	—(36)	—0.9

37.

s		″
+0.02=C′	—(37)	0.0
+0.03=D′	—(37)	+0.5
+0.06=E′	—(37)	0.0
+0.01=F′	—(37)	—0.8
—0.05=G′	—(37)	+1.5
+0.04=H′	—(37)	—0.1
—0.03=T′	—(37)	+0.1

38.

s		″
—0.07=C′	—(38)	—0.5
+0.04=D′	—(38)	—1.0
—0.03=E′	—(38)	+0.5
+0.02=F′	—(38)	+0.6
+0.01=G′	—(38)	+1.0
+0.05=H′	—(38)	—0.6
—0.02—T′	—(38)	+0.6

39.

s		″
+0.05=H′	—(39)	+0.9
+0.09=P	—(39)*	+1.5
—0.02=T′	—(39)*	+0.1
—0.06=V′	—(39)	+1.2

40.

s		″
—0.05=H′	—(40)	—0.1
+0.09=P	—(40)	—0.5
—0.02=T′	—(40)*	+0.1
+0.04=V′	—(40)	—0.8

41.

s		″
—0.10=H′	—(41)	+0.9
—0.06=P	—(41)	+2.5
—0.02=T′	—(41)	+0.1
—0.06=V′	—(41)	+2.2

42.

s		″
+0.01=F′	—(42)	
+0.04=G′	—(42)	—1.8
—0.05=O′	—(42)*	+1.4
+0.02=Q′	—(42)	+0.8
0.00=U′	—(42)	0.0
+0.04=W′	—(42)	—0.6
+0.04=X′	—(42)	+0.1
—0.05=Y′	—(42)	+1.3

43.

s		″
+0.05=F′	—(43)	—0.1
—0.02=G′	—(43)	—0.7
—0.01=O′	—(43)	—1.5
+0.11=Q′	—(43)*	—0.1
—0.01=U′	—(43)	0.0
+0.03=W′	—(43)	—0.5
—0.07=X′	—(43)	+0.2
—0.01=Y′	—(43)	+0.4

44, Dec. only.

″	
—0.2=A'_{-3}	—(44)
—0.8=A'_{-1}	—(44)
+0.7=A′	—(44)
—0.3=B′	—(44)
+0.6=C′	—(44)
+0.1=D′	—(44)

45.

s		″
+0.06=F′	—(45)	—1.0
0.00=L′	—(45)	0.0
—0.09=T′	—(45)	+0.1
+0.04=W′	—(45)	+0.6

46, Dec. only.

″	
0.0=F′	—(46)
0.0=L′	—(46)
—0.9=T′	—(46)
—1.4=W′	—(46)

47.

s		″
+0.11=F′	—(47)	+0.1
0.00=L′	—(47)	0.0
+0.11=T′	—(47)	+2.2
+0.24=W′	—(47)	+1.7

48.

s		″
—0.15=F′	—(48)	0.0

s		″
0.00=L′	—(48)	0.0
—0.09=T′	—(48)	—0.9
—0.11=W′	—(48)	—2.4

49.

s		″
+0.01=C′	—(49)	+1.9
+0.01=D′	—(49)	—0.7
+0.13=I″	—(49)	+0.7
+0.05=F′	—(49)	+1.7
+0.03=R″	—(49)	0.0
0.00=L′	—(49)	0.0
+0.08=I″I″	—(49)	+1.0
+0.09=W′	—(49)	+1.5
+0.06=L″L″	—(49)	+2.6
+0.04=Y′	—(49)	+0.2

50.

s		″
+0.11=C′	—(50)	—0.1
+0.11=D′	—(50)	+0.3
0.00=F′	—(50)	—1.3
—0.02=R″	—(50)	0.0
0.00=L′	—(50)	0.0
—0.02=I″I″	—(50)	0.0
—0.01=W′	—(50)	—0.5
+0.16=L″L″	—(50)	—1.4
+0.04=Y′	—(50)	—2.8

51.

s		″
+0.01=C′	—(51)	0.0
—0.18=D′	—(51)	+0.4
—0.07=I″	—(51)	—0.2
—0.10=F′	—(51)	—1.3
+0.13=R″	—(51)	+0.1
0.00=L′	—(51)	0.0
—0.10=E″E″	—(51)	+0.9
—0.12=I″I″	—(51)	+1.1
—0.16=W′	—(51)	—0.4
—0.09—L″L″	—(51)	—0.4
+0.05=Y′	—(51)	+0.3

52.

s		″
+0.01=C′	—(52)	+1.0
+0.02=D′	—(52)	+2.4
—0.02=I″	—(52)	—0.2
+0.10=F′	—(52)	+1.7
—0.07=R″	—(52)	+1.1
0.00=L′	—(52)	0.0
—0.04=E″E″	—(52)	+0.9
+0.08=I″I″	—(52)	+1.1
—0.11=W′	—(52)	+0.6
—0.04=L″L″	—(52)	+0.6
0.00=Y′	—(52)	—0.7

53.

s		″
+0.11=C′	—(53)	+0.5
+0.03=E″	—(53)	+0.4
—0.03=D′	—(53)	—0.1
—0.02=I″	—(53)	—0.6
—0.01=K″	—(53)	+0.9
+0.02=P″	—(53)	—0.3
—0.11=Q″	—(53)	—0.6

54.

s		″
+0.07=C′	—(54)	—0.3
+0.09=E″	—(54)	+0.6
—0.02=D′	—(54)	+0.1
—0.01=I″	—(54)	+0.6
—0.05=K″	—(54)	+0.1
+0.03=P″	—(54)	—0.1
—0.10=Q″	—(54)	—0.4

55.

s		″
—0.02=C′	—(55)	0.0
0.00=H″	—(55)	0.0
+0.03=L″	—(55)	+0.2
—0.01=P″	—(55)	+0.2

56.

s		″
—0.02=C′	—(56)	0.0
0.00=H″	—(56)	0.0
+0.23=L″	—(56)	+0.2
—0.01=P″	—(56)	+0.2

57.

s		″
—0.19=C′	—(57)	—0.1
+0.02=F″	—(57)	0.0
—0.11=K″	—(57)	—0.6
—0.06=G′	—(57)	+0.4
0.00=L′	—(57)	0.0

58.

s		″
—0.04=C′	—(58)	—0.1
—0.03=F″	—(58)	0.0
+0.09=K″	—(58)	—0.6
+0.09=G′	—(58)	—0.6
0.00=L′	—(58)	0.0

59.

s		″
—0.07=C′	—(59)	—0.1
—0.08=D″	—(59)*	+0.6
+0.10=I″	—(59)	—0.2
+0.05=R″	—(59)	0.0
+0.09=A″A″	—(59)	—0.7
—0.01=C″C″	—(59)	+0.2
0.00=I″I″	—(59)	+1.0

60.

s		″
—0.03=C′	—(60)	—0.5
+0.01=D″	—(60)*	+0.2
—0.11=I″	—(60)	+0.4
0.00=K_1″	—(60)	0.0
+0.14=R″	—(60)	+0.6
+0.03=A″A″	—(60)	—0.1
—0.12=C″C″	—(60)	—0.2
+0.14=I″I″	—(60)	+0.6

61.

s		″
—0.04=C′	—(61)	—1.1
—0.12=E″	—(61)	—0.1
—0.02=R″	—(61)	—0.9
0.00=U″	—(61)	0.0
—0.24=E″E″	—(61)	—2.1
—0.03=N″N″	—(61)	0.0
—0.17=O″O″	—(61)	+0.6

62.

s		″
+0.06=C′	—(62)	—1.1
=E″	—(62)	—1.1
—0.02=R″	—(62)*	—0.9
0.00=L′	—(62)	0.0
—0.04=E″E″	—(62)	—1.1
—0.08=N″N″	—(62)	—1.0
+0.03=O″O″	—(62)	+1.6

63.

s		″
+0.06=C′	—(63)	—0.5
+0.09=K″	—(63)	—0.1
—0.18=L″	—(63)	—0.3
+0.05=O′	—(63)	0.0
—0.13=Q′	—(63)*	—0.6
—0.08=L″L″	—(63)	—0.9
—0.05=Y′	—(63)	+0.8

64.

s		″
—0.03=P″	—(64)	0.0
+0.19=Q″	—(64)	+0.7
0.00=L′	—(64)	0.0
+0.17=C″C″	—(64)	0.0
+0.20=E″E″	—(64)	+1.7

65.

s		″
—0.03=P″	—(65)	0.0
+0.04=Q″	—(65)	+0.7
0.00=L′	—(65)	0.0
—0.03=C″C″	—(65)	0.0
+0.15=E″E″	—(65)	—0.3

66, AR. only.

s	
—0.03=P″	—(66)
+0.09=Q″	—(66)
0.00=L′	—(66)
+0.07=C″C″	—(66)*
+0.05=E″E″	—(66)

67.

s		″
+0.05=R″	−(67)	+0.4
+0.14=Y″	−(67)	0.0
−0.04=B″B″	−(67)	−0.4
−0.04=W′	−(67)	−0.5

68.

s		″
−0.03=R″	−(68)	−0.6
−0.14=Y″	−(68)	0.0
+0.03=B″B″	−(68)	+0.5
+0.03=W′	−(68)	+0.4

69, Dec. only.

″	
+0.5=N″	−(69)
−0.4=R″	−(69)
0.0=A″A″	−(69)
−0.4=H″H″	−(69)
−1.4=I″I″	−(69)
+1.1=K″K″	−(69)

70, Dec. only.

″	
−0.3=N″	−(70)
−0.2=R″	−(70)
+1.2=A″A″	−(70)
−0.2=H″H″	−(70)
−0.2=I″I″	−(70)
+0.3=K″K″	−(70)

71 b.

s		″
0.00=R″	−(71 b)	−0.5
−0.16=A″A″	−(71 b)	−0.2
−0.05=H″H″	−(71 b)	+0.6
−0.05=I″I″	−(71 b)	−0.4
−0.02=K″K″	−(71 b)	0.0
+0.04=N″N″	−(71 b)	+1.4
−0.05=O″O″	−(71 b)	0.0

71 a, AR. only.

s	
+0.05=R″	−(71 a)
+0.04=A″A″	−(71 a)
+0.05=H″H″	−(71 a)
0.00=I″I″	−(71 a)
−0.02=K″K″	−(71 a)
−0.01=N″N″	−(71 a)
−0.10=O″O″	−(71 a)

72.

s		″
0.00=A″$_{-1}$	−(72)	0.0
−0.01=A″	−(72)	+0.2
+0.02=B″	−(72)	−0.2
+0.03=E″	−(72)	+0.2
−0.02=I″	−(72)	−0.8
−0.01=M″	−(72)	+0.4
+0.07=N″	−(72)	
−0.12=R″	−(72)	+0.4
−0.05=F″F″	−(72)	+1.3
−0.02=I″I″	−(72)	−0.6
+0.06=L″L″	−(72)	+0.2
+0.03=O″O″	−(72)	0.0
−0.02=P″P″	−(72)	−0.3

73.

s		″
−0.06=A″$_{-1}$	−(73)	+0.5
+0.13=A″	−(73)	−0.3
−0.09=B″	−(73)	+1.3
+0.02=E″	−(73)	−0.3
+0.02=I″	−(73)	+0.7
−0.02=M″	−(73)	
+0.01=N″	−(73)	
−0.08=R″	−(73)	−0.1
−0.01=F″F″	−(73)	+0.8
−0.03=I″I″	−(73)	−0.1
−0.05=L″L″	−(73)	−0.3
+0.07=O″O″	−(73)	−0.5
+0.07=P″P″	−(73)	+0.2

74.

s		″
−0.10=A″$_{-1}$	−(74)	−0.2
−0.11=A″	−(74)	0.0
−0.08=B″	−(74)	+2.6
−0.07=E″	−(74)	+0.1
−0.02=I″	−(74)	0.0
−0.06=M″	−(74)	+0.2
−0.03=N″	−(74)	+0.4
−0.02=R″	−(74)	+0.2
−0.05=F″F″	−(74)	+1.1
−0.02=I″I″	−(74)	−0.8
+0.02=L″L″	−(74)	0.0
+0.13=O″O″	−(74)	−0.2
+0.08=P″P″	−(74)	−0.5

75.

s		″
+0.08=B″	−(75)	−1.0
−0.04=D″	−(75)	−0.8
−0.08=G″	−(75)	+1.5
−0.05=Z″	−(75)	0.0
0.00=G″G″	−(75)	+0.5
+0.02=L″L″	−(75)	+0.4
−0.05=X′	−(75)	−0.9

76.

s		″
+0.08=B″	−(76)	0.0
+0.11=D″	−(76)	−0.8
+0.07=G″	−(76)	−1.5
+0.05=Z″	−(76)	0.0
0.00=G″G″	−(76)	−0.5
+0.02=L″L″	−(76)	+0.4
+0.10=X′	−(76)	+0.1

77.

s		″
−0.01=M″	−(77)	−1.2
−0.13=N″	−(77)	−2.1
0.00=U″	−(77)	0.0
+0.05=P‴	−(77)	−2.2
−0.15=Q‴	−(77)	−2.3
−0.04=K″K″	−(77)	−1.6

78.

s		″
+0.13=M″	−(78)	−0.2
+0.06=N″	−(78)	−1.1
0.00=U″	−(78)	0.0
+0.14=P‴	−(78)	−0.2
+0.09=Q‴	−(78)	−0.3
+0.10=K″K″	−(78)	−0.6

79.

s		″
+0.02=A‴	−(79)	−1.4
+0.20=A″$_{-1}$	−(79)	+0.2
+0.18=B″	−(79)	−0.9
+0.16=C‴	−(79)	−0.2
+0.12=E‴	−(79)	−1.1
−0.02=F‴	−(79)	−0.9
+0.04=L‴	−(79)	−0.5
0.00=M‴	−(79)	−0.4
−0.04=P‴	−(79)	+0.6
+0.16=Q‴	−(79)	−0.4
+0.06=S‴	−(79)	+0.5
+0.13=V‴	−(79)	+1.0
−0.03=N″N″	−(79)	−0.5
−0.07=O″O″	−(79)	−0.8
−0.11=P″P″	−(79)	0.0

80.

s		″
+0.05=A‴	−(80)	−0.9
−0.01=A″$_{-1}$	−(80)	−0.3
+0.01=B″	−(80)	−1.4
−0.06=C‴	−(80)	−0.7
−0.05=E‴	−(80)	+0.4
+0.06=F‴	−(80)	−0.5
+0.02=L‴	−(80)	0.0
−0.02=M‴	−(80)	−0.9
−0.11=P‴	−(80)	+1.1
−0.01=Q‴	−(80)	−0.9
−0.01=S‴	−(80)	−1.0
+0.01=V‴	−(80)	+0.5
+0.06=N″N″	−(80)	0.0
+0.07=O″O″	−(80)	−0.3
−0.03=P″P″	−(80)	+0.5

81.

s		″
−0.03=A‴	−(81)	−0.2
−0.09=C‴	−(81)*	+1.0
−0.13=D‴	−(81)*	+1.1
−0.06=G‴	−(81)	−0.2

s		″
−0.12=H‴	−(81)	−0.4
−0.16=I‴	−(81)	−1.8
−0.08=O‴	−(81)	−0.5
−0.04=R‴	−(81)	−0.3
−0.08=T‴	−(81)	+0.3
0.00=U‴	−(81)	0.0
0.00=W‴	−(81)	0.0
+0.08=Y‴	−(81)	−0.9

82.

s		″
−0.02=A‴	−(82)	+0.3
+0.02=C‴	−(82)*	+0.5
+0.08=D‴	−(82)*	−1.4
+0.10=G‴	−(82)*	+0.3
−0.01=H‴	−(82)	−1.8
+0.05=I‴	−(82)	−1.3
+0.08=O‴	−(82)	−1.0
+0.07=R‴	−(82)	−1.8
+0.08=T‴	−(82)	−0.3

83.

s		″
−0.06=A‴	−(83)	+0.5
−0.10=B″	−(83)	−1.0
−0.02=C‴	−(83)	−0.3
−0.01=E‴	−(83)	−0.2
−0.04=L‴	−(83)	+0.5
−0.03=M‴	−(83)	−0.4
+0.08=Q‴	−(83)	−0.4
−0.07=S‴	−(83)	+0.5
−0.10=V‴	−(83)	−1.0
−0.10=Y‴	−(83)	−1.2

84.

s		″
+0.02=A‴	−(84)	+0.1
−0.07=B″	−(84)	−1.4
−0.04=C‴	−(84)	+0.3
−0.03=E‴	−(84)	−0.6
−0.01=L‴	−(84)	+0.1
+0.05=M‴	−(84)	−0.8
−0.04=Q‴	−(84)	+0.2
+0.01=S‴	−(84)	+0.1
+0.03=V‴	−(84)	−0.4
+0.13=Y‴	−(84)	−0.6

85.

s		″
+0.04=M‴	−(85)*	+0.4
−0.03=N″	−(85)	−0.7
0.00=U″	−(85)	0.0

86.

s		″
+0.04=M‴	−(86)*	+1.4
+0.07=N″	−(86)	+1.4
0.00=U″	−(86)	0.0

87.

s		″
−0.06=M‴	−(87)*	+0.4
−0.13=N″	−(87)	+1.3
0.00=U″	−(87)	0.0

88.

s		″
+0.09=M‴	−(88)*	+0.4
+0.02=N″	−(88)	+0.3
0.00=U″	−(88)	0.0

89.

s		″
−0.07=H‴	−(89)	−0.5
+0.09=I‴	−(89)	0.0
+0.04=M‴	−(89)	+0.2

90.

s		″
+0.08=H‴	−(90)	+1.5
+0.09=I‴	−(90)	+1.0
+0.04=M‴	−(90)	+0.2

91.

s		″
+0.18=H‴	−(91)	+1.5
−0.01=I‴	−(91)	+1.9
+0.04=M‴	−(91)	+0.2

92.

s		″
+0.03=H‴	−(92)	+1.5
−0.01=I‴	−(92)	+1.9
+0.04=M‴	−(92)	+0.2

93.

s		″
+0.05=I‴	−(93)	+0.4
−0.10=K‴	−(93)	+0.3
+0.10=N‴	−(93)	+0.4
−0.02=O‴	−(93)	+1.6

94.

s		″
+0.04=I‴	−(94)	+0.7
+0.09=K‴	−(94)	−0.4
−0.11=N‴	−(94)*	−0.3
+0.02=O‴	−(94)	+0.9

95.

s		″
+0.10=A‴	−(95)	+2.0
+0.06=D‴	−(95)	−0.4
+0.06=E‴	−(95)	+0.4
+0.02=F‴	−(95)	+0.6
−0.03=L‴	−(95)	−0.1
+0.03=M‴	−(95)	0.0
+0.01=O‴	−(95)	0.0
−0.01=Q‴	−(95)	−1.1
0.00=R‴	−(95)	+1.2
−0.03=V‴	−(95)	−0.3
0.00=X‴	−(95)*	0.0
−0.03=Y‴	−(95)	+1.6

96.

s		″
−0.14=A‴	−(96)	+1.2
−0.03=D‴	−(96)	+0.8
−0.18=E‴	−(96)	+1.6
−0.12=F‴	−(96)	+1.8
−0.02=L‴	−(96)	+0.1
+0.04=M‴	−(96)	+0.2
−0.04=R‴	−(96)	+2.4
−0.12=V‴	−(96)	−0.1
−0.12=Y‴	−(96)	+1.8

SECTION I. PART IV.

FINAL LEAST SQUARE EQUATIONS. AR.

Zone.	s
V.	$-0.24=A_{-1}+D+G+H+L+Q+V+W-8$ (V)
VI.	$-0.17=A_{-1}+D+G+H+L+Q+W-7$ (VI)
VII.	$+0.34=A_{-1}+D+G+H+L+Q+V+W-8$ (VII)
4.	$+0.05=A_{-2}+A_{-1}+A+C-4$ (4)
5.	$-0.10=A_{-2}+A_{-1}+A+C+D+F+G+K+Q-9$ (5)
6.	$+0.43=A_{-2}+A_{-1}+A+C+D+F+K+Q+W+Y+0.5Z-10.5$ (6)
7.	$-0.27=A_{-2}+A_{-1}+A+C+D+F+K+Q+W+Y+Z+BB-12$ (7)
8.	$-\ .21=C+D+E+F+K-5$ (8)
9.	$-\ .01=C+D+E+F+K-5$ (9)
10.	$+\ .01=C+D+0.5G+0.5H+0.5I+K-4.5$ (10)
11.	$-\ .34=C+D+0.5G+0.5H+0.5I+K-4.5$ (11)
12.	$+\ .32=C+D+0.5G+0.5H+0.5I+K-4.5$ (12)
13.	$+\ .13=A_{-2}+A_{-1}+A+C+D+F+K+L+Q+W+Y-11$ (13)
14.	$+\ .01=A_{-2}+A_{-1}+A+C+E+F+H+L+Q+W+Y-11$ (14)
15.	$-\ .02=F+G+0.5I+K+N+P+R-6.5$ (15)
16.	$+\ .06=F+0.5G+I+K+N+P+R-6.5$ (16)
17.	$-\ .05=F+0.5G+I+K+N+P+R-6.5$ (17)
18.	$-\ .21=K+M+Q+T+0.5V-4.5$ (18)
19.	$+\ .20=K+M+Q+T+V-5$ (19)
20.	$+\ .08=K+M+O+U+V_1-5$ (20)
21.	$-\ .11=K+M+O+U+0.5V_1-4.5$ (21)
22.	$+\ .08=Q+V+W+Y+Z+AA-6$ (22)
23.	$-\ .07=Q+V+W+Y+Z+AA-6$ (23)
24.	$-\ .24=A+B+C+E'+F'+K+T'+0.5X-7.5$ (24)
25.	$+\ .24=A+B+C+E'+F'+K+M+T'+X-9$ (25)
26.	$+\ .12=A+B'+C+E'+F'+K+T'+AA-8$ (26)
27.	$+\ .15=A+B'+B+C+E+N+P-7$ (27)
28a.	$-\ .17=H'+K+R'+T'+V'-5$ (28a)
28b.	$-\ .17=H'+K+R'+T'+V'-5$ (28b)
29.	$-\ .20=K+N+T'+X-4$ (29)
30.	$.00=K+N+T'+X-4$ (30)
31.	$+\ .18=A'_{-3}+A'_{-1}+A'+B'+C'+D'+E'+F'+H'+M'+T'+W'+Y'-13$ (31)
33.	$-\ .17=A'_{-3}+A'_{-1}+C'+D'+E'+F'+M'+T'-8$ (33)
34.	$+\ .06=A'_{-3}+A'_{-1}+C'+D'+E'+F'+M'+T'+W'+Y'-10$ (34)
35.	$.00=A'_{-3}+A'_{-2}+A'_{-1}+A'+B'-5$ (35)
36.	$+\ .05=A'_{-3}+A'_{-2}+A'_{-1}+A'+B'-5$ (36)
37.	$+\ .08=C'+D'+E'+F'+G'+H'+T'-7$ (37)
38.	$.00=C'+D'+E'+F'+G'+H'+T'-7$ (38)
39.	$+\ .02=H'+0.5P'+0.5T'+V'-3$ (39)
40.	$+\ .07=H'+P+0.5T'+V'-3.5$ (40)
41.	$-\ .24=H'+P+T'+V'-4$ (41)
42.	$+\ .08=F'+G'+0.5O'+Q'+U'+W'+X'+Y'-7.5$ (42)
43.	$+\ .01=F'+G'+O'+0.5Q'+U'+W'+X'+Y'-7.5$ (43)
45.	$+\ .01=F'+L'+T'+W'-4$ (45)
47.	$+\ .46=F'+L'+T'+W'-4$ (47)
48.	$-\ .35=F'+L'+T'+W'-4$ (48)
49.	$+\ .50=C'+D'+I''+F'+R''+L'+I''I''+W'+L''L''+Y'-10$ (49)
50.	$+\ .37=C'+D'+F'+R''+L'+I''I''+W'+L''L''+Y'-9$ (50)
51.	$-\ .63=C'+D'+I''+F'+R''+L'+E''E''+I''I''+W'+L''L''+Y'-11$ (51)

Zone. s

52. $-0.07=C'+D'+I''+F'+R''+L'+E''E''+I''I''+W'+L''L''+Y'-11$ (52)

53. $-.01=C'+E''+D'+I''+K''+P''+Q''-7$ (53)

54. $+.01=C'+E''+D'+I''+K''+P''+Q''-7$ (54)

55. $.00=C'+H''+L''+P''-4$ (55)

56. $+.20=C'+H''+L''+P''-4$ (56)

57. $-.34=C'+F''+K''+G'+L'-5$ (57)

58. $+.11=C'+F''+K''+G'+L'-5$ (58)

59. $+.12=C'+0.5D''+I''+R''+A''A''+C''C''+I''I''-6.5$ (59)

60. $+.05=C'+0.5D''+I''+K''_1+R''+A''A''+C''C''+I''I''-7.5$ (60)

61. $-.62=C'+E''+R''+U''+E''E''+N''N''+O''O''-7$ (61)

62. $-.04=C'+0.5R''+L'+E''E''+N''N''+O''O''-5.5$ (62)

63. $-.17=C'+K''+L''+O'+0.5Q'+L''L''+Y'-6.5$ (63)

64. $+.53=P''+Q''+L'+C''C''+E''E''-5$ (64)

65. $+.13=P''+Q''+L'+C''C''+E''E''-5$ (65)

66. $+.14=P''+Q''+L'+0.5C''C''+E''E''-4.5$ (66)

67. $+.11=R''+Y''+B''B''+W'-4$ (67)

68. $-.11=R''+Y''+B''B''+W'-4$ (68)

71b. $-.29=R''+A''A''+H''H''+I''I''+K''K''+N''N''+O''O''-7$ (71b)

71a. $+.01=R''+A''A''+H''H''+I''I''+K''K''+N''N''+O''O''-7$ (71a)

72. $-.04=A''_{-1}+A''+B''+E''+I''+M''+N''+R''+F''F''+I''I''+L''L''+O''O''+P''P''-13$ (72)

73. $-.02=A''_{-1}+A''+B''+E''+I''+M''+N''+R''+F''F''+I''I''+L''L''+O''O''+P''P''-13$ (73)

74. $-.33=A''_{-1}+A''+B''+E''+I''+M''+N''+R''+F''F''+I''I''+L''L''+O''O''+P''P''-13$ (74)

75. $-.12=B''+D''+G''+Z''+G''G''+L''L''+M''M''-7$ (75)

76. $+.43=B''+D''+G''+Z''+G''G''+L''L''+M''M''-7$ (76)

77. $-.28=M''+N''+U''+P'''+Q'''+K''K''-6$ (77)

78. $+.52=M''+N''+U''+P'''+Q'''+K''K''-6$ (78)

79. $+.80=A'''+A''_{-1}+B''+C'''+E'''+F'''+L'''+M'''+P'''+Q'''+S'''+V'''+N''N''+O''O''+P''P''-15$ (79)

80. $-.02=A'''+A''_{-1}+B''+C'''+E'''+F'''+L'''+M'''+P'''+Q'''+S'''+V'''+N''N''+O''O''\ P''P''-15$ (80)

81. $-.60=A'''+0.5C'''+0.5D'''+G'''+H'''+I'''+O'''+R'''+T'''+U'''+W'''+Y'''-11$ (81)

82. $+.35=A'''+0.5C'''+0.5D'''+0.5G'''+H'''+I'''+O'''+R'''+T'''-7.5$ (82)

83. $-.45=A'''+B''+C'''+E'''+L'''+M'''+Q'''+S'''+V'''+Y'''-10$ (83)

84. $+.05=A'''+B''+C'''+E'''+L'''+M'''+Q'''+S'''+V'''+Y'''-10$ (84)

85. $-.01=0.5M'''+N''+U''-2.5$ (85)

86. $+.09=0.5M'''+N''+U''-2.5$ (86)

87. $-.16=0.5M'''+N''+U''-2.5$ (87)

88. $+.06=0.5M'''+N''+U''-2.5$ (88)

89. $+.06=H'''+I'''+M'''-3$ (89)

90. $+.21=H'''+I'''+M'''-3$ (90)

91. $+.21=H'''+I'''+M'''-3$ (91)

92. $+0.06=H'''+I'''+M'''-3$ (92)

93. $+0.03=I'''+K'''+N'''+O'''-4$ (93)

94. $+0.10=I'''+K'''+0.5N'''+O'''-3.5$ (94)

95. $+0.18=A'''+D'''+E'''+F'''+L'''+M'''+O'''+Q'''+R'''+V'''+0.5X'''+Y'''-11.5$ (95)

96. $-0.73=A'''+D'''+E'''+F'''+L'''+M'''+R'''+V'''+Y'''-9$ (96)

Star.	s
A_{-2}	−0.01=6A_{-2}−(4)−(5)−(6)−(7)−(13)−(14)
A_{-1}	−0.05=9A_{-1}−(V)−(VI)−(VII)−(4)−(5)−(6)−(7)−(13)−(14)
A	+0.04=10A−(4)−(5)−(6)−(7)−(13)−(14)−(24)−(25)−(26)−(27)
B	−0.02=3B−(24)−(25)−(27)
C	0.00=15C−(4)−(5)−(6)−(7)−(8)−(9)−(10)−(11)−(12)−(13)−(14)−(24)−(25)−(26)−(27)
D	+0.08=12D−(V)−(VI)−(VII)−(5)−(6)−(7)−(8)−(9)−(10)−(11)−(12)−(13)
E	0.00=4E−(8)−(9)−(14)−(27)
F	+0.02=10F−(5)−(6)−(7)−(8)−(9)−(13)−(14)−(15)−(16)−(17)
G	+0.03=7.5G−(V)−(VI)−(VII)−(5)−0.5(10)−0.5(11)−0.5(12)−(15)−0.5(16)−0.5(17)
H	+0.00=5.5H−(V)−(VI)−(VII)−0.5(10)−0.5(11)−0.5(12)−(14)
I	0.00=4I−0.5(10)−0.5(11)−0.5(12)−0.5(15)−(16)−(17)
K	−0.16=23K−(5)−(6)−(7)−(8)−(9)−10)−(11)−(12)−(13)−(15)−(16)−(17)−(18)−(19)−(20)−(21)−(24)−(25)−(26)−(28a)−(28b)−(29)−(30)
L	0.00=5L−(V)−(VI)−(VII)−(13)−(14)
M	−0.01=5M−(18)−(19)−(20)−(21)−(25)
N	−0.01=6N−(15)−(16)−(17)−(27)−(29)−(30)
O	0.00=2O−(20)−(21)
P	+0.02=6.5P−(15)−(16)−(17)−(27)−0.5(39)−(40)−(41)
Q	+0.06=12Q−(V)−(VI)−(VII)−(5)−(6)−(7)−(13)−(14)−(18)−(19)−(22)−(23)
R	+0.01=3R−(15)−(16)−(17)
T	+0.01=2T−(18)−(19)
U	0.00=2U−(20)−(21)
V	+0.02=5.5V−(V)−(VII)−0.5(18)−(19)−(22)−(23)
$V_1=v_4$	+0.01=1.5V_1−(20)−0.5(21)
W	+0.02=9W−(V)−(VI)−(VII)−(6)−(7)−(13)−(14)−(22)−(23)
X	−0.01=3.5X−0.5(24)−(25)−(29)−(30)
Y	+0·02=6Y−(6)−(7)−(13)−(14)−(22)−(23)
Z	−0.04=3.5Z−0.5(6)−(7)−(22)−(23)
AA	+0.01=3AA−(22)−(23)−(26)
BB	0.00=BB−(7)
A'_{-3}	0.00=5A'_{-3}−(31)−(33)−(34)−(35)−(36)
A'_{-2}	0.00=2A'_{-2}−(35)−(36)
A'_{-1}	+0.03=5A'_{-1}−(31)−(33)−(34)−(35)−(36)
A′	0.00=3A′−(31)−(35)−(36)
B′	0.00=5B′−(26)−(27)−(31)−(35)−(36)
C′=C″	−0.06=20C′−(31)−(33)−(34)−(37)−(38)−(49)−(50)−(51)−(52)−(53)−(54)−(55)−(56)−(57)−(58)−(59)−(60)−(61)−(62)−(63)
D′	+0.02=11D′−(31)−(33)−(34)−(37)−(38)−(49)−(50)−(51)−(52)−(53)−(54)
E′	+0.02=8E′−(24)−(25)−(26)−(31)−(33)−(34)−(37)−(38)
F′	0.00=17F′−(24)−(25)−(26)−(31)−(33)−(34)−(37)−(38)−(42)−(43)−(45)−(47)−(48)−(49)−(50)−(51)−(52)
G′=O″	+0.01=6G′−(37)−(38)−(42)−(43)−(57)−(58)
H′	−0.19=8H′−(28a)−(28b)−(31)−(37)−(38)−(39)−(40)−(41)
L′=U″	0.00=20L′−(45)−(47)−(48)−(49)−(50)−(51)−(52)−(57)−(58)−(61)−(62)−(64)−(65)−(66)−(77)−(78)−(85)−(86)−(87)−(88)
M′	−0.01=3M′−(31)−(33)−(34)
O′	+0.02=2.5o′−0.5(42)−(43)−(63)
Q′	+0.01=2Q′−(42)−0.5(43)−0.5(63)
R′	+0.01=2R′−(28a)−(28b)
T′	−0.09=17T′−(24)−(25)−(26)−(28a)−(28b)−(29)−(30)−(31)−(33)−(34)−(37)−(38)−0.5(39)−0.5(40)−(41)−(45)−(47)−(48)
U′	−0.01=2U′−(42)−(43)
V′=S	−0.02=5V′−(28a)−(28b)−(39)−(40)−(41)
W′	−0.01=13W′−(31)−(34)−(42)−(43)−(45)−(47)−(48)−(49)−(50)−(51)−(52)−(67)−(68)

Star.	s
X′=M″M″	+0.02=4X′−(42)−(43)−(75)−(76)
Y′	+0.04=9Y′−(31)−(34)−(42)−(43)−(49)−(50)−(51)−(52)−(63)
A''_{-1}	+0.03=5A''_{-1}−(72)−(73)−(74)−(79)−(80)
A″	+0.01=3A″−(72)−(73)−(74)
B″=B‴	+0.03=9B″−(72)−(73)−(74)−(75)−(76)−(79)−(80)−(83)−(84)
D″	+0.03=3D″−0.5(59)−0.5(60)−(75)−(76)
E″	−0.02=6E″−(53)−(54)−(61)−(72)−(73)−(74)
F″	−0.01=2F″−(57)−(58)
G″	−0.01=2G″−(75)−(76)
H″	0.00=2H″−(55)−(56)
I″	−0.02=10I″−(49)−(51)−(52)−(53)−(54)−(59)−(60)−(72)−(73)−(74)
K″	+0.01=5K″−(53)−(54)−(57)−(58)−(63)
K''_1	0.00=K''_1−(60)
L″	+0.08=3L″−(55)−(56)−(63)
M″=M‴	+0.31=17M″−(72)−(73)−(74)−(77)−(78)−(79)−(80)−(83)−(84)−0.5(85) −0.5(86)−0.5(87)−0.5(88)−(89)−(90)−(91)−(92)−(95)−(96)
N″	−0.09=9N″−(72)−(73)−(74)−(77)−(78)−(85)−(86)−(87)−(88)
P″	−0.06=7P″−(53)−(54)−(55)−(56)−(64)−(65)−(66)
Q″	+0.11=5Q″−(53)−(54)−(64)−(65)−(66)
R″	+0.08=14.5R″−(49)−(50)−(51)−(52)−(59)−(60)−(61)−0.5(62)−(67)−(68) −(71b)−(71a)−(72)−(73)−(74)
Y″	0.00=2Y″−(67)−(68)
Z″	0.00=2Z″−(75)−(76)
A″A″	0.00=4A″A″−(59)−(60)−(71b)−(71a)
B″B″	−0.01=2B″B″−(67)−(68)
C″C″	+0.04=4.5C″C″−(59)−(60)−(64)−(65)−0.5(66)
E″E″	−0.02=7E″E″−(51)−(52)−(61)−(62)−(64)−(65)−(66)
G″G″	0.00=2G″G″−(75)−(76)
H″H″	0.00=2H″H″−(71b)−(71a)
I″I″	+0.04=11I″I″−(49)−(50)−(51)−(52)−(59)−(60)−(71b)−(71a)−(72)−(73) −(74)
K″K″	+0.02=4K″K″−(71b)−(71a)−(77)−(78)
L″L″	+0.08=10L″L″−(49)−(50)−(51)−(52)−(63)−(72)−(73)−(74)−(75)−(76)
N″N″	−0.05=6N″N″−(61)−(62)−(71b)−(71a)−(79)−(80)
O″O″	−0.06=9O″O″−(61)−(62)−(71b)−(71a)−(72)−(73)−(74)−(79)−(80)
P″P″	−0.01=5P″P″−(72)−(73)−(74)−(79)−(80)
A‴	−0.06=8A‴−(79)−(80)−(81)−(82)−(83)−(84)−(95)−(96)
C‴	+0.01=5C‴−(79)−(80)−0.5(81)−0.5(82)−(83)−(84)
D‴	+0.01=3D‴−0.5(81)−0.5(82)−(95)−(96)
E‴	−0.09=6E‴−(79)−(80)−(83)−(84)−(95)−(96)
F‴	−0.06=4F‴−(79)−(80)−(95)−(96)
G‴	−0.01=1.5G‴−(81)−0.5(82)
H‴	+0.09=6H‴−(81)−(82)−(89)−(90)−(91)−(92)
I‴	+0.14=8I‴−(81)−(82)−(89)−(90)−(91)−(92)−(93)−(94)
K‴	−0.01=2K‴−(93)−(94)
L‴	−0.04=6L‴−(79)−(80)−(83)−(84)−(95)−(96)
N‴	+0.05=1.5N‴−(93)−0.5(94)
O‴	+0.01=5O‴−(81)−(82)−(93)−(94)−(95)
P‴	+0.04=4P‴−(77)−(78)−(79)−(80)
Q‴	+0.01=10Q‴−(72)−(73)−(74)−(77)−(78)−(79)−(80)−(83)−(84)−(95)
R‴	−0.01=4R‴−(81)−(82)−(95)−(96)
S‴	0.01=4S‴−(79)−(80)−(83)−(84)
T‴	0.00=2T‴−(81)−(82)
U‴	0.00=U‴−(81)
V‴	−0.08=6V‴−(79)−(80)−(83)−(84)−(95)−(96)
W‴	0.00=W‴−(81)
X‴	0.00=0.5X‴−0.5(95)
Y‴	−0.04=5Y‴−(81)−(83)−(84)−(95)−(96)

Zone.	
I.	$-1''.2=D+L+W-3$ (I)
II.	$-0.3=D+L+W-3$ (II)
III.	$+1.3=A_{-1}+G+H+Q-4$ (III)
IV.	$+0.7=A_{-1}+G+H+Q+V-5$ (IV)
4.	$+0.3=A_{-2}+A_{-1}+A+C-4$ (4)
5.	$+3.2=A_{-2}+A_{-1}+A+C+D+F+K+Q-8$ (5)
6.	$+1.7=A_{-2}+A_{-1}+A+C+D+F+K+Q+W+Y+Z-11$ (6)
7.	$+5.0=A_{-2}+A_{-1}+A+C+D+F+K+Q+W+Y+Z+BB-12$ (7)
8.	$-0.9=C+D+E+F+K-5$ (8)
9.	$+2.6=C+D+E+F+K-5$ (9)
10.	$-1.8=C+D+G+H+I+K-6$ (10)
12.	$+2.9=C+D+G+H+I+K-6$ (12)
13.	$-1.4=A_{-2}+A_{-1}+A+C+D+F+K+L+Q+W+Y-11$ (13)
14.	$-4.1=A_{-2}+A_{-1}+A+C+E+F+H+L+Q+W+Y-11$ (14)
15.	$+0.5=F+G+I+K+N+P+R-7$ (15)
16.	$-4.1=F+G+I+K+N+P+R-7$ (16)
17.	$-5.9=F+G+I+K+N+P+R-7$ (17)
18.	$+1.4=K+M+Q+T+V-5$ (18)
19.	$+0.4=K+M+Q+T+V-5$ (19)
20.	$+0.3=K+M+O+U+V_1-5$ (20)
21.	$-0.7=K+M+O+U+V_1-5$ (21)
22.	$-2.8=Q+V+W+Y+Z+AA-6$ (22)
23.	$-1.4=Q+V+W+Y+Z+AA-6$ (23)
24.	$-3.9=A+B+C+E'+F'+K+T'+X-8$ (24)
25.	$-0.3=A+B+C+E'+F'+K+M+T'+X-9$ (25)
26.	$+7.0=A+B'+C+E'+K+T'+AA-7$ (26)
27.	$-1.2=A+B'+B+C+E+N+P-7$ (27)
27b.	$-1.3=H'+K+R'+T'+V'-5$ (27b)
27c.	$+1.7=H'+K+R'+T'+V'-5$ (27c)
27d.	$+1.3=K+N+T'+v'_3+X-5$ (27d)
27e.	$+0.2=K+N+T'+X-4$ (27e)
28a.	$-5.0=H'+K+R'+T'+V'-5$ (28a)
29.	$+0.5=K+N+T'+X-4$ (29)
30.	$+1.6=K+N+T'+X-4$ (30)
31.	$+1.9=A'_{-3}+A'_{-1}+A'+B'+D'+F'+M'+T'+W'+Y'-10$ (31)
32.	$-4.2=A'_{-3}+A'_{-1}+A'+E'+F'+M'+T'+W'+Y'-9$ (32)
33.	$+3.2=A'_{-3}+A'_{-1}+C'+D'+E'+F'+M'+T'-8$ (33)
34.	$-1.3=A'_{-3}+A'_{-1}+C'+D'+E'+F'+M'+T'+W'+Y'-10$ (34)
35.	$-0.2=A'_{-3}+A'_{-2}+A'_{-1}+A'+B'-5$ (35)
36.	$+0.3=A'_{-3}+A'_{-2}+A'_{-1}+A'+B'-5$ (36)
37.	$+1.2=C'+D'+E'+F'+G'+H'+T'-7$ (37)
38.	$+0.6=C'+D'+E'+F'+G'+H'+T'-7$ (38)
39.	$+3.7=H'+P+T'+V'-4$ (39)
40.	$-1.3=H'+P+T'+V'-4$ (40)
41.	$+5.7=H'+P+T'+V'-4$ (41)
42.	$+1.2=G'+O'+Q'+U'+W'+X'+Y'-7$ (42)
43.	$-2.3=F'+G'+O'+Q'+U'+W'+X'+Y'-8$ (43)
44.	$+0.1=A'_{-3}+A'_{-1}+A'+B'+C'+D'-6$ (44)
45.	$-0.3=F'+L'+T'+W'-4$ (45)
46.	$-2.3=F'+L'+T'+W'-4$ (46)
47.	$+4.0=F'+L'+T'+W'-4$ (47)
48.	$-3.3=F'+L'+T'+W'-4$ (48)
49.	$+8.9=C'+D'+I''+F'+R''+L'+I''I''+W'+L''L''+Y'-10$ (49)
50.	$-5.8=C'+D'+F'+R''+L'+I''I''+W'+L''L''+Y'-9$ (50)
51.	$+0.5=C'+D'+I''+F'+R''+L'+E''E''+I''I''+W'+L''L''+Y'-11$ (51)
52.	$+8.5=C'+D'+I''+F'+R''+L'+E''E''+I''I''+W'+L''L''+Y'-11$ (52)
53.	$+0.2=C'+E''+D'+I''+K''+P''+Q''-7$ (53)
54.	$+0.6=C'+E''+D'+I''+K''+P''+Q''-7$ (54)

Zone. ″

55. $+0.4 = C' + H'' + L'' + P'' - 4$ (55)
56. $+0.4 = C' + H'' + L'' + P'' - 4$ (56)
57. $-0.3 = C' + F'' + K'' + G' + L' - 5$ (57)
58. $-1.3 = C' + F'' + K'' + G' + L' - 5$ (58)
59. $+0.8 = C' + D'' + I'' + R'' + A''A'' + C''C'' + I''I'' - 7$ (59)
60. $+1.0 = C' + D'' + I'' + K''_{1} + R'' + A''A'' + C''C'' + I''I'' - 8$ (60)
61. $-3.6 = C' + E'' + R'' + U'' + E''E'' + N''N'' + O''O'' - 7$ (61)
62. $-3.6 = C' + E'' + R'' + L' + E''E'' + N''N'' + O''O'' - 7$ (62)
63. $-1.6 = C' + K'' + L'' + O' + Q' + L''L'' + Y' - 7$ (63)
64. $+2.4 = P'' + Q'' + L' + C''C'' + E''E'' - 5$ (64)
65. $+0.4 = P'' + Q'' + L' + C''C'' + E''E'' - 5$ (65)
67. $-0.5 = R'' + Y'' + B''B'' + W' - 4$ (67)
68. $+0.3 = R'' + Y'' + B''B'' + W' - 4$ (68)
69. $-0.6 = N'' + R'' + A''A'' + H''H'' + I''I'' + K''K'' - 6$ (69)
70. $+0.6 = N'' + R'' + A''A'' + H''H'' + I''I'' + K''K'' - 6$ (70)
71b. $+0.9 = R'' + A''A'' + H''H'' + I''I'' + K''K'' + N''N'' + O''O'' - 7$ (71b)
72. $+0.8 = A''_{-1} + A'' + B'' + E'' + I'' + M'' + R'' + F''F'' + I''I'' + L''L'' + O''O'' + P''P'' - 12$ (72)
73. $+1.9 = A''_{-1} + A'' + B'' + E'' + I'' + R'' + F''F'' + I''I'' + L''L'' + O''O'' + P''P'' - 11$ (73)
74. $+2.9 = A''_{-1} + A'' + B'' + E'' + I'' + M'' + N'' + R'' + F''F'' + I''I'' + L''L'' + O''O'' + P''P'' - 13$ (74)
75. $-0.3 = B'' + D'' + G'' + Z'' + G''G'' + L''L'' + M''M'' - 7$ (75)
76. $-2.3 = B'' + D'' + G'' + Z'' + G''G'' + L''L'' + M''M'' - 7$ (76)
77. $-9.4 = M'' + N'' + U'' + P''' + Q''' + K''K'' - 6$ (77)
78. $-2.4 = M'' + N'' + U'' + P''' + Q''' + K''K'' - 6$ (78)
79. $-4.8 = A''' + A''_{-1} + B'' + C''' + E''' + F''' + L''' + M''' + P''' + Q''' + S''' + V''' + N''N'' + O''O'' + P''P'' - 15$ (79)
80. $-4.4 = A''' + A''_{-1} + B'' + C''' + E''' + F''' + L''' + M''' + P''' + Q''' + S''' + V''' + N''N'' + O''O'' + P''P'' - 15$ (80)
81. $-1.9 = A''' + C''' + D''' + G''' + H''' + I''' + O''' + R''' + T''' + U''' + W''' + Y''' - 12$ (81)
82. $-6.5 = A''' + C''' + D''' + G''' + H''' + I''' + O''' + R''' + T''' - 9$ (82)
83. $-3.0 = A''' + B'' + C''' + E''' + L''' + M''' + Q''' + S''' + V''' + Y''' - 10$ (83)
84. $-3.0 = A''' + B''' + C''' + E''' + L''' + M''' + Q''' + S''' + V''' + Y''' - 10$ (84)
85. $-0.3 = M''' + N'' + U'' - 3$ (85)
86. $+2.8 = M''' + N'' + U'' - 3$ (86)
87. $+1.7 = M''' + N'' + U'' - 3$ (87)
88. $+0.7 = M''' + N'' + U'' - 3$ (88)
89. $-0.3 = H''' + I''' + M''' - 3$ (89)
90. $+2.7 = H''' + I''' + M''' - 3$ (90)
91. $+3.6 = H''' + I''' + M''' - 3$ (91)
92. $+3.6 = H''' + I''' + M''' - 3$ (92)
93. $+2.7 = I''' + K''' + N''' + O''' - 4$ (93)
94. $+0.9 = I''' + K''' + N''' + O''' - 4$ (94)
95. $+3.9 = A''' + D''' + E''' + F''' + L''' + M''' + O''' + Q''' + R''' + V''' + X''' + Y''' - 12$ (95)
96. $+9.8 = A''' + D''' + E''' + F''' + L''' + M''' + R''' + V''' + Y''' - 9$ (96)

Star.	″
A_{-2}	+0.1=6A_{-2}—(4)—(5)—(6)—(7)—(13)—(14)
A_{-1}	+0.3=8A_{-1}—(III)—(IV)—(4)—(5)—(6)—(7)—(13)—(14)
A	—0.3=10A—(4)—(5)—(6)—(7)—(13)—(14)—(24)—(25)—(26)—(27)
B	0.0=3B—(24)—(25)—(27)
C	+1.2=14C—(4)—(5)—(6)—(7)—(8)—(9)—(10)—(12)—(13)—(14)—(24)—(25)—(26)—(27)
D	—0.3=10D—(I)—(II)—(5)—(6)—(7)—(8)—(9)—(10)—(12)—(13)
E	+0.3=4E—(8)—(9)—(14)—(27)
F	—0.4=10F—(5)—(6)—(7)—(8)—(9)—(13)—(14)—(15)—(16)—(17)
G	—0.2=7G—(III)—(IV)—(10)—(12)—(15)—(16)—(17)
H	—0.1=5H—(III)—(IV)—(10)—(12)—(14)
I	—0.2=5I—(10)—(12)—(15)—(16)—(17)
K	+0.8=25K—(5)—(6)—(7)—(8)—(9)—(10)—(12)—(13)—(15)—(16)—(17)—(18)—(19)—(20)—(21)—(24)—(25)—(26)—(27b)—(27c)—(27d)—(27e)—(28a)—(29)—(30)
L	—0.4=4L—(I)—(II)—(13)—(14)
M	—0.1=5M—(18)—(19)—(20)—(21)—(25)
N	+0.4=8N—(15)—(16)—(17)—(27)—(27d)—(27e)—(29)—(30)
O	0.0=2O—(20)—(21)
S′=P	+0.1=7P—(15)—(16)—(17)—(27)—(39)—(40)—(41)
Q	+0.2=11Q—(III)—(IV)—(5)—(6)—(7)—(13)—(14)—(18)—(19)—(22)—(23)
R	0.0=3R—(15)—(16)—(17)
T	0.0=2T—(18)—(19)
U	0.0=2U—(20)—(21)
V	0.0=5V—(IV)—(18)—(19)—(22)—(23)
$V_1=v_4$	0.0=2V_1—(20)—(21)
W	—0.3=8W—(I)—(II)—(6)—(7)—(13)—(14)—(22)—(23)
X	+0.4=6X—(24)—(25)—(27d)—(27e)—(29)—(30)
Y	—0.3=6Y—(6)—(7)—(13)—(14)—(22)—(23)
Z	—0.3=4Z—(6)—(7)—(22)—(23)
AA	+0.2=3AA—(22)—(23)—(26)
BB	0.0=BB—(7)
A'_{-3}	0.0=7A'_{-3}—(31)—(32)—(33)—(34)—(35)—(36)—(44)
A'_{-2}	—0.1=2A'_{-2}—(35)—(36)
A'_{-1}	—0.2=7A'_{-1}—(31)—(32)—(33)—(34)—(35)—(36)—(44)
A′	+0.3=5A′—(31)—(32)—(35)—(36)—(44)
B′	0.0=6B′—(26)—(27)—(31)—(35)—(36)—(44)
C′=C″	+0.4=20C′—(33)—(34)—(37)—(38)—(44)—(49)—(50)—(51)—(52)—(53)—(54)—(55)—(56)—(57)—(58)—(59)—(60)—(61)—(62)—(63)
D′	+1.8=12D′—(31)—(33)—(34)—(37)—(38)—(44)—(49)—(50)—(51)—(52)—(53)—(54)
E′	+0.5=8E′—(24)—(25)—(26)—(32)—(33)—(34)—(37)—(38)
F′	+0.2=17F′—(24)—(25)—(31)—(32)—(33)—(34)—(37)—(38)—(43)—(45)—(46)—(47)—(48)—(49)—(50)—(51)—(52)
G′=O″	—0.2=6G′—(37)—(38)—(42)—(43)—(57)—(58)
H′	+0.4=8H′—(27b)—(27c)—(28a)—(37)—(38)—(39)—(40)—(41)
L′=U″	0.0=20L′—(45)—(46)—(47)—(48)—(49)—(50)—(51)—(52)—(57)—(58)—(61)—(62)—(64)—(65)—(77)—(78)—(85)—(86)—(87)—(88)
M′	+0.1=4M′—(31)—(32)—(33)—(34)
O′	—0.1=3O′—(42)—(43)—(63)
Q′	+0.1=3Q′—(42)—(43)—(63)
R′	0.0=3R′—(27b)—(27c)—(28a)
T′	+0.6=23T′—(24)—(25)—(26)—(27b)—(27c)—(27d)—(27e)—(28a)—(29)—(30)—(31)—(32)—(33)—(34)—(37)—(38)—(39)—(40)—(41)—(45)—(46)—(47)—(48)
U′	0.0=2U′—(42)—(43)
V′=S	0.0=6V′—(27b)—(27c)—(28a)—(39)—(40)—(41)
(v'_3)	0.0=(v'_3)—(27d)

Star.	″
W′	−0.7=15W′−(31)−(32)−(34)−(42)−(43)−(45)−(46)−(47)−(48)−(49)−(50)−(51)−(52)−(67)−(68)
X′=M″M″	−0.5=4X′−(42)−(43)−(75)−(76)
Y′	−0.7=10Y′−(31)−(32)−(34)−(42)−(43)−(49)−(50)−(51)−(52)−(63)
A''_{-1}	+0.2=5A''_{-1}−(72)−(73)−(74)−(79)−(80)
A″	−0.1=3A″−(72)−(73)−(74)
B″=B‴	−2.0=9B″−(72)−(73)−(74)−(75)−(76)−(79)−(80)−(83)−(84)
D″	−0.8=4D″−(59)−(60)−(75)−(76)
E″	−0.2=7E″−(53)−(54)−(61)−(62)−(72)−(73)−(74)
F″	0.0=2F″−(57)−(58)
G″	0.0=2G″−(75)−(76)
H″	0.0=2H″−(55)−(56)
I″	+0.4=10I″−(49)−(51)−(52)−(53)−(54)−(59)−(60)−(72)−(73)−(74)
K″	−0.3=5K″−(53)−(54)−(57)−(58)−(63)
K''_1	0.0=K''_1−(60)
L″	+0.1=3L″−(55)−(56)−(63)
M″=M‴	+0.3=18M″−(72)−(74)−(77)−(78)−(79)−(80)−(83)−(84)−(85)−(86)−(87)−(88)−(89)−(90)−(91)−(92)−(95)−(96)
N″	−0.3=9N″−(69)−(70)−(74)−(77)−(78)−(85)−(86)−(87)−(88)
P″	0.0=6P″−(53)−(54)−(55)−(56)−(64)−(65)
Q″	+0.4=4Q″−(53)−(54)−(64)−(65)
R″	−0.8=16R″−(49)−(50)−(51)−(52)−(59)−(60)−(61)−(62)−(67)−(68)−(69)−(70)−(71b)−(72)−(73)−(74)
Y″	0.0=2Y″−(67)−(68)
Z″	0.0=2Z″−(75)−(76)
A″A″	+0.2=5A″A″−(59)−(60)−(69)−(70)−(71b)
B″B″	+0.1=2B″B″−(67)−(68)
C″C″	0.0=4C″C″−(59)−(60)−(64)−(65)
E″E″	0.0=6E″E″−(51)−(52)−(61)−(62)−(64)−(65)
G″G″	0.0=2G″G″−(75)−(76)
H″H″	0.0=3H″H″−(69)−(70)−(71b)
I″I″	+1.3=12I″I″−(49)−(50)−(51)−(52)−(59)−(60)−(69)−(70)−(71b)−(72)−(73)−(74)
K″K″	−0.8=5K″K″−(69)−(70)−(71b)−(77)−(78)
L″L″	+1.2=10L″L″−(49)−(50)−(51)−(52)−(63)−(72)−(73)−(74)−(75)−(76)
N″N″	−0.1=5N″N″−(61)−(62)−(71b)−(79)−(80)
O″O″	+0.4=8O″O″−(61)−(62)−(71b)−(72)−(73)−(74)−(79)−(80)
P″P″	−0.1=5P″P″−(72)−(73)−(74)−(79)−(80)
A‴	+1.6=8A‴−(79)−(80)−(81)−(82)−(83)−(84)−(95)−(96)
C‴	+0.6=6C‴−(79)−(80)−(81)−(82)−(83)−(84)
D‴	+0.1=4D‴−(81)−(82)−(95)−(96)
E‴	+0.5=6E‴−(79)−(80)−(83)−(84)−(95)−(96)
F‴	+1.0=4F‴−(79)−(80)−(95)−(96)
G‴	+0.1=2G‴−(81)−(82)
H‴	+1.8=6H‴−(81)−(82)−(89)−(90)−(91)−(92)
I‴	+2.8=8I‴−(81)−(82)−(89)−(90)−(91)−(92)−(93)−(94)
K‴	−0.1=2K‴−(93)−(94)
L‴	+0.1=6L‴−(79)−(80)−(83)−(84)−(95)−(96)
N‴	+0.1=2N‴−(93)−(94)
O‴	+1.0=5O‴−(81)−(82)−(93)−(94)−(95)
P‴=D″D″	−0.7=4P‴−(77)−(78)−(79)−(80)
Q‴=F″F″	−2.0=10Q‴−(72)−(73)−(74)−(77)−(78)−(79)−(80)−(83)−(84)−(95)
R‴	+1.5=4R‴−(81)−(82)−(95)−(96)
S‴	+0.1=4S‴−(79)−(80)−(83)−(84)
T‴	0.0=2T‴−(81)−(82)
U‴	0.0=U‴−(81)
V‴	−0.3=6V‴−(79)−(80)−(83)−(84)−(95)−(96)
W‴	0.0=W‴−(81)
X‴	0.0=X‴−(95)
Y‴	+0.7=5Y‴−(81)−(83)−(84)−(95)−(96)

SECTION I. PART V.

FINAL RESULTS OF SOLUTION.

No. of Zone.	$\delta\xi$	$\delta\eta$	No. of Zone.	$\delta\xi$	$\delta\eta$	No. of Zone.	$\delta\xi$	$\delta\eta$
	s	″		s	″		s	″
I.		+0.46	28a.	+ .050	+0.95	62.	+ .010	+0.57
II.		+0.16	28b.	+ .050		63.	+ .039	+0.25
III.		−0.35	29.	+ .056	−0.07	64.	− .131	−0.65
IV.		−0.15	30.	+ .006	−0.35	65.	− .051	−0.25
V.	+0.031		31.	− .016	−0.24	66.	− .055	
VI.	+0.023		32.		+0.42	67.	− .027	+0.08
VII.	− .041		33.	+0.021	−0.44	68.	+ .028	−0.11
4.	− .018	−0.17	34.	− .007	+0.07	69.		+0.09
5.	+ .011	−0.46	35.	− .008	−0.02	70.		−0.11
6.	− .043	−0.19	36.	− .018	−0.09	71b.	+0.054	−0.06
7.	+ .023	−0.49	37.	− .010	−0.23	71a.	+0.011	
8.	+ .044	+0.20	38.	+ .002	−0.14	72.	+ .009	−0.06
9.	+ .004	−0.49	39.	+ .006	−1.05	73.	+ .008	−0.17
10.	.000	+0.35	40.	− .007	+0.19	74.	+ .032	−0.22
11.	+ .078		41.	+ .072	−1.55	75.	.000	+0.12
12.	− .069	−0.43	42.	− .011	−0.13	76.	− .079	+0.41
13.	− .013	+0.11	43.	− .001	+0.32	77.	+ .045	+1.75
14.	− .002	+0.37	44.		−0.07	78.	− .089	+0.58
15.	+ .009	+0.11	45.	− .005	+0.03	79.	− .053	+0.63
16.	− .002	+0.79	46.		+0.53	80.	+ .001	+0.60
17.	+ .015	+1.04	47.	− .118	−1.04	81.	+ .072	+0.59
18.	+ .046	−0.35	48.	+ .085	+0.78	82.	− .044	+1.19
19.	− .041	−0.15	49.	− .048	−0.96	83.	+ .049	+0.60
20.	− .018	−0.04	50.	− .039	+0.59	84.	− .001	+0.60
21.	+ .022	+0.16	51.	+ .058	−0.12	85.	.000	+0.07
22.	− .017	+0.46	52.	+ .008	−0.85	86.	− .040	−0.94
23.	+ .008	+0.23	53.	− .002	−0.14	87.	+ .060	−0.59
24.	+ .030	+0.48	54.	− .005	−0.19	88.	− .028	−0.25
25.	− .028	+0.02	55.	− .016	−0.22	89.	− .025	+0.12
26.	− .018	−1.07	56.	− .066	−0.22	90.	− .075	−0.88
27.	− .020	+0.20	57.	+ .075	+0.11	91.	− .075	−1.19
27b.		+0.21	58.	− .011	+0.31	92.	− .025	−1.19
27c.		−0.38	59.	− .020	−0.28	93.	− .015	−0.94
27d.		−0.27	60.	− .009	−0.30	94.	− .040	−0.49
27e.		0.00	61.	+ .095	+0.58	95.	− .011	−0.09
						96.	+0.089	−0.81

Star's Name.	$\delta \xi$	$\delta \eta$
	s	″
A_{-2}	−0.01	−0.12
A_{-1}	− .01	− .12
A	.00	− .15
B	− .01	+ .23
C	.00	− .02
D	+ .01	− .10
E	+ .01	+ .15
F	+ .01	+ .06
G	+ .01	+ .16
H	.00	− .06
I	+ .01	+ .33
K	.00	+ .02
L	.00	+ .18
M	.00	− .09
N	+ .01	+ .24
O	+ .01	+ .06
P	+ .01	− .02
Q	+ .01	− .06
R	+ .01	+ .64
T	+ .01	− .25
U	+ .01	+ .06
V	.00	+ .01
V_1	+ .01	+ .06
W	.00	+ .10
X	+ .01	+ .04
Y	.00	+ .03
Z	− .01	− .07
AA	− .01	− .06
BB	+ .02	− .49
A'_{-3}	− .01	− .05
A'_{-2}	− .02	− .10
A'_{-1}	.00	− .08
A′	− .02	+ .06
B′	− .02	− .21
C′ = C″	.00	− .06
D′	0.00	−0.07

Star's Name.	$\delta \xi$	$\delta \eta$
	s	″
E′	0.00	−0.05
F′	− .01	− .03
G′ = O″	+ .01	+ .01
H′	− .01	− .20
M′	− .01	.00
O′	+ .02	+ .11
Q′	+ .01	+ .18
R′	+ .06	+ .26
T′	.00	− .11
U′	− .01	+ .09
V′ = S	+ .03	− .27
v'_3		− .27
W′	− .01	− .08
X′ = M″M″	− .02	+ .05
Y′	.00	− .13
A''_{-1}	+ .01	+ .19
A″	+ .02	− .18
B″ = B‴	.00	+ .05
D″	− .03	− .21
E″	+ .02	+ .02
F″	+ .03	+ .21
G″	− .05	+ .26
H″	− .05	− .22
I″	.00	− .28
K″	+ .02	+ .01
K''_1	− .01	− .30
L″	.00	− .03
M″ = M‴	+ .01	− .05
N″	− .01	+ .01
P″	− .06	− .28
Q″	− .03	− .20
R″	+ .02	− .13
U″ = L′	− .01	+ .02
Y″	.00	+ .01
Z″	−0.04	+0.26

Star's Name.	$\delta \xi$	$\delta \eta$
	s	″
A″A″	+0.01	−0.09
B″B″	.00	+ .04
C″C″	− .05	− .37
E″E″	− .02	− .12
G″G″	− .04	+ .26
H″H″	+ .03	− .02
I″I″	+ .01	− .09
K″K″	+ .01	+ .29
L″L″	+ .01	+ .01
N″N″	+ .01	+ .44
O″O″	+ .01	+ .28
P″P″	.00	+ .13
A‴	+ .01	+ .61
C‴	+ .01	+ .80
D‴	+ .04	+ .24
E‴	.00	+ .34
F‴	− .01	+ .33
G‴	+ .03	+ .94
H‴	− .01	+ .07
I‴	− .01	.00
K‴	− .04	− .76
L‴	+ .01	+ .27
N‴	+ .01	− .66
O‴	.00	+ .25
P‴ = D″D″	− .01	+ .71
Q‴ = F″F″	.00	+ .22
R‴	+ .03	+ .59
S‴	.00	+ .63
T‴	+ .02	+ .89
U‴	+ .08	+ .59
V‴	.00	+ .20
W‴	+ .08	+ .60
X‴	− .01	− .0[illegible]
Y‴	0.00	+0.31

SECTION I. PART VI.

REVISION ZONES OF SMALL STARS NEAR θ ORIONIS.

Revision Zone 1, Jan. 4, 1858.

Name of Star.	Mag.	Reading of Scale.	Reduction.	Correct.	$\delta-\delta_0$	Principal Star.	U′—U.	Correction.	$\alpha-\alpha_0$
		′ ″	″	″	′ ″	m s	s	s	m s
a_1	14	39 30	—0.5	—1.1	39 28.4	A —1 30.87	+ 4.50	0.00	—1 26.37
a_2	17	36 10		—0.5	36 09.0	A	+11.00		—1 19.87
a_3	16	31 30		—0.2	31 29.3	B —1 8.38	— 1.25		—1 9.63
c_1	13	38 27		—0.9	38 25.6	C —1 1.88	+ 3.10		—0 58.78
f_1	15	30 50		—0.1	30 49.4	F —0 35.06	+ 0.25		—0 34.81
f_2	16	39 50		—1.0	39 48.5	F	+10.50		—0 24.56
f_3	14	31 45		—0.2	31 44.3	F	+15.00		—0 20.06
k_1	16.17	36 30		—0.7	36 28.8	K —0 10.87	+ 3.00		—0 7.87

Revision Zone 2, Jan. 7, 1858.

Name of Star.	Mag.	Reading of Scale.	Reduction.	Correct.	$\delta-\delta_0$	Principal Star.	U′—U.	Correction.	$\alpha-\alpha_0$
		′ ″	″	″	′ ″	m s	s	s	m s
k_1*	17	36 30	+0.6	—1.5	36 29.1	I —0 13.07	+ 5.00	0.00	—0 8.07
n_1	16.17	36 15		—1.4	36 14.2	M +0 04.03	+ 0.75		+0 4.78
n_2	16	34 13		—1.0	34 12.6	M	+ 1.15		+0 5.18
n_5	14.15	33 40		—0.8	33 39.8	M	+ 4.75		+0 8.78
o_1	16	32 15		—0.5	32 15.1	P +0 09.81	— 0.10		+0 9.71
n_3	17	32 20		—0.5	32 20.1	N +0 04.28	+ 2.00		+0 6.28

Revision Zone 3, Jan. 12, 1858.

Name of Star.	Mag.	Reading of Scale.	Reduction.	Correct.	$\delta-\delta_0$	Principal Star.	U′—U.	Correction.	$\alpha-\alpha_0$
		′ ″	″	″	′ ″	m s	s	s	m s
k_1	17	36 35	0.0	—1.2	36 33.8	I —0 13.07	+ 5.00	0.00	—0 8.07
n_1	16.17	36 20		—1.1	36 18.9	M +0 4.03	+ 1.00		+0 5.03
n_2	16	34 13		—0.8	34 12.2	M	+ 1.25		+0 5.28
n_3	16.17	32 30		—0.4	32 29.6	N +0 4.28	+ 2.35		+0 6.63
n_4	16	32 32		—0.4	32 31.6	N	+ 4.50		+0 8.78
n_5	14	33 37		—0.6	33 36.4	N	+ 4.65		+0 8.93
o_1	15	32 18		—0.4	32 17.6	O +0 9.08	+ 0.60		+0 9.68
p_1	16	37 20		—1.3	37 18.7	O	+ 2.60		+0 11.68
p_2	15	39 12		—1.6	39 10.4	O	+ 4.50		+0 13.58
p_3	15	34 2		—0.7	34 1.3	M +0 4.03	+11.00		+0 15.03
p_4	16.17	39 53		—1.8	39 51.2	M	+12.75		+0 16.78
r_1	16	34 30		—0.8	34 29.2	R +0 18.70	+ 3.60		+0 22.30
r_2	16	31 25		—0.2	31 24.8	S +0 25.98	0.00		+0 25.98
r_3	15.16	34 10		—0.7	34 9.3	S	0.00		+0 25.98
r_4	16	37 40		—1.3	37 38.7	S	0.00		+0 25.98
s_1	14.15	36 13		—1.1	36 11.9	S	+ 1.75		+0 27.73
t_1	16.17	36 20		—1.1	36 18.9	T +0 30.65	+ 3.40		+0 34.05
v_1	15.16	39 20		—1.6	39 18.4	V +0 35.29	+ 1.50		+0 36.79
v_2	15	30 35		—0.1	30 34.9	V	+ 6.00		+0 41.29

*** There seem to be one or two faint stars in the vicinity of k_1, but vision is very bad.**

Revision Zone 3, Jan. 12, 1858.

Name of Star.	Mag.	Reading of Scale.	Reduction.	Correct.	$\delta-\delta_0$	Principal Star.	U′—U.	Correction.	$\alpha-\alpha_0$
		′ ″	″	″	′ ″	m s	s	s	m s
v_3	16	35 58	0.0	−1.0	35 57.0	V +0 35.29	+10.25	0.00	+0 45.54
v_4	13	36 22		−1.1	36 20.9	V	+11.75		+0 47.04
v_5	17	34 0		−0.6	33 59.4	V	+12.00		+0 47.29
w_1	16.17	32 25		−0.4	32 24.6	W+0 53.42	+ 4.00		+0 57.42
x_1	15	38 33		−1.5	38 31.5	X +1 0.08	+ 1.50		+1 1.58
y_1	14	31 53		−0.3	31 52.7	Y +1 21.49	+ 2.60		+1 24.09
z_1	15	30 13		+0.1	30 13.1	Z +1 26.16	+ 8.50		+1 34.66
z_2	16	40 0		−1.7	39 58.3	Z	+13.50		+1 39.66
z_3	15					Z	+23.25		+1 49.41

Revision Zone 4, Jan. 14, 1858.

Name of Star.	Mag.	Reading of Scale.	Reduction.	Correct.	$\delta-\delta_0$	Principal Star.	U′—U.	Correction.	$\alpha-\alpha_0$
		′ ″	″	″	′ ″	m s	s	s	m s
a'_1	15	22 3	+1.0	−0.5	22 3.5	B′−1 23.03	− 8.65	0.00	−1 31.68
a'_2	15	29 0		−1.5	28 59.5	B′	− 8.65		−1 31.68
a'_3	17	23 20		−0.5	23 20.5	B′	− 0.50		−1 23.53
b'_1	16	29 30		−1.8	29 29.2	B′	+13.90		−1 9.13
d'_1	14.15	20 12		−0.1	20 12.9	D′−0 51.49	+ 8.50		−0 42.99
d'_2	15	20 30		−0.1	20 30.9	D′	+16.15		−0 35.34
g'_1	16.17	27 30		−1.4	27 29.6	H′−0 11.67	− 2.90		−0 14.57
h'_1	14	26 26		−1.2	26 25.8	H′	+ 6.50		−0 5.17
p'_1	16	20 50		−0.1	20 50.9	P′+0 0.77	+ 0.75		+0 1.52

Revision Zone 5, Jan. 14, 1858.

Name of Star.	Mag.	Reading of Scale.	Reduction.	Correct.	$\delta-\delta_0$	Principal Star.	U′—U.	Correction.	$\alpha-\alpha_0$
		′ ″	″	″	′ ″	m s	s	s	m s
ϕ	16	20 37	+1.0	−0.1	20 37.9	O′+0 00.55	+ 1.00	0.00	+0 1.55
ψ	17.18	20 47		−0.1	20 47.9	O′	+ 1.35		+0 1.90
	17	21 49		−0.3	21 49.7	O′	+ 0.20		+0 0.75
δ	13.14	22 26		−0.4	22 26.6	O′	+ 3.35		+0 3.90
ϵ	15	22 45		−0.5	22 45.5	O′	+ 2.25		+0 2.80
ζ	16	22 51		−0.5	22 51.5	O′	+ 1.50		+0 2.05
η	17	22 52		−0.5	22 52.5	O′	+ 5.75		+0 6.30
θ	15	26 49		−1.2	26 48.8	O′	+ 1.35		+0 1.90
ι	16	27 10		−1.3	27 9.7	O′	+ 1.50		+0 2.05
κ	16.17	28 21		−1.5	28 20.5	O′	− 1.35		−0 0.80
t'_1	17	24 18		−0.8	24 18.2	T′+0 14.51	+ 0.25		+0 14.76
t'_2	18	21 48		−0.3	21 48.7	T′	+ 4.50		+0 19.01

Revision Zone 6, Jan. 20, 1858.

Name of Star.	Mag.	Reading of Scale.	Reduction.	Correct.	$\delta-\delta_0$	Principal Star.	U′—U.	Correction.	$\alpha-\alpha_0$
		′ ″	″	″	′ ″	m s	s	s	m s
v'_2	16	25 32	+1.0	−1.1	25 31.9	V′ +0 25.98	+ 4.50	0.00	+0 30.48
v'_3	13	26 31		−1.3	26 30.7	V′	+ 5.00		+0 30.98
v'_1	17.18	22 49		−0.6	22 49.4	U′ +0 24.92	+ 4.50		+0 29.42
v'_4	17	27 0		−1.4	26 59.6	V′ +0 25.98	+ 8.75		+0 34.73
w'_1	14.15	25 58		−1.2	25 57.8	W′+0 42.08	+ 3.50		+0 45.58
w'_2	15	23 48		−0.7	23 48.3	W′	+12.25		+0 54.33
y'_1	16	29 40		−2.0	29 39.0	Y′ +1 22.83	+ 0.33		+1 23.16
y'_2	16.17	22 30		−0.4	22 30.6	Y′	+13.00		+1 35.83

Revision Zone 7, Feb. 12, 1858.

Name of Star.	Mag.	Reading of Scale.	Reduction.	Correct.	$\delta-\delta_0$	Principal Star.		U′—U.	Correction.	$\alpha-\alpha_0$
		′ ″	″	″	′ ″		m s	s	s	m s
a″₁	16	11 55	+3.1	−0.6	11 57.5	A″	−1 38.83	+ 8.00	0.00	−1 30.83
b″₁	15	17 5		−1.9	17 6.2	B″	−1 26.53	+17.75		−1 8.78
e″₁	17	12 15		−0.7	12 17.4	E″	−0 56.77	+ 1.25		−0 55.52
e″₂	15.16	18 53		−2.4	18 53.7	E″		+ 7.00		−0 49.77
g″₁	17	18 12		−2.2	18 12.9	G″	−0 48.27	+ 0.25		−0 48.02
i″₁*	15	20 20		−2.8	20 20.3	I″	−0 45.84	+ 2.7		−0 43.09
i″₂	14	14 44		−1.3	14 45.8	I″		+ 3.50		−0 42.34
i″₃	16	10 58		−0.3	11 0.8	I″		+ 5.25		−0 40.59
k″₁	13.14	16 33		−1.8	16 34.3	K″	−0 36.44	+ 0.10		−0 36.34
l″₁	16.17	19 10		−2.4	19 10.7	L″	−0 30.98	+ 5.60		−0 25.38

Revision Zone 8, March 10, 1858.

Name of Star.	Mag.	Reading of Scale.	Reduction.	Correct.	$\delta-\delta_0$	Principal Star.		U′—U.	Correction.	$\alpha-\alpha_0$
		′ ″	″	″	′ ″		m s	s	s	m s
o″₁	15	11 32	+2.4	−0.3	11 34.1	N″	−0 20.64	+ 1.25	0.00	−0 19.39
o″₂	15.16,16	19 30		−1.9	19 30.5	O″	−0 20.40	+ 2.00		−0 18.40
p″₁	15	13 17		−0.7	13 18.7	P″	−0 16.15	+ 3.12		−0 13.03
p″₂	17 or 17.18	13 25		−0.7	13 26.7	P″		+ 4.87		−0 11.28
q″₁	15.16	13 12		−0.7	13 13.7	R″	−0 6.32	− 0.62		−0 6.94
r″₂	16	17 0		−1.4	17 1.0	R″		+ 0.50		−0 5.82
r″₃	17	17 20		−1.5	17 20.9	R″		+ 1.25		−0 5.07
r″₁	15	19 37		−1.9	19 37.5	R″		+ 0.50		−0 5.82
r″₄	16.17	19 39		−1.9	19 39.5	R″		+ 2.25		−0 4.07
p	16	19 35		−1.9	19 35.5	V″	0 0.00	− 0.50		−0 0.50
ξ	16.17	20 25		−2.1	20 25.3	V″		− 1.12		−0 1.12
π	17	20 24		−2.1	20 24.3	V″		− 0.65		−0 0.65
r″₅	14	12 23		−0.5	12 24.9	R″	−0 6.32	+ 3.75		−0 2.57
β	15	19 38		−1.9	19 38.5	V″	0 0.00	+ 4.90		+0 4.90
γ	16	19 35		−1.9	19 35.5	V″		+ 5.50		+0 5.50
b″b″₁	14	17 42		−1.5	17 42.9	B″B″	+0 10.09	+ 0.10		+0 10.19
b″b″₂	17	13 1		−0.6	13 2.8	Z″	+0 7.96	+ 1.62		+0 9.58

Revision Zone 9, March 13, 1858.

Name of Star.	Mag.	Reading of Scale.	Reduction.	Correct.	$\delta-\delta_0$	Principal Star.		U′—U.	Correction.	$\alpha-\alpha_0$
		′ ″	″	″	′ ″		m s	s	s	m s
m	17.18	14 25	+2.5	−0.9	14 26.6	Y″	+0 6.51	+ 9.25	0.00	+0 15.76
n	17	14 10		−0.8	14 11.7	Y″		+12.00		+0 18.51
o	16	16 45		−1.4	16 46.1	Y″		+14.62		+0 21.13
p	15	16 25		−1.3	16 26.2	Y″		+18.00		+0 24.51
h″h″₁	17	15 53		−1.2	15 54.3	H″H″	+0 25.89	+ 0.50		+0 26 39
h″h″₂	13.14	11 21		−0.2	11 23.3	H″H″		+ 1.75		+0 27.64
i″i″₁	16	14 36		−0.9	14 37.6	I″I″	+0 34.32	+ 1.50		+0 35.82
i″i″₂†	17.18	18 0		−1.6	18 0.9	I″I″		+ 5.90		+0 40.22
i″i″₃	17	10 20		0.0	10 22.5	I″I″		+ 9.75		+0 44.07
k″k″₁	16.17	17 45		−1.5	17 46.0	K″K″	+0 53.44	+ 5.50		+0 58.94
l″l″₁	16	17 36		−1.5	17 37.0	L″L″	+1 10.88	+ 4.00		+1 14.88
m″m″₁	17	11 15		−0.1	11 17.4	L″L″		+ 8.00		+1 18.88

* Declination of i″₁ probably erroneous.

† There are perhaps two or more fainter stars between this star and I″I″. There is, in the region following I″I″ and north of it, a large number of very minute stars, too faint to be observed steadily, six or eight within 4′ of it, and apparently large numbers at the limit of vision.

Revision Zone 9, March 13, 1858.

Name of Star.	Mag.	Reading of Scale.	Reduction.	Correct.	$\delta-\delta_0$	Principal Star.	U′—U.	Correction.	$\alpha-\alpha_0$
		′ ″	″	″	′ ″	m s	s	s	m s
$m''m''_2$	17	17 20	+2.5	−1.4	17 21.1	L″L″ +1 10.88	+10.50	0.00	+1 21.38
$n''n''_1$	17	12 40		−0.4	12 42.1	O″O″ +1 32.22	− 1.25		+1 30.97
$o''o''_1$	16.17	19 30		−1.8	19 30.7	O″O″	+11.00		+1 43.22
$o''o''_2$	16	20 10		−2.0	20 10.5	O″O″	+14.50		+1 46.72

Revision Zone 10, March 18, 1858.

Name of Star.	Mag.	Reading of Scale.	Reduction.	Correct.	$\delta-\delta_0$	Principal Star.	U′—U.	Correction.	$\alpha-\alpha_0$
		′ ″	″	″	′ ″	m s	s	s	m s
a'''_1	16	8 25	+4.0	−0.8	8 28.2	B‴ −1 26.53	− 2.75	0.00	−1 29.28
d'''_1	16.17	7 15		−0.7	7 18.3	D‴ −1 20.23	+ 4.00		−1 16.23
d'''_2	17	6 33		−0.7	6 36.3	D‴	+ 6.25		−1 13.98
d'''_3	15	6 10		−0.6	6 13.4	D‴	+ 7.65		−1 12.58
f'''_1	16.17	9 30		−0.9	9 33.1	F‴ −0 59.09	+ 1.25		−0 57.84
f'''_2	17	9 10		−0.8	9 13.2	F‴	+ 1.50		−0 57.59
g'''_1	15.16	1 30		−0.2	1 33.8	G‴ − 0 55.56	+ 8.00		−0 47.56
i'''_1	17.18	7 0		−0.6	7 3.4	G‴	+16.00		−0 39.56
i'''_2	16.17	8 15		−0.7	8 18.3	G‴	+17.25		−0 38.31
i'''_3	16	2 0		−0.2	2 3.8	I‴ −0 40.50	+ 3.00		−0 37.50
K‴		3 2		−0.3	3 5.7	I‴	+ 5.00		−0 35.50
k'''_1	17	1 24		−0.2	1 27.8	I‴	+ 6.50		−0 34.00
k'''_2	16.17	4 10		−0.4	4 13.6	I‴	+11.00		−0 29.50
l'''_1	17	6 0		−0.5	6 3.5	I‴	+12.50		−0 28.00
l'''_2	16	1 10		−0.1	1 13.9	L‴ −0 28.73	+ 2.00		−0 26.73
m'''_1	14	0 50		−0.1	0 53.9	I‴ −0 40.50	+16.00		−0 24.50
m'''_2	17	1 30		−0.2	1 33.8	I‴	+21.25		−0 19.25
n'''_1	16	6 30		−0.5	6 33.5	N‴ −0 2.43	− 1.67		−0 4.10
n'''_{-2}	14					N‴	− 2.50		−0 4.93
n'''_1	16	−0 10		0.0	−0 6.0	N‴	0.00		−0 2.43
n'''_5		0 0		0.0	+0 4.0	N‴	+ 2.00		−0 0.43
n'''_3	17	3 3		−0.2	3 6.8	N‴	+ 1.08		−0 1.35
n'''_2	16	4 3		−0.3	4 6.7	N‴	+ 0.58		−0 1.85
n'''_4	15	4 3		−0.3	4 6.7	N‴	+ 1.83		−0 0.60
o'''_1	16	0 35		0.0	0 39.0	O‴ +0 0.73	+ 0.25		+0 0.98
o'''_2	14	0 40		0.0	0 44.0	O‴	+ 2.00		+0 2.73
o'''_3	16	8 40		+0.7	8 44.7	O‴	+ 4.75		+0 5.48
o'''_4	15.16	8 5		+0.6	8 9.6	O‴	+11.25		+0 11.98
q'''_1	16	7 30		−0.6	7 33.4	Q‴ +0 15.54	+10.25		+0 25.79
q'''_2	14.15	7 0		−0.5	7 3.5	Q‴	+12.25		+0 27.79
q'''_3	16.17	8 0		−0.6	8 3.4	O‴	+19.00		+0 34.54

Revision Zone 11, March 19, 1858.

Name of Star.	Mag.	Reading of Scale.	Reduction.	Correct.	$\delta-\delta_0$	Principal Star.	U′—U.	Correction.	$\alpha-\alpha_0$
		′ ″	″	″	′ ″	m s	s	s	m s
u‴	15.16	1 8	+3.8	−0.1	1 11.7	U‴ +0 51.08	+ 0.75	0.00	+0 51.83
v'''_1	16	1 50		−0.2	1 53.6	V‴ +0 59.52	+ 5.50		+1 5.02
v'''_2	16.17	1 10		−0.1	1 13.7	V‴	+13.00		+1 12.52
w'''_1	15	0 55		0.0	0 58.8	W‴ +1 15.53	+ 2.00		+1 17.53
w'''_2	14.15	8 40		−1.2	8 42.6	W‴	+10.75		+1 26.28

SECTION II. PART I.

ZONES NEAR c AND ι ORIONIS.

Zone I., Jan. 21, 1864. $\tau = -0^h\ 50^m$ $\theta = 22°$

Letter.	Mag.	Reading of Scale.	d		1864.0 δ—δ₀	Transit.	k		1864.0 α—α₀
		′ ″	′ ″	″	′ ″	h m s	h m s	s	m s
A	12	6 55	+19 53.9	−1.8	+26 47.1	4 36 58.6	−4 39 32.51	.00	−2 33.91
B	9	5 20		−1.4	25 12.5	37 1.0		+.01	−2 31.50
C	13	4 10		−1.1	24 2.8	37 8.7		+.01	−2 23.80
D	13	5 28		−1.4	25 20.5	37 45.7		.00	−1 46.81
E	10	2 20		−0.6	22 13.3	37 56.1		.00	−1 36.41
F	13	3 12		−0.8	23 5.1	38 14.7		.00	−1 17.81
G	13	5 35		−1.3	25 27.6	38 36.9		.00	−0 55.61
H	10.11	0 20		−0.1	20 13.8	38 49.4		.00	−0 43.11
I	13	3 28		−0.8	23 21.1	38 50.8		.00	−0 41.71
K	13	7 30		−1.8	27 22.1	38 59.7		.00	−0 32.81
L	9.10	−0 25		+0.1	19 29.0	39 33.3		.00	+0 0.79
M	9	9 19		−2.2	29 10.7	39 45.5		−.01	+0 12.98
N	12	5 30		−1.3	25 22.6	40 9.4		−.01	+0 36.88
O	10.11	7 32		−1.7	27 24.2	40 15.4		−.01	+0 42 88
P	9.10	0 1		0.0	19 54.9	40 25.9		.00	+0 53.39
Q	13	5 50		−1.3	25 42.6	41 4.1		−.01	+1 31.58
R	11.12	4 15		−0.9	24 8.0	41 5.6		−.01	+1 33.08
S	11	1 45		−0.3	21 38.6	4 41 31.8		−.01	+1 59.28

Zone II., 10′ n. of I. Jan. 21, 1864. $\tau = -0^h\ 40^m$ $\theta = 22°$

Letter.	Mag.	Reading of Scale.	d		1864.0 δ—δ₀	Transit.	k		1864.0 α—α₀
		′ ″	′ ″	″	′ ″	h m s	h m s	s	m s
A	14	1 30	+29 43.6	−0.4	+31 13.2	4 46 59.9	−4 48 48.96	+.01	−1 49.05
B	12	8 44		−2.1	38 25.5	47 32 3		.00	−1 16.66
C	13	6 50		−1.7	36 31.9	47 39.3		.00	−1 9.66
D	13	2 50		−0.7	32 32.9	47 41.4		.00	−1 7.56
E	7.8	5 16		−1.2	34 58.4	47 54.6		.00	−0 54.36
F	10	7 40		−1.8	37 21.8	48 26.9		.00	−0 22.06
G	8	9 44		−2.3	39 25.3	48 35.3		−.01	−0 13.67
H	10	1 50		−0.4	31 33.2	48 44.3		.00	−0 4.66
I	6	3 20		−0.8	33 2.8	48 53.8		.00	+0 4.84
K		−0 35		+0.1	29 8.7	49 2.1		.00	+0 13.14
L	7	2 20		−0.5	32 3.1	49 10.4		.00	+0 21.44
M	11	8 45		−2.0	38 26.6	49 34.0		−.01	+0 45.03
N	13	1 40		−0.3	31 23.3	50 30.4		−.01	+1 41.43
O	12.13	2 50		−0.6	32 33.0	51 10.6		−.01	+2 21.63
P	12.13	2 55		−0.6	32 38.0	51 12.2		−.01	+2 23.23
Q	8	5 20		−1.2	35 2.4	51 24.0		−.01	+2 35.03
R	11.12	6 50		−1.5	36 32.1	5 51 26.5		−.02	+2 37.52

NOTE. — The same letters in different zones of this Section represent in general different stars.

Zone III.=II. repeated. Jan. 21, 1864. $\tau=-0^h\ 31^m$ $\theta=+22°$

Letter.	Mag.	Reading of Scale.	d	1864.0 $\delta-\delta_0$	Transit.	k	1864.0 $\alpha-\alpha_0$
		′ ″	′ ″ ″	′ ″	h m s	h m s s	m s
A	14	1 33	+29 41.5 −0.4	+31 14.1	4 56 21.8	−4 58 10.38 +.01	−1 48.57
B		8 43	−2.1	38 22.4	56 53.9	.00	−1 16.48
C		6 50	−1.7	36 29.8	57 0.5	.00	−1 9.88
D		2 50	−0.7	32 30.8	57 2.9	.00	−1 7.48
E	9	5 18	−1.3	34 58.2	57 15.9	.00	−0 54.48
F	10.11	7 40	−1.8	37 19.7	57 48.2	.00	−0 22.18
G	9	9 44	−2.3	39 23.2	57 56.4	−.01	−0 13.99
H	6	3 27	−0.8	33 7.7	58 15.4	.00	+0 5.02
I		−0 33	+0.1	29 8.6	58 23.4	.00	+0 13.02
K	7	2 20	−0.5	32 1.0	58 31.8	.00	+0 21.42
L	13	5 50	−1.4	35 30.1	58 46.5	−.01	+0 36.11
M	11	8 45	−2.0	38 24.5	58 56.1	−.01	+0 45.71
N	14	1 40	−0.3	31 21.2	59 52.0	−.01	+1 41.61
O	12.13	2 55	−0.6	32 35.9	5 0 32.3	−.01	+2 21.91
P	12.13	3 4*	−0.6	32 44.9			
Q	8	5 20	−1.2	35 0.3	0 45.6	−.01	+2 35.21
R	11	6 55	−1.5	36 35.0	5 0 47.9	−.02	+2 37.50

Zone IV., 10′ n. of II. III. Jan. 21, 1864. $\tau=-0^h\ 18^m$ $\theta=+22°$

Letter.	Mag.	Reading of Scale.	d	1864.0 $\delta-\delta_0$	Transit.	k	1864.0 $\alpha-\alpha_0$
		′ ″	′ ″ ″	′ ″	h m s	h m s s	m s
A	11	8 40	+39 40.0 −2.2	+48 17.8	5 9 13.2	−5 11 36.99 +.01	−2 23.78
B	9.10	9 18	−2.3	48 55.7	9 14.9	.00	−2 22.09
C	10	5 17	−1.3	44 55.7	9 32.4	+.01	−2 4.58
D	12.13	1 12	−0.4	40 51.6	9 49.1	+.01	−1 47.88
E	13	3 24	−0.9	43 3.1	10 2.5	.00	−1 34.49
F	12.13	7 50	−1.9	47 28.1	10 30.6	.00	−1 6.39
G	13	5 10	−1.2	44 48.8	10 31.5	.00	−1 5.49
H	13	0 32†	−0.1	40 11.9	10 41.4	.00	−0 55.59
I	10.11	10 12	−2.5	49 49.5	10 46.2	.00	−0 50.79
K	9	−0 16	+0.1	39 24.1	11 23.3	.00	−0 13.69
L	11	0 20	−0.1	39 59.9	11 24.0	.00	−0 12.99
M	11	8 2	−1.9	47 40.1	11 31.4	.00	−0 5.59
N	11.12	2 38	−0.6	42 17.4	11 36.2	.00	−0 0.79
O	11	2 44	−0.6	42 23.4	11 37.1	.00	+0 0.11
P	12	7 30	−1.8	47 8.2	12 1.0	−.01	+0 24.00
Q	12	7 20	−1.8	46 58.2	12 14.6	−.01	+0 37.60
R	12.13	4 10	−0.9	43 49.1	13 11.9	−.01	+1 34 90
S	12.13	7 3	−1.6	46 41.4	13 20.7	−.01	+1 43.70
T	12	5 40	−1.3	45 18.7	13 22.2	−.01	+1 45.20
U	12	3 46	−0.8	43 25.2	13 40.2	−.01	+2 3.20
V	13	5 20	−1.2	44 58.8	5 13 49.9	−.01	+2 12.90

Zone V., 10′ n. of IV. Jan. 21, 1864. $\tau=-0^h\ 8^m$ $\theta=+22°$

Letter.	Mag.	Reading of Scale.	d	1864.0 $\delta-\delta_0$	Transit.	k	1864.0 $\alpha-\alpha_0$
		′ ″	′ ″ ″	′ ″	h m s	h m s s	m s
A	12	0 29	+49 46.1 −0.2	+50 14.9	5 20 0.2	−5 22 23.29 +.01	−2 23.08
B	9	−0 48	+0.1	48 58.2	20 1.6	+.01	−2 21.68
C	12.13	2 23	−0.7	52 8.4	20 3.4	+.01	−2 19.88
D	14	5 3	−1.3	54 47.8	5 20 19.7	+.01	−2 3.58

* Same as II. P. † Corrected from 0′40″ Jan. 27.

Zone V. 10′ n. of IV., Jan. 21, 1864. $\tau = -0^h\ 8^m$ $\theta = +22°$

Letter.	Mag.	Reading of Scale.	d	1864.0 $\delta-\delta_0$	Transit.	k	1864.0 $\alpha-\alpha_0$
		′ ″	′ ″ ″	′ ″	h m s	h m s s	m s
E	12.13	2 53	+49 46.1 −0.8	+52 38.3	5 20 32.0	−5 22 23.29 +.01	−1 51.28
F	13	8 20	−2.0	58 4.1	21 1.7	.00	−1 21.59
G	11	9 40	−2.4	59 23.7	21 7.3	.00	−1 15.99
H	13	1 48	−0.4	51 33.7	21 18.7	.00	−1 4.59
I	13	4 50	−1.2	54 34.9	21 22.4	.00	−1 0.89
K	12.13	3 10	−0.8	52 55.3	21 27.4	.00	−0 55.89
L	10	0 3	0.0	49 46.4	21 32.5	.00	−0 50.79
M	9	4 12	−1.0	53 57.1	21 36.4	.00	−0 46.89
N	9	10 0	−2.4	59 43.7	21 47.0	.00	−0 36.29
O	13	0 30	−0.1	50 16.0	21 58.8	.00	−0 24.49
P	10	1 57	−0.5	51 42.6	22 4.3	.00	−0 18.99
Q	7.8	3 58	−0.9	53 43.2	22 26.0	.00	+0 2.71
R	7.8	8 12	−1.9	57.56.2	22 26.0	.00	+0 2.71
S	9	8 17	−1.9	58 1.2	23 10.1	−.01	+0 46.80
T	12.13	9 30	−2.2	59 13.9	23 39.7	−.01	+1 16.40
U	12	8 10	−1.9	57 54.2	23 44.9	−.01	+1 21.60
V	11	8 18	−1.9	58 2.2	23 55.4	−.01	+1 32.10
W	13	8 4	−1.8	57 48.3	5 24 11.3	−.01	+1 48.00

Zone VI. 10′ n. of V., Jan. 21, 1864. $\tau = +0^h\ 3^m$ $\theta = +22°$

Letter.	Mag.	Reading of Scale.	d	1864.0 $\delta-\delta_0$	Transit.	k	1864.0 $\alpha-\alpha_0$
		′ ″	′ ″ ″	′ ″	h m s	h m s s	m s
A	13	6 10	+59 27.2 −1.5	+65 35.7	5 30 50.8	−5 33 5.53 +.01	−2 14.72
B	10	2 15	−0.6	61 41.6	31 3.7	+.01	−2 1.82
C	13	1 35	−0.4	61 1.8	31 23.7	+.01	−1 41.82
D	12	8 54	−2.2	68 19.0	31 39.3	.00	−1 26.23
E	11.12	5 56	−1.4	65 21.8	31 46.0	.00	−1 19.53
F	11.12	3 57	−1.0	63 23.2	31 56.5	.00	−1 9.03
G	11	5 7	−1.2	64 33.0	32 7.3	.00	−0 58.23
H	10	1 28	−0.4	60 54.8	32 12.4	.00	−0 53.13
I	8.9	0 16	−0.1	59 43.1	32 29.1	.00	−0 36.43
K	11 or 12	3 12	−0.7	62 38.5	32 41.2	.00	−0 24.33
L	9.10	2 36	−0.6	62 2.6	32 45.0	.00	−0 20.53
M	10	0 45	−0.2	60 12.0	32 49.8	.00	−0 15.73
N	8.9	2 8	−0.5	61 34.7	33 17.0	.00	+0 11.47
O	12	8 48	−2.0	68 13.2	33 37.4	−.01	+0 31.86
P	11.12	2 30	−0.5	61 56.7	33 59.2	−.01	+0 53.66
Q	11.12	1 37	−0.3	61 3.9	34 3.7	−.01	+0 58.16
R	12.13	−0 14	+0.1	59 13.3	34 22.2	.00	+1 16.67
S	12	7 38	−1.7	67 3.5	35 19.9	−.01	+2 14.36
T	11	4 6	−0.8	63 32.4	5 35 28.8	−.01	+2 23.26

Zone VII. same as VI., Jan. 21, 1864. $\tau = +0^h\ 14^m$ $\theta = +22°$

Letter.	Mag.	Reading of Scale.	d	1864.0 $\delta-\delta_0$	Transit.	k	1864.0 $\alpha-\alpha_0$
		′ ″	′ ″ ″	′ ″	h m s	h m s s	m s
A	12.13	8 50	+59 32.2 −2.2	+68 20.0	5 42 4.6	−5 43 31.05 .00	−1 26.45
B	11.12	5 50	−1.5	65 20.7	42 11.5	.00	−1 19.55
C	11	−0 11	0.0	59 21.2	42 15.1	+.01	−1 15.94
D	11.12	3 50	−1.0	63 21.2	42 21.7	.00	−1 9.35
E	12	2 11	−0.5	61 42.7	5 42 32.3	.00	−0 58.75

Zone VII. same as VI., Jan. 21, 1864. $\tau=+0^h\ 14^m$ $\theta=+22°$

Letter.	Mag.	Reading of Scale.	d		1864.0 $\delta-\delta_0$	Transit.	k		1864.0 $\alpha-\alpha_0$
		′ ″	′ ″	″	′ ″	h m s	h m s	s	m s
F	9.10	1 20*	+59 32.2	−0.3	+60 51.9	5 42 37.8	−5 43 31.05	.00	−0 53.25
G	8	0 12		−0.1	59 44.1	42 54.8		.00	−0 36.25
H	11	3 10		−0.7	62 41.5	43 6.7		.00	−0 24.35
I	9.10	2 30		−0.6	62 1.6	43 10.4		.00	−0 20.65
K	10	0 43		−0.2	60 15.0	43 15.4		.00	−0 15.65
L	13	5 50		−1.4	65 20.8	43 27.2		.00	−0 3.85
M		2 3		−0.5	61 34.7	5 43 42.5		.00	+0 11.45

Zone VIII. 6′ n. of VI., Jan. 21, 1864. $\tau=+0^h\ 54^m$ $\theta=+22°$

Letter.	Mag.	Reading of Scale.	d		1864.0 $\delta-\delta_0$	Transit.	k		1864.0 $\alpha-\alpha_0$
		′ ″	′ ″	″	′ ″	h m s	h m s	s	m s
A	12.13	9 25	+65 25.7	−2.4	+74 48.3	6 21 18.2	−6 23 49.08	+.01	−2 30.87
B	13	0 2		−0.1	65 27.6	21 34.5		+.01	−2 14.57
C	14	2 45		−0.7	68 10.0	21 43.7		+.01	−2 5.37
D	12	2 55		−0.8	68 19.9	22 22.7		+.01	−1 26.37
E	12	−0 6		0.0	65 19.7	22 29.6		+.01	−1 19.47
F	12	6 59		−1.7	72 23.0	22 36.6		.00	−1 12.48
G	12	3 14		−0.8	68 38.9	22 40.8		.00	−1 8.28
H	12	0 20		−0.1	65 45.6	22 40.8		.00	−1 8.28
I†	12	−0 50		+0.2	64 28.9				−0 58.32
K	13	−0 12		0.0	65 13.7	23 8.3		.00	−0 40.78
L	13	0 20		−0.1	65 45.6	23 10.0		.00	−0 39.08
M	12.13	0 20		−0.1	65 45.6	23 17.0		.00	−0 32.08
N	12	5 48		−1.3	71 12.4	23 31.1		.00	−0 17.98
O	13	7 30		−1.8	72 53.9	23 38.2		.00	−0 10.88
P	13	5 45		−1.3	71 9.4	23 45.9		.00	−0 3.18
Q	11	2 48		−0.6	68 13.1	24 20.8		.00	+0 31.72
R	13	1 15		−0.3	66 40.4	24 21+3		.00	+0 32.22
S	12	5 58		−1.4	71 22.3	24 55.9		.00	+1 6.82
T	13	6 11		−1.4	71 35.3	25 28.8		−.01	+1 39.71
U	13	4 50		−1.0	70 14.7	25 35.4		−.01	+1 46.31
V	14	5 32		−1.2	70 56.5	25 50.5		−.01	+2 1.41
W	12	1 40		−0.3	67 5.4	6 26 3.6		−.01	+2 14.51

Zone IX. 10′ n. of VIII., Jan. 21, 1864. $\tau=+1^h\ 7^m$ $\theta=+22°$

Letter.	Mag.	Reading of Scale.	d		1864.0 $\delta-\delta_0$	Transit.	k		1864.0 $\alpha-\alpha_0$
		′ ″	′ ″	″	′ ″	h m s	h m s	s	m s
A	12.13	−0 32	+75 22.1	0.0	+74 50.1	6 34 46.2	−6 37 17.05	+.01	−2 30.84
B	12.13	1 50		−0.5	77 11.6	34 49.6		+.01	−2 27.44
C	11.12	6 10		−1.5	81 30.6	35 10.1		+.01	−2 6.94
D	12	9 20		−2.3	84 39.8	35 11.9		+.01	−2 5.14
E	13	5 50		−1.4	81 10.7	36 3.8		.00	−1 13.25
F	12.13	5 50		−1.4	81 10.7	36 8.1		.00	−1 8.95
G	12.13	1 0		−0.3	76 21.8	36 20.0		.00	−0 57.05
H	12.13	1 4		−0.3	76 25.8	36 21.6		.00	−0 55.45
I	12	9 10		−2.2	84 29.9	36 44.4		.00	−0 32.65
K	9	1 16		−0.3	76 37.8	36 50.8		.00	−0 26.25
L	13	−0 18		+0.1	75 4.2	37 14.4		.00	−0 2.65
M	12.13	5 55		−1.4	81 15.7	37 28.3		.00	+0 11.25
N	12.13	−0 11		+0.1	75 11.2	6 37 47.4		.00	+0 30.35

* Corrected from 0′20″ by revision.

† The position of I is taken from observation, March 3.

Zone IX. 10′ n. of VIII., Jan 21, 1864. $\tau = +1^h\ 7^m$ $\theta = +22°$

Letter.	Mag.	Reading of Scale.	d		1864.0 $\delta - \delta_0$	Transit.	k		1864.0 $\alpha - \alpha_0$
		′ ″	′ ″	″	′ ″	h m s	h m s	s	m s
O	12.13	4 20	+75 22.1	−0.9	+79 41.2	6 38 55.6	−6 37 17.05	−.01	+1 38.54
P	11	9 23		−2.2	84 42.9	38 57.1		−.01	+1 40.04
Q	12	2 54		−0.6	78 15.5	6 39 19.3		−.01	+2 2.24

Zone X,* Jan. 25, 1864. $\tau = +1^h\ 22^m$ $\theta = +35°$

Letter.	Mag.	Reading of Scale.	d		1864.0 $\delta - \delta_0$	Transit.	k		1864.0 $\alpha - \alpha_0$
		′ ″	′ ″	″	′ ″	h m s	h m s	s	m s
A	11	8 20	−29 58.9	−1.4	−21 40.3	4 4 59.2	−4 8 5.55	+.00	−3 6.35
B	10	4 20		−0.8	−25 39.7	5 7.2		+.01	−2 58.34
C	12	10 20		−1.7	−19 40.6	5 41.3		.00	−2 24.25
D	12.13	9 0		−1.5	−21 0.4	5 45.7		.00	−2 19.85
E	13	8 50		−1.5	−21 10.4	5 47.1		.00	−2 18.45
F	10	7 45		−1.3	−22 15.2	6 20.2		.00	−1 45.35
G	10	7 20		−1.2	−22 40.1	6 39.3		.00	−1 26.25
H	13	1 20		−0.2	−28 39.1	7 10.7		.00	−0 54.85
I		7 28		−1.2	−22 32.1	7 22.8		.00	−0 42.75
K		10 20		−1.7	−19 40.6	7 23.7		−.01	−0 41.86
L	11	1 14		−0.2	−28 45.1	7 38.9		.00	−0 26.65
M	9	1 12		−0.2	−28 47.1	7 58.0		.00	−0 7.55
N	8	3 43		−0.6	−26 16.5	8 4.4		.00	−0 1.15
O	12	2 15		−0.3	−27 44.2	8 12.4		.00	+0 6.85
P	10	2 12		−0.3	−27 47.2	8 16.0		.00	+0 10.45
Q	10	5 11		−0.8	−24 48.7	8 16.7		−.01	+0 11.14
R	8	7 25		−1.1	−22 35.0	9 19.7		−.01	+1 14.14
S	12	−0 26		+0.1	−30 24.8	9 23.3		.00	+1 17.75
T	9.10	1 15		−0.1	−28 44.0	9 47.4		−.01	+1 41.84
U	12	4 20		−0.6	−25 39.5	9 51.2		−.01	+1 45.64
V	13	8 8		−1.2	−21 52.1	10 4.9		−.02	+1 59.33
W	11	5 14		−0.7	−24 45.6	10 13.5		−.01	+2 7.94
X	10.11	0 46		0.0	−29 12.9	10 25.3		−.01	+2 19.74
Y	12.13	1 13		−0.1	−28 46.0	4 10 30.1		−.01	+2 24.54

Zone XI. same as X., Jan. 25, 1864. $\tau = -1^h\ 14^m$ $\theta = +35°$

Letter.	Mag.	Reading of Scale.	d		1864.0 $\delta - \delta_0$	Transit.	k		1864.0 $\alpha - \alpha_0$
		′ ″	′ ″	″	′ ″	h m s	h m s	s	m s
A	11.12	8 20	−29 59.2	−1.4	−21 40.6	4 12 26.3	−4 15 33.16	.00	−3 6.86
B	10	4 20		−0.8	−25 40.0	12 34.6		+.01	−2 58.55
C	12	10 20		−1.7	−19 40.9	13 8.9		.00	−2 24.26
D	12.13	9 0		−1.5	−21 0.7	13 12.9		.00	−2 20.26
E	12.13	8 50		−1.5	−21 10.7	13 14.5		.00	−2 18.66
F	9.10	7 48		−1.3	−22 12.5	13 47.6		.00	−1 45.56
G	9.10	7 23		−1.2	−22 37.4	14 6.9		.00	−1 26.26
H	13	1 25		−0.2	−28 34.4	14 37.7		.00	−0 55.46
I	9.10	7 25		−1.2	−22 35.4	14 50.4		.00	−0 42.76
K		10 20		−1.7	−19 40.9	14 51.4		−.01	−0 41.77
L	11	1 14		−0.2	−28 45.4	15 6.4		.00	−0 26.76
M	8.9	1 12		−0.2	−28 47.4	15 25.6		.00	−0 7.56
N	8	3 44		−0.6	−26 15.8	15 32.1		.00	−0 1.06
O	11	2 13		−0.3	−27 46.5	15 40.0		.00	+0 6.84
P	10	2 13		−0.3	−27 46.5	4 15 43.4		.00	+0 10.24

* This is the northernmost of the zones about ι Orionis.

Zone XI. same as X., Jan. 25, 1864, $\tau = +1^h\ 14^m$ $\theta = +35°$

Letter.	Mag.	Reading of Scale.	d	1864.0 δ−δ₀	Transit.	k	1864.0 α−α₀
		′ ″	′ ″ ″	′ ″	h m s	h m s s	m s
Q	13	2 20	−29 59.2−0.3	−27 39.5	4 16 2.9	−4 15 33.16 .00	+0 29.74
R	8	7 20	−1.1	−22 40.3	16 47.3	−.01	+1 14.13
S		−0 30	+0.1	−30 29.1	16 50.7	−.01	+1 17.53
T	9	1 15	−0.1	−28 44.3	17 14.8	−.01	+1 41.63
U	12.13	8 10	−1.2	−21 50.4	17 32.2	−.02	+1 59.02
V	11	5 17	−0.7	−24 42.9	17 40.9	−.01	+2 7.73
W	10	0 44	0.0	−29 15.2	17 52.8	−.01	+2 19.63
X	12	1 16	−0.1	−28 43.3	4 17 57.8	−.01	+2 24.63

Zone XII. 10′ s. of X. XI., Jan. 25, 1864. $\tau = -1^h\ 02^m$ $\theta = +35°$

Letter.	Mag.	Reading of Scale.	d	δ−δ₀	Transit.	k	α−α₀
		′ ″	′ ″ ″	′ ″	h m s	h m s s	m s
A	13	−0 27	−40 5.0−0.1	−40 32.1	4 25 6.2	−4 28 42.22+.01	−3 36.01
B	12	10 10	−1.8	−29 56.8	25 16.8	.00	−3 25.42
C	13	7 0	−1.2	−33 6.2	25 35.2	.00	−3 7.02
D	13	4 30	−0.8	−35 35.8	25 46.8	+.01	−2 55.41
E	12	1 13	−0.3	−38 52.3	26 3.4	+.01	−2 38.81
F	10	3 17	−0.6	−36 48.6	26 5.8	+.01	−2 36.41
G	13	2 45	−0.5	−37 20.5	26 32.2	+.01	−2 10.01
H	13	5 44	−1.0	−34 22.0	26 35.8	.00	−2 6.42
I	12.13	7 7	−1.2	−32 59.2	26 44.7	.00	−1 57.52
K	12.13	6 40	−1.1	−33 26.1	26 48.0	.00	−1 54.22
L	10	0 10	−0.1	−39 55.1	26 55.6	+.01	−1 46.61
M	12	5 58	−1.0	−34 8.0	27 14.0	.00	−1 28.22
N	13	8 15	−1.4	−31 51.4	27 30.6	.00	−1 11.62
O	13	3 18	−0.5	−36 47.5	27 55.9	.00	−0 46.32
P	10.11	3 50	−0.6	−36 15.6	28 11.6	.00	−0 30.62
Q	9.10	3 28	−0.6	−36 37.6	28 16.2	.00	−0 26.02
R	9	2 58	−0.5	−37 7.5	28 16.6	.00	−0 25.62
S	7.8	2 48	−0.5	−37 17.5	28 28.6	.00	−0 13.62
T	7.8	3 20	−0.5	−36 45.5	28 30.4	.00	−0 11.82
U	8*	7 50	−1.3	−32 16.3	28 38.6	−.01	−0 3.63
V	7.8*	1 27	−0.2	−38 38.2	28 43.5	.00	+0 1.28
W	5*	8 54	−1.4	−31 12.4	28 53.4	−.01	+0 11.17
X	9	2 48	−0.4	−37 17.4	29 13.9	.00	+0 31.68
Y	11	2 50	−0.4	−37 15.4	29 35.9	−.01	+0 53.67
Z	12.13	1 12	−0.1	−38 53.1	29 50.2	−.01	+1 7.97
AA	12	5 55	−0.9	−34 10.9	29 55.8	−.01	+1 13.57
BB		−0 13	+0.1	−40 17.9	30 3.7	−.01	+1 21.47
CC	11	2 57	−0.4	−37 8.4	30 49.6	−.01	+2 7.37
DD	8	7 35	−1.1	−32 31.1	30 54.8	−.01	+2 12.57
EE	12	1 23	−0.1	−38 42.1	4 31 5.0	−.01	+2 22.77

Zone XIII. revision of part of XII., Jan. 25, 1864. $\tau = -0^h\ 53^m$ $\theta = +35°$

Letter.	Mag.	Reading of Scale.	d	δ−δ₀	Transit.	k	α−α₀
		′ ″	′ ″ ″	′ ″	h m s	h m s s	m s
A	11	3 50	−40 5.0−0.6	−36 15.6	4 35 0.0	−4 35 30.76 .00	−0 30.76
B	10	3 25	−0.6	−36 40.6	35 4.7	.00	−0 26.06
C	9	2 59	−0.5	−37 6.5	35 5.3	.00	−0 25.46
D	12	5 18	−0.9	−34 47.9			

* Magnitude supplied March 14.

Zone XIII. revision of part of XII., Jan. 25, 1864. $\tau = -0^h\ 53^m$ $\theta = +35°$

Letter.	Mag.	Reading of Scale.	d		1864.0 $\delta-\delta_0$	Transit.	k		1864.0 $\alpha-\alpha_0$
		′ ″	′ ″	″	′ ″	h m s	h m s	s	m s
E	7	2 48	−40 5.0	−0.4	−37 17.4	4 35 17.3	−4 35 30.76	.00	−0 13.46
F	6.7	3 17		−0.5	−36 48.5				
G	9	7 50		−1.3	−32 16.3	35 26.9		−.01	−0 3.87
H	9	1 29		−0.2	−38 36.2	35 32.0		.00	+0 1.24
I		8 53		−1.4	−31 13.4	35 41.9		−.01	+0 11.13
K		2 48		−0.4	−37 17.4	4 36 2.4		.00	+0 31.64

Zone XIV. 10′ s. of XII. XIII., Jan. 25, 1864. $\tau = -0^h\ 47^m$ $\theta = +35°$

Letter.	Mag.	Reading of Scale.	d		1864.0 $\delta-\delta_0$	Transit.	k		1864.0 $\alpha-\alpha_0$
		′ ″	′ ″	″	′ ″	h m s	h m s	s	m s
A	12.13	0 13	−50 4.6	−0.1	−49 51.7	4 39 6.9	−4 41 53.35	+.01	−2 46.44
B	12	4 14		−0.8	−45 51.4	39 25.2		+.01	−2 28.14
C	12.13	−0 24		0.0	−50 28.6	39 31.0		+.01	−2 22.34
D	12	5 59*		−1.1	−44 6.7	39 38.8		.00	−2 14.55
E	10	6 28		−1.1	−43 37.7	39 54.7		.00	−1 58.65
F	9.10	10 10		−1.7	−39 56.3	40 6.7		.00	−1 46.65
G	13	5 2		−0.9	−45 3.5	40 37.8		.00	−1 15.55
H	12	7 40		−1.3	−42 25.9	40 41.8		.00	−1 11.55
I	13	8 40		−1.4	−41 26.0	41 4.8		.00	−0 48.55
K	12	2 50		−0.4	−47 15.0	41 7.6		.00	−0 45.75
L	12	7 20		−1.2	−42 45.8	41 8.7		.00	−0 44.65
M	12.13	6 7		−1.0	−43 58.6	41 47.7		.00	−0 5.65
N	11	5 13		−0.8	−44 52.4	41 54.5		.00	+0 1.15
O	9	8 32		−1.4	−41 34.0	42 13.9		−.01	+0 20.54
P	13	1 20		−0.2	−48 44.8	42 28.2		.00	+0 34.85
Q	13	1 20		−0.1	−48 44.7	42 51.4		.00	+0 58.05
R	13	−0 20		+0.1	−50 24.5	42 57.8		.00	+1 4.45
S	11.12	5 28		−0.8	−44 37.4	43 4.7		−.01	+1 11.34
T		9 50		−1.5	−40 16.1	43 14.7		−.01	+1 21.34
U	12	9 40		−1.5	−40 26.1	43 22.8		−.01	+1 29.44
V	9.10	4 30		−0.6	−45 35.2	43 38.4		−.01	+1 45.04
W	11	7 33		−1.1	−42 32.7	4 43 49.2		−.01	+1 55.84
β†		9 58				4 50 36.			
γ†		6 56				51 25.			
χ†		1 40				51 49.5			
δ†		1 30				53 3.7			
χ†		8 58				4 54 3.			

Zone XV. 10′ s. of XIV., Jan. 25, 1864. $\tau = -0^h\ 25^m$ $\theta = +35°$

Letter.	Mag.	Reading of Scale.	d		1864.0 $\delta-\delta_0$	Transit.	k		1864.0 $\alpha-\alpha_0$
		′ ″	′ ″	″	′ ″	h m s	h m s	s	m s
A	9	−1 0	−60 3.6	0.0	−61 3.6	5 2 2.5	−5 4 59.83	+.01	−2 57.32
B	8.9	−0 46		0.0	−60 49.6	2 5.4		+.01	−2 54.42
C	10	5 10		−1.0	−54 54.6	2 12.8		+.01	−2 47.02
D		4 20		−0.8	−55 44.4	2 15.7		+.01	−2 44.12
E		−0 12		−0.1	−60 15.7	2 18.3		+.01	−2 41.52
F	13	7 0		−1.2	−53 4.8	2 45.0		.00	−2 14.83
G	12.13	0 0		−0.1	−60 3.7	2 49.9		+.01	−2 9.92
H	12	0 5		−0.1	−59 58.7	5 3 3.8		+.01	−1 56.02

* Scale reading for D corrected from 4′59″ to 5′59″ March 12.

† These five stars not observed on the chronograph.

Zone XV. 10′ s. of XIV., Jan. 25, 1864. $\tau = -0^h\,25^m$ $\theta = +35°$

Letter.	Mag.	Reading of Scale.	d	1864.0 $\delta - \delta_0$	Transit.	k	1864.0 $\alpha - \alpha_0$
		′ ″	′ ″ ″	′ ″	h m s	h m s s	m s
I	12	5 58	−60 3.6 −1.1	−54 6.7	5 3 5.9	−5 4 59.83 .00	−1 53.93
K		0 13	0.0	−59 50.6	3 23.3	+.01	−1 36.52
L	13	4 14	−0.7	−55 50.3	3 42.5	.00	−1 17.33
M	12	3 23	−0.6	−56 41.2	3 44.9	.00	−1 14.93
N	12	4 47	−0.8	−55 17.4	4 44.0	.00	−0 15.83
O	9	6 13	−1.0	−53 51.6	4 50.4	.00	−0 9.43
P	10	9 4	−1.5	−51 1.1	4 55.0	−.01	−0 4.84
Q	10	1 30	−0.2	−58 33.8	5 21.1	.00	+0 21.27
R	13	7 30	−1.2	−52 34.8	5 42.1	−.01	+0 42.26
S	9	7 2	−1.1	−53 2.7	5 45.6	−.01	+0 45.76
T	10.11	2 45	−0.4	−57 19.0	5 50.8	−.01	+0 50.96
U	10	3 50	−0.6	−56 14.2	5 52.5	−.01	+0 52.66
V	10.11	−0 20	+0.1	−60 23.5	6 1.9	.00	+1 2.07
W	11	6 2	−0.9	−54 2.5	6 11.4	−.01	+1 11.56
X	13	4 12	−0.6	−55 52.2	6 18.8	−.01	+1 18.96
Y	10.11	9 3	−1.4	−51 2.0	7 15.7	−.01	+2 15.86
Z	10	3 40	−0.5	−56 24.1	7 20.6	−.01	+2 20.76
AA	10.11	0 30	0.0	−59 33.6	7 21.4	−.01	+2 21.56
BB	8	0 27	0.0	−59 36.6	5 7 48.4	−.01	+2 48.56

Zone XVI. 10′ s. of XV., Jan. 25, 1864. $\tau = -0^h\,15^m$ $\theta = +35°$

Letter.	Mag.	Reading of Scale.	d	1864.0 $\delta - \delta_0$	Transit.	k	1864.0 $\alpha - \alpha_0$
		′ ″	′ ″ ″	′ ″	h m s	h m s s	m s
A	9	9 3	−70 4.2 −1.6	−61 2.8	5 12 29.1	−5 15 26.25 +.01	−2 57.14
B	12	−0 30	0.0	−70 34.2	12 30.3	+.01	−2 55.94
C	8	9 18	−1.6	−60 47.8	12 31.9	+.01	−2 54.34
D	12	9 49	−1.7	−60 16.9	12 44.6	+.01	−2 41.64
E	12	3 49	−0.7	−66 15.9	12 55.4	+.01	−2 30.84
F	9	3 14	−0.6	−66 50.8	13 4.9	+.01	−2 21.34
G	8.9	2 41	−0.5	−67 23.7	13 5.3	+.01	−2 20.94
H	13	2 45	−0.5	−67 19.7	13 17.5	+.01	−2 8.74
I	10	8 30	−1.5	−61 35.7	13 29.7	.00	−1 56.55
K	9	1 11	−0.2	−68 53.4	13 39.3	+.01	−1 46.94
L	9	10 12	−1.7	−59 53.9	13 49.7	.00	−1 36.55
M	12	5 0	−0.9	−65 5.1	13 59.4	.00	−1 26.85
N	10	1 55	−0.4	−68 9.6	14 24.5	.00	−1 1.75
O	12	7 40	−1.3	−62 25.5	14 27.5	.00	−0 58.75
P	13	7 40	−1.3	−62 25.5	14 41.0	.00	−0 45.25
Q	9	−0 27	−0.1	−70 31.1	15 21.6	.00	−0 4.65
R	13	6 45	−1.1	−63 20.3	15 54.3	−.01	+0 28.04
S	12	2 0	−0.3	−68 4.5	16 0.8	.00	+0 34.55
T	10	9 55	−1.6	−60 10.8	16 19.9	−.01	+0 53.64
U	10	9 42	−1.6	−60 23.8	16 28.2	−.01	+1 1.94
V	11	6 38	−1.0	−63 27.2	16 35.4	−.01	+1 9.14
W	8	6 24	−1.0	−63 41.2	16 44.1	−.01	+1 17.84
X	10	8 20	−1.3	−61 45.5	16 50.9	−.01	+1 24.64
Y	12	0 14	0.0	−69 50.2	17 13.6	−.01	+1 47.34
Z	10.11	7 38	−1.1	−62 27.3	17 31.5	−.01	+2 5.24
AA	10	−0 10	+0.1	−70 14.1	17 54.9	−.01	+2 28.64
BB	9.10	5 17	−0.7	−64 47.9	5 18 2.6	−.01	+2 36.34

Zone XVII. 10′ s. of XVI., Jan. 25, 1864. $\tau = -0^h\ 2^m$ $\theta = +35°$

Letter.	Mag.	Reading of Scale.	d		1864.0 $\delta-\delta_0$	Transit.	k		1864.0 $\alpha-\alpha_0$
		′ ″	′ ″	″	′ ″	h m s	h m s	s	m s
A	12	0 45	−80 14.1	−0.3	−79 29.4	5 25 20.8	−5 28 34. 6	+.01	−3 13.25
B	8	0 27		−0.2	−79 47.3	25 28.9		+.01	−3 5.15
C*	12	9 18		−1.6	−70 57.7	25 38.3		+.01	−2 55.75
D	10	−0 15		−0.1	−80 29.2	25 44.4		+.01	−2 49.65
E	13	4 28		−0.8	−75 46.9	25 52.5		+.01	−2 41.55
F	12.13	2 50		−0.6	−77 24.7	26 4.6		+.01	−2 29.45
G	13	1 30		−0.3	−78 44.4	26 5.3		+.01	−2 28.75
H	13	6 20		−1.1	−73 55.2	26 16.8		+.01	−2 17.25
I	12.13	7 47		−1.3	−72 28.4	26 19.9		+.01	−2 14.15
K	12.13	5 47		−1.0	−74 28.1	26 33.6		+.01	−2 0.45
L	12	4 59		−0.9	−75 16.0	26 39.6		+.01	−1 54.45
M	12	3 37		−0.6	−76 37.7	26 42.6		+.01	−1 51.45
N	10	0 22		−0.1	−79 52.2	26 45.4		+.01	−1 48.65
O	9	6 20		−1.1	−73 55.2	27 13.7		.00	−1 20.36
P	11	0 20		−0.1	−79 54.2	27 34.6		.00	−0 59.46
Q	10	4 12		−0.7	−76 2.8	27 43.0		.00	−0 51.06
R	13	4 3		−0.7	−76 11.8	27 54.8		.00	−0 39.26
S	13	8 50		−1.5	−71 25.6	28 2.4		.00	−0 31.66
T	12	5 42		−0.9	−74 33.0	28 4.3		.00	−0 29.76
U	12	−0 3		0.0	−80 17.1	28 20.6		.00	−0 13.46
V	8.9†	9 43		−1.6	−70 32.7	28 29.6		.00	−0 4.46
W	11	8 50		−1.5	−71 25.6	28 32.4		.00	−0 1.66
X	12	2 47		−0.5	−77 27.6	28 43.5		.00	+0 9.44
Y	13	6 12		−1.0	−74 3.1	28 54.7		.00	+0 20.64
Z	9.10	−0 30		+0.1	−80 44.0	29 2.7		.00	+0 28.64
AA	9	2 2		−0.3	−78 12.4	29 8.0		.00	+0 33.94
BB	10	4 58		−0.8	−75 16.9	29 35.5		−.01	+1 1.43
CC	9.10	0 49		−0.1	−79 25.2	29 47.3		−.01	+1 13.23
DD	9	8 59		−1.4	−71 16.5	30 34.9		−.01	+2 0.83
EE	10	10 6		−1.5	−70 9.6	31 3.0		−.01	+2 28.93
FF	9	1 32		−0.1	−78 42.2	31 8.9		−.01	+2 34.83
GG	8	0 45		0.0	−79 29.1	5 31 12.7		− 01	+2 38.63

Zone XVIII. 10′ s. of XVII., Jan. 25, 1864. $\tau = +0^h\ 9^m$ $\theta = +35°$

Letter.	Mag.	Reading of Scale.	d		1864.0 $\delta-\delta_0$	Transit.	k		1864.0 $\alpha-\alpha_0$
		′ ″	′ ″	″	′ ″	h m s	h m s	s	m s
A		9 49	−90 15.8	−1.7	−80 28.5	5 36 6.1	−5 38 55.77	+.01	−2 49.66
B	9	0 18		−0.1	−89 57.9	36 23.1		+.01	−2 32.66
C	12	8 48		−1.5	−81 29.3	36 29.2		+.01	−2 26.56
D	10.11	3 20		−0.6	−86 56.4	36 43.6		+.01	−2 12.16
E	13	5 28		−1.0	−84 48.8	36 52.6		+.01	−2 3.16
F	13	9 50		−1.7	−80 27.5	36 58.7		.00	−1 57.07
G	10.11	4 12		−0.7	−86 4.5	37 2.6		+.01	−1 53.16
H	10.11	7 23		−1.3	−82 54.1	37 13.8		.00	−1 41.97
I	12	8 40		−1.5	−81 37.3	37 36.5		.00	−1 19.27
K	11	−0 10		0.0	−90 25.8	37 37.6		+.01	−1 18.16
L	8.9	3 6		−0.5	−87 10.3	37 49.2		.00	−1 6.57
M	9.10	1 42		−0.3	−88 34.1	37 56.5		.00	−0 59.27
N	12	−0 50		+0.1	−91 5.7	38 3.0		.00	−0 52.77
O	12	5 30		−0.9	−84 46.7	38 13.3		.00	−0 42.47
P	12	−0 29		+0.1	−90 44.7	5 38 21.6		.00	−0 34.17

* C does not accord with B of XVI. in Dec. By observations March 14, the Dec. of C in Zone XVII. is 70 32.5, and agrees with XVI. B.

† Magnitude supplied March 14.

Zone XVIII. 10′ s. of XVII., Jan. 25, 1864. $\tau = +0^h\ 9^m$ $\theta = +35°$

Letter.	Mag.	Reading of Scale.	d		1864.0 $\delta - \delta_0$	Transit.	k		1864.0 $\alpha - \alpha_0$
		′ ″	′ ″	″	′ ″	h m s	h m s	s	m s
Q	12	10 0	−90 15.8	−1.7	−80 17.5	5 38 42.4	−5 38 55.77	.00	−0 13.37
R	13	1 40		−0.3	−88 36.1	38 54.2		.00	−0 1.57
S	12	−0 34		+0.1	−90 49.7	39 8.3		.00	+0 12.53
T	12	2 12		−0.3	−88 4.1	39 17.3		.00	+0 21.53
U	12	8 28		−1.4	−81 49.2	39 20.4		−.01	+0 24.62
V	10	9 34		−1.5	−80 43.3	39 24.6		−.01	+0 28.82
W	10	3 30*		−0.5	−86 46.3	39 45.2		−.01	+0 49.42
X	11.12	4 13		−0.7	−86 3.5	39 55.6		−.01	+0 59.82
Y	12	4 28		−0.7	−85 48.5	40 31.7		−.01	+1 35.92
Z	12	8 4		−1.3	−82 13.1	40 32.3		−.01	+1 36.52
AA	12	5 40		−0.8	−84 36.6	41 29.4		−.01	+2 33.62
BB	11.12	6 40		−1.1	−83 36.9	41 38.7		−.01	+2 42.92
CC	13	0 10		+0.1	−90 5.7	5 41 43.7		−.01	+2 47.92

* Scale reading altered from 2′30″ to 3′30″ March 28.

SECTION II. PART II.

POSITIONS OF ZERO STARS FOR THE REDUCTION OF ZONES NEAR c AND ι ORIONIS, REFERRED TO θ ORIONIS. MEAN EQUINOX, 1864.0

Star.	Mag.	$\alpha - \alpha_0$	Weight.	$\delta - \delta_0$	Weight.
		m s		′ ″	
W. 614*		−3 6.08		+72 38.2	
W. 609†	8	−3 6.05	2	−79 47.4	2
W. 617	8	−2 54.57	2	−60 47.1	2
W. 628	9	−2 32.61	2	−89 55.7	2
W. 632 = Str. 593	9	−2 21.18	5	−67 30.7	2
W. 633	9	−2 20.91	2	−66 48.6	2
W. 642	9	−1 46.93	2	−68 57.1	2
a		−1 36.52	3	+22 13.2	3
W. 647	9	−1 36.26	2	−59 53.2	2
W. 655	9	−1 20.47	2	−64 52.0	2
β		−1 17.67	3	+23 4.1	3
W. 660	9	−1 7.54	2	+61 1.0	2
W. 661	8	−1 6.21	2	−87 11.2	2
W. 670	7	−0 54.36	3	+34 56.6	3
1		−0 53.87	2	−22 32.9	3
W. 676 = Str. 597	9	−0 46.85	5	+53 59.8	5
H‴		−0 41.82	3	−19 38.8	3
v		−0 41.75	3	+23 13.8	3
Str. 602	7.5	−0 36.34	4	+59 43.0	4
Str. 603	8.2	−0 25.55	4	−37 6.1	4
W. 690	8	−0 13.97	2	+39 22.6	2
2		−0 12.52	1	−26 15.3	3
Mädler, 801		−0 11.51	2	−36 50.0	2
W. 697	8	−0 4.63	2	−70 30.6	2
φ		−0 1.13	2	−26 17.3	3
3		−0 0.12	2	−24 53.9	3
I. L		+0 0.73	3	+19 29.8	3
I. L		+0 0.98	2	+19 27.5	3
W. 698	7.8	+0 2.59	3	+53 44.3	3
W. 700 = Str. 605	7.8	+0 2.77	5	+57 56.0	5
c′ Orionis		+0 4.90		+33 5.9	5
ι Orionis		+0 11.14	7	−31 12.5	7
W. 706 = Str. 607	7.8	+0 11.61	5	+61 34.4	5
15‡		+0 13.78	2	+29 10.0	3
c² Orionis		+0 21.37	5	+32 2.1	5
W. 712	6	+0 21.42	2	+32 4.4	2

* This declination is about 8″ in error.

† The value of $\alpha - \alpha_0$ was determined by the meridian circle of this observatory, and the value of $\delta - \delta_0$ includes an observation made here. Weisse's AR. is wrong 14ˢ.

‡ The AR. of this star is about 0.ˢ8 in error.

Star.	Mag.	$\alpha-\alpha_0$	Weight.	$\delta-\delta_0$	Weight.
		m s		′ ″	
W. 715	8.9	+0 31.79	2	−37 13.6	2
W. 716	8	+0 34.64	2	−53 1.6	2
16		+0 43.29	1	+27 23.5	3
W. 721*	8	+0 46.84	5	+58 0.6	5
W		+0 53.47	3	+19 56.2	3
W. 734 = Str. 611	7	+1 21.30	5	−40 15.1	5
W. 741	9	+1 41.78	2	−28 45.1	2
W. 756	6.7	+2 12.35	2	−32 33.7	2
W. 773	7	+2 35.24	3	+35 1.0	3
χ		+2 36.47	1	−26 2.3	3
W. 780	7.8	+2 54.83	3	+77 30.2	3

* This position is from Bessel, combined with our own observations.

The following positions derived from W. 614 in AR., and W. 780 in AR. and Dec., have been used to reduce Zones VIII. and IX.; each with weight = 2

ζ	−2 31.04	+74 49.0
ε	−1 26.28	
ε′	−1 19.26	
L	−0 2.39	+75 6.0
N	+0 30.22	+75 11.0

SECTION II. PART III.

ZONES NEAR c AND ι ORIONIS. SUPPLEMENTARY STARS.

Supplement to Zone I., Jan. 27, 1864.

Name of Star.	Mag.	Reading of Scale.	Reduction.	Correct-tion.	$\delta-\delta_0$	Principal Star.	U′—U.	Correc-tion.	$\alpha-\alpha_0$
		′ ″	′ ″	″	′ ″	m s	s	s	m s
c_1	13	2 0	+19 53.9	−0.7	+21 53.2	−2 23.80	+ 4.7	+0.01	−2 19.09
e_1	16	4 45		−1.2	+24 37.7	−1 36.41	+ 9.5	−0.01	−1 26.92
e_2	13	−0 24		−0.1	+19 29.8		+10.0	+0.01	−1 26.40
f_1	16	5 55		−1.4	+25 47.5	−1 17.81	0.0	−0.01	−1 17.82
f_2	17	1 10		−0.4	+21 3.5		+ 4.5	+0.01	−1 13.30
g_1	15	6 20		−1.4	+26 12.5	−0 55.61	+ 3.5	0.00	−0 52.11
g_2	17	2 10		−0.6	+22 3.3		+ 6.0	+0.01	−0 49.60
i_1	15	4 40		−1.1	+24 32.8	−0 41.71	+ 4.0	0.00	−0 37.71
i_2	14	3 32		−0.9	+23 25.0		+ 6.0	0.00	−0 35.71
l_{-1}	13	2 59		−0.8	+22 52.1	+0 0.79	− 2.7	−0.01	−0 1.92
l_1	13	10 20		−2.2	+30 11.7		+ 3.0	−0.04	+0 3.75
l_2	13.14	8 30		−1.9	+28 22.0		+10.0	−0.03	+0 10.76
m_1	17	−0 8		−0.2	+19 45.7	+0 12.98	+ 3.0	+0.03	+0 16.01
m_2	14	2 18		−0.6	+22 11.3		+ 5.7	+0.03	+0 18.71
m_3	17	6 10		−1.4	+26 2.5		+ 8.0	+0.01	+0 20.99
m_4	17	6 30		−1.5	+26 22.4		+11.0	+0.01	+0 23.99
n_1	14	−0 40		−0.1	+19 13.8	+0 36.88	0.0	+0.02	+0 36.90
n_2	16	6 40		−1.5	+26 32.4		+ 2.0	0.00	+0 38.88
o_1	13	9 45		−2.1	+29 36.8	+0 42.88	+ 1.0	−0.01	+0 43.87
o_2	13	4 0		−1.0	+23 52.9		+ 4.3	+0.01	+0 47.19
r_1	16	3 55		−0.9	+23 48.0	+1 33.08	+ 2.5	0.00	+1 35.58
r_2	15	6 45		−1.5	+26 37.4		+ 3.0	−0.01	+1 36.07
r_3	16	0 45		−0.3	+20 38.6		+ 5.2	+0.01	+1 38.29
r_4	15	−0 5		−0.1	+19 48.8		+ 5.7	+0.02	+1 38.80
r_5	14	7 30		−1.6	+27 22.3		+20.5	−0.01	+1 53.57
r_6	17	6 55		−1.5	+26 47.4		+21.5	−0.01	+1 54.57
s_1	12	7 25		−1.6	+27 17.3	+1 59.28	+ 5.0	−0.02	+2 4.26
s_2*	14	−0 45		0.0	+19 8.9		+ 8.0	+0.01	+2 7.29
s_4	16	4 40		−1.0	+24 32.9		+18.0	−0.01	+2 17.27
s_5	8	7 55		−1.7	+27 47.2		+20.0	−0.02	+2 19.26

Supplement to Zone II., Jan. 27, 1864.

Name of Star.	Mag.	Reading of Scale.	Reduction.	Correct-tion.	$\delta-\delta_0$	Principal Star.	U′—U.	Correc-tion.	$\alpha-\alpha_0$
		′ ″	′ ″	″	′ ″	m s	s	s	m s
a_{-1}	15	2 16	+29 43.6	−0.7	+31 58.9	−1 49.05	− 0.2	0.00	−1 49.25
a_1	16	3 50		−1.0	+33 32.6		+12.0	−0.01	−1 37.06
c_1	14	9 0		−2.0	+38 41.6	−1 9.66	−14.0	−0.01	−0 55.67
d_1	13.14	3 0		−0.8	+32 42.8	−1 7.56	+ 2.0	0.00	−1 5.56

* s_2 corrected from 8′45″ Feb. 29.

Supplement to Zone II., Jan. 27, 1864.

Name of Star.	Mag.	Reading of Scale.	Reduction.	Correction.	$\delta-\delta_0$	Principal Star.	U′—U	Correction.	$\alpha-\alpha_0$
		′ ″	′ ″	″	′ ″	m s	s	s	m s
e_{-1}	15	9 50	+29 43.6	—2.2	+39 31.4	—0 54.36	— 2.7	—0.02	—0 57.08
e_1	13	6 20		—1.5	+36 2.1		0.0	0.00	—0 54.36
e_2	17	3 40		—1.0	+33 22.6		+ 6.0	+0.01	—0 48.35
e_3	13	5 14		—1.3	+34 56.3		+18.0	0.00	—0 36.36
e_4	17	5 58		—1.4	+35 40.2		+23.5	0.00	—0 30.86
e_5	15	5 34		—1.3	+35 16.3		+27.0	0.00	—0 27.36
e_6	16	5 10		—1.2	+34 52.4		+31.5	0.00	—0 22.86
f_{-1}	15	10 0		—2.2	+39 41.4	—0 22.06	— 5.7	—0.01	—0 27.77
f_1	15	4 14		—1.0	+33 56.6		+ 7.0	+0.01	—0 15.05
f_2	13	7 2		—1.6	+36 44.0		+ 8.5	0.00	—0 13.56
g_1	17	9 10		—2.0	+38 51.6	—0 13.67	+ 7.0	0.00	—0 6.67
h_1	12.13	8 58		—2.0	+38 39.6	—0 4.66	+ 1.5	—0.02	—0 3.18
h_2	17	7 55		—1.8	+37 36.8		+ 1.7	—0.02	—0 2.98
h_3	17	8 48		—1.9	+38 29.7		+ 4.0	—0.02	—0 0.68
h_4	13	0 30		—0.3	+30 13.3		+ 8.5	0.00	+0 3.84
h_5	15	2 55		—0.8	+32 37.8		+ 8.5	0.00	+0 3.84
i_{-0}	13	10 20		—2.3	+40 1.3	+0 4.84	0.0	—0.02	+0 4.82
i_1	13	6 12		—1.4	+35 54.2		+ 4.0	—0.01	+0 8.83
i_2	15	8 40		—1.9	+38 21.7		+ 4.5	—0.02	+0 9.32
i_3	16	6 25		—1.5	+36 7.1		+ 4.8	—0.01	+0 9.63
i_4	15	9 50		—2.2	+39 31.4		+ 9.5	—0.02	+0 14.32
i_5	9	3 30		—0.9	+33 12.7		+10.3	0.00	+0 15.14
l_1	17	8 10		—1.8	+37 51.8	+0 21.44	+ 5.5	—0.02	+0 26.92
l_2	16	7 40		—1.7	+37 21.9		+ 8.5	—0.02	+0 29.92
l_3	14	2 35		—0.7	+32 17.9		+11.5	0.00	+0 32.94
l_4	17	6 15		—1.4	+35 57.2		+13.5	—0.01	+0 34.93
l_5	14	5 48		—1.3	+35 30.3		+15.0	—0.01	+0 36.43
l_6	17	3 40		—0.9	+33 22.7		+21.8	—0.01	+0 43.23
l_7	14	—0 10		—0.1	+29 33.5		+22.5	+0.01	+0 43.95
l_8	15	3 35		—0.9	+33 17.7		+24.0	—0.01	+0 45.43
m_1	14	9 10		—2.0	+38 51.6	+0 45.03	+ 0.7	0.00	+0 45.73
m_2	13	2 40		—0.7	+32 22.9		+11.0	+0.02	+0 56.05
m_3	15	9 0		—2.0	+38 41.6		+16.0	0.00	+1 1.03
p_1	15	7 20		—1.6	+37 2.0	+2 23.23	+ 1.8	—0.01	+2 25.02
r_1	12	1 25		—0.4	+31 8.2	+2 37.52	+ 0.5	+0.02	+2 38.04

Supplement to Zone IV., Jan. 28, 1864.

Name of Star.	Mag.	Reading of Scale.	Reduction.	Correction.	$\delta-\delta_0$	Principal Star.	U′—U	Correction.	$\alpha-\alpha_0$
		′ ″	′ ″	″	′ ″	m s	s	s	m s
b_1	16	8 50	+39 40.0	—1.9	+48 28.1	—2 22.09	+11.5	0.00	—2 10.59
b_2	16	6 30		—1.5	+46 8.5		+14.0	+0.01	—2 8.08
c_1	16	4 40		—1.2	+44 18.8	—2 4.58	+ 8.5	0.00	—1 56.08
c_2	15.16	9 4		—1.9	+48 42.1		+ 9.0	—0.01	—1 55.59
c_3	15	9 40		—2.0	+49 18.0		+15.0	—0.01	—1 49.59
d_1	16	—0 30		—0.5	+39 9.5	—1 47.88	+ 3.7	0.00	—1 44.18
d_2	17	9 30		—1.9	+49 8.1		+ 7.0	—0.02	—1 40.90
e_0	17	7 50		—1.7	+47 28.3	—1 34.49	0.0	—0.01	—1 34.50
e_1	17	3 45		—1.1	+43 23.9		+ 4.0	0.00	—1 30.49
e_2	16	4 48		—1.2	+44 26.8		+ 6.0	0.00	—1 28.49
e_3	15	3 58		—1.1	+43 36.9		+23.5	0.00	—1 10.99
h_{-1}	15	—0 12		—0.5	+39 27.5	—0 55.59	— 2.0	0.00	—0 57.59
i_{-1}	15	9 50		—1.9	+49 28.1	—0 50.79	— 7.0	0.00	—0 57.79

* Clouded suddenly ; continued on Feb. 12.

Supplement to Zone IV., continued Feb. 12, 1864.

Name of Star.	Mag.	Reading of Scale.	Reduction.	Correction.	$\delta-\delta_0$	Principal Star.	U′—U	Correction.	$\alpha-\alpha_0$
		′ ″	′ ″	″	′ ″	m s	s	s	m s
i_1	17	2 20	+39 40.0	−0.8	+41 59.2	−0 50.79	+ 7.5	+0.03	−0 43.26
i_2	16	3 10		−0.9	+42 49.1		+12.5	+0.02	−0 38.27
i_3	14	0 15		−0.5	+39 54.5		+17.0	+0.03	−0 33.76
i_4	15	7 20		−1.6	+46 58.4		+17.0	+0.01	−0 33.78
i_5	15	8 8		−1.7	+47 46.3		+22.5	+0.01	−0 28.28
i_6	17	1 40		−0.7	+41 19.3		+22.0	+0.03	−0 28.76
i_7	15	−0 8		−0.4	+39 31.6		+23.5	+0.03	−0 27.26
i_8	16	3 28		−1.0	+43 7.0		+24.7	+0.02	−0 26.07
i_9	16	10 3		−2.0	+49 41.0		+36.5	0.00	−0 14.29
{ k_1*	12	6 28		−1.4	+46 6.6	−0 13.69	+ 0.7	−0.02	−0 13.01
{ l_1	12	6 28		−1.4	+46 6.6	−0 12.99	+ 0.3	−0.02	−0 12.71
l_2	12	4 30		−1.1	+44 8.9		+ 2.3	0.01	−0 10.70
m_1	11	−1 0		−0.2	+38 39.8	−0 5.59	+ 2.2	+0.03	−0 3.36
m_2	14	10 15		−2.0	+49 53.0		+ 2.5	−0.01	−0 3.10
m_3	15	3 9		−0.9	+42 48.1		+ 3.7	+0.02	−0 1.87
o_1	14	6 45		−1.5	+46 23.5	+0 0.11	+ 0.6	−0.01	+0 0.70
o_2	16	1 30		−0.6	+41 9.4		+ 4.2	0.00	+0 4.31
o_3	14	0 16		−0.4	+39 55.6		+ 5.0	+0.01	+0 5.12
o_4	15	−0 18		−0.3	+39 21.7		+15.0	+0.01	+0 15.12
o_5	16	10 15		−2.0	+49 53.0		+18.0	−0.03	+0 18.08
p_1	13	1 0		−0.5	+40 39.5	+0 24.00	+ 0.5	+0.02	+0 24.52
p_2	16	2 15		−0.7	+41 54.3		+ 8.0	+0.02	+0 32.02
q_1	13	7 23		−1.6	+47 1.4	+0 37.60	+ 2.0	0.00	+0 39.60
q_2	15	9 40		−1.9	+49 18.1		+ 5.5	−0.01	+0 43.09
q_3	12.13	−0 45		−0.2	+38 54.8		+ 8.5	+0.03	+0 46.13
s_0	16	5 48		−1.3	+45 26.7	+1 43.70	0.0	0.00	+1 43.70
u_{-1}	15	3 12		−0.8	+42 51.2	+2 3.20	− 2.7	0.00	+2 0.50

Supplement to Zone V., Feb. 12, 1864.

Name of Star.	Mag.	Reading of Scale.	Reduction.	Correction.	$\delta-\delta_0$	Principal Star.	U′—U	Correction.	$\alpha-\alpha_0$
		′ ″	′ ″	″	′ ″	m s	s	s	m s
c_1	16	0 30	+49 46.1	−0.5	+50 15.6	−2 19.88	+ 0.7	+0.01	−2 19.17
c_2	16	2 58		−0.9	+52 43.2		+ 2.3	0.00	−2 17.58
c_3	15	3 0		−0.9	+52 45.2		+ 9.5	0.00	−2 10.38
d_{-1}	16	8 10		−1.8	+57 54.3	−2 3.58	− 1.0	−0.01	−2 4.59
d_1	17	6 35		−1.6	+56 19.5		+ 6.7	−0.01	−1 56.89
d_2	17	7 30		−1.7	+57 14.4		+10.0	−0.01	−1 53.59
e_1	16	8 30		−1.9	+58 14.2	−1 51.28	+ 0.7	−0.02	−1 50.60
e_2	15	−0 32		−0.3	+49 13.8		+ 1.0	+0.01	−1 50.27
e_3	17	5 50		−1.4	+55 34.7		+ 4.5	−0.01	−1 46.79
e_4	15	0 47		−0.5	+50 32.6		+ 7.0	+0.01	−1 44.27
e_5	15	6 10		−1.5	+55 54.6		+ 8.7	−0.01	−1 42.59
f_{-1}	14	8 3		−1.8	+57 47.3	−1 21.59	− 9.0	0.00	−1 30.59
g_1	14	6 14		−1.5	+55 58.6	−1 15.99	+ 6.7	+0.01	−1 9.28
g_2	15	5 0		−1.3	+54 44.8		+ 7.0	+0.02	−1 8.97
h_0	15	4 16		−1.1	+54 1.0	−1 4.59	0.0	−0.01	−1 4.60
i_1	17	4 40		−1.2	+54 24.9	−1 0.89	+ 1.5	0.00	−0 59.39
i_2	14	−0 15		−0.3	+49 30.8		+ 3.5	+0.02	−0 57.37
k_0	17	8 40		−1.9	+58 24.2	−0 55.89	0.0	−0.02	−0 55.91
k_1	14	4 25		−1.1	+54 10.0		+ 1.7	0.00	−0 54.19
m_{-2}	10.11	8 43		−1.9	+58 27.2	−0 46.89	− 3.0	−0.02	−0 49.91

* Same star.

† N and O are nebulous.

Supplement to Zone V., Feb. 12, 1864.

Name of Star.	Mag.	Reading of Scale.	Reduction.	Correction.	$\delta-\delta_0$	Principal Star.	U′—U	Correction.	$\alpha-\alpha_0$
		′ ″	′ ″	″	′ ″	m s	s	s	m s
m_{-1}	11	8 43	+49 46.1	−1.9	+58 27.2	−0 46.89	− 1.7	−0.02	−0 48.61
m_1	13	6 3		−1.4	+55 47.7		+ 3.0	−0.01	−0.43.90
m_2	13.14	8 28		−1.9	+58 12.2		+ 3.0	−0.01	−0 43.90
n_{-2}	13	9 0		−2.0	+58 44.1	−0 36.29	− 1.5	0.00	−0 37.79
n_{-1}	11.12	8 38		−1.9	+58 22.2		− 0.5	+0.01	−0 36.78
n_1	16	7 15		−1.6	+56 59.5		+ 1.5	+0.01	−0 34.78
n_2	16	6 40		−1.5	+56 24.6		+ 1.5	+0.01	−0 34.78
n_3	12	6 13		−1.4	+55 57.7		+ 1.7	+0.01	−0 34.58
n_4	14	4 28		−1.1	+54 13.0		+ 3.7	+0.02	−0 32.57
n_5	15	4 30		−1.1	+54 15.0		+10.7	+0.02	−0 25.57
p_0	17	8 7		−1.8	+57 51.3	−0 18.99	0.0	−0.02	−0 19.01
p_1	16	−0 5		−0.3	+49 40.8		+ 1.8	+0.01	−0 17.18
p_2	13	4 25		−1.1	+54 10.0		+ 4.0	−0.01	−0 15.00
p_3	13	5 43		−1.3	+55 27.8		+ 8.5	−0.01	−0 10.50
p_4	16	5 50		−1.4	+55 34.7		+10.0	−0.01	−0 9.00
p_5	16	0 3		−0.3	+49 48.8		+15.5	+0.01	−0 3.48
p_6	13	4 15		−1.1	+54 0.0		+17.5	−0.01	−0 1.50
r_1	13	8 12		−1.8	+57 56.3	−0 2.71	+ 5.0	0.00	+0 7.71
r_2	16	10 15		−2.1	+59 59.0		+12.5	−0.01	+0 15.20
r_3	16	0 2		−0.3	+49 47.8		+15.0	+0.03	+0 17.74
r_4	13	8 33		−1.8	+58 17.3		+17.5	0.00	+0 20.21
r_5	16	5 58		−1.4	+55 42.7		+17.8	+0.01	+0 20.52
r_6	15	8 50		−1.9	+58 34.2		+21.0	0.00	+0 23.71
r_7	17	7 28		−1.7	+57 12.4		+25.0	0.00	+0 27.71
r_8	17	9 10		−1.9	+58 54.2		+27.0	−0.01	+0 29.70
r_9	17	8 40		−1.9	+58 24.2		+30.0	0.00	+0 32.71
s_{-1}	15	−0 30		−0.2	+49 15.9	+0 46.80	− 3.7	+0.03	+0 43.13
s_{-2}	17	8 22		−1.8	+58 6.3		− 4.0	0.00	+0 42.80
s_{-3}	16	6 6		−1.4	+55 44.7		− 4.0	+0.01	+0 42.81
s_{-4}	16	1 40		−0.5	+51 25.6		− 8.5	+0.02	+0 38.32
s_{-5}	14	3 40		−0.9	+53 25.2		−13.5	+0.02	+0 33.32
s_1	14	10 12		−2.1	+59 56.0		+ 8.0	−0.01	+0 54.79
s_2	15	9 12		−1.9	+58 56.2		+11.0	0.00	+0 57.80
s_3	17	7 20		−1.6	+57 4.5		+11.5	0.00	+0 58.30
s_4	17	7 8		−1.5	+56 52.6		+20.7	0.00	+1 7.50
s_5	17	6 45		−1.5	+56 29.6		+23.5	0.00	+1 10.30
s_6	17	6 45		−1.5	+56 29.6		+26.5	0.00	+1 13.30
t_1	16	6 50		−1.5	+56 34.6	+1 16.40	+ 3.0	+0.01	+1 19.41
t_2	15	9 40		−2.0	+59 24.1		+ 4.7	0.00	+1 21.10
u_1	16	8 0		−1.7	+57 44.4	+1 21.60	+ 4.0	0.00	+1 25.60

Supplement to Zone VI., Feb. 27, 1864.

Name of Star.	Mag.	Reading of Scale.	Reduction.	Correction.	$\delta-\delta_0$	Principal Star.	U′—U	Correction.	$\alpha-\alpha_0$
		′ ″	′ ″	″	′ ″	m s	s	s	m s
a_1	17	4 40	+59 27.2	−1.2	+64 6.0	−2 14.72	+ 6.0	+0.01	−2 8.71
a_2	13	9 0		−2.2	+68 25.0		+ 9.2	−0.01	−2 5.53
b_{-1}	14	2 45		−0.8	+62 11.4	−2 1.82	− 0.5	0.00	−2 2.32
b_1	13	6 45		−1.7	+66 10.5		+ 3.25	−0.02	−1 58.59
c_{-2}	16	7 42		−1.9	+67 7.3	−1 41.82	− 3.0	−0.02	−1 44.84
c_{-1}	14	8 42		−2.1	+68 7.1		− 2.0	−0.02	−1 43.84
c_1	14	8 30		−2.0	+67 55.2		+ 1.5	−0.02	−1 40.34

Supplement to Zone VI., Feb. 27, 1864.

Name of Star.	Mag.	Reading of Scale.	Reduction.	Correction.	$\delta-\delta_0$	Principal Star.	U′—U	Correction.	$\alpha-\alpha_0$
		′ ″	′ ″	″	′ ″	m s	s	s	m s
c_2	16	10 0	+59 27.2	−2.4	+69 24.8	−1 41.82	+ 4.0	−0.03	−1 37.85
c_3	17	5 10		−1.3	+64 35.9		+ 7.25	−0.01	−1 34.58
d_{-1}	17	7 50		−1.9	+67 15.3	−1 26.23	− 1.12	0.00	−1 27.35
d_1	16	4 40		−1.2	+64 6.0		+ 1.0	+0.01	−1 25.22
d_2	14	7 10		−1.7	+66 35.5		+ 1.75	+0.01	−1 24.47
d_3	16	7 45		−1.9	+67 10.3		+ 2.75	0.00	−1 23.48
e_1	15	6 43		−1.6	+66 8.6	−1 19.53	+ 3.0	0.00	−1 16.53
e_2	11	−0 7		−0.1	+59 20.1		+ 3.75	+0.02	−1 15.76
f_0	14	7 30		−1.8	+66 55.4	−1 9.03	0.0	−0.01	−1 9.04
f_1	11	9 18*		−2.2	+68 43.0		+ 0.88	−0.02	−1 8.17
f_2	12.13	6 22		−1.5	+65 47.7		+ 1.1	−0.01	−1 7.94
f_3	14	8 48		−2.1	+68 13.1		+ 6.25	−0.02	−1 2.80
h_{-4}	15	2 30		−0.7	+61 56.5	−0 53.13	− 7.25	0.00	−1 0.38
h_{-3}	12	2 12		−0.6	+61 38.6		− 5.5	0.00	−0 58.63
h_{-2}	13	3 20		−0.9	+62 46.3		− 4.5	−0.01	−0 57.64
h_{-1}	16	5 35		−1.4	+65 0.8		− 2.5	−0.01	−0 55.64
h_0	17	6 5		−1.5	+65 30.7		0.0	−0.02	−0 53.15
h_1	14	5 47		−1.4	+65 12.8		+ 1.5	−0.02	−0 51.65
h_2	14	2 11		−0.6	+61 37.6		+ 8.0	0.00	−0 45.13
h_3	13	5 52		−1.5	+65 17.7		+12.5	−0.02	−0 40.65
h_4	12.13	6 25		−1.6	+65 50.6		+14.5	−0.02	−0 38.65
i_{-2}	17	2 15		−0.6	+61 41.6	−0 36.43	− 2.5	−0.01	−0 38.94
i_{-1}	14	−0 41		0.0	+58 46.2		− 2.0	0.00	−0 38.43
i_1	13	6 24		−1.6	+65 49.6		+ 4.5	−0.02	−0 31.95
i_2	17	6 40		−1.6	+66 5.6		+ 7.0	−0.02	−0 29.45
i_3	15	5 25		−1.3	+64 50.9		+ 7.75	−0.02	−0 28.70
i_4	17	4 5		−1.0	+63 31.2		+ 9.5	−0.01	−0 26.94
i_5	17	5 10		−1.2	+64 36.0		+ 9.75	−0.02	−0 26.75
k_{-3}	17	2 12		−0.6	+61 38.6	−0 24.33	− 3.25	0.00	−0 27.58
k_{-2}	16	2 12		−0.6	+61 38.6		− 1.5	0.00	−0 25.83
k_{-1}	16.17	0 40		−0.3	+60 6.9		− 1.25	+0.01	−0 25.57
k_1	15	4 35		−1.1	+64 1.1		+ 2.0	−0.01	−0 22.34
k_2	15	3 20		−0.8	+62 46.4		+ 3.25	0.00	−0 21.08
m_1	15	4 50		−1.2	+64 16.0	−0 15.73	+ 0.85	−0.01	−0 14.89
m_2	13.14	5 7		−1.2	+64 33.0		+ 2.75	−0.02	−0 13.00
m_3	15	1 40		−0.5	+61 6.7		+ 6.5	0.00	−0 9.23
m_4	16	3 22		−0.8	+62 48.4		+ 8.0	−0.01	−0 7.74
m_5	13	6 0		−1.4	+65 25.8		+12.0	−0.02	−0 3.75
m_6	13.14	10 5		−2.3	+69 29.9		+16.0	−0.03	+0 0.24
n_{-3}	16	6 45		−1.6	+66 10.6	+0 11.47	− 7.75	−0.02	+0 3.70
n_{-2}	17	6 55		−1.6	+66 20.6		− 7.75†	−0.02	+0 3.70
n_{-1}	14	8 11		−1.9	+67 36.3		− 5.0	−0.02	+0 6.45
n_1	15	4 0		−1.0	+63 26.2		+ 1.85	−0.01	+0 13 31
n_2	16	0 32		−0.2	+59 59.0		+ 3.75	+0.01	+0 15.23
n_3	14	6 32		−1.5	+65 57.7		+15.15	−0.02	+0 26.60
n_4	17	7 7		−1.6	+66 32.6		+19.5	−0.02	+0 30.95
o_1	13	7 22		−1.7	+66 47.5	+0 31.86	− 0.5	+0.01	+0 32.37
p_{-2}	15	0 58		−0.3	+60 24.9	+0 53.66	− 9.5	+0.01	+0 44.17
p_{-1}	16	9 30		−2.2	+68 55.0		− 5.0	−0.02	+0 48.64
q_{-1}	16	0 25		−0.2	+59 52.0	+0 58.16	− 2.75	0.00	+0 55.41
q_0	16	−0 30		+0.1	+58 57.3		0.0	+0.01	+0 58.17

* Small stars of 16 and 17 Mag. are so numerous near the middle of this zone, that a few of the faintest may have been omitted.

† Same AR.

Supplement to Zone VI., Feb. 27, 1864.

Name of Star.	Mag.	Reading of Scale.	Reduction.	Correction.	$\delta-\delta_0$	Principal Star.	U′—U	Correction.	$\alpha-\alpha_0$
		′ ″	′ ″	″	′ ″	m s	s	s	m s
q_1	15	10 5	+59 27.2	−2.3	+69 29.9	+0 58.16	+ 2.5	−0.03	+1 0.63
q_2	15	5 58		−1.4	+65 23.8		+13.5	−0.02	+1 11.64
q_3	15	7 45		−1.8	+67 10.4		+15.5	−0.02	+1 13.64
r_1	15	4 35		−1.1	+64 1.1	+1 16.67	+ 2.5	−0.02	+1 19.15
r_2	16	0 0		−0.1	+59 27.1		+ 4.5	0.00	+1 21.17
r_3	16	10 15		−2.3	+69 39.9		+14.0	−0.04	+1 30.63
r_4	15	2 58		−0.7	+62 24.5		+33.0	−0.01	+1 49.66
r_5	15	8 15		−1.8	+67 40.4		+43.5	−0.03	+2 0.14
s_{-3}*	17	6 10		−1.4	+65 35.8	+2 14.36	−21.0	+0.01	+1 53.37
s_{-2}	14	1 15		−0.3	+60 41.9		−16.5	+0.02	+1 57.84
s_{-1}	16	5 28		−1.2	+64 54.0		−15.0	+0.01	+1 59.37
t_1	16	8 58		−2.0	+68 23.2	+2 23.26	+ 0.75	−0.02	+2 23.99

Supplement to Zone VIII., Feb. 27, 1864.

Name of Star.	Mag.	Reading of Scale.	Reduction.	Correction.	$\delta-\delta_0$	Principal Star.	U′—U	Correction.	$\alpha-\alpha_0$
		′ ″	′ ″	″	′ ″	m s	s	s	m s
b_1	17	6 50	+65 25.7	−1.7	+72 14.0	−2 14.57	+ 1.0	−0.02	−2 13.59
c_{-1}	15	5 18		−1.3	+70 42.4	−2 5.37	− 1.0	−0.01	−2 6.38
c_1	15	9 20		−2.2	+74 43.5		+ 3.0	−0.02	−2 2.39
c_2	16	0 40		−0.3	+66 5.4		+ 7.0	+0.01	−1 58.36
d_{-2}	17	5 10		−1.3	+70 34.4	−1 26.37	−16.5	−0.01	−1 42.88
d_{-1}	17	6 30		−1.6	+71 54.1		−11.0	−0.01	−1 37.38
f_0	16	8 20		−2.0	+73 43.7	−1 12.48	0.0	0.00	−1 12.48
f_1	16	7 29		−1.8	+72 52.9		+ 1.75	0.00	−1 10.73
g_{-1}	16	1 18		−0.4	+66 43.3	−1 8.28	− 1.0	+0.01	−1 9.27
g_1	17	4 30		−1.1	+69 54.6		+ 2.0	0.00	−1 6.28
g_2	15	2 43		−0.7	+68 8.0		+ 5.0	0.00	−1 3.28
g_3	17	7 0		−1.7	+72 24.0		+ 8.5	−0.01	−0 59.79
g_4†	17	9 4		−2.1	+74 27.6		+15.35	−0.02	−0 52.95
g_5	16	8 2		−1.9	+73 25.8		+24.5	−0.02	−0 43.80
l_1	16	7 30		−1.8	+72 53.9	−0 39.08	+ 6.5	−0.02	−0 32.60
m_0	17	10 20		−2.4	+75 43.3	−0 32 08	0.0	−0.03	−0 32.11
n_{-2}	17	6 50		−1.6	+72 14.1	−0 17.98	− 5.75	0.00	−0 23.73
n_{-1}	17	9 20		−2.1	+74 43.6		− 3.0	−0.01	−0 20.99
o_{-1}	15.16	5 10		−1.2	+70 34.5	−0 10.88	− 2.0	+0.01	−0 12.87
p_1	14	9 30		−2.2	+74 53.5	−0 3.18	+ 0.5	−0.01	−0 2.69
p_2	15	3 48		−0.9	+69 12.8		+ 4.25	+0.01	+0 1.08
p_3	16	5 25		−1.3	+70 49.4		+11.75	0.00	+0 8.57
p_4	17	6 18		−1.5	+71 42.2		+12.5	0.00	+0 9.32
p_5	14	4 18		−1.0	+69 42.7		+18.0	0.00	+0 14.82
p_6	17	8 20		−1.9	+73 43.8		+19.5	−0.01	+0 16.31
p_7	16	7 50		−1.8	+73 13.9		+20.0	−0.01	+0 16.81
q_{-2}	13	9 40		−2.2	+75 3.5	+0 31.72	− 2.0	−0.02	+0 29.70
q_{-1}	16	9 30		−2.2	+74 53.5		− 0.5	−0.02	+0 31.20
q_0‡	14	1 15		−0.3	+66 40.4		0.0	0.00	+0 31.72
q_1§	14	4 45		−1.2	+70 9.5		+ 5.0	−0.01	+0 36.71
q_2§	17	7 32		−1.7	+72 56.0		+12.5	−0.01	+0 44.21
q_3§	15	7 15		−1.6	+72 39.1		+15.0	−0.01	+0 46.71
s_{-2}	16	3 30		−0.8	+68 54.9	+1 6.82	−14.0	+0.01	+0 52.83
s_{-1}	15	4 0		−0.9	+69 24.8		− 7.0	+0.01	+0 59.83
s_1	16	8 55		−2.0	+74 18.7		+17.75	−0.01	+1 24.56
t_{-3}	16	4 2		−0.9	+69 26.8	+1 39.71	− 9.5	+0.01	+1 30.22

* s_{-3} precedes r_5. † Follows I. ‡ Same as R. § Follows R.

Supplement to Zone VIII., Feb. 27, 1864.

Name of Star.	Mag.	Reading of Scale.	Reduction.	Correction.	$\delta-\delta_0$	Principal Star.	U′—U	Correction.	$\alpha-\alpha_0$
		′ ″	′ ″	″	′ ″	m s	s	s	m s
t_{-2}	15	9 58	+65 25.7	−2.3	+75 21.5	+1 39.71	− 2.5	−0.01	+1 37.20
t_{-1}	16	9 10		−2.0	+74 33.7		− 0.75	−0.01	+1 38.95
t_0	17	7 0		−1.6	+72 24.1		0.00	0.00	+1 39.71
v_1	17	5 20		−1.2	+70 44.5	+2 1.41	+ 2.5	0.00	+2 3.91
v_2	16	4 40		−1.0	+70 4.7		+ 9.5	0.00	+2 10.91
v_3	17	6 55		−1.6	+72 19.1		+10.00	0.00	+2 11.41

Supplement to Zone IX., Feb. 29, 1864.

Name of Star.	Mag.	Reading of Scale.	Reduction.	Correction.	$\delta-\delta_0$	Principal Star.	U′—U	Correction.	$\alpha-\alpha_0$
		′ ″	′ ″	″	′ ″	m s	s	s	m s
d_1	17	8 45	+75 22.1	−2.1	+84 5.0	−2 5.14	+ 0.5	0.00	−2 4.64
d_2	13	6 9		−1.5	+81 29.6		+ 1.38	+0.01	−2 3.75
c_1*	15.16	−0 40		0.0	+74 42.1	−2 6.94	+ 4.75	+0.02	−2 2.17
c_2	16.17	3 55		−1.1	+79 16.0		+ 6.0	+0.01	−2 0.93
c_3†	17	4 55		−1.3	+80 15.8		+ 8.5	0.00	−1 58.44
c_4†	15	10 12		−2.3	+85 31.8		+14.5	−0.01	−1 52.45
c_5†	17	6 50		−1.7	+82 10.4		+14.5	0.00	−1 52.44
c_6†	16	3 25		−1.0	+78 46.1		+19.0	+0.01	−1 47.93
c_7	17	3 20		−1.0	+78 41.1		+21.0	+0.01	−1 45.93
c_8	17	8 15		−2.0	+83 35.1		+24.5	−0.01	−1 42.45
c_9	14	2 10		−0.7	+77 31.4		+26.0	+0.01	−1 40.93
c_{10}‡	17	5 10		−1.3	+80 30.8		+27.0	0.00	−1 39.94
c_{11}‡	17	7 10		−1.7	+82 30.4		+30.5	−0.01	−1 36.45
c_{12}‡	17	6 48		−1.6	+82 8.5		+31.5	0.00	−1 35.44
c_{13}‡	17	10 10		−2.3	+85 29.8		+36.0	−0.02	−1 30.96
c_{14}‡	17	7 0		−1.7	+82 20.4		+37.25	−0.01	−1 29.70
c_{15}‡	17	8 17		−2.0	+83 37.1		+38.75	−0.01	−1 28.20
c_{16}‡	16.17	3 10		−0.9	+78 31.2		+40.50	+0.01	−1 26.43
c_{17}‡	17	2 30		−0.8	+77 51.3		+46.0	+0.01	−1 20.93
e_{-1}§	14	8 12		−1.9	+83 32.2	−1 13.25	− 0.75	−0.01	−1 14.01
e_{-2}	17	3 28		−0.9	+78 49.2		− 0.88	+0.01	−1 14.12
e_{-3}	17	3 12		−0.8	+78 33.3		− 0.88	+0.01	−1 14.12
f_1	17	9 15		−2.1	+84 35.0	−1 8.95	+ 3.5	−0.01	−1 5.46
f_2	17	0 10		−0.2	+75 31.9		+ 5.5	+0.02	−1 3.43
f_3	17	0 12		−0.2	+75 33.9		+ 6.5	+0.02	−1 2.43
f_4	15.16	8 45		−2.0	+84 5.1		+ 6.5	−0.01	−1 2.46
f_5	17	2 15		−0.7	+77 36.4		+ 8.5	+0.01	−1 0.44
h_0	14	9 4		−2.1	+84 24.0	−0 55.45	0.0	−0.03	−0 55.48
h_1‖	13	1 10		−0.4	+76 31.7		+ 1.25	0.00	−0 54.20
h_2‖	13	10 20		−2.3	+85 39.8		+ 1.0	−0.03	−0 54.48
h_3¶	15	8 45		−2.0	+84 5.1		+ 2.25	−0.03	−0 53.23
h_4¶	14	−0 55		0.0	+74 27.1		+ 2.0	+0.01	−0 53.44
h_5¶	17	7 10		−1.7	+82 30.4		+ 2.0	−0.02	−0 53.47
h_6**	17	7 40		−1.8	+83 0.3		+ 2.0	−0.02	−0 53.47
h_7	17	0 42		−0.4	+76 3.7		+ 5.0	0.00	−0 50.45
h_8	17	7 5		−1.7	+82 25.4		+ 5.0	−0.02	−0 50.47

* No star precedes C within 8^s.=Limit 2^m 15^s.

† The number of small stars gives almost a nebulous ground to the field.

‡ These stars were determined from c_9. The difference of AR. between c_9 and F was found to equal 32″25.

§ e_{-1} is nebulous. ‖ About same AR. ¶ Same AR. ** Same AR. as h_4.

Supplement to Zone IX., Feb. 29, 1864.

Name of Star.	Mag.	Reading of Scale.	Reduction.	Correction.	δ—δ₀	Principal Star.	U′—U	Correction.	α—α₀
		′ ″	′ ″	″	′ ″	m s	s	s	m s
i_{-4}	17	8 42	+75 22.1	−2.0	+84 2.1	−0 32.65	− 7.75	0.00	−0 40.40
i_{-3}	14	10 18		−2.3	+85 37.8		− 7.75	0.00	−0 40.40
i_{-2}	16.17	3 15		−0.8	+78 36.3		− 3.5	+0.02	−0 36.13
i_{-1}	17	3 15		−0.8	+78 36.3		− 2.5	+0.02	−0 35.13
k_{-1}	17	0 0		−0.2	+75 21.9	−0 26.25	− 4.25	0.00	−0 30.50
k_1	17	8 15		−1.8	+83 35.3		+ 0.25	−0.02	−0 26.02
k_2	16	−0 50		0.0	+74 32.1		+ 6.0	+0.01	−0 20.24
k_3	15	4 55		−1.2	+80 15.9		+ 7.75	−0.01	−0 18.51
k_4	16	10 0		−2.3	+85 19.8		+ 8.0	−0.03	−0 18.28
k_5	14	7 18		−1.7	+82 38.4		+ 9.25	−0.02	−0 17.02
k_6	17	8 10		−1.8	+83 30.3		+10.5	−0.02	−0 15.77
k_7	14	7 45		−1.8	+83 5.3		+12.5	−0.02	−0 13.77
k_8	16	5 17		−1.2	+80 37.9		+13.5	−0.01	−0 12.76
k_9	16.17	5 10		−1.2	+80 30.9		+14.5	−0.01	−0 11.76
k_{10}	16	7 40		−1.8	+83 0.3		+20.5	−0.02	−0 5.77
l_{-1}*	17	2 30		−0.7	+77 51.4	−0 2.65	− 3.5	−0.01	−0 6.16
m_{-2}	17	2 25		−0.7	+77 46.4	+0 11.25	− 5.75	+0.01	+0 5.51
m_{-1}	16	7 50		−1.8	+83 10.3		− 1.25	−0.01	+0 9.99
m_1	17	8 30		−1.9	+83 50.2		+ 0.75	−0.01	+0 11.99
m_2	12	10 10		−2.2	+85 29.9		+ 1.0	−0.01	+0 12.24
m_3	16	6 44		−1.5	+82 4.6		+ 9.5	0.00	+0 20.75
n_1	15.16	−0 20		−0.1	+75 2.0	+0 30.35	+ 1.5	0.00	+0 31.85
n_2	16	0 40		−0.3	+76 1.8		+ 1.75	0.00	+0 32.10
n_3	16	0 50		−0.3	+76 11.8		+ 5.5	0.00	+0 35.85
n_4	17	6 50		−1.6	+82 10.5		+ 6.5	−0.02	+0 36.83
n_5	15	5 53		−1.4	+81 13.7		+ 8.25	−0.02	+0 38.58
n_6	16	6 0		−1.4	+81 20.7		+26.25	−0.02	+0 56.58
n_7	17	5 50		−1.4	+81 10.7		+43.0	−0.02	+1 13.33
n_8	17	5 35		−1.3	+80 55.8		+56.0	−0.02	+1 26.33
o_{-2}	17	6 50		−1.5	+82 10.6	+1 38.54	− 3.5	−0.01	+1 35.03
o_{-1}	15	−0 8		−0.1	+75 14.0		− 1.0	+0.01	+1 37.55
o_1	15	6 4		−1.3	+81 24.8		+ 0.62	−0.01	+1 39.15
p_{-1}	16	−0 45		+0.1	+74 37.2	+1 40.04	− 0.75	+0.03	+1 39.32
o_2†	17	5 12		−1.1	+80 33.0	+1 38.54	+ 4.0	0.00	+1 42.54
o_3	17	3 20		−0.8	+78 41.3		+13.0	0.00	+1 51.54
q_{-2}	17	5 50		−1.3	+81 10.8	+2 2.24	− 4.75	−0.01	+1 57.48
q_{-1}	16	8 45		−1.9	+84 5.2		− 0.5	−0.02	+2 1.72
q_1	16.17	6 35		−1.4	+81 55.7		+ 1.5	−0.01	+2 3.73
q_2	16	6 10		−1.3	+81 30.8		+ 2.25	−0.01	+2 4.48
q_3	15	9 11		−2.0	+84 31.1		+ 5.5	−0.02	+2 7.72
q_4	17	8 10		−1.7	+83 30.4		+10.0	−0.02	+2 12.22
q_5	17	7 55		−1.7	+83 15.4		+10.75	−0.02	+2 12.97
q_6	13	6 42		−1.5	+82 2.6		+11.25	−0.01	+2 13.48
q_7	15	4 40		−1.0	+80 1.1		+15.0	−0.01	+2 17.23
q_8	15	4 0		−0.9	+79 21.2		+16.5	0.00	+2 18.74

Supplement to Zone X., Feb. 29, 1864.

Name of Star.	Mag.	Reading of Scale.	Reduction.	Correction.	δ—δ₀	Principal Star.	U′—U	Correction.	α—α₀
		′ ″	′ ″	″	′ ″	m s	s	s	m s
e_1	15	1 15	−29 58.9	−0.1	−28 44.0	−2 18.45	+ 3.0	+0.02	−2 15.43
g_1	16	8 20		−1.5	−21 40.4	−1 26.25	+ 3.25	0.00	−1 23.00

* The star L is 2″ or 3″ wrong. The declinations here were right by M.

† o_2 follows p.

Supplement to Zone X., Feb. 29, 1864.

Name of Star.	Mag.	Reading of Scale.	Reduction.	Correction.	$\delta - \delta_0$	Principal Star.	U′—U	Correction.	$\alpha - \alpha_0$
		′ ″	′ ″	″	′ ″	m s	s	s	m s
g_2	17	2 45	−29 58.9	−0.4	−27 14.3	−1 26.25	+ 9.75	+0.01	−1 16.49
h_{-1}	17	6 25		−1.1	−23 35.0	−0 54.85	− 5.25	−0.01	−1 0.11
i_{-1}	16	6 2		−1.0	−23 57.9	−0 42.75	− 4.0	0.00	−0 46.75
i_1*	15	8 0		−1.4	−22 0.3		+ 8.5	0.00	−0 34.25
i_2*	14	9 30		−1.7	−20 30.6		+15.0	−0.01	−0 27.76
i_3*	15	6 40		−1.2	−23 20.1		+23.0	0.00	−0 19.75
l_{-2}	13	3 0		−0.4	−26 59.3	−0 26.65	− 8.5	0.00	−0 35.15
l_{-1}	16	1 55		−0.2	−28 4.1		− 2.0	0.00	−0 28.65
i_4	17	8 50		−1.6	−21 10.5	−0 42.75	+27.0	−0.01	−0 15.76
m_1	14	6 47		−1.2	−23 13.1	−0 7.55	+ 2.5	−0.01	−0 5.06
m_2	15	10 10		−1.8	−19 50.7		+ 4.0	−0.02	−0 3.57
m_3	14	10 10		−1.8	−19 50.7		+ 5.75	−0.02	−0 1.82
n_1	14	1 30		−0.1	−28 29.0	−0 1.15	+ 1.0	+0.01	−0 0.14
q_1	17	4 40		−0.7	−25 19.6	+0 11.14	+ 0.5	0.00	+0 11.64
q_2	14	4 38		−0.7	−25 21.6		+ 1.0	0.00	+0 12.14
q_3	15	3 40		−0.5	−26 19.4		+ 2.75	0.00	+0 13.89
q_4	15	4 40		−0.7	−25 19.6		+ 3.5	0.00	+0 14.64
q_5	17	2 0		−0.2	−27 59.1		+ 4.0	+0.01	+0 15.15
q_6	15	2 15		−0.3	−27 44.2		+19.0	+0.00	+0 30.14
q_7	16	3 50		−0.6	−26 9.5		+21.5	0.00	+0 32.64
q_8	15	1 40		−0.1	−28 19.0		+24.75	0.00	+0 35.89
q_9	16	5 10		−0.8	−24 49.7		+36.25	0.00	+0 47.39
r_{-2}	16	8 50		−1.5	−21 10.4	+1 14.14	− 6.0	0.00	+1 8.14
r_{-1}†	17	6 50		−1.1	−23 10.0		− 2.5	0.00	+1 11.64
s_1	14	−0 40		+0.3	−30 38.6	+1 17.75	+ 7.0	0.00	+1 24.75
r_1‡	16	7 50		−1.4	−22 10.3	+1 14.14	+18.5	0.00	+1 32.64
r_2	16	4 10		−0.6	−25 49.5		+20.5	0.00	+1 34.64
u_1	16	1 40		−0.1	−28 19.0	+1 45.64	+ 4.5	0.00	+1 50.14
u_2	17	0 40		+0.1	−29 18.8		+ 6.75	+0.01	+1 52.40
u_3	16.17	2 0		−0.2	−27 59.1		+ 8.75	0.00	+1 54.39
v_1	16	9 30		−1.6	−20 30.5	+1 59.33	+ 6.0	0.00	+2 5.33
w_0	16	4 27		−0 6	−25 32.5	+2 7.94	0.0	0.00	+2 7.94
w_1	16	3 40		−0.5	−26 19.4		+ 7.5	0.00	+2 15.44

Supplement to Zone XII., March 2, 1864.

Name of Star.	Mag.	Reading of Scale.	Reduction.	Correction.	$\delta - \delta_0$	Principal Star.	U′—U	Correction.	$\alpha - \alpha_0$
		′ ″	′ ″	″	′ ″	m s	s	s	m s
c_{-1}	17	9 0	−40 5.0	−1.7	−31 6.7	−3 7.02	− 4.75	0.00	−3 11.77
d_{-1}	16	5 0		−0.9	−35 5.9	−2 55.41	− 2.75	0.00	−2 58.16
e_{-2}	16	10 10		−2.0	−29 57.0	−2 38.81	− 9.5	−0.01	−2 48.32
e_{-1}	16	8 15		−1.6	−31 51.6		− 1.0	−0.01	−2 39.82
f_{-1}	15	9 40		−1.9	−30 26.9	−2 36.41	− 1.5	−0.01	−2 37.92
f_1	17	8 20		−1.6	−31 46.6		+ 1.0	0.00	−2 35.41
f_2	16	0 40		0.0	−39 25.0		+ 9.5	0.00	−2 26.91
f_3	17	8 0		−1.5	−32 6.5		+13.5	0.00	−2 22.91
i_0	17	−0 10		+0.1	−40 14.9	−1 57.52	0.0	+0.01	−1 57.51
l_1	17	4 0		−0.7	−36 5.7	−1 46.61	+ 4.5	0.00	−1 42.11
l_2	16	10 20		−2.0	−29 47.0		+ 8.5	−0.01	−1 38.12
m_1	17	2 40		−0.4	−37 25.4	−1 28.22	+ 1.25	0.00	−1 26.97
n_{-1}	17	9 40		−1.8	−30 26.8	−1 11.62	− 3.0	0.00	−1 14.62
n_1	16.17	7 28		−1.4	−32 38.4		+ 4.75	0.00	−1 6.87
o_{-1}	15	2 10		−0.3	−37 55.3	−0 46.32	− 3.25	0.00	−0 49.57

* Follows K also. † Several other very faint stars near R and in the nebulosity. ‡ Follows S also.

Supplement to Zone XII., March 3, 1864. Continued from March 2.

Name of Star.	Mag.	Reading of Scale.	Reduction.	Correction.	$\delta-\delta_0$	Principal Star.	U′—U	Correction.	$a-a_0$
		′ ″	′ ″	″	′ ″	m s	s	s	m s
p_{-2}	16	5 38	−40 5.0	−0.9	−34 27.9	−0 30.62	− 5.25	−0.01	−0 35.88
p_{-1}	16	4 50		−0.8	−35 15.8		− 0.5	0.00	−0 31.12
p_1	14	4 27		−0.7	−35 38.7		+ 2.35	0.00	−0 28.27
p_2	13.14	5 22		−0.9	−34 43.9		+ 4.0	−0.01	−0 26.63
r_1	14	1 0		0.0	−39 5.0	−0 25.62	+ 1.5	+0.01	−0 24.11
r_2	11.12	5 20		−0.9	−34 45.9		+ 4.0	−0.01	−0 21.63
r_3	10.11	1 18		−0.1	−38 47.1		+ 5.25	+0.01	−0 20.36
s_{-1}	17	8 50		−1.6	−31 16.6	−0 13.62	− 3.5	−0.02	−0 17.14
t_1	9.10	4 25		−0.7	−35 40.7	−0 11.82	+ 1.0	0.00	−0 10.82
t_2	12	3 20		−0.5	−36 45.5		+ 3.5	0.00	−0 8.32
t_3	17	5 5		−0.8	−35 0.8		+ 6.5	−0.01	−0 5.33
t_4	10	1 28		−0.1	−38 37.1		+ 6.5	+0.01	−0 5.31
v_1	15	5 8		−0.8	−34 57.8	+0 1.28	+ 1.5	−0.01	+0 2.77
v_2	15	3 50		−0.6	−36 15.6		+ 8.5	−0.01	+0 9.77
v_3	16	5 45		−1.0	−34 21.0		+ 9.0	−0.01	+0 10.27
w_1	9	8 39		−1.6	−31 27.6	+0 11.17	+ 0.5	0.00	+0 11.67
w_2	10	8 44		−1.6	−31 22.6		+ 3.25	0.00	+0 14.42
x_{-2}	16	0 26		+0.1	−39 38.9	+0 31.68	−16.50	+0.01	+0 15.19
x_{-1}	12	2 0		−0.2	−38 5.2		−13.50	0.00	+0 18.18
z_{-1}	15	−0 29		+0.3	−40 33.7	+1 7.97	− 6.50	+0.01	+1 1.48
aa_1	13	9 44		−1.7	−30 22.7	+1 13.57	+ 4.5	−0.01	+1 18.06
aa_2	13	9 24		−1.7	−30 42.7		+16.0	−0.01	+1 29.56
bb_1	14	1 24		−0.1	−38 41.1	+1 21.47	+ 2.5	−0.01	+1 23.96
bb_2	13	−0 20		+0.3	−40 24.7		+ 7.75	0.00	+1 29.22
cc_1	17	0 34		+0.2	−39 30.8	+2 7.37	+ 1.75	+0.01	+2 9.13

Supplement to Zone XIV., March 12, 1864.

Name of Star.	Mag.	Reading of Scale.	Reduction.	Correction.	$\delta-\delta_0$	Principal Star.	U′—U	Correction.	$a-a_0$
		′ ″	′ ″	″	′ ″	m s	s	s	m s
d_1	15	2 55	−50 4.6	−0.6	−47 10.2	−2 14.55	+ 0.5	+0.01	−2 14.04
d_2	15	4 57		−0.9	−45 8.5		+ 1.35	0.00	−2 13.20
d_3	16	2 40		−0.5	−47 25.1		+ 7.75	+0.01	−2 6.79
d_4	16	0 59		−0.2	−49 5.8		+ 8.0	+0.01	−2 6.54
e_{-1}	16	10 0		−1.8	−40 6.4	−1 58.65	− 0.75	−0.01	−1 59.41
e_1	16	−0 20		0.0	−50 24.6		+ 1.0	+0.02	−1 57.63
e_2	17	4 20		−0.8	−45 45.4		+ 4.0	0.00	−1 54.65
e_3	15	1 12		−0.2	−48 52.8		+ 7.0	+0.01	−1 51.64
e_4	12	3 43		−0.7	−46 22.3		+12.0	+0.01	−1 46.64
e_5*	16	6 30		−1.2	−43 35.8		+20.0	0.00	−1 38.65
e_6	16	2 35		−0.5	−47 30.1		+23.75	+0.01	−1 34.89
h_1	17	4 30		−0.8	−45 35.4	−1 11.55	+ 8.0	+0.01	−1 3.54
k_{-1}	17	2 10		−0.3	−47 54.9	−0 45.75	− 2.0	0.00	−0 47.75
k_1	17	4 15		−0.7	−45 50.3		+ 0.75	0.00	−0 45.00
l_1	17	7 50		−1.4	−42 16.0	−0 44.65	+ 6.0	0.00	−0 38.65
l_2	16.17	7 10		−1.2	−42 55.8		+ 8.75	0.00	−0 35.90
m_{-2}†	16.17	1 0		+0.2	−49 4.4	−0 5.65	−11.0	+0.02	−0 16.63
m_{-1}	15	3 30		−0.6	−46 35.2		− 0.75	+0.01	−0 6.39
m_0	15	8 28		−1.4	−41 38.0		0.0	−0.01	−0 5.66
m_1	10	−0 57		+0.2	−51 1.4		+ 1.0	+0.02	−0 4.63
m_2	13.14	8 16		−1.4	−41 50.0		+ 5.75	−0.01	+0 0.09
n_1	12	2 48		−0.4	−47 17.0	+0 1.15	+ 0.5	+0.01	+0 1.66

* Follows F which is hard to observe.

† m_{-2} Scale reading altered from —1′0″ which was found incorrect.

Supplement to Zone XIV., March 12, 1864.

Name of Star.	Mag.	Reading of Scale.	Reduction.	Correction.	$\delta-\delta_0$	Principal Star.	U′—U	Correction.	$\alpha-\alpha_0$
		′ ″	′ ″	″	′ ″	m s	s	s	m s
n_2	13	8 45	−50 4.6	−1.5	−41 21.1	+0 1.15	+ 2.38	−0.01	+0 3.52
q_{-2}	16	−0 50		+0.2	−50 54.4	+0 58.05	− 5.75	+0.01	+0 52.31
q_{-1}	16.17	3 0		−0.4	−47 5.0		− 2.25	0.00	+0 55.80
s_{-1}	16	9 32		−1.5	−40 34.1	+1 11.34	− 9.75	−0.01	+1 1.58
w_{-1}	16	7 5		−1.1	−43 0.7	+1 55.84	− 0.62	0.00	+1 55.22
w_1	17	6 45		−1.0	−43 20.6		+ 2.0	0.00	+1 57.84
w_2	10	−0 52		+0.3	−50 56.3		+20.0	+0.02	+2 15.86

Supplement to Zone XV., March 12, 1864.

Name of Star.	Mag.	Reading of Scale.	Reduction.	Correction.	$\delta-\delta_0$	Principal Star.	U′—U	Correction.	$\alpha-\alpha_0$
		′ ″	′ ″	″	′ ″	m s	s	s	m s
f_1	14	2 48	−60 3.6	−0.5	−57 16.1	−2 14.83	+ 0.75	+0.01	−2 14.07
g_{-1}	16	0 35		−0.2	−59 28.8	−2 9.92	− 4.75	0.00	−2 14.67
g_1	13.14	2 12		−0.4	−57 52.0		+ 2.5	0.00	−2 7.42
g_2	15	0 15		−0.1	−59 48.7		+ 7.5	0.00	−2 2.42
h_{-3}	16	0 35		−0.2	−59 28.8	−1 56.02	− 4.75	0.00	−2 0.77
h_{-2}	16	−0 20		0.0	−60 23.6		− 4.5	0.00	−2 0.52
h_{-1}	15	1 28		−0.3	−58 35.9		− 2.0	0.00	−1 58.02
i_{-3}	16	6 0		−1.1	−54 4.7	−1 53.93	− 5.0	0.00	−1 58.93
i_{-2}	15	9 40		−1.7	−50 25.3		− 3.75	−0.01	−1 57.69
i_{-1}	15	6 45		−1.2	−53 19.8		− 2.5	0.00	−1 56.43
i_0	17	2 45		−0.5	−57 19.1		0.0	+0.01	−1 53.92
i_1	15	0 3		−0.1	−60 0.7		+ 4.0	+0.01	−1 49.92
k_{-2}	16	9 5		−1.5	−51 0.1	−1 36.52	− 1.75	−0.02	−1 38.29
k_{-1}	16	0 50		−0.2	−59 13.8		− 1.25	0.00	−1 37.77
k_1	15	2 45		−0.5	−57 19.1		+ 1.75	0.00	−1 34.77
k_2	16	0 25		−0.1	−59 38.7		+ 3.75	0.00	−1 32.77
k_3	14	2 15		−0.4	−57 49.0		+10.5	0.00	−1 26.02
k_4	16.17	2 30		−0.5	−57 34.1		+12.0	−0.01	−1 24.53
m_0	13	8 0		−1.3	−52 4.9	−1 14.93	0.0	−0.01	−1 14.94
m_1	17	2 15		−0.4	−57 49.0		+ 7.5	0.00	−1 7.43
m_2	17	0 45		−0.2	−59 18.8		+13.0	0.00	−1 1.93

Supplement to Zone XV., completed March 14, 1864.

Name of Star.	Mag.	Reading of Scale.	Reduction.	Correction.	$\delta-\delta_0$	Principal Star.	U′—U	Correction.	$\alpha-\alpha_0$
		′ ″	′ ″	″	′ ″	m s	s	s	m s
n_1	15	1 35	−60 3.6	−0.1	−58 28.7	−0 15.83	+ 2.0	0.00	−0 13.83
n_2	15	8 2		−1.5	−52 3.1		+ 3.5	0.00	−0 12.33
p_1	15	3 58		−0.6	−56 6.2	−0 4.84	+ 4.5	0.00	−0 0.34
s_1	14	7 20		−1.2	−52 44.8	+0 45.76	+ 0.5	0.00	+0 46.26
u_1	11	−0 7		+0.3	−60 10.3	+0 52.66	+ 1.0	0.00	+0 53.66
u_2	15	3 35		−0.5	−56 29.1		+ 4.5	0.00	+0 57.16
v_{-1}	16	6 0		−1.0	−54 4.6	+1 2.07	− 1.0	0.00	+1 1.07
w_{-1}	15	9 35		−1.7	−50 30.3	+1 11.56	− 7.0	0.00	+1 4.56

Supplement to Zone XVI., March 14, 1864.

Name of Star.	Mag.	Reading of Scale.	Reduction.	Correction.	$\delta-\delta_0$	Principal Star.	U′—U	Correction.	$\alpha-\alpha_0$
		′ ″	′ ″	″	′ ″	m s	s	s	m s
h_1	13.14	4 12	−70 4.2	−0.7	−65 52.9	−2 8.74	+ 8.0	0.00	−2 0.74
h_2	14	5 40		−1.0	−64 25.2		+10.0	0.00	−1 58.74

Supplement to Zone XVI., March 14, 1864.

Name of Star.	Mag.	Reading of Scale.	Reduction.	Correction.	$\delta-\delta_0$	Principal Star.	U′—U.	Correction.	$\alpha-\alpha_0$
		′ ″	′ ″	″	′ ″	m s	s	s	m s
i_{-2}	13	10 10	−70 4.2	−1.9	−59 56.1	−1 56.55	−13.5	0.00	−2 10.05
i_{-1}	14	8 42		−1.7	−61 23.9		− 1.5	0.00	−1 58.05
i_0	15	1 45		−0.3	−68 19.5		0.0	0.00	−1 56.55
i_1	13	10 12		−1.9	−59 54.1		+ 0.5	0.00	−1 56.05
i_2	12	7 52		−1.5	−62 13.7		+ 4.5	0.00	−1 52.05
i_3	14	4 40		−0.8	−65 25.0		+ 5.25	0.00	−1 51.30
i_4	14	5 48		−1.1	−64 17.3		+ 6.5	0.00	−1 50.05
k_1	11	3 16		−0.5	−66 48.7	−1 46.94	+ 1.12	0.00	−1 45.82
m_{-1}	15	8 47		−1.6	−61 18.8	−1 26.85	− 2.75	0.00	−1 29.60
m_1	15	2 37		−0.4	−67 27.6		+13.0	0.00	−1 13.85
p_1	14	6 15		−1.1	−63 50.3	−0 45.25	+ 5.0	0.00	−0 40.25
u_{-1}	13	6 0		−1.0	−64 5.2	+1 1.94	− 1.85	0.00	+1 0.09
w_1	13	3 20		−0.4	−66 44.6	+1 17.84	+ 1.75	0.00	+1 19.59
z_1	14	0 30		+0.1	−69 34.1	+2 5.24	+ 7.25	−0.01	+2 12.48

Supplement to Zone XVII., March 28, 1864.

Name of Star.	Mag.	Reading of Scale.	Reduction.	Correction.	$\delta-\delta_0$	Principal Star.	U′—U.	Correction.	$\alpha-\alpha_0$
		′ ″	′ ″	″	′ ″	m s	s	s	m s
h_0	16	−0 33	−80 14.1	0.0	−80 47.1	−2 17.25	0.0	0.00	−2 17.25
h_1	15.16	0 33		−0.1	−79 41.2		+ 0.50	0.00	−2 16.75
i_1	17	8 30		−1.5	−71 45.6	−2 14.15	+ 2.00	0.00	−2 12.15
i_2	15.16	3 35		−0.7	−76 39.8		+ 6.50	0.00	−2 7.65
k_1	14	−0 12		0.0	−80 26.1	−2 0.45	+ 3.00	0.00	−1 57.45
n_{-1}	14	−0 14		0.0	−80 28.1	−1 48.65	− 0.75	0.00	−1 49.40
n_1	15	8 20		−1.5	−71 55.6		+ 6 50	0.00	−1 42.15
n_2	16	4 10		−0.7	−76 4.8		+17.50	0.00	−1 31.15
o_{-2}	17	8 0		−1.4	−72 15.5	−1 20.36	−14.50	0.00	−1 34.86
o_{-1}	16.17	7 45		−1.3	−72 30.4		− 8.00	0.00	−1 28.36
o_1	16	0 18		−0.1	−79 56.2		+ 5.00	0.00	−1 15.36
q_1	17	−0 10		0.0	−80 24.1	−0 51.06	+ 0.25	0.00	−0 50.81
q_2	16.17	0 0		0.0	−80 14.1		+ 4.00	0.00	−0 47.06
r_{-1}	16	4 40		−0.8	−75 34.9	−0 39.26	− 3.75	0.00	−0 43.01
r_1	14	8 04		−1.3	−72 11.4		+ 2.75	0.00	−0 36.51
w_1	13	7 50		−1.3	−72 25.4	−0 1.66	+ 1.75	0.00	+0 0.09
aa_1	16	5 32		−0.9	−74 43.0	+0 33.94	+11.25	0.00	+0 45.19
aa_2	16.17	6 50		−1.1	−73 25.2		+11.50	0.00	+0 45.44
aa_3	15	0 55		−0.2	−79 19.3		+11.50	0.00	+0 45.44
aa_4	14.15	1 0		−0.2	−79 14.3		+13.75	0.00	+0 47.69
aa_5	16.17	2 40		−0.4	−77 34.5		+14.25	0.00	+0 48.19
dd_{-1}	16	8 30		−1.3	−71 45.4	+2 0.83	− 3.25	0.00	+1 57.58
dd_1	15	7 37		−1.1	−72 38.2		+ 3.25	0.00	+2 4.08
dd_2	16	8 05		−1.2	−72 10.3		+14.50	0.00	+2 15.33

Supplement to Zone XVIII., March 28, 1864.

Name of Star.	Mag.	Reading of Scale.	Reduction.	Correction.	$\delta-\delta_0$	Principal Star.	U′—U.	Correction.	$\alpha-\alpha_0$
		′ ″	′ ″	″	′ ″	m s	s	s	m s
d_{-1}	11	8 15	−90 15.8	−1.3	−82 2.1	−2 12.16	− 0.50	+0.02	−2 12.64
d_{-2}	16	5 12		−0.9	−85 4.7		− 2.50	+0.01	−2 14.65
d_1	16	−0 20		−0.2	−90 36.0		+ 5.50	−0.01	−2 6.67
e_1	15	8 10		−1.3	−82 7.1	−2 3.16	+ 1.00	+0.01	−2 2.15
e_2	13	−0 50		−0.1	−91 5.9		+ 4.00	−0.02	−1 59.18

Supplement to Zone XVIII., March 28, 1864.

Name of Star.	Mag.	Reading of Scale.	Reduction.	Correction.	$\delta-\delta_0$	Principal Star.	U′—U	Correction.	$\alpha-\alpha_0$
		′ ″	′ ″	″	′ ″	m s	s	s	m s
g_{-1}	16	8 15	+90 15.8	−1.3	−82 2.1	−1 53.16	− 9.00	+0.01	−2 2.15
g_1	16	4 42		−0.9	−85 34.7		+ 0.15	0.00	−1 53.01
g_2*	15	10 0		−1.6	−80 17.4		+ 4.00	+0.02	−1 49.14
h_0	14	8 02		−1.3	−82 15.1	−1 41.97	0.00	0.00	−1 41.97
h_1	15	5 58		−1.0	−84 18.8		+ 5.50	0.00	−1 36.47
h_2	17	4 40		−0.9	−85 36.7		+ 8.00	−0.01	−1 33.98
h_3	17	6 04		−1.0	−84 12.8		+11.00	−0.01	−1 30.98
k_1	16	0 40		−0.4	−89 36.2	−1 18.16	+ 1.00	0.00	−1 17.16
l_{-1}	15	2 50		−0.6	−87 26.4	−1 6.57	− 6.25	0.00	−1 12.82
m_1	14	1 15		−0.4	−89 1.2	−0 59.27	+ 2.25	0.00	−0 57.02
o_{-2}	15	−0 20		−0.2	−90 36.0	−0 42.47	− 3.75	−0.02	−0 46.24
o_{-1}	15	2 20		−0.5	−87 56.3		− 3.50	−0.01	−0 45.98
o_1	16	5 25		−0 9	−84 51.7		+ 1.75	0.00	−0 40.72
o_2	15	1 40		−0.5	−88 36.3		+ 7.50	−0.01	−0 34.98
t_{-5}	17	4 40		−0.7	−85 36.5	+0 21.53	−15.50	+0.01	+0 6.04
t_{-4}	15	7 40		−0.9	−82 36.7		−13.00	+0.02	+0 8.55
t_{-3}	17	5 50		−0.9	−84 26.7		− 9.50	+0.02	+0 12.05
t_{-2}	17	8 25		−1.1	−81 51.9		− 3.00	+0.03	+0 18.56
t_{-1}	13	1 05		−0.4	−89 11.2		− 0.15	0.00	+0 21.38
t_1	10	0 0		−0.4	−90 16.2		+ 4.00	−0.01	+0 25.52
x_{-2}	16	6 47		−0.9	−83 29.7	+0 59.82	−15.00	+0.02	+0 44.84
x_{-1}	16	7 10		−0.9	−83 6.7		− 7.75	+0.02	+0 52.09
x_1	16.17	5 0		−0.8	−85 16.6		+16.00	+0.01	+1 15.83
y_{-1}	16	4 20		−0.7	−85 56.5	+1 35.92	− 4.50	0.00	+1 31.42
aa_1	16	3 20		−0.7	−86 56.5	+2 33.62	+ 1.25	−0.02	+2 34.85
aa_2	10	−1 0		−0.7	−91 16.5		+ 4.75	−0.07	+2 38.30

* g_2 has a brighter companion on the edge of the field.

SECTION III. PART I.

CATALOGUE OF STARS NEAR ι, θ AND c ORIONIS, FROM OBSERVATIONS MADE AT THE OBSERVATORY OF HARVARD COLLEGE. 1857–1864.

THE POSITIONS ARE REFERRED TO θ' ORIONIS AS THE ORIGIN.

No.	Herschel's and Struve's No.	Letter. G. P. Bond.	Letter. Herschel.	Number. W. C. Bond.	Letter. Lassell.	Letter. Liapunoff.	Mag. by Argelander's Scale.	Eq. 1857.0 $\alpha-\alpha_0$ In Time.	Eq. 1857.0 $\alpha-\alpha_0$ In sec. of Arc.	Prec. Cöef.	Eq. 1857.0 $\delta-\delta_0$ In sec. of Arc.	Prec. Cöef.
								m s	″	″	″	″
1		XII. A					11.5	−3 35.89	−3238.3	−0.26	−2434.1	+0.28
2		XII. B					10.8	−3 25.33	−3079.9	−0.20	−1798.7	+0.26
3		XVII. A					10.8	−3 13.02	−2895.3	−0.49	−4771.1	+0.25
4		XII. c_{-1}					14.8	−3 11.67	−2875.1	−0.21	−1868.4	+0.25
5		XII. C					11.5	−3 6.92	−2803.8	−0.22	−1987.9	+0.24
6		X. A; XI. A					10.3	−3 6.53	−2798.0	−0.15	−1302.1	+0.24
7		XVII. B					8.6	−3 4.92	−2773.8	−0.49	−4789.0	+0.24
8		X. B; XI. B					9.7	−2 58.36	−2675.4	−0.18	−1541.4	+0.23
9		XII. d_{-1}					13.9	−2 58.05	−2670.7	−0.23	−2107.5	+0.23
10		XV. A; XVI. A					9.2	−2 57.05	−2655.7	−0.38	−3664.8	+0.22
11		XVI. B; XVII. C					11.2	−2 55.64	−2634.6	−0.44	−4234.9	+0.22
12		XII. D					11.5	−2 55.30	−2629.5	−0.23	−2137.3	+0.22
13		XV. B; XVI. C					8.7	−2 54.20	−2613.0	−0.38	−3650.2	+0.22
14		XVII. D; XVIII. A					9.5	−2 49.42	−2541.3	−0.50	−4830.3	+0.21
15		XII. e_{-2}					13.9	−2 48.22	−2523.3	−0.20	−1798.5	+0.21
16		XV. C					9.7	−2 46.85	−2502.8	−0.35	−3296.1	+0.21
17		XIV. A					11.0	−2 46.29	−2494.4	−0.32	−2993.2	+0.21
18		XV. D						−2 43.95	−2459.3	−0.35	−3345.8	+0.20
19		XV. E; XVI. D					10.8	−2 41.40	−2421.0	−0.38	−3617.7	+0.20
20		XVII. E					11.5	−2 41.33	−2419.9	−0.47	−4548.3	+0.20
21		XII. e_{-1}					13.9	−2 39.72	−2395.8	−0.21	−1913.0	+0.20
22		XII. E					11.2	−2 38.69	−2380.3	−0.25	−2333.6	+0.20
23		XII. f_{-1}					13.1	−2 37.82	−2367.3	−0.20	−1828.3	+0 20
24		XII. F					9.5	−2 36.29	−2344.4	−0.24	−2210.0	+0.19
25		XII. f_1					14.8	−2 35.31	−2329.7	−0.21	−1907.9	+0.19
26		I. A					10.8	−2 33.97	−2309.5	+0.13	+1605.8	+0.19
27		XVIII. B					8.9	−2 32.40	−2286.0	−0.55	−5399.2	+0.19
28		I. B					9.2	−2 31.55	−2273.3	+0.12	+1511.2	+0.19
29		VIII. A; IX. A					11.0	−2 31.04	−2265.6	+0.41	+4487.9	+0.19
30		XVI. E					10.8	−2 30.65	−2259.7	−0.41	−3977.2	+0.19
31		XVII. F					11.0	−2 29.23	−2238.4	−0.48	−4646.0	+0.18
32		XVII. G					11.5	−2 28.52	−2227.8	−0.49	−4725.7	+0.18
33		XIV. B					10.8	−2 28.00	−2220.0	−0.29	−2752.7	+0.18
34		IX. B					11.0	−2 27.63	−2214.5	+0.42	+4630.3	+0.18
35		A'_{-3}					10.4	−2 27.22	−2208.3	+0.02	+ 454.6	+0.18

No.	Herschel's and Struve's No.	Letter. G. P. Bond.	Letter. Herschel.	Number. W. C. Bond.	Letter. Lassell.	Letter. Liapunoff.	Mag. by Argelander's Scale.	Eq. 1857.0 $\alpha-\alpha_0$ In Time.	Eq. 1857.0 $\alpha-\alpha_0$ In sec. of Arc.	Prec. Cöef.	Eq. 1857.0 $\delta-\delta_0$ In sec. of Arc.	Prec. Cöef.
								m s	″	″	″	″
36		XII. f_2					13.9	−2 26.79	−2201.8	−0.26	−2366.3	+0.18
37		XVIII. C					10.3	−2 26.33	−2194.9	−0.50	−4890.6	+0.18
38		A_{-2}					10.2	−2 24.26	−2163.9	+0.05	+ 812.2	+0.18
39		X. C; XI. C					10.8	−2 24.19	−2162.8	−0.14	−1182.0	+0.18
40		IV. A					10.2	−2 23.90	−2158.5	+0.25	+2896.6	+0.18
41		I. C					11.5	−2 23.85	−2157.8	+0.11	+1441.6	+0.18
42		V. A					10.8	−2 23.20	−2148.0	+0.27	+3013.7	+0.18
43		XII. f_3					14.8	−2 22.81	−2142.1	−0.21	−1927.7	+0.17
44		XIV. C					11.0	−2 22.19	−2132.8	−0.32	−3029.8	+0.17
45*		II. a_{-3}					14.8	−2 22.12	−2131.8	+0.16	+1909.8	+0.17
46		IV. B; V. B					9.3	−2 22.00	−2130.0	+0.26	+2935.7	+0.17
47		XVI. F					9.2	−2 21.15	−2117.2	−0.42	−4012.0	+0.17
48		XVI. G					8.8	−2 20.75	−2111.2	−0.42	−4044.9	+0.17
49		A'_{-2}					10.8	−2 20.04	−2100.6	+0.02	+ 511.3	+0.17
50		V. C					11.0	−2 20.01	−2100.1	+0.28	+3127.2	+0.17
51		X. D; XI. D					11.0	−2 19.99	−2099.8	−0.15	−1261.7	+0.17
52		V. c_1					13.9	−2 19.29	−2089.4	+0.27	+3014.4	+0.17
53*		II. a_{-2}					14.8	−2 19.14	−2087.1	+0.18	+2140.9	+0.17
54		I. c_1					11.5	−2 19.14	−2087.1	+0.10	+1312.0	+0.17
55		X. E; XI. E					11.3	−2 18.48	−2077.2	−0.15	−1271.7	+0.17
56		V. c_2					13.9	−2 17.71	−2065.6	+0.28	+3162.0	+0.17
57		XVII. H					11.5	−2 17.03	−2055.5	−0.46	−4436.4	+0.17
58		XVII. h_0					13.9	−2 17.02	−2055.3	−0.50	−4848.3	+0.17
59		XVII. h_1					13.3	−2 16.52	−2047.8	−0.49	−4782.4	+0.17
60		X. e_1					13.1	−2 15.34	−2030.1	−0.19	−1725.2	+0.16
61*		VI. A; VIII. B					11.5	−2 14.81	−2022.1	+0.35	+3930.5	+0.16
62		XV. F					11.5	−2 14.67	−2020.1	−0.34	−3185.9	+0.16
63		XV g_{-1}					13.9	−2 14.49	−2017.4	−0.37	−3569.9	+0.16
64		XIV. D					10.8	−2 14.41	−2016.2	−0.28	−2647.8	+0.16
65		XVIII. d_{-2}					13.9	−2 14.41	−2016.1	−0.52	−5105.8	+0.16
66*		$(a''_{-1})_{-1}$					10.8	−2 14.11	−2011.6	−0.05	− 210.1	+0.16
67		XVII. I					11.0	−2 13.94	−2009.1	−0.45	−4349.5	+0.16
68		XV. f_1					12.3	−2 13.90	−2008.5	−0.36	−3437.2	+0.16
69		XIV. d_1					13.1	−2 13.90	−2008.5	−0.30	−2831.3	+0.16
70		VIII. b_1					14.8	−2 13.77	−2006.6	+0.39	+4332.9	+0.16
71		XIV. d_2					13.1	−2 13.06	−1995.9	−0.29	−2709.6	+0.16
72*		$(a'_{-1})_{-1}$					13.1	−2 12.63	−1989.5	−0.01	+ 198.8	+0.16
73		XVIII. d_{-1}					10.2	−2 12.41	−1986.1	−0.51	−4923.2	+0.16
74		XVII. i_1					14.8	−2 11.94	−1979.1	−0.45	−4306.7	+0.16
75		XVIII. D					9.8	−2 11.91	−1978.7	−0.53	−5217.5	+0.16
76		A‴					10.3	−2 10.71	−1960.7	−0.09	− 702.2	+0.16
77		IV. b_1					13.9	−2 10.71	−1960.6	+0.26	+2907.0	+0.16
78		V. c_3					13.1	−2 10.51	−1957.6	+0.28	+3164.1	+0.16
79		XII. G					11.5	−2 9.89	−1948.4	−0.24	−2241.6	+0.16
80		XV. G; XVI. i_{-2}					11.2	−2 9.81	−1947.1	−0.38	−3601.0	+0.16

No.	Herschel's and Struve's No.	Letter. G. P. Bond.	Letter. Herschel.	Number. W. C. Bond.	Letter. Lassell.	Letter. Liapunoff.	Mag. by Argelander's Scale.	Eq. 1857.0 $\alpha-\alpha_0$ In Time.	Eq. 1857.0 $\alpha-\alpha_0$ In sec. of Arc.	Prec. Cöef.	Eq. 1857.0 $\delta-\delta_0$ In sec. of Arc.	Prec. Cöef.
								m s	″	″	″	″
81	1	A$'_{-1}$	Z				9.7	−2 9.13	−1937.0	−0.01	+ 150.9	+0.16
82		VI. a_1					14.8	−2 8.87	−1933.0	+0.35	+3844.9	+0.15
83		XVI. H					11.5	−2 8.55	−1928.2	−0.42	−4040.8	+0.15
84		IV. b_2					13.9	−2 8.19	−1922.9	+0.24	+2767.4	+0.15
85		XVII. i_2					13.3	−2 7.43	−1911.4	−0.47	−4600.9	+0.15
86		XV. g_1					11.7	−2 7.25	−1908.7	−0.36	−3473.1	+0.15
87		IX. C					10.4	−2 7.15	−1907.2	+0.45	+4889.6	+0.15
88		XIV d_3					13.9	−2 6.65	−1899.7	−0.30	−2846.2	+0.15
89		VIII. c_{-1}					13.1	−2 6.56	−1898.4	+0.39	+4241.4	+0.15
90		XVIII. d_1					13.9	−2 6.41	−1896.2	−0.56	−5437.1	+0.15
91		XIV. d_4					13.9	−2 6.39	−1895.9	−0.31	−2946.9	+0.15
92		XII. H					11.5	−2 6.31	−1894.7	−0.23	−2063.1	+0.15
93*		VIII. C; VI. a_2					11.9	−2 5.62	−1884.3	+0.37	+4096.5	+0.15
94		IX. D					10.8	−2 5.35	−1880.3	+0.47	+5078.8	+0.15
95		IX. d_1					14.8	−2 4.85	−1872.8	+0.46	+5044.0	+0.15
96		V. d_{-1}					13.9	−2 4.73	−1871.0	+0.31	+3473.3	+0.15
97		IV. C					9.7	−2 4.69	−1870.3	+0.24	+2694.7	+0.15
98		IX. d_2					11.5	−2 3.96	−1859.4	+0.45	+4888.6	+0.15
99		V. D					12.3	−2 3.71	−1855.7	+0.29	+3286.8	+0.15
100		XVIII. E					11.9	−2 2.91	−1843.7	−0.52	−5089.8	+0.15
101		VI. b_{-1}					12.3	−2 2.47	−1837.1	+0.34	+3730.4	+0.15
102		VIII. c_1; IX. c_1					13.1	−2 2.47	−1837.0	+0.41	+4481.8	+0.15
103		XV. g_2					13.1	−2 2.25	−1833.7	−0.37	−3589.7	+0.14
104		VI. B					9.4	−2 1.97	−1829.6	+0.33	+3700.6	+0.14
105		XVIII. g_{-1}; XVIII. e_3					13.5	−2 1.91	−1828.7	−0.51	−4925.6	+0.14
106		IX. c_2					14.2	−2 1.13	−1817.0	+0.44	+4755.0	+0.14
107		XV. h_{-3}					13.9	−2 0.59	−1808.9	−0.37	−3569.8	+0.14
108		XVI. h_1					11.7	−2 0.55	−1808.2	−0.41	−3953 9	+0.14
109		XV. h_{-2}					13.9	−2 0.34	−1805.1	−0.38	−3624.6	+0.14
110		XVII. K					11.0	−2 0.23	−1803.5	−0.46	−4469.1	+0.14
111		XIV. e_{-1}					13.9	−1 59.29	−1789.3	−0.26	−2407.4	+0.14
112*		a$'_{-2}$					12.3	−1 59.12	−1786.8	−0.01	+ 189.5	+0.14
113		XVIII. e_2					11.5	−1 58.92	−1783.8	−0.56	−5466.9	+0.14
114		XV. i_{-3}					13.9	−1 58.77	−1781.5	−0.34	−3245.7	+0.14
115		IX. c_3					14.8	−1 58.65	−1779.7	+0.44	+4814.8	+0.14
116		VI. b_1; VIII. c_2					11.5	−1 58.64	−1779.6	+0.36	+3967.0	+0.14
117		XVI. h_2					12.3	−1 58.55	−1778.3	−0.40	−3866.2	+0.14
118		XIV. E					9.7	−1 58:52	−1777.8	−0.28	−2618.7	+0.14
119		XVI. i_{-1}					12.3	−1 57.87	−1768.0	−0.39	−3684.9	+0.14
120		XV. h_{-1}					13.1	−1 57.85	−1767.7	−0.37	−3516.9	+0.14
121		XIV. e_1; XV. i_{-2}					13.5	−1 57.51	−1762.6	−0.32	−3025.9	+0 14
122*		a_{-3}					10.8	−1 57.43	−1761.5	+0.07	+ 999.4	+0.14
123		XII. I.					11.7	−1 57.42	−1761.3	−0.22	−1980.2	+0.14
124		XII. i_0					14.8	−1 57.39	−1760.8	−0.26	−2415.9	+0.14
125		XVIII. F; XVII. k_1					11.9	−1 57.03	−1755.5	−0.50	−4827.8	+0.14

No.	Herschel's and Struve's No.	Letter. G. P. Bond.	Letter. Herschel.	Number. W. C. Bond.	Letter. Lassell.	Letter. Liapunoff.	Mag. by Argelander's Scale.	Eq. 1857.0 $\alpha-\alpha_0$ In Time.	Eq. 1857.0 $\alpha-\alpha_0$ In sec. of Arc.	Prec. Cöef.	Eq. 1857.0 $\delta-\delta_0$ In sec. of Arc.	Prec. Cöef.
								m s	″	″	″	″
126		V. d_1					14.8	−1 57.03	−1755.4	+0.30	+3378.5	+0.14
127		XVI. I					9.5	−1 56.37	−1745.5	−0.39	−3696.7	+0.14
128		XVI. i_0					13.1	−1 56.35	−1745.3	−0.43	−4100.5	+0.14
129		XV. i_{-1}					13.1	−1 56.27	−1744.1	−0.34	−3200.8	+0.14
130		IV. c_1					13.9	−1 56.19	−1742.8	+0.23	+2657.9	+0.14
131*		XV. H; XVI. i_1					11.2	−1 55.86	−1737.9	−0.38	−3597.4	+0.14
132		IV. c_2					13.3	−1 55.71	−1735.6	+0.26	+2921.2	+0.14
133		XIV. e_2					14.8	−1 54.51	−1717.7	−0.29	−2746.3	+0.13
134		A''_{-1}					10.2	−1 54.35	−1715.3	−0.06	− 312.4	+0.13
135		XVII. L					10.8	−1 54.23	−1713.5	−0.47	−4516.9	+0.13
136		XII. K					11.3	−1 54.11	−1711.7	−0.22	−2007.0	+0.13
137		XV. I					10.8	−1 53.77	−1706.5	−0.34	−3247.6	+0.13
138		XV. i_0					14.8	−1 53.75	−1706.3	−0.36	−3440.0	+0.13
139		V. d_2					14.8	−1 53.73	−1706.0	+0.31	+3433.5	+0.13
140		XVIII. G					9.8	−1 52.91	−1693.7	−0.53	−5165.4	+0.13
141		XVIII. g_1					13.9	−1 52.77	−1691.5	−0.53	−5135.6	+0.13
142		IX. c_4					13.1	−1 52.67	−1690.0	+0.47	+5130.9	+0.13
143		IX. c_5					14.8	−1 52.65	−1689.7	+0.45	+4929.5	+0.13
144*		a'_{-1}					11.5	−1 52.53	−1688.0	0.00	+ 322.	+0.13
145		A′					11.0	−1 51.98	−1679.7	−0.02	+ 106.6	+0.13
146		XVI. i_2					10.8	−1 51.87	−1678.0	−0.39	−3734.6	+0.13
147		XIV. e_3					13.1	−1 51.49	−1672.4	−0.31	−2933.7	+0.13
148		V. E					11.0	−1 51.41	−1671.1	+0.28	+3157.4	+0.13
149		XVII. M					10.8	−1 51.23	−1668.4	−0.47	−4598.6	+0.13
150		XVI. i_3					12.3	−1 51.11	−1666.6	−0.41	−3925.9	+0.13
151		V. e_1					13.9	−1 50.75	−1661.2	+0.31	+3493.3	+0.13
152*		a_{-2}					10.2	−1 50.53	−1658.0	+0.06	+ 887.1	+0.13
153		IV. c_3; V. e_2					13.1	−1 50.05	−1650.7	+0.26	+2955.0	+0.13
154		XVI. i_4					12.3	−1 49.86	−1647.9	−0.40	−3858.2	+0.13
155		XV. i_1					13.1	−1 49.75	−1646.2	−0.38	−3601.6	+0.13
156*		II. a_{-1}					13.1	−1 49.47	−1642.1	+0.16	+1920.1	+0.13
157		XVII. n_{-1}					12.3	−1 49.17	−1637.5	−0.50	−4829.0	+0.13
158*		XVIII. g_2					13.1	−1 48.91	−1633.6	−0.50	−4818.3	+0.13
159		II. A; III. A					13.1	−1 48.88	−1633.2	+0.16	+1872.8	+0.13
160*		XVII. N					9.7	−1 48.42	−1626.3	−0.49	−4793.1	+0.13
161		IX. c_6					13.9	−1 48.13	−1621.9	+0.43	+4725.2	+0.12
162		IV. D					11.0	−1 47.97	−1619.6	+0.21	+2450.7	+0.12
163		V. e_3					14.8	−1 46.93	−1603.9	+0.30	+3333.9	+0.12
164		I. D					11.5	−1 46.87	−1603.1	+0.12	+1519.7	+0.12
165		XVI. K					8.9	−1 46.74	−1601.1	−0.43	−4134.3	+0.12
166		XII. L; XIV. F					9.4	−1 46.51	−1597.6	−0.26	−2396.6	+0.12
167		XIV. e_4					10.8	−1 46.50	−1597.5	−0.29	−2783.2	+0.12
168		IX. c_7					14.8	−1 46.13	−1591.9	+0.43	+4720.3	+0.12
169		XVI. k_1					10.2	−1 45.62	−1584.3	−0.42	−4009.6	+0.12
170		X. F; XI. F					9.4	−1 45.38	−1580.7	−0.16	−1334.7	+0.12

No.	Herschel's and Struve's No.	Letter. G. P. Bond.	Letter. Herschel.	Number. W. C. Bond.	Letter. Lassell.	Letter. Liapunoff.	Mag. by Argelander's Scale.	Eq. 1857.0 $\alpha - \alpha_0$ In Time.	Eq. 1857.0 $\alpha - \alpha_0$ In sec. of Arc.	Prec. Cöef.	Eq. 1857.0 $\delta - \delta_0$ In sec. of Arc.	Prec. Cöef.
								m s	″	″	″	″
171		VI. c_{-2}					13.9	−1 45.01	−1575.1	+0.36	+4026.5	+0.12
172		V. e_4					13.1	−1 44.39	−1565.9	+0.27	+3031.8	+0.12
173		IV. d_1					13.9	−1 44.27	−1564.1	+0.20	+2348.7	+0.12
174		VI. c_{-1}					12.3	−1 44.01	−1560.2	+0.37	+4086.3	+0.12
175*		a_{-1}					13.9	−1 43.80	−1557.0	+0.08	+1120.0	+0.12
176		A_{-1}					10.8	−1 43.36	−1550 4	+0.08	+1083.1	+0.12
177		VIII. d_{-2}					14.8	−1 43.06	−1545.9	+0.38	+4233.6	+0.12
178*		b''_{-1}					11.5	−1 43.05	−1545.7	+0.03	+ 565.4	+0.12
179		V. e_5					13.1	−1 42.73	−1540.9	+0.30	+3353.8	+0.12
180		IX. c_8					14.8	−1 42.66	−1539.9	+0.46	+5014.3	+0.12
181		XII. l_1					14.8	−1 42.00	−1530.0	−0.24	−2166.5	+0.12
182		VI. C					11.5	−1 41.97	−1529.6	+0.33	+3661.0	+0.12
183		XVII. n_1					13.1	−1 41.94	−1529.1	−0.45	−4316.4	+0.12
184		XVIII. h_0					12.3	−1 41.73	−1526.0	−0.51	−4935.9	+0.12
185		XVIII. H					9.8	−1 41.73	−1526.0	−0.51	−4974.9	+0.12
186		IX. c_9					12.3	−1 41.13	−1516.9	+0.42	+4650.6	+0.11
187		IV. d_2					14.8	−1 41.02	−1515.3	+0.26	+2947.3	+0.11
188		VI. c_1					12.3	−1 40.51	−1507.6	+0.37	+4074.4	+0.11
189		IX. c_{10}					14.8	−1 40.15	−1502.2	+0.44	+4830.0	+0.11
190		A″					10.7	−1 38.83	−1482.5	−0.05	− 230.2	+0.11
191		XIV. e_5					13.9	−1 38.52	−1477.8	−0.28	−2616.6	+0.11
192		XV. k_{-2}					13.9	−1 38.14	−1472.1	−0.33	−3060.9	+0.11
193		XII. l_2					13.9	−1 38.03	−1470.4	−0.20	−1787.8	+0.11
194		VI. c_2					13.9	−1 38.03	−1470.4	+0.38	+4164.0	+0.11
195		XV. k_{-1}					13.9	−1 37.59	−1463.9	−0.37	−3554.6	+0.11
196		VIII. d_{-1}					14.8	−1 37.56	−1463.4	+0.39	+4313.3	+0.11
197		II. a_1					13.9	−1 37.14	−1457.1	+0.17	+2011.8	+0.11
198		IX. c_{11}					14.8	−1 36.66	−1449.9	+0.45	+4949.6	+0.11
199		I. E					9.7	−1 36.45	−1446.8	+0.10	+1332.5	+0.11
200		XV. K; XVI. L					9.2	−1 36.36	−1445.4	−0.38	−3593.0	+0.11
201		XVIII. h_1					13.1	−1 36.23	−1443.4	−0.52	−5059.6	+0 11
202		IX. c_{12}					14.8	−1 35.65	−1434.7	+0.45	+4927.8	+0.11
203		XIV. e_6					13.9	−1 34.75	−1421.2	−0.30	−2850.8	+0.11
204		VI. c_3					14.8	−1 34.74	−1421.1	+0.35	+3875.2	+0.11
205		XVII. o_{-2}					14.8	−1 34.65	−1419.8	−0.45	−4336.2	+0.11
206		IV. e_0					14.8	−1 34.61	−1419.2	+0.25	+2847.6	+0.11
207		XV. k_1					13.1	−1 34.60	−1419.0	−0.36	−3439.8	+0.11
208		IV. E					11.5	−1 34.59	−1418.9	+0.22	+2582.4	+0.11
209		XVIII. h_2					14.8	−1 33.73	−1406.0	−0.53	−5137.4	+0.10
210		XV. k_2					13.9	−1 32.59	−1388.9	−0.38	−3579.4	+0.10
211		a'_1					13.1	−1 31.68	−1375.2	−0.02	+ 123.5	+0.10
212		a'_2					13.1	−1 31.68	−1375.2	+0.03	+ 539.5	+0.10
213		IX. c_{13}					14.8	−1 31.18	−1367.7	+0.47	+5129.1	+0.10
214		XVII. n_2					13.9	−1 30.93	−1364.0	−0.47	−4565.5	+0.10
215		A					9.8	−1 30.87	−1363.1	+0.05	+ 806.6	+0.10

No.	Herschel's and Struve's No.	Letter. G. P. Bond.	Letter. Herschel.	Number. W. C. Bond.	Letter. Lassell.	Letter. Liapunoff.	Mag. by Argelander's Scale.	Eq. 1857.0 $\alpha-\alpha_0$ In Time.	Eq. 1857.0 $\alpha-\alpha_0$ In sec. of Arc.	Prec. Cöef.	Eq. 1857.0 $\delta-\delta_0$ In sec. of Arc.	Prec. Cöef.
								m s	″	″	″	″
216		a''_1					13.9	−1 30.83	−1362.5	−0.07	− 482.5	+0.10
217		XVIII. h_3					14.8	−1 30.74	−1361.1	−0.52	−5053.5	+0.10
218		V. f_{-1}					12.3	−1 30.73	−1361.0	+0.31	+3466.6	+0.10
219		IV. e_1					14.8	−1 30.59	−1358.9	+0.23	+2603.2	+0.10
220		IX. c_{14}					14.8	−1 29.91	−1348.6	+0.45	+4939.7	+0.10
221		XVI. m_{-1}					13.1	−1 29.42	−1341.3	−0.39	−3679.5	+0.10
222		a'''_1					13.9	−1 29.28	−1339.2	−0.10	− 691.8	+0.10
223		IV. e_2					13.9	−1 28.60	−1329.0	+0.23	+2666.1	+0.10
224		IX. c_{15}					14.8	−1 28.41	−1326.2	+0.46	+5016.4	+0.10
225		XVII. o_{-1}					14.2	−1 28.15	−1322.2	−0.45	−4351.1	+0.10
226		XII. M					10.8	−1 28.11	−1321.7	−0.23	−2048.7	+0.10
227		VI. d_{-1}					14.8	−1 27.52	−1312.8	+0.36	+4034.6	+0.09
228		I. e_1					13.9	−1 26.97	−1304.6	+0.12	+1477.1	+0.09
229		XII. m_1					14.8	−1 26.85	−1302.8	−0.25	−2246.1	+0.09
230*		XVI. M					10.8	−1 26.66	−1299.9	−0.41	−3905.8	+0.09
231		IX. c_{16}					14.2	−1 26.63	−1299.4	+0.43	+4710.6	+0.09
232	2	B″; B‴	(χ)				10.3	−1 26.53	−1298.0	−0.08	− 581.6	+0.09
233		VI. D; VII. A; VIII. D					10.8	−1 26.52	−1297.8	+0.37	+4099.0	+0.09
234		I. e_2; a_1					11.8	−1 26.41	−1296.1	+0.09	+1168.8	+0.09
235		X. G; XI. G					9.4	−1 26.18	−1292.7	−0.16	−1359.4	+0.09
236		XV. k_3					13.3	−1 25.85	−1287.7	−0.37	−3469.6	+0.09
237		VI. d_1					13.9	−1 25.38	−1280.7	+0.35	+3845.4	+0.09
238		VI. d_2					12.3	−1 24.64	−1269.6	+0.36	+3994.9	+0.09
239		XV. k_4					14.2	−1 24.36	−1265.4	−0.36	−3454.7	+0.09
240		VI. d_3					13.9	−1 23.65	−1254.7	+0.36	+4029.7	+0.09
241		a'_3					14.8	−1 23.53	−1253.0	−0.01	+ 200.5	+0.09
242		B′					11.2	−1 23.03	−1245.5	+0.01	+ 391.8	+0.09
243		X. g_1					13.9	−1 22.93	−1243.9	−0.15	−1301.0	+0.09
244		V. F					11.5	−1 21.73	−1226.0	+0.31	+3483.5	+0.09
245		IX. c_{17}					14.8	−1 21.13	−1216.9	+0.43	+4670.7	+0.09
246	3	C‴	(μ)				9.3	−1 20.50	−1207.5	−0.10	− 793.9	+0.09
247	4	D‴	(ν)				9.6	−1 20.23	−1203.5	−0.12	− 944.2	+0.08
248		XVII. O					9.2	−1 20.15	−1202.2	−0.46	−4435.8	+0.08
249		a_2					14.8	−1 19.87	−1198.1	+0.07	+ 969.0	+0.08
250		VI. E; VII. B; VIII. E					10.4	−1 19.68	−1195.2	+0.35	+3920.1	+0.08
251		XVIII. I					10.7	−1 19.03	−1185.5	−0.50	−4897.9	+0.08
252		XVIII. K					10.0	−1 17.90	−1168.5	−0.56	−5426.4	+0.08
253		I. f_1					13.9	−1 17.87	−1168.1	+0.12	+1546.9	+0.08
254		I. F					11.5	−1 17.86	−1167.9	+0.11	+1384.5	+0.08
255		XV. L					11.5	−1 17.17	−1157.5	−0.35	−3350.9	+0.08
256		XVIII. k_1					13.9	−1 16.90	−1153.5	−0.55	−5376.8	+0.08
257		VI. e_1					13.1	−1 16.69	−1150.4	+0.36	+3968.1	+0.08
258		II. B; III. B					10.5	−1 16.66	−1149.9	+0.20	+2303.4	+0.08
259		X. g_2					14.8	−1 16.40	−1146.0	−0.19	−1634.9	+0.08
260		d'''_1					14.2	−1 16.23	−1143.5	−0.10	− 761.7	+0.08

No.	Herschel's and Struve's No.	Letter. G. P. Bond.	Letter. Herschel.	Number. W. C. Bond.	Letter. Lassell.	Letter. Liapunoff.	Mag. by Argelander's Scale.	Eq. 1857.0 $\alpha-\alpha_0$ In Time.	Eq. 1857.0 $\alpha-\alpha_0$ In sec. of Arc.	Prec. Cöef.	Eq. 1857.0 $\delta-\delta_0$ In sec. of Arc.	Prec. Cöef.
								m s	″	″	″	″
261		V. G; VII. C; VI. e_2					10.2	−1 16.05	−1140.7	+0.32	+3561.1	+0.08
262		XIV. G					11.5	−1 15.41	−1131.2	−0.29	−2704.0	+0.08
263		XVII. o_1					13.9	−1 15.13	−1126.9	−0.50	−4796.7	+0.08
264		XV. m_0					11.5	−1 14.79	−1121.8	−0.33	−3125.4	+0.08
265		XV. M					10.8	−1 14.76	−1121.4	−0.36	−3401.7	+0.08
266		XII. n_{-1}					14.8	−1 14.53	−1117.9	−0.21	−1827.3	+0.08
267		IX. e_{-3}					14.8	−1 14.32	−1114.8	+0.43	+4712.8	+0.08
268		IX. e_{-2}					14.8	−1 14.32	−1114.8	+0.43	+4728.7	+0.08
269*		IX. e_{-1}					12.3	−1 14.22	−1113.3	+0.46	+5011.7	+0.08
270		d'''_2					14.8	−1 13.98	−1109.7	−0.11	− 803.7	+0.08
271		XVI. m_1					13.1	−1 13.65	−1104.8	−0.42	−4048.1	+0.07
272		IX. E					11.5	−1 13.46	−1101.9	+0.44	+4870.2	+0.07
273		I. f_2					14.8	−1 13.34	−1100.1	+0.10	+1263.0	+0.07
274		VIII. f_0					13.9	−1 12.67	−1090.0	+0.40	+4423.2	+0.07
275		VIII. F					10.8	−1 12.66	−1089.9	+0.39	+4342.5	+0.07
276		d'''_3					13.1	−1 12.58	−1088.7	−0.11	− 826.6	+0.07
277		XVIII. l_{-1}					13.1	−1 12.57	−1088.5	−0.54	−5246.9	+0.07
278		XII. N					11.5	−1 11.52	−1072.8	−0.21	−1911.9	+0.07
279		XIV. H					10.8	−1 11.42	−1071.3	−0.28	−2546.4	+0.07
280		IV. e_3					13.1	−1 11.09	−1066.4	+0.23	+2616.4	+0.07
281		VIII. f_1					13.9	−1 10.91	−1063.7	+0.40	+4372.4	+0.07
282		II. C; III. C					11.5	−1 9.85	−1047.8	+0.19	+2190.3	+0.07
283		a_3					13.9	−1 9.63	−1044.5	+0.04	+ 689.3	+0.07
284		V. g_1					12.3	−1 9.42	−1041.3	+0.30	+3358.1	+0.07
285		VI. F; VII. D					10.4	−1 9.35	−1040.2	+0.34	+3801.7	+0.07
286*		VI. f_0; VIII. g_{-1}					13.1	−1 9.32	−1039.8	+0.36	+4008.8	+0.07
287		IX. F					11.0	−1 9.16	−1037.4	+0.44	+4870.2	+0.07
288		b'_1					13.9	−1 9.13	−1037.0	+0.03	+ 569.2	+0.07
289		V. g_2					13.1	−1 9.11	−1036.6	+0.29	+3284.3	+0.07
290		b''_1					13.1	−1 8.78	−1031.7	−0.04	− 173.8	+0.07
291		VIII. G; VI. f_1					10.4	−1 8.40	−1026.0	+0.37	+4120.4	+0.07
292		B					11.2	−1 8.38	−1025.7	+0.05	+ 771.8	+0.07
293		VIII. H; VI. f_2					11.0	−1 8.28	−1024.2	+0.36	+3946.2	+0.07
294*		II. D; III. D					11.2	−1 7.59	−1013.9	+0.16	+1951.4	+0.07
295		XV. m_1					14.8	−1 7.26	−1008.9	−0.37	−3469.5	+0.07
296		XII. n_1					14.2	−1 6.77	−1001.5	−0.22	−1958.9	+0.06
297		IV. F					11.0	−1 6.51	− 997.6	+0.25	+2847.7	+0.06
298		VIII. g_1					14.8	−1 6.46	− 996.9	+0.38	+4194.2	+0.06
299		XVIII. L					8.9	−1 6.32	− 994.8	−0.54	−5230.8	+0.06
300		IX. f_1					14.8	−1 5.67	− 985.1	+0.46	+5074.6	+0.06
301		II. d_1					11.7	−1 5.63	− 984.5	+0.16	+1962.4	+0.06
302		IV. G					11.5	−1 5.60	− 984.0	+0.23	+2688.4	+0.06
303	5	C′; C″	Q			w	9.9	−1 5.28	− 979.2	−0.03	+ 13.9	+0.06
304		V. h_0					13.1	−1 4.73	− 971.0	+0.29	+3240.6	+0.06
305		V. H					11.5	−1 4.71	− 970.7	+0.27	+3093.3	+0.06

No.	Herschel's and Struve's No.	Letter. G. P. Bond.	Letter. Herschel.	Number. W. C. Bond.	Letter. Lassell.	Letter. Liapunoff.	Mag. by Argelander's Scale.	Eq. 1857.0 $\alpha-\alpha_0$ In Time.	Eq. 1857.0 $\alpha-\alpha_0$ In sec. of Arc.	Prec. Cöef.	Eq. 1857.0 $\delta-\delta_0$ In sec. of Arc.	Prec. Cöef.
								m s	"	"	"	"
306		IX. f_2					14.8	−1 3.62	−954.3	+0.41	+4531.5	+0.06
307		XIV. h_1					14.8	−1 3.40	−951.0	−0.30	−2735.8	+0.06
308		VI. f_3; VIII. g_2					12.3	−1 3.21	−948.2	+0.37	+4090.1	+0.06
309		IX. f_4					13.3	−1 2.67	−940.1	+0.46	+5044.7	+0.06
310		IX. f_3					14.8	−1 2.62	−939.3	+0.41	+4533.5	+0.06
311	6	C	(ζ)				10.7	−1 1.88	−928.2	+0.04	+ 681.9	+0.06
312		XV. m_2					14.8	−1 1.76	−926.4	−0.37	−3559.2	+0.06
313		XVI. N					9.7	−1 1.55	−923.3	−0.43	−4090.0	+0.06
314	8	D''				$\beta_{\prime\prime}$	11.4	−1 1.54	−923.1	−0.06	− 307.8	+0.06
315	7	E'''	(ξ)				10.2	−1 1.42	−921.3	−0.11	− 825.4	+0.06
316		V. I					11.5	−1 1.03	−915.4	+0.29	+3274.5	+0.06
317		IX. f_5					14.8	−1 0.64	−909.6	+0.42	+4656.0	+0.06
318		VI. h_{-4}					13.1	−1 0.53	−908.0	+0.33	+3716.1	+0.06
319		X. h_{-1}					14.8	−1 0.03	−900.5	−0.17	−1415.4	+0.06
320		VIII. g_3					14.8	−0 59.97	−899.6	+0.39	+4343.6	+0.06
321		V. i_1					14.8	−0 59.53	−892.9	+0.29	+3264.5	+0.05
322		XVII. P					10.2	−0 59.23	−888.5	−0.50	−4794.6	+0.05
323	9	F'''					10.7	−0 59.09	−886.4	−0.11	− 816.6	+0.05
324		XVIII. M					9.3	−0 59.02	−885.3	−0.55	−5314.5	+0.05
325		VII. E; VI. h_{-3}					10.8	−0 58.84	−882.6	+0.33	+3700.3	+0.05
326		c_1					11.5	−0 58.78	−881.7	+0.08	+1105.6	+0.05
327		XVI. O					10.8	−0 58.57	−878.5	−0.39	−3745.9	+0.05
328*		VI. G; VIII. I					10.4	−0 58.43	−876.5	+0.35	+3870.6	+0.05
329		f'''_1					14.2	−0 57.84	−867.6	−0.09	− 626.9	+0.05
330		VI. h_{-2}					11.5	−0 57.80	−867.0	+0.34	+3765.9	+0.05
331		V. i_2; IV. i_{-1}					12.7	−0 57.70	−865.5	+0.26	+2969.1	+0.05
332		f'''_2					14.8	−0 57.59	−863.9	−0.09	− 646.8	+0.05
333		II. e_{-1}; IV. h_{-1}					13.1	−0 57.43	−861.4	+0.20	+2369.1	+0.05
334		IX. G					11.0	−0 57.24	−858.6	+0.42	+4581.4	+0.05
335	10	E''	π			s	10.9	−0 56.77	−851.6	−0.05	− 239.0	+0.05
336		XVIII. m_1					12.3	−0 56.77	−851.5	−0.55	−5341.6	+0.05
337		V. k_0					14.8	−0 56.05	−840.8	+0.31	+3503.9	+0.05
338		V. K					11.0	−0 56.02	−840.3	+0.28	+3175.0	+0.05
339*		e''_{-1}					14.8	−0 55.98	−839.7	−0.07	− 446.6	+0.05
340		VI. h_{-1}					13.9	−0 55.80	−837.0	+0.35	+3900.5	+0.05
341		II. c_1					12.3	−0 55.76	−836.4	+0.20	+2321.3	+0.05
342		IX. h_0					12.3	−0 55.69	−835.4	+0.46	+5063.7	+0.05
343		IV. H					11.5	−0 55.69	−835.3	+0.21	+2411.6	+0.05
344		I. G					11.5	−0 55.67	−835.0	+0.12	+1527.3	+0.05
345		IX. H					11.0	−0 55.64	−834.6	+0.42	+4585.5	+0.05
346	11	G'''	(π)				10.7	−0 55.56	−833.4	−0.12	− 947.1	+0.05
347*	15	e''_1					14.8	−0 55.29	−829.4	−0.07	− 462.6	+0.05
348		X. H; XI. H					11.5	−0 55.07	−826.0	−0.20	−1717.1	+0.05
349		IX. h_2					11.5	−0 54.70	−820.5	+0.47	+5139.5	+0.05
350		II. E; III. E					8.6	−0 54.50	−817.5	+0.18	+2098.0	+0.05

No.	Herschel's and Struve's No.	Letter. G. P. Bond.	Letter. Herschel.	Number. W. C. Bond.	Letter. Lassell.	Letter. Liapunoff.	Mag. by Argelander's Scale.	Eq. 1857.0 $\alpha-\alpha_0$ In Time.	Eq. 1857.0 $\alpha-\alpha_0$ In sec. of Arc.	Prec. Cöef.	Eq. 1855.0 $\delta-\delta_0$ In sec. of Arc.	Prec. Cöef.
								m s	″	″	″	″
351		II. e_1					11.5	−0 54.45	−816.7	+0.18	+2161.8	+0.05
352		IX. h_1					11.5	−0 54.39	−815.9	+0.42	+4591.4	+0.05
353		V. k_1					12.3	−0 54.32	−814.8	+0.29	+3249.7	+0.05
354		IX. h_5					14.8	−0 53.68	−805.2	+0.45	+4950.1	+0.05
355		IX. h_6					14.8	−0 53.68	−805.2	+0.46	+4980.0	+0.05
356		IX. h_3					13.1	−0 53.45	−801.7	+0.46	+5044.8	+0.05
357*		VIII. g_4; IX. h_4					13.4	−0 53.38	−800.7	+0.41	+4467.0	+0.05
358		VI. H; VII. F					9.2	−0 53.34	−800.1	+0.33	+3653.0	+0.05
359		VI. h_0					14.8	−0 53.31	−799.7	+0.35	+3930.4	+0.05
360		XVIII. N					10.8	−0 52.51	−787.7	−0.56	−5466.0	+0.04
361		I. g_1					13.1	−0 52.17	−782.5	+0.13	+1572.2	+0.04
362		VI. h_1					12.3	−0 51.81	−777.2	+0.35	+3912.5	+0.04
363	12	D′	R		54	v	10.7	−0 51.49	−772.4	−0.02	+ 66.3	+0.04
364		IV. I; V. L					9.6	−0 50.91	−763.7	+0.26	+2987.7	+0.04
365		XVII. Q					9.7	−0 50.84	−762.6	−0.47	−4563.1	+0.04
366		IX. h_8					14.8	−0 50.68	−760.2	+0.45	+4945.1	+0.04
367		IX. h_7					14.8	−0 50.64	−759.6	+0.41	+4563.4	+0.04
368		XVII. q_1					14.8	−0 50.58	−758.7	−0.50	−4824.4	+0.04
369		V. m_{-2}					9.9	−0 50.05	−750.8	+0.31	+3506.9	+0.04
370	13	e''_2				$u_{'''}$	13.3	−0 49.77	−746.6	−0.03	− 66.3	+0.04
371		I. g_2					14.8	−0 49.65	−744.7	+0.10	+1323.0	+0.04
372		XII. o_{-1}					13.1	−0 49.45	−741.8	−0.25	−2275.6	+0.04
373	14	F″				$u_{'}$	12.0	−0 48.86	−732.9	−0.03	− 70.5	+0.04
374		V. m_{-1}					10.2	−0 48.75	−731.3	+0.31	+3506.9	+0.04
375		II. e_2					14.8	−0 48.43	−726.4	+0.17	+2002.3	+0.04
376		XIV. I					11.5	−0 48.43	−726.4	−0.27	−2486.3	+0.04
377	16	G″				$p_{'}$	11.3	−0 48.27	−724.1	−0.08	− 525.0	+0.04
378		g''_1					14.8	−0 48.02	−720.3	−0.04	− 107.1	+0.04
379		XIV. k_{-1}					14.8	−0 47.61	−714.1	−0.31	−2875.2	+0.04
380		g'''_1					13.3	−0 47.56	−713.4	−0.14	−1106.2	+0.04
381		V. M					8.9	−0 47.02	−705.3	+0.29	+3236.9	+0.04
382	17	H″			50	u	11.7	−0 46.90	−703.5	−0.03	− 45.0	+0.04
383		XVII. q_2					14.2	−0 46.83	−702.4	−0.50	−4814.4	+0.04
384		X. i_{-1}					13.9	−0 46.67	−700.1	−0.17	−1438.2	+0.04
385		XII. O					11.5	−0 46.21	−693.1	−0.24	−2207.8	+0.04
386		XVIII. o_{-2}					13.1	−0 45.98	−689.7	−0.56	−5436.3	+0.04
387	18	I″	t		44	r	10.4	−0 45.84	−687.6	−0.05	− 252.2	+0.04
388		XVIII. o_{-1}					13.1	−0 45.73	−685.9	−0.54	−5276.5	+0.04
389		XIV. K					10.8	−0 45.61	−684.1	−0.30	−2835.2	+0.03
390		VI. h_2					12.3	−0 45.28	−679.2	+0.33	+3697.4	+0.03
391		XVI. P					11.5	−0 45.07	−676.0	−0.39	−3745.7	+0.03
392		XIV. k_1					14.8	−0 44.86	−672.9	−0.30	−2750.5	+0.03
393		XIV. L					10.8	−0 44.52	−667.8	−0.28	−2566.0	+0.03
394		V. m_2					11.7	−0 44.05	−660.7	+0.31	+3492.0	+0.03
395		V. m_1					11.5	−0 44.04	−660.6	+0.30	+3347.5	+0.03

No.	Herschel's and Struve's No.	Letter. G. P. Bond.	Letter. Herschel.	Number. W. C. Bond.	Letter. Lassell.*	Letter. Liapunoff.	Mag. by Argelander's Scale.	Eq. 1857.0 $\alpha-\alpha_0$ In Time.	Eq. 1857.0 $\alpha-\alpha_0$ In sec. of Arc.	Prec. Cöef.	Eq. 1857.0 $\delta-\delta_0$ In sec. of Arc.	Prec. Cöef.
								m s	″	″	″	″
396*		VIII. g_5					13.9	−0 43.99	−659.8	+0.40	+4405.6	+0.03
397		IV. i_1					14.8	−0 43.36	−650.4	+0.22	+2519.0	+0.03
398*		D; I. H					9.5	−0 43.12	−646.8	+0.09	+1211.4	+0.03
399*	19	d'_1; i''_1			52	$u_{\prime\prime}$	12.5	−0 42.99	−644.9	−0.03	+ 15.4	+0.03
400		XVII. r_{-1}					13.9	−0 42.79	−641.9	−0.47	−4535.1	+0.03
401	20	X. I; XI. I					9.3	−0 42.68	−640.2	−0.16	−1353.9	+0.03
402	21	i''_2			41	$a_{\prime\prime}$	12.3	−0 42.34	−635.1	−0.06	− 314.2	+0.03
403		XVIII. O					10.8	−0 42.23	−633.4	−0.52	−5086.9	+0.03
404*	23	H'''; X. K; XI. K					9.4	−0 41.77	−626.6	−0.14	−1181.1	+0.03
405		I. I					11.5	−0 41.76	−626.4	+0.11	+1400.9	+0.03
406		VIII. K; VI. h_3					10.8	−0 40.88	−613.2	+0.35	+3915.5	+0.03
407		IX. i_{-3}					12.3	−0 40.62	−609.3	+0.47	+5137.6	+0.03
408		IX. i_{-4}					14.8	−0 40.61	−609.2	+0.46	+5041.9	+0.03
409	22	i''_3				$q_{\prime}$	13.9	−0 40.59	−608.9	−0.08	− 539.2	+0.03
410	24	I'''					9.1	−0 40.50	−607.5	−0.14	−1103.0	+0.03
411		XVIII. o_1					13.9	−0 40.48	−607.2	−0.52	−5091.9	+0.03
412		XVI. p_1					12.3	−0 40.06	−601.0	−0.40	−3830.5	+0.03
413		i'''_1					15.0	−0 39.56	−593.4	−0.10	− 776.6	+0.03
414		VI. i_{-2}					14.8	−0 39.09	−586.4	+0.33	+3701.4	+0.03
415		XVII. R					11.5	−0 39.04	−585.6	−0.47	−4572.0	+0.03
416		VIII. L; VI. h_4					11.2	−0 39.03	−585.4	+0.36	+3947.9	+0.02
417		XIV. l_1					14.8	−0 38.52	−577.8	−0.28	−2536.2	+0.02
418		IV. i_2					13.9	−0 38.37	−575.6	+0.22	+2568.9	+0.02
419		i'''_2					14.2	−0 38.31	−574.7	−0.10	− 701.7	+0.02
420		V. n_{-2}; VI. i_{-1}					11.9	−0 38.25	−573.8	+0.31	+3525.0	+0.02
421		I. i_1					13.1	−0 37.76	−566.4	+0.11	+1472.6	+0.02
422		i'''_3					13.9	−0 37.50	−562.5	−0.13	−1076.2	+0.02
423	25	E	(σ)				10.8	−0 37.22	−558.3	+0.05	+ 790.8	+0.02
424		V. n_{-1}					10.4	−0 36.93	−553.9	+0.31	+3502.1	+0.02
425*		V. N; VI. I; VII. G					8.7	−0 36.47	−547.1	+0.32	+3583.5	+0.02
426		II. e_3					11.5	−0 36.44	−546.6	+0.18	+2096.2	+0.02
427	27	K''	r		47	t	10.7	−0 36.44	−546.6	−0.03	− 71.1	+0.02
428		IX. i_{-2}					14.2	−0 36.33	−544.9	+0.43	+4716.2	+0.02
429		XVII. r_1					12.3	−0 36.30	−544.5	−0.45	−4331.5	+0.02
430	26	K''_1; k''_1			42	$r_{\prime}$	11.7	−0 36.15	−542.2	−0.05	− 206.8	+0.02
431		XII. p_{-2}					13.9	−0 35.77	−536.6	−0.23	−2068.0	+0.02
432		XIV. l_2					14.2	−0 35.77	−536.6	−0.28	−2575.9	+0.02
433		I. i_2					12.3	−0 35.76	−536.4	+0.11	+1404.9	+0.02
434	29	K'''					11.5	−0 35.34	−530.1	−0.13	−1014.6	+0.02
435	30	d'_2				$t_{\prime\prime}$	13.1	−0 35.34	−530.1	−0.03	+ 30.9	+0.02
436		IX. i_{-1}					14.8	−0 35.33	−529.9	+0.43	+4716.2	+0.02
437		X. l_{-2}					11.5	−0 35.06	−526.0	−0.19	−1619.4	+0.02
438	31	F	(υ)				9.4	−0 35.06	−525.9	+0.07	+ 963.4	+0.02
439		V. n_1					13.9	−0 34.92	−523.8	+0.30	+3419.4	+0.02
440		V. n_2					13.9	−0 34.92	−523.8	+0.30	+3384.5	+0.02

* Lassell's numbers are also included in this column.

No.	Herschel's and Struve's No.	Letter. G. P. Bond.	Letter. Herschel.	Number. W. C. Bond.	Letter. Lassell.	Letter. Liapunoff.	Mag. by Argelander's Scale.	Eq. 1857.0 $\alpha-\alpha_0$ In Time.	Eq. 1857.0 $\alpha-\alpha_0$ In sec. of Arc.	Prec. Cöef.	Eq. 1857.0 $\delta-\delta_0$ In sec. of Arc.	Prec. Cöef.
								m s	″	″	″	″
441		XVIII. o_2					13.1	−0 34.73	−520.9	−0.55	−5316.4	+0.02
442		V. n_3					10.8	−0 34.72	−520.8	+0.30	+3357.6	+0.02
443*	28	f_1					13.1	−0 34.64	−519.6	+0.04	+ 649.4	+0.02
444		X. i_1					13.1	−0 34.18	−512.7	−0.16	−1320.4	+0.02
445		k'''_1					14.8	−0 34.00	−510.0	−0.14	−1112.2	+0.02
446		XVIII. P					10.8	−0 33.91	−508.7	−0.56	−5444.8	+0.02
447		IV. i_4					13.1	−0 33.89	−508.4	+0.24	+2818.3	+0.02
448		IV. i_3					12.3	−0 33.85	−507.8	+0.20	+2394.4	+0.02
449	32	E′	ξ	1	55	z	10.5	−0 33.03	−495.5	0.00	+ 290.3	+0.02
450		I. K					11.5	−0 32.87	−493.1	+0.13	+1642.0	+0.02
451		IX. I					10.3	−0 32.87	−493.0	+0.46	+5069.8	+0.02
452		VIII. l_1					13.9	−0 32.79	−491.8	+0.40	+4373.8	+0.02
453		V. n_4					12.3	−0 32.71	−490.6	+0.29	+3252.9	+0.02
454		VIII m_0					14.8	−0 32.30	−484.5	+0.41	+4543.2	+0.01
455		VIII. M; VI. i_1					11.2	−0 32.18	−482.7	+0.36	+3947.5	+0.01
456		XVII. S					11.5	−0 31.45	−471.8	−0.44	−4285.7	+0.01
457		XII. p_{-1}					13.9	−0 31.01	−465.2	−0.24	−2115.9	+0.01
458	33	L″		2	46	$t_{,}$	11.2	−0 30.98	−464.7	−0.04	− 107.8	+0.01
459		II. e_4					14.8	−0 30.94	−464.2	+0.18	+2140.1	+0.01
460		IX. k_{-1}					14.8	−0 30.69	−460.4	+0.41	+4521.8	+0.01
461		XII. P; XIII. A					9.9	−0 30.58	−458.7	−0.24	−2175.7	+0.01
462		VI. i_2					14.8	−0 29.61	−444.2	+0.36	+3965.5	+0.01
463		XVII. T					10.8	−0 29.55	−443.2	−0.46	−4473.1	+0.01
464		k'''_2					14.2	−0 29.50	−442.5	−0.12	− 946.4	+0.01
465		VI. i_3					13.1	−0 28.86	−432.9	+0.35	+3890.8	+0.01
466		IV. i_6					14.8	−0 28.86	−432.9	+0.21	+2479.2	+0.01
467	34	L‴	M			ρ	8.7	−0 28.73	−431.0	−0.09	− 658.6	+0.01
468		X. l_{-1}					13.9	−0 28.56	−428.4	−0.19	−1684.2	+0.01
469		IV. i_5					13.1	−0 28.39	−425.9	+0.25	+2866.2	+0.01
470		XII. p_1					12.3	−0 28.16	−422.4	−0.24	−2138.8	+0.01
471		l'''_1					14.8	−0 28.00	−420.0	−0.11	− 836.5	+0.01
472		VI. k_{-3}					14.8	−0 27.73	−416.0	+0.33	+3698.5	+0.01
473		X. i_2					12.3	−0 27.69	−415.4	−0.15	−1230.7	+0.01
474*		II. f_{-1}; IV. i_7					13.1	−0 27.61	−414.2	+0.20	+2376.4	+0.01
475		II. e_5					13.1	−0 27.44	−411.6	+0.18	+2116.2	+0.01
476		VI. i_4					14.8	−0 27.10	−406.5	+0.34	+3811.2	+0.01
477		VI. i_5					14.8	−0 26.91	−403.7	+0.35	+3876.0	+0.01
478		l'''_2					13.9	−0 26.73	−401.0	−0.14	−1126.1	+0.01
479	35	F′	o	3	56	y	10.0	−0 26.69	−400.4	−0.02	+ 272.3	+0.01
480		X. L; XI. L					10.2	−0 26.61	−399.2	−0.20	−1725.3	+0.01
481		XII. p_2					11.7	−0 26.52	−397.8	−0.23	−2083.9	+0 01
482		IX. K					8.7	−0 26.45	−396.7	+0.42	+4597.8	+0.01
483		IX. k_1					14.8	−0 26.23	−393.5	+0.46	+5015.3	+0.01
484		IV. i_8					13.9	−0 26.17	−392.6	+0.22	+2587.0	+0.01
485		VI. k_{-2}					13.9	−0 25.98	−389.8	+0.33	+3698.6	+0.01

No.	Herschel's and Struve's No.	Letter. G. P. Bond.	Letter. Herschel.	Number. W. C. Bond.	Letter. Lassell.	Letter. Liapunoff.	Mag. by Argelander's Scale.	Eq. 1857.0 $\alpha-\alpha_0$ In Time.	Eq. 1857.0 $\alpha-\alpha_0$ In sec. of Arc.	Prec. Cöef.	Eq. 1857.0 $\delta-\delta_0$ In sec. of Arc.	Prec. Cöef.
								m s	″	″	″	″
486		XII. Q; XIII. B					9.4	−0 25.93	−388.9	−0.24	−2199.1	0.00
487		VI. k_{-1}					14.2	−0 25.72	−385.8	+0.32	+3606.9	0.00
488		V. n_5					13.1	−0 25.71	−385.6	+0.29	+3255.0	0.00
489		XII. R; XIII. C					9.0	−0 25.43	−381.4	−0.25	−2227.0	0.00
490	36	l''_1			48	$l_{,}$	14.2	−0 25.38	−380.7	−0.03	− 49.3	0.00
491		V. O					11.5	−0 24.61	−369.2	+0.26	+3016.0	0.00
492		f_2					13.9	−0 24.56	−368.4	+0.09	+1188.5	0.00
493		m'''_1					12.3	−0 24.50	−367.5	−0.14	−1146.1	0.00
494		VI. K; VII. H					10.0	−0 24.49	−367.4	+0.34	+3760.0	0.00
495		XII. r_1					12.3	−0 23.99	−359.9	−0.26	−2345.0	0.00
496		VIII. n_{-2}					14.8	−0 23.91	−358.7	+0.39	+4334.1	0.00
497	37	M″; M‴	K		37	q	9.9	−0 23.74	−356.1	−0.09	− 592.2	0.00
498		II. e_6					13.9	−0 22.94	−344.1	+0.18	+2092.4	0.00
499		VI. k_1					13.1	−0 22.50	−337.5	+0.34	+3841.1	0.00
500	C 1	II. F; III. F					9.4	−0 22.21	−333.1	+0.19	+2240.7	0.00
501		XIII. D; XII. r_2					10.4	−0 21.52	−322.8	−0.23	−2086.9	0.00
502		VI. k_2					13.1	−0 21.24	−318.6	+0.34	+3766.4	0.00
503		VIII. n_{-1}; IX. k_2					14.4	−0 20.81	−312.1	+0.41	+4477.8	0.00
504		VI. L; VII. I					9.0	−0 20.75	−311.2	+0.33	+3722.1	0.00
505	40	N″	N		38	p	9.6	−0 20.64	−309.6	−0.07	− 424.7	0.00
506	38	G′; O″		4	51	x	11.3	−0 20.40	−306.0	−0.03	+ 5.6	0.00
507		XII. r_3					9.9	−0 20.24	−303.6	−0.26	−2327.1	0.00
508	42	f_3					12.3	−0 20.06	−300.9	+0.04	+ 704.3	0.00
509		X. i_3					13.1	−0 19.67	−295.1	−0.17	−1400.1	0.00
510	43	o''_1			36	$q_{,,}$	13.1	−0 19.39	−290.9	−0.08	− 505.9	0.00
511		m'''_2					14.8	−0 19.25	−288.8	−0.14	−1106.2	0.00
512		V. p_0					14.8	−0 19.15	−287.3	+0.31	+3471.3	0.00
513		V. P					9.5	−0 19.12	−286.8	+0.27	+3102.6	0.00
514		IX. k_3					13.1	−0 18.71	−280.7	+0.44	+4815.9	−0.01
515		IX. k_4					13.9	−0 18.50	−277.5	+0.47	+5119.8	−0.01
516	41	o''_2			49	$x_{,,}$	13.5	−0 18.40	−276.0	−0.03	− 29.5	−0.01
517		VIII. N					10.7	−0 18.16	−272.4	+0.39	+4272.4	−0.01
518		V. p_1					13.9	−0 17.30	−259.5	+0.26	+2980.9	−0.01
519		IX. k_5					12.3	−0 17.23	−258.5	+0.45	+4958.5	−0.01
520		XII. s_{-1}					14.8	−0 17.04	−255.6	−0.21	−1876.5	−0.01
521		G					10.2	−0 16.77	−251.6	+0.08	+1094.3	−0.01
522		XIV. m_{-2}					14.2	−0 16.48	−247.2	−0.32	−2944.3	−0.01
523	45	P″	τ	5	40	l	10.1	−0 16.15	−242.3	−0.04	− 116.0	−0.01
524*	44	g'_2			53	$x_{,}$	12.5	−0 16.09	−241.4	−0.03	+ 16.8	−0.01
525		IX. k_6					14.8	−0 15.98	−239.7	+0.46	+5010.4	−0.01
526		VI. M; VII. K					9.4	−0 15.84	−237.6	+0.32	+3613.6	−0.01
527		X. i_4					14.8	−0 15.69	−235.3	−0.15	−1270.4	−0.01
528		XV. N					10.3	−0 15.67	−235.0	−0.35	−3317.3	−0.01
529		V. p_2					11.5	−0 15.13	−227.0	+0.29	+3250.1	−0.01
530	C 3	II. f_1					13.1	−0 15.13	−226.9	+0.17	+2036.7	−0.01

No.	Herschel's and Struve's No.	Letter. G. P. Bond.	Letter. Herschel.	Number. W. C. Bond.	Letter. Lassell.	Letter. Liapunoff.	Mag. by Argelander's Scale.	Eq. 1857.0 $\alpha-\alpha_0$ In Time.	Eq. 1857.0 $\alpha-\alpha_0$ In sec. of Arc.	Prec. Cöef.	Eq. 1857.0 $\delta-\delta_0$ In sec. of Arc.	Prec. Cöef.
								m s	″	″	″	″
531		VI. m_1					13.1	−0 15.05	−225.8	+0.35	+3856.1	−0.01
532	46	g'_1					14.2	−0 14.57	−218.6	+0.02	+ 449.6	−0.01
533		IV. i_9					13.9	−0 14.41	−216.2	+0.26	+2981.1	−0.01
534		H					10.3	−0 14.01	−210.2	+0.08	+1062.5	−0.01
535		IX. k_7					12.3	−0 13.98	−209.7	+0.45	+4985.4	−0.01
536*	C 4	II. G; III. G; IV. K					8.7	−0 13.88	−208.2	+0.20	+2364.3	−0.01
537		XV. n_1					13.1	−0 13.66	−204.9	−0.37	−3508.6	−0.01
538	C 2	II. f_2					11.5	−0 13.65	−204.7	+0.19	+2204.1	−0.01
539		XII. S; XIII. E					8.0	−0 13.43	−201.4	−0.25	−2237.4	−0.01
540		XVII. U; XVIII. Q					10.8	−0 13.18	−197.8	−0.50	−4817.2	−0.01
541		VI. m_2					11.7	−0 13.16	−197.4	+0.35	+3873.1	−0.01
542	C 5	IV. L					10.2	−0 13.08	−196.2	+0.20	+2400.0	−0.01
543	I	I				λ,	10.3	−0 13.07	−196.1	+0.06	+ 909.4	−0.01
544		VIII. o_{-1}					13.3	−0 13.05	−195.7	+0.38	+4234.6	−0.01
545	47	p''_1			34	$n_{,,}$	13.1	−0 13.03	−195.5	−0.07	− 401.3	−0.01
546		IX. k_8					13.9	−0 12.97	−194.5	+0.44	+4838.0	−0.01
547		IV. k_1; IV. l_1					10.8	−0 12.97	−194.5	+0.24	+2766.7	−0.01
548		XV. n_2					13.1	−0 12.17	−182.6	−0.33	−3123.0	−0.01
549		IX. k_9					14.2	−0 11.97	−179.5	+0.44	+4831.0	−0.01
550		XII. T; XIII. F					7.6	−0 11.71	−175.6	−0.24	−2206.9	−0.01
551*	48	H′	h	6.8	57	*j*	10.1	−0 11.67	−175.1	+0.02	+ 510.8	−0.01
552		p''_2		7			14.9	−0 11.28	−169.2	−0.07	− 393.3	−0.02
553		VIII. O					11.5	−0 11.07	−166.0	+0.40	+4374.0	−0.02
554	49	K	ι		58	μ,	9.0	−0 10.87	−163.1	+0.04	+ 666.0	−0.02
555		IV. l_2					10.8	−0 10.81	−162.1	+0.23	+2649.0	−0.02
556		XII. t_1					9.3	−0 10.71	−160.7	−0.24	−2140.6	−0.02
557		V. p_3					11.5	−0 10.64	−159.6	+0.29	+3327.9	−0.02
558	50	Q″		9	39	ν	10.7	−0 10.59	−158.9	−0.04	− 118.6	−0.02
559		VI. m_3					13.1	−0 9.38	−140.7	+0.33	+3666.8	−0.02
560		XV. O					9.3	−0 9.27	−139.1	−0.34	−3231.5	−0.02
561		V. p_4					13.9	−0 9.14	−137.1	+0.29	+3334.8	−0.02
562		XII. t_2					10.8	−0 8.21	−123.1	−0.24	−2205.4	−0.02
563*		k_1					14.4	−0 8.00	−120.0	+0.07	+ 990.6	−0.02
564		VI. m_4					13.9	−0 7.90	−118.5	+0.34	+3768.5	−0.02
565		X. M; XI. M					9.1	−0 7.46	−112.0	−0.20	−1727.1	−0.02
566	52	q''_1			32	$p_{,,}$	13.3	−0 6.94	−104.1	−0.07	− 406.3	−0.02
567*	51	μ		10			13.9	−0 6.85	−102.8	−0.03	− 8.3	−0.02
568		II. g_1					14.8	−0 6.76	−101.4	+0.20	+2331.8	−0.02
569		IX. l_{-1}					14.8	−0 6.36	− 95.4	+0.42	+4671.6	−0.02
570	53	R″	σ	13	33	*n*	9.4	−0 6.32	− 94.8	−0.06	− 273.2	−0.02
571		XIV. m_{-1}					13.1	−0 6.25	− 93.8	−0.30	−2795.0	−0.02
572		IX. k_{10}					13.9	−0 5.98	− 89.7	+0.45	+4980.5	−0.02
573	54	r''_2		12	35	$n_{,}$	13.9	−0 5.82	− 87.3	−0.05	− 179.0	−0.02
574		IV. M					9.5	−0 5.71	− 85.6	+0.25	+2859.9	−0.02
575*	57	r''_1		11	45		11.9	−0 5.65	− 84.8	−0.03	− 22.3	−0.02

No.	Herschel's and Struve's No.	Letter. G. P. Bond.	Letter. Herschel.	Number. W. C. Bond.	Letter. Lassell.	Letter. Liapunoff.	Mag. by Argelander's Scale.	Eq. 1857.0 $\alpha-\alpha_0$ In Time.	Eq. 1857.0 $\alpha-\alpha_0$ In sec. of Arc.	Prec. Cöef.	Eq. 1857.0 $\delta-\delta_0$ In sec. of Arc.	Prec. Cöef.
								m s	″	″	″	″
576		XIV. m$_0$					13.1	−0 5.53	−83.0	−0.27	−2497.8	−0.02
577		XIV. M					11.0	−0 5.52	−82.8	−0.29	−2638.4	−0.02
578		XII. t$_3$					14.8	−0 5.22	−78.3	−0.23	−2100.6	−0.02
579		XII. t$_4$					9.7	−0 5.19	−77.9	−0.26	−2316.9	−0.02
580	56	h$'_1$		42	59	*i*$_,$	12.3	−0 5.17	−77.6	+0.01	+ 385.8	−0.02
581	ad 54	r$''_3$					14.2	−0 5.07	−76.1	−0.05	− 159.1	−0.02
582		X. m$_1$					12.3	−0 4.99	−74.8	−0.16	−1392.9	−0.03
583*	58	n$'''_{-2}$					11.5	−0 4.93	−74.0	−0.12	− 914.	−0.02
584	C 6	II. H		14			9.4	−0 4.73	−71.0	+0.16	+1893.4	−0.03
585		XV. P; XIV. m$_1$					9.7	−0 4.58	−68.7	−0.33	−3061.0	−0.03
586		XVI. Q; XVII. V					8.9	−0 4.35	−65.3	−0.44	−4231.7	−0.03
587	60	n$'''_{-1}$					13.9	−0 4.10	−61.5	−0.11	− 806.5	−0.03
588		VII. L; VI. m$_5$					11.5	−0 3.96	−59.4	+0.35	+3923.5	−0.03
589*	57*	r$''_4$		15*			12.7	−0 3.81	−57.2	−0.03	− 20.4	−0.03
590		XII. U; XIII. G					8.7	−0 3.65	−54.7	−0.22	−1936.1	−0.03
591		X. m$_2$					13.1	−0 3.50	−52.5	−0.15	−1190.5	−0.03
592		IV. m$_2$; V. p$_5$					13.1	−0 3.41	−51.2	+0.26	+2991.1	−0.03
593	C 7	II. h$_1$; IV. m$_1$					10.6	−0 3.36	−50.4	+0.20	+2319.9	−0.03
594		VIII. P					11.5	−0 3.36	−50.4	+0.39	+4269.6	−0.03
595*		ν		15	43 *e*		13.9	−0 3.13	−46.9	−0.03	− 15.0	−0.03
596	C 10	II. h$_2$					14.8	−0 3.07	−46.0	+0.19	+2257.0	−0.03
597		IX. L; VIII. p$_1$					12.0	−0 2.86	−42.9	+0.41	+4499.0	−0.03
598	62	r$''_5$			31	$\rho_{,,}$	12.3	−0 2.57	−38.6	−0.07	− 455.1	−0.03
599	59	N$'''$					11.8	−0 2.43	−36.5	−0.12	− 974.1	−0.03
600	61	n$'''_1$					13.9	−0 2.43	−36.5	−0.15	−1206.0	−0.03
601*							15.6	−0 2.4	−36.	−0.03	− 31.	−0.03
602*		ν'					14.3	−0 2.20	−33.0	−0.03	− 67.5	−0.03
603		IV. m$_3$					13.1	−0 1.97	−29.6	+0.22	+2568.3	−0.03
604		I. l$_{-1}$					11.5	−0 1.97	−29.5	+0.10	+1372.3	−0.03
605		n$'''_2$					13.9	−0 1.85	−27.8	−0.12	− 953.3	−0.03
606		X. m$_3$					12.3	−0 1.75	−26.3	−0.15	−1190.5	−0.03
607		V. p$_6$					11.5	−0 1.63	−24.5	+0.29	+3240.2	−0.03
608*		ν''			*f*		14.3	−0 1.58	−23.7	−0.03	− 18.0	−0.03
609		XVII. W					10.1	−0 1.45	−21.8	−0.45	−4285.4	−0.03
610		n$'''_3$					14.8	−0 1.35	−20.3	−0.13	−1013.2	−0.03
611		XVIII. R					11.5	−0 1.32	−19.8	−0.55	−5315.9	−0.03
612*		ξ		16	*i*		13.5	−0 1.09	−16.4	−0.03	+ 24.6	−0.03
613		X. N; XI. N					8.6	−0 1.02	−15.3	−0.18	−1575.9	−0.03
614	C 8	IV. N					9.9	−0 0.89	−13.4	+0.22	+2537.6	−0.03
615	66	κ					14.2	−0 0.80	−12.0	+0.02	+ 500.5	−0.03
616	C 11	II. h$_3$					14.8	−0 0.77	−11.6	+0.20	+2309.9	−0.03
617*	64	I$'$	γ'			*bb*$_,$		−0 0.71	−10.7	0.00	+ 12.9	0.00
618*		π		19	*h*		13.1	−0 0.69	−10.4	−0.03	+ 24.6	−0.03
619*	65	K$'$	γ	17		*b*		−0 0.66	−10.0	0.00	+ 8.7	0.00
620	63	n$'''_4$					13.1	−0 0.60	− 9.0	−0.12	− 953.3	−0.03

No.	Herschel's and Struve's No.	Letter. G. P. Bond.	Letter. Herschel.	Number. W. C. Bond.	Letter. Lassell.	Letter. Liapunoff.	Mag. by Argelander's Scale.	Eq. 1857.0 $\alpha-\alpha_0$ In Time.	Eq. 1857.0 $\alpha-\alpha_0$ In sec. of Arc.	Prec. Cöef.	Eq. 1857.0 $\delta-\delta_0$ In sec. of Arc.	Prec. Cöef.
								m s	″	″	″	″
621*	ad II.						15.6	—0 0.53	— 8.	—0.03	— 36.	—0.03
622*	II.	ρ		18			12.7	—0 0.50	— 7.5	—0.03	— 27.8	—0.03
623	68	n'''_5					13.9	—0 0.43	— 6.5	—0.15	—1196.0	—0.03
624*	67	U″; L′	δ	21		*d*		—0 0.33	— 5.0	0.00	+ 16.1	0.00
625	ad II.				*d*		15.6	—0 0.27	— 4.	—0.03	— 28.	—0.03
626		XV. p_1					13.1	—0 0.17	— 2.6	—0.36	—3366.0	—0.03
627		X. n_1					12.3	—0 0.05	— 0.7	—0.20	—1708.8	—0.03
628	69	M′	a	22		*a*		0 0.00	0.0	0.00	0.0	0.00
629	C 9	IV. O					9.7	+0 0.01	+ 0.1	+0.22	+2543.6	—0.03
630		VI. m_6					11.7	+0 0.07	+ 1.0	—0.38	+4170.1	—0.03
631*		τ					14.3	+0 0.20	+ 3.	—0.03	— 42.	—0.03
632		XIV. m_2					11:7	+0 0.22	+ 3.3	—0.27	—2509.8	—0.03
633*	71	N′	a'					+0 0.23	+ 3.5	0.00	— 2.1	0.00
634		XVII. w_1					11.5	+0 0.30	+ 4.5	—0.45	—4345.2	—0.03
635	70	O′		23	2	*i*	10.5	+0 0.55	+ 8.3	—0.02	+ 98.3	—0.03
636*		σ		24			13.3	+0 0.56	+ 8.4	—0.03	— 8.7	—0.03
637		IV. o_1					12.3	+0 0.59	+ 8.8	+0.24	+2783.7	—0.03
638*		L; I. L					9.3	+0 0.70	+10.6	+0.09	+1171.2	—0.03
639	74	O‴	B			$\tau_{,}$	11.1	+0 0.73	+11.0	—0.12	— 951.5	—0.03
640*	73	P′	β	25		*c*		+0 0.77	+11.5	0.00	+ 6.8	0.00
641*	III.	o'_1					14.8	+0 0.79	+11.9	—0.02	+ 111.2	—0.03
642*		υ					15.6	+0 0.87	+13.	—0.03	+ 48.	—0.03
643		VIII. p_2					13.1	+0 0.91	+13.6	+0.37	+4153.0	—0.03
644	72	o'''_1					13.9	+0 0.98	+14.7	—0.14	—1161.0	—0.03
645		XIV. N					10.2	+0 1.29	+19.3	—0.29	—2692.2	—0.03
646		XII. V; XIII. H					8.6	+0 1.38	+20.7	—0.26	—2317.0	—0.03
647*	75	ϕ; p'_1		26	9 *l*		12.1	+0 1.51	+22.6	—0.03	+ 38.0	—0.03
648*		χ					14.3	+0 1.61	+24.2	—0.03	— 8.7	—0.03
649		XIV. n_1					10.8	+0 1.80	+27.0	—0.31	—2836.8	—0.04
650	79	θ		29	1	$k_{,}$	13.1	+0 1.90	+28.5	+0.01	+ 408.8	—0.04
651*	ad 75	ψ		27			13.1	+0 1.96	+29.4	—0.03	+ 47.8	—0.04
652*	76	ζ	y′	32		$f_{,,}$	13.9	+0 2.01	+30.2	—0.01	+ 171.6	—0.04
653	83	ι		28		$k_{,,}$	13.9	+0 2.05	+30.8	+0.01	+ 429.7	—0.04
654*	78	ω		31			12.3	+0 2.21	+33.2	—0.03	+ 10.0	—0.04
655		V. R					8.0	+0 2.57	+38.5	+0.31	+3476.5	—0.04
656		V. Q					8.1	+0 2.58	+38.7	+0.28	+3223.5	—0.04
657*	80	ϵ	y″	33	4	$f_{,}$	13.1	+0 2.64	+39.6	—0.01	+ 165.2	—0.04
658	77	o'''_2					12.3	+0 2.73	+41.0	—0.14	—1156.0	—0.04
659		XII. v_1					13.1	+0 2.88	+43.2	—0.23	—2097.6	—0.04
660		VI. n_{-3}					13.9	+0 3.53	+53.0	+0.36	+3970.9	—0.04
661		VI. n_{-2}					14.8	+0 3.53	+53.0	+0.36	+3980.9	—0.04
662		XIV. n_2					11.5	+0 3.65	+54.7	—0.27	—2480.8	—0.04
663*	84	δ	w	37		$g_{,}$	11.7	+0 3.70	+55.5	—0.02	+ 147.1	—0.04
664	C 14	I. l_1; II. h_4		35			11.5	+0 3.73	+55.9	+0.15	+1812.8	—0.04
665	C 12	II. h_5		36			13.1	+0 3.77	+56.5	+0.16	+1958.1	—0.04

No.	Herschel's and Struve's No.	Letter. G. P. Bond.	Letter. Herschel.	Number. W. C. Bond.	Letter. Lassell.	Letter. Liapunoff.	Mag. by Argelander's Scale.	Eq. 1857.0 $\alpha-\alpha_0$ In Time.	Eq. 1857.0 $\alpha-\alpha_0$ In sec. of Arc.	Prec. Cöef.	Eq. 1857.0 $\delta-\delta_0$ In sec. of Arc.	Prec. Cöef.
								m s	″	″	″	″
666*	81	x''_1		30			13.9	+0 3.98	+ 59.7	−0.05	− 195.8	−0.04
667	85	M	J			$\kappa_{\prime}$	9.4	+0 4.03	+ 60.5	+0.05	+ 848.9	−0 04
668		IV. o_2					13.9	+0 4.21	+ 63.2	+0.21	+2469.7	−0.04
669	87	Q′	ν	39	10	*k*	9.8	+0 4.22	+ 63.3	−0.02	+ 100.0	−0.04
670	86	N	d	38	3	$\nu_{\prime}$	10.8	+0 4.28	+ 64.2	+0.04	+ 673.2	−0.04
671*	88	β		41	18	$e_{\prime\prime}$	11.5	+0 4.64	+ 69.6	−0.03	− 24.4	−0.04
672	C 15	II. I; III. H		40			6.0	+0 4.85	+ 72.8	+0.17	+1985.5	−0.04
673	C 13	II. i_0; IV. o_3					11.9	+0 4.87	+ 73.1	+0.20	+2398.7	−0.04
674		n_1					14.2	+0 4.91	+ 73.6	+0.07	+ 976.5	−0.04
675*		ω'					15.2	+0 4.97	+ 74.5	−0.04	− 93.4	−0.04
676*	ad 88	γ		43	*k*		13.1	+0 5.23	+ 78.5	−0.03	− 27.6	−0.04
677*	ad 81	x''_2		34			14.8	+0 5.24	+ 78.6	−0.05	− 201.4	−0.04
678*		n_2					13.9	+0 5.28	+ 79.2	+0.06	+ 852.2	−0.04
679		IX. m_{-2}					14.8	+0 5.31	+ 79.7	+0.42	+4666.7	−0.04
680	92	o'''_3					13.9	+0 5.48	+ 82.2	−0.10	− 675.3	−0.04
681*	89	η	z			$e_{\prime}$	14.8	+0 6.02	+ 90.3	−0.01	+ 173.2	−0.04
682		VI. n_{-1}					12.3	+0 6.28	+ 94.2	+0.37	+4056.6	−0.04
683		XVIII. t_{-5}					14.8	+0 6.29	+ 94.3	−0.53	−5136.2	−0.04
684*	90	n_3		56			14.5	+0 6.45	+ 96.8	+0 04	+ 744.8	−0.04
685	93	Y″	ϵ	45	26	*e*	8.3	+0 6.51	+ 97.7	−0.04	− 95.0	−0.04
686*	91			44			15.6	+0 6.67	+100.	−0.03	− 39.	−0.04
687		X. O; XI. O					10.3	+0 6.93	+104.0	−0.19	−1665.1	−0.04
688*							15.6	+0 7.07	+106.	−0.03	− 18.	−0.04
689		V. r_1					11.5	+0 7.57	+113.5	+0.31	+3476.6	−0.04
690	95	Z″			30	$\rho_{\prime}$	10.3	+0 7.96	+119.4	−0.07	− 443.7	−0.04
691		VIII. p_3					13.9	+0 8.39	+125.9	+0.38	+4249.7	−0.05
692	C 16	II. i_1		46			11.5	+0 8.75	+131.2	+0.18	+2154.5	−0.05
693	94	n_4		58			13.9	+0 8.78	+131.7	+0.04	+ 751.6	−0.05
694		XVIII. t_{-4}					13.1	+0 8.79	+131.8	−0.51	−4956.4	−0.05
695	97	n_5					12.5	+0 8.85	+132.8	+0.05	+ 818.1	−0.05
696	98	O					11.5	+0 9.08	+136.2	+0.06	+ 886.3	−0.05
697		VIII. p_4					14.8	+0 9.14	+137.1	+0.39	+4302.5	−0.05
698	C 18	II. i_2					13.1	+0 9.23	+138.5	+0.19	+2302.0	−0.05
699	C 17	II. i_3					13.9	+0 9.55	+143.2	+0.18	+2167.4	−0.05
700	102	R′	e	47	6	$\alpha_{\prime}$	11.5	+0 9.56	+143.4	+0.02	+ 492.7	−0.05
701		$b''b''_2$					14.8	+0 9.58	+143.7	−0.07	− 417.2	−0.05
702		XVII. X					10.1	+0 9.67	+145.0	−0.48	−4647.3	−0.05
703	96	o_1		59			13.9	+0 9.69	+145.4	+0.04	+ 736.4	−0.05
704		IX. m_{-1}					13.9	+0 9.78	+146.7	+0.46	+4990.6	−0.05
705	99	P; S′	c	48	5	$\beta_{\prime}$	11.5	+0 9.81	+147.2	+0.03	+ 611.2	−0.05
706		XII. v_2					13.1	+0 9.88	+148.2	−0.24	−2175.3	−0.05
707	103	$A''A''$	ψ	49	27	σ	11.2	+0 10.08	+151.2	−0.05	− 253.5	−0.05
708	101	$B''B''$	ζ	50	23	*f*	9.6	+0 10.09	+151.4	−0.04	− 98.5	−0.05
709*	100	$b''b''_1$	G	51		μ	12.3	+0 10.19	+152.9	−0.04	− 136.4	−0.05
710		XII. v_3					13.9	+0 10.38	+155.6	−0.23	−2060.7	−0.05

No.	Herschel's and Struve's No.	Letter. G. P. Bond.	Letter. Herschel.	Number. W. C. Bond.	Letter. Lassell.	Letter. Liapunoff.	Mag. by Argelander's Scale.	Eq. 1857.0 $\alpha-\alpha_0$ In Time.	Eq. 1857.0 $\alpha-\alpha_0$ In sec. of Arc.	Prec. Cöef.	Eq. 1857.0 $\delta-\delta_0$ In sec. of Arc.	Prec. Cöef.
								m s	″	″	″	″
711		X. P; XI. P					9.5	+0 10.43	+156.5	−0.19	−1666.5	−0.05
712	C 19	I. l_2		52			11.7	+0 10.70	+160.5	+0.14	+1702.3	−0.05
713		IX. M					10.9	+0 11.04	+165.6	+0.44	+4876.0	−0.05
714		X. Q					9.4	+0 11.22	+168.3	−0.17	−1488.4	−0.05
715		XII. W; XIII. I		54			5.0	+0 11.25	+168.7	−0.21	−1872.6	−0.05
716*		VI. N; VII. M					8.3	+0 11.31	+169.6	+0.33	+3695.0	−0.05
717*		XII. w_0					9.2	+0 11.65	+174.7	−0.21	−1883.0	−0.05
718		p_1					13.9	+0 11.68	+175.2	+0.07	+1038.7	−0.05
719		X. q_1					14.8	+0 11.72	+175.8	−0.18	−1519.3	−0.05
720		XII. w_1					9.2	+0 11.77	+176.5	−0.21	−1887.3	−0.05
721		IX. m_1					14.8	+0 11.78	+176.7	+0.46	+5030.5	−0.05
722	105	o'''_4					13.3	+0 11.98	+179.7	−0.10	− 710.4	−0.05
723		IX. m_2					10.8	+0 12.02	+180.3	+0.47	+5130.3	−0.05
724	104	C″C″	λ	55	25	*h*	10.5	+0 12.22	+183.3	−0.05	− 176.0	−0.05
725		X. q_2					12.3	+0 12.22	+183.3	−0.18	−1521.3	−0.05
726		XVIII. t_{-3}					14.8	+0 12.29	+184.4	−0.52	−5066.4	−0.05
727		XVIII. S					10.8	+0 12.79	+191.8	−0.56	−5449.4	−0.05
728	C 20	I. M; II. K; III. I		57			9.0	+0 12.98	+194.7	+0.14	+1749.7	−0.05
729		VI. n_1					13.1	+0 13.15	+197.3	+0.34	+3806.6	−0.05
730		p_2					13.1	+0 13.58	+203.7	+0.08	+1150.4	−0.05
731		X. q_3					13.1	+0 13.97	+209.6	−0.18	−1579.0	−0.05
732	106	P‴; D″D″		63		$\zeta_{,,}$	11.5	+0 13.98	+209.7	−0.09	− 570.4	−0.05
733	C 21	II. i_4					13.1	+0 14.23	+213.4	+0.20	+2371.8	−0.05
734	108	T′	μ	60	8	*a*	9.0	+0 14.51	+217.7	+0.01	+ 443.8	−0.05
735		XII. w_2					9.7	+0 14.52	+217.8	−0.21	−1882.2	−0.05
736		VIII. p_5					12.3	+0 14.65	+219.7	+0.38	+4183.1	−0.05
737		t'_1; q'_1					15.0	+0 14.68	+220.2	0.00	+ 266.1	−0.05
738*	109	X. q_4			12		13.1	+0 14.72	+220.8	−0.18	−1519.2	−0.05
739		IV. o_4					13.1	+0 15.03	+225.4	+0.20	+2362.1	−0.06
740	107	p_3					13.1	+0 15.03	+225.5	+0.05	+ 841.3	−0.06
741	110	E″E″	η	61	19	*g*	10.0	+0 15.06	+225.9	−0.04	− 110.5	−0.06
742	C 22	II. i_5		62			9.2	+0 15.06	+226.0	+0.16	+1993.1	−0.06
743		V. r_2; VI. n_2					13.9	+0 15.07	+226.0	+0.32	+3599.4	−0.06
744		X. q_5					14.8	+0 15.24	+228.6	−0.19	−1678.7	−0.06
745		XII. x_{-2}					13.9	+0 15.31	+229.7	−0.26	−2378.5	−0.06
746	111	Q‴; F″F″	E	64	29	$\zeta_{,}$	10.8	+0 15.54	+233.1	−0.09	− 583.8	−0.06
747		m					15.0	+0 15.76	+236.4	−0.06	− 333.4	−0.06
748		VIII. p_6					14.8	+0 16.12	+241.9	+0.40	+4424.2	−0.06
749		p_4; I. m_1					14.5	+0 16.38	+245.7	+0.09	+1188.6	−0.06
750	112	G″G″		65	28	$\sigma_{,}$	10.8	+0 16.56	+248.4	−0.08	− 467.1	−0.06
751		VIII. p_7					13.9	+0 16.63	+249.4	+0.40	+4394.3	−0.06
752		Q					9.7	+0 16.90	+253.5	+0.07	+1028.4	−0.06
753		IV. o_5; V. r_3					13 9	+0 17.79	+266.8	+0.26	+2990.8	−0.06
754		XII. x_{-1}					10.8	+0 18.30	+274.5	−0.25	−2284.8	−0.06
755		n					14.8	+0 18.51	+277.7	−0.06	− 348.3	−0.06

No.	Herschel's and Struve's No.	Letter. G. P. Bond.	Letter. Herschel.	Number. W. C. Bond.	Letter. Lassell.	Letter. Liapunoff.	Mag. by Argelander's Scale.	Eq. 1857.0 $\alpha-\alpha_0$ In Time.	Eq. 1857.0 $\alpha-\alpha_0$ In sec. of Arc.	Prec. Cöef.	Eq. 1857.0 $\delta-\delta_0$ In sec. of Arc.	Prec. Cöef.
								m s	″	″	″	″
756		I. m_2					12.3	+0 18.66	+280.0	+0.10	+1331.7	−0.06
757	113	R	b	66	7	$\delta_,$	10.0	+0 18.70	+280.5	+0.04	+ 666.1	−0.06
758		XVIII. t_{-2}					14.8	+0 18.80	+281.9	−0.51	−4911.5	−0.06
759*	114	t'_2		67			15.6	+0 19.01	+285.2	−0.02	+ 108.7	−0.06
760		V. r_4					11.5	+0 20.07	+301.0	+0.31	+3497.7	−0.06
761		V. r_5					13.9	+0 20.38	+305.7	+0.29	+3343.1	−0.06
762*	115	s'''_{-1}					14.8	+0 20.54	+308.1	−0.11	− 848.7	−0.06
763		IX. m_3					13.9	+0 20.54	+308.1	+0.45	+4925.0	−0.06
764		XIV. O					9.2	+0 20.67	+310.0	−0.27	−2493.6	−0.06
765		XVII. Y					11.5	+0 20.86	+312.8	−0.46	−4442.7	−0.06
766		I. m_3					14.8	+0 20.93	+314.0	+0.12	+1562.9	−0.06
767	IV.	o		69*	20	$\gamma_{,,,}$	13.9	+0 21.13	+317.0	−0.05	− 193.9	−0.06
768	C 23	II. L; III. K		70			7.8	+0 21.36	+320.4	+0.16	+1922.5	−0.06
769		XV. Q					9.7	+0 21.44	+321.6	−0.37	−3513.4	−0.06
770		XVIII. t_{-1}					11.5	+0 21.63	+324.5	−0.55	−5350.8	−0.06
771		XVIII. T					10.8	+0 21.78	+326.7	−0.54	−5283.7	−0.06
772		r_1					13.9	+0 22.30	+334.5	+0.06	+ 869.2	−0.06
773		V. r_6					13.1	+0 23.57	+353.5	+0.31	+3514.7	−0.07
774		IV. P					10.8	+0 23.89	+358.3	+0.25	+2828.7	−0.07
775		I. m_4					14.8	+0 23.93	+359.0	+0.12	+1582.9	−0.07
776*	116	u'_{-1}					16.4	+0 24.2	+363.	+0.01	+ 380.	−0.07
777	C 24	IV. p_1					11.5	+0 24.42	+366.4	+0.21	+2440.0	−0.07
778*	117	p	p	75	21	$\gamma_{,,}$	13.1	+0 24.45	+366.7	−0.05	− 216.0	−0.07
779*		s_{-1}					15.6	+0 24.67	+370.	+0.05	+ 864.	−0.07
780		XVIII. U					10.8	+0 24.86	+372.9	−0.51	−4908.7	−0.07
781	120	U′	χ	76	15	β	10.8	+0 24.92	+373.8	−0.01	+ 195.5	−0.07
782		XVIII. t_1					9.7	+0 25.78	+386.7	−0.56	−5415.7	−0.07
783	122	q'''_1		81*			13.9	+0 25.79	+386.9	−0.10	− 746.6	−0.07
784	123	H″H″	k	78	24	$\gamma_,$	10.8	+0 25.89	+388.4	−0.06	− 286.0	−0.07
785	124	V′; S	a	79	11	$\epsilon_,$	10.8	+0 25.98	+389.7	+0.03	+ 587.2	−0.07
786	118	r_2					13.9	+0 25.98	+389.7	+0.04	+ 684.8	−0.07
787	119	r_3					13.3	+0 25.98	+389.7	+0.05	+ 849.3	−0.07
788		r_4					13.9	+0 25.98	+389.7	+0.07	+1058.7	−0.07
789	121	$h''h''_1$	q				14.8	+0 26.39	+395.9	−0.05	− 245.7	−0.07
790		VI. n_3					12.3	+0 26.43	+396.5	+0.36	+3958.2	−0.07
791		II. l_1					14.8	+0 26.83	+402.5	+0.19	+2272.3	−0.07
792		V. r_7					14.8	+0 27.57	+413.5	+0.30	+3432.9	−0.07
793	126	$h''h''_2$	l	80	26 B	$\sigma_{,,}$	11.7	+0 27.64	+414.6	−0.08	− 516.7	−0.07
794		s_1					12.5	+0 27.73	+416.0	+0.07	+ 971.9	−0.07
795	125	q'''_2		84		$\dot{\xi}_,$	12.5	+0 27.79	+416.9	−0.11	− 776.5	−0.07
796		XVI. R					11.5	+0 28.23	+423.4	−0.40	−3799.8	−0.07
797*		v'_1		85			15.0	+0 28.49	+427.4	−0.03	+ 172.7	−0.07
798*		XVII. z_0					11.5	+0 28.82	+432.3	−0.50	−4831.3	−0.07
799		XVII. Z; XVIII. V					9.3	+0 28.96	+434.4	−0.50	−4843.1	−0.07
800		V. r_8					14.8	+0 29.55	+443.3	+0.31	+3534.7	−0.08

No.	Herschel's and Struve's No.	Letter. G. P. Bond.	Letter. Herschel.	Number. W. C. Bond.	Letter. Lassell.	Letter. Liapunoff.	Mag. by Argelander's Scale.	Eq. 1857.0 $\alpha-\alpha_0$ In Time.	Eq. 1857.0 $\alpha-\alpha_0$ In sec. of Arc.	Prec. Cöef.	Eq. 1857.0 $\delta-\delta_0$ In sec. of Arc.	Prec. Cöef.
								m s	″	″	″	″
801*		(i)		86			13.1	+0 29.67	+445.	−0.06	− 282.	−0.08
802		II. l_2					13.9	+0 29.83	+447.5	+0.19	+2242.4	−0.08
803		IX. N; VIII. q_{-2}					11.4	+0 29.83	+447.5	+0.41	+4507.9	−0.08
804		XI. Q; X. q_6					12.3	+0 30.03	+450.4	−0.19	−1661.3	−0.08
805	128	v'_2		83			13.9	+0 30.48	+457.2	0.00	+ 331.9	−0.08
806	127	T					11.7	+0 30.65	+459.8	+0.05	+ 780.1	−0.08
807		VI. n_4					14.8	+0 30.79	+461.8	+0.36	+3993.1	−0.08
808*	129	v'_3		82	13	$\beta_{'''}$	11.9	+0 30.98	+464.7	+0.01	+ 391.2	−0.08
809		VIII. q_{-1}					13.9	+0 31.01	+465.2	+0.41	+4494.0	−0.08
810		VI. O; VIII. Q					10.2	+0 31.62	+474.3	+0.37	+4093.7	−0.08
811		IX. n_1					13.3	+0 31.66	+474.9	+0.41	+4502.6	−0.08
812		XII. X; XIII. K					9.2	+0 31.77	+476.6	−0.25	−2236.9	−0.08
813		IX. n_2					13.9	+0 31.91	+478.6	+0.41	+4562.4	−0.08
814		IV. p_2					13.9	+0 31.92	+478.8	+0.21	+2514.9	−0.08
815		VIII. R; VI. o_1					11.5	+0 32.13	+481.9	+0.36	+4004.5	−0.08
816		V. r_9					14.8	+0 32.57	+488.5	+0.31	+3504.8	−0.08
817		X. q_7					13.9	+0 32.73	+490.9	−0.18	−1568.9	−0.08
818	C 25	II. l_3					12.3	+0 32.87	+493.0	+0.16	+1938.5	−0.08
819		V. s_{-5}					12.3	+0 33.19	+497.8	+0.28	+3205.8	−0.08
820		t_1					14.2	+0 34.05	+510.8	+0.07	+ 978.9	−0.08
821		XVII. AA					9.2	+0 34.17	+512.5	−0.49	−4691.8	−0.08
822	133	I''I''	θ	87	22	γ	10.7	+0 34.32	+514.8	−0.06	− 306.0	−0.08
823	132	R'''	(*a*)				10.7	+0 34.47	+517.1	−0.13	−1019.0	−0.08
824		U					12.1	+0 34.53	+518.0	+0.06	+ 922.2	−0.08
825	131	q'''_3		88			14.2	+0 34.54	+518.1	−0.10	− 716.6	−0.08
826	130	v'_4					14.8	+0 34.73	+521.0	+0.01	+ 419.6	−0.08
827		XVI. S					10.8	+0 34.75	+521.2	−0.43	−4083.9	−0.08
828		II. l_4					14.8	+0 34.85	+522.7	+0.18	+2157.8	−0.08
829		XIV. P					11.5	+0 34.99	+524.9	−0.31	−2924.2	−0.08
830		V					11.0	+0 35.29	+529.4	+0.08	+1107.2	−0.08
831		IX. n_3					13.9	+0 35.66	+534.9	+0.41	+4572.4	−0.08
832	134	$i''i''_1$		89			13.9	+0 35.82	+537.3	−0.06	− 322.4	−0.08
833		X. q_8					13.1	+0 35.98	+539.7	−0.20	−1698.4	−0.08
834		III. L; II. l_5		90			11.9	+0 36.19	+542.8	+0.18	+2130.8	−0.08
835		VIII. q_1					12.3	+0 36.53	+548.0	+0.38	+4210.1	−0.09
836		IX. n_4					14.8	+0 36.62	+549.3	+0.45	+4931.1	−0.09
837		I. N					10.8	+0 36.83	+552.4	+0.12	+1523.2	−0.09
838		v_1; I. n_1					12.8	+0 36.83	+552.4	+0.08	+1156.4	−0.09
839		IV. Q					10.5	+0 37.49	+562.3	+0.24	+2818.8	−0.09
840*		$(i''i'')_2$					15.6	+0 37.53	+563.	−0.04	− 171.	−0.09
841		V. s_{-4}					13.9	+0 38.19	+572.9	+0.27	+3086.2	−0.09
842		IX. n_5					13.1	+0 38.37	+575.6	+0.44	+4874.3	−0.09
843	135	S'''	A	91		ζ	8.6	+0 38.54	+578.1	−0.11	− 853.6	−0.09
844		I. n_2					13.9	+0 38.82	+582.3	+0.13	+1593.0	−0.09
845		IV. q_1					11.5	+0 39.49	+592.3	+0.24	+2822.0	−0.09

No.	Herschel's and Struve's No.	Letter. G. P. Bond.	Letter. Herschel.	Number. W. C. Bond.	Letter. Lassell.	Letter. Liapunoff.	Mag. by Argelander's Scale.	Eq. 1857.0 $\alpha-\alpha_0$ In Time.	Eq. 1857.0 $\alpha-\alpha_0$ In sec. of Arc.	Prec. Cöef.	Eq. 1857.0 $\delta-\delta_0$ In sec. of Arc.	Prec. Cöef.
								m s	″	″	″	″
846		$i''i''_2$					15.0	+0 40.22	+603.3	−0.04	− 119.1	−0.09
847		v_2					13.1	+0 41.29	+619.4	+0.03	+ 634.9	−0.09
848	136	W′	κ	93	16	*m*	9.9	+0 42.08	+631.2	−0.02	+ 60.2	−0.09
849		XV. R					11.5	+0 42.42	+636.3	−0.34	−3154.1	−0.09
850		V. s_{-2}					14.8	+0 42.66	+639.8	+0.31	+3487.0	−0.09
851		V. s_{-3}					13.8	+0 42.67	+640.1	+0.30	+3345.4	−0 09
852		I. O					9.8	+0 42.82	+642.3	+0.13	+1644.9	−0.09
853		IV. q_2; V. s_{-1}					13.1	+0 42.99	+644.8	+0.26	+2957.7	−0.09
854		II. l_6					14.8	+0 43.15	+647.3	+0.17	+2003.4	−0.10
855	138	T‴	(β)				11.0	+0 43.60	+654.0	−0.13	− 989.3	−0.10
856		I. o_1; II. l_7					11.9	+0 43.84	+657.6	+0.14	+1775.8	−0.10
857		VI. p_{-2}					13.1	+0 44.02	+660.3	+0.32	+3625.6	−0.10
858		VIII. q_2					14.8	+0 44.03	+660.4	+0.39	+4376.7	−0.10
859		$i''i''_3$					14.8	+0 44.07	+661.1	−0.09	− 577.5	−0.10
860		XVIII. x_{-2}					13.9	+0 45.08	+676.2	−0.52	−5009.0	−0.10
861		II. M; III. M					10.2	+0 45.28	+679.2	+0.20	+2306.2	−0.10
862	C 26	II. l_8					13.1	+0 45.35	+680.3	+0.17	+1998.4	−0.10
863*	137	w'_1		92	14		12.5	+0 45.39	+680.9	+0.01	+ 357.8	−0.10
864		XVII. aa_1					13.9	+0 45.41	+681.1	−0.47	−4482.3	−0.10
865		v_3					13.9	+0 45.54	+683.1	+0.06	+ 957.0	−0.10
866		XVII. aa_2					14.2	+0 45.65	+684.8	−0.46	−4404.5	−0.10
867		XVII. aa_3					13.1	+0 45.67	+685.0	−0.49	−4758.6	−0.10
868		II. m_1; IV. q_3					11.7	+0 45.84	+687.6	+0.20	+2333.9	−0.10
869		XV. S					8.9	+0 45.92	+688.8	−0.34	−3182.0	−0.10
870		XV. s_1					12.3	+0 46.42	+696.3	−0.34	−3164.1	−0.10
871		VIII. q_3					13.1	+0 46.53	+697.9	+0.39	+4359.8	−0.10
872		V. S					8.9	+0 46.66	+699.8	+0.31	+3481.9	−0.10
873*		V_1; v_4					11.9	+0 47.13	+707.0	+0.07	+ 981.7	−0.10
874		I. o_4					11.5	+0 47.14	+707.1	+0.11	+1433.6	−0.10
875		v_5					14.8	+0 47.29	+709.4	+0.05	+ 839.4	−0.10
876		X. q_9					13.9	+0 47.47	+712.1	−0.18	−1489.0	−0.10
877		XVII. aa_4					12.5	+0 47.92	+718.8	−0.49	−4753.6	−0.10
878		XVII. aa_5					14.2	+0 48.41	+726.2	−0.48	−4653.8	−0.10
879		VI. p_{-1}					13.9	+0 48.47	+727.0	+0.37	+4135.7	−0.10
880*		XVIII. W					9.7	+0 49.67	+745.0	−0.54	−5205.6	−0.10
881	139	U‴	(γ)				10.5	+0 51.08	+766.2	−0.14	−1108.0	−0.11
882		XV. T					9.8	+0 51.13	+766.9	−0.37	−3438.3	−0.11
883	140	u'''_1					13.3	+0 51.83	+777.5	−0.14	−1128.3	−0.11
884		XVIII. x_{-1}					13.9	+0 52.33	+784.9	−0.52	−4986.0	−0.11
885		XIV. q_{-2}					13.9	+0 52.46	+786.9	−0.33	−3053.7	−0.11
886		VIII. s_{-2}					13.9	+0 52.66	+789.9	+0.37	+4135.7	−0.11
887		XV. U					9.5	+0 52.83	+792.4	−0.36	−3373.4	−0.11
888*		W; I. P					9.2	+0 53.40	+801.0	+0.09	+1198.6	−0.11
889	142	K″K″	(λ)	94	17	$\delta_{\prime\prime}$	11.3	+0 53.44	+801.6	−0.06	− 258.2	−0.11
890		VI. P					10.0	+0 53.51	+802.6	+0.33	+3717.5	−0.11

No.	Herschel's and Struve's No.	Letter. G. P. Bond.	Letter. Herschel.	Number. W. C. Bond.	Letter. Lassell.	Letter. Liapunoff.	Mag. by Argelander's Scale.	Eq. 1857.0 $\alpha-\alpha_0$ In Time.	Eq. 1857.0 $\alpha-\alpha_0$ In sec. of Arc.	Prec. Cöef.	Eq. 1857.0 $\delta-\delta_0$ In sec. of Arc.	Prec. Cöef.
								m s	"	"	"	"
891		XII. Y					10.2	+0 53.79	+ 806.8	−0.25	−2234.6	−0.11
892		XVI. T; XV. u_1					10.0	+0 53.83	+ 807.4	−0.38	−3609.8	−0.11
893	141	w'_2					13.1	+0 54.33	+ 815.0	−0.01	+ 228.3	−0.11
894		V. s_1; VI. q_{-1}					13.1	+0 54.95	+ 824.3	+0.32	+3594.8	−0.11
895		XIV. q_{-1}					14.2	+0 55.94	+ 839.1	−0.31	−2824.2	−0.11
896		II. m_2					11.5	+0 55.98	+ 839.6	+0.16	+1943.7	−0.11
897		IX. n_6					13.9	+0 56.37	+ 845.6	+0.44	+4881.5	−0.11
898		XV. u_2					13.1	+0 57.33	+ 859.9	−0.36	−3388.3	−0.12
899		w_1					14.2	+0 57.42	+ 861.3	+0.04	+ 744.6	−0.12
900		V. s_2; VI. q_0					13.5	+0 57.84	+ 867.6	+0.31	+3537.6	−0.12
901		VI. Q					10.0	+0 58.01	+ 870.1	+0.33	+3664.7	−0.12
902		V. s_3					14.8	+0 58.16	+ 872.4	+0.30	+3425.3	−0.12
903		XIV. Q					11.5	+0 58.19	+ 872.9	−0.31	−2923.9	−0.12
904		$k''k''_1$					14.2	+0 58.94	+ 884.1	−0.04	− 134.0	−0.12
905	143	V'''	X			τ	7.8	+0 59.52	+ 892.8	−0.12	− 918.3	−0.12
906		VI. q_1; VIII. s_{-1}					13.1	+1 0.05	+ 900.8	+0.38	+4168.2	−0.12
907		XVIII. X					10.4	+1 0.07	+ 901.0	−0.53	−5162.7	−0.12
908		X					13.2	+1 0.08	+ 901.2	+0.04	+ 714.9	−0.12
909		XVI. u_{-1}					11.5	+1 0.28	+ 904.2	−0.40	−3844.4	−0.12
910		II. m_3					13.1	+1 0.94	+ 914.1	+0.20	+2322.4	−0.12
911		XV. v_{-1}					13.9	+1 1.23	+ 918.4	−0.35	−3243.8	−0.12
912		x_1					13.1	+1 1.58	+ 923.7	+0.08	+1111.5	−0.12
913		XVII. BB					9.7	+1 1.65	+ 924.7	−0.47	−4516.1	−0.12
914		XII. z_{-1}; XIV. s_{-1}					13.4	+1 1.65	+ 924.8	−0.27	−2433.1	−0.12
915		XV. V; XVI. U					9.8	+1 2.18	+ 932.7	−0.38	−3622.8	−0.12
916		XIV. R; XV. w_{-1}					12.3	+1 4.66	+ 969.9	−0.32	−3026.5	−0.13
917		v'''_1					13.9	+1 5.02	+ 975.3	−0.14	−1086.4	−0.13
918		VIII. S					10.8	+1 6.64	+ 999.6	+0.39	+4283.2	−0.13
919		V. s_4					14.8	+1 7.36	+1010.4	+0.30	+3413.5	−0.13
920		XII. Z					11.0	+1 8.09	+1021.3	−0.26	−2332.2	−0.13
921		X. r_{-2}					13.9	+1 8 21	+1023.2	−0.15	−1269.5	−0.13
922		XVI. V					10.2	+1 9.33	+1039.9	−0.40	−3806.3	−0.13
923		V. s_5					14.8	+1 10.16	+1052.4	+0.30	+3390.5	−0.13
924	145	L''L''	S	95		δ	10.0	+1 10.88	+1063.2	−0.03	− 53.3	−0.13
925		XIV. S					10.4	+1 11.48	+1072.1	−0.29	−2676.5	−0.14
926		VI. q_2					13.1	+1 11.48	+1072.1	+0.35	+3924.8	−0.14
927	144	X. r_{-1}					14.8	+1 11.72	+1075.8	−0.17	−1389.1	−0.14
928		XV. W					10.5	+1 11.72	+1075.8	−0.35	−3241.6	−0.14
929		v'''_2					14.2	+1 12.52	+1087.8	−0.14	−1126.3	−0 14
930		IX. n_7					14.8	+1 13.12	+1096.9	+0.44	+4871.7	−0.14
931		V. s_6					14 8	+1 13.16	+1097.4	+0.30	+3390.6	−0.14
932		XVII. CC					9.3	+1 13.46	+1101.9	−0.50	−4764.2	−0.14
933		VI. q_3					13.1	+1 13.47	+1102.1	+0.36	+4031.4	−0.14
934		XII. AA					10.8	+1 13.67	+1105.1	−0.23	−2049.9	−0.14
935	146	X. R; XI. R					8.6	+1 14.21	+1113.1	−0.16	−1356.6	−0.14

No.	Herschel's and Struve's No.	Letter. G. P. Bond.	Letter. Herschel.	Number. W. C. Bond.	Letter. Lassell.	Letter. Liapunoff.	Mag. by Argelander's Scale.	Eq. 1857.0 $\alpha-\alpha_0$ In Time.	$\alpha-\alpha_0$ In sec. of Arc.	Prec. Cöef.	Eq. 1857.0 $\delta-\delta_0$ In sec. of Arc.	Prec. Cöef.
								m s	″	″	″	″
936		$l''l''_1$					13.9	+1 14.88	+1123.2	−0.04	− 143.0	−0.14
937		W‴					10.2	+1 15.53	+1133.0	−0.15	−1166.9	−0.14
938	147	X′; M″M″	T	96		ε	10.8	+1 15.77	+1136.6	−0.03	− 7.2	−0.14
939		XVIII. x_1					14.2	+1 16.07	+1141.1	−0.53	−5115.6	−0.14
940		V. T; VI. R					11.0	+1 16.39	+1145.8	+0.31	+3554.6	−0.14
941		w'''_1					13.1	+1 17.53	+1163.0	−0.14	−1141.2	−0.14
942		X. S; XI. S; XII. aa_1					11.5	+1 17.88	+1168.2	−0.21	−1824.5	−0.15
943		XVI. W					8.6	+1 18.03	+1170.4	−0.40	−3820.2	−0.15
944		$m''m''_1$					14.8	+1 18.88	+1183.2	−0.08	− 522.6	−0.15
945		VI. r_1					13.1	+1 18.99	+1184.8	+0.34	+3842.1	−0.15
946		XV. X					11.5	+1 19.13	+1186.9	−0.36	−3351.2	−0.15
947		V. t_1					13.9	+1 19.27	+1189.1	+0.30	+3395.6	−0.15
948		XVI. w_1					11.5	+1 19.79	+1196.8	−0.42	−4003.6	−0.15
949		XII. bb_{-1}					9.8	+1 20.98	+1214.7	−0.27	−2412.2	−0.15
950*		V. t_2; VI. r_2					13.5	+1 20.99	+1214.8	+0.32	+3566.6	−0.15
951		$m''m''_2$					14.8	+1 21.38	+1220.7	−0.05	− 158.9	−0.15
952		V. U					10.8	+1 21.45	+1221.8	+0.31	+3475.3	−0.15
953	148	Y	V				9.3	+1 21.49	+1222.4	+0.06	+ 951.0	−0.15
954		XII. BB; XIV. T					8.4	+1 21.53	+1222.9	−0.27	−2416.0	−0.15
955	149	N″N″	Y				10.5	+1 22.03	+1230.5	−0.06	− 295.3	−0.15
956	150	Y′	U				10.2	+1 22.83	+1242.5	−0.01	+ 227.9	−0.15
957		y'_1					13.9	+1 23.16	+1247.4	+0.03	+ 579.0	−0.15
958		XII. bb					12.3	+1 24.08	+1261.2	−0.26	−2320.0	−0.15
959*		y_1					12.3	+1 24.09	+1261.4	+0.04	+ 712.7	−0.15
960		VIII. s					13.9	+1 24.37	+1265.6	+0.40	+4459.8	−0.16
961		XVI. X					9.7	+1 24.82	+1272.3	−0.39	−3704.4	−0.16
962		X. s_1					12.3	+1 24.85	+1272.7	−0.21	−1837.5	−0.16
963		V. u_1					13.9	+1 25.46	+1281.9	+0.31	+3465.5	−0.16
964		IX. n_8					14.8	+1 26.13	+1291.9	+0.44	+4856.9	−0.16
965		Z					11.0	+1 26.16	+1292.4	+0.07	+1016.9	−0.16
966		w'''_2					12.5	+1 26.28	+1294.2	−0.10	− 677.4	−0.16
967		XIV. U; XII. bb_2					11.2	+1 29.45	+1341.8	−0.27	−2424.3	−0.16
968*		XII. aa_2					11.5	+1 29.65	+1344.8	−0.21	−1841.6	−0.16
969*		VIII. t_{-3}; VI. r_3					13.9	+1 30.25	+1353.7	+0.38	+4174.4	−0.16
970		$n''n''_1$					14.8	+1 30.97	+1364.6	−0.07	− 437.9	−0.16
971		I. Q					12.7	+1 31.53	+1372.9	+0.12	+1543.8	−0.16
972		XVIII. y_{-1}					13.9	+1 31.67	+1375.0	−0.53	−5155.4	−0.17
973		V. V					10.2	+1 31.95	+1379.3	+0.31	+3483.4	−0.17
974		O″O″					10.6	+1 32.22	+1383.3	−0.06	− 318.0	−0.17
975		X. r_1					13.9	+1 32.71	+1390.7	−0.16	−1329.1	−0.17
976		I. R					10.1	+1 33.03	+1395.4	+0.11	+1449.2	−0.17
977		z_1					13.1	+1 34.66	+1419.9	+0.03	+ 613.1	−0.17
978		X. r_2					13.9	+1 34.72	+1420.8	−0.18	−1548.3	−0.17
979		IV. R					11.3	+1 34.79	+1421.9	+0.23	+2630.3	−0.17
980		IX. o					14.8	+1 34.82	+1422.3	+0.45	+4931.8	−0.17

No.	Herschel's and Struve's No.	Letter. G. P. Bond.	Letter. Herschel.	Number. W. C. Bond.	Letter. Lassell.	Letter. Liapunoff.	Mag. by Argelander's Scale.	Eq. 1857.0 $\alpha-\alpha_0$ In Time.	Eq. 1857.0 $\alpha-\alpha_0$ In sec. of Arc.	Prec. Cöef.	Eq. 1857.0 $\delta-\delta_0$ In sec. of Arc.	Prec. Cöef.
								m s	″	″	″	″
981		I. r_1					13.9	+1 35.53	+1432.9	+0.11	+1429.2	−0.17
982		y'_2					14.2	+1 35.83	+1437.5	−0.03	+ 150.6	−0.17
983		I. r_2					13.1	+1 36.01	+1440.2	+0.13	+1598.6	−0.17
984		XVIII. Y					10.8	+1 36.17	+1442.5	−0.53	−5147.3	−0.17
985		XVIII. Z					10.8	+1 36.75	+1451.3	−0.51	−4931.9	−0.17
986		VIII. t_{-2}; IX. o_{-1}					13.1	+1 37.19	+1457.8	+0.41	+4518.9	−0.17
987		I. r_3					13.9	+1 38.25	+1473.7	+0.09	+1239.8	−0.17
988		IX. O					10.9	+1 38.34	+1475.1	+0.43	+4782.4	−0.18
989		I. r_4					13.1	+1 38.76	+1481.4	+0.09	+1190.0	−0.18
990		IX. o_1					13.1	+1 38.94	+1484.1	+0.44	+4886.0	−0.18
991		VIII. t_{-1}; IX. p_{-1}					13.9	+1 38.95	+1484.2	+0.41	+4477.7	−0.18
992		VIII. t_0					14.8	+1 39.53	+1492.9	+0.39	+4345.3	−0.18
993		VIII. T					11.5	+1 39.53	+1492.9	+0.39	+4296.5	−0.18
994		z_2					13.9	+1 39.66	+1494.9	+0.09	+1198.3	−0.18
995		IX. P					10.0	+1 39.83	+1497.4	+0.46	+5084.1	−0.18
996		II. N; III. N					11.9	+1 41.45	+1521.7	+0.15	+1883.5	−0.18
997		X. T; XI. T					9.2	+1 41.83	+1527.4	−0.20	−1722.9	−0.18
998		IX. o_2					14.8	+1 42.33	+1535.0	+0.44	+4834.3	−0.18
999		$o''o''_1$					14.2	+1 43.22	+1548.3	−0.03	− 29.3	−0.18
1000		IV. S					11.0	+1 43.59	+1553.8	+0.24	+2802.7	−0.18
1001		IV. s_0					13.9	+1 43.59	+1553.8	+0.24	+2728.0	−0.18
1002		IV. T					10.3	+1 45.09	+1576.3	+0.23	+2720.0	−0.19
1003		XIV. V					9.3	+1 45.17	+1577.6	−0.30	−2733.9	−0.19
1004*		y'''_{-1}					11.5	+1 45.67	+1585.0	−0.14	−1073.7	−0.19
1005		X. U					11.5	+1 45.72	+1585.8	−0.18	−1538.2	−0.19
1006		VIII. U					11.5	+1 46.13	+1592.0	+0.38	+4216.0	−0.19
1007		X'''					11.0	+1 46.68	+1600.2	−0.11	− 764.4	−0.19
1008		$o''o''_2$					13.9	+1 46.72	+1600.8	−0.03	+ 10.5	−0.19
1009		XVI. Y					10.8	+1 47.54	+1613.1	−0.44	−4188.9	−0.19
1010		V. W					11.5	+1 47.85	+1617.8	+0.31	+3469.6	−0.19
1011*		z_3; aa_{-1}					13.1	+1 49.41	+1641.2	+0.05	+ 818.	−0.19
1012		VI. r_4					13.1	+1 49.51	+1642.6	+0.33	+3745.8	−0.19
1013		X. u_1					13.9	+1 50.23	+1653.4	−0.20	−1697.6	−0.19
1014		IX. o_3					14.8	+1 51.34	+1670.1	+0.43	+4722.7	−0.19
1015		AA					10.8	+1 52.01	+1680.2	+0.05	+ 788.9	−0.19
1016		X. u_2					14.8	+1 52.49	+1687.4	−0.20	−1757.4	−0.20
1017		VI. s_{-3}					14.8	+1 53.21	+1698.1	+0.35	+3937.2	−0.20
1018		I. r_5					12.3	+1 53.51	+1702.6	+0.13	+1643.7	−0.20
1019		X. u_3					14.2	+1 54.48	+1717.2	−0.20	−1677.7	−0.20
1020		I. r_6					14.8	+1 54.51	+1717.7	+0.13	+1608.8	−0.20
1021		XIV. w_{-1}					13.9	+1 55.35	+1730.2	−0.28	−2579.3	−0.20
1022		XIV. W					10.0	+1 55.97	+1739.5	−0.28	−2551.3	−0.20
1023*		aa_1					13.1	+1 56.73	+1751.	+0.06	+ 969.	−0.20
1024		IX. q_{-2}					14.8	+1 57.27	+1759.1	+0.44	+4872.2	−0.20
1025		VI. s_{-2}					12.8	+1 57.73	+1765.9	+0.32	+3643.3	−0.20

No.	Herschel's and Struve's No.	Letter. G. P. Bond.	Letter. Herschel.	Number. W. C. Bond.	Letter. Lassell.	Letter. Liapunoff.	Mag. by Argelander's Scale.	Eq. 1857.0 $\alpha-\alpha_0$ In Time.	Eq. 1857.0 $\alpha-\alpha_0$ In sec. of Arc.	Prec. Cöef.	Eq. 1857.0 $\delta-\delta_0$ In sec. of Arc.	Prec. Cöef.
								m s	″	″	″	″
1026		XVII. dd$_{-1}$					13.9	+1 57.79	+1766.8	−0.45	−4304.0	−0.20
1027		XIV. w$_1$					14.8	+1 57.97	+1769.6	−0.28	−2599.2	−0.20
1028		Y‴					10.4	+1 58.17	+1772.6	−0.13	−1001.7	−0.20
1029*		aa$_2$					13.1	+1 58.40	+1776.	+0.05	+ 844.	−0.20
1030		VI. s$_{-1}$					13.9	+1 59.21	+1788.1	+0.35	+3895.4	−0.21
1031		I. S					10.0	+1 59.23	+1788.5	+0.10	+1300.0	−0.21
1032		X. V; XI. U					11.3	+1 59.25	+1788.7	−0.16	−1309.8	−0.21
1033		VI. r$_5$					13.1	+1 59.97	+1799.5	+0.36	+4061.8	−0.21
1034		IV. u$_{-1}$					13.1	+2 0.40	+1806.0	+0.22	+2572.6	−0.21
1035*		y‴$_1$					12.3	+2 0.67	+1810.0	−0.14	−1087.7	−0.21
1036		XVII. DD					9.3	+2 1.03	+1815.5	−0.45	−4275.1	−0.21
1037		VIII. V					12.3	+2 1.23	+1818.5	+0.38	+4258.0	−0.21
1038		IX. q$_{-1}$					13.9	+2 1.51	+1822.6	+0.46	+5046.7	−0.21
1039		IX. Q					10.5	+2 2.04	+1830.6	+0.43	+4697.0	−0.21
1040		IV. U					10.3	+2 3.09	+1846.4	+0.22	+2606.7	−0.21
1041		IX. q$_1$					14.2	+2 3.52	+1852.8	+0.45	+4917.2	−0.21
1042		VIII. v$_1$					14.8	+2 3.73	+1856.0	+0.38	+4246.0	−0.21
1043		P″P″					10.9	+2 4.16	+1862.4	−0.08	− 512.4	−0.21
1044*		I. s$_1$					11.5	+2 4.21	+1863.1	+0.13	+1640.3	−0.21
1045		IX. q$_2$					13.9	+2 4.27	+1864.1	+0.44	+4892.3	−0.21
1046		XVII. dd$_1$					13.1	+2 4.29	+1864.3	−0.46	−4356.7	−0.21
1047*		aa$_3$					15.6	+2 4.53	+1868.	+0.08	+1112.	−0.21
1048		X. v$_1$					13.9	+2 5.40	+1881.0	−0.15	−1229.0	−0.21
1049		XVI. Z					9.9	+2 5.42	+1881.3	−0.40	−3745.8	−0.21
1050*		aa$_4$					15.6	+2 6.47	+1897.	+0.07	+1084.	−0.22
1051*		(aa)$_6$; (y)′$_1$					11.5	+2 6.91	+1903.7	+0.02	+ 539.5	−0.22
1052*		XII. CC					10.1	+2 7.48	+1912.2	−0.25	−2226.9	−0.22
1053		aa$_5$; I. s$_2$					13.1	+2 7.37	+1910.6	+0.08	+1152.7	−0.22
1054		IX. q$_3$					13.1	+2 7.51	+1912.6	+0.46	+5072.6	−0.22
1055		X. W; XI. V					10.2	+2 7.91	+1918.7	−0.18	−1482.7	−0.22
1056		X. w$_0$					13.9	+2 8.02	+1920.3	−0.18	−1531.0	−0.22
1057		XII. cc$_1$					14.8	+2 9.25	+1938.7	−0.26	−2369.3	−0.22
1058*		I. s$_3$					13.9	+2 9.69	+1945.3	+0.14	+1753.5	−0.22
1059		VIII. v$_2$					13.9	+2 10.73	+1961.0	+0.38	+4206.3	−0.22
1060		VIII. v$_3$					14.8	+2 11.23	+1968.4	+0.39	+4340.7	−0.22
1061*		aa$_6$					12.3	+2 11.50	+1972.5	+0.05	+ 881.5	−0.22
1062		IX. q$_4$					14.8	+2 12.01	+1980.1	+0.46	+5012.0	−0.22
1063		XII. DD					8.4	+2 12.67	+1990.1	−0.22	−1949.5	−0.22
1064		XVI. z$_1$					12.3	+2 12.68	+1990.2	−0.44	−4172.5	−0.22
1065		IX. q$_5$					14.8	+2 12.76	+1991.4	+0.45	+4997.0	−0.22
1066		IV. V					11.7	+2 12.79	+1991.9	+0.23	+2700.4	−0.22
1067*		(y)′$_2$; p″p″$_1$					12.3	+2 13.24	+1998.6	−0.03	+ 5.1	−0.23
1068		IX. q$_6$					11.5	+2 13.27	+1999.1	+0.45	+4924.2	−0.23
1069		VI. S; VIII. W					10.8	+2 14.27	+2014.0	+0.36	+4026.0	−0.23
1070		(y)′$_3$					10.2	+2 14.83	+2022.4	−0.01	+ 340.2	−0.23

No.	Herschel's and Struve's No.	Letter. G. P. Bond.	Letter. Herschel.	Number. W. C. Bond.	Letter. Lassell.	Letter. Liapunoff.	Mag. by Argelander's Scale.	Eq. 1857.0 $\alpha-\alpha_0$ In Time.	Eq. 1857.0 $\alpha-\alpha_0$ In sec. of Arc.	Prec. Cöef.	Eq. 1857.0 $\delta-\delta_0$ In sec. of Arc.	Prec. Cöef.
								m s	″	″	″	″
1071		X. w_1					13.9	+2 15.53	+2032.9	−0.19	−1577.8	−0.23
1072		XVII. dd_2					13.9	+2 15.54	+2033.1	−0.45	−4328.7	−0.23
1073		XV. Y; XIV. w_2					9.8	+2 16.01	+2040.2	−0.33	−3057.6	−0.23
1074		IX. q_7					13.1	+2 17.03	+2055.4	+0.44	+4802.7	−0.23
1075		I. s_4					13.9	+2 17.22	+2058.3	+0.11	+1474.5	−0.23
1076		IX. q_8					13.1	+2 18.54	+2078.1	+0.43	+4762.8	−0.23
1077*		I. s'_3; I. s_5					8.6	+2 18.96	+2084.4	+0.12	+1550.4	−0.23
1078		X. X; XI. W					9.8	+2 19.77	+2096.6	−0.20	−1752.4	−0.23
1079		XV. Z					9.7	+2 20.93	+2113.9	−0.36	−3382.5	−0.24
1080		II. O; III. O					11.0	+2 21.69	+2125.4	+0.16	+1956.1	−0.24
1081		XV. AA					9.9	+2 21.73	+2126.0	−0.38	−3572.0	−0.24
1082		BB					10.4	+2 22.31	+2134.7	+0.08	+1115.9	−0.24
1083		XII. EE					10.8	+2 22.89	+2143.3	−0.26	−2320.4	−0.24
1084		VI. T					10.1	+2 23.10	+2146.5	+0.34	+3814.1	−0.24
1085		II. P; III. P					11.0	+2 23.15	+2147.3	+0.16	+1963.2	−0.24
1086		VI. t_1					13.9	+2 23.82	+2157.3	+0.37	+4104.9	−0.24
1087		X. Y; XI. X					11.0	+2 24.67	+2170.1	−0.20	−1723.0	−0.24
1088		II. p_1					13.1	+2 24.93	+2174.0	+0.18	+2223.7	−0.24
1089		XVI. AA; XVII. EE					9.7	+2 28.99	+2234.9	−0.44	−4210.1	−0.24
1090		XVIII. AA					10.8	+2 33.87	+2308.0	−0.53	−5074.8	−0.25
1091		II. Q; III. Q					8.6	+2 35.04	+2325.6	+0.17	+2103.2	−0.26
1092		XVII. FF					8.9	+2 35.06	+2325.9	−0.49	−4720.4	−0.26
1093		XVIII. aa_1					13.9	+2 35.10	+2326.5	−0.54	−5214.7	−0.26
1094		XVI. BB					9.3	+2 36.53	+2348.0	−0.41	−3886.1	−0.26
1095		II. R; III. R					10.3	+2 37.43	+2361.4	+0.18	+2195.3	−0.26
1096		II. r_1					10.8	+2 37.97	+2369.5	+0.15	+1870.0	−0.26
1097		XVIII. aa_2					9.7	+2 38.56	+2378.4	−0.57	−5474.7	−0.26
1098		XVII. GG					8.4	+2 38.86	+2382.9	−0.50	−4767.3	−0.26
1099		XVIII. BB					10.4	+2 43.16	+2447.4	−0.52	−5015.0	−0.27
1100		XVIII. CC					11.5	+2 48.18	+2522.7	−0.56	−5403.8	−0.28
1101		XV. BB					8.6	+2 48.74	+2531.1	−0.38	−3574.7	−0.28

SECTION III. PART II.

DIFFERENTIAL CATALOGUE: COMPARISONS WITH OTHER AUTHORITIES.

No. G. P. B.	No. Hersch. and Struve.	Letter. G. P. Bond.	No. W. C. B.	No. Ll.	Struve and Liap. —Bond.		Herschel. —Bond.		W. C. Bond. —Bond.		Lassell. —Bond.	
					$\alpha-\alpha_0$	$\delta-\delta_0$	$\alpha-\alpha_0$	$\delta-\delta_0$	$\alpha-\alpha_0$	$\delta-\delta_0$	$\alpha-\alpha_0$	$\delta-\delta_0$
					″	″	″	″	″	″	″	″
81	1	A'_{-1}					−1.2	−3.7				
232	2	B''; B'''					−22.1	+3.7				
246	3	C'''					−14.0	+4.4				
247	4	D'''					−7.9	+3.8				
303	5	C'; C''			−0.3	−1.8	−3.9	+0.5				
311	6	C					−2.5	−4.8				
314	8	D''			−3.7	−1.4	−2.1	−1.9				
315	7	E'''					−4.9	+3.8				
323	9	F'''					−8.3	+1.3				
335	10	E''			−1.5	−1.8	−7.4	−4.8				
346	11	G'''					−1.5	+3.4				
347*	15	e''_1			+2.8	+1.4	+93.0	−3.8				
363	12	D'		54	0.0	−1.6	−0.5	−3.9			−8.1	+4.4
370	13	e''_2			−5.7	−2.2	−11.5	+12.8				
373	14	F''			+0.2	−1.1	−2.7	+13.8				
377	16	G''			−1.8	+0.6	+8.5	−45.5				
382	17	H''		50	−3.3	−1.0	−5.1	+6.6			−2.6	+2.4
387	18	I''		44	−1.8	−0.3	−9.4	+0.5			+0.5	+8.3
399	19	d'_1; i''_1		52	−5.0	−0.5	−15.7	+2.0			+1.9	−5.0
401	20	X. I; XI. I					−17.0	−25.6				
402	21	i''_2		41	+0.2	+2.6	+8.4	+13.0			−3.6	+13.7
404	23	H'''; X. K; XI. K					+14.8	−18.1				
409	22	i''_3			−4.3	+1.0	−7.7	−8.9				
410	24	I'''					+22.7	−20.4				
423	25	E					−4.7	−4.2				
427	27	K''		47	−1.9	+0.4	−7.5	+0.8			+0.7	−0.6
430	26	K''_1; k''_1		42	−4.2	−0.8	−15.3	+12.7			+3.3	−0.7
434	29	K'''					−21.5	−21.3				
435	30	d'_2			−3.7	+1.2	−6.0	+17.1				
438	31	F					+12.8	−3.6				
443	28	f_1					−31.6	−2.5				
449	32	E'	1	55	+0.1	−0.8	−1.0	−5.2	−3.8	−0.7	+0.4	−3.8
458	33	L''	2	46	−3.8	−0.2	−10.1	+8.0	+0.8	−6.6	−0.9	−5.8
467	34	L'''			−3.7	−0.7	−8.8	+2.6				
479	35	F'	3	56	+0.5	+0.4	−5.0	−2.1	0.0	−0.9	−1.7	−1.8

No. G. P. B.	No. Hersch. and Struve.	Letter. G. P. Bond.	No. W. C. B.	No. Ll.	Struve and Liap. —Bond. $\alpha-\alpha_0$	$\delta-\delta_0$	Herschel. —Bond. $\alpha-\alpha_0$	$\delta-\delta_0$	W. C. Bond. —Bond. $\alpha-\alpha_0$	$\delta-\delta_0$	Lassell. —Bond. $\alpha-\alpha_0$	$\delta-\delta_0$
					″	″	″	″	″	″	″	″
490	36	l''_1		48	− 5.5	+ 3.0	−14.4	+14.0			−3.9	− 1.3
497	37	M''; M'''		37	− 3.3	+ 0.2	−10.2	+ 0.7			−1.3	+ 8.9
500*	C 1	II. F; III. F					−20.6	+ 1.5				
505	40	N''		38	− 3.2	+ 0.5	+ 2.2	+ 2.5			−9.4	+ 3.8
506	38	G'; O''	4	51	− 2.0	− 0.5	−12.6	+ 9.1	− 1.0	− 0.3	+0.5	− 5.6
508	42	f_3					+10.7	−11.1				
510	43	o''_1		36	− 3.5	+ 8.4	− 0.2	+18.8			−1.2	+22.5
516	41	o''_2		49	− 7.6	− 1.9	−21.6	+ 5.0			+4.0	− 6.6
523	45	P''	5	40	− 2.0	− 0.7	− 3.0	− 0.3	− 0.4	− 2.1	−0.8	− 2.0
524	44	g'_2		53	− 6.9	− 1.2	−20.2	+10.4			−6.7	− 8.2
530	C 3	II. f_1					+15.3	+ 8.6				
532	46	g'_1					− 3.0	+21.2				
536	C 4	II. G; III. G; IV. K					+ 4.6	− 1.1				
538	C 2	II. f_2					− 9.5	−13.5				
542	C 5	IV. L					+10.8	+ 5.3				
543*	I	I			+12.1	− 0.1						
545	47	p''_1		34	− 3.0	+ 1.6	+ 9.6	+20.5			+1.7	+ 3.4
551*	48	H'	6; 8	57	− 1.0	+ 0.1	− 6.0	− 1.3	+11.2	+ 2.1	+1.5	− 8.8
552*		p''_2	7						+ 4.8	+ 7.9		
554	49	K		58	+ 0.1	0.0	− 1.1	− 2.8			+1.4	− 2.4
558	50	Q''	9	39	− 2.2	− 0.4	+ 5.1	− 5.9	− 0.9	− 1.7	−5.7	+ 1.8
566	52	q''_1		32	− 4.0	+10.3	+ 5.2	+39.0			−2.2	+ 7.9
567*	51	μ	10				+ 1.7	− 2.9	+ 1.5	−15.5		
570	53	R''	13	33	− 1.5	+ 1.1	+ 0.6	+ 2.1	+ 7.2	− 0.4	+2.7	0.0
573	54	r''_2	12	35	− 1.4	+ 2.3	− 3.7	+ 6.0	− 1.1	− 2.2	+0.9	+ 3.8
575*	57	r''_1	11	45	− 0.7	− 1.9	+15.2	+20.0	− 5.7	− 0.5	0.0	+ 0.4
580*	56	h'_1	42	59	− 1.3	+ 1.2	− 6.2	+12.5	+ 2.2	− 2.1	−8.8	− 5.5
581	ad 54	r''_3										
583	58	n'''_{-2}					+ 7.1	− 2.2				
584	C 6	II. H	14				+ 1.5	+ 8.2	+ 1.1	+ 3.9		
587	60	n'''_{-1}					+24.8	+36.5				
589*	57	r''_4	15		− 0.4	− 2.2	−12.4	+17.9	+ 4.9	+15.1		
593	C 7	II. h_1; IV. m_1					− 6.1	−28.0				
595		ν	15	43					− 5.4	+ 9.7	+6.2	0.0
596	C 10	II. h_2					−29.8	−16.8				
598	62	r''_5		31	+ 0.6	+ 1.7	+10.2	+ 7.5			−1.0	+ 5.6
599	59	N'''					− 0.4	+ 7.2				
600	61	n'''_1					+ 0.5	+29.9				
612*		ξ	16	*i*	− 2.	− 3.			+ 4.1	− 2.9		
614	C 8	IV. N					−23.2	−21.4				
615	66	κ					+ 3.4	+ 3.1				
616	C 11	II. h_3					+13.6	−31.3				
617	64	I'			+ 0.3	− 0.6	+ 0.2	− 0.4				
618*		π	19	*h*	− 2.	− 2.			+ 4.1	− 5.9		
619	65	K'	17		+ 0.5	− 0.3	− 0.5	0.0	+ 0.7	− 0.3		

No. G. P. B.	No. Hersch. and Struve.	Letter. G. P. Bond.	No. W. C. B.	No. Ll.	Struve and Liap. —Bond. $\alpha-\alpha_0$	Struve and Liap. —Bond. $\delta-\delta_0$	Herschel. —Bond. $\alpha-\alpha_0$	Herschel. —Bond. $\delta-\delta_0$	W. C. Bond. —Bond. $\alpha-\alpha_0$	W. C. Bond. —Bond. $\delta-\delta$	Lassell. —Bond. $\alpha-\alpha_0$	Lassell. —Bond. $\delta_0-\delta_0$
					″	″	″	″	″	″	″	″
620	63	n'''_4					− 3.9	+ 7.3				
621*	ad II.				+2.	+5.						
622	II.	ρ	18		+0.2	+0.2			+ 0.2	+13.7		
623	68	n'''_5					+ 2.0	+30.7				
624	67	U″; L′	21		+0.3	−0.2	− 1.0	0.0	+ 0.2	− 0.9		
628	C 9	IV. O					−29.2	− 4.1				
633	71	N′			−0.6	+0.2	+ 1.0	+ 0.9				
635	70	O′	23	2	−2.1	−0.2	− 4.2	− 3.7	− 3.4	− 0.6	+0.7	−1.9
636		σ	24						+ 0.3	+ 0.4		
639	74	O‴			−8.8	−1.6	− 1.4	+ 4.3				
640	73	P′	25		+0.5	−0.2	+ 0.5	− 0.1	+ 1.1	− 0.7		
641	III.	o'_1			−4.4	−4.0						
644	72	o'''_1					−10.0	+43.8				
647	75	ϕ; p'_1	26	9	−1.3	+1.2	− 0.7	+ 3.5	− 7.2	−12.8	+1.7	+3.4
650	79	θ	29	1	−3.1	−1.9	+ 9.2	− 9.3	− 2.1	− 1.7	−7.4	−8.6
651*	ad 75	ψ	27		−3.	−5.			−12.5	−20.7		
652*	76	ζ	32	4	+0.2	+0.3	+ 1.1	− 6.0	0.0	− 2.3	+2.2	−5.9
653	83	ι	28		+3.1	+2.8	+17.4	− 8.3	−10.2	+ 2.9		
654	78	ω	31		+1.3	−0.3	− 0.8	+ 9.6	− 4.5	+ 0.6		
657	80	ϵ	33	4	−3.7	−0.8	− 0.8	− 3.7	− 4.0	− 5.2	−7.2	+0.5
658	77	o'''_2					−10.8	+46.3				
663	84	δ	37		−1.7	−1.0	+ 1.1	− 7.7	+ 0.1	− 0.2		
664	C 14	I. l_1; II. h_4	35				+18.6	+ 0.4	− 6.4	− 6.3		
665	C 12	II. h_5	36				− 1.3	− 7.9	− 6.6	−64.1		
666	81	x''_1	30		−2.9	−0.1	−20.2	+39.2	−32.5	+ 5.2		
667	85	M			+1.0	+1.4	− 1.0	− 3.4				
669	87	Q′	39	10	−1.4	0.0	+ 0.8	− 1.7	− 1.4	− 1.6	−2.0	−0.4
670	86	N	38	3	−0.2	+0.3	− 4.9	−27.7	− 2.8	+ 1.6	−0.3	−4.4
671	88	β	41	18	−0.6	+0.1	+ 3.3	+ 0.5	+ 3.6	−12.8	−2.7	−0.2
672	C 15	II. I; III. H	40				+ 5.1	+ 2.0	+ 1.3	+ 3.1		
673	C 13	II. i_0; IV. o_3					− 9.6	−27.8				
676*	ad 88	γ	43		−4.	0.			− 2.3	−10.8		
677*	ad 81	x''_2	34						−41.7	+ 8.7		
680	92	o'''_3					+17.8	+ 1.8				
681	89	η			−7.9	0.0	− 6.5	− 9.1				
684*	90	n_3	56				− 7.5	−16.4	+93.6	−14.1		
685	93	Y″; θ^2	45	26	+0.1	+1.0	+ 5.0	+ 1.4	− 0.6	+ 0.2	−1.2	+1.6
686*	91		44				− 3.1	−19.0	−20.3	− 1.4		
690	95	Z″		30	−2.9	+1.3	+ 5.2	− 2.2			−5.3	+3.0
692	C 16	II. i_1	46				−24.6	+13.6	− 3.7	− 2.1		
693*	94	n_4	58				−16.9	−24.8	+68.7	−16.0		
695	97	n_5					− 1.3	−20.6				
696	98	O					− 1.5	−15.7				
698	C 18	II. i_2					+ 8.8	− 0.2				
699	C 17	II. i_3					−23.1	+16.6				

No. G. P. B.	No. Hersch. and Struve.	Letter. G. P. Bond.	No. W. C. B.	No. Ll.	Struve and Liap. —Bond. $\alpha-\alpha_0$	$\delta-\delta_0$	Herschel. —Bond. $\alpha-\alpha_0$	$\delta-\delta_0$	W. C. Bond. —Bond. $\alpha-\alpha_0$	$\delta-\delta_0$	Lassell. —Bond. $\alpha-\alpha_0$	$\delta-\delta_0$
					″	″	″	″	″	″	″	″
700	102	R′	47	6	−0.8	+0.6	+13.0	−10.7	−2.7	+1.5	+1.9	−4.9
703*	96	o_1	59				−17.1	−21.4	+63.0	−6.6		
705	99	P; S′	48	5	0.0	+0.7	−7.1	−5.3	−3.2	+2.4	+2.4	−0.2
707	103	A″A″	49	27	−1.6	+2.4	+8.3	+19.0	−2.7	+1.2	−0.6	+5.2
708	101	B″B″	50	23	−0.6	+2.5	−0.7	+2.5	−1.3	+2.2	+0.6	+1.1
709	100	$b''b''_1$	51		−3.2	+2.0	−2.2	+3.6	−2.4	+1.9		
712	C 19	I. l_2	52				+6.8	−28.8	−4.2	−4.7		
715		XII. W; XIII. I	54						+0.4	+2.2		
722	105	o'''_4					+4.3	+16.4				
724	104	C″C″	55	25	−1.5	+1.3	−1.3	0.0	−3.5	+0.1	−4.0	+5.3
728	C 20	I. M; II. K; III. I	57				+8.6	+4.6	+1.4	−2.7		
732	106	P″; D″D″	63		−1.4	+2.7	−9.0	+5.6	+24.5	+4.9		
733	C 21	II. i_4					−5.9	+24.7				
734	108	T′	60	8	+0.6	+0.7	−1.5	−1.7	−2.0	+2.1	−6.0	−3.8
737	109	q'_1; t'_1		12			−4.2	+2.9			−4.8	+6.5
740	107	p_3					−22.0	−18.5				
741*	110	E″E″	61	19	0.0	+0.3	−1.7	+0.4	+1.0	−1.7	−0.3	+3.0
742	C 22	II. i_5	62				+9.4	+9.6	+2.1	−8.2		
746	111	Q‴; F″F″	64	29	−2.1	+1.5	−2.4	+1.5	+9.1	−1.7	−5.3	+4.5
750	112	G″G″	65	28	−3.4	+2.2	+3.5	+11.2	−4.4	+2.6	−8.0	+9.9
757	113	R	66	7	+2.2	+2.8	−5.2	−3.7	−2.8	−3.3	−1.4	−7.4
759	114	t'_2	67				+14.4	+18.7	−0.2	+0.5		
762	115	s'''_{-1}					+10.7	+34.3				
767	IV.	o	69*	20					−8.6	+69.8	+2.9	+9.6
768	C 23	II. L; III. K	70				+9.3	+6.0	+1.4	+4.3		
776	116	u'_{-1}					−8.8	+3.5				
777	C 24	IV. p_1					−2.7	+1.4				
778	117	p	75	21	−2.8	+3.0	−1.7	−22.2	+8.2	+2.1	−5.8	+2.5
781	120	U′	76	15	−2.2	+0.7	+2.5	0.0	−0.6	−0.8	−4.8	−0.5
783	122	q'''_1	81*				−1.9	+3.2	+73.8	−47.8		
784	123	H″H″	78	24	−2.3	+1.8	−1.1	−7.5	−3.5	+2.8	−6.4	+1.3
785	124	V′; S	79	11	−0.4	+0.6	+0.9	+0.5	+15.6	+8.2	−3.0	−2.8
786	118	r_2					−18.4	+5.1				
787	119	r_3					−18.2	−4.8				
789	121	$h''h''_1$					−15.9	−12.7				
793	126	$h''h''_2$	80	26 B	+4.7	+2.3	−2.2	+2.0	+8.5	−7.7	−0.4	+12.9
795*	125	q'''_2	84		−6.4	−2.5	−5.1	+10.7	+57.5	−23.9		
797		v'_1	85						−3.3	−4.7		
801		(i)	86						−30.8	−25.5		
805	128	v'_2	83		−2.5	+2.0	−17.7	+35.3	−1.3	−1.2		
806	127	T					−40.3	−29.1				
808	129	v'_3	82	13	−3.7	+1.1	−19.0	+18.6	+0.4	−0.2	−3.6	−3.6
818	C 25	II. l_3					+16.7	+41.5				
822	133	I″I″	87	22	−1.8	+2.1	+1.5	+1.0	+5.1	+2.5	−5.6	+4.0
823	132	R‴					−12.7	+2.3				

No. G. P. B.	No. Hersch. and Struve.	Letter. G. P. Bond.	No. W. C. B.	No. Ll.	Struve and Liap. —Bond. $\alpha-\alpha_0$	Struve and Liap. —Bond. $\delta-\delta_0$	Herschel. —Bond. $\alpha-\alpha_0$	Herschel. —Bond. $\delta-\delta_0$	W. C. Bond. —Bond. $\alpha-\alpha_0$	W. C. Bond. —Bond. $\delta-\delta_0$	Lassell. —Bond. $\alpha-\alpha_0$	Lassell. —Bond. $\delta-\delta_0$
					″	″	″	″	″	″	″	″
825	131	q'''_3	88				−17.6	+24.7	+ 8.0	−55.1		
826	130	v'_4					−19.8	+ 9.4				
832	134	$i''i''_1$	89				− 4.5	+ 1.7	− 6.8	+14.7		
834		III. L; II. l_5	90						+ 5.8	− 5.5		
843	135	S‴	91		−1.5	+1.8	− 7.3	+ 3.5	+ 2.2	−23.5		
848	136	W′	93	16	−0.3	+2.9	+ 2.9	+ 0.8	− 1.2	− 2.2	− 2.9	+5.1
855	138	T‴					−13.1	+ 3.2				
862	C 26	II. l_8					− 2.6	− 2.2				
863*	137	w'_1	92	14			−44.7	+ 9.3	−84.2	− 2.9	−11.7	+5.6
881	139	U‴					− 5.5	+ 7.7				
883	140	u'''_1					+ 1.2	+11.2				
889	142	K″K″	94	17	−0.8	+3.1	+ 4.2	−10.9	− 1.8	− 1.2	−14.7	+2.8
893	141	w'_2					−27.7	+36.9				
905	143	V‴			−0.2	+1.2	+ 0.3	+ 3.2				
924	145	L″L″	95		−0.8	+1.5	+ 2.7	− 0.8	− 1.7	− 4.7		
927	144	X. r_{-1}					−29.2	− 3.4				
935	146	X. R; XI. R					−31.8	+ 0.1				
938	147	X′; M″M″	96		−1.4	−0.1	+ 8.8	+ 9.4	− 0.5	− 3.1		
953	148	Y					+ 2.8	−19.2				
955	149	N″N″					− 0.2	+ 2.8				
956	150	Y′					+11.3	− 6.1				

SECTION III. PART III.

(A.) NOTES TO THE GENERAL CATALOGUE.

The assignment of weights in most of the following cases was made by Professor Bond upon his own judgment; and in the cases of those stars of which positions are given in the following pages, additional to those obtained by the process of Section I. and II., the combination has been made according to a manuscript list in his own handwriting, correcting a few manifest errors. [S.]

45, 53 were determined in short revision zones like those of Section I. Part 6.

61. The observed declinations of this star differ 8.″1, but their mean was found very exact by an observation Feb. 25, 1864.

66, 72. Like 45.

93. The declinations of this star observed in Zone VIII., and the Supplement to VI., differ 15.″0. The simple mean has been taken.

112. Determined as 45, etc. This was combined with a chart position, 1862, March 27th.

The positions are

Revision Zone,	−1794.″5	+188.″9
Chart,	−1779.	+190.

122. A chart position −1766.″4 +997.″1, combined with a position from a short revision zone −1756.″7 +1001.″6.

131. In this as in other similar cases, I have assumed that an observation in one of the "revision" zones should have equal weight with one in one of the principal zones. These latter observations were not made so deliberately as the former, although the chronograph was used for them. [S.]

144. From a chart, 1862, March 27th.

152. From short revision zones in 1864, $\alpha - \alpha^{\circ} = -1655.''3$ $\delta - \delta^{\circ} = +887.1$.

It was also observed Feb. 7, 1863, its difference from 176 being found −7.ˢ35 and 3′16″, giving −1660.″7 +887.″1. The mean of the two determinations was taken.

156. From a short extension zone.

158. Probably identical with 157.

160. This is probably the star mentioned in the foot-note to p. 69.

175. "Entered on charts 1862, March 27th. Observed Feb. 7, 1863, chart 25."

The position is given in Prof. Bond's papers from the chart.

178. Observed Feb. 7, 1863; found to differ from 232 by −16.ˢ5 and −53″.

The position should be 20′ farther south, viz., in δ −634.″6.

230. The star Weisse, V. 655, should be in this neighborhood, but is not found.

269. Nebulous.

286. The two declinations of this star differ 12.″1.

294. Double; the preceding is given in the catalogue.

339. Its position and distance from 347 were estimated as 329° and 20″, from whence the position in the catalogue was derived; but a value −463.″7 was adopted for the declination of the comparison star when the calculations were made.

347. On April 7, 1864, this star was compared in AR. with 314, 377, 387; the differences were respectively, +6.ˢ5 −6.ˢ9 −9.ˢ9, and the resulting, $\alpha-\alpha_0$ reduced to 1857.0 = − 829.″7. In the same manner on April 8th, +6.ˢ0 −6.ˢ5 were obtained for its differences from 314 and 377, hence $\alpha-\alpha_0$ = − 827.″4. The value of the same coördinate from the observation on page 43 is −832.″8. These three results were combined with the weights 3, 2, 1, respectively.

357, 396. These stars' catalogue-positions do not correspond with their nomenclature, as they follow not only 291, but also 293 and 328.

398. This star's position in Section I. depends on 12 and 10 observations, and has received three times the weight of that from Section II.

399. There is also a micrometrical measure of the difference of declination between this star and 624, which places this one 0.″4 farther north, and hence at 16.″4 (see note to 624).

404. The position of this star from Section I., depending on six observations, has received a weight of 3; that from Section II., from two observations, a weight = 2.

425. A close double star. The position and distance were estimated by G. P. B., as 300° and 2.″5. The star is Struve, 743 of the Mensurae Micrometricae.

443. The AR. of this star was observed 1858, Jan. 4th, page 41; and on April 7, 1864, it was found to precede 554 by 23.ˢ6. The two results for 1857.0 are −34.ˢ81 and −34.ˢ47, whose mean has been taken.

474. The two determinations of this star's declination differ 9.″8.

524. Certain estimates and chart-positions for this star are

	AR.	Dec.
1859, Feb. 22, Estimate,	−232.″5	+24.″
1862, March 19, Chart,	−249.	+23.5
1863, Jan. 31, Chart,	−235.	+24.

These, as appears by a note in Prof. Bond's handwriting, have been rejected for declination, and the result obtained December 19, 1857, viz., 524−624 = 0.″0, has been combined with an observation made with the scale on April 4, 1864, 524−506 = 12″. The combination gives

	″
From 624 $\delta-\delta_0$ =	16.0
From 506	17.6
Mean	16.8

The AR. of this star depends upon the differences −0.ˢ05 and +4.ˢ1 from 523 and 506, which give us −16.ˢ20 and −16.ˢ30 respectively, combined with the observations just mentioned.

536. There is a note to this star in Zone IV. "Companion 11th, −0′14″."

563. See note on p. 41.

567. The position in the catalogue is from two diagrams of March 10, 1859, and one of Jan. 19, 1863.

575. Two measures (made April 15, 1864) of this star's difference of declination from the principal star

in the trapezium give $\Delta\delta = -21.''62$ and $-23.''35$ respectively. The former has received a weight 2, and the two thus combined with the observation on page 43 give the result in the catalogue.

Its right ascension depends on observations made

1858, March 10th, page 43,	$\alpha-\alpha_0 =$	$-87.''3$
1864, April 14th, three observations by transits		$-81.7 = -5.^s45$
" April 16th, Micrometer-measure		-85.62
" " " " "		-85.07

With the weights 1, 2, 3, 6, assigned to these observations by G. P. B., the catalogue position is the result.

583. The declination of this star is from a diagram.

589. This star's AR. is derived from the observation on page 43, to which a weight = 1 has been assigned; and two measures of its difference of AR. from 575, made April 15, 1864. These give $589-575 = +29.''48$ (wt. 2) and $+27.''48$ (wt. 1) respectively, and we thus have —

	$\alpha-\alpha_0$	wt.
Page 43,	$-61.''0$	1
April 15, 1864,	-55.32	2
" " "	-57.32	1

The mean result is as in the catalogue.

For declination, a single observation April 15, 1864, gives $589-575 = +2.''0$, or 589 $\delta-\delta_0 = -20.''3$. This is combined with the result on page 43.

595. "A very small star. Position by two diagrams, March 10, 1852." (G. P. B.)

601. 1858, March 10. "Besides the above, (671,676,) very faint stars were suspected in the following positions." (G. P. B.)

$\alpha-\alpha_0$	$\delta-\delta_0$
$-36''$	$-31''$
-8	-36
-4	-28
$+100$	-39
$+106$	$-18.$

I have retained them in the catalogue, as the manuscript containing them had been examined by Prof. Bond. [S.]

602, 608. Seen certainly Feb. 10, 1863; positions by diagram of that date.

	$\alpha-\alpha_0$	$\delta-\delta_0$	
612, 618. Mean of	$-16.''0$	$+24.''0$	from numerous diagrams.
	-11.0	$+25.0$	
and of	-16.8	$+25.3$	from page 43.
	-9.8	$+24.3$	

617, 619, 624, 633, 640. These with 628 are the six stars of the trapezium, and do not occur, with the exception of 624, in the zones; but were determined by micrometric observations in connection with 628, made December 14, 1857.

Denoting the star No. 633 by $a_{,}$ and the remainder by Liapunoff's letters, we have the following observations.

A. Of Position and Distance.

	Position.	No. Obs.	Dist.	No. Obs.	No. in Series.
Star $a_{,}$ from a	120° 16′	5	4.″01	4	1
b " a	310 26	3	13.07	4	2
$b_{,}$ " b	349 40	3	4.05	4	3
d " b	32 12	2	9.10	4	4
d " c	298 56	2	18.86	4	5

B. Of Differences of Declination.

		No. Obs.	No. in Series.
Diff. Dec. d and $b_{,}$	2.95	4	6
d " b	7.46	4	7
d " c	9.44	4	8
d " a	16.25	4	9
d " $a_{,}$	18.28	4	10

The values of these stars' coördinates were now assumed as follows:—

		$\delta-\delta_0$		$(\alpha-\alpha_0)\cos\delta$
$a_{,}$	ξ^{iv}	− 2.″02	η^{iv}	+ 3.″46
b	ξ	+ 8.44	η	− 9.45
$b_{,}$	ξ'''	+12.30	η'''	−10.39
c	ξ'	+ 6.65	η'	+11.97
d	ξ''	+15.92	η''	− 4.71

These, except those of the first line, were taken from Liapunoff.

No. in Series.	Computed Angle.	Computed Distance.	C—O Angle.	C—O Dist.	Δ dp
1.	120° 17′	4.″01	+ 1′	0.″00	+0.″00
2.	311 46	12.67	+ 80	−0.40	+0.29
3.	346 19	3.97	−201	−0.08	−0.23
4.	32 22	8.86	+ 10	−0.24	+0.03
5.	299 4	19.09	+ 8	+0.23	+0.04

	Comp'd Diff. Dec.	C—O
6.	− 3.″62	−0.″67
7.	+ 7.48	+0.02
8.	+ 9.27	−0.17
9.	+15.92	−0.33
10.	−17.94	+0.34

The equations to be solved by least squares, are these.

		wt.
$0 = 0.''00 - 0.86\, d\xi^{iv}$	$-0.50\, d\eta^{iv}$	1.25
$= +0.29 + 0.75\, d\xi$	$+0.67\, d\eta$	0.75
$= -0.23 + 0.24(d\xi''' - d\xi)$	$+0.97(d\eta''' - d\eta)$	0.75
$= +0.03 - 0.54(d\xi'' - d\xi)$	$+0.84(d\eta'' - d\eta)$	0.50
$= +0.04 + 0.86(d\xi'' - d\xi')$	$+0.49(d\eta'' - d\eta')$	0.50

			wt.
$0 =$	$0''.00 - 0.50\, d\xi^{iv}$	$+0.86\, d\eta^{iv}$	1.
$=$	$-\ 0.40 + 0.67\, d\xi$	$-0.75\, d\eta$	1.
$=$	$-\ 0.08 + 0.97(d\xi''' - d\xi)$	$-0.24(d\eta''' - d\eta)$	1.
$=$	$-\ 0.24 + 0.84(d\xi'' - d\xi)$	$+0.54(d\eta'' - d\eta)$	1.
$=$	$+\ 0.23 + 0.49(d\xi'' - d\xi')$	$-0.86(d\eta'' - d\eta')$	1.

		wt.
$0 =$	$-\ 0.67 + d\xi''' - d\xi''$	1.
$=$	$+\ 0.02 + d\xi'' - d\xi$	1.
$=$	$-\ 0.17 + d\xi'' - d\xi'$	1.
$=$	$-\ 0.33 + d\xi''$	1.
$=$	$+\ 0.34 + d\xi^{iv} - d\xi''$	1.

And the results finally obtained,

$d\xi = +0''.22$	$d\eta = -0''.46$	$\xi = +\ 8''.66$	$\eta = -\ 9''.91$	$\eta \sec\delta = -\ 9''.96$
$d\xi' = +0.17$	$d\eta' = -0.49$	$\xi' = +\ 6.82$	$\eta' = +11.48$	$\eta' \sec\delta = +11.53$
$d\xi'' = +0.23$	$d\eta'' = -0.27$	$\xi'' = +16.15$	$\eta'' = -\ 4.98$	$\eta'' \sec\delta = -\ 5.00$
$d\xi''' = +0.62$	$d\eta''' = -0.23$	$\xi''' = +12.92$	$\eta''' = -10.62$	$\eta''' \sec\delta = -10.67$
$d\xi^{iv} = -0.05$	$d\eta^{iv} = 0.00$	$\xi^{iv} = -\ 2.07$	$\eta^{iv} = +\ 3.46$	$\eta^{iv} \sec\delta = +\ 3.48$

as in the catalogue.

621, 625. See note to 601.

622. The difference of declination between this star and θ was obtained on 1863, Feb. 7th, by four observations, which read thus:—

	r		″
	58.25		27.0
	58.13		28.1
	63.79		27.3
	63.84		27.8
Coincidence,	61.002	Mean,	27.55

Two observations of *double* distance give 29.″8 and 27.″8, employing the previous value of the coincidence. These received each half weight, and the value 27.″8 in the catalogue was obtained.

The right-ascension was derived from a chart by G. P. B. himself.

624. See note to 617. The value 16.″0 of the declination of this star has been used throughout, as a zero of reference for all the zones about θ Orionis. The brightest star of the trapezium (No. 628) was observed but 3 and 4 times in the zones, whereas 624 was 20 times.

631. A very faint star. The place depends upon diagrams of 1850 March 5, 1863 Jan. 31, Feb. 7.

633. See note to 617.

636. From various diagrams.

638. The observations of this star in the zones near θ and in those near c Orionis, give the following values.

	$a - a_0$	No. Obs.	$\delta - \delta_0$	No. Obs.
Zones near θ	$+0^s.69$	5	$+1172''.2$	4
" " c	$+0.75$	1	$+1169.2$	1

The former have had double weight.

640. See 617.

641. Two determinations. On April 7th, 1864, it was estimated to be 15″ distant from 635, in the direction of 652; this would give

$\alpha - \alpha_0$ +12.″6 $\delta - \delta_0$ +112.″7; the observation on page 42, Rev. Zone, 5,
gives +11.″2 +109.″7, and the mean of these two has been taken.

642. From a very carefully executed diagram, March 5, 1850; the definition was then most admirable.

647. The position of the star p'_1 page 42, is not given in the catalogue. Prof. Bond says, "It is probably an imperfect observation of ϕ, and should be rejected." It will be noticed that the position of page 42 is nearer in declination to the star ψ = No. 651. I think that when the declination was observed the latter fainter star was seen on the scale.

The position of 647 is derived from the following sources.

1. Its angle of position from the two stars 624 and 635 was measured 1863, Jan. 31st.

Angle from U″ (624) 51° 58′ (3 obs.)
" " O′ (635) 166 3 (3 obs.)

These results give us $\alpha - \alpha_0$ +23.″33 $\delta - \delta_0$ +38.″06

2. Its difference of AR. from θ Orionis (628) was measured on April 16, 1864; $\alpha - \alpha_0$ = 22.″65; of declination 38.″33 on the same date, and 37.″62 on Jan. 31, 1863.

3. Its difference of declination from 624 was found on Jan. 31, 1863, to be +21.″94, or +37.″94 from 628, employing as usual 16.″0 for 624.

4. The star occurs in Revision Zone 5, $\alpha - \alpha_0$ +23.″25 $\delta - \delta_0$ +37.″9.

The adopted value of $\alpha - \alpha_0$ is slightly erroneous, owing to the position from No. 1 being taken as +22.″17. The values might be thus combined:—

	$\alpha - \alpha_0$	$\delta - \delta_0$	wt.
From 624 and 635	+23.″33	+38.″06	1
" 628	+22.65	+38.33	1
" 628		+37.62	1
" 624		+37.94	1
" Revision Zone,	+23.25	+37.9	½
Mean,	+23.1	+37.98	

648. From diagrams 1859 March 10, 1860 Jan. 16, 1863 Feb. 7, with estimate.

651. This star is distant from 647 12.″4 in angle of position 38° 40′ by chart and diagrams; the employment of the former star's catalogue place, makes that of this one +30.″4 +47.″7; while from page 42 we derive +28.″5 +47.″9; the mean of which occurs in the catalogue. It is Struve's ad 75, and on G. P. Bond's authority I have identified it with W. C. Bond, No. 27, with some hesitation. I am inclined to think the *observation* of 27 is erroneous, but there can be no doubt that G. P. Bond's 651 was seen by the elder Bond.

652, 657, 663, 681. The coördinates of these stars with reference to 669 were determined in December, 1857. Observations of differences of AR. (by transits) were made on the 14th of that month, and of differences of declination on the 19th; which give (employing for 669 its place +63.″3 +100.″0 as in the catalogue)

	$\alpha - \alpha_0$	$\delta - \delta_0$
652	+29.″5	+171.″6
657	+37.1	+165.1

	$a-a_0$	$\delta-\delta_0$
	$''$	$''$
663	+52.1	+147.3
681	+85.8	+173.6

These have been combined with the observations of 1858, Jan. 14th (page 42).

As here given, the right ascensions have been corrected by 0.″3, since their combination with those just alluded to, and I have left the catalogue places as they stood before the correction was made.

In combining the declinations, those of 1857, Dec. 19th, have received double weight, because they are measures with the filar micrometer.

654. This is the variable Herschel 78; see Sect. IV. Part II. The right ascension of this star was determined by micrometrical measurement from θ Orionis, April 16th, 1864; its declination depends on numerous diagrams.

666, 677.

1857, Dec. 19th, 666 precedes 741 11.^{s}00; is south of it 85.″34 by micrometer.

1858, March 10th, 666 follows 570 10. 50

1864, March 24th, 666 follows 570 10. 15

1857, Dec. 19th, 677 precedes 685 1. 50

1858, March 10th, 677 follows 570 11. 75

1857, Dec. 19th, 677 is south of 666 by 5″ on the scale.

1864, March 24th, estimated position and distance of 677 from 666, 198° (an evident error for 108°) and 20″. Hence difference AR. Dec. +19.″0 −6.″2.

We have then

		$a-a_0$	wt.	$\delta-\delta_0$
		$''$		$''$
666.	1857, Dec. 19,	+60.9	1	−195.84
	1858, March 10,	+62.7	2	
	1864, March 24,	+57.4	3	
	Adopted,	+59.7		−195.8
677.	1857, Dec. 19,	+75.2	1	
	1858, March 10,	+81.5	1	
		+78.4	2	
	From 666,	+78.7	3	−201.4
	Adopted,	+78.6		−201.4

671. 1857, Dec. 14th. Angle of position and distance from 624, 118° 52′ and 83.″94; hence $a-a_0$ $=+68.''86$ and $\delta-\delta_0=-24.''52$; employing as usual −5.″0 and +16.″0 for the coördinates of 624.

1858, March 10th, page 43, we find $a-a_0=+73.''5$ $\delta-\delta_0=-21.''5$.

1864, April 16, by direct measurement, $a-a_0=+69.''09$ $\delta-\delta_0=-25.''67$.

Combining these values with the weights for AR. 3, 1, 3, and for Dec. 2, 1, 2, we obtain the numbers in the catalogue.

675. From a diagram of February 7, 1863.

676. The angle of position of this star from 671 was twice estimated as 113° 30′ and 109° 30′ by the con-

figuration relatively to the trapezium. This was combined with their *relative* situation, from page 43, to furnish the differences between them, +8.″9 and −3.″2, actually employed in the catalogue.

678. A chart-value for this star's declination +857″ was originally combined with the result here given from page 41. I suppose the latter to be the more correct.

681. See 652.

684. In identifying this star with Herschel 90, which otherwise is not in this catalogue, it has been assumed that the Δ NPD of that astronomer has the wrong sign.

686, 688. See 601. No. 686 is marked by G. P. B. as "not on chart"; and Herschel 91, with which it is probably identical, is most likely a condensation of nebulous matter, as O. Struve observes in his article, "Ueber das von Herrn W. Lassell in Malta aufgestellte Spiegelteleskop," in Volume III. of the "Mélanges Mathématiques et Astronomiques."

709. The difference of declination between this star and 741 was micrometrically determined on December 19, 1857, to be −25.″64; giving as $\delta - \delta_0$ −136.″14. This received a weight of 3 in combination with the value on page 43.

716. Double. Companion distant 4″, in position 50°; magnitude 10.2. See also Struve C. G. 607, Mensurae micrometricae, page 91.

717. Position derived from 715, by estimate of position and distance, 150° and 12″.

738. A very faint star; three charts give +219″ +274″, and the observation on page 42 +221.″4 +258.″2; the mean has been with some doubt inserted in the catalogues. It was seen 1862 March 19th, 1863 Jan. 19th, 1864 Jan. 25th, besides 1858 Jan. 14th.

759. "The place of this very faint star by observation in regular course, 1858, Jan. 14th, is confirmed by diagrams and observations Jan. 23, 1863"; the value thus obtained being $\alpha - \alpha_0 = +284''$, $\delta - \delta_0$ +108″.

762. Note in Equatorial day-book, April 7, 1864. "H. 115 was seen with difficulty, 18.ˢ0 preceding S‴ (No. 843), and 5″ north of it." The catalogue position is derived from this observation, and it appears not to have been previously visible at Cambridge, possibly from variability.

776. "Two readings from projection of place, on tracing from Herschel's chart, give

$\alpha - \alpha_0$	$\delta - \delta_0$
+360″	+385″
+366	+374
+363	+380."

G. P. B. has taken the mean here without the tenths, as the observations are but estimates.

778. This star occurs in Revision Zone 9, March 13, 1858, and was also observed on March 24, 1864, being found to precede 784 by 1.ˢ5, and by the scale to be 1′ 8″ north of it. Consequently we have

	s	′ ″
1858, March 13,	+24.51	−3 33.8
1864, March 24,	+24.39	−3 38.1
	+24.45	−3 36.0

779. By chart, 1862, March 28th.

797. By chart, April 7, 1864, found to be 57″ distant from 781 in the angle of position 110°; hence $\Delta\alpha$ +53.″8 $\Delta\delta$ −19.″5; the position being thus +427.″4 +176.″0, which latter combined with +169.″4 from Revision Zone 6, Jan. 20, 1858, gives +172.″7. The AR. from the Revision Zone appears to be 1ˢ in error.

798. Derived from 799 by the angle of position and distance.

801. This is a new star, and *variable*, near also to the variable, No. 822.

808. See pages 8, 40, 42, for the declination of this star.

840. From a chart, Feb. 7, 1863; see note on page 43.

863. For the right ascension of this star an observation, made April 7, 1864, 863—734 =30.ˢ70, was combined with that made on Jan. 20, 1858. The resulting $\alpha-\alpha_0$ are

1858, Jan. 20,	+45.ˢ58
1864, April 7,	+45. 21
	+45. 395

873. The position from zones 20, 21 has received a weight = 2; that from page 42 a weight = 1.

880. Double. Comp. 10ᵐ.2; 3″ dist.

888. The position from Zone I. page 45, has received a weight = 1; that from pages 23 and 40 combined, depending on 9 and 8 observations, a weight = 3.

932. Annular nebula.

949. From three estimates of distance, 8″, 9″, 10″, and one estimate of angle of position, 295°, with reference to 954.

957. Confirmed by a chart, March 28, 1862, which gives $\alpha-\alpha_0$ +1249″, $\delta-\delta_0$ +577″.

968. Nomenclature not in regular order, as the star was more conveniently referred to 934.

1004, 1035. Compared on Jan. 23, 1863, with 1028, as follows:—

	$\Delta\alpha$	$\Delta\delta$
1004—1028	—12.ˢ50	—1′12″
1035—1028	+ 2. 50	—1 26

1011. Declination from a diagram, March 28, 1862; right ascension from page 42.

1023, 1029, 1047, 1050. From chart, March 28, 1862.

1035. See 1004.

1044, 1058. From brief extension zones.

1051. Observed February 7, 1863; 1051—1015 = +15.ˢ00 —4′10.″, which gives

	$\alpha-\alpha_0$	$\delta-\delta_0$
	+2ᵐ 7.ˢ01=1905.″15	+538.″9
An extension zone,	1902.27	+540.0

The mean of these is given in the catalogue.

1053. Observed in two extension zones, and on page 57.

The three positions, reduced to 1864, give

	m s	′ ″
Extension Zone near θ Orionis,	+2 7.53	+19 11.8
" " " c Orionis,	7.79	10.0
Page 57,	7.29	8.9
Mean 1864.0,	+2 7.537	+19 10.23
Precession,	— 0.037	+ 1.54
1857.0,	+2 7.50	+19 11.8

1061. From an extension zone near θ Orionis.

1067. From two extension zones near θ Orionis: which give respectively

(1864.0)	$+2^m\ 13.^s08$	$-1.''1$
	$+2\ 13.37$	$+8.0$

The declination does not appear especially secure.

1077. From a brief extension zone, combined with p. 57; the result of the former (1864.0) is $+2^m\ 18.^s77$ $+25'50.''5$. The declination of p. 57 is $2'$ in error by my own observation.

(B.) NOTES TO THE DIFFERENTIAL CATALOGUE.

I have given in what follows some marginal notes, in Prof. Bond's handwriting, to a copy of Struve's Memoir. These are the statements mentioned as G. P. B's. B. denotes here W. C. B.

347. Herschel 15 was identified with this star by Struve, and there appears to be no other star to which it can be referred.

500. In this and other similar cases, Nos. 530, 536, 538, etc., I have made the comparison after referring Herschel's positions of these stars (Cape Results, page 12) to G. P. B.'s, by the mean difference of all the stars near c Orionis.

543. There is some large error in Liapunoff's determination of this star.

551. "B. 6, 8, probably identical" (Struve). Confirmed by G. P. B.

552. B. 7. Not seen by Struve. Supposed by him to be identical with 580 = H. 56.

567, 575. B. 10 and 11. Supposed by Struve to be identical = H. 57. G. P. B. says, "B. 10 and B. 11 are distinct stars, μ and r_1''." I have assumed that Herschel 57 is composed of 575 and 589, indistinctly seen. Struve suspects his H. 57, which is No. 589, to be variable. Probably in Struve's Memoir, page 93, Bond 10 should be called H. 51, and Bond 15, H. 57.

580. B. 42. Wrong sign in AR.

589. B. 15 called by Struve = Ll. 43 = H. 51, probably an error for H. 57. Lassell 43 is nearer to 595 than to 589; so also is W. C. B. 15.

612, 622, 618, 636. W. C. B. 16, 18, 19, 20, 24. Struve says, "None of these stars, situated in the immediate vicinity of the trapezium, have been recognized by me, and they are in no other catalogue. The existence of these five stars appears to me very doubtful." G. P. B. says, "All these stars I have often seen, excepting B. 20, unless that be v" (642).

647, 651. B. 26 and 27. Struve says, probably identical. He has since seen them separately at Malta. B. 27 is his ad 75. See note to 651 on p. 105.

666, 677. B. 30 and 34 Struve says, are identical = H. 81. He has since seen them separately, and B. 34 is his ad 81.

676, 686. B. 43, 44. Struve says, "probably identical = H. 91." B. 44 is probably the same object as H. 91, and Struve has since seen B. 43 at Malta; it is his ad 88.

684, 693, 703. B. 56, 58, 59. G. P. B. says, "H. in Struve's catalogue, is wrong both in AR. and Dec.

for these three stars 15″ to 20″. B. 60″. to 100″ in AR. Struve's catalogue has wrong sign for Dec. of H. 90, and it is missing on his chart."

732, 746. B. 63, 64. Struve supposes B. 72 and 74 identical with 63 and 64. G. P. B. says, "probably explained rightly."

759. B. 67. = H. 114. "B. is right. H. 20″ in error." G. P. B.

767. B. 69, 73. "Not in Harvard Zones." G. P. B. I have assumed the identity of 767 and B. 69.

B. 71. Not found in Harvard Zones.

Struve V = B. 77. G. P. B. says, "I doubt the existence of this star independently of t'_2 (H. 114)."

783, 795. B. 81, 84. Struve says, "identical." G. P. B. says, "These are two distinct stars, H. 122 and 125." There is, however, an error of 1′ in B.'s position.

797. B. 85. Struve says, "B. 85 does not exist in the heavens." Its position agrees very closely with the Harvard Zones.

801. B. 86. Struve says, "Probably identical with B. 78." These are distinct stars. G. P. B.

825. B. 88. Struve says, it "does not exist in the heavens." It is H. 131. Error of about 25″ in Herschel, of about 55″ in B.

864. B. 92. Error of 40″ in Herschel, 80″ in Bond.

SECTION IV. PART I.

ON THE MAGNITUDES OF THE STARS CONTAINED IN THE PRECEDING CATALOGUE.

THE difficulties which attend the proper investigation of this subject are very great; and it was Prof. Bond's desire and intention, which he only gave up very reluctantly, to submit the whole matter to a very thorough discussion. But time and strength failed him to accomplish this object to his own satisfaction; and the preliminary results for magnitude which precede, are those which are given in the papers entrusted to my care. They accompany the stars' positions, as well in the partial catalogues which were subjected to his own revision, as in the copy of the final catalogue which he did not see completed, but which was made according to his own directions.

I have, however, omitted from the printed catalogue the magnitudes of the stars *a* and *d* of the trapezium, as they were manifestly given too faint; and, as will be seen by the comparisons which come a few pages later, it would perhaps have been better to do the same with a few others.

I shall begin by giving an account of the method of sequences which Prof. Bond employed, to connect the magnitudes of neighboring stars, and of the method in which I myself reduced these in 1864, under his immediate direction.

The sequences are thus arranged in the original manuscripts.

The stars' names are written in the order of their magnitude.

Upon the *same* line are given stars which differ but slightly.

Upon *consecutive* lines those which differ more largely.

And in some cases it is stated that the difference between two stars on consecutive lines is larger than usual.

The following method was suggested by myself for the reduction of such observations, and carried out.

The slight difference between two stars on the same line was called 1 grade; the greater difference between two stars on consecutive lines, 2 grades, and where noted as especially large, 3 grades. To the brightest star observed in one group or sequence of stars was assigned the magnitude 0 grades; to the next brightest, the number of grades by which it differed from the previous one, and so on.

For example, I give below the first sequence observed, 1858, Jan. 4th.

K, much the brightest.
F. D. A.
H. G. I.
C. E.
B.
e_1
f_3

I inferred from this, that if g represented the number of grades by which each star differed from K, the value of g for the star F would be 3, as the name F occurs first on the second line, and K is called much brighter. Again, g for D would be 4, for A 5, as these occur in succession on the same line; for H, it would be 7, and so forth, as in the following reduction.

Half grades occur where the difference between two stars is considered doubtful.

It now becomes necessary to ascertain what these numbers g represent. In the first place the stars selected as the brightest in each sequence being not identical in different cases, for the magnitudes on any ordinary scale each sequence will be represented for any case by

$$m = x + yg$$

Where x is a number varying from one sequence to another, denoting the magnitude of the star for which $g = 0$, or the brightest star in the sequence; while y is the ratio of a grade to a magnitude, which may or may not vary from one sequence to another. I assumed, by Prof. Bond's direction, that y was constant for the whole series of observations, and equal to 0.25; a previous determination of the value of y from the materials contained below, by the method of least squares, having shown that in general the variation of y from 0.25 exceeded but slightly its probable error.

In other words, the assumption was that a variation of $0.^{m}25$ on G. P. Bond's scale was equivalent to the difference between the magnitudes of two stars upon the same line; that the difference between two stars in succeeding lines was equal to $0.^{m}50$ (or $0.^{m}75$ in case of very large differences), and these assumptions best harmonized the observations of sequences with those of magnitudes made directly.

This process has in general given results corresponding quite nearly to the magnitudes estimated directly, though there are one or two cases of discrepancy among the brighter stars.

The following tables give the sequences, reduced in the first place to grades, and secondly to magnitudes; g being the grade, m the observed magnitude.

1858, Jan. 4.

Star's Name.	g	m	m′
K	0	9.0	8.8
F	3	10.0	9.5
D	4	10.0	9.8
A	5	10.3	10.0
H	7	11.0	10.5
G	8	11.3	10.8
I	9	11.0	11.0
C	11	11.0	11.5
E	12	11.0	11.8
B	14	12.0	12.3
e_1	16	12.3	12.8
f_3	18		13.3

1858, Jan. 7.

Star's Name.	g	m	m′
K	0	7.3	8.1
L	3	9.0	8.9
Q	5	9.0	9.4
R	6	9.3	9.6
M	7.5	10.0	10.0
S	9.5	11.3	10.5
N	10.5	11.3	10.7
V	12.5	12.0	11.2
P	13.5	11.3	11.5
O	14.5	12.0	11.7
T	16.5	12.0	12.2
U	17.5	12.0	12.5

1858, Jan. 20, Sequence I.

Star's Name.	g	m	m′
C′	0	10.5	10.3
F′	2	10.5	10.8
H′	3	12.0	11.0
E′	5	12.0	11.5
D′	6	11.5	11.8
A′	8	12.2	12.3
B′	9	12.3	12.5
G′	11	12.2	13.0

1858, Jan. 20, Sequence II.

Star's Name.	g	m	m′
M′	0	6.3	8.2
T′	2	7.3	8.7
P′	3	8.3	8.9
K′	4	9.0	9.2
L′	6	10.0	9.7
Q′	8	11.0	10.2
F′	9	10.5	10.4
H′	10	12.0	10.7
E′	12	12.0	11.2
O′	13	12.0	11.4
I′	14	12.0	11.7
G′	16	12.2	12.2
S′	17	12.0	12.4
R′	19	13.0	12.9

1858, Feb. 8, Sequence I.

Star's Name.	g	m	m′
C″	0	11.0	10.8
B″	1	11.0	11.1
I″	3	11.7	11.6
A″	4	12.0	11.8
E″	5	12.3	12.1
G″	7	12.3	12.6
D″	8	13.0	12.8
H″	10	13.0	13.3
F″	11	13.3	13.6

1858, Feb. 8, Sequence II.

Star's Name.	g	m	m′
R″	0	9.0	9.7
N″	1	9.3	10.0
M″	3	10.3	10.5
I″	5	11.3	11.0
P″	6	11.3	11.2
K″	8	12.0	11.7
O″	10	12.3	12.2
L″	11	13.0	12.5
Q″	12	13.0	12.7

1858, March 20, Sequence I.

Star's Name.	g	m	m′
S‴	0	7.3	7.9
L‴	1	8.3	8.2
I‴	4	9.3	8.9
C‴	5	9.0	9.2
H‴	6	9.3	9.4
D‴	8	10.3	9.9
M‴	9	11.0	10.2

Sequence II. of this date consists of but two stars, and cannot therefore give results of any value; the numbers are

Star's Name.	g	m
V‴	0	7.0
S‴	1	7.3

1858, March 20, Sequence III.

Star's Name.	g	m	m′
F‴	0	11.3	11.2
U‴	1	12.0	11.5
R‴	2	12.0	11.7
T‴	3	11.3	12.0
O‴	4	12.0	12.2

1858, March 20, Sequence I. Continued.

Star's Name.	g	m	m′
B′′′	10	11.0	10.4
E′′′	12	11.0	10.9
A′′′	13	11.0	11.2
G′′′	15	12.0	11.7
F′′′	17	12.0	12.2
O′′′	19	12.0	12.7
P′′′	20	12.3	12.9
N′′′	22	13.3	13.4

All the preceding sequences were reduced by the formula $m' = m'^{\circ} + 0.25\,g$ where m'° is the value of m' for $g = 0$; and the sum of the deviations $m' - m$ is 0, except so far as owing to decimals, beyond the first, of a magnitude. The values of m were always estimated on the same night, for the foregoing sequences.

The sequences which follow differ nowise from those that precede, excepting that the value of m° below were taken from a preliminary catalogue of magnitudes, obtained from the Zones of Section I.; and were not estimated the same night that the values of g were.

1858, Jan. 8, Sequence I.				**1858, Jan. 8, Sequence II.**				**1858, Jan. 8, Sequence III.**			
Star's Name.	g	m°	m′	Star's Name.	g	m°	m′	Star's Name.	g	m°	m′
K	0.	9.3	10.1	L	0.	10.3	9.9	M	0.	11.3	11.7
L	2.0	10.3	10.6	Y	0.5	10.0	10.1	Z	1.	12.3	11.9
Q	2.5	10.3	10.7	W	1.5	10.3	10.3	N	2.	12.0	12.2
M	4.5	11.3	11.2	Q	2.5	10.3	10.6	B	2.	12.3	12.2
R	5.5	12.0	11.4								
N	7.5	12.0	11.9								
S	7.5	12.3	11.9								
V	8.5	13.0	12.2								
O	10.5	13.3	12.7								
P	11.5	12.3	12.9								
T	12.5	13.3	13.2								
U	14.5	13.0	13.7								

1858, Jan. 12.

Star's Name.	g	m°	m′	Star's Name.	g	m°	m′	Star's Name.	g	m°	m′
K	0	9.3	7.7	M	13	11.3	10.9	N	22	12.0	13.2
W	2	10.3	8.2	G	14	11.3	11.2	V	23	13.0	13.4
Y	3	10.0	8.4	I	15	11.0	11.4	S	24	12.3	13.7
L	4	10.3	8.7	H	16	12.0	11.7	T	26	13.3	14.2
Q	6	10.3	9.2	C	17	11.3	11.9	O	27	13.3	14.4
F	7	10.3	9.4	E	18	12.0	12.2	P	28	12.3	14.7
D	8	11.0	9.7	Z	20	12.3	12.7	U	29	14.0	14.9
A	10	10.3	10.2	B	21	12.3	12.9	X	30	13.0	15.2
R	11	12.0	10.4								

1858, Feb. 12,* Sequence I.

Star's Name.	g	m°	m′
Y″	0.	6.0	7.6
V″	1.0	6.0	7.8
X″	5.0	7.0	8.8
T″	3.0	8.0	8.3
R″	7.0	10.3	9.3
N″	9.0	10.0	9.8
U″	9.5	10.3	9.9
M″	11.5	10.3	10.4
P″	11.5	10.3	10.4
S″	14.5	12.0	11.2
Z″	14.0	13.0	11.1
Q″	13.5	12.0	10.9
W″	15.5	12.0	11.4

1858, Feb. 12, Sequence II.

Star's Name.	g	m°	m′
B″B″	0.	8.0	9.9
E″E″	2.0	10.3	10.4
F″F″	4.0	12.0	10.9
C″C″	6.0	11.0	11.4
I″I″	6.0	11.0	11.4
H″H″	8.5	12.3	12.0
G″G″	8.0	13.0	11.9
A″A″	10.5	12.3	12.5
D″D″	11.5	13.0	12.7

1858, Feb. 12, Sequence III.

Star's Name.	g	m°	m′
E″E″	0.	10.3	10.6
L″L″	0.	10.0	10.6
O″O″	4.	12.0	11.6
C″C″	3.	11.0	11.4
N″N″	3.	12.0	11.4
M″M″	2.	12.0	11.1
H″H″	6.	12.3	12.1
I″I″	6.	11.0	12.1
P″P″	6.	12.3	12.1
K″K″	8.	13.0	12.6

The numbers m' in the previous discussion were brought together for each star included in it, by myself; and on the sheet containing them is this remark, written in red ink, to call attention to it, by Prof. Bond himself. "These results have been adopted for the final catalogue," under the heading, in my own hand, "Final magnitudes (from Sequences) of Prof. Bond's Scale."

For other stars the adopted magnitudes on Prof. Bond's scale were those directly estimated by himself, and given in the preceding pages, and to a slight extent in the following cases. In order to reduce these magnitudes to greater accordance with the common (Argelander's) scale, he observed a number of stars in the latter author's Sternverzeichniss, near the equator, and in the constellation Orion and vicinity, as near as possible to the nebula; and these observations now follow. In these I acted as his amanuensis, and selected for him, as he sat at the telescope, a region which could be easily found by neighboring stars, and which was north of — 2° declination; consequently Prof. Bond was unaware of the magnitudes of the catalogue stars, to which I directed him as they came in their order of AR., and could make an estimate very free from previous knowledge. In connection with this, he reobserved some stars in his own nebula-catalogue, thus nearly diminishing any effect from the altitude of the nebula, or the state of the atmosphere.

* Some of these Sequences are written in inverted order in the journal of observations, and are thus a little confused. They have been printed as in the sheets which passed under Prof. Bond's inspection.

1864, March 18.

Observations began at about 7^h Sid. ended at $8^h\ 15^m$. The region of Sternverzeichniss compared, was near AR. of *6′* Orionis, and about 4° higher in altitude.

Stars near ε Orionis compared with magnitudes in Sternverzeichniss.

G. P. B's Mag.	Dec. — 1° No. in Sternv.	Sternv. Mag.
8.9	973	9.3
7.8	987	8.1
11	989	9.4
8.9	990	8.4
9	991	9.0
10.11	993	9.5
6 about	1004	6.2
9	1006	9.3
10 faint	1011	9.5
9.10	1029	9.5
9	1040	8.9
10.11	1041	9.5
8.9	1055	9.2

Zone VI. of Preceding Book, near c Orionis.

Est. Mag.	Star.
9	B
12	D
11	E
11.12	F
12	G
8.9	H
8 double	I as one star.
10	K
8	L
9	M
7.8	N
10.11	O
10.11	P
10.11	Q
12	S
10.11	T

Stars near ι Orionis. Zone XII.

Est. Mag.	Star.
10	F
9.10	L
12	M
10	P
9.10	Q
8.9	R
7.8	S

Stars near ι Orionis. Zone XII.

Est. Mag.	Star.
7	T
8	U
5	W
9	X
11	Y
12	AA
7.8	BB
10.11	CC
7.8	DD
12	EE

Telescope set back upon Sternv. near δ Orionis, 4° higher altitude than nebula.

Est. Mag. G. P. B.	Dec. — 0° Sternv. No.	Mag. fr. Sternv.
8	982	8.0
9 9.10	991	9.3
10	995	9.4
9 9.10	997	9.1
9.10	998	9.3
9	1002	8.9
8	1007	8.6
10	1010	9.5
10	1014	9.5
9	1016	9.3
8.9	1018	9.2
9.10	1022	9.4
9	1028	8.9
10	1030	9.5
9.10	1040	9.5
9.10	1043	9.5
9.10	1048	9.5
9.10	1050	9.5
10	1053	9.5
10 } 10 }	1055	9.5
9.10	1066	9.4
10	1069	9.5
9.10	1071	9.5
9.10	1072	9.5
10	1078	9.5

Orion Zones (near nebula). Old Zones near 6′ Orionis.

Est. Mag.	Star.	1858. Mag.
9.10	A	10
12	B	12

1864, March 18. Continued.

Orion Zones (near nebula). Old Zones near θ' Orionis.

Est. Mag.	Star.	1858. Mag.	Est. Mag.	Star.	1858. Mag.
10.11	C	11	10.11	I''	11.12
10	D	10	11.12	K''	12
10	E	11	12.13	L''	13
9.10 9	F	10	10	M'''	10.11 11
10	G	11.12	9	N''	9.10
10	H	11	10	P''	11.12
10	I	11.12	12	Q''	12.13
8	K	8	8	R''	9
9.10	L	9.10	7	Y''	6
9.10	M	10	13	Z''	13 12.13
11.12	N	11.12	12.13	A''A''	12.13
13	O	12	8 7.8	B''B''	7.8 8
12.13	P	11.12	9.10	C''C''	11
9	Q	9.10	12.13	D''D''	12.13
9.10	R	10	9.10	F''F''	
12	S	11.12	8.9	E''E''	9.10
12	T	12	10	I''I''	11.12
13	U	12	10	N''N''	11.12
7.8	T'	7.8	11	O''O'	11.12
9	A'_{-1}	10	8.9 9	C'''	9
12	A'	12 12.13	9.10 10	D'''	10.11
12.13	B'	12.13	10	E'''	11
10.11	C'	10.11 11	12	F'''	12
10.11	D'	11.12 12	11	G'''	12
10.11	E'	12	9	H'''	9.10
10	F'	10.11 11	9	I'''	9.10
12	G'	12 12.13	13.14	K'''	13
10.11	H'	12	8.9 8	L'''	8.9
10.11	O'	12	10.11	M'''	10.11 11
9.10	Q'	11	13	N'''	13.14
11	R'	13	11.12	O'''	12
7.8	T'	7.8	12.13	P'''	12.13
11	U'	12	11	R'''	12
9	W'	10	7.8 8	S'''	7.8
12	E''	12.13	11.12	T'''	11.12
13.14	F''	13.14	12	U'''	12
13	G''	12.13	7.8	V'''	7
12.13	H''	13	13	W''	13

Left off on account of clouds and a south-westerly haze gathering, which must, at the low altitude of the nebula, have had a very sensible effect on brightness of the stars. Had intended to return again to Sternverzeichniss. The strong moonlight, and latterly the suspicion of clouds and haze, diminished confidence in above estimates. Power used, 141. Illuminating lamp on, and other circumstances as usual when magnitudes have been observed.

1864, March 21.

Magnitudes from Sternverzeichniss. Bright moonlight, quite clear, definition not good; estimates made with a good deal of hesitation.

Est. Mag.	No. fr. Sternv. +2°	Mag. fr. Sternv.	Est. Mag.	No. fr. Sternv. +2°	Mag. fr. Sternv.
9	804	9.0	9.10	788	9.5
9.10	805	9.3	9	786	8.8
11	812*	9.3	9.10	784	9.5
9	801	9.2	10	780	9.5
6.7	800	5.0	9	774	9.3
10 }	795	9.5	7	773	7.2
9.10 }			9	772	9.2
9	799	9.0	8.9	767	9.3
10.11 }	790	9.2	9.10 9 }	766	8.8
10 }			9.10 9 }		

The stars observed were at about the altitude of the nebula.

This series from $7^h\ 20^m$ to $7^h\ 37^m$ Sid. time.

Stars in Nebula.

Mag.	Star of Old Or. Zone.	Mag. 1858.
11	E′	12
9.10	F′	10.10 11
9.10	C′	10.11 11
10.11	D′	11.12 12
12	G′	12 12.13
10	H′	12
11	O′	12
9.10	Q′	11
11.12	R′	13
10	Y′	11.12
11	N	11.12
12	P	11.12
9	M	10
8.9	L	9.10
7.8 8	K	8
10	I	11.12

These are the stars of the Orion Zones themselves; we now go on to Zones about c Orionis.

Zone I.

Star.	Mag.
M	8.9
N	12
O	10
P	8
Q	16
R	10.11
S	10

Zone II.

Star.	Mag.
H	9
G	7.8
F	8.9
E	7.8
D	12
C	13
B	11
A	17 barely vis.

Zone IV.

Star.	Mag.
K	8 7.8
L	11
M	8.9
N	9.10
O	9
P	12
Q	11
R	13
S	12
T	10.11
U	10.11
V	13.14

Zone V.

Star.	Mag.
L	9.10
M	7.8 8
N	7.8
O	13
P	9.10

* Another star (10th) is 2′ N.

1864, March 21. Continued.

Zone V.

Star.	Mag.
Q	7.8
R	7
S	8

Zone VI.

Star.	Mag.
D	11.12
E	11
F	11.12
G	11
H	8.9
I	8 7.8
K	10
L	8.9
M	9
N	7.8
O	11
P	10
Q	10

Zone VIII.

Star.	Mag.
F	12
G	11
H	11
K	11
L	not seen.
M	12
N	11.12
O	13
P	13
Q	10

Zone IX.

Star.	Mag.
I	10.11
K	7.8
L	13.14
M	12
N	13
O	12
P	10
Q	11

Zones near θ Orionis.

Star.	Mag.	Mag. 1858.
M‴	10.11	10.11 11
N″	8.9	9.10
P″	9	11.12

Zones near θ Orionis.

Star.	Mag.	Mag. 1858.
R″	8	9
B″B″	7.8	7.8 8
C″C″	9.10	11
E″E″	8	9.10
F″F″	10	
I″I″	10.11	11.12
N″N″	10.11	11.12
O″O″	10.11 11	11.12
Q‴	9.10	
D‴	9.10	10.11
E‴	9.10	11
F‴	11.12	12
G‴	11.12	12
H‴	9	9.10
I‴	8.9	9.10
K‴	13	13
L‴	8	8.9

Zone X. near ι Orionis.

Star.	Mag.
L	11
M	9
N	8
O	10.11
P	9
Q	9
R	8
S	14
T	9
U	14 just visible.
W	11

Zone XII.

Star.	Mag.
Comp. ι Or.	9
U	8
T	6
S	7
R	8.9
Q	9
P	10
O	not visible.
N	" "
M	12
L	9.10
K	13
I	14
F	9
E	13

1864, March 21. Continued.

Zone XIV.

Star.	Mag.
G	not seen.
H	" "
N	11
O	9
T	7
V	9.10
W	10

Zone XV.

Star.	Mag.
S	8
T	10
U	9.10
V	10
W	12
Q	10
P	10
O	9
N	10.11
K	9

Zone XVI.

Star.	Mag.
K	8
I	9.10
G	8.9
H	barely visible.
F	9
E	12
D	12
C	8.9
B	13
A	9

Zone XVII.

Star.	Mag.
V	8
W	10.11
X	9.10
Y	not seen.
Z	9
AA	9
BB	10
CC	9.10
DD	9.10
EE	10
FF	8
GG	7.8

Zone XVIII.

Star.	Mag.
A	9.10
B	8
C	10.11
D	10
E	14
F	not seen.
G	10
H	10
I	11.12
K	10
L	8 7.8
M	9

Ended this set $8^h\ 47^m$.

Stars about 12′ n. of ε Orionis.

Sternv. of Arg. —1°	Est. Mag.	Mag. fr. Sternv.
965	8	8.7
966	9	9.0
964	9	9.3
958	9.10	9.5
957	9	9.3
953	7	8.3
949	7	7.7
942	8	9.1
940	10	9.5
936	9	9.2
924	10	9.5
974	8	9.0
981	10.11*	9.5
986	9.10	9.5
988	7.8	8.6
987	7	8.1
989	10.11	9.4
990	7.8	8.4
991	8	9.0
993	9.10	9.5
992	9	9.5

Finished at $9^h\ 2^m$.

Zones near θ Orionis, $9^h\ 12^m$ Sid. T.

Star.	Est. Mag.	Mag. 1858.
M‴	9.10	10.11 11
R″	8. 7.8	9
F″F″	10	
I″I″	10	11.12
N″N″	10.11	11.12

* Two stars of this magnitude make up —1° 981.

1864, March 21. Continued.

Star.	Est. Mag.	Mag. 1858.	Star.	Est. Mag.	Mag. 1858.
Y′	10	11.12	T′	7	7.8
F′	10	10.11 11	K	8 7.8	8
H‴	9 8.9	9.10	M	10	10
I‴	8.9	9.10	Q	9	9.10
C‴	8	9	I	9.10	11.12
D‴	10 9.10	10.11	Star* ½ (c^1 to c^2 Or.)	8 7.8	
E‴	10 9.10	11			

Continued clear (perfectly so to appearance), with bright moonlight, moon nearly full.

Stars ordinarily estimated as 13 are near limit of vision to-night at low altitude of nebula. Those noted as 13—14 and 14 are the faintest visible to-night.

* This is N.

REMARKS.

The plan proposed in the foregoing observations was to observe a certain number of stars in Argelander's Sternverzeichniss (without knowledge of his magnitudes), at near the altitudes of the nebula, and then to observe a few in each zone of the nebula, afterwards returning to the Sternverzeichniss. The sky was so full of light, that there was not much confidence felt in the estimate of magnitudes; but I should anticipate that the 12th magnitudes would be pretty consistent, as the illuminated scale was employed as usual. The fainter stars, called 15th, 16th, and 17th, in zones about θ, c and ι Orionis, were invisible, probably also the 14th.

N. B. The identity of system of magnitudes here, as well as in zones about θ, c and ι Orionis, has not been intentionally deviated from. G. P. B.

The previous observations were projected on squared paper, under Professor Bond's direction; the paper containing the curves states that $15^m.0$ was adopted as the limit of usual visibility on Argelander's scale, with the 14 inch refractor; this corresponds to the 17.18 magnitude on G. P. B.'s scale.

The notation 10.11 11 and the like denotes that the magnitude was estimated as between 10.11 and 11, or other numbers so used. See the bottom of page 122. The following table and introduction were written out for press, Nov. 29, 1864; and its results have been used in the subsequent reductions of magnitudes in the partial catalogues, which were completed in Prof. Bond's lifetime. I do not know why the column "G. P. B. in 1857–8" was not used at all, and I have therefore not ventured to use them myself, but have left the magnitudes as they stand in the partial catalogues, excepting when two or more observations of the same star were combined.

Results of Preliminary Reduction of Observations upon the Magnitudes of Stars contained in the Zones about θ′, c *and* ι *Orionis, compared with Argelander's Sternverzeichniss, for the purpose of ascertaining the correction to Argelander's Scale.*

The present reduction is not regarded as final, although probably correct, considerably within the limits of the errors of the determination, and will be used for engraving the Star Charts.

G. P. B. in 1864.	Arg. Stv.	G. P. B. in 1857–58.	G. P. B. in 1864.	Arg. Stv.	G. P. B. in 1857–8.	G. P. B. in 1864.	Arg. Stv.	G. P. B. in 1857–58.	G. P. B. in 1864.	Arg. Stv.	G. P. B. in 1857–58.
m	m	m	m	m	m	m	m	m	m	m	m
7.0	7.8	7.1	10.0	9.7	11.0	13.0	11.5	12.8	16.0	13.9	15.1
.1	.9	.2	.1	.8	.1	.1	.6	.8	.1	14.0	.2
.2	8.0	.3	.2	.8	.1	.2	.7	.8	.2	.1	.4
.3	.1	.4	.3	.9	.2	.3	.7	.9	.3	.2	.5
.4	.2	.5	.4	.9	.3	.4	.8	.9	.4	.3	.7
7.5	8.3	7.7	10.5	10.0	11.4	13.5	11.9	13.0	16.5	14.3	15.9
.6	.3	.8	.6	.0	.5	.6	12.0	.0	.6	.4	16.0
.7	.4	.9	.7	.1	.6	.7	.1	.1	.7	.5	.2
.8	.5	8.0	.8	.1	.6	.8	.1	.1	.8	.6	.4
.9	.6	.1	.9	.2	.7	.9	.2	.2	.9	.7	.5
8.0	8.6	8.3	11.0	10.2	11.8	14.0	12.3	13.3	17.0	14.8	16.7
.1	.7	.4	.1	.3	.8	.1	.4	.3	.1	.8	.9
.2	.7	.6	.2	.3	.9	.2	.5	.3	.2	.9	17.1
.3	.8	.7	.3	.4	.9	.3	.5	.4	.3	15.0	.3
.4	.9	.8	.4	.5	12.0	.4	.6	.5	.4	.1	.4
8.5	8.9	9.0	11.5	10.5	12.0	14.5	12.7	13.6	17.5	15.2	17.6
.6	9.0	.2	.6	.6	.1	.6	.8	.6	.6	.3	.9
.7	.0	.3	.7	.7	.2	.7	.8	.7	.7	.4	18.1
.8	.1	.5	.8	.7	.2	.8	.9	.7	.8	.4	.2
.9	.1	.6	.9	.8	.3	.9	13.0	.8	.9	.5	18.4
9.0	9.2	9.7	12.0	10.8	12.3	15.0	13.1	13.9	18.0	15.6	18.6
.1	.2	.9	.1	.9	.3	.1	.2	14.0			
.2	.3	10.0	.2	11.0	.4	.2	.2	.1			
.3	.3	.1	.3	.0	.4	.3	.3	.2			
.4	.4	.3	.4	.1	.5	.4	.4	.3			
9.5	9.4	10.4	12.5	11.2	12.5	15.5	13.5	14.4			
.6	.4	.5	.6	.3	.6	.6	.6	.5			
.7	.5	.6	.7	.3	.6	.7	.7	.7			
.8	.6	.7	.8	.4	.7	.8	.7	.8			
.9	.6	.9	.9	.5	.7	.9	.8	.9			

Reduction to decimals of Mag. was effected, as in the following example,—

9.10 = 9.3	10.9 = 9.7
9.10 9 = 9.2	10.9 10 = 9.8
9.10 10 = 9.5	10.9 9 = 9.5

The magnitudes given on pages 1–22, I suppose to be only provisional; but as they sometimes differ rather more than is desirable from the values in the General Catalogue, especially in the case of the brighter stars, I give them below, in all cases where a star occurs in the zones near θ Orionis, as brighter than the 10th magnitude. The values here given appear for these stars to be the best.

No. of Star in General Catalogue.	Letter. G. P. B.	Magnitude. G. P. B.	Magnitude Reduced.*	Magnitude. Gen. Cat.
81	A'_{-1}	10.8	9.3	9.7
246	C‴	8.3	8.6	9.3
467	L‴	8.3	8.6	8.7
554	K	9.3	9.0	9.0
570	R″	9.9	9.2	9.4
624	U″=L′	9.7	9.2	9.6
628	M′	6.		
685	Y″	6.		8.3
708	B″B″	8.	8.5	9.6
734	T′	8.	8.5	9.0
845	S‴	7.1	7.8	8.6
850	W′	9.9	9.2	9.9
907	V‴	6.5		7.8
926	L″L″	9.9	9.2	10.0

* By table on previous page.

SECTION IV. PART II.

ON THE VARIABLE STARS IN THE GREAT NEBULA OF ORION.

It is proper to begin this Part with the observations relating to the star Herschel 78 (No. 654 of the Catalogue in the present work), which I have been able to collect from the manuscripts of W. C. Bond and G. P. Bond.

The variability of this star was alluded to in Prof. W. C. Bond's Memoir on the Nebula, read before the American Academy, in April, 1848, and published, as it appears, in that year. The statement is the following. "His [Sir John Herschel's] star No. 75 is well seen, but No. 78, to which the same magnitude is given in his table, has not been seen steadily by me. Indeed, the observations upon it at different times have been so contradictory, that I could only account for the discrepancies by supposing it to be a variable star of short period."

The foot-notes to the observations are the editor's; so also the remarks in brackets.

1848, Feb. 26. Sir J. H.'s star just following β in trapezium, we only see by glimpses. W. C. B.

1848, March 22. It was uncommonly fine seeing just as twilight ended; I saw many stars in the neighborhood of the trapezium, and Sir John Herschel's star,* the sixth star, stood out boldly. W. C. B.

1850, Feb. 1, 7 P. M. The star 78 of Sir John Herschel's Catalogue is steadily seen considerably above the line joining the two following stars of the trapezium, making an angle of about 15° or 20°. W. C. B.

1850, Feb. 7, 9.30 P. M. Looked at nebula in Orion; very fine vision; the small stars easily seen, but the Herschel star has disappeared: x a very small star is seen as well as Lassell's stars, steadily, but of "h" there is not a suspicion; under the circumstances there can be no doubt that it is variable with a short period. I remember having seen it in January (or December), when it was so distinct that I never once thought of its being the star we had tried so hard to see when preparing the engraving, until father told me it was the "h" star. G. P. B. [A diagram accompanies this; $x = 636$ G. P. B. $h =$ Herschel 78.]

1850, Feb. 11, 12, 15, 16. Star † not seen, but the seeing is bad. G. P. B.

1850, Feb. 13. The variable star near θ' Orionis (78) was not visible to-night. Good seeing. W. C. B.

1850, March 2. Bad seeing, but the variable star is now visible whenever the air is not violently disturbed, and it is easier to see than the star n ‡ s. of a. For a fortnight past the moon has been too bright for certainly deciding whether the star had reappeared or not, but it was only within a few days that we have suspected its reappearance. G. P. B.

* I suppose this note refers to the known sixth star of the trapezium.

† This is written on the same page, and immediately below the last note.

‡ This appears to be W. C. B. 24 = No. 636 of the present work.

1850, March 4. Good seeing. Saw the Herschel star No. 78 very distinctly; it can even be seen when looking steadily at it. It is now brighter than No. 24, a little less than No. 26, and about equal to No. 27 (which seems to have been estimated too low). The new star near No. 23 * is seen easily. I should call No. 78 = 17 Mag. G. P. B.

1850, March 5. I can scarcely recall an evening of more admirable definition; "*h*" † is brighter, I think, than on last evening. It is now decidedly brighter than *n*, and almost equal to *m*, but very much brighter than *p*, which is a very small star. It is not easy to say whether "*h*" or *m* is brightest. One new star is seen to-night near *m* and *n*; its place is indicated by *b*, and there seem to be others near it. Another new star is seen at *a*. G. P. B.

1850, March 8. Bad seeing. Can see No. 78, but I fancy it fainter than when last seen, as it is less than *m*; but the air is disturbed. G. P. B.

1850, March 9. Pretty good definition, but through clouds, consequently it is difficult to see the small stars. No. 78 is now *decidedly* less than *m*, less even than *n*, at least it is not so easy to see as *n*. G. P. B.

1850, March 15. Variable star in Orion still in sight. G. P. B.

1850, March 29. Cannot see No. 78, but when I looked it was getting low; could see *m*, and just see *n*, but not *p*. (See March 5.) G. P. B.

1850, Oct. 31. "*h*" star not visible, seeing not good.

1850, Dec. 27, 11 P. M. The variable star in Orion not visible. It cannot be so bright as either of the stars ‡ 1, 2, 3, or 4. Vision hardly more than tolerably good. G. P. B.

1851, Jan. 23, 10 P. M. Looked at Orion; fine seeing; Herschel star *not* visible. G. P. B.

1851, Feb. 3. Looked at Orion; very fine seeing, and the "*h*" star (variable) certainly not to be seen, though we easily see the Lassell stars, etc. G. P. B.

1852, Feb. 16, 9 P. M. Looked at Orion nebula. Saw a star 18.19 Mag. in place of *x*. Sir J. Herschel's variable star? [G. P. B. who wrote this note, afterwards found this to be the star.]

1857, Feb. 23 § [with diagram]. While examining *a*, saw *k* by indirect vision; could not see it for a long time even by putting *a* and *e* out of the field. Saw it however at last, and have no doubt of its existence. There seemed now and then to be something near *x*. S. C.

1858, March 10. A diagram, drawn by G. P. B. gives the position and name of "*h*." See next note.

1859, Feb. 23. At the time when *z* ‖ was visible, though faint, no trace of a star could be discerned at "*h*," where, on March 10, 1858, was a star brighter than *z*. Seen also in Feb. 1857, by Mr. Coolidge. G. P. B.

1859, March 4. "*h*" not seen. G. P. B.

1859, March 5.¶ There are three stars near the trapezium; *a* is at $2\frac{2}{3}$ the distance (3^d-2^d) east, and at

* The references here are to W. C. B.'s Catalogue, except for H. 78.

† By the diagram accompanying this observation, *p* is easily recognized as W. C. B. 24 = G. P. B. 636; *m* and *n* as 647 and 651, H. 75 and Str. ad 75; the other two new stars here indicated are 631 and 642. The name "*h*" is used throughout for Herschel 78.

‡ 1 = 647, 2 = 651, 3 = 612, 4 = 618, from the diagram in the Observing Book.

§ This note by the late Major Sidney Coolidge, U. S. A., a very careful and accurate observer, is a valuable one. The diagram which he gives makes *k* = Struve's II. (see a few pages on), and *x* = Herschel 78.

‖ G. P. B. 636, is *z* by diagram; "*h*" as usual is Herschel 78.

¶ The diagram shows that the expression "distance (3d—2d)" means the distance from *d* to *c* of the trapezium; by this and the diagram itself, *a* is possibly Herschel 78. The 1st star is *a* of the trapezium; and *m* is the point where the line from *a* of the trapezium to the unknown *b*, meets the line from *d* to *c* of the trapezium. With these explanations Mr. Coolidge's *c* is found to be 622, Struve's II. and his *b* 647 = Herschel 75.

$\frac{2}{3}$ the distance north of it; $m-b$ is twice the distance $1^{st}-m$; it has a reddish tinge; c was seen once, its distance from 1st is not quite twice the distance $1^{st}-2^{d}$. S. C.

1859, March 10. Of "*h*" not a trace, though its vicinity was often explored; it must be nearly 1 mag. or more $< z$.* G. P. B.

1860, Jan. 20. Under tolerably good definition; I could not see the "*h*" star, when *k* was plainly in sight. Once or twice, when the definition was best, I fancied that it was seen by glimpses, but could obtain no confirmation. G. P. B.

1860, Jan. 23. There is no possibility of deciding on the question of the visibility of the "*h*" star, as the definition is not sufficiently good to admit of seeing the sixth star of the trapezium. G. P. B.

1860, Jan. 25. Cannot see "*h*" or even *k*, but the definition is not good. G. P. B.

1860, Feb. 12. Under good definition saw the small star *k*, but could not see "*h*," though by glimpses a brightness in nebula, possibly a star, could be pretty frequently suspected. G. P. B.

1860, Feb. 21,† 6^h Sid. T. [Diagram] drawn rapidly at instrument. *k* was seen at first certainly, afterwards by glimpses, as was "*h*"; "*h*" is whitish. S. C. Could see nothing of "*h*" at 7 P. M.; *k* just discernible. G. P. B.

1860, Feb. 21, 10 ½ P. M. [Another diagram by S. C. contains both the stars in question; "*h*" and G. P. B.'s *k*.]

[There is a little discrepancy of nomenclature between the two observers; S. C. oftenest saw Str. II., G. P. B. his own No. 636 = W. C. B. 24.]

1861, Jan. 28. Nearly clear sky; definition pretty good; could see *k* by glimpses, but no suspicion of "*h*.' G. P. B. [The following notes are all by G. P. B.]

1861, Feb. 13. Good definition; slight haze in sky; *k* easily seen, no trace of "*h*."

1861, March 20. No sign of "*h*," though *k* and many small stars were seen.

1862, Jan. 31. I cannot see the "*h*" star, nor make out *k* with certainty.

1862, Feb. 18.† The variable "*h*" star was not seen though looked for expressly.

1862, March 27. Good definition; no trace of "*h*" star.

1863, Jan. 30. No sign of ω. So also on Jan. 23d, see p. 129.

1863, Feb. 7. υ and ω ("*h*" star), both frequently sought for, but could not be seen.

1863, Dec. 29. "*h*" star not seen.

1863, Dec. 30. The variable "*h*" star looked for and not seen.

1864, Jan. 3, Sunday. $\omega =$ "*h*" star, $5^h\ 30^m$ Sid. time. Clear, definition tolerably good. Saw immediately on putting the eye to the telescope that the old variable "*h*" has reappeared (see also April 16). It is quite conspicuous $= k\ (\rho)$.

The sudden reappearance of ω ("*h*" star) is highly interesting. It was looked for carefully on the 29th and 30th, and although at one moment there was a suspicion (as often before) of a star, or brightish knot of nebulosity, it did not stand close scrutiny, whereas now the star is conspicuous, scarcely inferior to ϕ and about $= \rho$.

1864, Jan. 4, $5^h\ 20^m$ Sid. time. Saw the "*h*" star (ω) for a few moments only, between clouds, about as bright as ρ, not quite $= \phi$, certainly seen; clouded suddenly.

Was absent until 14th, and on 16th could not see "*h*" star, when ρ could be seen, though not always on account of atmospheric disturbance. Seen in April.

* Here z is No. 636. $= k$ of 1860, Jan. 20 and 25.

† Observation made at A. Clark's, Esq., with his 18½ inch object glass, now belonging to the Dearborn Observatory of Chicago.

1864, Jan. 17. Variable "*h*" not visible with utmost attention, and must be $< \xi$ or π.

1864, Jan. 19. No "*h*" star.

1864, Jan. 20. No sign of "*h*" star, or that near $I''I''$, but vision is bad.

1864, Jan. 22. Have not seen "*h*" star since 4th, though always looked for.

1864, Jan. 25. Vision is sufficiently good to bring out σ, ξ, π, ψ, etc., but "h" star and 'new star' near $I''I''$ are not in sight.

1864, Jan. 27. Glimpses of σ, and vague suspicions of "h" not confirmed.

1864, Feb. 3. σ is easily seen, but no sign of "h" star.

1864, March 14. No "h" star since Jan. 4th, though always looked for.

1864, March 15. ω ("h" star) not visible.

1864, March 17. "h" (ω) looked for; not seen.

1864, April 14. Looked for ω (H.); 78 not seen.

1864, April 16. "h" star (H. 78, ω) is in full view; fainter than ϕ, $\omega < \phi$ 0.7 1.0.
Measured diff. AR. H. 78 $=$ ω from θ'. Not difficult.

1864, April 16, continued. H. 78, which reappeared to-night, was last seen Jan. 4th. I have been in the constant practice of looking for it, but to-night having no time to lose, as I wished to make use of the few minutes between daylight and the last moment when Orion could be sufficiently well seen, to measure the distances of some of the small stars, and in so doing, while observing ϕ saw H. 78 without having it at all in mind.

It appears from the preceding observations, that without a doubt, Herschel 78 was seen on the following dates, 1850 Feb. 1, March 2, March 4, March 5, March 8, March 9, March 15, 1858 March 10, 1864 Jan. 3, Jan. 4, April 16; that it was not seen 1850 Feb. 7, Feb. 13, 1851 Jan. 23, Feb. 3, 1859 Feb. 23, March 10, 1861 Jan. 28, Feb. 13, March 20, 1862 Feb. 18, 1863 Jan. 23, Feb. 7, Dec. 29, Dec. 30, 1864 Jan. 17, Jan. 19, Jan. 25, Feb. 3, March 15, March 17, April 14, under circumstances which seem to leave no doubt as to its invisibility from faintness; and that it disappeared between the dates 1850 Feb. 1, and Feb. 7, reappearing again before March 2d, and continued visible until the 15th of the same month; that between Dec. 30, 1863, and Jan. 3, 1864, it reappeared, disappearing before the 17th of the latter month; and that, invisible April 14, 1864, it was plainly seen on April 16th.

Otto Struve's observations on this star, similarly treated, indicate decided visibility on the dates 1857 March 24, 1858 Feb. 23, Oct. 28, 1859 Feb. 24, 1862 March 6; and decided invisibility, 1856 Oct. 30, 1857 March 18, Sept. 2, Sept. 24, Oct. 24, 1858 Feb. 1, 1859 Feb. 27, Feb. 28, 1860 March 5, 1861 March 9, March 15, March 27, April 4, Sept. 27, 1862 March 21.

Struve's set of observations is for the most part at dates not very near to those made at Cambridge; it is, however, noticeable that, while the star was invisible at Cambridge, 1859, Feb. 23, it was visible at Pulcova, Feb. 24, and again disappeared before the 27th of the same month. There are no indications by which we can derive the period from these observations with any degree of certainty.

Naturally connected with the star just mentioned, are other fainter stars in and near the Huyghenian region; those especially which O. Struve has suspected to be variable. To give a more definite idea of these stars, I have extracted from the General Catalogue the numbers and letters according to Prof. Bond, the number according to Herschel as given by Struve, the magnitude according to Prof. Bond's reduction to Argelander's scale, and the positions of the General Catalogue itself; and have appended the names of other authorities upon which the visibility of the star rests. They have not all been seen at Pulcova, nor should this be expected; for the nebula at Cambridge attains the altitude of 42° on the meridian, but of only 25° at Pulcova; so that the extinction of light is necessarily much more powerful at the latter place; added to this is the circumstance that the climate of Pulcova is still more severe than that of New England,* and that consequently the winter nights are generally of little use for observation there on account of the bad definition.

No. G. P. B.	Letter. G. P. B.	No. Struve.	Mag.	$\alpha-\alpha_0$	$\delta-\delta_0$	
567	μ		13.9	$-102''.8$	$-\ \ 8''.3$	W. C. B. 10
575	r''_1	51	11.9	$-\ 84.8$	$-\ 22.3$	
589	r''_4	57	12.7	$-\ 57.2$	$-\ 20.4$	
595	ν		13.9	$-\ 46.9$	$-\ 15.0$	Lassell's e
602	ν'		14.3	$-\ 33.0$	$-\ 67.5$	
608	ν''		14.3	$-\ 23.7$	$-\ 18.0$	Lassell's f
612	ξ		13.5	$-\ 16.4$	$+\ 24.6$	Lassell's i
618	π		13.1	$-\ 10.4$	$+\ 26.6$	Lassell's h
622	ρ	II.	12.7	$-\ \ 7.5$	$-\ 27.8$	
625		ad II.	15.6	$-\ \ 4.$	$-\ 28.$	Lassell's d
631	τ		14.3	$+\ \ 3.$	$-\ 42.$	Lassell
636	σ		13.3	$+\ \ 8.4$	$-\ \ 8.7$	Lassell, W. C. B. 24
641	o'_1	III.	14.8	$+\ 11.9$	$+111.2$	
642	υ		15.6	$+\ 13.$	$+\ 48.$	
647	ϕ; p'_1	75	12.1	$+\ 22.6$	$+\ 38.0$	Lassell's 9 1
648	χ		14.3	$+\ 24.2$	$-\ \ 8.7$	Lassell
651	ψ	ad 75	13.1	$+\ 29.4$	$+\ 47.8$	
654	ω	78	12.3	$+\ 33.2$	$+\ 10.0$	
671	β	88	11.5	$+\ 69.6$	$-\ 24.4$	
676	γ	ad 88	13.1	$+\ 78.5$	$-\ 27.6$	

The stars 602 and 642 were seen on the 7th of February, 1863, and the 5th of March, 1850, respectively, each time by Prof. G. P. Bond, and in each case the star is stated to be a new one and to be certainly seen.

The stars from the list given above which have been most carefully examined, are

* The climate of Cambridge approximates quite nearly to that of Schwerin or Königsberg.

Nos. 575, 589, 622, 647, 671, as well as H. 78 = No. 654, which has been considered in the previous pages.

The following notices of these stars were extracted from the journal of observations, beginning from the date when this question appears to have been taken up by Prof. G. P. Bond, and include all that I have been able to find. It will be seen from a portion of the notes, that he considered their physical variability to be at least questionable, and that the appearance of change which they undoubtedly present is due for the most part to the effect of atmospheric disturbance upon the nebula in which they are situated.

After this series I have given such scattered notes as were found relating to earlier dates.

1863, Jan. 17. The atmosphere being very much disturbed, not a star was visible in the bright part of the nebula, excepting the trapezium. The influence of bad definition seems for the most part far greater on a small star involved in nebulosity, than on one in open space, as is natural.

1863, Jan. 18. o'_1 (Struve's III) not seen.

1863, Jan. 19. Looked carefully for all the small stars near the trapezium, but could see only those marked 1—11; all distinct, although the sky was not clear, nor the definition satisfactory. 6th star of trapezium distinct, however.

Numbers in pencil give order of brightness; 1, 2, easily seen; 7, 8, 9, 10, rather difficult; no sign of "*h*" star, 10=11 nearly.

[The pencil numbers are applied as follows, as the diagram upon which they are written shows,—

1=No. 575 H. 51* according to Struve.	5=No. 622 Struve's II.	9=No. 612 Lassell's i
2=No. 671 H. 88	6=No. 676 " ad 88	10=No. 636 (σ)
3=No. 647 H. 75	7=No. 651 " ad 75	11=No. 567 (μ, H. 51).]
4=No. 589 H. 57 according to Struve.	8=No. 618 Lassell's h	

1863, Jan. 23. The little star ρ, quite easily seen; much brighter than σ, which was difficult.

[On the accompanying diagram, which is on a separate paper inserted in the book, and I suppose prepared with the star's places, beforehand, are traced the positions of a number of stars marked "not seen." Those seen are

$$\mu,\ r''_1,\ r''_4,\ \xi,\ \pi,\ \iota,\ \phi,\ \psi,\ \sigma,\ \rho,$$

of which ξ, π, σ, are marked difficult, and those not seen are ν, τ, χ, ω.]

1863, Jan. 30. At $2^h\,23^m$ by Chro. 236† ϕ visible.
" 2 28 " " " ρ "
r''_1 and r''_4
" 2 34 " " " σ easy, though faint.

Was impressed with the fact that before any of the nebulosity came in sight the star β was already distinct at $2^h\ 20^m$, and must be far brighter than $b''b''_1$, [H. 100] which came in sight at $2^h\ 28^m$, at the same time with

* On account of the close agreement in position, I have called No. 567 H. 51.

† Usually within 1^m of Sidereal time.

ρ and γ; $\phi < \beta$, but ϕ was seen at $2^h\ 25^m$ (Chro. 236) r''_1 and r''_4 at $2^h\ 30^m$; at $2^h\ 34^m$ σ was easy to discern; ψ, ξ, and π, were also seen, although the haze and moonlight were detrimental. At no time could the star t'_2 be discerned. No sign of ω, μ, ν, τ, χ, υ; looked for, but not seen, though the sky was full of light and moon-milkiness.

1863, Jan. 31. Comparison of magnitudes.

$$\beta > r''_2\ \sigma = \xi \text{ and } \pi;\ r''_1 > \epsilon;\ r''_2 < r''_1 \text{ a little.}$$
$$\phi > \epsilon;\ \beta > \epsilon;\ \rho = \gamma;\ r''_4 > \rho \text{ a little, } \phi > r''_2.$$

1863, Feb. 2. At times the small stars of the bright region clearly seen, but only for a moment. It is remarkable how much more their visibility is affected by atmospheric disturbance, than is that of others less bright, situated in dark regions.

Thus frequently every star will disappear in the bright triangle, excepting those of the trapezium; but the moment the images become tranquil, β, ϕ, r''_1, r''_4, ρ, γ, ψ, ξ, π, and σ, will come in sight distinctly. Three or four of the brightest are then far more intense than other stars constantly in sight in other quarters.

1863, Feb. 7. τ certainly seen; its distance from ρ is a v. little ($1''$) less than dist. of longest side of trapezium, and line $\rho\ \tau$ makes an angle of $100°$ with $\rho\ \sigma$.

1863, Feb. 14. A fine clear evening; excellent definition; clouded suddenly after 7 P. M.

Chro.* 424 $3^h\ 31^m\ 30^s$. β in sight (preceding of pair s. f. θ) $< Q''$, but $= A''A''$ [H. 103.] $=$ 13th.

$3^h\ 50^m$. r''_1, r''_4, ρ, σ, ϕ, ψ, β, γ, ξ, π, μ, all in sight; broad twilight.

Of t'_2 no sign till long after. New stars ν' and τ by glimpses; they are $< \sigma$.

Order of Mag. $\beta > \phi$ a little, just perceptibly.
$\phi > r''_1$ or $\phi = r''_4$
$\rho > r''_1$ decidedly; $\rho = r''_4$; $\rho > \psi$ decidedly.
$\gamma = \pi$.
$\psi = \pi$.
$\pi > \xi$ v. little. $\sigma < \xi$.
$\mu = \sigma$ or $\mu < \sigma$; neither μ nor σ are difficult, though requiring attention and effort.

1863, Feb. 16. $3^h\ 58^m$ (by 424); β, r''_1 and ϕ, nearly equal, r''_4 a v. little $> \rho$; ψ and γ not visible, bad definition. Commenced observing as usual shortly after twilight; vision rather disturbed, and it clouded up after 7 P. M. Comparisons in strong twilight seem to be the most decisive; as it gets dark the disturbing effect of neighboring stars, or of the masses of nebulosity, is very apparent.

1863, Feb. 17. When quite dark every star in bright nebula about the trapezium was visible only by glimpses, and commonly undistinguishable, although sky was quite clear and no moon. Their light is very easily confounded with nebulosity, by atmospheric disturbances. Yet the very faint star t'_2 was always conspicuous and seen as well as ever.

1863, Feb. 20. Order of brightness. $\beta > r''_1$ very little; $r''_1 > \phi$, $\phi > \rho$, $\rho = r''_4$, $\rho > \xi$ and π decidedly.

1863, Feb. 23. EC.† $4^h\ 35^m$. $\beta > \phi\ 0^m.2$, $\phi > r''_1\ 0^m.1$, $\rho < r''_1\ 0^m.2$, $\rho > r''_4\ 0^m.1$.

1863, Feb. 25. β, ϕ, and r''_1, are so nearly equal that without noticeable error their succession in brightness might be reversed.

$$\beta > \phi,\ \phi = r''_1,\ \rho < r''_1\ 0^m.1,\ \rho > r''_4\ 0^m.1.$$

I have perceived no evident change since January, except such as vision accounted for; ψ and γ are plainer

* 1^m fast of Sidereal time. † Electric clock; $1^m\ 33^s$ fast of Sidereal time.

than usual, considering the vision, which is not particularly good. They are usually as difficult as ξ and π. I always discern σ when atmosphere is not very bad; also π and ξ.

1863, Feb. 27. (After 4^h 38^m 30^s E.C.) Order of brightness β, ϕ, r''_1, r''_4, ρ as usual, but ρ is more affected by the bad seeing.

1863, March 2. 5^h 5^m. ρ is quite as bright as $b''b''_1$ [H. 100.]
$\beta >$ A''A'' a very little; $\beta > \phi$ just perceptibly, $\phi = r''_1$.
$r''_1 > r''_4$ decidedly, $\rho > r''_4$ or $\rho = r''_4$.
$\gamma = \psi$, ξ and $\pi < \psi$, $\sigma = \mu$, both rather difficult.
$\delta = \beta$.

1863, March 4. Vision very bad, at 5^h 45^m (by 236). A''A'' [H. 103] and b''b''' [H. 100] even, seen without difficulty, while no trace of any star in bright nebula, from β downward, can be seen by a glimpse even after long watching.

1863, March 9. β, ϕ, and r''_1, now and then in sight of usual relative magnitudes; β barely $> \phi$ and r''_1, $\phi = r''_1$.

1863, March 11. Under not good definition, saw the stars of the nebula as follows: $\beta > \phi$ just perceptibly, $\phi = r''_1$, $r''_4 < r''_1$ very decidedly, say $0^m.4$, $r''_1 > \rho$ $0^m.1$, $\sigma < \rho$ $0^m.2$, ψ and γ about as faint as σ.

1863, March 12. No star visible in bright regions except trapezium, yet A''A'' is plain, and $b''b''_1$ distinguished without difficulty; δ, ϵ, and ζ, are all invisible. The effect of the bad vision is to fill up the dark regions with dim diffuse light.

1863, March 16. Small stars of nebula. $\beta > \phi$ $0^m.1$, $r''_1 < \phi$ $0^m.1$. This comparison made in haze, and not to be relied upon. Vision not suitable for the faint stars.

1863, March 17. Looked at faint stars near trapezium; β seems, as usual, a very little brighter than ϕ or r''_1. I do not see that the latter has changed certainly, though I fancied it a little fainter than usual with reference to r''_4.

1863, March 18. β, ϕ, etc., well seen, of usual relative brightness; $\beta > \phi$ just perceptibly, $\phi = r''_1$, $r''_1 > r''_4$ $0^m.3$, r''_4 may be a little $> \rho$, say $0^m.05$.

1863, March 19. [After 6^h 54^m Sid. time.] From atmospheric disturbance β not always visible, but $b''b''_1$ always easily seen.

1863, March 22. β, ϕ, r''_1, r''_4, and ρ, as usual; r''_1 nearly as bright as β. Later, ξ, π, ψ, γ, of usual order of brightness.

1863, March 27. Sid. time 7^h 8^m, was surprised to see β, ϕ, r''_1, ρ, and r''_4, all easily, before the form of bright triangles of nebulosity could be clearly made out. 7^h 14^m, σ is easily seen as bright as ϵ; $\rho = b''b''_1$.

[After 7^h 40^m.] Order of magnitude, β, ϕ, and r''_1, nearly equal; β a little the largest; ρ may be $> r''_4$ $0^m.05$, σ fainter than r''_4.

1863, March 30. β to ρ as usual for brightness. [After 7^h 30^m Sid. time.]

1863, April 1. β, ϕ, ρ, as usual. [After 8^h 30^m Sid. time.]

1863, April 9. β and r''_1 as usual.

1863, April 14. $\beta > \phi > r''_1$ just perceptible, but probably owing to confusion of images, nebula being low.

1863, Oct. 15. After 2^h 12^m Sid. time. Could not see the new star of 1863, Feb. 20th, and March 2d, although companion of I''I'' is very plain, as are also the small stars about the trapezium. It is certainly a variable. Near trapezium ρ at least as bright as Jan. and Feb. last.

1863, Dec. 10. Sid. time $4^h\,50^m - 5^h\,10^m$. No star excepting four of trapezium is to be seen in R. Huygeniana; worst character of definition.

1864, Jan. 3. [The stars ϕ and ρ are compared with H. 78.] At times the vision was excellent, bringing out σ, ξ, and π.

1864, Jan. 4. [The stars ϕ and ρ are again compared with H. 78.]

1864, Jan. 16. $\beta = r''_1$, $r''_1 > \phi$ perhaps. These three not very unequal, ρ decidedly smaller. The vision was always disturbed.

1864, Jan. 17. $\beta > r''_1$, but just perceptibly; $r''_1 =$ or $> \phi$; $\rho < \phi$ decidedly, but is next [?] brightest; ξ and π just in sight, σ not visible.

1864, Jan. 20. $r''_1 = \beta$, or $r''_1 > \beta$ perhaps; $\phi > \beta$ a little; others seen only by glimpses.

1864, Jan. 25. Vision is sufficiently good to bring out σ, ξ, π, ψ, etc.

1864, Jan. 27. Small stars of trapezium of usual order of brightness, β, r''_1, ϕ, etc.; glimpses of σ.

1864, Feb. 3. Fine definition. $7^h\,30^m$ Sid. time, σ is easily seen. In the stars of the Huyghenian region I do not recognize any (unless probably σ is fainter) which have changed since last year. Thus β, r''_1, and ϕ are the brightest; when vision is best, r''_1 seems to gain slightly on the others, but in general β is $> r''_1$ a very little, but $\beta > \phi$ pretty decidedly; ρ not difficult, but decidedly $< r''_1$ or ϕ; σ I do not see unless vision is above the average (Vide Feb. 25, 1863).

1864, Feb. 12. Clear, fine definition. As usual when vision is best, the small stars of the trapezium region β, ϕ, r''_1, do not differ much in brightness; the latter is critically placed on preceding edge of bright edge of strong light; σ easily seen.

1864, Feb. 19. Clear, bright moonlight, vision tolerably good; δ [H. 84] is conspicuously brighter to-night than ρ or ϵ.

1864, Feb. 29. I always look at small stars about trapezium β, ϕ, r''_1, etc., but can recognize no decided changes not accounted for by atmosphere; σ was readily seen, and possibly may be brighter.

1864, March 14. ρ is seen of usual brightness, β, r''_1, and ϕ, as usual.

1864, March 15. $\beta > r''_1$ slightly? $r''_1 > \phi$ not much; ρ readily seen as usual; σ considerably $< \phi$; ξ and π not very difficult. On the whole, with the exception of ω, I can recognize no instance of evident variation in these stars.

1864, March 17. Notwithstanding bright moonlight, the small stars near θ' [Orionis] such as σ, ξ, π, ψ, ρ, are readily seen; ρ is much $> \sigma$, and generally seen with ease.

1864, March 24. $6^h\,56^m$ Sid. time. β visible $= A''A''$ under bad definition; is near boundary of darker region n. of it, enveloping trapezium. $6^h\,58^m$, ϕ not difficult, but $> r''_1$ now first in sight.

1864, March 28. $7^h\,17^m$ Sid. time. β in sight, and ϕ a very little fainter. $7^h\,19^m$, r''_1 in sight, it is fainter than ϕ.

1864, April 7. $8^h\,9^m$ Sid. time. β easily seen, and $= A''A''$, also $\phi < \beta$ $0^g.3$. $8^h\,15^m$, ρ in sight; $8^h\,21^m$, comp. of β easily seen; this and ρ are brighter than n_3, n_4, by far, as neither of these is yet in sight; $\rho > o_1 > n_5$. $8^h\,31^m$, the small comp. of O' [Str. III.] easily seen, and is $= \eta$, a little $< \rho$.

1864, April 14. $8^h\,41^m$ Sid. time. β seen certain; $8^h\,43^m$, $A''A''$ not yet in sight, carefully looked for; $8^h\,47^m$ ϕ and r''_1 easily seen; $8^h\,47^m$ $\beta > A''A''$ $0^g.7$; $8^h\,52^m$ γ comp. of β seen, and $b''b''_1$ $b''b''_1 > \gamma$: $8^h\,55^m$ ρ, images undulatory; looked for $\omega =$ H. 78, not seen.

Of late I see β sooner than ϕ, and ϕ sooner than r''_1, though this is undoubtedly the effect of low altitude and worse definition of nebula. Precisely the same has been noticed when definition has been unusually bad.

So far as I can now recall, I have seen nothing to lead me to think either of the stars of Huyghenian region, from β downward, to be variable, excepting H. 74 [78] $=\omega$.

[1864, April 15, occurs a measure of r''_1.]

1864, April 16. $8^h\ 56^m$ Sid. time. β, A″A″, D″D″, G″G″, all nearly equal; [afterwards ϕ is measured.]

Besides the observations just given, there occur in the observing books for the years between 1847 and 1862, allusions to the same stars, or some of them. They are often in the shape of notes to little diagrams, whose scale is generally determined, together with the orientation, by the trapezium, which is always given. And it is in consequence not often difficult to identify the stars meant, though sometimes a little uncertainty arises, from the fact that there was no exact catalogue of them all in existence when the diagrams were drawn. One object, in fact, for which the researches whose results are contained in the present work was undertaken, was to give as nearly as possible all the stars visible with the Cambridge refractor, in the central portions of the nebula, under the most favorable circumstances which came for a series of years. The journals of observations, with the 21 foot refractor at Cambridge, — numbering 55 volumes, — contain each about 96 pages in 4to; and their dates extend from July 1847, when the telescope was erected, until the present time. From these books I have extracted all the scattered notes relating to these small stars, until the date when the regular observations upon Struve's supposed variables begin. When no statement to the contrary is made, the observations are by Prof. G. P. Bond, otherwise by Prof. W. C. Bond [W. C. B.] and the late Major Sidney Coolidge [S. C.], of the 16th U. S. Infantry, — who fell, gallantly fighting for his country, in the battle of Chickamauga, — a very conscientious observer, and one who has left valuable results of his labors at this observatory.

The remarks enclosed in brackets [ ] are mine, added when the original seemed to require them.

1847, Dec. 6, 11 P. M. [Note by W. C. B.] [The star marked] 2 is on the edge of the nebula; *d* is just on the edge of the small bright spot.

[A diagram accompanies this note, in which are given the positions of stars near the trapezium, including the six of the trapezium itself. The star 2 is manifestly Struve's H. 51, and H. 57 is also given in nearly its proper place, as well as Struve II. in a nearly right line with the two preceding, but too far north. Herschel 88 appears double, the following being of course Struve's ad 88, Herschel 75, =W. C. B. 26, also double; Struve's ad 75, =W. C. B. 27, being the companion, which is the star *d* also alluded to. This last pair also is too near the trapezium by about 10″. The trapezium is in fact on too large a scale in proportion to the rest of the drawing, which includes θ^2 Orionis, and the bright star H. 101. Upon the drawing are also given Lassell's *i* and *h*.]

[Same date, same observer.] No. 1 annular [micrometer; power, 103, see Annals I. 1, xliv.] the 6th star is steadily seen with this.

1848, Jan. 12. [A diagram and note by W. C. B.; from which I make out with certainty only the following remarks.] The following of this pair [H. 88, ad 88] is the smallest, they point to *a* [*a* of the trapezium?], and this is in a line with ϵ and 13, and they are a little below from *a* to 5. [I suppose 5 to be H. 110 ϵ is θ^2 Orionis, 13, H. 101.]

1848, Feb. 7. [W. C. B.] I saw the small star off the n. f. cape this evening; it is very minute on a dark ground in the channel. I called George, who saw it certain; it must be two magnitudes less than Herschel's 18th magnitude. [This is probably H. 114.]

1850, Feb. 1, 7^h. [Note by W. C. B.] Found a new star in the square below the trapezium, situated a (*a*) 19 mag.; both these stars [H. 78 has just been alluded to] must be variable; *a* could not have been overlooked if it had been as bright last year [1848?] as it now is. [The star *a* is by diagram and description Struve's III.]

1850, Feb. 7, 9^h 30^m. [By G. P. B.] Looked at nebula in Orion, very fine vision, all the small stars easily seen, but the Herschel star [H. 78] has disappeared; *x*, a very small star is seen, as well as Lassell's stars, steadily. [*x* is No. 636 G. P. B.]

1850, March 2. [The star No. 636 is alluded to as seen by G. P. B., identified by diagram.]

1850, March 4. [G. P. B. compares H. 78 with 24, 26, and 27 [W. C. B.], and remarks that No. 27 seems to have been estimated too low, and that the new star [Str. III.] near No. 23 [H. 70] is seen easily.

It must be remembered that G. P. B. did not begin making a catalogue of stars in the nebula, till 1857.]

1850, March 5. [G. P. B.] One new star is seen to-night near *m* and *n* [H. 75 and S. ad 75; the new star is in the neighborhood of the knot of nebula remarked by Dr. Gyldèn; see* Mélanges Mathématiques et Astronomiques, III. 501]; its place is indicated by *b*, and there seem to be others near it; another new star is seen at *a*; these stars I have no doubt belong to the nebula, for at the moment when they become visible the vision is distracted by others just on the verge of being seen as separate stars. [On the accompanying diagram appear Struve's II., the new star at *a*, (Struve's ad II.), $p =$ No. 636 G. P. B., and H. 78; H. 75 ad 75, and the new star $b =$ G. P. B. 642.]

[1850, March 9 and 10. *m* and *n* were seen by G. P. B.] March 11. Very good definition; the small stars steadily seen [*n* and *p* mentioned by names]. March 12, *p* not seen. March 29, *m* and *n* seen, *p* not. [As on previous dates in 1850, $m =$ H. 75, $n =$ ad 75, $p =$ No. 636 G. P. B.]

[1850, Dec. 27. A diagram by G. P. B. gives Lassell's *i* and *h*, probably; though perhaps *h* and the star seen March 5th; together with H. 75 and ad 75.]

[1851, Feb. 3. A diagram by G. P. B. gives *i*, *h*, and G. P. B. 636.]

1857, Feb. 23. [S. C.] The diagram was drawn at the instrument. While examining *a*, saw *k* by indirect vision; could not see it for a long time even by putting *a* [θ' Orionis max.] and *e* [θ^2 Orionis] out of the field. Saw it however at last, and have no doubt of its existence.

* The passage here cited (in an article of Winnecke's) is "Dr. Gyldèn, welcher an den Beobachtungen Theil nahm, sah alle diese Sterne [ad 81, ad 54, ad 75, *i*, *h*,] ebenfalls auf das Bestimmteste. Ein von ihm bemerktes Object, welchem die Coordinaten $\Delta\,\alpha = 0''$ $\Delta\,\delta = +40''$ zukommen, und das anfänglich für einen Stern gehalten wurde, erwies sich bei sorgsamer Betrachtung als ein Nebelknoten. In der Umgegend desselben schienen übrigens bisweilen sehr schwache Sternchen aufzublitzen, so dass dort möglicher Weise ein Haufen äusserst feiner Sterne vorhanden ist." 186.

[Note by G. P. B. in ink, evidently written later.] N. B. That *k* is the star (5) [II.] reported "invisible" by M. O. Struve, Feb. 21, 1857.

1857, Feb. 25. [S. C.] Nothing could be seen of star *k*.

1857, March 6. [S. C. Gives a diagram with *k*, seen "by glimpses."]

[In October, 1857, begin the observations for the General Catalogue.]

[1857, Dec. 14. The star β = Struve's H. 51, is measured.]

[1858, Jan. 14. There is an observation with the partially illuminated scale on Struve's III. ϕ, ψ.]

[1858, March 10. The stars r''_1, r''_4, ρ = Struve's II, ξ, π, β, γ, were observed in like manner, though in both these cases the estimates may have been made by alternating the illumination by which the scale was seen, with the dark field in which the star was visible.]

1859, Jan. 19. All stars of trapezium easily seen. [A diagram contains No. 636 also.]

1859, Feb. 22. Star *z* is quite plain in the position given below relatively to *a* and *c*. [Star *z* is 636 by diagram.]

1859, March 5. [S. C.] The definition is indifferent; there are three stars near the trapezium. [These are Struve's II. marked *c*, H. 75 marked *b*, and H. 78? marked *a*.]

1859, March 6. [S. C.] [The same stars *a*, *b*, *c*.] *a* was seen with comparative ease [H. 78?]; *a* inclines to red; *b* is also reddish and harder to make out; *c* could occasionally be seen.

1859, March 10. [G. P. B.] Though slightly hazy and with the moon near, the vision was admirable. In the brighter part of the nebula near the trapezium, by glimpses many new stars were suggested; *y* [648, χ] was certainly seen and located; it is much fainter than *z* [636, σ]; of *h* [H. 78] not a trace, though its vicinity was often explored.

[Upon the diagram which accompanies this, are represented also the stars afterwards denoted by μ, and ν, the stars r''_1, r''_4 (Struve's H. 51 and H. 57), and Struve's II., as also Lassell's *i* and *h*.]

[1860, Jan. 16th, occurs a diagram on which the stars 636, σ, and 648, χ, are given; the latter is stated to be very faint, but seen with certainty; the former to be easily seen. Jan. 20th, a diagram on which is marked a star *k*, which is probably the same as No. 636.]

1860, Jan. 25. G. P. B. Cannot see *h* or even *k*, but the definition is not good. [*k* is here No. 636.]

[A diagram of 1860, Feb. 12th, contains No. 636, called *k*.]

[A diagram of 1860, Feb. 21st, by S. C., contains a star marked *k*, but this is nearer Struve's II.]

Same date [G. P. B.]; could see nothing of *h* at 7 P. M.; *k* just discernible.

[Same date; another diagram by S. C. contains *k* in nearly the place of No. 636.]

1860, Feb. 23. [S. C.] The only small stars near trapezium that are visible are *k* and *m*, *m* [H. 78] is easily seen, *k* is certainly seen at moments for four or five seconds steadily. [*k* is here G. P. B. 636, by diagram.]

1861, Jan. 28. *k* seen by glimpses.

1861, Feb. 13. *k* seen easily.

1862, Jan. 31. Vision is passably good. I cannot see the *h* star, nor make out *k* with certainty.

These earlier observations decide nothing with respect to the stars r''_1 and r''_4, or Struve's H. 51 and H. 57; as those stars are always visible both with the Cambridge and Pulcova refractors, except in exceedingly disturbed states of the atmosphere.

The star Struve II. is not always seen with certainty at Cambridge, but this may be owing to its situation, which renders it often a difficult object, and to the state of the atmosphere.

The star Struve III. is especially noted as a new one, on Feb. 1, 1850; my own interpretation of the diagram of that date was made before I found a note by G. P. B. on page 113 of Struve's Memoir, which indicated the same thing. This star is undoubtedly variable, but owing to its distance from the trapezium, was not always looked for.

The star H. 75 does not occur so constantly on the earlier diagrams as I should have supposed; my own experience with it, which began very lately, does not indicate that it is often invisible here. It is, however, sometimes invisible under unfavorable atmospheric circumstances.

The small stars not contained in Struve's Memoir, "Observations de la Grande Nebulese d'Orion, faites a Cazan et a Poulkova," but which have since 1862 been seen at Malta and Pulcova, have also been suspected of variability; and about this the earlier notes give but little that is decisive. The star ad 75 is often mentioned, generally with H. 75, and both these stars were well known objects here, so that their absence under favorable circumstances would have been noticed; ad 88, more distant from the trapezium, was sometimes seen; on 1847, Dec. 6th, 1848, Jan. 12th, 1858, March 10th.

The star W. C. B. 24 = G. P. B. 636, σ, has not been seen at Pulcova. It is often mentioned in these earlier observations; as *certainly* seen on the dates 1850, Feb. 7th, March 2d, March 5th, March 11th; it is not mentioned March 10th, and was not seen March 12th; not mentioned 1850, Dec. 27th, but seen again 1851, Feb. 3d.

On 1858, March 10th, the same star was observed in regular course; it was seen again 1859, Jan. 19th, Feb. 22d, March 10th, 1860, Jan. 16th, Jan. 20th, probably, but not Jan. 25th, owing, as is stated, to bad definition; seen again Feb. 12th, Feb. 21st, just discernible, Feb. 23d, S. C. saw it "by moments for 4 or 5 seconds steadily." On 1861, Jan. 28th, it was seen by glimpses; Feb. 13th, easily; not seen 1862, Jan. 31st. These observations might be considered as indications of variability, were it not that those made later in connection with the other stars near the trapezium, and which are given on pages 129 to 133, afford another, and quite different explanation of the phenomena in many of these cases.

With regard to the three stars, two of which Lassell denotes by *i* and *h*, and the third is called by Struve ad 75, it may be proper to remark that I find no suspicion expressed by Professor Bond that they are variable. They have not been always seen or at least always noticed in these earlier observations.

All these objects deserve prolonged accurate study with high optical power, and I am inclined to think that the interest of the subject, and the abundant means now at the command of astronomers, will accomplish such a result.

A pair of stars some distance from the Huyghenian region, have shown marked variations. These are denoted in the present catalogue as F″F″ and I″I″; they are Herschel's Nos. 111 and 133, given as of the 10th and 9.10th magnitudes respectively. Liapunoff assigns them both to the 8th magnitude. The star F″F″ is not in a region especially interesting for its nebulosity; its coördinates are $\alpha - \alpha_0 + 15.^s54$, and $\delta - \delta_0 = -583.''8$; I″I″ however, follows the Messierian branch closely; its coördinates are $\alpha - \alpha_0 = + 34^s.32$, and $\delta - \delta_0 = - 306.''0$. Of these two, the latter (H. 133) exhibits the most decided and considerable variability; and a change in it, noticed Jan. 17, 1863, led to a careful examination of other stars, with respect to the same phenomenon. Of the number suspected, however, only F″F″ (H. 111) was confirmed as a variable, together with H. 133. The following are the observations of H. 133, made before it was suspected.

On 1857, December 4th, 7th, and 10th, I″I″ was noted as of the 12th magnitude on G. P. B.'s Zone Scale, 9 times; twice as the 11th, once as the 11.12th. On 1858, Feb. 12th, in the observations of sequences of magnitudes it was placed below H. 145, H. 104, H. 147, and H. 149, but above H. 112, H. 123, H. 103, H. 106, and H. 142; hence on Liapunoff's scale it would be about 9.5; substantially equal to its magnitude in December, 1857. On March 4th, 1858, it was noted as of the 11.12 magnitude on G. P. B.'s scale.

On March 20th, 1861, I noted it myself as of the 10th magnitude. My own scale is essentially that of Argelander's Sternverzeichniss, calling, however, the fainter stars there given as of the 10th magnitude, as Argelander himself suggests in the first paragraph of the Introduction to vol. iv. of the Bonn observations.

On March 17th, 1862, H. 111 was noted upon a chart of the stars as fainter than Herschel 149, and also fainter than the star following H. 149, termed O″O″ in the present catalogue. But on Jan. 17th, 1863, it was noted as brighter than these same stars, and also brighter than H. 45.

On Jan. 25th, 1863, it was noted as brighter than H. 37 and H. 45; on Jan. 31st, fainter than 45, but brighter than 37; Feb. 2d, fainter than either; Feb. 10th, brighter than 45. The magnitudes of the stars with which both H. 111 and H. 133 have been compared, are, upon Prof. Bond's scale of 1858, the following. This scale differs, it is true, from that of Argelander, as well as from Liapunoff's, but the relative magnitudes are well estimated.

No. Herschel.	G. P. B.'s Letter.	Mag. G. P. B. 1858.	Mag by Sequences.	Mag. Liapunoff.	Mag. Herschel.
123	H″H″	12.3	12.0	10	12
10	E″	12.0	12.1	9	11
—	O″O″	11.7	11.6	—	—
32	E′	11.7	11.4	8.9	11
18	I″	11.7	11.3	9	10.11
149	N″N″	11.6	11.4	8.9	10
35	F′	11.2	10.6	8.9	10
37	M‴	11.3	10.2	9	10.11
7	E‴	11.2	10.9	8	10
45	P″	10.6	10.8	9	11
4	D‴	10.6	9.9	8	9.10

The suspicions about F″F″ = H. 111 arose about the same time, and it will be well to give the observations of the pair together; so that it is necessary here to mention, that H. 111 was on December 10th, 1857, estimated as 12th, but on March 4th, 1858, as 10.11; and on Jan. 25th, 1863, it appeared brighter than H. 37, and less than H. 7; on Jan. 31st, brighter than H. 45; Feb. 2d, brighter than H. 37, and H. 45.

Comparisons were now made in February, March, and April, 1863, between these two suspected variables and others of the previous list; the differences were expressed in decimals of a magnitude, with however, the intention, as appears from the following note of Prof. Bond, that $0^m.1$ should express a decided difference of magnitude, or 1 grade, according to Argelander's idea.

"In the comparisons made subsequent to Feb. 13th, 0.1, 0.2 signify, 1st, 0.1 = A perceptible gradation of light, as small as will constitute a difference of magnitude certainly recognizable.

0.2 = A strongly marked difference of magnitude, which strikes the eye at once without special effort of attention.

The higher numbers, 0.3 0.4, indicate still larger differences, but beyond 0.3 or perhaps even beyond 0.2, I have little confidence in the numbers. The distinction between 0.0, 0.1, 0.2, I think are to be depended on.

Two stars differing by 0.2 I should regard as separated by an amount larger than any ordinary errors of observation would account for. That is if the stars were between the 9th mag. and the 12th, and compared directly with each other.

Two stars really equal might — though rarely — be thought to differ 0.1 = 1 grade.

In the course of observations I soon became sensible of an inconsistency between my estimates of 1 grade, and the $0^m.1$ (one tenth mag.) with which notation I began recording the 1 grade. But when the differences of compared stars exceeded 3 grades,

I found myself resorting to my previously formed habit of distinguishing '*magnitudes*' as in my zones. Thus where the differences are stated as 0.1, 0.2, 0.3, I think they may be safely treated as quantities pretty clearly defined and consistent; above these not much reliance can be placed on them as to consistency of scale. After some practice —say about March 1st—by 2 grades, I should have designated the difference between my zone 11th and 12th mag."

Under these circumstances I have not thought it necessary to print *in extenso* the observations of these months in 1863, as it would be difficult to reduce them to a fixed scale; but with the help of the following little table, which is derived partially from the results already given and partly from the determination made by grades in the winter of 1863 and 1864, I have obtained such approximations to the magnitude of these two stars as could be obtained by interpolation; noticing especially all cases where the usual magnitudes were not confirmed, or where variations were suspected from night to night.

The magnitudes here given, and consequently employed throughout for this pair of stars, are on Professor Bond's scale of 1858; which can be reduced to that of Argelander, at least approximately, by the table on page 122.

	Mag. finally adopted.
H″H″	12.2
E″	12.0
O″O″	11.6
E′	11.5
I″	11.3
N″N″	11.3
F′	11.0
M‴	10.8
E‴	10.8
P″	10.8
D‴	10.3

On 1863, Feb. 13th, I″I″ was called fainter than P″ by $0^m.4$; brighter than F″F″ by $0^m.4$; F″F″ brighter than H″H″ by $0^m.2$.

On 1863, Feb. 15th, I″I″ was $0^m.1$ fainter than F″F″ which is equal to M‴ = $10^m.8$.

On 1863, Feb. 17th, I″I″ was very nearly equal to P″; here about $10^m.8$.

On 1863, Feb. 18th, 20th, 21st, 23d, no great change is noticed; though F″F″ is gradually growing brighter, so far as the indications are decisive.

On 1863, Feb. 25th, it appears probable that F″F″ has increased; being now greater than M‴, than E‴, than P″; while I″I″ is estimated as equal to N″N″, and as standing half-way between E′ and F′. Its magnitude may be estimated as $11^{m}.4$.

The 27th of the same month, this is confirmed; F″F″ is, compared with M‴, $0^{m}.5$ brighter; with E‴, $0^{m}.3$ brighter; with P″, $0^{m}.2$ brighter; while I″I″ is now less than N″N″ and even than O″O″, equal to E′, and 1 magnitude less than F″F″. A note by Prof. Bond reads thus: "I″I″ looks quite faint, and F″F″ as decidedly brighter than usual."

February 28th, F″F″ seems about to have retained its magnitude; on this date, March 2d, and March 4th, it is still brighter than the stars E‴, M‴, P″. But on March 5th it sank to a magnitude between M‴ and E‴, that is $10^{m}.8$.

On the same dates I″I″ had the following relative magnitudes.

		m
Feb. 28th,		
	$>$ E′	0.1
	$>$ O″O″	0.1
	$<$ N″N″	0.1
March 2d,*		
	$<$ N″N″	0.2
	N″N″ $>$ O″O″	0.3
March 2d,		
	$= \frac{1}{2}(E' + F')$	
March 4th,		
	$>$ O″O″	0.07
	$<$ N″N″	0.25
	$<$ P″	0.4
	$<$ M‴	0.4
March 5th,		
	$=$ N″N″	
	$>$ O″O″	0.3
March 5th,*		
	$>$ N″N″	0.15

On March 11th, no change in F″F″ from the 5th; but with some fluctuations which appear partly to arise from the difficulties of observation, it retained nearly the magnitude $10^{m}.8$ on the 11th, 12th, 13th, 15th, 16th, 17th, 18th, 19th, 20th; on the 22d I find the note "F″F″ has certainly diminished," and by interpolation the following observations are reduced, –

* By T. H. S.

$$\begin{array}{lll} F''F'' & < M''' & \overset{m}{0}.05 \\ & < E''' & 0.1 \\ & = N''N'' & \\ & > N''N'' & 0.05 \end{array}$$

from which its magnitude appears to be $11^{m}.2$.

The other star (I″I″) was observed by G. P. B. as follows : —

1863, March 11th,

$$\begin{array}{lll} I''I'' & < N''N'' & \overset{m}{0}.4 \\ & < O''O'' & 0.15 \\ & > H''H'' & 0.2 \\ & & 0.1 \end{array}$$

Two comparisons with P″ on this date are respectively $1^{m}.0$ and $0^{m}.6$ of G. P. B.'s scale ; thus, $11^{m}.6$.

T. H. S.* also compared it with N″N″ and O″O″ making it respectively $0^{m}.25$ and $0^{m}.1$ fainter.

1863, March 12th, (by G. P. B.)

$$\begin{array}{lll} I''I'' & < O''O'' & \overset{m}{0}.2 \\ & < N''N'' & 0.4 \\ & < E'' & 0.05 \\ & < I'' & 0.2 \\ & > H''H'' & 0.2 \end{array}$$

1863, March 13th,

$$\begin{array}{lll} I''I'' & < O''O'' & 0.15 \\ & < E' & 0.15 \end{array}$$

"In a perfectly clear sky, quite dark, can barely discern I″I″ in finder [34‴ aperture] ; E‴, M‴, P″, N″N″, E′, O″O″, are all comparatively easy."

1863, March 14, "I″I″ is very faint ; is too faint to be compared with F″F″ or P″ or even with N″N″."

$$\begin{array}{lll} I''I'' & < M''M'' & \overset{m}{0}.2 \\ & < O''O'' & 0.3 \\ & > H''H'' & 0.1 \\ & < E' & 0.3 \end{array}$$

1863, March 15th,

$$\begin{array}{lll} I''I'' & > H''H'' & 0.15 \\ & < O''O'' & 0.3 \end{array}$$

* The present editor.

"I″I″ is scarcely brighter than either of the stars [H. 95, H. 106, H. 112,] in the triangle near F″F″ of which its companion [H. 106] is the apex; was decidedly less than E″; I″I″ = D″ [H. 8]."

1863, March 16th,

I″I″	=	M″M″	
	<	O″O″	0.1
	=	E″ or < E″	0.05
	<	E′	0.15
	>	D″	0.1

"I″I″ is certainly brighter than on last evening, relatively to H″H″ by 1 grade = $0^m.1$, but it is too faint to be compared with N″N″, P″, or M‴."

1863, March 17th,

			m
I″I″	<	O″O″	0.05
	<	I″	0.15
	>	E″	0.1
	=	O″O″	
	<	M‴	0.3

1863, March 18th,

I″I″	<	P″	0.4
	<	O″O″	0.1
	=	E″	
	=	M″M″	
	>	H″H″	0.2

1863, March 18th, T. H. S.

I″I″	=	H″H″	
	<	N″N″	0.3
	<	O″O″	0.15

These make I″I″ relatively fainter than Prof. Bond's observations.

1863, March 19th, G. P. B.

			m
I″I″	<	E′	0.15
	>	H″H″	0.05
	=	E″	
	<	O″O″	0.2
	=	H″H″	
	>	M″M″	0.05
	<	O″O″	0.1
	>	E″	0.02
	>	H″H″	0.1

1863, March 20th, G. P. B.

$$
\begin{aligned}
I''I'' &< O''O'' && 0^{m}.05 \\
&> E'' && 0.05 \\
&> M''M'' && 0.05 \\
&> H''H'' && 0.2 \\
&< N''N'' && 0.3
\end{aligned}
$$

1863, March 20th, T. H. S.

$$
\begin{aligned}
I''I'' &> H''H'' && 0^{m}.25 \\
&< N''N'' && 0.3 \\
&< \tfrac{1}{2}(H''H''+N''N'') && 0.05 \\
&< O''O'' && 0.3
\end{aligned}
$$

1863, March 22d, G. P. B.

$$
\begin{aligned}
I''I'' &< O''O'' && 0^{m}.05 \\
&< E' && 0.05 \\
&< E'' && 0.02 \\
&< E''' && 0.1 \\
&< M''' && 0.05 \\
&> H''H'' && 0.25
\end{aligned}
$$

On 1863, March 23d, it was found by careful comparisons that the two variables were equally bright, or at least that the difference between them was very slight. The separate observations of each give

$$
\begin{aligned}
I''I'' &> O''O'' && 0^{m}.1 \\
&= N''N'' && \\
&> I'' && 0.05 \\
F''F'' &< M''' && 0.1 \\
&< E''' && 0.15 \\
&> I'' && 0.05
\end{aligned}
$$

On 1863, March 26th, the following observations were taken:—

$$
\begin{aligned}
I''I'' &< P'' && 0^{m}.1 \\
&> N''N'' && 0.1 \\
&> E' && 0.2 \\
&> I'' && 0.2
\end{aligned}
$$

"$I''I''$ much increased, as sky grew darker the superiority of $I''I''$ to $N''N''$ was very evident."

$$
\begin{aligned}
F''F'' &< M''' && 0^{m}.1 \\
&< P'' && 0.1 \\
&> I''I'' && 0.05
\end{aligned}
$$

On 1863, March 27th, the observations of the two stars, by G. P. B. are as follows:—

			m
F″F″	>	I″I″	0.1
	=	I″I″	
	<	M‴	0.3
	<	E‴	0.1
I″I″	>	N″N″	0.15
	>	I″	0.2
	<	E‴	0.1

On the same date, the observations of T. H. S. give (not very consistently)

			m
I″I″	>	N″N″	0.15
	<	F″F″	0.05
F″F″	<	M‴	0.2
	<	N″N″	0.1
	>	O″O″	0.1

Upon March 29th, F″F″ was noted to be very much diminished. We have

			m
F″F″	<	E″	0.15
	>	H″H″	0.12
	<	O″O″	0.3
	<	E′	0.2

"Too faint to be compared with M‴ by grades, or even with I″I″ safely."

			m
*I″I″	>	I″	0.1
	>	E′	0.15
	>	N″N″	0.1

March 30th ("I″I″ evidently fainter to-night").

			m
I″I″	<	N″N″	0.1
	>	O″O″	0.1
	<	I″	0.02
	>	E′	0.1
F″F″	<	O″O″	0.12
	<	I″I″	0.2
	>	H″H″	0.15
	<	E″	0.05

* Probably no change since 27th.

Observations on both stars were made by both observers April 1st.

By T. H. S.

$I''I''$	$<$	$N''N''$	$0^{m}.15$
	$>$	$O''O''$	0.1
$F''F''$	$<$	M'''	0.3
	$<$	E'''	0.2
	$<$	$N''N''$	0.1
	$>$	$I''I''$	0.1

By G. P. B.

$I''I''$	$<$	$N''N''$	0.15
	$>$	$O''O''$	0.1
	$<$	I''	0.15
	$>$	E''	0.2
$F''F''$	$<$	$N''N''$	0.2
	$>$	$O''O''$	0.05
	$<$	$I''I''$	0.08

"F″F″ is too small for comparison with M‴, though it is perhaps brighter than on 30th."

April 2d, by G. P. B.

$I''I''$	$>$	$O''O''$	$0^{m}.05$
	$<$	$N''N''$	0.15
	$<$	I''	0.05
	$=$	E'	
$F''F''$	$=$	$O''O''$	
	$=$	$I''I''$	

"F″F″ is much too faint to be compared with M‴."

April 3d, by G. P. B.

$I''I''$	$<$	$N''N''$	$0^{m}.1$
	$>$	$O''O''$	0.1
$F''F''$	$=$	$I''I''$	

April 13th, by G. P. B.

$I''I''$	$=$	M'''	
	$>$	$N''N''$	0.2
	$<$	P''	0.05

"I″I″ has become quite bright."

$F''F''$	$<$	$I''I''$	0.25
	$<$	M'''	0.15

April 14th, by G. P. B.

$I''I''$	$>$	P''	0.08

	$> N''N''$	$0^m.3$
	$> M'''$	0.05
$F''F''$	$< N''N''$	0.05
	$< M'''$	0.2

This completes the series for the season of 1862–63; the following *résumé* of the results is sufficient to give a distinct idea of the variability of the two stars in question. The reductions have been made in cases where the stars were compared both with brighter and fainter comparison stars, as the decimals of a magnitude given above are not referred to the scale to which the magnitudes of comparison stars are, but are quantities more nearly like the grades of the next season. I would remark here that some of the casual variations in the table are probably owing to the difficulty of estimating the magnitudes of stars involved in nebulosity, especially in nights of inferior definition.

The observations before Feb. 28th, have also, where practicable, been reduced.

Date, 1863.	Magnitude of $F''F''$ = H. 111.	Magnitude of $I''I''$ = H. 133.	Remarks.
Feb. 13	11.9	11.4	
15	10.8	11.0 :	
17		10.8	
23	10.9		
25		11.3	H. 111 brighter than $10^m.8$
28		11.4	“ “ “
March 2		11.4	“ “ “
4		11.5	“ “ “
5	10.8	11.3	
11		11.9	H. 111 nearly 10^m8.
12		11.9	“ “
14		12.0	“ “
15		12.0	“ “
16			“ “
17		11.7	“ “
18		11.8	“ “
19		12.0	“ “
20		11.8	“ “
22	11.2	11.5	
23	11.2	11.3	
26	10.9	11.0	
27	11.1	11.1	
29	12.0		
30	11.9	11.4	
April 1	11.4	11.5	
2	11.5	11.5	
3	11.4	11.4	
13		10.9	

There would seem to be no room for reasonable doubt with respect to the variability of this pair of stars; as careful comparisons of the kind here given are the best means for deciding such a question. They were made largely in strong twilight; so that the disturbing influence of the nebula itself was reduced to a minimum.

In the next season Professor Bond determined to abandon the notation by decimals of a magnitude, which had proved a source of some inequality, and to substitute that by grades. He also omitted to use a number of the comparison stars previously given. The following are the magnitudes of the stars now employed, expressed in grades, as derived from comparisons among themselves; estimated from $O''O''$ as a zero.

Letter of Star.	Relative Magnitude.	Letter of Star.	Relative Magnitude.
$O''O''$	0.0	F'	1.7
I''	1.1	M'''	1.9
$N''N''$	1.1	E'''	2.5
P''	1.7	D'''	2.6

These numbers are tolerably consistent with the former ones. In the first of the tables which follow, are given the observations themselves of the two variables, compared with the stars just given; and the results for the variables of the individual comparisons. In the second table come the results for each evening, from all comparison stars, and as a control the observed differences of the two variables *inter se.* What these results represent on the scale of 1858, may be approximately ascertained by comparing the adopted magnitudes of the comparison stars in grades with the adopted magnitudes on that scale.

Date. 1863.	Star comp'd with F''F''.	F''F''— Comp. Star.	Rel. Mag. of F''F''.	Star comp'd with I''I''.	I''I''— Comp. Star.	Rel. Mag. of I''I''.
		Grades.	Grades.		Grades.	Grades.
Oct. 15.	M'''	+2.0	3.9	M'''	−1.0	0.9
	E'''	+1.0	3.5	N''N''	+0.5	1.6
				F'	+0.5	2.2
Dec. 7.	E'''	0.0	2.5	N''N''	−0.5	0.6
	M'''	+2.0	3.9	O''O''	+1.0	1.0
	M'''	+1.0	2.9	M'''	−0.7	1.2
	N''N''	+2.0	3.1	M'''	+0.5	2.4
	F'	+2.0	3.7	M'''	0.0	1.9
				P''	+0.5	2.2
				F'	−0.5	1.2
Dec. 8.	M'''	+1.0	2.9	N''N''	+1.0	2.1
	N''N''	+2.0	3.1	M'''	−0.5	1.4
	E'''	+0.5	3.0	M'''	0.0	1.9
				F'	0.0	1.7
				M'''	+0.5	2.4
				I''	+1.0	2.1
Dec. 9.	M'''	+1.5	3.4	M'''	+0.5	2.4
	E'''	0.0	2.5	N''N''	+0.7	1.8
	E'''	0.0	2.5	O''O''	+2.0	2.0
				I''	+2.0	3.1
Dec. 10.	M'''	+1.0	2.9	M'''	−0.7	1.2
	E'''	−0.5	2.0	I''	+1.0	2.1
Dec. 15.	M'''	+2.0	3.9	M'''	+0.5	2.4
	E'''	+0.7	3.2	N''N''	+1.0	2.1
				I''	+2.5	3.6
				M'''	+0.7	2.6
				F'	0.0	1.7
Dec. 16.	E'''	+0.7	3.2	M'''	+0.5	2.4
	M'''	+2.5	4.4	N''N''	+1.0	2.1
				F'	0.0	1.7
Dec. 29.	E'''	+0.7	3.2	M'''	+0.5	2.4
	M'''	+2.0	3.9	F'	+0.7	2.4
				N''N''	+1.3	2.4
Dec. 30.	M'''	+2.0	3.9	M'''	+0.5	2.4
	E'''	+0.5	3.0	F'	+0.7	2.4
				N''N''	+1.5	2.6
				I''	+2.0	3.1
				E'''	−0.3	2.2

Date. 1864.	Star comp'd with F″F″.	F″F″— Comp. Star.	Rel. Mag. of F″F″.	Star comp'd with I″I″.	I″I″— Comp. Star.	Rel. Mag. of I″I″.
		Grades.	Grades.		Grades.	Grades.
Jan. 3.	E‴	+0.7	3.2	M‴	+1.0	2.9
	E‴	+0.5	3.0	N″N″	+0.7	1.8
				N″N″	+1.3	2.4
				F′	+0.5	2.2
				M‴	+0.5	2.4
Jan. 16.	M‴	+3.0	4.9	M‴	−1.0	0.9
	E‴	+0.7	3.2	I″	+1.7	2.8
	D‴	+1.0	3.6	I″	+1.5	2.6
				N″N″	0.0	1.1
				I″	+0.7	1.8
				F′	0.0	1.7
				N″N″	+0.7	1.8
Jan. 17.	M‴	+2.0	3.9	N″N″	−0.5	0.6
	E‴	+1.0	3.5	O″O″	+0.7	0.7
				M‴	−1.5	0.4
				N″N″	−0.3	0.8
				I″	+1.0:	2.1
Jan. 19.	M‴	+0.5	2.4	N″N″	+0.3	1.4
	E‴	−0.3	2.2	M‴	−1.0	0.9
Jan. 20.	M‴	−0.7	1.2	M‴	−1.5	0.4
	M‴	0.0	1.9	N″N″	−0.3	0.8
	E‴	−1.0	1.5	O″O″	+1.0	1.0
				I″	+0.7	1.8
				I″	+0.3	1.4
Jan. 21.	M‴	−0.7	1.2	N″N″	0.0	1.1
				N″N″	−0.3	0.8
				M‴	−2.0	−0.1
				I″	+0.7	1.8
				N″N″	−0.3	0.8
Jan. 22.	M‴	+1.0	2.9	N″N″	0.0	1.1
				N″N″	−0.3	0.8
				M‴	−1.0	0.9
Jan. 25.	M‴	+1.7	3.6	N″N″	−0.7	0.4
	E‴	+0.5	3.0	O″O″	+0.5	0.5
	D‴	0.0	2.6	I″	−0.5	0.6
				M‴	−3.0	−1.1
				N″N″	−1.0	0.1
				O″O″	+0.3	0.3
				I″	+0.3	1.4
Jan. 26.	M‴	+1.7	3.6	O″O″	0.0	0.0
	E‴	+0.5	3.0	I″	−0.5	0.6
				N″N″	−1.7	−0.6
				N″N″	−2.0	−0.9

Date. 1864.	Star comp'd with F″F″.	F″F″— Comp. Star.	Rel. Mag. of F″F″.	Star comp'd with I″I″.	I″I″— Comp. Star.	Rel. Mag. of I″I″.
		Grades.	Grades.		Grades.	Grades.
Jan. 27.	M‴	+2.0	3.9	O″O″	−0.5	−0.5
				O″O″	−0.5	−0.5
				N″N″	−1.7	−0.6
				I″	−0.5	0.6
Feb. 3.	E‴	+1.0	3.5	O″O″	+0.3	0.3
	M‴	+2.0 :	3.9	N″N″	−0.7	0.4
	M‴	+2.3	4.2	I″	+0.5	1.6
Feb. 5.	D‴	0.0	2.6	N″N″	0.0	1.1
	E‴	+0.5	3.0	M‴	−1.0	0.9
				N″N″	+0.5	1.6
Feb. 8.	E‴	+0.5	3.0	M‴	−1.5	0.4
	M‴	+2.0	3.9	N″N″	+1.0	2.1
Feb. 9.	E‴	+1.5	4.0	M‴	−1.0	0.9
	M‴	+2.0	3.9	N″N″	0.0	1.1
				N″N″	+0.3	1.4
Feb. 10.	M‴	+2.0	3.9	N″N″	0.0	1.1
	E‴	+1.7	4.2	M‴	−1.0	0.9
	D‴	+1.5	4.1	N″N″	0.0	1.1
Feb. 11.	M‴	+2.0	3.9	N″N″	0.0	1.1
				N″N″	+0.3	1.4
Feb. 12.	M‴	+1.7	3.6	N″N″	+0.7	1.8
	M‴	+2.5	4.4	M‴	−1.5	0.4
	E‴	+1.7	4.2			
Feb. 16.	D‴	+1.0	3.6	N″N″	+1.7	2.8
	M‴	+2.0	3.9	M‴	−0.5	1.4
Feb. 18.	M‴	+3.0	4.9	N″N″	+2.5	3.6
	E‴	+1.5	4.0	M‴	0.0	1.9
Feb. 19.	E‴	+1.7	4.2	E‴	−1.0	1.5
				N″N″	+2.5	3.6
				M‴	+0.7	2.6
Feb. 23.	M‴	+2.5	4.4	M‴	+1.0	2.9
	E‴	+1.5	4.0	N″N″	+2.0	3.1
Feb. 25.	D‴	+0.5	3.1	N″N″	+1.5	2.6
				M‴	+0.3	2.2
Feb. 26.	D‴	+1.0	3.6	N″N″	+1.0	2.1
	M‴	+2.5	4.4	M‴	+0.5	2.4

Date. 1864.	Star comp'd with F″F″.	F″F″— Comp. Star.	Rel. Mag. of F″F″.	Star comp'd with I″I″.	I″I″— Comp. Star.	Rel. Mag. of I″I″.
		Grades.	Grades.		Grades.	Grades.
Feb. 27.				M‴	+1.0	2.9
March 2.	E‴	+2.5	5.0	N″N″	+3.0	4.1
	D‴	+2.0	4.6	M‴	+1.5	3.4
March 3.	E‴	+2.0	4.5	N″N″	+3.0	4.1
				M‴	+1.3	3.2
				E‴	+0.5	3.0
				D‴	+0.5	3.1
March 9.	M‴	+2.5 :	4.4	M‴	+2.5 :	4.4
				D‴	+0.5	3.1
				E‴	+0.7	3.2
March 12.	M‴	+2.0	3.9	N″N″	−0.5	0.6
				M‴	−2.0	−0.1
				O″O″	+0.5	0.5
March 14.	M‴	+3.0	4.9	N″N″	−1.0	0.1
				O″O″	+1.0	1.0
				I″	−0.7	0.4
March 15.				N″N″	+0.7	1.8
				I″	+1.5	2.6
March 16.	M‴	+2.5	4.4	N″N″	+2.5	3.6
	E‴	+1.5	4.0	M‴	−0.3	1.6
March 17.	M‴	+2.5	4.4	N″N″	+0.7	1.8
	M‴	+3.0	4.9	M‴	−1.7	0.2
March 18.	M‴	+3.0	4.9	N″N″	−0.5	0.6
				O″O″	+1.0	1.0
				M‴	−2.5	−0.6
March 19.	P″	0.0	1.7	N″N″	0.0	1.1
	M‴	+2.5	4.2			
March 21.	M‴	+1.3	3.2	N″N″	+2.0	3.1
				M‴	−0.3	1.6
				M‴	0.0	1.9
March 24.	M‴	+3.0	4.9	M‴	−1.0	0.9
	E‴	+1.7	4.2	M‴	−0.7	1.2
				N″N″	+3.0	4.1
March 28.	M‴	+2.5	4.4	M‴	0.0*	1.9
				N″N″	+2.5	3.6
April 4.	M‴	+2.0	3.9	P″	−1.3	0.4
				M‴	−0.7	1.2
				N″N″	+1.0	2.1

* Two observations.

Date. 1864.	Star comp'd with F″F″.	F″F″— Comp. Star.	Rel. Mag. of F″F″.	Star comp'd with I″I″.	I″I″— Comp. Star.	Rel. Mag. of I″I″.
		Grades.	Grades.		Grades.	Grades.
April 7.	M‴	+2.5	4.4	N″N″	+1.0	2.1
				M‴	+1.0	2.9
				N″N″	+1.5	2.6
April 8.				M‴	+0.5	2.4
April 9.				M‴	+2.0	3.9

Date. 1863.	Relative Mag. F″F″.	Relative Mag. I″I″.	Observed Diff. F″F″—I″I″.
	Grades.	Grades.	Grades.
Oct. 15.	3.7	1.6	2.0
Dec. 7.	3.2	1.5	2.5
8.	3.0	1.9	1.2
9.	2.8	2.3	1.0
10.	2.4	1.6	1.5
15.	3.6	2.5	1.0
16.	3.8	2.1	2.0
29.	3.6	2.4	1.7
30.	3.4	2.5	1.0
1864.			
Jan. 3.	3.1	2.3	1.0
16.	3.9	1.8	
17.	3.7	0.6	
19.	2.3	1.2	
20.	1.5	1.1	1.0
21.	1.2	0.9	1.3
22.	2.9	0.9	2.0
25.	3.1	0.3	
26.	3.3	−0.2	
27.	3.9	−0.2	
Feb. 3.	3.9	0.8	
5.	2.8	1.2	
8.	3.4	1.2	
9.	4.0	1.1	
10.	4.1	1.0	

Date. 1864.	Relative Mag. F″F″.	Relative Mag. I″I″.	Observed Diff. F″F″—I″I″.
	Grades.	Grades.	Grades.
Feb. 11.	3.9	1.2	
12.	4.1	1.1	3.5
16.	3.8	2.1	
18.	4.4	2.8	
19.	4.2	2.6	2.5
23.	4.2	3.0	2.0
25.	3.1	2.4	
26.	4.0	2.2	
27.		2.9	1.7
March 2.	4.8	3.8	2.0
3.	4.5	3.3	1.7
9.	4.4*	3.2	0.0
12.	3.9	0.3	
14.	4.9	0.5	
15.		2.2	
16.	4.2	2.6	
17.	4.6	1.0	
18.	4.9	0.3	
19.	3.0	1.1	
21.	3.2	2.2	
24.	4.6	2.1	
28.	4.4	2.5	
April 4.	3.9	1.2	
7.	4.4	2.5	3.0
8.		2.4	1.0
9.		3.9	1.3

It will be seen from the foregoing observations that the general results of 1862–63, with reference to these stars, were confirmed in the following season. I add a few observations of my own, made by Prof. Bond's direction, which may therefore be considered as in some degree a continuation of his series. I find upon examination that 1 grade of my scale represented very nearly $0^g.44$ of Prof. Bond's, so that we appear to have proceeded upon moderately different ideas of what a grade should be. The reductions are effected from the same series of magnitudes of the comparison stars as

* Difference too great for safe comparison.

have already been employed, multiplying, however, the quantities in the columns F″F″ – Comp. star, I″I″ – Comp. star, by 0.44 before using them.

I have preferred this course for uniformity's sake, as it gave nearly the same result as a reduction, using my own values of the relative magnitudes of the stars, and with about the same probable errors. The magnitudes expressed in my own grades are:—

O″O″	0.0
N″N″	1.2
M‴	3.0
E‴	4.0
P″	4.5
D‴	5.2

but are derived from too few observations. The series was interrupted by Prof. Bond's decease, in February, 1865, and the pressing duties consequent upon that sad event.

Date. 1865.	Star comp'd with F″F″.	F″F″— Comp. Star.	Rel. Mag. of F″F″.	Star comp'd with I″I″.	I″I″— Comp. Star.	Rel. Mag. of I″I″.
		Grades.	Grades.		Grades.	Grades.
Jan. 11.	M‴	+3.0	3.2	M‴	+2.0	2.8
	P″	+3.0	3.0	P″	+1.0	2.1
	D‴	+1.0	3.0	D‴	−1.0	2.2
				N″N″	+3.0	2.4
Jan. 12.	M‴	+1.0	2.3	M‴	+1.0	2.3
	E‴	+2.0	3.4	D‴	−1.5	1.9
	D‴	−1.0	2.2	E‴	+1.0	2.9
				N″N″	+1.5	1.8
Jan. 16.	M‴	+1.0	2.3	D‴	0.0	2.6
	D‴	−1.5	1.9	M‴	+1.5	2.6
				N″N″	+1.0	1.5
				N″N″	+1.5	1.8
Jan. 18.	M‴	−0.5	1.7	M‴	+1.0	2.3
	D‴	−1.5	1.9	D‴	−3.0	1.3
	E‴	−2.0	1.6	E‴	−1.5	1.8
	N″N″	+1.0	1.5	N″N″	+1.5	1.8
Jan. 20.	M‴	+0.5	2.1	M‴	−0.5	1.7
	E‴	−0.5	2.3	E‴	−2.0	1.6
				N″N″	+1.5	1.8
				O″O″	+1.0	0.4
Jan. 20, bis.	M‴	+1.0	2.3	N″N″	+0.5	1.3
	N″N″	+1.0	1.5	M‴	+0.5	2.1
	E‴	−2.0	1.6	E‴*	+2.0	(3.4)
Jan. 25.	M‴	+0.5	2.1	N″N″	−1.5	0.4
	E‴	−1.0	2.1	O″O″	−1.0	−0.4
				H″H″	+3.0	

* This should probably be —2.0, as F″F″ is distinctly stated to be brighter than I″I″ 1 grade. T. H. S.

Date. 1865.	Relative Mag. F″F″.	Relative Mag. I″I″.	Observed Diff. F″F″—I″I″.
	Grades.	Grades.	Grades.
Jan. 11.	3.1	2.4	
12.	2.6	2.2	0.4
16.	2.1	2.1	0.0
18.	1.7	1.8	0.4
20.	2.2	1.4	
20, bis.	1.8	1.7	
25.	2.1	0.0	

It may, I think, be assumed as unquestionable, that the stars H. 111 and H. 133 are variable. It would, however, be premature to attempt to determine their periods and the curves of brilliancy; which latter appear to be of a somewhat irregular nature, not unlike other instances of the same class of phenomena.

Other stars have been also suspected; the most probably variable of them is No. 801. It is identical in all probability with W. C. B. 86, and has therefore not been seen at Pulcova, as the latter is considered by Struve to be identical with W. C. B. 78. It was first seen by G. P. B. at Cambridge, Feb. 20th, 1863, and on March 6th, March 11th, March 22d, of that year was easily seen; March 17th it was not certainly visible, and on March 27th and 30th it was not visible at all.

A note by Prof. Bond, written later than March 17th, 1864, states that this star is certainly variable, "no trace of it [was seen] Jan. 3d, 1864, under definition showing the companion of I″I″ [H. 134, No. 832 of the General Catalogue] bright; not visible Feb. 3d, 1864, under fine definition; not visible in 1864 up to March 17th"; and on the latter date a note states that "these two stars [H. 78 and No. 801 General Catalogue] have been regularly sought for on every night when atmosphere has not been very much disturbed," and there is no note that No. 801 has been seen since March 22d, 1863. The variability of this star is only possibly questionable, because it is situated in the comparatively dense nebulosity of the Messierian branch. The nebulosity has, as has been abundantly noticed in the preceding pages, a tendency to obliterate faint stars when the atmosphere is disturbed. This however does not account for its invisibility Jan. 3d, and Feb. 3d, 1864.

SECTION V.

PHYSICAL OBSERVATIONS OF THE NEBULA IN ORION.

THE structure of this — the most remarkable of the nebulae known to astronomers in Europe and North America — is of very great importance in its bearing upon the nebular hypothesis. The observations made here were largely embodied in drawings, from which the engraving of the nebula which accompanies this work was executed; but, besides, there are a great many allusions in the journals of observation to physical characteristics, which have been extracted and are here given; the observations previous to 1858, however, as a portion and a continuation of Prof. W. C. Bond's, were omitted here, to be given afterwards in Appendix I. As in the last section, the notices within brackets were supplied by the editor, as were also the foot-notes.

1858, Feb. 8. In the region preceding the trapezium, I was struck with the apparent increase of small stars in the parts where the nebulosity was brightest. I have before received a similar impression in the northern zones for the region preceding the nebula.

1859, Feb. 21, 8 P. M. Clear, but only tolerably good vision; Chart I. Commenced drawing the nebula in chalk, on the charts prepared by Mr. Hall, of all the stars on a black ground. There is a remarkable dark region extending from AR. — 0^{m} 35^{s} Dec. + 11′ 20″ rel. to θ' to the Dec. + 26′ about, at which point bright nebulosity again appears. I have not thus far, since commencing in 1857, made any attempt to compare our results with previous observations.

1859, Feb. 22. The region about F″, H″, etc., has an appearance of resolvability.

1859, Feb. 23. Drew again on Chart I. There is a very large area north of the nebula, in which there is but little if any nebulosity. Under the best definition no star is to be seen within the trapezium. The edge *ab* [the S. E. edge of the Huyghenian region] seems to be just perceptibly (more strongly seen March 4th) brighter than the region within it; *w* is the brightest part of the nebula, *x* is very black, *y* bright. There is a curved bright outline as at *pq*.

1859, March 2. I think that there is often a disposition of the nebulosity to *shrink* from the immediate contact or neighborhood of some stars, and as evident a tendency to condense about others. I am disposed to think that, in general, where the stars are most numerous, there is also an increase of nebulosity.

1859, March 4. The moisture condensed on object-glass. The nebulosity preceding trapezium, distant say 1′ at *w*, and towards a point half-way from *a* to *q* of Feb. 23d, is the brightest of all. Several stars about the trapezium as A, etc., are decidedly nebulous, or have nebulous attachments. This would be highly important to verify. (Subsequently verified, vide March 10th.) [A is here the star θ^2 Orionis = Y″.]

1859, March 10. Admirable definition. Moon six days old in neighborhood. Two features were very noticeable. 1st. The great number (probably at least 20 could have been located) of very small stars in the region south of the three bright stars Y″, B″B″, and E″E″, [H. 93, H. 101, H. 110,] as far as P‴ and F″F″ and $h''h''_2$, and between this and the Great Proboscis. Another feature exhibited, and several times noted on other dates, was the nebulous appendages of the stars Y″, B″B″, and E″E″, as above represented; that of Y″ is very evident. On a scrap of paper I have indicated the region which has most bright stars, and it is remarkable that it is just where the small stars appeared in the greatest number on the 10th. That is in the region s. f. θ, and preceding the Great Proboscis, and is occupied with a bright diffuse nebulosity, mostly very evenly graduated. In the *black* regions near and n. f. θ, but a single very faint star could be seen.

1859, March 23. 236* = 6^h 57^m, a patch of light just visible 40″ preceding the centre of trapezium and a little south of the centre. 236 = 7^h 00^m, a patch of light 1′30″ south of centre of trapezium and following 1′. Looked at the nebula in early twilight, and found that the bright part surrounding the trapezium is outlined as in my water-color drawing, and as described Feb. 23.

[1860, Jan. 16th, occurs a diagram in which the bright portions of the nebula about 1′ preceding the trapezium are marked as probably resolvable under very fine definition.]

1860, Feb. 21. 6^h Sid. T. The spiral whirl of the nebula is readily distinguished when attention is once directed to this feature.

1860, Dec. 6. Have more than once noticed that there is a tendency in the nebulosity to dispose itself about the groups of bright stars in preference to other regions.

1861, Jan. 2. Thought I could recognize at least nine distinct instances where the nebulosity of Orion was expressly associated with bright stars or groups, sufficiently marked to arrest attention when directed to this particular aspect. There are included within a region 30′ on each side of trapezium, from ι Orionis, a nebulous group, and a group north of trapezium 1′ more distant north of it than ι is south, and in nearly the same AR.

1861, Jan. 28. Nearly clear sky. Definition pretty good. Now that attention is directed to the fact, the impression of a diverging spiral-like radiation of wisps of nebulosity from the bright region near the trapezium is very certain. There are one or two such near *a* [about 1′ preceding trapezium, 0.′5 south]. The three groups of stars constituting the stars of the sword-handle of Orion, ι, θ, and [c,] are foci of nebulous regions. The northernmost has about it a great mass of nebulosity which brightens up around several of the stars.

1861, Jan. 31. [Some differences of AR. are given among four stars, of which (c) is c^1 Orionis, (d) is c^2 Orionis, (a) is 536 of the present catalogue, and (b) 614 or 629. This region is brightly nebulous, especially about these four stars, and in the field preceding (c) and (d).] About (a) the accumulation of nebulosity is especially marked. Connection can be traced between this nebulosity and that about the trapezium. There is a decided accumulation of nebulosity about a star of 9th Mag. 11′ south of the trapezium, and 35 seconds of time preceding it [probably H. 34 = 467 of the General Catalogue]. All the region for several fields around ι Orionis is nebulous. Definition good. Eye-piece and object-glass clear. It is remarkable that about each group composing "sword-handle" of Orion, there is an accumulation of nebulosity quite evident even in the finder. See Obs. Feb. 21, 1863. There are, besides, several instances of a similar preference to the neighborhood of individual stars. The whirl-like arrangement of the drifts of nebulosity about the region of trapezium is very striking to-night.

1861, Feb. 6. Clear, with tolerably good definition. When the attention is given expressly to the dark

* Chronometer nearly indicating Sidereal time.

striae or furrows, their continuity becomes much more distinct, and the radiation more than ever evident. Those numbered [in a diagram accompanying] are clearly recognized as tending to give the wisp-like aspect and whirl of the brighter streaks of nebulosity. No. 3 is about 12′ preceding T′ [H. 108], and is 1′ north of it. This streak tends a little to the south of T′, above it is 4 tending to 2. This region about 2, 3, 4, and 5, has a twisted net-like aspect. 12 is remarkable, reaches half way from apex of nebula towards n. I propose to draw the dark features of the nebula with as little regard as possible to the bright regions, and vice versa. There is a field having only one or two stars visible, and those of the smallest visible, 15′ prec. ι Orionis while about it are grouped a brilliant assemblage of all grades of brightness.

1861, Feb. 8. Very cold, [6° Fahrenheit,] and bad definition; not free from haze. One edge of the two dark channels (of the lower of them) passes at x.

1861, Feb. 12. After warm rain, cleared suddenly; not good definition, but clear. Object-glass covered with moisture twice. Dark channels y and x distinctly traced as represented.

1861, Feb. 13. Good definition. Slight haze in sky. Examined the dark wreath-like openings, and traced a large number on chart M (tracing linen); was more than ever convinced of their whirl-like disposition. The whole bright region n. p. trapezium, shows the tendency strongly, and I suspect a radiation in the bright regions, lying to the n. p. side of line joining the stars L″, N″, and R″. The region about T′ has an indication of like tendency.

N. B. That line from ι Orionis to c Orionis, and to northernmost stars of sword of Orion, is an axis of light (and stars) radiating from trapezium.

1861 [no date]. The examination of spiral structure was begun systematically about Feb. 6, in the Regio Godiniana, and proceeded regularly on successive nights round in the p., s. p., s. f., and f. quadrants.

1861, Feb. 25. Quite hazy. In finder [aperture 34 lines] the general indication of radiation of light from nebula is towards A, B, C, and D, noticeable in finder. [I cannot identify these points very exactly; A is about 12′ from R. Huygeniana in AR. — 3′ ±; B is upon the Proboscis Major; C is about 9′ north of the nebula Mairanni; D is perhaps $\alpha - 8'$, $\delta + 4'$.]

1861, Feb. 26. A fine clear night, with good definition. The best opportunity for viewing the nebula that has offered this winter. Set upon the nebula in strong twilight, and prepared for examining the quarter s. p. the trapezium with reference to the disposition of the wreaths or streaks of nebulosity. On previous evenings I have examined the n., n. p., p., and s. p., portions in succession. From the star R″ a narrow brush or ray extended towards Q″ as indicated on the chart M. in yellow. From E″E″ there is a broad diffused mass of nebulosity extending towards F″F″ not regularly curved from E″E″. From A″A″ towards trapezium the radiations seem to cross each other, forming a net-work, as it were, reticulated. Indeed, elsewhere, in many instances the wreaths, like smoke from wet weeds, grass, or hay thrown on coals, seem to intertwine in a way quite difficult to draw. On turning to the region bounded by the following stars, viz: $h''h''_2$, K″K″, I″I″, H″H″, W′, U′, and the trapezium, it was with great interest that I was able to trace, under a fine sky, very evident narrow wreaths especially crossing the line I″I″ to W′. All of them could not be accurately located in the sketch on chart M; but the general aspect of long narrow wisp-like bright streaks, "Streifen," some of them 10′ or 15′ long, alternating and intertwined darker channels were very plainly seen and could not be overlooked after the attention had once been arrested upon the feature. Between K″K″ (H″H″?*) and W′ are as many as four dark streaks, and three or four, if not more bright wisps. The latter are rather more readily distinguished than the darker tracings.

* Note in parenthesis by G. P. B.

The evidence afforded by the fine view of this evening leaves no doubt whatever of the structure of the nebula. It is impossible now to see it in any other aspect than as a maze of radiating spiral-like wreaths of nebulosity or filaments, tentacles; the centre of the vertex being about the trapezium. I think that the brightest region of all, that surrounding the trapezium, looked at in the light afforded by this disposition of the more distant nebulosity, shows itself to be nothing else than an intricate convolution of similar wreaths. T. H. S., who looked at the nebula, compared it to an appearance of auroral rays streaming up and out from the trapezium in an ascending spiral. The trapezium at the vertex, like a watch-spring, with its outer turn horizontal and centre drawn down, seen from above. H. P. T.* compared radiation to smoke from burning weeds when wet.

N. B. The character of the convolutions can be represented best by dropping a little ink in a tumbler of water; after the ink has diffused itself for eight or ten minutes the resemblance is exact, as to the aspect of the filaments. Earlier, they are too well defined and too narrow and threadlike.

N. B. The star R$''$ is to be added to those having *brushes* of nebulosity appended on their southern sides.

1861, Feb. 28. A few glimpses between clouds. Noticed an interesting feature in parallel preceding trapezium. The bright striae from H$''$, (K$''$?), 12th, can be traced as indicated on chart M. to the star under M (as in margin) in whirl towards trapezium. [The star under M (M$'$?) appears to be O$'$ = H. 70.]

1861, March 4. Definition good; sky not always free from clouds. The striae between I$''$I$''$, H$''$H$''$, W$'$, and U$'$ well seen. Traced connection of light from d$'_2$ as it passes between F$'$ and G$'$ across the sweep of the curve (from G$'$ to below F$'$) in continuation with a large bright mass directed from trapezium towards E. I also trace more alternating bands from C towards T.

1861, March 5. A very clear night. Definition is not better than usual, though good. A dark channel passes from F$'$ between D$'$ and H$''$, and can be traced 12$'$ beyond C$'$, and D$'$ and is quite distinct. The dark channel directed towards the small star nearest n. f. F$'$ nearly in Dec. of c Orionis, has well-defined straight parallel edges, clean cut. I draw only the bright and dark striae which are to be superimposed on the general shading of the nebula. The sudden sweep of the striae in the region n. p. E$'$ is well and certainly recognized.

1861, March 7. Very cold, but vision quite good, considering the low temperature. One of the clearest of winter nights. Noticed another instance of nebulosity radiating to southward from star of 6th (?) Mag. located on map M. near stars g_1, g_2, g_3, and 12$'$ south of apex of light. This is perhaps a continuation of nebulosity from H$''$H$''$, h$''$h$''_2$, but the star is too bright to trace the connection. Dark channels of irregular width and tortuous, run up on either side of the nebulosity, trending from this star a little to left of ι Orionis. The dark canal was traced from t$'_2$ past W$'$ to almost as far beyond W$'$ as W$'$ is from trapezium.

1861, March 11. Night perfectly clear, definition very fine. Traced the terminus of bright nebulosity on following side of trapezium towards K$''$K$''$. The brightness continues without much diminution to line from I$''$I$''$ to W$'$. It is plain that the termination of the brighter nebulosity on the n. p. side of the nebula is the bright mass indicated on chart M. which sweeps around E$'$ to D$'$.

The opening south of E$'$, F$'$, to D$'$, H$''$, K$''$, is not at all *dark*, it is rather indicated by its *outlines* than by deficiency of light. Whereas beyond the sweep preceding D$'$, E$'$, the falling off is abrupt and decided. There is, however, a distinct continuation of the nebulosity of the region s. f. E to just two fields (24$'$) preceding S, where it sweeps up.

The long wreaths from the region P$''$, Q$''$, to M$'$ particularly well seen and delineated on chart M. There

* H. P. Tuttle, Esq., now Assistant Paymaster U. S. N.

is plainly a long stria made up of several filaments, having a general origin near the bright mass close preceding M′, which extends almost to the parallel of L″. The most interesting observation of the evening was the aspect in which the whole bright region close about the trapezium presented itself as a mass of *wisps*, curling out *towards the eye*, and to the left in short tufts curving back. I associate plainly into one group of wisps all the bright regions on the preceding sidė of the trapezium, ėven the abrupt vertical ray β [the preceding or W. boundary of the Huygenian region] is clearly connected as a whirl from the ray, shooting out more faintly towards P″.

1861, March 20. Was satisfied of the whirl character of bright masses s. p. trapezium.

1862, Jan. 31. Notice that both c^1 and c^2 Orionis are decidedly nebulous, *i. e.* have well marked aggregations of nebulosity about them. There are also two other stars within 4′ or 5′ of them having a similar character. The spirality is most distinct in the region preceding trapezium, and in same Dec. or near it, say 3′ either side.

1862, Feb. 18. Clark's 18½ inch object-glass. Good definition, calm and clear. Temp. about 32° [Fahrenheit]. Obtained a very fine view of the nebula with the great object-glass. The cirrus-like filaments sweeping outward from the Regio Huygeniana perfectly distinct, far more so, indeed, than with our 15-inch object glass, and entirely unquestionable as to their wisp-like curve or *spirality* in the sense attached to this term by Lord Rosse. The filaments were more numerous, distinct, and cirrus-like than I have before seen them. In the R. Gentiliana, Derhamiana, and Picardiana, this disposition was very strikingly evident. The R. Huygėniana was resolved into a confused assemblage of wispy masses verging on resolution. The region south of R. Subnebulosa towards, and following Sinus Gentilii, abounds with very small stars. The nebulous light shooting southward from the bright stars in the R. Subnebulosa, more especially the preceding one, are very plain, also the nebulosity of the nebulous star s. of R. Gentiliana has same tendency very plainly seen. As the result of this view I was entirely confirmed in my previous impressions of the spiral structure of this nebula.

1862, March 26. [Group of stars AR. 0′ Dec.—16′ noted as strongly nebulous.]

1862, March 27. N. B. That nebulosity tends to aggregate about the groups of small stars. That the *smallest* stars in the bright masses of nebulosity about the trapezium are easily seen in strong twilight, and before others in darker regions come into sight, although when the sky becomes dark the latter are much more easily seen. This shows that the small stars near the trapezium are really much brighter than they appear to be, their light being commonly overpowered by that of the nebula. This fact is important as evidence of a clustering of stars about the bright nebulous regions.

1863, Jan. 18. Bridge over Sinus Magnus easily seen, with brightening up in the middle.

1863, Jan. 19. The bright ridge (in parallel in Sinus Magnus) was distinct.

1863, Jan. 25. The dark streaks from the vicinity of R″ towards L″ are very remarkable. Region about H″ has a strong appearance of resolution.

1863, Feb. 7. $3^h\ 1^m$. Outline of nebulosity prec. tr. just visible. [Later in evening] in the field following in Dec. of L‴, bringing L‴ to preceding edge of field, are fifty stars, large and small. All this region, and thence to H″H″ and between N‴ and Y″ is strewed thick with very small stars which could be kept steadily in view. There are hundreds visible, not on charts. North prec. T′ are great quantities of stars; evidently the mass of stars is in line of AR. of θ'. But the field following T has many very small stars. The large starless areas two or three fields following S‴ are remarkable in contrast. I several times found fields destitute of the minutest star under the finest definition. 2½ fields following the bright stars S‴ and V‴ is a field without the slightest trace of a star, under *superb definition.* Not a trace, on the meridian. Field 12′, and much of the neighborhood, has but two or three stars. The area destitute of stars (contiguous) is probably at least equal to two fields. No nebulosity in same area. Have rarely, if ever before, known such fine definition to continue uninterruptedly for so long an interval. First viewed the nebula in strong twilight, but was annoyed by moisture condensing on the lenses of the eye-piece, which seldom happens; but vision was very fine. Later in evening went back to telescope at 7 P. M. and stayed until 11^h; during the whole time definition of the first quality.

1863, Feb. 16. As haze gathered, compared the finished drawing of 1859, with nebula about the trapezium, and found it unexpectedly exact; the only corrections were: —

1st. Bridge of Sinus Magnus should incline more to hour-circle.

2d. The Promontory (in parallel of Q′) is rather too bright in general color.

1863, Feb. 21. Looked in Comet-seeker at the three stars of "Sword-handle" of Orion; the clustering of small stars about each is indubitable, and as evident as is the aggregation of nebulosity about the same centres.

1863, Feb. 23. Marked on copy for Mr. Watts to engrave several corrections of drawings derived from combination of all sketches. Light in region n. p. trapezium falls off pretty suddenly, about in AR. of $\frac{1}{2}$ (P″ + Q″). Next examined with Shimmin eye-piece, power 90, as follows. Noticed particularly, that at least part of the effect of "reticulation" is caused by the two wisps x and y, the former bright. The latter fainter but *certain*, proceeding from the bright triangular mass z. These two are intersected by the wisp streaming off from θ^2. [Here x and y is the wisp running off from the neighborhood of the s. apex of the R. Huygeniana, towards the s. f. quadrant; y is parallel to x.] Noted sweep of the large convolutions n. p. trapezium, about c, especially s. p., and was well satisfied with drawing. The annular nebula near ι Orionis is a very beautiful object (a cluster?) star or stellar nucleus, is about as bright as O′. This is the nebula H. IV. 33.

1863, Feb. 28. Noticed that the sweep of nebula in W. C. B.'s engraving, in region n. p. trapezium, represented as a gulf, may have been intended to express the rather sudden falling off of brighter light from the Huygeniana Region, to which it bears considerable resemblance.

1863, Feb. (no day stated.) The drawing of 1859 was compared in details with nebula in end of Feb. 1863, and no change of any prominent feature could be recognized. Mr. Watts's copy of nebula was corrected in a few details, I think Feb. 23d, not later than 25th.

1863, Dec. 7. I always look at *Sinus Magnus*, etc., for change of feature, but was never satisfied of any not accounted for by change of atmospheric condition. The region close s. p. trapezium shows pretty plainly the wispy structure. The wisp a is narrower than in our engraving [this appears to be the wisp bounding the R. Huygeniana on the preceding side].

1864, Jan. 3. Looked particularly at clustering of stars, and am persuaded that they are associated with the aggregations of nebulosity about ι, θ, and c Orionis. The nebulosity about c is quite bright, involving a large group of stars.

1864, Jan. 20. A chart of all the stars from 2° north to 1° south of θ Orionis, would illustrate the aggregation of stars about nebula effectively.

1864, Jan. 26. F″F″ has a strong wisp of nebulosity running off to s. f. side, and should be perhaps reckoned among the stars with attachments. I find not a large, and by no means very faint mass of nebulosity, sweeps from the region preceding D‴ (H. 4), and 35^s at earliest preceding H‴ (H. 23) to ι Orionis, besides that from Messierian branch.

1864, Feb. 3. Under fine definition the third field following Piazzi [star] (= V‴) is *starless*. The wisps shooting southward from Y″, B″B″, and E″E″, finely seen; that from B″B″ certain, but less distinct, owing perhaps to confusion with b″b″ [$b''b''_1$?]. That from E″E″ I thought tended to sweep round towards direction of A″A″. As usual under fine definition, the region between Y″ and F″F″ seems to have many very faint stars which have not been recorded, though with time they might be; δ [H. 84] near Q′ [H. 87] is just = ε [H. 80]; both are in narrow dark stripe, with bright nebulosity near at hand on all sides. The nebulosity from θ, sweeps up both on preceding and following sides to ι Orionis; from the "Cam" [the nebula of Mairan] it tends rather to the following side of c Orionis. The annular nebula, H. IV. 33, is a beautiful object; central star *vivid* and opening, remarkably outlined, coming close to star on n. p. side. This nebula is connected by diffuse light with ι Orionis.

1864, Feb. 26. Clear, but definition much disturbed by fine very rapid undulations. Can trace connection of nebula c Orionis with that of θ with entire certainty, the junction being best pronounced by way of region north, and a little preceding T′. By moving the telescope fixed in declination with [of] faintest part of connecting nebulosity, and then several degrees away in AR., I could readily recognize, in sweeping past, the region of this connection by the nebulous light filling the field. Both the above and the connection recognized Feb. 27, were confirmed perfectly, March 2. (See drawing on large chart, March 3 and 9.)

1864, Feb. 27. Going to the telescope at 5^h 30^m Sid. Nebula on meridian, sky clear, and twilight not quite vanished; could very easily connect the nebula c Orionis with θ. Nearly in direction of AR. the mass tending to preceding side of meridian of θ. This nebulosity extends at least 20′ north of c and as much preceding; the limits becoming diffuse and blending with stars.

1864, March 2. At 6^h 10^m. Under a very fine and clear sky, with good definition, explored and drew on chart portion of nebula about c Orionis, using power 146. A double connection with θ′ Orionis was clearly made out, one on the preceding side of T′ towards F′, the other running in a direction n. p. to the north side of Cam. The dawning of light in approaching the great nebula, is sensible very much outside (north) of the engraving. Light shades very gradually, and on it are superimposed as it were the "Cam" [nebula Mairanni] and other features.

1864, March 3. Micrometer taken off [to apply a Huygenian eye-piece of low power]. Atmosphere of perfect transparency, and every condition favorable, vision quite good as to definition also.

(1.) Power, 90. Light may be discerned 448 beats = 224^s preceding ι Orionis in same parallel, and a full field s. and s. f. ι Orionis, shading insensibly; field = 161^s about (see below; field = 36.′3).

(2.) Messierian Branch sweeps finely to point, say 8′ n. p. ι Orionis, connecting strongly with nebulosity sweeping over from D‴ encircling dark area, of which centre is in AR. of E‴, and say 8′ s. of it. The sweep of nebulosity on both sides clearly defining this dark area, and blending away insensibly to s. and s. p., is one of the grandest features of the nebula. By Obs. March 9, field = 36′.

(3.) Can discern light one field preceding, and in parallel of E‴, and 8′ fol. S‴ (= V‴?) (2d of Piazzi's stars) very faint.

(4.) A pretty strong body of light, in breadth = ⅔ extreme length of "Cam," extends ½ field in parallel of S′ (of Cam) fol. it.

(5.) Light extends strongly, one field preceding, and in parallel of θ′, and thence suddenly fainter, and very faint, say ⅔ field following.

(6). Faintly to one field preceding (30° north of) F′.

(7.) I think a faint nebulous ground is still discernible as far as $1\frac{1}{2}$ to $1\frac{1}{4}$ field prec. c^1 Orionis, and in parallel and with entire certainty $\frac{8}{10}$ field following it.

(8.) Quite decided light to $1\frac{1}{4}$ field n. 45° fol. c^1. Faint $\frac{8}{10}$ of field due north of c^1 and $1\frac{1}{4}$ field in direction prec. 45° north (see 11). See observations Feb. 27. Nebulosity at least 20′ north of c^1, and as much preceding, but both sky and power then used were less favorable.

(9.) The association of nebulosity about stars near c Orionis, is very remarkable. There are four striking examples, including c^1, not reckoning c^2. The most remarkable is in $\delta - \delta_0 + 39.'5$ $\alpha - \alpha_0 = -4'$ rough, double star 7th mag.

(10.) Revised drawing of c^1 Orionis made last night, and with a few corrections made it entirely satistory.

(11.) The coarse cluster straggling round a centre in $\delta - \delta_0 + 60'$ $\alpha - \alpha_0 - 10^m$, gives so much light that I cannot decide whether it is enveloped in nebulosity from c^1.

Notes which follow were written out on the 4th, using rough memoranda, put down on 3d.

(1 *b.*) Power, 90. Obs. 4th. Measured *apparent* field of view $= 58°$. Hence the low power eye-piece, has a power of 90 roughly.

Low power eye-piece, March 7, has a power $= 91$ by another method.

(2 *b.*) The grand sweep of nebulosity from D''' on one side, and Messierian Branch on the other, bright and well defined on the boundary of the dark region enclosed by it south of E''', is a fine feature, brought out with far greater distinctness to-night, with the low power, than I had before seen it. To the southward and s. p. side light shades off insensibly, without decided feature, excepting the nebulosity about ι Orionis, and some of the neighboring stars. The light to s. p. side of D''' associates itself so decidedly with that, it should be classed among examples of that feature. The whorl of the nebulous wreaths from prec. side of the Huyghenian Region is admirably displayed, rising like smoke of grass or weeds in a bonfire, in filaments as it were, and ascending past D''', bending round to a little north of ι Orionis, which, however, it envelopes in its fainter southerly expansion.

The dark area south of E''' is remarkable; did not record its diameter, but it must have been 8′ or 10′.

(5 *b.*) A great diffusion of light n. p. θ diminishing rather suddenly, but continuing faintly, as indicated by (6).

(7 *b.*) I found it very difficult to fix the limit of light on prec. side of c'_1 Orionis, beyond about a field from that star. About Dec. + 34′ and AR. −30′ to 35′, is a region of comparatively few small stars; I almost suspected diffuse faint light in approaching this region. The connection of c Orionis with great nebula, is perhaps most decided *with the low power* in the diffused light extending s. p. from c. Though with the higher power, that by way of the Cam [nebula Mairanni] may be easiest discerned. Evidently the mass of light from nebula c Orionis sweeps off southerly, and especially s. prec.

(9 *b.*) The association of nebulosity with stars AR. + 1′ Dec. + 33′ = c^1 Orionis, AR. −5′ Dec. + 39′ (very strongly), AR. 0′ Dec. = + 42′, and AR. + 2′, Dec. + 36′ is here referred to; all these stars are coarsely double.

(10 *b.*) On revision, was well satisfied with last night's sketch; made some additions and continued light s. p. c to its connection with great nebula about AR. −20′ Dec. + 14′. It seems that the immediate neighborhood of the great nebula, especially the following side, and that of ι Orionis have considerable areas, comparatively clear of small stars, while at greater distances, as for instance, far north of c^1 Orionis, the ground of the sky assumes the aspect of the milky way. It is as though there had been a process of absorption into

the nebula. (Purity of sky throughout unexceptionable, and definition quite good.) Had made arrangements for drawing the nebula to-night, and put the low power eye-piece (that given by Mr. Shimmin), and the object-glass in condition for observing with least obstruction of light. Fortunately, the night was one of the finest we ever experienced, for the object particularly required, to trace the utmost limits of the nebula, and the connection of ι and c Orionis with θ Orionis. I do not know that I ever saw the *tout ensemble* of the nebula to such advantage. To ascertain the limits of faint light, the telescope was moved rather quickly from a distance of several fields, until the dawning of the light could be discerned, the operation being often repeated, three or four times, until the limit was definitely determined on. I think that in most cases there was perceptible, though barely visible, faint light outside of the point taken.

1864, March 9. A fine clear night; completed drawing of nebula about ι Orionis under very favorable circumstances, the sky being clear, dark, and tolerably quiet. Using Shimmin eye-piece, power, = 90, the fine curve from S‴ to region preceding D‴ is composed of quite bright nebulosity, contrasting strongly with the gulf or bay enclosed by it, having its centre about H‴ and I‴. With this low power and large field, combined with the Messierian Branch whence it issues at S‴, and with the bright masses in AR.—14′, Dec. —4′ (from θ′) it forms decidedly the grandest feature of the nebula. To the south of this (with the exception of the brighter, very diffuse nebulosity surrounding the nebulous star ι, and the strongly nebulous, coarsely double star s. p. ι Orionis in AR. —3′, Dec. —37′) can trace no feature, all is a diffuse nebulosity shading off by insensible gradations. There is a good deal of complication to s. and s. p. (especially) of S‴, the latter itself belonging decidedly to the class of stars having nebulous appendages, the connection of nebulosity being perfectly decisive. A rather faint and irregular, but certainly recognized, bridge of light traverses the dark bay, passing the neighborhood of H‴, I‴, to S‴. It does not much affect the general darkness of the bay. The star N‴ [H. 58] n. p. O‴, about AR. —2′, Dec. —15′, as I have often before noticed, is very strongly nebulous, the light connecting with R″. Traced very clearly the upward (southerly) sweep of nebulosity in region far preceding (and n. p.) θ′. Especially a remarkable dark vortex in AR.—16′, Dec. + 6′ from θ′, and wispy convolutions stretching s. p. over region about AR. —24′, Dec. + 5′, forming last of principal light on this side of nebula; beyond are no details. From this vicinity a strong mass of diffuse light connects with c[1] Orionis. The s. f. sweep of nebulosity over region between I″I″ and O″O″ well seen; this part of nebula more regular in the curvature of wisps. N. B. The presence of fewer stars of all grades, is it an absence of centre of disturbance? The limit of light (not faintest) well pronounced, passes 1′ to n. of N″N″ and O″O″.

Recognized the *three* communications between c Orionis and θ′; the middle one is least evident of the three. Tried to fix a limit of nebulous light from c Orionis in direction of the coarse cluster n. p. it. The nebulosity certainly penetrates the south side of cluster, say to Dec. + 56′, but the number and brightness of stars as the nebulosity grows fainter, make it impossible to follow it through the cluster with certainty, though it can be traced farther on either side preceding or following cluster. (N. B. March 3. Traced to + 62′ with ease and with finest possible sky.)

With Comet-seeker it appeared that the three star-groups about ι, θ′, and c Orionis, were projected upon a ground comparatively less massed with milky-way stars than the more distant regions; at least one might fancy that the three groups were supplied somewhat at the expense of contiguous territory. Thus, if we suppose ι, θ, and c, grouped as in drawing, the region surrounding, to a distance represented by dotted lines, has fewer stars, or less of *star-dust* than that outside of it. The middle star, especially, of Belt of Orion, has an evident aggregation of stars about it. [After these observations the filar micrometer was replaced.]

1864, March 19. As usual, the portion of R. Huygeniana prec. θ' Orionis is brightest, and in very early twilight the bright region is defined so as to extend borders of Sinus Magnus to preceding side of θ'. Have often remarked this before.

1864, March 24. Struve's remark (Observations Great Nebula Orion, p. 110,) as to ratio of distances from H. 110 to Proboscis and θ^2 Orionis, through 110 to Huyghenian Region, is not confirmed. Proportions are as in my drawings and engravings. I find that in Herschel's drawing at the Cape of Good Hope, the cusp near star p (H. 117), might readily be sketched in attempting to represent its present aspect. The darkness to north of H″H″, where is also a recession of the edge from the star, gives, though less decidedly, a similar effect with H.'s drawing; the nebulosity touching p or nearly so. See Struve, p. 110. I do not think his view, as to a change, sustained in this feature. Struve, p. 110, remarks change in Prom. Herschelii that it no longer has its cusp or apex at star 126. Found that this, as represented by Herschel, would answer well for present aspect of nebula, without a better atmosphere than we have to-night.

1864, March 28. 236,* 7^h 17^m. Brightest of nebulosity just preceding θ' at distance 35″ is visible over area of 0.′5 by side.

7^h 19^m. Outline of Huyghenian region is readily distinguished, especially near Y″. The light near ϕ is a little, but not much fainter than that prec. θ'.

7^h 25^m. The s. p. margin of Huyghenian region is evidently less clearly defined than s. f. edge, and fully explains Herschel's having given it less expression. His engraving, however, is far too indefinite here.

7^h 31^m. Compared carefully the bright masses of Huyghenian region with engraving, and found an excellent agreement.

Comparing bright spots of Huyghenian R., was much pleased at excellent agreement of engraving with nebula. The only correction suggested by this examination was, that the Sinus Lamontii should have been a very little darker; yet it is not darker than the dark space first met in crossing from it to Y″, nor than that immediately n. f. the trapezium. It is conspicuous, rather from its size than its intensity. Were the trapezium stars removed, I think that region would be much darker than Sinus Lamontii. I was confident of tracing continuation of the decided margin of s. p. edge of Huyghenian R. across Sinus Lamontii. The engraving does not represent Huyghenian R. proportionally too bright. Nebula minima, seen as usual; has two centres of somewhat brighter light. The sweep of Messierian Branch on preceding side, objected to in Herschel's and W. C. Bond's engravings, by Struve, at the points near star $h''h''_2$ (Prom. Herschelii) and near p, I have examined, and find, that as a general expression, my engraving is exact; but that south of $h''h''_2$ there is less decision, and, as shown in some of my old drawings, a fringe of nebulosity on the southern sweep, which might account, imperfectly seen, for Struve's remark that $h''h''_2$ was 20″ north of Prom. Herschelii. In engraving the wisp from trapezium across S. Lamontii is rather too bright.

1864, April 7. Liapunoff's remark as to relative brightness of R. Picardiana and Messierian Branch [is not confirmed]. I find the former immediately north of R. Huygeniana, by far the brightest.

The remarks, in Annual Report of the Royal Astr. Society, of Messrs. Stone and Carpenter, as to "Squareness of outline" of S. Magnus [are also not confirmed]. I find on comparison no correction needed in my engraving, excepting, perhaps, a very little less definiteness of outline, and rather more light between the "Bridge" and the part of Sinus Magnus towards the trapezium.

1864, April 9. Atmosphere very much disturbed. Comparing the Sinus Magnus on my engraving, and remark of Messrs. Stone and Carpenter as to squareness of this feature, I think that I perceive that under

* Nearly Sidereal time.

very bad definition, and the reduced light of the nebula, owing to its low altitude, there is more resemblance to their description, and to Sir J. Herschel's drawing in *this particular,* than when the nebula is seen to advantage. There is a loss of precision in northern edge, although the terminus near ϕ is quite sharp. In bad definition, with feeble light, the dark channel which passes δ, ϵ, and ζ, might be confounded, near Q′, with the S. Magnus, and so explain Sir J. Herschel's figure.

1864, April 14. Just before it clouded, I prepared to sketch the brighter parts of the nebula, and noticed that the outline of Huyghenian R., the part strongly expressed on my engraving, is distinctly recognizable, and certainly, even in the part between AR. of Y″ and B″B″ at origin of Messierian Branch. The latter does not blend here with R. Huygeniana, nor does the light sweep out in definite curve, nearly so far in the parallel a little s. of θ', and towards following side, as is represented in Sir J. Herschel's drawing, but its limit even falls short of AR. of B″B″ as in my engraving. In fact no part of the bright region shades off more indefinitely than is shown in my drawing, excepting it be about x, where the bright wisps of R. Gentil. originate, but even this is not noticeable. (April 15, seems to be outline at x as well, and a narrow darker space separates R. Huygeniana from origin of wisps 2″ or 3″ broad.) I had begun an outline noticing that ay was more readily traced than northern limit of S. Magnus, and that general effect of prec. margin of S. Magnus is that of a nearly straight, and *sharply* defined edge inclining to s. f. direction. More nearly straight than I have represented it in the engraving, where the bright mass in n. p. corner of S. Magnus is perhaps a little too bright, and too distinctly tending to s. p. direction. These appearances were noted hastily, but without thinking at the time of Messrs. Stone and Carpenter's remarks, in Report of Council of Astr. Soc.; but the correction here noticed would rather increase the *squareness* of preceding part of S. Magnus. [x is the preceding point of the Huyghenian Region; ay is the southern boundary of the Sinus Magnus.]

1864, April 15. Sky at first not clear; but improved, though it clouded again towards close of observations. Vision at first disturbed; improved towards middle of observations.

At $8^h\ 50^m\ 45^s$, by Chro. 236. Can distinguish brighter masses of nebulosity especially preceding the trapezium.

At $8^h\ 56^m$. Light preceding θ in parallel, may be traced $4^s.3$. Centre of brightest part $2^s.25$ [preceding θ], brighter mass is about 50″ long.; axis [in] s. f. direction inclined 25° to hour circle. Brightest part [is] about in parallel of θ.

The south shore of Cape* is south of Q′ 19.″2. Herschel's engraving has this limit north of Q′.

[G. P. B. allows 3″ for the half breadth of the "Cape," and this, with the declination of Q′, (H. 87) 100.″0, from page 85, gives for the centre of "Cape" the declination + 83.″8.]

[The difference of declination between θ' Orionis and the north point of bright mass which reaches southward to ϕ (H. 75) is now measured, 69.″8.]

It is about 15″ s. of s. face of Cape; there seems to be an error of 10″ in the engraving, which has $\Delta\,\delta = 81''$.

The limit of general terminus of Huyghenian Region from θ' [measured in declination]. $\Delta\,\delta = 53.''3$.

Following edge of light mass about (n. of) ϕ [measured] in AR. from θ'; $\Delta\,\alpha = 25.''6$. This edge is nearly in direction of AR. but very slightly inclines to s. f. direction. No other part of preceding edge of S. Magnus is so near to θ' in AR. My engraving is here to be corrected as to this mass axis.

* By "Cape," is meant promontory forming the southern boundary of S. Magnus; this edge trends a little n. of parallel towards ϑ in AR. [Note by G. P. B.]

Also the n. p. corner of S. Magnus is to be filled in with more light. Noticed separation of wisps at origin from point x, April 14th, of Huyghenian Region; wisps emanating from x should be broader and more diffuse than in engraving, and brighter.

More light in S. Gentilii, by diffusing the masses there somewhat. The end of Cape at C_8 is well placed in AR. on engraving; greater than AR. of B″B″ [+151.″4, page 85, No. 708]. From C_8 the boundary of light is continuous, across interval between AR. of Q′ and O′, although dark canal from ϵ, ζ, and δ, is traced as intersecting it without by any means obliterating the light.

The least definite edge of bright light of Huyghenian R. is on the side n. and n. p. θ', though even here the light fades quite suddenly, so that line n. of Fig. A should be made a little more pronounced in engraving. The bays μ, μ', etc. (Fig. 3 of Herschel's engraving), do not exist apart from possible slight deficiencies of light which are not now to be recognized.

Star ϕ [H. 75] is far within nebulosity, not at its edge, as Herschel has it. His limit π π' of nebulosity north of Q′ is far too much north of it. The "nebula minima," he places in parallel of Q′, when it is really far south of it

For the above I compared Herschel's engraving, reversed by reflection so as to present it as seen through refractor.

From the foregoing observations it may be gathered that in all probability not only the stars of the trapezium (the multiple star θ' Orionis), but also many in the neighborhood, are physically connected with the nebula. This is especially true of the groups, which, to the naked eye, form ι, θ, c, Orionis. For we see that each of these groups is accompanied by a nebula. We have three nebulae in a nearly right line; the probability that mere chance should have superposed on these nebulae three bright star-groups, also nearly in a right line, and the extremes equidistant from the middle one, is very small indeed. The conclusion that the star-groups and nebulae are physically connected, each group with its nebula, and the three systems of stars together, is much strengthened by the manifest connection of the three nebulae *inter se.* Sir John Herschel, it is true, was not able, with his 18-inch reflector, even at the Cape of Good Hope, to trace the connection; but he carefully avoided deciding the point, as the fact appeared to him extremely probable, and as he had traced the nebulous connection between θ and ι Orionis. The previous observations, as well as those made by Prof. W. C. Bond, confirm the connection of all three nebulae in the most decided manner; the luminous circle whose centre is near the star Herschel 34, and which encloses the principal great nebula, and nearly touches ι Orionis, connects ι and θ; while θ and c are joined by three filaments of light extremely attenuated.

The name "Corona Herschelii," is suggested, in accordance with what are believed to have been Prof. Bond's wishes, for the luminous circle between ι and θ Orionis; as Sir John Herschel first traced a portion of it (I suppose the continuation of the Messierian Branch), and as Prof. Bond's view of the whole subject accords so exactly with the wise suggestion or surmise of Sir William Herschel. The following quotations from these two great authorities, will be read with interest.

"Fifth Class. Very large nebulae. V. 35. Diffused m. nebulosity, extending over no less than 10 degrees of P. D., and many degrees of R. A. It is of very different brightness, and in general, extremely F. and difficult to be perceived. Most probably the nebulosities of the 28th, 30th, 31st, 33d, 34th, and 38th, of this class, are connected together, and form an immense stratum of far distant stars, to which must also belong the nebula in Orion." Sir W. H., in *Philosophical Transactions* for 1789, p. 249.

Again. "Table of extensive diffused nebulosity, No. 24. Visible and unequally bright nebulosity R. A.=5^h 28^m 31^s P. D.=94° 22′, Par.=1° 48′, Mer.=2° 32′, Area in square degrees=4.6. I am pretty sure this joins to the great nebula in Orion." Sir W. H. in *Philosophical Transactions* for 1811, p. 276.

"Although I have not succeeded in tracing any nebulous connection between this nebula (c Orionis) and the great one about θ Orionis, yet as their distance is not much more than half a degree, it not improbably forms part of one great nebulous system extending southwards, through and beyond that nebula as far as ι Orionis, up to which star a pretty conspicuous branch of the great nebula runs. More powerful telescopes than mine must decide this point." Sir J. H. in "Results of Astronomical Observations made during the years 1834, 35, 36, 37, 38, at the Cape of Good Hope," p. 11.

The connection of stars and nebulae is rendered additionally probable from the absence of nebulosity and of small stars in the same field; from the nebulous wisps attached to bright stars, and perhaps from the discovery by the Professors Bond, and by O. Struve, of variable stars among those of the General Catalogue, especially near θ Orionis.

Portions of the nebula have been seen to approximate to resolvability; and in one place both Prof. G. P. Bond and Dr. Gyldèn appear to have detected a cluster of very faint stars, which were visible but by glimpses. This is especially interesting in connection with the researches of Huggins, on the nebula, as viewed by the spectroscope, and other such researches now in progress. The limits of visible nebulosity, as sketched by Professor Bond on a chart of all the brighter stars in the General Catalogue, are the following:—

AR.	Dec.	AR.	Dec.	AR.	Dec.
0′	−68′	−39′	+21′	+31′	+33′
−19	−60	−46	+33	+22	+ 7
−53	−31	−31	+66	+23	−15
−51	−14	+ 1	+62	+23	−44
−36	0	+28	+69	+20	−56

SECTION VI.

ON THE SPIRAL CHARACTER OF THE GREAT NEBULA IN ORION.

Extracted from the "Proceedings of the American Academy," Vol. V. pp. 227–230.

[Prof. G. P. Bond exhibited a drawing of the great nebula surrounding the star θ Orionis, representing its appearance in the twenty-three foot refractor of the Observatory of Harvard College.]

The feature to which attention was particularly directed was the spiral structure of the principal masses of light, or, more correctly, the tendency to an arrangement in elongated wisps or whirls, sweeping outward from the bright region of the trapezium. A disposition of the nebulosity in some localities to radiate from the vicinity of the trapezium, noticed in the Memoir published by Prof. W. C. Bond in 1848, has repeatedly attracted attention in subsequent years. The idea of a spiral character in the radiations had even been suggested, without however presenting itself definitely to the mind as the true conception of the leading features of the nebula.

During the past winter, opportunities were taken to review the whole region, with particular reference to this peculiarity; attention being given exclusively to the arrangement of the diverging wisps of nebulosity, and the alternating dark spaces by which they are separated from each other. A particular scrutiny of the latter was of considerable assistance in tracing the fainter convolutions. The form and disposition of the whirls were thus defined by two independent processes, the nebula being first sketched as a bright object on a dark ground, and, again, its darker openings and channels as dark objects on a white ground.

The quarter designated in Herschel's chart* as the Regio Godiniana, was first explored. The nebulosity was here resolved into an assemblage of three or four long wisps, interlaced with each other, or crossed by offsets; these were ultimately traced from a point near the northern margin of the Sinus Magnus, over the whole length of

* Mem. Astr. Soc., Vol. II.

the Regio Picardiana and the Regio Godiniana, forming a sweep of 120°. After passing the well defined northern boundary of the last-named region, and beyond Herschel's stars o and ξ, these wreaths bend rather suddenly, and tend towards the south-preceding direction. Indications of their presence in this quarter are imperfectly suggested in Lassell's and in Sir J. Herschel's latest drawing. From this point feeble traces exist for 10′ or 15′ in a south-preceding direction. Their course over the R. Picardiana gives a decidedly reticulated aspect to the whole region; but, though bright, they are here so closely intertwined and connected by offsets, that it is a matter of no little difficulty to gain a clear comprehension of their proper relations. The complexity of the details is further increased by several offshoots from this quarter, which cross over into the adjacent R. Derhamiana; still the general effect is easily recognized.

From the southern corner of the Regio Picardiana, and from those parts of R. Derhamiana and R. Huygeniana which lie near the trapezium on its north-preceding and preceding sides, a number of narrow and bright branches diverge, their extremities tending also to the south-preceding direction. Some of these cross the R. Gentiliana and seem to merge together, forming a nebulous mass, which can be followed through an arc of 10′ or 15′. Others, which are less curved, originate near the Sinus Gentilii; these are narrow and somewhat tortuous.

It is to be noticed that the initial direction of the wreaths (Nebelstreifen) changes continuously from an angle of position of 330° on the northern margin of the Sinus Magnus, to one of 220°, or less, at the S. Gentilii, and the sweep of the curve correspondingly diminishes, so that throughout the whole nebulous region preceding the sharply defined apex of the R. Huygeniana, the extremities of the filaments have a pretty uniform tendency in the angle of position 220°. As soon, however, as we pass to the fields on the following side of the apex, a change is immediately apparent, the ultimate direction being about in the angle 160°. The principal group of wisps results from the resolution of the R. Messieriana, and the region between the trapezium and the Proboscis Minor, including both these features, into four or five distinct wreaths, having a common initial direction in the angle of position 110°. The very bright nebulosity lying between the S. Gentilii, the trapezium, and the R. Subnebulosa, cannot be resolved into a regular structure, but three or four condensed spots, constituting the most brilliant part of the nebula close on the south-preceding side of the trapezium, are plainly distinguished as tufts or curled offsets from a prominent wisp of light which extends from its origin, near the trapezium, across the R. Gentiliana.

The general aspect of the greater part of the nebula is therefore that of an assemblage of curved wisps of luminous matter, which, branching outward from a common origin in the bright masses in the vicinity of the trapezium, sweep towards a southerly direction, on either side of an axis passing through the apex of the Regio Huygeniana, nearly in the angle of position 180°. About twenty of these convolutions have been distinctly traced, while others giving a like impression are too faint or too intricate to be subjected to precise description. It may therefore be properly classed among "the spiral nebulae," under the definition given by their first discoverer, Lord Rosse; including in the term all objects in which a curvilinear arrangement, not consisting of regular reëntering curves, may be detected.

That the existence of this feature in the great nebula of Orion should have hitherto escaped notice, after the many careful scrutinies to which it has been subjected, with the help of the largest instruments and the most skilful observers, may seem scarcely credible; a few words of explanation on this point will not therefore be amiss. It is to be ascribed partly to the confusing effect produced by the crossing and intersection of the principal striae and of their offsets, which the eye cannot unravel without the aid of some clue to their mutual relation and significance; and partly also to the faintness of some of the details, which are, nevertheless, very essential features in a correct apprehension of its structure, supplying, as they do, what would otherwise appear as breaks of continuity, and assisting materially in the recognition of a principle of regularity pervading the whole structure. Until the law of relation and continuity in the several parts of such an object is entertained in the mind, it must remain an incoherent, confused assemblage of material, having no orderly or connected arrangement.

The change from the previous notion of its configuration is not more considerable than that which took place with reference to the celebrated nebula 51 Messier, in which the original discovery of the spiral arrangement was made. This object had been subjected to a careful examination and description by both the Herschels, but neither their drawings nor descriptions furnished the slightest intimation of a spiral structure. It deserves particular notice, too, that there was no want of sufficient optical power to exhibit the appearance in question; for the spirality of 51 Messier is seen with perfect distinctness in a refractor of 15 inches' aperture, and must certainly be within reach of the twenty-foot Herschelian reflectors. Nor can it for a moment be thought that the earlier observations and delineations were in any proper sense erroneous. They were simply made at a great disadvantage, in the absence of a clear conception of the general plan of structure presented in the object. Some of the details indispensable to its recognition, being only faintly presented, were overlooked, or, appearing by mere suggestions and glimpses of vision, they conveyed an erroneous impression; in this way the mutual relation of the various parts came to be entirely misconceived. The missing links were supplied by the larger optical power of Lord Rosse's telescope, too plainly not to insure notice; and the nebula then presented itself under a totally different aspect. Instances of similar revelations, completely at variance with previous conjectures, have indeed so often occurred in the history of astronomical discovery, that the process ought to be regarded as the ordinary rule, rather than as an unusual exception.

[The notes of the observations referred to will be found in the text of the previous Section, at the dates 1860, Feb. 21st, 1861, Jan. 28th, Feb. 6th (when the regular examination mentioned in the early part of the present Section began), Feb. 13th, Feb. 25th, Feb. 26th, Feb. 28th, March 5th, March 7th, March 11th.

The paper bears date March 12th, 1861, but underwent some revision after that, before it was finally printed. In 1862, observations upon Feb. 18th, made with the 18½ inch object-glass, by Alvan Clark, now belonging to the Dearborn Observatory of Chicago, and at that time temporarily mounted in Cambridgeport, confirmed the views the author had previously expressed. There are also some notes bearing on this point, 1864, March 9th.]

APPENDIX I.

OBSERVATIONS BY PROF. W. C. BOND.

The following pages contain such measures mostly of right ascension and declination directly made in 1847–48, of the stars in the nebula, as I have been able to find in the observing books. They have been reduced to arc, employing the value 9.″800 of a revolution of the micrometer screw, omitting, however, the small correction for temperature, which does not appear to be very certain, and also the refraction correction, which cannot be strictly computed, as the time is not given. These corrections amount at most to a very few tenths of a second, and are not essential here.

After the observations for position, come some physical notes on the nebula, which have also been extracted from the observing books, for dates between 1847 and the commencement of Prof. G. P. Bond's systematic course of observations, the main subject of the present book.

OBSERVATIONS OF STARS IN THE NEBULA. 1847–48.

DIRECT MEASURES OF DIFFERENCES OF RIGHT-ASCENSION AND DECLINATION.

No. W. C. B.	Date.	$\alpha-\alpha_0$	$\delta-\delta_0$	No. W. C. B.	Date.	$\alpha-\alpha_0$	$\delta-\delta_0$
		″	″			″	″
1	1847, Nov. 26.		+283.8	5	1847, Nov. 30.		− 116.5
	1848, Feb. 29.	−499.2	288.5		1848, Feb. 29.	−242.1	118.2
2	1848, Feb. 29.	−463.4	−114.5	(530*)	1848, Jan. 7.		+2037.7
3	1847, Nov. 5.		+268.2	(543)	1848, Jan. 3.	−197.9	
	26.		268.4				
	1848, Feb. 29.	−400.2	271.3	8	1847, Nov. 26.		+ 510.1
					Dec. 29.	−178.4	
4	1847, Nov. 26.	−308.6	+ 4.5		1848, Jan. 3.	178.9	516.0
	1848, Feb. 29.	304.8	6.0		Feb. 21.	172.7	511.2
5	1847, Nov. 26.	−242.4	−117.1	(554)	1847, Dec. 29.	−161.5	

* The numbers in parentheses are G. P. B.'s, for stars not in W. C. B.'s printed catalogue.

No. W. C. B.	Date.	$\alpha-\alpha_0$	$\delta-\delta_0$
(554)	1848, Jan. 3.	−162″.0	+ 671″.9
	Feb. 21.	162.5	
9	1847, Nov. 26.	−156.7	− 119.0
	Nov. 30.		120.4
	1848, Feb. 29.	−162.1	121.2
11	1847, Dec. 29.		− 21.9
12	1848, Feb. 29.	− 88.0	− 180.9
13	1847, Nov. 26.	− 93.3	− 272.8
	30.		271.9
	1848, Feb. 29.	94.2	273.3
42	1847, Dec. 29.	− 82.4	
	1848, Jan. 3.	80.5	+ 386.4
	Feb. 21.	73.8	382.1
14	1848, Jan. 7.	− 73.1	+1896.1
	Feb. 23.	69.7	1897.2
17	1847, Nov. 26.	− 9.9	+ 7.7
	30.	9.2	8.8
21	1847, Nov. 5.	− 2.7	+ 17.0
	26.	3.8	16.1
	30.	4.8	16.0
23	1847, Nov. 5.	+ 7.0	+ 97.2
	26.	4.9	99.8
	30.		98.2
	Dec. 24.		98.7
25	1847, Nov. 5.	+ 12.7	+ 8.1
	26.	11.8	5.5
	30.	12.9	6.4
28	1848, Jan. 3.	+ 31.5	+ 434.1
	Feb. 21.	21.7	431.3
29	1847, Dec. 29.	+ 27.0	
	1848, Jan. 3.	25.1	+ 409.6
	Feb. 21.	27.9	405.6
32	1847, Nov. 5.	+ 32.0	+ 166.1
	26.	30.3	171.1
	30.		168.7
33	1847, Nov. 5.	+ 42.9	+ 160.9
	26.	35.7	161.7
	30.		159.3
35	1848, Jan. 7.	+ 57.2	+1809.6
	Feb. 23.	53.9	1811.5

No. W. C. B.	Date.	$\alpha-\alpha_0$	$\delta-\delta_0$
36	1848, Feb. 23.	+ 49″.6	+1955″.3
(667)	1848, Jan. 3.	+ 59.0	
37	1847, Nov. 5.	+ 65.4	+ 150.1
	26.	55.4	148.4
	30.		146.6
38	1847, Dec. 29.	+ 63.7	+ 670.3
	1848, Jan. 3.	62.7	
	Feb. 21.	62.5	673.4
39	1847, Nov. 5.	+ 66.7	+ 102.0
	26.	61.8	99.8
	30.		98.2
	Dec. 24.		98.7
40	1848, Jan. 7.	+ 74.1	+1985.9
	Feb. 23.	74.1	1986.8
41	1847, Dec. 24.	+ 66.2	− 34.8
43	1847, Dec. 24.	+ 74.3	− 36.8
45	1847, Nov. 5.	+ 95.8	− 93.1
	26.	97.5	93.6
	30.		93.8
	Dec. 6.		95.5
(690)	1847, Dec. 24.	+116.8	− 442.6
46	1848, Jan. 7.	+128.3	
	Feb. 23.	127.4	+2154.2
(695)	1848, Jan. 3.	+130.2	+ 819.2
(696)	1848, Jan. 3.	+134.8	
59	1848, Jan. 3.	+140.8	+ 740.8
47	1847, Nov. 26.	+140.3	+ 495.0
	Dec. 24.	140.7	
	29.	140.2	492.9
	1848, Jan. 3.	144.0	493.4
	Feb. 21.	142.1	492.6
48	1847, Dec. 29.	+146.2	+ 611.0
	1848, Jan. 3.	142.6	613.9
	Feb. 21.	145.0	612.3
49	1847, Nov. 5.	+149.3	− 251.1
	26.	148.5	252.4
	30.		252.4
50	1847, Nov. 5.	+149.5	− 93.1

No. W. C. B.	Date.	$\alpha-\alpha_0$	$\delta-\delta_0$
50	1847, Nov. 26.	+150″.4	− 95″.4
	30.		95.6
	Dec. 6.		96.4
51	1847, Nov. 5.		− 122.0
	26.	+150.8	135.6
	30.		135.6
	Dec. 29.	149.1	135.0
52	1848, Jan. 7.	+157.4	+1700.4
	Feb. 23.	157.6	1701.5
55	1847, Nov. 5.	+182.0	− 175.7
	26.	178.4	175.3
	30.		175.4
57	1848, Jan. 7.	+200.8	+1749.4
	Feb. 23.	198.3	1750.0
60	1847, Nov. 26.	+212.7	+ 446.3
	Dec. 24.	218.7	443.4
	29.	218.4	
	1848, March 1.	218.1	445.6
(740)	1848, Jan. 3.	+217.2	+ 841.6
61	1847, Nov. 5.	+227.3	− 110.0
	26.	225.9	107.4
	30.		111.3
	Dec. 6.		111.2
	24.	223.7	110.2
	1848, March 1.	224.1	110.0
	24.	225.1	111.6
62	1848, Jan. 7.	+226.9	+1991.8
	Feb. 23.	228.2	1991.4
63	1847, Dec. 24.	+216.3	− 566.9
	29.	221.1	
64	1847, Dec. 24.	+242.7	− 585.5
65	1847, Dec. 24.	+243.9	− 464.6
	29.	243.8	
66	1847, Dec. 29.	+283.5	+ 668.3
	1848, Jan. 3.	279.3	666.9
	Feb. 21.	279.9	668.1
67	1847, Dec. 29.	+284.5	
68	1848, Feb. 23.	+306.1	+2099.8

No. W. C. B.	Date.	$\alpha-\alpha_0$	$\delta-\delta_0$
70	1848, Jan. 7.	+ 318″.8	+1924″.8
	Feb. 23.	321.8	1926.3
75	1847, Dec. 29.	+ 362.9	
76	1847, Dec. 24.	+ 371.6	+ 197.6
	24.	370.6	195.2
	29.	367.5	
	29.	365.7	194.7
	1848, Jan. 3.	372.8	
	3.	370.5	195.5
	March 1.	371.5	195.3
78	1847, Dec. 24.	+ 385.0	− 283.4
	29.		287.2
79	1847, Dec. 29.	+ 387.1	+ 588.6
	1848, Jan. 3.	387.1	585.3
80	1847, Dec. 24.	+ 422.5	− 516.2
	1848, Mch. 24.	416.5	520.9
82	1847, Dec. 24.	+ 457.5	+ 391.7
	29.	461.3	
	1848, Jan. 3.	460.1	386.4
	March 1.	462.1	392.1
(806)	1848, Jan. 3.	+ 464.4	
83	1848, Jan. 3.	+ 453.1	+ 329.9
	March 1.	454.0	331.7
87	1847, Dec. 24.	+ 513.9	− 304.2
	29.		306.9
	1848, March 1.		306.0
90	1848, Feb. 23.	+ 548.3	+2128.2
91	1848, Mch. 24.	+ 570.6	− 855.4
92	1848, March 1.	+ 665.8	+ 356.2
93	1847, Dec. 24.	+ 628.2	+ 63.0
	29.		60.1
	1848, March 1.		59.8
94	1848, March 1.	+ 798.4	− 258.2
(893)	1848, March 1.	+ 809.0	+ 232.5
95	1848, March 1.	+1059.9	− 56.6
96	1848, March 1.	+1134.6	− 8.7

Besides the above, there are a few measures of angles of position and distance, made October 10th, 12th, 15th. These, however, are not given in original, in the observing books, but simply as results; they are passed over here, as they appear to refer entirely to brighter and well-known stars, and some errors exist in the copy, which alone I have seen. It is probable that they were not used in the preparation of the Memoir.

The faint stars in the vicinity of the trapezium are not among them, and appear to have been given in the catalogue from estimates.

The remarks about the physical appearance of the nebula are in many cases interesting, and some extracts from them, including all where I have been able to make out with certainty the portion of the nebula observed, are now given.

1847, Dec. 6. The brighter parts are constantly sparkling with points of light, seen in favorable moments. No. 6, plain eye-piece, breaks the nebula into cumulus clouds.

1847, Dec. 24. [Some measures were made of points in the nebula; the following are definitely given as results.] From a to the edge of the preceding portion of the brightest part of the nebula, Δ AR. $10.^r03$ [$=98.''2$]. Breadth of the brighter portion of the nebula in the parallel of declination of a, $12.^r64$ [$=123.''9$]. Difference of declination of a and lower coast of the bay n. f. $8.^r29$ [$=81.''2$]; as a is almost exactly on the parallel of upper coast of this bay, this $8.^r29$ is the breadth. [The coördinates of the terminus of the brighter part of the nebula apparently above are found to be $\alpha-\alpha_0=0.''0$, $\delta-\delta_0=-157.''6$; and those of the bottom of the bay south preceding, $-77.''3$, $-75.''6$.]

The fixed wire on a, in a parallel of declination, runs along the upper edge of the opening following the trapezium, past No. 5 [H. 110].

1848, Jan. 4. The preceding boundary of the horn is very sharply defined all the distance from below 24 (H. 123) to above 25 (H. 126); the turn towards 25 is firmly outlined and beautifully curved, equally above and below; the following side of the horn is by no means so well defined, but flies off in wisps of cirrus. The branch below is of the same character; this branch originates in the stronger light. There is undoubtedly considerable nebulosity enveloping the stars ϵ, 4, 5, 6, and 7, but the sudden increased density of the "horn" is very marked and clearly defined.

The cam-shaped portion which is connected with the main in the s. p. direction from * 21 [H. 108], forms on the following side rather more than a right angle, terminating abruptly at 29 n. f., [H. 124?] the light is somewhat condensed to 21, but not strongly.

The brighter portions of the nebula are broken up into cumulus.

The northern nebula which we have brought in to-night is connected with the main by the preceding route, sweeping round with faint light. We do *not* resolve the nebula to-night; the stars are broad and blotty; thermometer falling.

1848, Jan. 12. The brighter portion seems full of points of light at times. I have the utmost confidence in treating these minute points.

1848, Jan. 17. [Diagram by W. C. B.; marginal note by G. P. B.] "This diagram shows terminal angle of R. Huygheniana in AR. $150''$."

1848, Feb. 7. From the star at the termination of the Messierian Branch the nebulosity radiates upward (south) and in the preceding direction, terminating in a cloudy appearance. The nebula passes on the following side of ι Orionis.

1848, March 22. It was uncommonly fine seeing just as twilight ended. I saw many stars in the neighborhood of the trapezium of Orion, and Sir John Herschel's star, the sixth star, stood out boldly. It is all but certain that this part of the nebula is composed of clusters of stars.

The first tendency of s. p. side from south cape is at a right angle to the s. f. side, provided we limit the region to the brightest portion; but the light is pretty strong in a southern direction, so that by taking the whole extent of the cape we lessen the angle to about 70 or 75 degrees.

There are other remarks with these, which are more or less obscure; and I am apprehensive that I have not, through my slight acquaintance with the nebula, got the sense of all which might be understood by a more competent editor. Some of Prof. W. C. Bond's notes upon the nebula are apparently not now to be found; at least I so explain the deficiency in places of the preceding observations of position, as compared with his catalogue. Yet in many cases, the catalogue positions may have been inserted by estimation, or perhaps have even been taken from the chart.

In connection with Prof. W. C. Bond's observations, I placed his original drawing (or what appears very certainly to be such) in the hands of Mr. Watts, to reëngrave. The difficulty of his task has been great, and the variations between his engraving and the former one considerable; but, as might be expected from so skilful and conscientious an artist (known to astronomers by the engravings in Vol. III. of these Annals), the present edition is much more accurate than the former. His principal care has been to render with fidelity the nebulous parts, any deviation in which would affect the appearance more sensibly than inaccuracies in the magnitudes of the stars, which are of less importance in the drawing.

Some very faint details in Prof. G. P. Bond's drawing, as, for instance, the very slight nebulosity about 2′ preceding H. 136 = No. 848 of the present catalogue, are, with the somewhat greater optical power of the Chicago equatorial, more conspicuous; and I am led to the conclusion that Prof. W. C. Bond's drawing represents the nebula as seen in a very advantageous state of the atmosphere; while the later and more elaborate engraving of Prof. G. P. Bond certainly represents it as seen on usual good nights, with great precision.

APPENDIX II.

ON THE ERRORS OF THE EQUATORIAL AND MICROMETER.

THE observations in the body of the work were made for the most part with the mica scale micrometer, (See Vol. I. Pt. II. of these Annals, p. iv,) or with the filar-micrometer, and in either case the errors in the position of the instrument with respect to the pole, and the flexure of different parts, produce corresponding errors in the zero of the position circle, which must be eliminated.

The formula for the correction of these errors counted positive when denoting a positive correction to the angle of position, is (for declination circle preceding)

$$\gamma \sin (\tau - \theta) \sec \delta + i_1 \sec \delta - (c + e \cos \phi \sin \tau) \tan \delta + \psi (\sin \phi \cos \delta - \cos \phi \sin \delta \cos \tau).$$

Here γ is the distance of the pole of the instrument from the pole of the heavens.

θ is the hour-angle of the pole of the instrument.

τ the hour-angle of the object observed, or rather of the middle point between the two objects observed.

δ is the declination of the same point.

$i_1 = i - \varepsilon \sin \phi$.

i is the complement of the angle between the hour and declination axes.

ε is the flexure of the declination axis.

ϕ is the latitude of the place.

c is the collimation of the instrument.

e is the flexure of the tube.

ψ is the angle by which the tube of the telescope, rigidly fixed to the declination axis, tends to rotate by the flexure of that axis.

This is of course different from the flexure of the declination axis itself.

The formula above given is essentially Bessel's; it may be found on p. 394 of Chauvenet's Manual of Spherical and Practical Astronomy, Vol. II., with the exception of the last term, which is derived from p. 430 of the same volume.

The sense in which each of these quantities is to be taken, is as Chauvenet gives them, namely:—

The hour-angle θ and τ are considered as increasing with a motion of the points to which they refer from the meridian above pole towards the west, and are counted from the meridian above pole.

The flexure ε of the declination axis is considered positive when the declination circle is depressed, and e is positive when the object end, as is natural, is depressed through flexure by a larger angle than the eye-end.

The angle i is considered positive when the pole of the declination axis is in north declination referred to the equator of the instrument.

The latitude ϕ, and declination δ, are of course positive when north; c is considered positive when, for declination circle preceding, the observed hour-angle is too small.

In the present series of observations the zero of position was always determined by the stars in the nebula, at times varying but two or three hours from those of observation.

Consequently all the parts of this correction to the angle of position which are independent of τ will disappear or be eliminated at once; and the remainder of the terms depending on e and ψ, being multiplied by sin δ or tan δ, about -0.1, will be greatly diminished. They may certainly be neglected, as observations of declinations (see below) indicate a value of e not exceeding $2'$ or $3'$, nor do observations of the zero of position on different sides of the meridian show a value of ψ greater than $3'$.

So that $0.1\ e$ and $0.1\ \psi$ will be much less than $1'$, and may therefore be neglected in the zero of position.

The quantity γ is in practice separated into two components,

$$\xi = \gamma \cos \theta. \qquad \eta = \gamma \sin \theta.$$

Here ξ represents the elevation of the pole of the instrument above the celestial pole, and η its distance in a westerly direction from the same point; and ξ and η are separately determined. In May and June, 1864, observations for the instrumental constants were carefully made by Professor Bond. They are of the following kinds.

To determine $i - \varepsilon \sin \phi$, η, ε, right ascensions of stars at culmination, or more properly differences of instrumental right ascensions of pairs of stars near the zenith and south horizon were observed. These observations were made May 27th, 1864, and are as follows, in four sets.

Letter.	Numbers of Stars, B. A. C.	Observed Diff. AR.	Computed Diff. AR.	δ	Telescope.	Diff. AR. Comp'd—Obs'd.
		m s	m s	° ′		s
A	4808	− 7 26.0	− 7 34.3	+30 58	W	−8.3
	4842			−37 12		
B	4916	+ 6 59.5	+ 6 59.1	−33 18	E	−0.4
	4943			+39 48		
C	4958	− 7 44.6	− 7 45.4	+40 56	E	−0.8
	4996			−35 35		
D	5084	−10 52.1	−10 58.5	+37 51	W	−6.4
	5151			−29 20		

The differences of AR. are always obtained by subtracting the chronometer time of culmination of the southern star, or the computed AR. of the south star, from the same quantities relative to the northern star.

The equation between the observed hour-angle of an object, the errors of the equatorial instrument and its true hour-angle, is for circle preceding

$$\tau = t + x - \eta \tan\phi - \gamma \sin(\tau - \theta) \tan\delta + c \sec\delta - i \tan\delta$$
$$+ \epsilon(\sin\phi \tan\delta + \cos\phi \cos\tau) + e \cos\phi \sec\delta \sin\tau.$$

Where t is the observed hour-angle, x the negative reading of the hour-circle when the pole of the declination axis is 90° w. of the meridian. Making $t = 0$, as in the above observations, and T the Sidereal time, we have, if α be AR.,

$$T - \alpha = x - \eta \tan\phi + \eta \tan\delta + c \sec\delta - (i - \epsilon \sin\phi) \tan\delta + \epsilon \cos\phi.$$

This formula holds for telescope east of the meridian, above pole; for the opposite position, or west of the meridian,

$$T - \alpha = x - \eta \tan\phi + \eta \tan\delta - c \sec\delta + (i - \epsilon \sin\phi) \tan\delta - \epsilon \cos\phi.$$

Denote now by T_n, T_s, α_n, α_s etc., the above quantities relative to *north* and *south* stars respectively. We shall then subtract the equation for a south star from that for a north star; the circle being supposed to precede, or the telescope to be east of the meridian.

This gives, for telescope east,

$$T_n - T_s - (\alpha_n - \alpha_s) = (\eta - i + \epsilon \sin\phi)(\tan\delta_n - \tan\delta_s) + c(\sec\delta_n - \sec\delta_s).$$

For telescope west,

$$T_n - T_s - (\alpha_n - \alpha_s) = (\eta + i - \epsilon \sin\phi)(\tan\delta_n - \tan\delta_s) - c(\sec\delta_n - \sec\delta_s).$$

Our four equations are then

$$+\overset{s}{8}.3 = 1.36\,(\eta + i - \epsilon \sin\phi) + 0.09c$$
$$+0.4 = 1.49\,(\eta - i + \epsilon \sin\phi) + 0.11c$$
$$+0.8 = 1.58\,(\eta - i + \epsilon \sin\phi) + 0.09c$$
$$+6.4 = 1.34\,(\eta + i - \epsilon \sin\phi) - 0.12c.$$

The combination of the first pair of these equations with each other gives

$$\eta = 3.^s19 - 0.07c \qquad i - \epsilon \sin\phi = 2^s.92 + 0.00c.$$

The second pair in like manner gives

$$\eta = 2.^s65 + 0.02c \qquad i - \epsilon \sin\phi = 2^s.14 + 0.07c.$$

The differences between these values are rather larger than was perhaps to be expected. Their means are

$$\eta = 2.^s92 - 0.03c \qquad i - \epsilon \sin\phi = 2.^s53 + 0.04c.$$

In order now to determine ε itself, we must go back to the observed culminations expressed in Sidereal time, as compared with the computed right-ascensions.

These are

	T	α	T—α	
	h m s	h m s	s	
B. A. C. 4808	14 26 19.4	14 26 1.0	+18.4	W.
" 4842	33 45.4	33 35.3	+10.1	W.
" 4916	47 51.9	47 27.7	+24.2	E.
" 4943	54 51.4	54 26.8	+24.6	E.
" 4958	57 16.6	56 52.4	+24.2	E.
" 4996	15 5 1.2	15 4 37.8	+23.4	E.
" 5084	19 43.1	19 24.1	+19.0	W.
" 5151	30 35.2	30 22.6	+12.6	W.

The equations between these and the instrumental corrections are then, if $x—\eta$ tan ϕ be called Δt

$$\begin{aligned}
\overset{s}{18.4} &= \Delta t + 0.60\eta - 1.17c + 0.60\ (i-\epsilon \sin \phi) - \epsilon \cos \phi \\
10.1 &= \Delta t - 0.76\eta - 1.26c - 0.76\ (i-\epsilon \sin \phi) - \epsilon \cos \phi \\
24.2 &= \Delta t - 0.66\eta + 1.20c + 0.66\ (i-\epsilon \sin \phi) + \epsilon \cos \phi \\
24.6 &= \Delta t + 0.83\eta + 1.30c - 0.83\ (i-\epsilon \sin \phi) + \epsilon \cos \phi \\
24.2 &= \Delta t + 0.87\eta + 1.32c - 0.87\ (i-\epsilon \sin \phi) + \epsilon \cos \phi \\
23.4 &= \Delta t - 0.72\eta + 1.23c + 0.72\ (i-\epsilon \sin \phi) + \epsilon \cos \phi \\
19.0 &= \Delta t + 0.78\eta - 1.27c + 0.78\ (i-\epsilon \sin \phi) - \epsilon \cos \phi \\
12.6 &= \Delta t - 0.56\eta - 1.15c - 0.56\ (i-\epsilon \sin \phi) - \epsilon \cos \phi.
\end{aligned}$$

Substituting the former values of η, and $i - \varepsilon \sin \phi$, we obtain

$$\begin{aligned}
\overset{s}{15.13} &= \Delta t - 1.16c - \epsilon \cos \phi \\
14.24 &= \Delta t - 1.27c - \epsilon \cos \phi \\
24.46 &= \Delta t + 1.25c + \epsilon \cos \phi \\
24.28 &= \Delta t + 1.24c + \epsilon \cos \phi \\
23.86 &= \Delta t + 1.26c + \epsilon \cos \phi \\
23.68 &= \Delta t + 1.28c + \epsilon \cos \phi \\
14.75 &= \Delta t - 1.26c - \epsilon \cos \phi \\
15.65 &= \Delta t - 1.16c - \epsilon \cos \phi.
\end{aligned}$$

We now obtain from these equations, by taking the mean of the first two and last two, for one set, and of the 3d, 4th, 5th, 6th, for another set,

$$\begin{aligned}
\overset{s}{14.94} &= \Delta t - 1.21c - \epsilon \cos \phi \\
24.07 &= \Delta t + 1.26c + \epsilon \operatorname{co} \phi
\end{aligned}$$

Hence

$$\begin{aligned}
\Delta t &= \overset{s}{19.51} + 0.02c \\
\epsilon \cos \phi &= 4.56 - 1.24c.
\end{aligned}$$

We have now to obtain an approximate value of c, which is readily found by observations of Polaris in reversed positions of the instrument, at culmination. For if we subtract from the formula for $T-\alpha$, circle preceding, its value for circle following, we shall have, denoting by $T'-T$ the difference in Sidereal time between the culminations, considered positive, when the one above taken with circle preceding would be later,

$$T'-T=2\ c\ \sec\delta-2\ (i-\epsilon\sin\phi)\tan\delta+2\ \epsilon\cos\phi,$$

and hence

$$\frac{(T'-T)\cos\delta}{2}=c-(i-\epsilon\sin\phi)\sin\delta+\epsilon\cos\phi\cos\delta.$$

For Polaris, $\cos\delta$ is very small, and $\sin\delta$ nearly 1; so that the equation thus formed is an advantageous one to combine with those previously obtained.

Observations made on June 2d give

$$T'-T=5^{m}\ 24.^{s}5$$
$$4\ 49.0$$

approximately, and hence, using $\delta=88°\ 34'\ 57''$,

$$c-(i-\epsilon\sin\phi)+0.025\ \epsilon\cos\phi=3.^{s}79,$$

and, as $i-\varepsilon\sin\phi=2.^{s}53+0.04c$, $\varepsilon\cos\phi=4.^{s}56-1.24c$.

$$0.93c=6.^{s}21$$
$$c=6.7\ =1.'7.$$

The following declinations, arranged by Prof. Bond, give now sufficient evidence that ξ and η for other dates are not large enough to affect sensibly the zero of position, and that the instrument is mounted with great stability. They were made mostly in connection with filar-micrometer comparisons.

Positions of stars observed in Declination, with the Great Refractor. To find position of instrumental axis.

The declinations will be more suitable for discussion, because, as the circles were usually read after the last transits over the wires in AR., the eye-piece will have a constant collimation error in declination, in whatever part of the slide it was left.

The last adjustment of the axis of the Great Refractor was made Aug. 19, 1847.

Cor. ref. = − [1.756] cot. $(\psi + \delta)$ in seconds.

June 12, 1848. Obs. of collimation in Dec. For tel. *east* of Pier, cor. for coll. in Dec. = + 31″ = 0′.5.

Date.	Name.	τ in time.	Mean Dec. Beg. of year.	Cor. to ap. pl.	App. Dec.	Reading for Dec.	Ref. Cor.	Obs. Dec.	C—O.
		h m	° ′	′	° ′	° ′	′	° ′	′
1847, Oct. 11.	Star comp.	+ 5 7.5	+36 48.1	+0.3	+36 48.4	+36 48.5	−0.8	+36 47.7	+0.7
Oct. 11.	η Herculis	+ 5 11.4	+39 13.0	+0.3	+39 13.3	+39 13.4	−0.8	+39 12.6	+0.7
Nov. 5.	B. A. C. 181	+ 2 42.2	+39 51.0	+0.5	+39 51.5	+39 53.0	−0.2	+39 52.8	−1.3
1848, June 12.	Capella	+ 0 56.6	+45 50.1	−0.2	+45 49.9	+45 50.5	0.0	+45 50.5	−0.6
1849, May 19.	χ Aurigae	+ 7 31.8	+29 32.9	−0.1	+29 32.8	+29 38.0	−6.5	+29 31.5	+1.3
May 19.	ϵ Geminorum	+ 7 8.1	+25 16.5	−0.1	+25 16.4	+25 20.5	−6.0	+25 14.5	+1.9
1850, July 8.	B. A. C. 4694	+ 2 17.2	+31 34.3	+0.2	+31 34.5	+31 34.2	−0.4	+31 33.8	+0.7
July 17.	H. C. 25380	+ 3 7.1	+ 5 52.3	+0.1	+ 5 52.4	+ 5 53.8	−1.0	+ 5 52.8	−0.4
July 25.	B. A. C. 4494	+ 4 38.2	−15 11.7	−0.1	−15 11.8	−15 3.2	−7.7	−15 10.9	+0.9
Aug. 29.	Arg. 125	− 5 22.6	+58 4.3	−0.2	+58 4.1	+58 4.3	−0.4	+58 3.9	+0.2
Sept. 8, 9.	Arg. 174	− 4 54.9	+54 22.3	−0.2	+54 22.1	+54 23.0	−0.3	+54 22.7	−0.6
1852, May 18.	Comp. star	− 5 6.6	+74 36.1	0.0	+74 36.1	+74 35.5	0.0	+74 35.5	+0.6
June 14.	β Urs. Maj.	+ 5 24.3	+56 10.5	+0.2	+56 10.7	+57 11.7	−0.4	+57 11.3	−0.6
June 14.	Star	+ 5 16.7	+56 14.1	+0.2	+56 14.3	+56 15.3	−0.4	+56 14.9	−0.6
Aug. 18.	Star	− 3 18.4	+23 22.9	+0.1	+23 23.0	+23 24.8	−0.6	+23 24.2	−1.2
Sept. 1, 2.	Star	− 4 21.1	+20 33.0	−0.2	+20 32.8	+20 36.0	−1.0	+20 35.0	−2.2
Sept. 2, 3.	Star	− 6 6.	+20 35.8	−0.2	+20 35.6	+20 36.0	−2.6	+20 33.4	+2.1
Sept. 15.	Star	− 5 10.2	+64 50.1	0.0	+64 50.1	+64 50.0	−0.2	+64 49 8	+0.3
Oct. 16.	Gr. 2037	+ 7 55.8	+80 6.3	+0.1	+80 6.4	+80 6.5	−0.7	+80 5.8	+0.6
Oct. 22.	Comp. star	+10 20.9	+76 18.0	+0.1	+76 18.1	+76 18.2	−1.5	+76 16.7	+1.4
1853, March 10.	Comp. star	+ 1 58.3	− 7 22.4	−0.2	− 7 22.6	− 7 22.0	−1.3	− 7 23.3	+0.7
March 12.	Rigel	− 0 0.	− 8 22.5	−0.2	− 8 22.7	− 8 21.4	−1.2	− 8 22.6	−0.1†
March 14.	Star	+ 2 46.	− 1 42.6	−0.1	− 1 42.7	− 1 42.0	−1.3	− 1 43.3	+0.6
March 14.	Lal. 9068	+ 3 15.2	− 7 22.3	−0.1	− 7 22.4	− 7 22.0	−1.7	− 7 23.7	+1.3
March 14.	Rigel	+ 2 54.9	− 8 22.5	−0.2	− 8 22.7	− 8 20.4	−1.6	− 8 22.0	−0.7
March 18.	Comp.*	+ 4 53.0	+ 2 7.8	−0.1	+ 2 7.7	+ 2 11.0	−2.5	+ 2 8.5	−0.8
March 22.	W. 720.	+ 3 28.1	+ 5 3.0	−0.1	+ 5 2.9	+ 5 3.2	−1.2	+ 5 2.0	+0.9§

1853, March 29.	Comp. *	+4 17.5	+ 8 54.2	−0.1	+ 8 54.1	+ 8 56.3	−1.5	+ 8 54.8	−0.7
Dec. 14.	B. A. C. 409	+5 27.4	+36 56.8	+0.4	+36 57.2	+36 58.5	−1.0	+36 57.5	−0.3
1854, Jan. 2.	Comet	+0 5.5	+19 5.6	0.0	+19 5.6	+19 6.0	−0.4	+19 5.6	0.0
April 2.	α Arietis	+6 24.0	+22 46.2	0.0	+22 46.2	+22 49.0	−3.1	+22 46.0	+0.2
April 3.	Comet	+5 40.1	+17 1.8	0.0	+17 1.8	+17 4.8	−2.3	+17 2.0	−0.2
April 4.	Comp. *	+6 10.1	+16 51.3	0.0	+16 51.3	+16 55.0	−3.7	+16 51.3	0.0
June 27.	Comet	+9 2.9	+60 17.1	+0.1	+60 17.2	+60 19.0	−2.2	+60 16.9	+0.3
June 28.	Comet	+9 4.2	+59 37.1	+0.1	+59 37.2	+59 38.0	−2.4	+60 35.9	+1.3
1857, Sept. 18.	η Bootis	+6 12.1	+19 7.1	−0.1	+19 7.0	+19 10.3	−3.1	+19 7.2	−0.2
Sept. 18.	B. Z. 460	+6 16.6	+20 50.3	−0.1	+20 50.2	+20 54.2	−3.1	+20 57.1	−0.9
Sept. 23.	W. 1071	−3 34.9	− 3 59.9	+0.1	−35 59.8	− 3 55.3	−1.7	− 3 57.0	+2.8
Nov. 11.	Oeltz. 15894	+8 52.8	+54 48.8	−0.2	+54 48.6	+54 52.0	−3.0	+54 49.0	−0.4
Nov. 13.	Oeltz. 15894	+9 0.3	+54 48.8	−0.2	+54 48.6	+54 52.0	−3.0	+54 49.0	−0.4
Nov. 13.	Oeltz. 16583	+8 52.8	+52 16.0	−0.1	+52 15.9	+52 19.0	−3.4	+52 15.5	+0.4
Nov. 13.	η Urs. Maj.	+8 44.0	+54 29.4	−0.5	+54 28.9	+54 31.0	−2.7	+54 28.3	+0.6
Nov. 14.	Oeltz. 16583	+8 19.0	+52 16.0	−0.1	+52 15.9	+52 18.0	−2.5	+52 15.5	+0.4
Nov. 17.		+6 45.8	+44 44.8	0.0	+44 44.8	+44 45.0	−1.5	+44 43.5	+1.3
Nov. 20.		+4 35.7	+38 23.3	+0.1	+38 23.4	+38 25.0	−0.5	+38 24.5	−1.1
1858, Jan. 4.		+5 12.0	+39 49.3	+0.2	+39 49.5	+39 50.2	−0.7	+39 49.5	0.0
Jan. 7.	Comp. * a′	+3 7.0	+37 2.9	+0.2	+37 3.1	+37 3.9	−0.3	+37 3.6	−0.5
Jan. 8.	Comp. * A″	+2 40.0	+35 59.9	+0.2	+36 0.1	+36 1.0	−0.3	+36 0.7	−0.6
Jan. 12.	Comp. * a‴	+2 31.3	+32 11.7	+0.2	+32 11.9	+32 12.5	−0.3	+32 12.2	−0.3
Feb. 6.	Comp. * a	+2 0.6	+ 5 59.8	+0.1	+ 5 59.9	+ 6 2.0	−0.8	+ 6 1.2	−1.3
March 3.	B. A. C. 1115	+2 48.9	−17 56.4	0.0	−17 56.4	−17 53.7	−2.4	−17 56.2	+0.2
May 3.	Comp. * ..	+4 0.5	+35 8.8	+0.1	+35 8.9	+35 10.0	−0.5	+35 9.5	−0.6
May 12.	Comp. * a	+3 14.7	+37 39.7	0.0	+37 39.7	+37 41.0	−0.3	+37 40.7	−1.0
May 12.	Comp. * a	+4 8.9	+37 37.7	0.0	+37 39.7	+37 33.3	−0.5	+37 32.8	+6.9
May 12.	B. A. C. 3728	+4 8.3	+34 58.8	0.0	+34 58.8	+34 52.6	−0.5	+34 52.1	+6.7
June 28.	Comet V.	+7 9.3	+26 27.6	0.0	+26 27.6	+26 34.0	−5.5	+26 28.5	−0.9
July 8.	Comp. * A	+6 58.6	+26 33.5	0.0	+26 33.5	+26 38.0	−4.6	+26 33.4	+0.1
July 13.	Comp. * L	+7 6.4	+27 48.8	0.0	+27 48.8	+27 54.0	−4.3	+27 49.7	−0.9
July 13.	μ Leonis	+7 3.5	+26 40.5	0.0	+26 40.5	+26 46.0	−4.5	+26 41.5	−1.0
July 15.	μ Leonis	+5 57.3	+26 40.5	0.0	+26 40.5	+26 44.0	−1.8	+26 42.2	−1.7
July 15.	Comp. * A	+6 20.4	+27 41.9	0.0	+27 41.9	+27 45.0	−2.3	+27 42.7	−0.8
Aug. 19.	Comet V.	+7 51.0	+32 8.5	−0.1	+32 8.4	+32 18.0	−8.7	+32 9.3	−0.9
Sept. 6, 7.	Comp. * a	−1 21.4	+45 13.7	+0.1	+45 13.8	+45 14.3	0.0	+45 14.3	−0.5
Sept. 6, 7.	Capella	−1 44.6	+45 50.8	+0.1	+45 50.9	+45 51.0	0.0	+45 51.3	−0.4
Sept. 6, 7.	α^2 Geminorum	−4 33.6	+32 11.9	0.0	+32 11.9	+32 12.6	−0.7	+32 11.9	0.0
Sept. 7, 8.	Comet	−1 21.1	+45 19.3	0.0	+45 19.3	+45 20.0	0.0	+45 20.0	−0.7
Sept. 8.	Comet V.	+7 42.7	+35 32.0	−0.1	+35 31.9	+35 37.0	−4.8	+35 32.2	−0.3
Sept. 8, 9.	Capella	−1 36.6	+45 50.8	+0.1	+45 50.9	+45 52.3	0.0	+45 52.3	−1.4
Sept. 8, 9.	Comp.* m	−1 51.3	+44 46.9	+0.1	+44 47.0	+44 44.0	0.0	+44 44.0	+3.0

† 4 obs. § 2 obs.

Besides the foregoing, the following observations were made, on June 2d, 1864, especially for the purpose of determining the values of ξ and e, as well as $\Delta\,\delta$, the index error of the declination circle.

Star.	Hour-angle West.	Comp'd Declination.	App. Obs'd Declination.	Circle.	Refraction.	Instrumental Declination.	C—O
	h m	° ′ ″	° ′ ″		′ ″	° ′ ″	″
Polaris	11 49.1	88 34 57	88 34 44	Prec.	−1 6	88 33 38	+79
Polaris	12 15.2		36 34	Foll.	−1 6	35 28	−31
α Cassiopeiæ	13 19.8	55 47 23	55 53 6	Foll.	−5 15	55 47 51	−28
α^2 Libræ	23 39.6	−15 28 38	−15 26 42	Prec.	−1 32	−15 28 14	−24
β Ursæ Minoris	23 57.7	74 42 43	74 43 39	Foll.	+ 37	74 44 16	−93
Polaris	0 4.0	88 34 57	88 35 20	Foll.	+1 0	88 36 20	−83
Polaris	0 15.8		88 33 32	Prec.	+1 0	88 34 32	+25

The formula with which these are to be compared, is

$$C-O = \pm\,\Delta\,\delta - \xi\cos\tau - \eta\sin\tau - e\cos(\delta+\psi)\frac{\sin\phi}{\cos\psi}$$

where as before, $\tan\psi = \cot\phi\cos\tau$, and the upper sign holds good for circle preceding, and the lower for circle following.

The equations thus formed are

$$\begin{array}{rlr}
+79'' = & \Delta\,\delta + 1.00\,\xi - 0.75e - 0.05\eta & n = +81'' \\
-31 = & -\Delta\,\delta + 1.00\,\xi - 0.75e + 0.07\eta & -34 \\
-28 = & -\Delta\,\delta + 0.94\,\xi - 0.95e + 0.34\eta & -42 \\
-24 = & \Delta\,\delta - 1.00\,\xi - 0.85e + 0.09\eta & -28 \\
-93 = & -\Delta\,\delta - 1.00\,\xi + 0.54e + 0.01\eta & -93 \\
-83 = & -\Delta\,\delta - 1.00\,\xi + 0.72e - 0.02\eta & -82 \\
+25 = & \Delta\,\delta - 1.00\,\xi + 0.72e - 0.07\eta & +28
\end{array}$$

(Substituting for η its value $2.^s92 - 0.03c = 2.^s72 = 40.''8$, and thus eliminating it, the numbers n take the place of the known terms in these equations.)

The final equations by the method of least squares, omitting fractions of seconds, are

$$\begin{array}{rl}
+332'' = & 7.00\,\Delta\,\delta - 0.94\,\xi - 0.44e \\
+183 = & -0.94\,\Delta\,\delta + 6.88\,\xi - 3.52e \\
-61 = & -0.44\,\Delta\,\delta - 3.52\,\xi + 4.08e
\end{array}$$

whence

$$\Delta\,\delta = 57''$$
$$\xi = 53$$
$$e = 37$$

It does not seem necessary to pursue the investigation farther, except to compare these values with observation. Substituting them in the equations, we have

	C—O
Polaris	$-1''$
Polaris	-2
α Cassiopeiæ	0
α^2 Libræ	-1
β Ursæ Minoris	-3
Polaris	$+1$
Polaris	-3

The probable error of one observation is thus about $\pm$ 1.″7.

For the declination of the nebula $-$ 5.°5, the correction of the angle of position is

$$1.00\ \xi \sin\tau - 1.00\ \eta \cos\tau \pm i, \pm 0.10c \pm 0.10e \cos\phi \sin\tau.$$

And for a change in τ only, without a reversal of the instrument, it does not vary by an amount capable of producing a sensible effect on these micrometric observations

The coefficient ψ depending on the rigidity of the connection between the tube and the declination axis, was found on June 3d to be about 2.′4, the angle of position of any object being apparently too large when the circle precedes. This again for $\delta = -$ 5.°5 becomes of sensible influence only when the zero is determined in opposite positions of the instrument with respect to the declination axis; which is not the case for the Orion Zones.

2. The stability of the instrument in right-ascension is readily tested by the accordance among themselves of the separate zones of Part I.; but it is desirable to show that there is no tendency to a general change in right-ascension. For this end, I have derived from the tables of reduction of the zones given in Vol. I. Part II. of these Annals, the values of x', which include such a general change, if supposed to exist, the clock-rate, and the variation for 1^h of α of the terms

$$-f - g \sin(G + \alpha) \tan\delta - h \sin(H + \alpha) \sec\delta.$$

This last variation, as δ is, *in maximo* $+20'$, and h, H are approximately 19.″5 and $270° - \odot$, becomes, in time

$$-h \cos(H + \alpha) \sin 1° = -0.^s34 \sin(\alpha - \odot),$$

and we can substitute for it nearly

$$-0.^s34 \sin t,$$

where t is the mean time of transit reduced to arc.

We then apply to the numbers x', the quantity

$$0.^s34 \sin t + r,$$

where r is the hourly rate of the clock, which does not appear to have exceeded $0^s.1$.

The observations were mostly made before 10^h m. t., and we can be sure that any constant change of the equatorial in AR. in 1^h can only amount to much less than $0^s.5$. The resulting error therefore for 2^m 15^s difference of AR. from θ Orionis will not exceed 0.″3.

Values of x':

Zone.	x'	Zone.	x'	Zone.	x'	Zone.	x'
	s		s		s		s
1	−0.28	18	−0.17	33	−0.34	48	−0.45
2	−0.23	19	−0.16	34	−0.28	49	−0.21
3	−0.48	20	−0.33	35	−0.26	50	−0.04
4	−0.50	21	−0.34	36	−0.68	51	−0.29
5	−0.60	22	−0.67	37	−0.94	52	−0.20
6	−0.28	23	−0.80	38	−0.21	53	−0.18
7	−0.35	24	−0.46	39	−0.40	54	−0.25
8	−0.31	25	−0.63	40	−0.28	55	−0.35
9	−0.15	26	−0.27	41	−0.43	56	−0.24
10	+0.08	27	−0.50	42	−0.27	57	−0.21
11	−0.05	28	−0.72	43	−1.32	58	−0.26
13	−0.05	29	−0.85	44	−0.46	59	−0.21
14	−0.25	30	−0.24	45	−0.47	60	+0.07
15	−0.10	31	−0.41	46	−0.68	61	−1.26
16	−0.43	32	−0.92	47	−0.67	62	−1.36
17	−0.35						

3. The value of one division of the mica scale employed in most of the observations of Sections I. and II. was derived from transits of Polaris and its companion; and the same means were employed to assure freedom from error in the subdivision into minutes of arc. The introduction to the second part of the first volume of these Annals contains an account of these observations. During the last few years various experiments have been made to deduce the correction of this value. These, however, did not produce very accordant results, although the mean correction was exceedingly small; so that it could not be guaranteed as a sensible amount, nor could they be considered as a considerable addition to the previous discussion.

It has seemed useful to compare Liapunoff's and G. P. Bond's declinations in such an order that the discrepancy, if any, between their numerical expressions for the same arc might be manifest. The results of this comparison are contained for the principal stars, denoted by capital letters in Sect. I., in the following table. The columns Decl. L. and L − B contain respectively Liapunoff's declination, and the amount by which this is farther north than G. P. Bond's; the latter extracted from Section III. Part II., the Differential Catalogue.

No. Herschel.	Decl. L.	L—B.	No. Herschel.	Decl. L.	L—B.	No. Herschel.	Decl. L.	L—B.	No. Herschel.	Decl. L.	L—B.
74	−952″.9	−1″.6	145	− 50″.9	+1″.5	53	−272″.0	+1″.1	48	+511″.0	+0″.1
143	916.3	+1.2	17	46.3	−1.0	142	254.3	+3.1	124	588.3	+0.6
135	851.2	+1.8	147	− 6.3	−0.1	18	252.8	−0.3	99	612.2	+0.7
34	659.4	−0.7	38	+ 5.1	−0.5	103	250.8	+2.5	49	666.1	0.0
37	592.0	+0.2	5	11.7	−1.8	10	241.2	−1.8	113	669.3	+2.8
111	−581.9	+1.5	30	+ 32.0	+1.2	26	−207.7	−0.8	86	+673.8	+0.3
106	567.4	+2.7	136	63.7	+2.9	104	174.4	+1.3	85	850.6	+1.4
16	524.7	+0.6	2	64.4	−1.6	50	118.9	−0.4	I.*	909.4	−0.1
112	464.5	+2.2	70	98.3	−0.2	45	116.6	−0.7			
95	442.1	+1.3	87	100.3	0.0	110	109.8	+0.3			
40	−424.2	+0.5	120	+196.7	+0.7	33	−108.1	−0.2			
47	399.6	+1.6	35	272.6	+0.4	101	95.7	+2.5			
8	309.6	−1.4	32	289.4	−0.8	93	93.7	+1.0			
133	303.3	+2.1	108	444.8	+0.7	14	71.9	−1.1			
123	283.7	+1.8	102	493.6	+0.6	27	70.8	+0.4			

These differences are in the majority of cases positive. I have taken the means of each 8 or 9 successive ones as follows.

No. of Stars.	Mean Dec. Liapunoff.	Mean L—B.
9	−679″	+0″.88
9	375	+1.09
9	158	+0.29
9	− 32	−0.04
9	+225	+0.30
8	685	+0.72
53	− 61	+0.54

These numbers do not show any sensible dependence on the declination of the star relative to θ' Orionis; the cause of the constant difference 0.″54 it is not easy to trace, and especially does it not seem likely to arise from the Cambridge observations, which are purely differential in character.

* Struve's notation.

www.ingramcontent.com/pod-product-compliance
Lightning Source LLC
LaVergne TN
LVHW020058110826
845151LV00001B/18

* 9 7 8 1 4 2 5 5 4 5 2 7 7 *